ENVIRONMENTAL GEOLOGY

Ecology, Resource and Hazard Management

2nd Edition

To The
People Of Tomorrow
Whose Future
We Are Selling
To
Finance Our Present

Mourn not the dead
But rather mourn the apathetic throng
The cowed and meek
Who see the world's great anguish and its wrong
And dare not speak

— Ralph Chaplin

Preface to the Second Edition

I wrote the book "*Environmental Geology: Indian Context*" at a time of the Indian history when the persons at the helm of affairs of the Government of India, and the academics and researchers all over India were imbued with engaging and energizing spirit of friendship for nature. The scientists were coming out with revealing and seminal works on the health and well-being of the environment of the Indian landmass.

Times have changed — changed radically in the last two decades. The waves of globalization have swept the world. The attendant storms of "liberalization" have overwhelmed the decision makers and the powerful sectors of the society. Our nation seems to be in the stranglehold of the insatiable urge to become rich at any cost, unmindful of the health and safety of the natural environment. The consequence is that in order to finance its present, this generation is selling the future of the generations to come. This is true everywhere in the world, more so in India.

Compared to the earth's biological capacity of 15.7 hectares per person, the peoples' environmental demand is more than 21.9 hectares per person, implying that we are living far beyond what nature offers us. Frightening is the rate of the consumption of resources, which is faster than the galloping rate of growth of the population, nearly half of which is hungry.

Since the systems of honest monitoring, guidance, supervision or inspection at every level of governance have collapsed, there is no way to check or contain the reckless overuse as well as misuse of the precious and finite resources the nature has offered us — land, soils, forests, water and minerals.

"Those who enjoy money power and who are able to manipulate the system of the governance" (borrowing the words of Justice G.S. Singhvi and Justice A.K. Ganguli, *The Hindu*, February 2, 2012) rule the roost in India. Under such a situation as this, while the corporate world and unscrupulous elements of the society are being offered natural assets at the throw-away costs, the people at large are deprived of the common property resources and therefore have turned indifferent to their natural surrounding and resources. The consequence is there to see — there are no compassionate stakeholders and no genuine custodians of the natural environment of India. So, everything is going wrong — the protective cover of natural forests is diminishing, the rate of loss of land is increasing, the springs are drying up and discharges of rivers dwindling, the groundwater table is going down, the rivers are getting increasingly poisoned, and the air is becoming progressively unbreathable. Today the common man feels that his life is becoming more and more difficult, whatever the Gross Domestic Product data indicate.

I cannot help feeling that I am part of the system that failed to redeem the pledge to strive for preserving the health of India's environment. I must confess my guilt for not having done enough.

When I first wrote this book thirty years ago, I had hoped that in the revised edition I will portray a beautiful India, a healthy India. Alas, I am presenting a more dismal, more disheartening picture.

I recall what a celebrated lawyer of India once said — we are the people who turn our heads and pretend that we do not see anything going wrong, and we shamelessly pretend not to have heard people cry.

The stark reality is that we do not mind the wilting or even death of the dreams of the generation that would replace us tomorrow.

I cannot help reciting the proverb of the Cree tribe of the Red Indians in North America —

> "Only after the last tree has fallen
> Only after the last river has been poisoned
> Only after the last fish has been caught
> Only then will be realized that money cannot be eaten".

However, looking at the eyes of the youths and the children of India, I feel hopeful that the people of tomorrow will sooner or later bring India's environment back to its health.

K.S. Valdiya
Jawaharlal Nehru Centre for Advanced Scientific Research
Bangalore – 560 064

Preface to the First Edition

This book is addressed primarily to students both at the undergraduate and postgraduate levels, and is written with the objective of creating an awareness for preserving the natural environment, conserving the earth's resources, coping with geological hazards, and participating in technological and industrial development without disturbing the ecological balance. However, it will also be of interest to the general public including engineers, architects, teachers and planners. The book makes an attempt to weave the diverse components of the ecology into the fabric of geology, bringing out the interlinkages and interdependence between various environmental factors. It highlights the various environmental stresses the land in India is being subjected to and the resultant deterioration in the quality of life, elucidating not only the natural and anthropogenic causes of the degradation of land and its soil carpet, but also the implications of resource depletion. The strategies used for the scientific management of land, soil, water and mineral resources are discussed, as also measures to prevent their destruction, degradation and misuse, and the means by which the damaged parts of the landscape and the earth's resources can be restored are suggested.

The book dwells on the occurrence of natural hazards resulting from quickened earth processes and suggests ways and means of coping with them, including mitigating or abating the risks.

Focusing attention on the interaction between man and earth, and elucidating the impact of irrational developmental activities on the stability and survival of the web of ecosystems, the section on engineering geology and geotechnics offers a strategy and methodology of development without destruction of the environment and harm to the resource base.

Explaining the causes, sources and dimensions of pollution and waste production, alternative options for energy generation without threat to the landscape, atmosphere and water bodies are suggested.

The book was written during my tenure as a Pitambar Pant National Environment Fellow in 1982-84. I am deeply grateful to the Department of Environment, Government of India, for the fellowship and the very generous assistance that came with it. To Ramesh Upadhyay, Kewalanand Lohni, Sukesh Bartariya, Dungar Singh Negi and Vishwanath Ghoshal I owe a debt of gratitude for the help rendered.

The work has benefited considerably from critical reviews by Prof. J. S. Singh, Dr. V. M. Meher-Homji, Dr. M. Deshayes, Prof. Mohammad Shafi, Prof. R. Vaidyanathan, Dr. R. K. Gupta, Shri D. C. Das, Dr. A. K. Dey (Late), Dr. N. G. K. Nair, Prof. B. B. S. Singhal, Dr. B. D. Pathak, Prof. Ishwar Prakash, Dr. R. K. Bhandari, Dr. C. R. Krishna Murti, Dr. S. Z. Qasim, Dr. Anna Mani, Prof. K. C. Sahu,

Prof. B. P. Ghildiyal, Prof. J. D. Ives, Prof. A. K. Saha, Col. N. Mohan Rao, Shri V. K. Raina, Prof. K. N. Khattri, Dr. H. K. Gupta, Prof. R. S. Yeats, Dr. Martin J. Haigh, Dr. S. V. Chitale, Dr. D. K. Ghosh, Shri Gyaneshwar Pant, and Shri R. S. Dixit. I am deeply grateful to them.

K.S. Valdiya
Kumaun University
Nainital 263002
Uttarakhand

Acknowledgements

This book was revised in the ashram-like edifying ambience of Jawaharlal Nehru Centre for Advanced Scientific Research. I do not know proper words to express my profound feelings of gratitude to JNC for the exceptionally warm hospitality and magnanimously generous help of all kinds for pursuing this work for the benefit of students, teachers and researchers.

I am extremely thankful to Shri Ratan Tata, Chairman, Tata Sons, without whose intervention, the revision of an old 1987 vintage book could not have been undertaken.

Simanta Borah, Sushil Gupta and Sohan Gaur of McGraw Hill Education (India) stimulated and inspired me to update the work and give it the look that it has. I am thankful to them.

The photographs sent by Professor K.N. Ganeshaiah (University of Agricultural Sciences, Bangalore), Professor B.N. Goswami (Indian Institute of Tropical Meteorology, Pune), Dr C.P. Rajendran (Indian Institute of Science, Bangalore), Dr N.P. Kurien (Centre for Earth Science Studies, Thiruvananthapuram), Professor M.P.S. Bisht (HNB Garhwal University, Srinagar), Professor M.S. Sisodia (JN Vyas University, Jodhpur), Dr V.C. Tewari (Wadia Institute of Himalaya Geology, Dehradun), and Dr J. Bhatt (Haldwani) portray various natural events affecting India. I owe them a debt of gratitude.

But for the tremendous help from Abhinav Khandka, Jaishri Sanwal and Sameer Valdiya in gathering through the internet a wealth of crucial data for the many tables in the book, I would have been toiling for months.

Universities Press, Hyderabad very kindly accorded permission to reproduce a number of diagrams from the author's book "*Geology, Environment and Society*" first published in 2004.

I am beholden to all those who helped me one way or the other.

K.S. Valdiya

Contents

Part-IV. Natural Hazards

Part-VI. Pollution and Energy

Part-VII. Alternative Sources of Energy

Preamble: Dimensions of Environmental Stress

OVER-USED EARTH

"The earth is more beautiful than useful and deserves better to be admired than used. Unfortunately, there are very few who discern beauty in the earth's landscape and virtue in conserving the earth's resources" and preserving its ecological integrity. In the eyes of the vast majority of people there is just greed for the varied resources of the earth. Consequently, the land of our country is in a distressingly sad te—it has been plundered, ravaged, defiled and polluted for obvious reasons. It has been swamped by growing numbers of people and livestock. **The tragedy of explosive population growth is manifest in the deterioration in the quality of life in spite of increasing developmental activities.** While one quarter of the world's seven billion-strong population is consuming two-thirds of the world's resources, one half gets just enough to stay alive in utter misery. More people means more land for living and production, more resources, (Fig. P 1) and more resultant wastes. One-sixth or 17.4% of the world's population—that is 1.21 billion—live in India striving for survival and fast eroding the resource base by exploiting the renewable resources beyond regeneration, covering the landscape with wastes, polluting the waters of rivers as well as exposing the population to environmentally-induced maladies. The various developmental projects involving engineering work have triggered unintended changes in the osystem that threaten the very benefits for which the projects were conceived and launched (Fig. P2). Over and above the backlash of these developmental activities are the hazards and disasters that nature genders by way of famine, floods, earthquakes and drought each year.

THREATS FROM THREE QUARTERS

e devastation of the natural environment and its floral and faunal domains far outweighs the hievements of man in the scientific and technological fields. The threat has come from three quarters: (i) the hungry and deprived people, unaware of and indeed unconcerned with the implications of denuding forests, damaging soils and mounting pressure on the fragile ecosystem, (ii) the rich and powerful segment of the population which manipulate technology and run industries which are defacing d defiling the environment, and (iii) the builders of roads, big dams, heavy engineering structures, etc. who destabilize the slopes in mountainous terrains leading to a variety of unintended but harmful changes.

Additionally, the pattern of overconsumption by the affluent section of the population is causing rapid pletion of basic natural resources, resulting in the denial of the benefits of burgeoning technology to the poor and the less fortunate. What Gilford Pinchot (1865–1946) said of the Americans applies equally well to the Indian society. "**The American Colossus was fiercely intent on appropriating an exploiting the riches of the richest continent—grasping with both hands, reaping where he had not sown, wasting what he thought would last for ever**." Indira Gandhi (1982) summed up the situation very cogently—"the rich out of greed and the poor out of need have been reckless in plundering earth's assets."

Fig. P.1 The tragedy of explosive population growth and the increasing demand for resources and food. (With the permission of *Ecoforum*, 7(1), 1982)

Fig. P.2 The blind application of technology in development programmes. (From *Ecoforum*, 7(1), 1982

So bad is the situation today in India that Justice G.S. Singhvi and Justice A.K. Ganguli of the Supreme Court had to remind the government that **the natural resources are vested with the government as a matter of trust in the name of the people of India, and it is the solemn duty of the state to protect the national interest, and the natural resources must always be used in the interest of the country** (*The Hindu*, February 2, 2012).

In spite of the concern expressed repeatedly by world leaders, who periodically assemble in many places all over the world, and despite the setting up of a Global Environmental Monitoring System together with Departments of Environment in various countries, people at large have remained indifferent to the needs of conserving resources and preserving the integrity of the environment. Even governmental agencies despite loud protestations have not put into practice the credo of sustainable development through conservation and without destruction. The consequence of this indifference has been the exhaustion of the earth's bearing capacity and a decline in the quality of life.

ECOLOGICAL FOOTPRINT

The measure of the capacity of productive land and available water to support the needs of people and absorb the wastes they generate and pollution they cause is called the ***ecological footprint*** (Miller, 2007). The ***per capital ecological footprint*** is the average ecological footprint of an individual in an area (Figure P 3). Ten years ago in 2002, the ecological footprint of all the peoples of the world was nearly 39% higher than the earth's ecological capacity (Miller, 2007). Today it must be far more than 50%. In other words, **the once bountiful earth is now unable to replenish its renewable resources and absorb the colossal wastes and pollutants we generate. It is but natural that we are overusing and degrading our croplands, forests and groundwater and other renewable resources faster than the nature can replenish them**. It would be hard to believe that at present the United States, the European Union, Japan, China and India collectively use quite more than 74% of the earth's ecological capacity (Miller, 2007). The ecological footprint of India is far more than twice as large as its domestic ecological capacity.

Our energy need is growing at a galloping pace. However, when we compare the consumption of total primary energy 9969 kg oil equivalent by the USA, 5160 kg oil equivalent by France, 1700 kg oil equivalent by Japan, and 832 kg oil equivalent by China, India's consumption was only 325 kg oil equivalent (Narayanan, 2004). It would be much higher now. And yet India is nowhere near the point of meeting its energy needs.

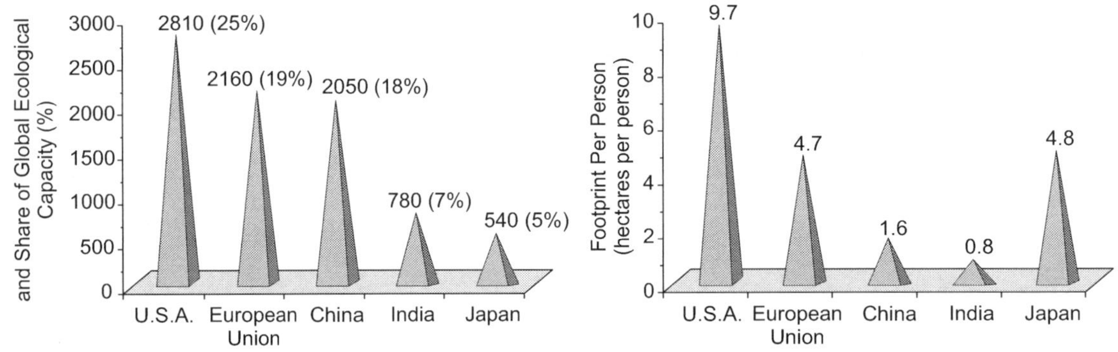

Fig. P3. Ecological footprint is the measure of the capacity of productive land and its available water to support the needs of people and to absorb the waste generated and pollution caused. (Based on: Miller, 2007, *People and Environment*, Cengage Learning India, New Delhi, 696 p.).

COMMON PROPERTY RESOURCES

Air, water, soil, animal and plant species, scenic beauty of the landscape are collectedly owned by the society at large — owned by all and accessible to everybody. These constitute the ***common property***

resources (Bawa et al, 2011). The tragedy is that the governments, the industries and even the people over-use and misuse them recklessly — damaging the very base of the resources, leading to what Hardin (1968, 1985) described as ***tragedy of commons***. The resulting damage — due to some individuals appropriating all or maximum benefits and gains, and the losses and pains shared by all members of t user groups — eventually culminates in the failure of the market, often economic collapse as we are witnessing today.

HISTORY OF ENVIRONMENTAL DEGRADATION

Ever since the human race took to agriculture, the symptoms of deterioration of the environment have been apparent. The reclamation of grasslands for cultivation initiated an unending cycle of erosion of the soil. The farmlands were expanded at the cost of forests, leading inevitably to loss of protective vegetal cover, reduction in the number of plant species in the cultivated tracts, thus rendering them vulnerable to disease and invasion of undesirable species. The domestication of animals and restricting their grazing to the lands of people's choice have enlarged the extent of deforestation and land degradation. Nearly 1000 animals and 25,000 plant species all over the world are threatened with extinction.

The destruction of forests in the developing countries has been so rapid and so alarming that only 0.2 hectare per capita forest was left in Asia and 0.5 ha/capita in Africa (compared to 2 to 2.7 ha/capita in the developed world some thirty years ago). India in the early 1980s was losing forest cover at the rate of 0.161 million ha/year or 441 ha/day. Today the situation is slightly better, *if one goes by the area covered by man-made plantations also*. The result of this depredation of forests is the expansion of vulnerable dryland (Fig. P 4).

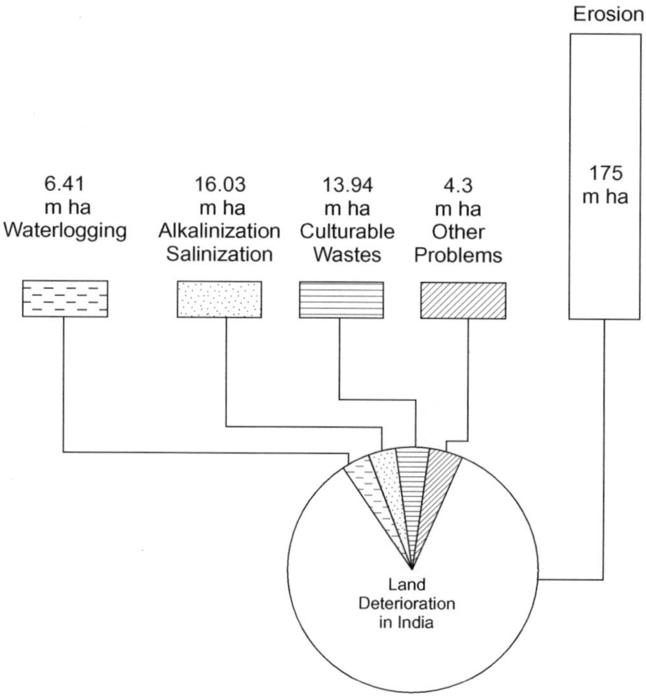

Fig. P4. Quite a large parts of Indian land is suffering from erosion, water logging, salinization and other form of degradation. (Based data from Anon, 2010, *Degraded And Wastelands of India*, ICA New Delhi, 158 p.)

Before the intervention of man, the sediments carried to the seas by rivers amounted to about 8,435 million tonnes per annum; it rose to 21,768 m t/yr after he started modifying the landscape. The rivers of the Indian subcontinent are today transporting every year to the ocean more than 1600 million tonnes of sediments containing precious nutrients, thus rapidly eroding the land. The result of mismanagement of land is that more than 167.57 million ha of Indian land are affected seriously by erosion — over 100 million ha of non-forest land including mined areas having become seriously degraded. The situation is extremely bad in almost all parts of Asia, as is evident from the fact that the Asian lands contribute three-fourths of the total sediments delivered to the oceans by rivers at the rate of 342 t/km^2/yr.

More than 60,000 km^2 of land all over the world is annually falling victim to the processes of sertification, and 43% of the South Asian and 49% of the African areas are subject to recurrent droughts. Excessive irrigation without proper drainage has destroyed the productivity of farmlands in many parts of the world. Thirty years ago about 35% of the irrigated farmlands of the world were flicted with soil sickness (alkalinity and salinity)—between 2,00,000 and 3,00,000 ha were then lost each year in this manner. Today more than 16.03 million ha of the farmlands in India seriously suffer from this bane of over-irrigation, and another 6.41 million ha are waterlogged and thus unproductive (Fig. P4).

The congregation of people with their livestock in villages and towns has added a new dimension to environmental distress—the generation of wastes and proliferation of pollution, with the attendant spread of diseases. In the later stages of evolution, urbanization considerably aggravated the environmental stresses. The hunger of cities for resources of all kinds has been growing phenomenally—food and raw material as much as huge amounts of fuel, energy and water. In return these urban centres pollute the atmosphere and contaminate rivers and streams. As a result of unchecked growth of population ere is an uninterrupted, unending flight of people from rural to urban centres, so that the towns and cities have become conglomerations of slums. In the absence of environmental planning and resource management, the slums in most developing countries including India have become expanding loci of arcity, inflation, economic destitution, unemployment and stagnation of sociocultural deterioration.

Technological development has not succeeded in turning the tide of distress stemming from the phenomenon of unchecked population growth in the developing countries, while the life styles of the rich people in the developed world and the needs of the poor of the developing countries have been making heavy and ever-escalating demands on the dwindling and depleting resources of the world.

"TYRANNY OF SMALL DECISIONS"

The stress build-up to nearly the threshold value is a result of what may be called the "tyranny of small decisions" (a phrase used by Alfred E. Kahn) — forests have been destroyed to feed paper factories and make furniture, wooden structures, etc., more than half a million hectares of forests in India have been to submerged in the waters of man-made lakes since 1953; roads have been constructed in geologically sensitive and ecologically vulnerable areas, often to satisfy the needs and whims of petty-minded people; industries and mines are permitted to dump their wastes and pollutants in water courses in order increase the profit margin of the privileged powerful; construction of buildings is being allowed in areas exposed to natural hazards so that people can live in close congregation; livestock population is allowed to multiply and animals let loose to defoliate the fragile forests because the "sentimental attachment' for these animals is too strong, and so on. Undoubtedly, **much of the present planning in our country has resulted from small decisions taken to meet political expediencies or to realize immediate objectives without taking into consideration the wider impacts of such short-sighted**

decisions. In the global perspective "when people in half the world are dying from too little to eat and people in the other half are dying from too much to eat", as Sydney Harris states, "there has to be something basically wrong in the global system of distributions." This cannot be more true than in the case of India where the population is sharply split into the many hungry and deprived and the few **pampered affluents. One segment of the population is threatened with disease and death while the oth** commands the whole system and luxuriates in plenty.

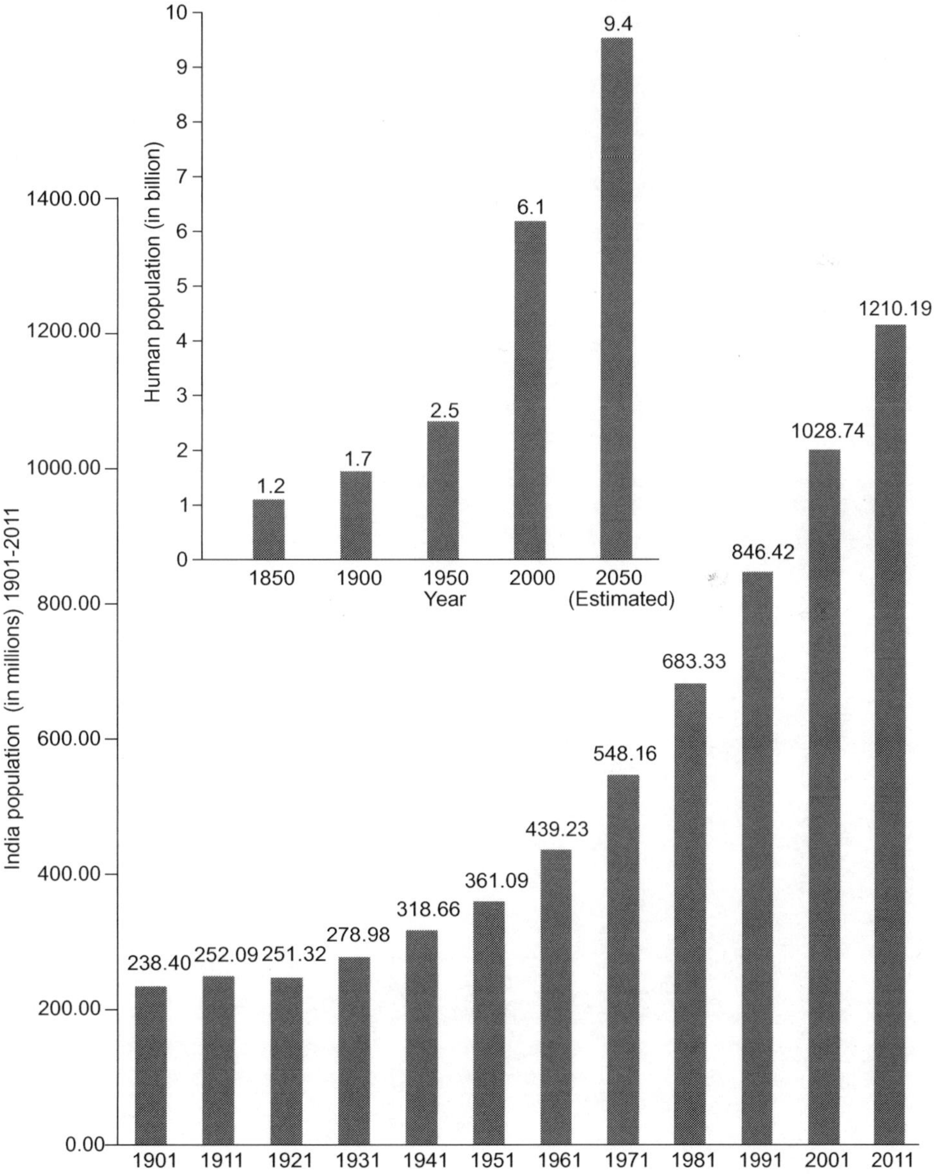

Fig. P5. The explosive population growth in the world and in India. The country has to take care of additional 27.8 million every year. (From: Census Report, 2011)

TRAGEDY OF POPULATION EXPLOSION

More than 1.21 billion people—about 17.5% of the world population—live in India having merely 2.4% of the earth's land. By A.D. 2025 India will have 1.46 billion people to bring up and support. The birth rate of the Indian population is alarmingly high—one birth every 34 minutes, or 2019 babies every day, or 1.76 million every year. But then, millions die every year so that the net yearly addition to the population is 27.8 million (Fig. P 5). To meet the needs of these additional millions the nation has to muster stupendous amounts of **additional** food-grains, cloth, houses, schools and jobs. In other words, all that this country is additionally growing, producing and building goes to meet the needs of the people who are annually adding to the population at the rate of 1.7%. There is naturally little, if at all, improvement in the quality of life of the vast majority of the people. **To quote John Galbraith, "economically speaking, we have to be running to stay where we are**." India is today a very poor country in the world in terms of per capita gross national products, social security and quality of life.

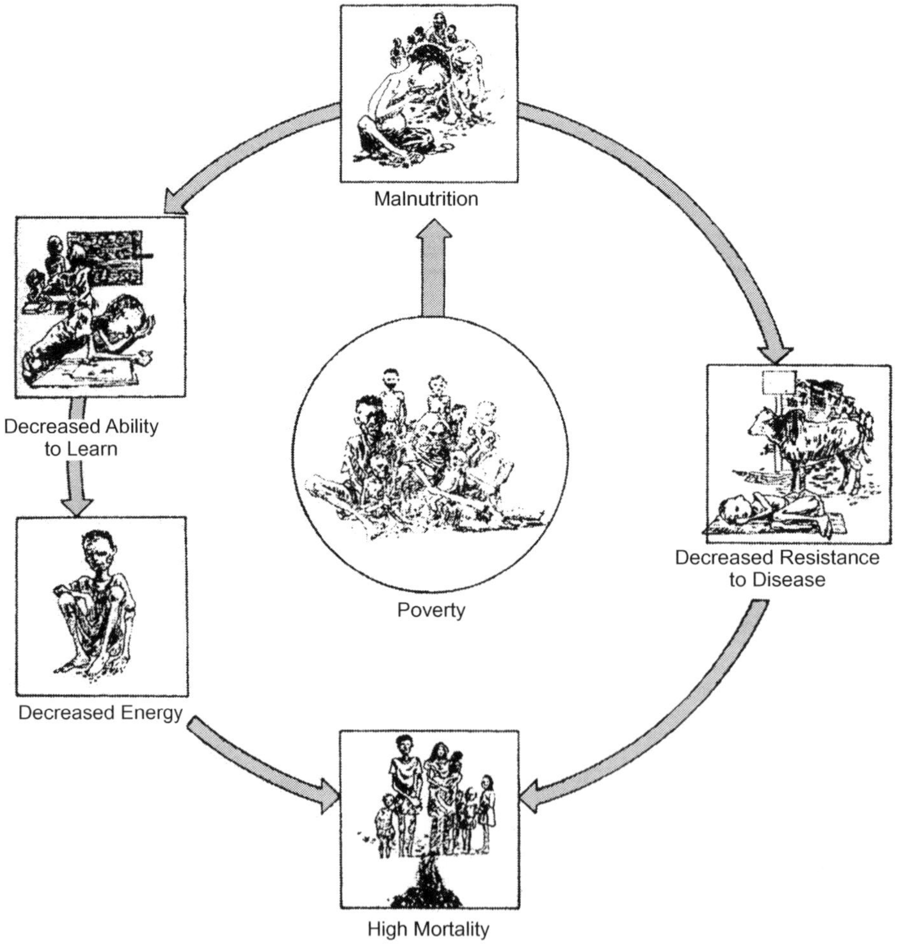

Fig. P6. The vicious cycle of destitution, disease and death. Nearly two-thirds of the population of India is hungry and suffering from malnutrition. (Modified after G.T. Miller and P. Armstrong, 1982).

Caught in the vicious cycle of destitution, exploitation and disease, the poor (Fig. P 6) do not think in terms of conserving and preserving for tomorrow, for theirs is a day-to-day struggle to survive. **When the rich do not care, and the powerful look away, the deprived cannot be expected to think of preserving the environment**. The destruction of our resource base, and the degradation of our environment will therefore continue at a galloping pace unless the growth of the population is reduced to the zero level.

Coupled with the distress caused by the burden of exploding population, there is absolute apathy, if not flippant disregard, for making use of the land in accordance with its capabilities and there is la of *popular* concerted efforts to tap alternative sources of energy in order to reduce the pressure on fast depleting resources, and to diminish pollution. **The main cause of the environmental crisis is that the development of the country is being sought without endeavouring to curb or minimize destruction of the resource-base and environment.** Decisions are taken without first ascertaining their place in t larger framework of long-term planning.

Alfred Whitehead once said: "When you have understood all about the sun and all about t atmosphere, and all about the rotation of the earth, you may still miss the radiance of the sunset." In other words, the world today is in the grip of an environmental crisis.

1 Spectrum of Environmental Geology

OBJECTIVES

In his eternal quest for improving the quality of life and satisfying his growing needs, man has used and abused his physical environment. He has been reluctant to share the resources of the earth with other organisms, even with fellow humans living in his own environment. Unmindful of the consequences of tampering with the delicate balance in ecological systems that constitute the web of his environment, man has induced changes that now pose a threat to his own well-being and even survival in some parts of the world. Although he has braved the inescapable natural hazards arising from the dynamic behaviour of the evolving earth, the changes or processes inadvertently initiated as a result of various developmental projects to improve his world are proving potentially hazardous to his world. An ironic situation has thus developed—**there is decreasing happiness for many in the midst of increasing comfort for a few**.

Environmental geology deals with the interrelationship of the various earth processes, their consequences d the activities of man who forms the central figure in the environmental scheme of things. The discipline embraces co-ordinated and integrated studies and the application of geology for the improvement and preservation of the environment through a more judicious use of natural resources, and by ensuring safeguards against degradation of the land and contamination of the air and water. **Environmental geology is thus a mission-oriented and problem-solving discipline**, as Coates (1981) puts it. The guiding ethics is that a better quality of life can be realized through use of the land in accordance with its capabilities, better management of water resources and judicious utilization of mineral and energy resources without dangering their bases. This implies socialization of the earth sciences. The motto of an environmental geologist in the words of John O'Leary and Peter Flawn (1970) should be:

Thou shalt not tear up the land,

Thou shalt not contaminate the water,

Thou shalt not pollute the air,

Thou shalt provide to Society the resources it needs.

SCOPE

ological factors greatly influence man and his activities. **The discipline of environmental geology embraces the whole gamut of human use of the earth—the prediction, assessment and combating of**

natural hazards, and the application of technology and engineering for the development of society without destruction of the environment. A geologist is well equipped to evaluate the capabilities of the earth for providing water, minerals and other vital resources, to select appropriate sites for construction of buildings, dams and bridges, to choose suitable places for the disposal of wastes and pollutants, to identify land capable of supporting agriculture, agroforestry, industries, urban centres, etc., and to ensure a harmonious relationship between man and his environment. T seven major areas in which the role of earth scientists are of vital importance (Fig. 1.1) are identified the following sections.

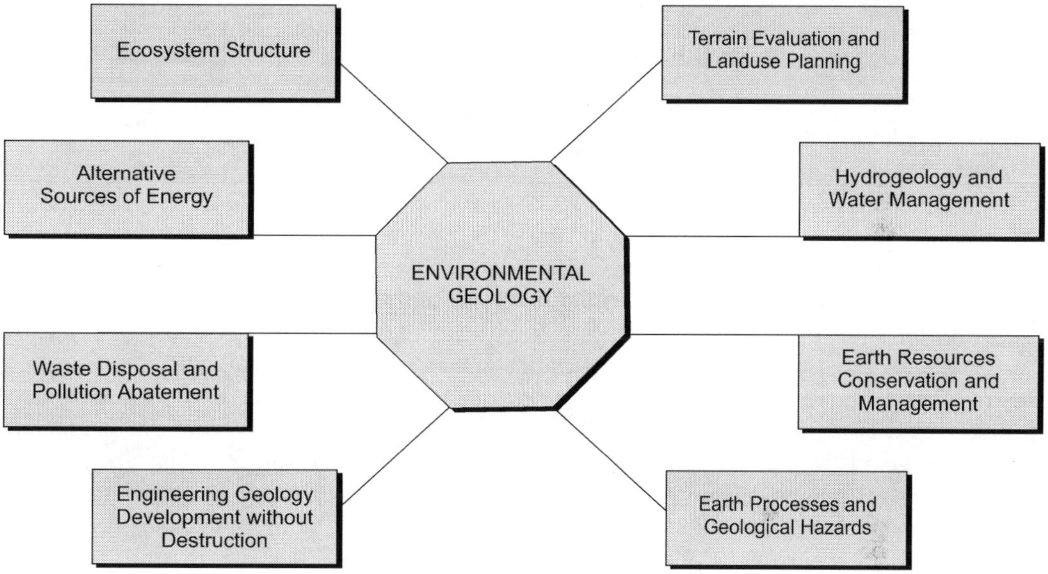

Fig. 1.1. The gamut of environmental geology

Ecosystem Structure

Man is a part of nature and geology is concerned with the study of nature, its evolution, dynamics and its resources. It is to the advantage of man to live in harmony with nature and share the limited resources of the earth with his fellow human beings and other organisms. He needs to realize the danger of actions that initiate changes that counter the normal evolution of the environment, and needs to exercise caution in embarking on ventures that have a direct bearing on the ecological balance. The sages of Vedic tim perceived interlinkages between the living and inanimate entities such as mountains, rivers, wind and sun which form the matrix of the environment. The dislocation of any one of these links would have impact on the other entities.

Appreciating the fact that the biological communities and their non-living environments function together as ***ecological systems or ecosystems,*** and realizing that man is totally dependent on primary producers (plants), he should be willing to share ungrudgingly the earth's resources with other beings—plants, animals and birds. This calls for an understanding of the mechanisms of various ecosystems, t eds and capabilities of producers (mainly plants), consumers (chiefly man) and decomposers (such as cteria) which form the components of the ecosystems. The study of the relationship between organisms in their original habitat or environment and their functional processes is the subject of ecology.

The most important element of ecological studies is the conservation of living resources in order to ensure a continuous yield of useful plants and animals and materials by establishing a balanced cycle harvest and renewal (Odum, 1959). This implies scientific management of forests, pasture farms, wildlife and livestock.

Mahatma Gandhi spelled out the essence of ecology in the following words, "In nature there is enough for everyone's need, but too little for everyone's greed."

errain Evaluation and Landuse Planning

Over geological ages, various geologic processes have shaped the terrain as it is today. Organisms have ade considerable modifications of the terrain in their use of it in accordance with their needs, but not necessarily in compatibility with the capabilities of the land and availability of resources. In this process man has caused extensive topographic changes and in some places even distortions.

The first step in planning for proper use of the terrain is the analysis of geomorphic condition, the nature and extent of weathering of rocks and development for soil cover, the slope condition, and the drainage network. Secondly, the suitability of sites for various kinds of uses is determined on the basis of analysis of the situation and likely changes that will take place as a result of construction activities, farming, grazing, deforestation, mining, industrialization, and urban expansion, etc.

The traditional "bhumi puja" of the Hindus performed before the initiation of construction activities mphasizes the desirability of ensuring the integrity of the land. The only way of maintaining the egrity of the land is by using it in accordance with its capability and carrying capacity. The primary requisite for landuse planning is thus the analysis of its capability on the basis of the study of rocks forming the foundation, the nature of soil constituting the mantle, the stability of the slopes, the quality d quantity of water stored underground or flowing on the surface and the nature of the vegetal carpet growing on the land.

Hydrogeology and Water Management

Water is not only the most effective and powerful agent of geomorphic changes, it is also the sustainer of life of all living beings. Availability of water is therefore the mainstay of the ecosystems of living ing. In Vedic literature a river is described as "jahnvi", or a life-giving and life-supporting system. e invocation in Hindu rituals to the rivers Ganga, Yamuna, Saraswati, Godavari, Narmada and Kaveri speaks volumes of the respectful place the water-systems occupied in the Indian society.

Since this resource of the required quality is available only in a limited quantity.—just 2.7% of the total water on earth—and in limited areas, its judicious use without despoiling the purity, freshness and source, and the strict avoidance of waste are essential elements of water-management planning. e qualitative and quantitative appraisal of surface and groundwater resources and the analysis of conditions suitable for movement and accumulation of this precious commodity is thus an essential subject of environmental geology. Related to this is the stress that the cultural and industrial activities of man have imposed on the water-balance and on the resource base itself—the problem of drying up of springs and wells, excess of water content in soils, the pollution of water, etc.

The geologist is required to provide, and is indeed capable of, gathering information for the planni and development of safe, adequate and continual supply of water in places that will serve the population. Hydrogeology is thus an essential component of environmental geology.

Earth Resources—Conservation and Management

Soils and minerals are the two categories of earth resources that are rapidly depleting and are non-renewable. Soils support the whole system of the living world through the supply of nutrients and moisture. The two are so intricately interwoven that many types of soils have become the livi foundation of ecosystems. Realizing that it takes a very long time and requires specific sets of clima and physicochemical conditions to produce even a thin veneer of soil, it is our primary concern to preserve that soil carpet, and save it from deterioration due to mismanagement. The subject of soil scien (pedology) involves understanding the processes that form the soils, their textural and compositional characters; and their capacities for supporting different types of plants and different uses such as construction of structures, roads, disposal of wastes, agriculture, horticulture, etc. The environmen geologist is thus required to identify the soils that can support different activities, and carry water within their pore-spaces and have the requisite strength to withstand the loads of the artificial structures bei built on them.

Minerals constitute a precious heritage of the geological past, playing a very crucial role in the destiny of man. They are inseparably intertwined with human society which consumes enormo quantities of minerals of diverse types, unearthing and excavating from different parts of the earth and thus modifying the topography drastically. Since the minerals are wasting assets, the assessment of the potential of existing reserves and planning their exploration and extraction in a manner that would ensure their availability for future generations are the two essential elements of resource management. Also related to this exercise are the endeavours to find new locations of mineral deposits, and developme of techniques of extraction that will ensure minimal or no pollution, degradation and erosion. T exploration of mineral resources requires an understanding of the nature, origin and evolution of the earth's crust and the relationship among different tectonic regimes—magmatism, metamorphism and metallogeny—the subject matter of economic geology. Thus an environmental geologist is also economic geologist.

Earth Processes and Geological Hazards

Any natural process that changes suddenly may have harmful effects on the community, and thus pose a hazardous threat. Environmental management planning incorporates a programme of identifying the nature and extent of the hazards, the factors which promote, initiate and accelerate the sudden harmful events such as landslides, floods, avalanches, etc., and of monitoring change that threaten the safety survival of the biotic communities, and predicting the time, location and magnitude of hazards so that hazard-preparedness and emergency operation plans can be formulated. An evaluation of the risks likely to be involved is also undertaken.

The hazards from such natural phenomena as earthquakes, volcanic eruptions and cyclones a inevitable and inescapable, and can be combated with effective preparedness plans. However, the incidence of human-induced changes such as landslides, subsidence of ground, floods and desertificati can certainly be minimized and their impacts abated by desisting from ill-advised tampering with the natural order or balance. The analysis of the condition of slopes, lithology, rock-structures and clima provide solutions for the minimization of hazards. Geologists who recognize the conditions that trigger hazardous events, can offer plans to cope with and combat such hazards.

Engineering Geology—Development without Destruction

Man has been effecting major changes in the terrain in his pursuit of a higher standard of living or a better quality of life, by developing cultural features, building utility and residential structures, improving mmunication and transportation, meeting energy requirements, augmenting irrigation, etc. The side effects and backlash of these development programmes have in many cases proved counterproductive. On the basis of terrain condition, stability of slopes, and endurance of the ground to withstand anticipated load or stress, the engineering geologist selects a site where there will he no damage, or the anticipated anges will be minimal, and recommends means to counter the harmful side-effects. Thus the principal aim of the engineering geologist is to ensure development without destruction, without jeopardizing the stability and integrity of the ecological balance and without despoiling the nature's beauty.

aste Disposal and Pollution Abatement

Man is a producer of colossal amounts of wastes and pollutants through varied activities in agriculture d industry and through his over-consumptive lifestyle in urban centres. The growing volume of wastes eatens not only to deface the landscape but also to pollute the atmosphere and the water supply. The vironmental geologist is required to determine the factors influencing suitability of sites for disposal of wastes and pollutants, and identify causes of pollution, particularly of water moving underground.

The fast depleting deposits of energy-minerals and the need to minimize pollution from conventional methods of energy generation have resulted in a keen demand for alternative or renewable sources of energy such as the sun, winds, tides, biomass etc.

Knowledgeable of the principles and practices of meteorology, an environmental geologist is well equipped to provide guidance on selecting suitable sites for tapping the pollution-free and inexhaustible sources of energy.

SUMMARY

e geologist plays a very crucial role in understanding, preserving, ameliorating and restoring the natural environment. He has adequate comprehension of the dynamics and mechanisms of the various earth processes that not only give rise to mineral deposits, form diverse soil types, permit movement of surface and underground waters, but also shape terrain morphology and fashion landscapes.

The knowledge of the character and capabilities of the terrain including the conditions of soils and eir stability permits scientifically sound and ecologically appropriate use of the land for maximum nefit with minimum damage.

A geohydrologist explores the ground to provide society with life-supporting water of requisite ality in adequate quantity. Knowing the problems related to underground water distribution and flow, and the sickness of waterlogged soil, the geohydrologist suggests remedial measures and techniques for augmenting supply of water on the surface as well as underground.

In the management of mineral resources, the economic geologist helps the society by ensuring predictable and plentiful supply of minerals, which constitute the foundation of all industry. Not only does he identify locations of deposits and suggest methodologies of extraction without harm to the environment, but also suggests ways and means of conserving the non-renewable, fast depleting resources.

As a practitioner of engineering geology and geotechnics, the environmental geologist provides insight into ground conditions, their stability and capability to bear loads and stresses, and identifi sites suitable for buildings, roads, bridges, dams and utility structures. He is thus an active but critical participant in all programmes of development without destruction.

He has also to determine the factors that influence the safety and stability of structures constructed man. The knowledge of the dynamics of earth processes equips the geologist well to evaluate the haza potential of an area, and plan strategies for mitigating risks to life and property.

The disposal of wastes and pollutants is another area in which the geologist is competent to provi guidance in selecting sites which would not create problems of poisoning the soils and polluting the waters.

Environmental geology is thus an integrated application of geology for the benefit of man an his living and inanimate world.

2 Ecological Perspective of Environment

***Some cynics say that ecology is a fad.
If it is, it will be the last fad.***

—G. T. Mller and P. Armstrong (1982)

MEANING OF ECOLOGY

Ecology is the science of the living environment concerned with the relationship between organisms and their habitats and the functional processes of organisms on land and in water. The word is derived from the Greek word ***oikos*** meaning a place to live. Ecology thus deals with the analysis and interpretation of the mutual relationship between organisms and their environment.

The two structural units of nature — the living component comprising biotic communities, and the non-living or abiotic component — embracing soil, water and air — function inseparably in the biosphere ***ecological systems*** or simply ***ecosystems***, such as forests, grasslands, farms, islands, deserts, lakes, ers, lagoons, etc. The driving forces in the ecosystems are the energy flow and the biogeochemical cycling involving carbon, oxygen, nitrogen, phosphorus, and other macro- and micronutrients like manganese, boron, copper, iron, etc. The interaction of the two components involves transformation of matter and energy into new forms and their transfer along the food chain ensuring stability for the system (Odum, 1959).

BIOTIC COMMUNITIES

Food Chain and Interdependence

There are two main groups of organisms in the biotic communities. One group is of self-nourishing organisms (***automorphs***) which utilize simple inorganic substances and build up complex substances in the presence of light energy from the sun, e.g. green plants ranging from algae and diatoms to giant trees. These are the producers (Fig. 2.1). The other group consists of consumers or ***heteromorphs*** (animals) and decomposers (bacteria and fungi), which being incapable of producing their own food, utilize the organic build-up of plants or devour one another, or rearrange complex substances to yield simpler usable substances. The producers, consumers and decomposers are associated together with inter-dependency in a functional unit with abiotic components to form the ecosystem. The transfer of energy from one trophic level to another constitutes the *food chain* (Fig. 2.2). The **food chains** link all

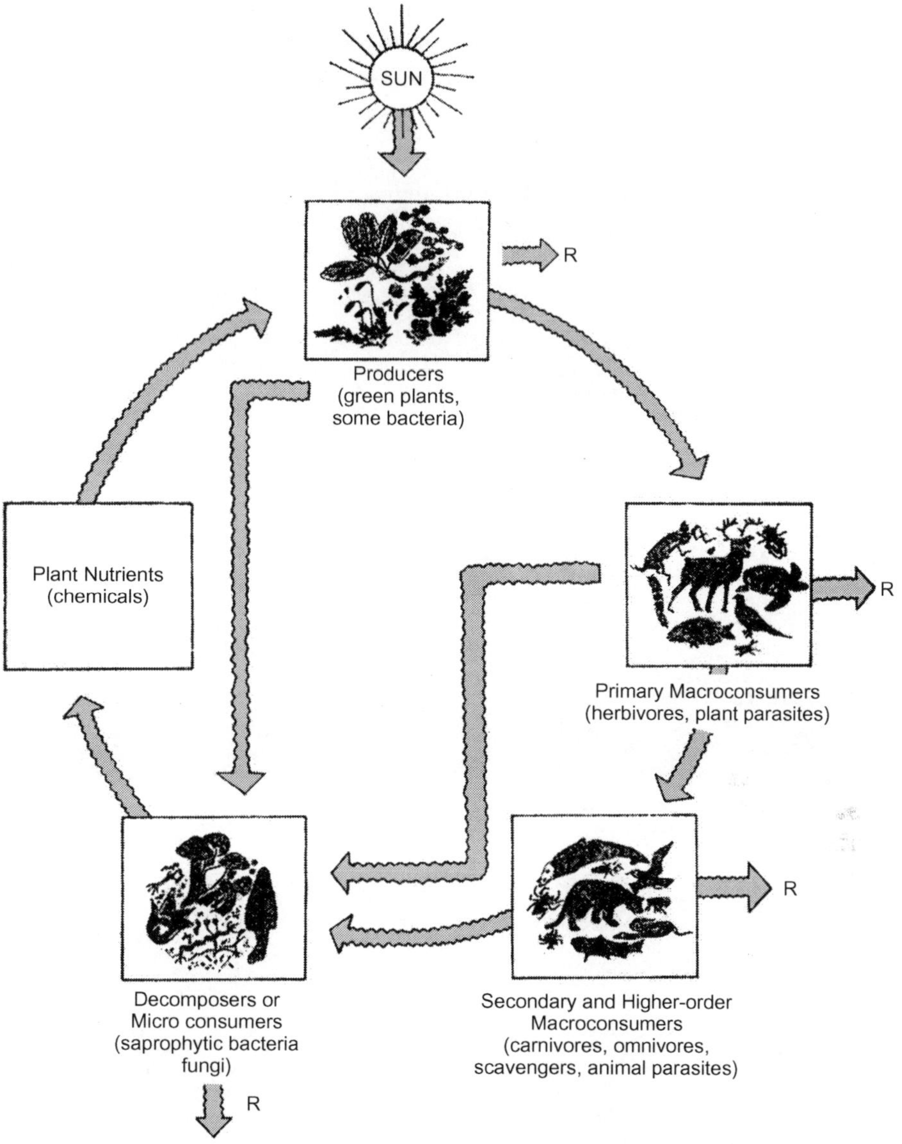

Fig. 2.1. The three members of the biotic community (producers, consumers and decomposers) are interdependently linked with the abiotic components to form a functional unit called ecosystem (Based on G.T. Miller and P. Armstrong, 1982).

other living beings to green plants. **Humans occupy the highest place in the chain, which means that man is inescapably and helplessly dependent on the other components of the system**. This is reflected in the following passage from the "Yajurved" (36/18):

दृते दृह मा मित्रस्य मा चक्षुषा सर्वाणि भूतानि समीक्षन्ताम्
मित्रस्याऽहं चक्षुषा सर्वाणि भूतानि समीक्षे
मित्रस्य चक्षुषा समीक्षामहे ।
यजुर्वेद (36/18)

"May all beings look on me as friend
May I look on all beings as friends
May we all look on one another with friendship

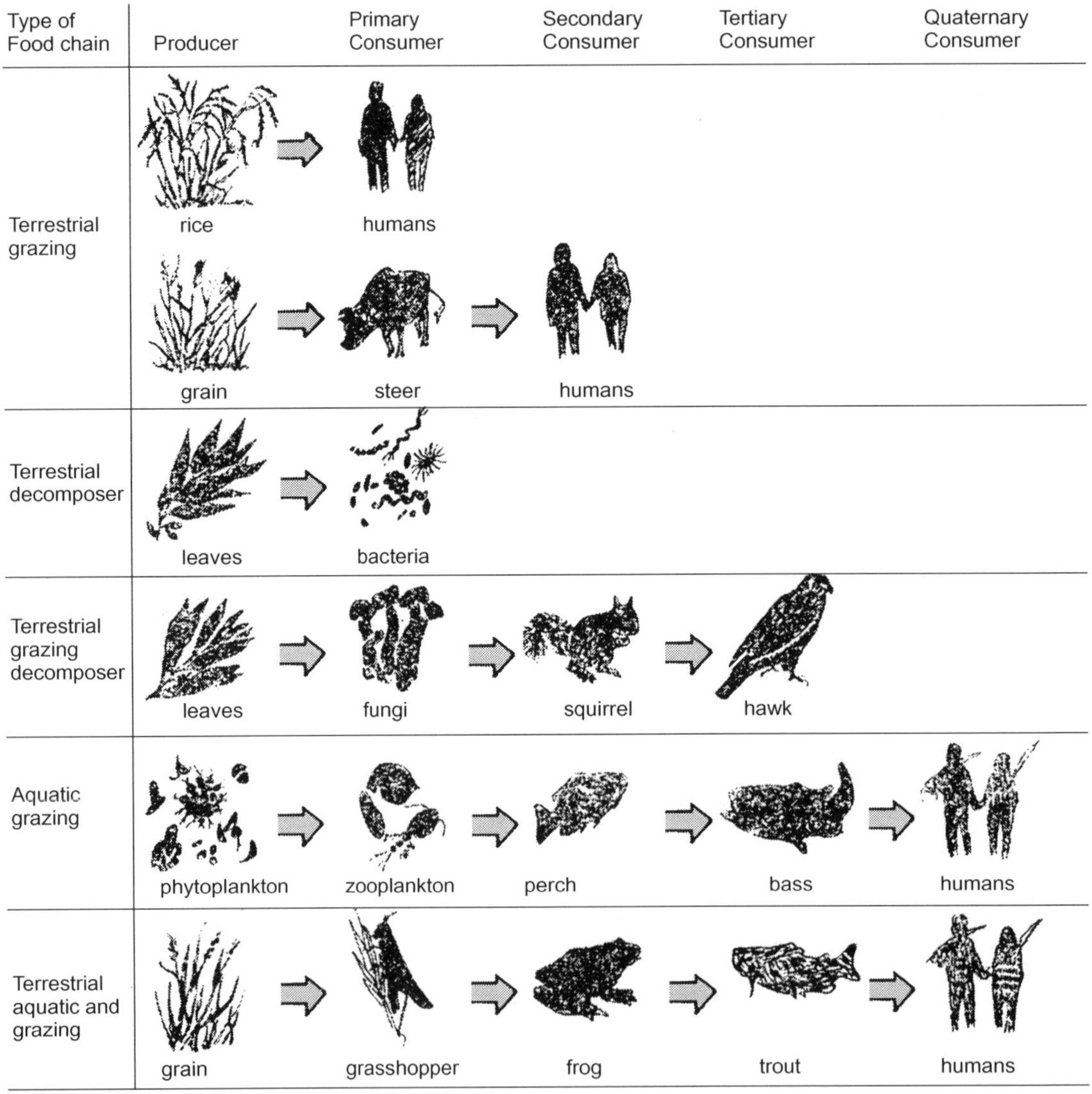

Fig. 2.2. Man occupies the highest place in the food chain, which links all living beings to green plants, the primary producers. (After G.T. Miller and P. Armstrong, 1982, *Living in the Environment*, Wadsworth, Belmont, 491 p. (Fig. 5.5)).

Stratification in Biotic Communities

Biotic communities exhibit stratifications. The upper sunlit zone is dominated by birds and se nourishing producers, the lower level by the consumers and decomposers, and the several intermediate strata provide habitat to a wide variety of organisms. The greater the stratification the larger would the habitats available and the lesser would be the interspecific competition, so that much larger numb of species would utilize energy effectively. Taking the example of a forest (Fig. 2.3) the soil suppo the decomposers, the middle stratum comprises herbs, shrubs and shorter trees, and the upper stratum is made up of taller trees. Each of these strata has its own community of animals, birds and insects. Lack of adaptability hinders, if not prevents, mobility from stratum to stratum—station to station. **Only humans have greater adaptability, and therefore are capable of living in a variety of habitats and strata. This is one of the reasons for man's pre-eminence in the organic world**.

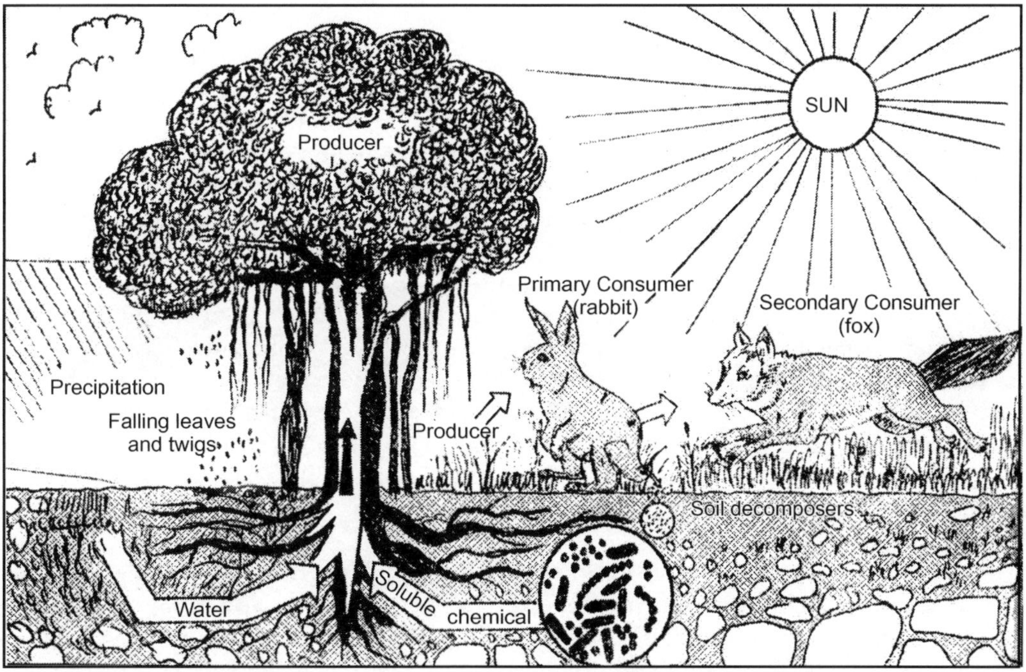

Fig. 2.3. Stratification in a forest ecosystem. Each stratum has its own community of creatures, and with the ecosystem the biotic community is interdependent. (Based on G.T. Miller and P. Armstrong 1982).

Community Stability

The welfare and survival of an organism depends on the integrity of the community. **Any modificati of the habitat changes the character of the community, throwing the organisms to stress**. For example, the lowering of water level of a pool is a modification in the habitat, and this change cur breeding of mosquitoes far more effectively than direct poisoning. Likewise, the introduction of a species into an alien habitat causes harm to both the species and the ecosystem, for the new system may not have predators or parasites and may thus grow uncontrollably to become a pest. The *Lantana* shrub ought from Australia to build decorative garden hedges in India has become a difficult weed, ravaging ricultural fields and forests in the Lesser Himalayan terrain in Kumaun and elsewhere in India.

Miller and Armstrong (1982) sum up this phenomenon succinctly: "**the survival of the fittest does not mean survival of the strongest or the most aggressive; it simply means that the species best able to adapt to changing circumstances has the best chance to survive and produce successful offsprings.**"

The welfare of man — who rides high on the food chain and depends on plants and other animals for food — depends on the stability of the biotic community. Therefore, any modification that adversely affects the ecosystem will also affect his life, welfare, culture and his very existence.

Species Diversity

Diversity in terms of the type and number of species living in a habitat is a characteristic feature of any biotic community or ecosystem. If the species diversity is limited in the desert ecosystem, it is understandably greatest in the rain forests. **The greater the specific (or genetic) diversity, the greater ll be the potentiality of the community for adapting to changing or fluctuating conditions, and the stabler would be the ecosystem. The larger number of species is an insurance against decline or extinction** as some of the species are bound to survive disease, poisoning (e.g. from spraying), destruction or invasion, and they pass on their resistance to their progeny. This is why there is a need for preserving the genetic variety in our various ecosystems such as the rainforests of the Silent Valley of e Sahyadri in Kerala and the Valley of Flowers in the Badarinath area in Uttarakhand Himalaya.

The Sahyadri mountain domain is one of the world's top 34 hotspot of biodiversity, with 4000 plant ecies (of which 1500 endemic including those of species and medicinal), very large number (120) of ique or endemic (14) mammals, reptiles and amphibians, and more than 40 ethnically distinct groups people (Bawa et al, 2007). Indeed like in the eastern Himalayan biodiversity hotspot, there are more an 30% of India's total plant species here in the Sahyadri. One might describe it as a veritable mine valuable genes. It is disconcerting to know that the Sahyadri has lost 75% of its original forest cover awa et al, 2007).

On the other hand, **ecosystems with a limited number of species such as the monocultures of pines, eucalyptus, poplars, food crops, etc. are very vulnerable to diseases and other external stresses ch as fire, flood and pollution**. This is one of the reasons why ecologists discourage monocultural antations of pines in the Himalayan region or eucalyptus in the plains and plateaus.

There is another, though selfish, reason for maintaining genetic diversity. Since it cannot be predicted which organism or plant species will in future become more important or useful as an alternative source of food, medicine or the like, it is advisable to preserve as **gene pools** all or most of the species in their tural habitats in the biosphere reserves (Miller and Armstrong, 1982). At present, the emphasis is on a few high-yielding, fast-growing varieties of crops. The rate at which the natural and semi natural vegetation containing the wild relatives of crops is disappearing is a cause of concern as no gene-pool will be left for future hybridization programmes. In case of an epidemic, the entire population of new varieties could be wiped out, leaving the agriculture high and dry and the evolution of new varieties would not be possible in the absence of wild or semi-wild stock.

ANTHROPOGENIC CHANGES IN ECOSYSTEMS

By developing an adaptability to live and survive in different ecological situations (strata), **man has become the most powerful member of the biotic community. But he remains the most selfish of the creatures as he is most inconsiderate towards other members of the biotic communities, and is very destructive in his own habit and predilections** (Fig. 2.4). For his own ends, **man has been effecting the most drastic modification in the composition of the ecosystems—by lopping and defoliati trees, by spraying insecticides on shrubs and pesticides on vegetables, by discharging pollutants into rivers, lakes and air, by over-irrigating agricultural farms, by impounding flowing waters behin dams, by introducing species in alien habitats, and so on. Admittedly, many of the adverse changes were inadvertent; nonetheless the consequences were far-reaching and severe.** Take the example of the reckless killing of crocodiles and hippopotamuses in African rivers. Digging deep in the rivers, these animals stir up sediments and keep the rivers free of clogging and enable plant growth, so that rivers are able to support large populations of fish and other organisms. The large-scale killing of crocodiles a hippopotamuses led to disappearance of fish and consequent destitution of the people living on them. T impoundment of the Nile River behind the Aswan Dam had brought about a proliferation of the snail that harbours a dreadful parasite blood-fluke (flatworm) causing the debilitating disease schistosomiasis ov the vast irrigated terrain in Egypt. This "small act" had brought half the population of Egypt to the bri of a mortal disease (Fahim, 1980).

Fig. 2.4. Though man has become the most powerful member of the biotic community, he remains the most selfish of all creatures, the most inconsiderate to his fellow beings and destructive in his behavior. (Fro G.T. Miller and P. Armstrong, 1982, *Living in the Environment*, Wadsworth, Belmont, 491 p., Fig. 3.3)

A recent study brought out an example of danger to food chain due to our own doing. There has been precipitous fall in the production of honey by bees (from 300 tonnes down to 50 tonnes), and decrea in the yield of date palm, lemon, papaya, '*kesar*' and mango in Kachchh. Almost similar is the case Bengal. This is the consequence of decline in the population of bees which play very crucial role in the cross-pollination of crop plants and nut trees. The bees died due to the use of pesticides like clothianidin, meonicotinoids, etc. (*Deccan Herald*, January 1, 2011). Nearly 7 million hectares of farmland in t most fertile plains of Punjab, Rajasthan, Haryana, Uttar Pradesh, Andhra Pradesh and Tamil Nadu, ha gone nearly out of production following irrigation beyond the receptivity of the soil in the semi-arid tracts. The water table having risen dangerously close to the surface, salts and alkalies have moved up and accumulated in the soils, damaging them forever.

How the water-lily or water-hyacinth, a plant brought into India as a decorative plant, is playing havoc with the water courses and water bodies in large parts of the country is a common knowledge. **ere is yet another menace creeping up the slopes of the Himalayan mountains. The dwindling and** local elimination of the highly desirable oak trees has resulted in the increase in population of pine trees. Although the pine is a commercially very useful plant, ecologically it is quite inadequate and irrelevant in the mountains, or it is unequal to the task of stabilizing the slopes and preventing erosion, and stimulating recharge of groundwater, and nurturing species diversity. **The elimination of the biologically mature trees has led to the decline of forests in all parts of the country**.

DYNAMICS OF THE HUMAN POPULATION

Explosively Fast Growth

As already stated, man has achieved a pre-eminent and predominant position through his uncanny **aptability to a wide variety of ecosystems, and his ability to manipulate the ecosystem to his benefit.** The result has been an **unchecked growth of human population** (Fig. 2.5). The current annual rate

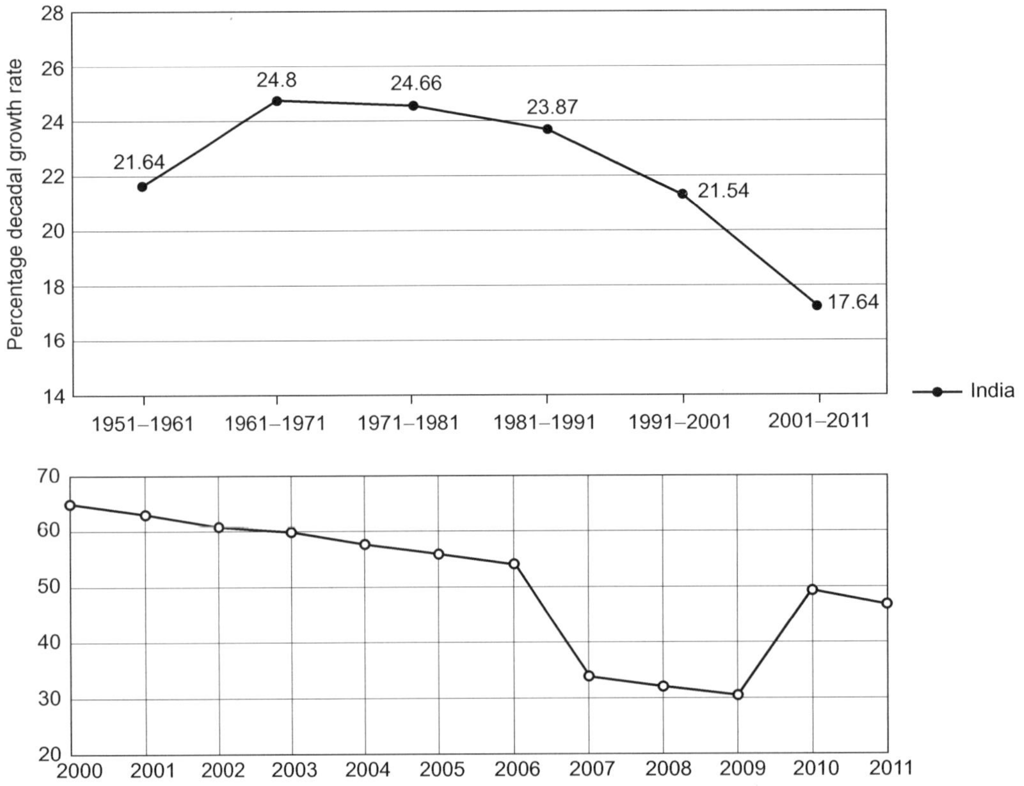

Fig. 2.5. *Upper*: Percentage decadal growth of the population of India. (From: Report of Registrar General of Census, 2011) *Lower*: The rate of infant mortality (per thousand birth) continues to be high — 47.57 infant per 1000 live birth (in 2011) indicating the poor state of health of the nation. (*Aquastat* Survey, 2010).

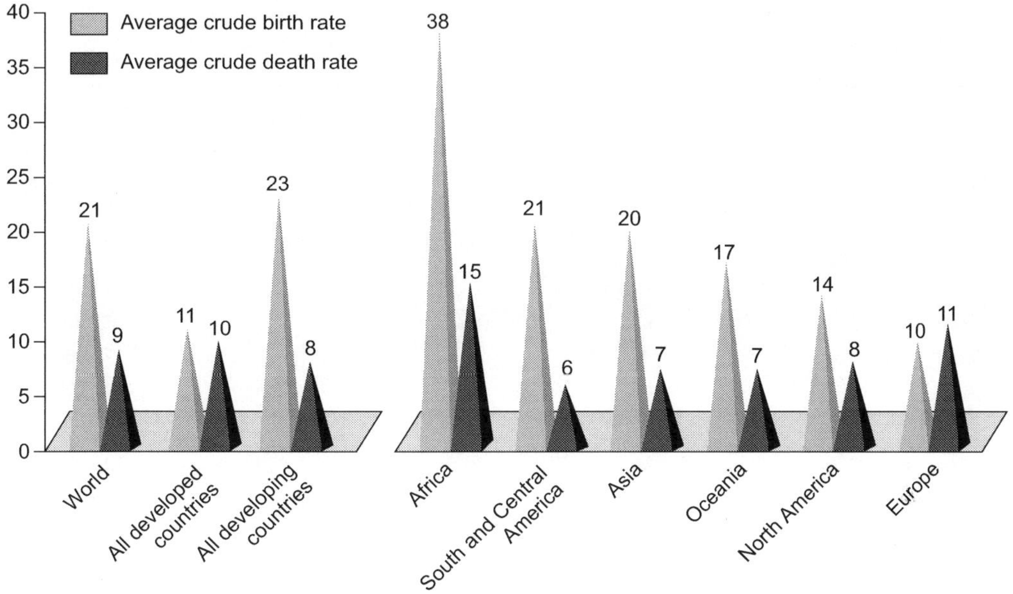

Fig. 2.6. Average crude birth and death rates in developed and developing countries of the World. (Based on: Tyler Miller, 2007. People and Environment, Cengage Learning India, New Delhi, 696 p.)

of increase of the world population is 1.1% (Bawa et al, 2007) and of India 1.764% (Census Repo 2011). With its finite resources and diminished **carrying capacity, the earth is called upon to support 19,71,200 extra persons every day** (Fig. 2.6). The present 7 billion strong global population is project to be about 9.3 billion by A.D. 2050. The growth rate of the Indian population 1.764% is very high (s Fig. P.5). **Every year India has to take the responsibility of feeding, clothing and sheltering some 27.8 million additional persons** (Fig. 2.7).

About 17.4% of the world population is densely packed in the 2.4% geographic area that is in Ind The population of India was 350 million in 1949 and today it is more than 1210 million. In 1981, of t more than 500 million people crowding the world, about 37% were below the age of 15, who had th the potentiality of procreating as prolifically as their parents did. Today on an average 382 persons li a crowded life in one square kilometre land.

If this growth rate is allowed to remain unchanged, Indians will number 1690 million in A.D. 20 when the population is expected to stabilize (Jain, 2011). **The galloping rate of population growth is not only the primary cause of poverty and hunger, but also is a dreadful threat to all ecosystems that the biosphere supports. The limits of endurance have been reached and our ecosystems cannot withstand further invasion and exploitation, and certainly cannot support more people than the earth already has**. This is evident from the fact that about half the world population is below the poverty line, about one-tenth suffers from hunger, and are absolutely poor (Fig. P. 6).

ble 2.1. Distribution of Population, Density and Decadal Growth Rate of Population: 2011 (Census Report, 2011).

State/Union Territory	Total Population	Density (per sq. km)	Decadal growth rate
ammu & Kashmir	12,548,926	124	23.71
imachal Pradesh	6,856,509	123	12.81
unjab	27,704,236	550	13.73
Chandigarh	1,054,686	9,252	17.10
ttarakhand	10,116,752	189	19.17
aryana	25,343,081	573	19.90
NCT of Delhi	16,753,235	11,297	20.96
ajasthan	68,621,012	201	21.44
ttar Pradesh	199,581,477	828	20.09
ihar	103,804,637	1,102	25.07
kkim	607,688	86	12.36
Arunachal Pradesh	1,382,611	17	25.92
Nagaland	1,980,602	119	-0.47
anipur	2,721,756	122	18.65
zoram	1,091,014	52	22.78
ripura	3,671,032	350	14.75
eghalaya	2,964,007	132	27.82
Assam	31,169,272	397	16.93
West Bengal	91,347,736	1,029	13.93
harkhand	32,966,238	414	22.34
Odisha	41,947,736	269	13.97
Chhattisgarh	25,540,196	189	22.59
adhya Pradesh	72,597,565	236	20.30
ujarat	60,383,628	308	19.17
aman & Diu	242,911	2,169	53.54
adra & Nagar Haveli	342,853	698	55.50
aharashtra	112,372,972	365	15.99
Andhra Pradesh	84,665,533	308	11.10
arnataka	61,130,704	319	15.67
oa	1,457,723	395	8.17
akshadweep	64,429	2,013	6.23
erala	33,387,677	859	4.86
amil Nadu	72,138,958	555	15.60
uducherry	1,244,464	2,598	27.72
Andaman & Nicobar Islands	399,944	46	6.68
otal	1,210,93,442	382	17.64

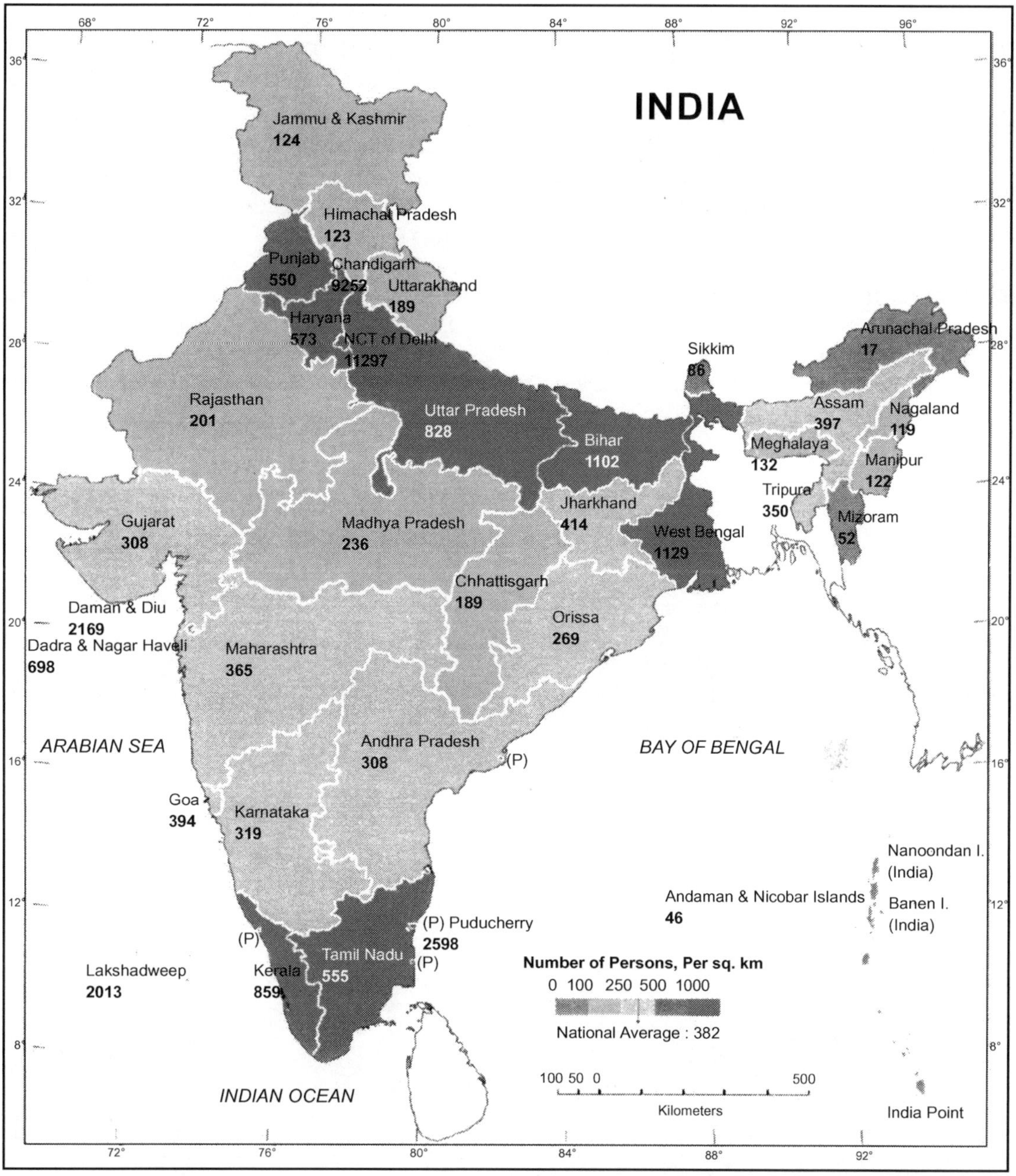

Density of Population, India, States and Union Territories: 2011

Fig. 2.7. The average density of the population in India is 382 persons per square kilometre. India is thus supporting and sustaining nearly 17% of the world's population. Bihar is the most densely populated state (1102 persons/km^2) and Arunachal Pradesh the least (17 persons/km^2) (The Census Report, 2011).

(See colour Fig. 2.7)

ungry India

cording to recent Report of the United Nations Development Programme (UNDP), if we look at the present rate of changes, **the number of the poor will be 279 million or 22.1% of the total Indian population by A.D. 2015, 40% of the children will remain underweight because of malnutrition, d the infant mortality will be 7% — far short of the target of 2.7% (*New Indian Express*, ptember 13, 2010). The sad fact is that nearly 230 million people (410 million according to Jason Burke, *The Hindu*, July 15, 2010) of the total population of 1210 million today remain hungry ery day (Fig. 2.8), and the infant mortality in 2010 was 38.3 per thousand** (Kounteya Sinha, *The Times of India*, February 2, 2012). **The report of the National Sample Survey-2012 brings out the stark reality that two-third population of India is hungry — in terms of malnutrition**. In the year 07 per person total cereal consumption was 174 kg in India, compared to 289 kg in China and 889 kg USA (Usha Patnaik, *The Hindu*, September 13, 2010). **Prime Minister's Advisory Council member G.K. Chadha admitted that India is a home to the world's largest food-insecure population with more than 200 million hungry people** (*The Hindu*, October 17, 2008).

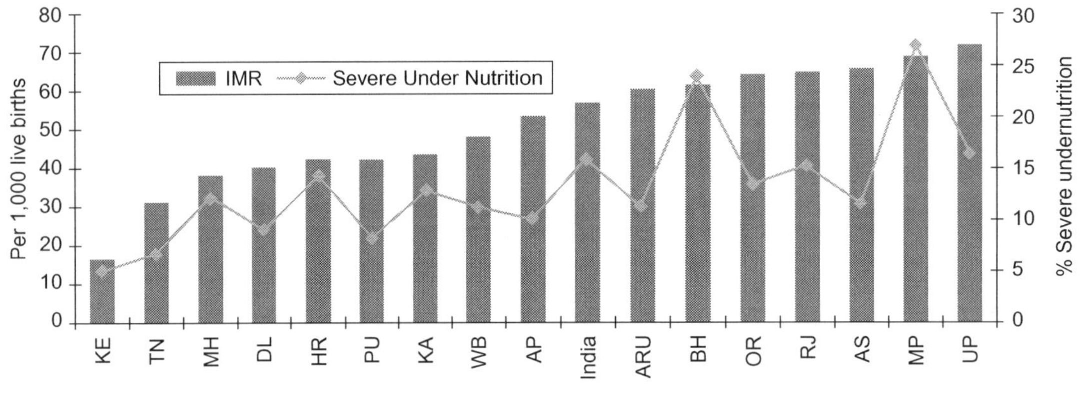

Source: Refrence 2.13

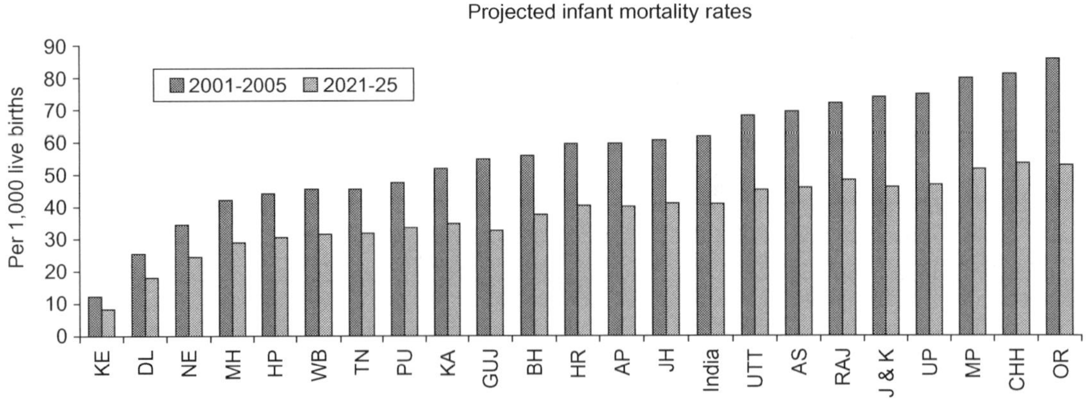

Fig. 2.8. Infant mortality rate (IMR) and severely undernourished children under five years in different states of India. (National Family Health Survey, www.nfshindi.org, 2007; Central Bureau of Health Intelligence-India, www.cbhididghs.nic.in, 2007).

"In terms of quality of life India scored 66th place in the 2008 Global Hunger List (HDI) of countries" — worst than nearly 25 sub-Sharan African countries and all of South Asia, except Banglade (Gargi Parsai, *The Hindu*, October 17, 2008).

The rate of infant mortality in 2011 was 47.57 infant per 1000 live births speaks poorly of the hea of the nation.

It is more than obvious that despite India's increase in Gross Domestic Product (GDP), its ranking in Human Development Index (HDI) is frightening low (0.519 in 2010) (*Aquaslat* Survey, 2010). In oth words, **there is vulgarly disproportionate increase in the affluence of the rich and powerful peop driven by greed and overambition at the expense of the poor**. It is the poor for whom the health of the environment matters very much, but the environment has deteriorated so much that the poor people have lost this very mean of their sustainance. **It is the poor people who have been uprooted from their forest habitations to make way for mining by corporates and multinationals, and forced to leave their lands and homes and give up livelihood due to building of large dams for water for irrigation and electricity primarily benefiting the rich people**.

These poor people are constrained to destroy the very source from which they scrape their living by using the ecosystems beyond their carrying capacity. With the resource bases severely endangere the only way of halting the inexorable march towards a collapse of the biosphere is to rigorously check population growth by means of stringent measures. Lincoln Day's words are relevant here, "Reprodu tion is a private act, but it is not a private affair: it has far-reaching social consequences." **India has no option but to cry halt to the growth of its population**.

In the Indian Himalayan region the population was nearly 47 million — 4% of the Indian populati in 2011. There was no much difference in the decadal growth in the Indian Himalaya (17.34%). Wh the population density in the country is 382 persor/km^2, that of Indian Himalaya (including its Bhabh and Tarai belts) is 181 person/km^2, according to 2011 Census Report. The density of population in t mountainous part of the Himalayan states would be far less than 181 person/km^2.

PRESERVING GENE POOLS AND CONSERVING ENDANGERED SPECIES

Concern for Wildlife in Indian Mythology

Genetic diversity, it may be recapitulated, imparts stability to the ecosystem and ensures its integrity. Genetic diversity cannot be maintained without preserving the gene pools — "the irreplaceable capita as Dilip Dey (1977) called them—in undisturbed conditions in biosphere reserves. Those species who existence is threatened due to the demands of humans must be saved and multiplied in the sanctuaries or wildlife parks.

Hindu mythology brings home a universal concern for the wildlife. All gods and deities have th respective "vaahans" or vehicles, either animal or bird: Vishnu flies on Garud (the eagle), Shiv rides Nandi (the bull), Lakshmi is mounted on the owl, Saraswati on the swan, Durgaa on the tiger, and so on. The association of gods with their 'vaahans' speaks eloquently for the concern or respect for wildlife in ancient India. The tender regard shown to Naag (snake), Kaakbhushundi (crow), Hanumaan (monkey Ganesh (elephant) and the respect for cows indicate the niches such animals and birds occupied in Hindu philosophy. The Rigved (7/54/1) exhorts —

शं नो भव द्विपदे शं चतुष्पदे

May bipeds and quadrupeds live in happiness.

ndangered Species

most 12% of the world's remaining bird species, 25% mammals, and 31% amphibians are eatened with extinction (Bawa et al., 2011). Likewise, the endemic plant species of eastern malaya (including Meghalaya) and Western Ghat are facing serious threats of being wiped out. As matter of fact, 65% of the primary forest habitats in tropical Asia has been lost due to deforestation andit et al., 2007).

India presently harbours about 15,000 species of higher plants and 80,000 species of animals, nstituting respectively 5.2 % and 4.3 % of all the known plant and animal species found on the rth (Gadgil, 1980). About 134 species of plants (99 from the Himalayan region) and 29 species of ammals, 29 types of birds and 3 species of reptiles are threatened, the greatest erosion of diversity curring in the drier tracts (Gadgil and Meher-Homji, 1983). There is a considerable decline of biota under the woodlands of the *Acacia-Anogeissus latifolia* series of the Maharashtra plateau, and of the *Manilkara-Chloroxylon* series of the eastern coast. An inhabitant of these ecosystems, the 'cheeta', nished some eight decades ago (Fig. 2.9) and the wolf, the Great Indian Bustard, and the floricans have become very rare. The common village bird sparrow is now not seen in cities and many big towns. e population of vulture is fast diminishing. Many of the evergreen forests in the Western Ghat have t their original architecture (Gadgil and Meher-Homji, 1983). The lion-tailed monkey inhabiting the rain forests of South India is on the verge of extinction, and the Nilgiri tahr, Malabar civet and 'Nilgai' in the south-western coastal belt are endangered.

The tiger (*Panthera tigris*) population had dwindled from 40,000 at the turn of the century to a few thousand today. The tigers are in great danger of being wiped out by ruthless poachers and plunderers today they are less than 40% of what was their numbers in 1995 (Bawa et al, 2011). The 'gharial' or crocodile and Gangetic dolphins of the Sundarbans are threatened with extinction. The numbers of lizards, snakes, crocodiles, tortoises and turtles in the rivers of the Indo-Gangetic Plains have greatly diminished. The one-horned rhinos (*Rhinoceros unicornis*) that once roamed the vast Indo-angetic floodplains extending from Pakistani Punjab to Assam is today restricted to just small areas in southeastern Nepal and Assam (Fig. 2.9). Likewise, the once abundant vulture (*Gyps indicus*, *Gyps bengalensis*) are dying at an alarmingly fast rate — and some are on the verge of extermination owing eating dead bodies of animals treated by veterinary drug diclofenc (Prakash et al, 2003).

The cheeta (*Acinonyx jubatus*) has vanished from India. In the Himalayan region, the snow leopard *Panthera uncia*), fishing cat (*Felis viverrina*) and lynx (*Felis lynx*) are threatened animals (Lamba, 77). Eleven species of squirrels and two of marmots face danger of extinction. Five out of eight species of deer have become rare—these are 'bharad' (blue sheep or *Pseudois nayaur*), 'ghural' (*Nemorhedus goral* or goat-antelope), 'sanzari' (Manipuri antelope), 'hangul' (*Cervus elaphus* or Kashmiri stag), 'barasingha' (*Cervus duvauceli*) and muskdeer (*Moschus moschiferus*). In lakes and rivers the mahseer d trout fish have dwindled considerably.

Deforestation has destroyed the habitats of birds and threatened the survival of many species of birds. The pink-headed duck (*Netta caryophyllacea*), which lived in northeastern India and Myanmar as last seen in 1935 (Bawa et al., 2007). The great pied hornbill (*Buceros bicornii komras*), bamboo rtridge, Himalayan Bearded Vulture (*Gypeatus barbatus aureous*), horned pheasant (*Tragopanme lenocephalus*), mountain quail (*Ophyrsia superciliosa*) and crimson horned pheasant (*Trago pansatyra*), monal pheasant (*Lophophorus impejanus*), chir pheasant (*Catreus wallichi*), the Great Indian Burtard of jasthan, are among the endangered species of birds.

Table 2.2. Numbers of Species Threatened with Extinction in India in Major Groups of Animals and Plan and Some Key Families and Orders of India. (From Bawa et al., 2007)

Group	Approximate no. of species	No. of species threatened	% of species Threatened
Fish	748	4	<1
Amphibians	209	3	1
Reptiles	390	16	4
Birds	926	73	7
Mammals	316	75	23
Gymnosperms	64	N/A	N/A
Angiosperms	15,000	785	5

Table 2.3. Threatened Species of Vascular Plants in South Asian Countries that have been Evaluated in R List Assessments. (From Bawa et al., 2007)

Country	Total No. of species	No. of species threatened	% of species threatened
Bangladesh	5000	12	0.20
Bhutan	5468	7	0.13
India	16000	248	1.55
Nepal	6973	7	0.10
Pakistan	4950	2	0.04
Sri Lanka	3314	280	8.45

Significantly, the extinction of birds and animals is directly related to the increase in the hum population as noticed thirty years ago (Fig. 2.10). Undoubtedly, the extermination of rich wildlife India is due to the demands of the fast multiplying hungry population.

Wildlife Sanctuaries

As no generation of people has the right to rob future generations of their natural heritage, the trusteeship of wildlife has to be accepted and attempts made to protect it in sanctuaries and biosphere reserves. In India there are more than 195 sanctuaries and 20 national parks includi Dachigam (Kashmir), Corbett (Uttarakhand), Dudhwa (Uttar Pradesh), Kaziranga (Assam), Simlip (Odisha), Kanha (Madhya Pradesh), Sariska and Bharatpur (Rajasthan), Gir (Gujarat), Pakhal (Andh Pradesh), and Bandipur (Karnataka).

Livestock Population

While wildlife has been declining, the number of domesticated animals has increased phenomenal along with the human population. **India with just 2.2% of geographic area supports 13% or 173 million heads of the cattle (*FAO Livestock Report*, 2007) of the world population of cattle — 50 buffaloes, 18% goats—which exceeds 400 million units.** In the Uttarakhand Himalaya the problem

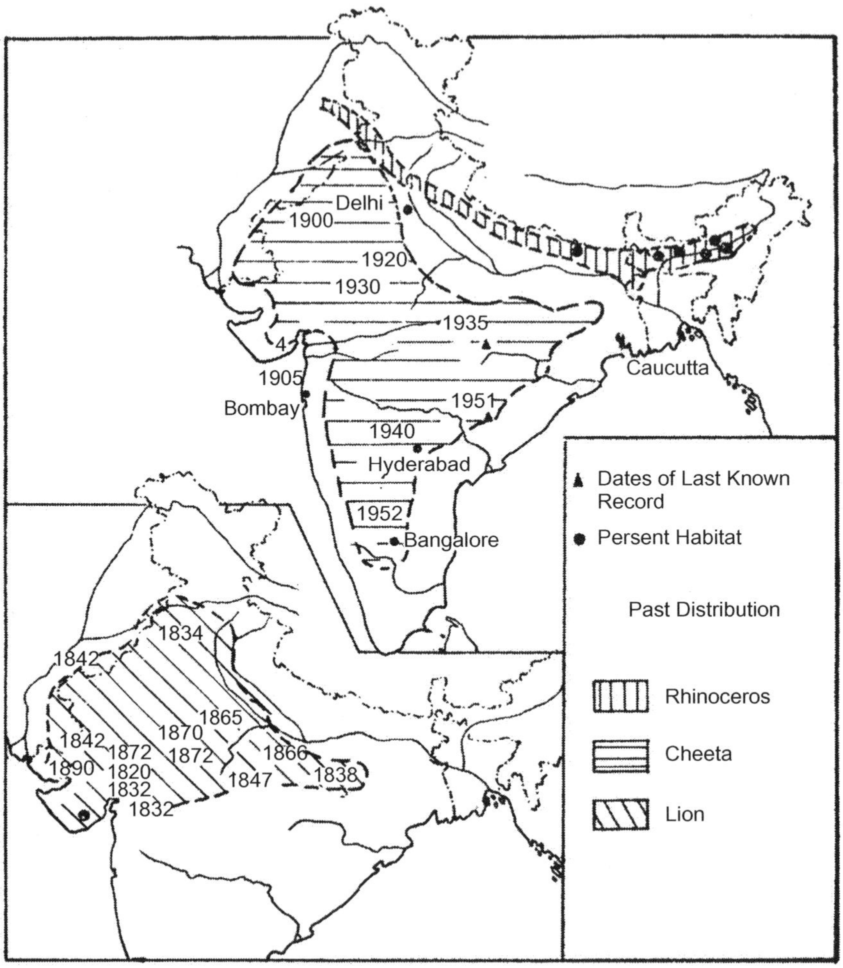

ed upon Survey of India map with the permission of the Surveyor General of India. The important waters of India extend into the sea to distance of twelve nautical miles measured from the appropriate the base time. Responsibility for the correctness of integral details shown the maps rests with the publisher. Government of India Copyright © 1988.

Fig. 2.9. Past and present distribution of a few threatened animals. The figures indicate the year when they were last seen. (Based on M. Gadgil, 1980).

ite serious. In Kumaun division the exponential growth rate of cattle population was 0.18% per year d per unit cattle had only 0.52 to 0.6 ha of the geographic land for grazing (Ashish, 1979), and in arhwal region only 0.2 ha (R.K. Gupta, 1983) as against the minimal need of 5 to 8 ha (ideally 15 ha). Kumaun the capacity of grazing lands diminished and the grass stock depleted at the rate of 5% per ar (Shah, 1981). The situation today remains more or less the same. Unless the population of livestock is reduced drastically there will not be enough forage left to feed them and no forests or pastures to keep e Himalayan domain green. **The attempt to preserve wildlife has to be coupled with an even more vigorous drive to reduce the population of livestock throughout the country.**

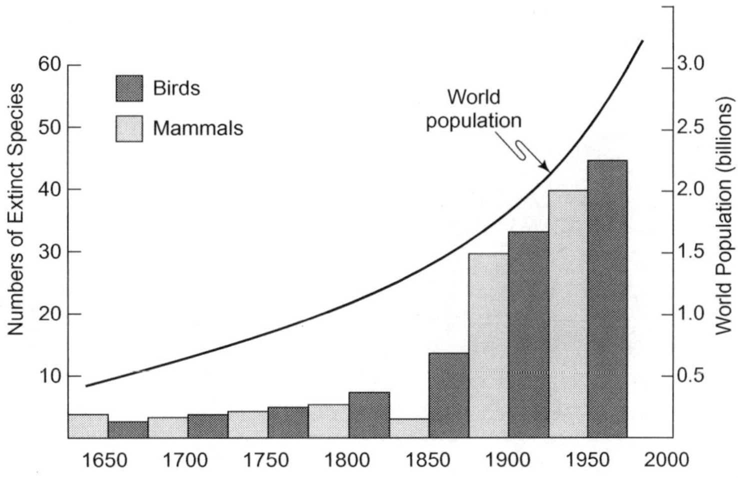

Fig. 2.10. The growth of human population is paralleled by the increase in the extinction rate of birds a animals. (From V. Ziswiler, 1967, *Extinct and Vanishing Species*, Springer-Verlag, New York).

FORESTS AND FOREST MANAGEMENT

Venerable Trees

The Hindus in ancient India held tress (like the 'peepal' and banyan) in great veneration and worshipp them to express their appreciation. The tribals all over India continue to venerate the forests, protecting them zealously around their shrines and temples. Not only do the trees provide shelter to a great variety of wildlife in the many strata of ecosystems, they also provide exceedingly useful commodities such as food, fruit, fodder and fuel. By intercepting raindrops on their foliage, trees protect the soil from erosion. Through their roots thrust deep into the ground, they stabilize mountain slopes against failure. They are primarily instrumental in the recharge of groundwater by promoting infiltration of rainwat and are responsible for the reduction in runoff and consequent floods. **One single tree is capable of cooling its surroundings equal to five average air conditioners operating 20 hours a day, abo 93 m^3 of woodland can reduce the noise level by 8 decibels, and one hectare of forest is capable of absorbing carbon dioxide and fumes generated by 20 cars** (Coates, 1981).

The forests moderate the climate and control local weather. The trees intercept solar radiation and reduce windspeed. The rainfall in the Amazon basin in South America is stimulated by dense forests in that region (Allen, 1980). The presence of forest decreases air temperature and decreases vapo pressure leading to increased condensation and rainfall (Meher-Homji, 1982). The evapotranspirati rate is lower in vegetation types having smoother canopy surface (such as of the monocultural plantations) than over those with irregular surface characteristics of the multistoried forests (Meher-Homji, 1982). The replacement of multistratal forests with single-storey pine forests, for examp may alter evapotranspiration, eventually affecting the rainfall. **Forests clearance, doubtless, caus perceptible decline in rainfall.**

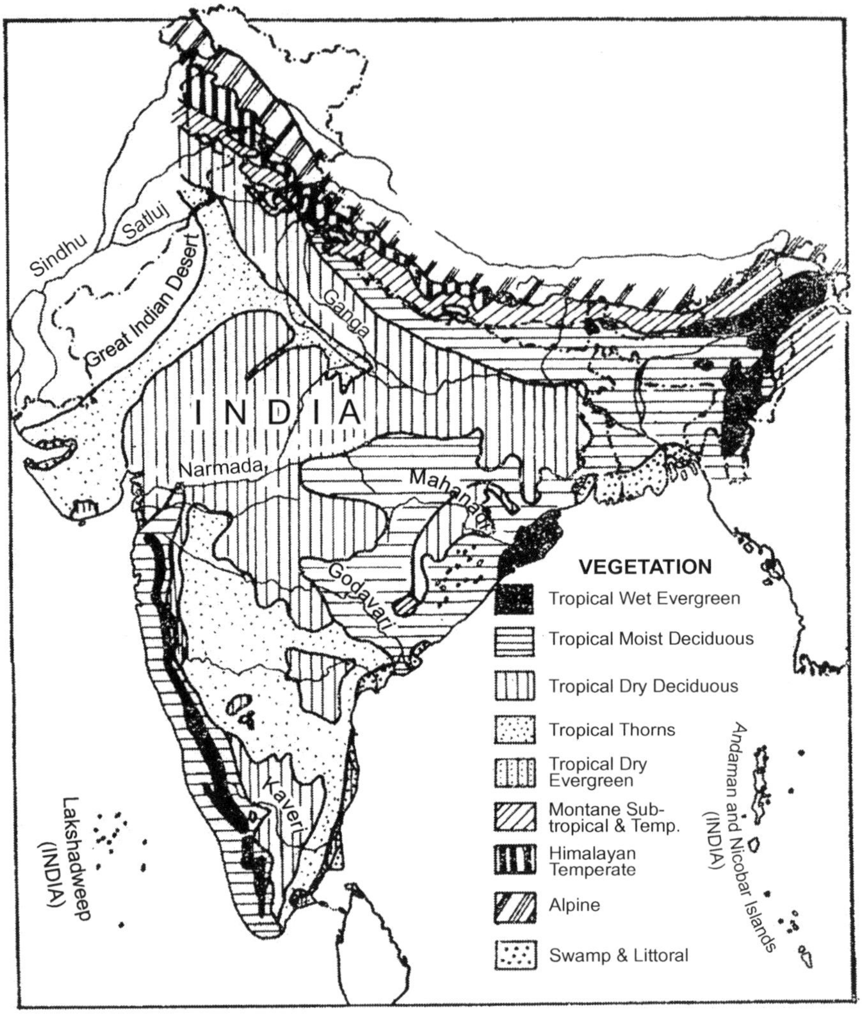

Fig. 2.11. Types of forests in India. (Based on *Frank School Atlas*, 1982, Frank Brothers, Delhi 10 p.)

Forest Fires

rest fires pose very serious threat to forests, practically all over India, but particularly in the Lesser malayan terrain of Uttarakhand and Himachal Pradesh, where pine forests predominate. Nearly 53% India's forests is periodically threatened by wild fires (Singh and Kushwaha, 2008). The February 12 forest fires ravaged very large swathes of Nagarahole National Park (> 500 ha) and the Bandipur Tiger Reserve in southern India, and in the national park in the Shimoga district (*The Hindu*, February 29,

2012). Primarily due to negligence by and sometimes mischief of the humans, our forests are loosing n only their spread but also their ecological health. Analysis of remote sensing-based time series analysis shows that in the southern part of the Aravali in Rajasthan recurrent forest fires burnt nearly 53,023 ha 2005, 44,681 ha in 2006, 57,689 ha in 2007, 89,655 ha in 2008, 1,99,837 ha in 2009 and 1,44,816 ha 2010, respectively accounting for 6.8%, 5.6%, 7.3%, 11.1%, 23% and 17.6% forest areas of Rajasth (Harishankar and Sudhakar Reddy, 2012). The forest fires are due to human activities. Gone are the da when the departments of forests used to have **fire lines** of sufficient width developed within the requis distances, and have them cleared of by biomass before the onset of summer. Now such fire lines a few and far between, and are seldom, if at all, cleared in time. Moreover, due to shortage of staff, the extent of the "beats" of the forest guards are unmanageably large for single individuals, particular when conflagrations take place almost simultaneously. Not considering them to be the stakeholde the alienated villagers or tribals around the forests do not assist the forest department to extinguish conflagration.

The situation becomes dismal during the summers year after year in Uttarakhand when sever thousands hectares of forests are grievously damaged, and often within the close reach of the superior headquarters of the forest department. The annual toll of forests from fire in Uttarakhand is 2408 (*Dainik Jagaran*, May 3, 2012). Quite often the villagers themselves are culprits.

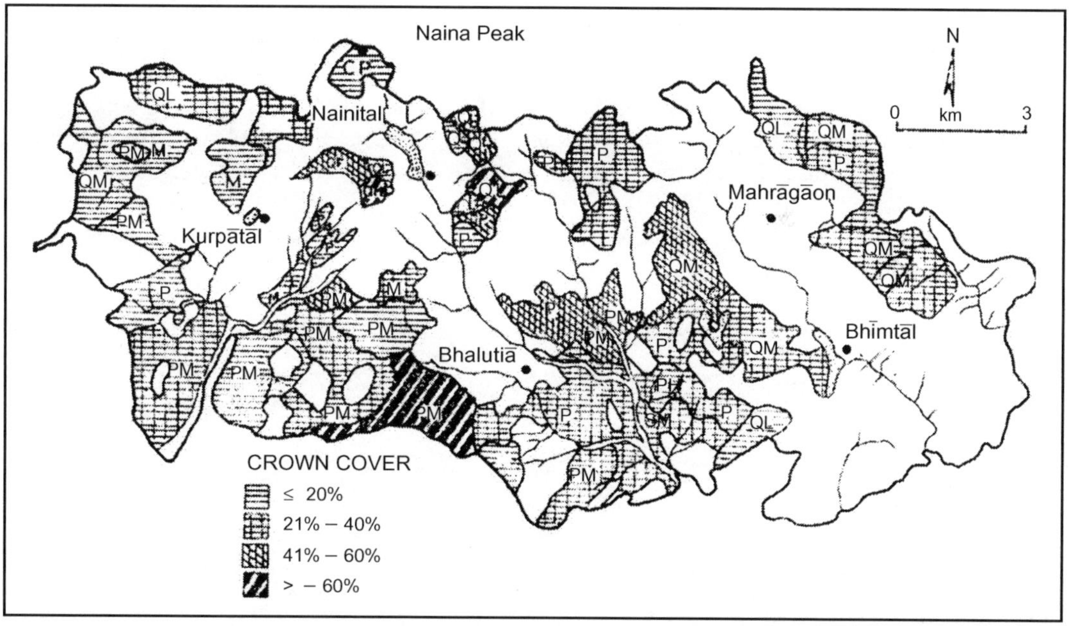

P = Pinus roxburghii, PM = Pine-mixed broadleaf, QL = Quercus leucotrichophora, QF = Quercus floribunda, QM = Mixed oak, CP = Cypressus torulosa, M = Mixed broadleaf forest, SM = Mixed sal forest, Unshaded = Agriculture and Deforested slopes

Fig. 2.12. Forest types and crown cover in the Nainital–Bhimtal areas, Kumaun Himalaya. (From J.S. Singh, U. Pandey and A.K. Tiwari, 1984, *Ambio*, 13(2): 80–87)

Forest Cover

uite less than 20% of the geographic land surface of the world is today covered with forests covering arly 4633 million ha (*FAO Global Forest Assessment*, 2008). In 1982 only 14.10% *of the Indian* ndmass had the shelter of the tree canopy, compared to 32% of the U.S.A., 41% of the erstwhile S.S.R., and 69% of Japan (Anon, 1983). Today the situation is a little better. According to the Forest rvey of India Report, 2009, the cover of ***dense*** forests was 2.82% of the geographic area of India able 2.2 and 2.3). A glance at Tables 2.3 and 2.5 will bring out the reality of the situation. In India day, of the total forest and tree cover of 7,82,871 km^2 (23.81%), only 6,92,027 km^2 or 21.05% is really natural forests. Of this **the good forest cover only 83,471 km^2 (2.54%) area of the country** (Forest rvey of India Report, 2011).

In the early eighties of the last century although in Kumaun region (Uttarakhand Himalaya) was said to ve forests on its 33% geographic area, actually only 1.63% of the tract had a tree canopy cover of more an 60% (Rawat et al., 1983). The landsat imagery interpretation indicated that only 28.7% of the Central malayan sector had then remained forested, and that only 4.4% of the area had forests with greater than % crown density (J. S. Singh et al., 1984). In other words, much of the forest cover was then in a very degraded, defoliated condition. The situation elsewhere in India then was not much better.

The 2011 report of the Forest Survey of India, Dehradun, showed that the India's forest cover declined 367 km^2 between 2007 and 2009, the largest decrease (550 km^2) taking place in northeast India Manipur, Nagaland, Arunachal Pradesh, Meghalaya), although in Punjab, Jharkhand, Tamil Nadu and jasthan the area of the social forestry increased by 380 km^2 (*The Times of India*, January 2012).

As already stated **out of the total of 692,207 km^2 of forests, only 83,471 km^2 have canopy cover of more than 40%** (Forest Survey of India Report, 2011). As shown in Table 2.8, the percentage of rrently threatened frontier forests ranges from 57% for India to 100% in Bangladesh and Bhutan awa et al., 2011).

It is disconcerting to realize that massive deforestation, though unreported, is going on everywhere in the Himalaya.

Table 2.4. Forest and Tree Cover of India in 2011 (Report, Forest Survey of India, 2011)

Class	Area (km^2)	% of Geographical area
Forest Cover		
(a) Very dense forest	83,471	2.54
(b) Moderately dense forest	320,736	9.76
(c) Open Forest	287,820	8.75
Total Forest Cover (including mangroves)	**692,027**	**21.05**
Tree cover	90,844	2.76
Total Forest and Tree cover	**7,82,871**	**23.81**
Scrub	42,177	1.28
Non-forest	2,553,059	77.67
Total geographical area	**3,287,263**	**100.00**

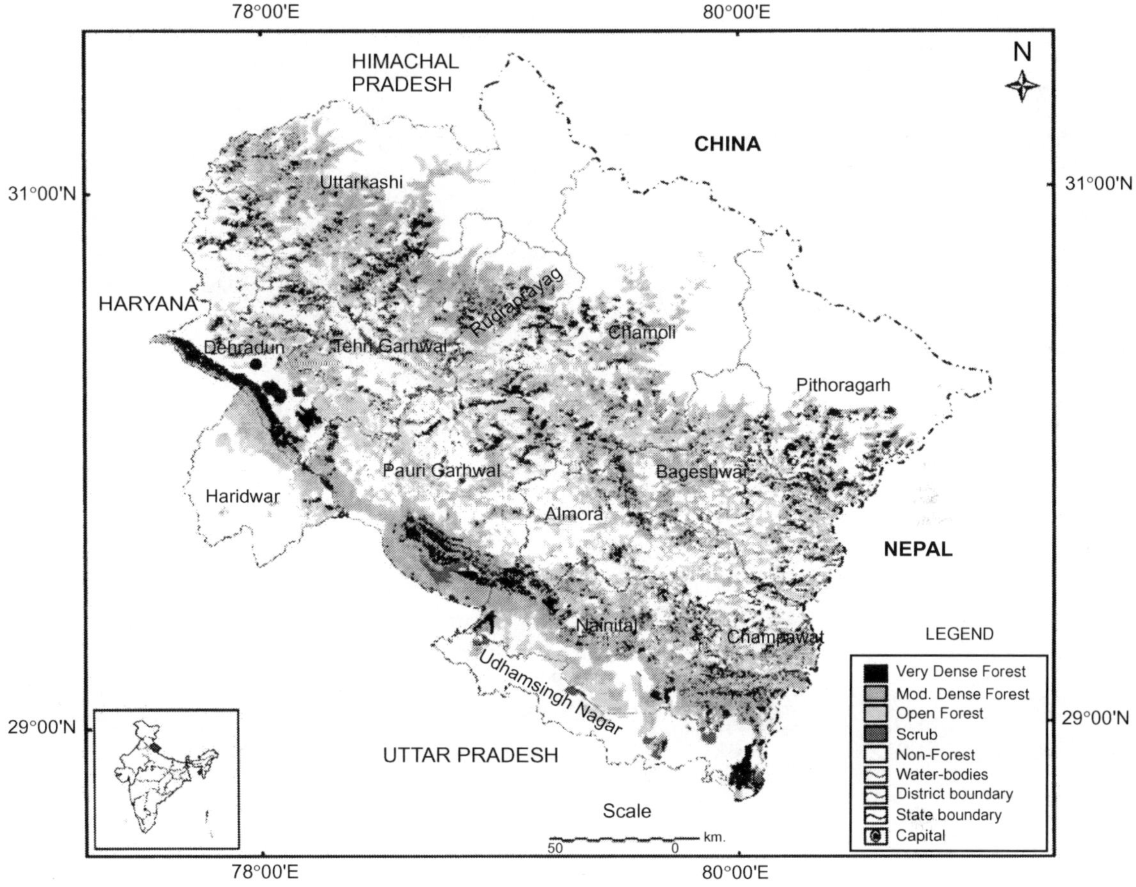

Fig. 2.13. Forest cover in Uttarakhand. (From *Forest Survey of India-2011 Report*) (See colour Fig. 2.13)

Forest Productivity and Extraction

In the early 1980s the growing stock of Indian forests, estimated variously between 1126 and 26 million m^3, was roughly 2400 million m^3 (Bowander, 1982). The annual rate of increment was ve low—0.5 to 0.61 m^3/ha compared to the then global rate of 2.0m^3/ha/yr. Thus the annual increment forests was of the order of 16 to 34 million m^3 (Bowander, 1982).

The reasons for the low productivity of the Indian forests was incredibly low investment (R 800 million compared to an earning of Rs 40,000 million in 1981), low pricing and unjustifiab subsidy conceded to the industries, excessive extraction bordering on exploitation without husbanding, and abusive management.

Table 2.5. Extent of Forest Cover in the Indian Landmass and the Volume of Growing Stock (Forest Survey India Report, 2011).

State and Union Territories	Geographical Area (km^2)	Recorded Forest Area (km^2)	Volume of Growing Stock (m^3)		
			In Forest	In TOF	Total
Andhra Pradesh	275069	63814	255.082	115.683	370.765
Arunachal Pradesh	83743	51540	492.689	74.516	567.205
Assam	78438	26832	173.494	41.336	214.830
Bihar	94163	6473	35.186	47.195	82.381
Chhattisgarh	135191	59772	334.381	70.069	404.450
Delhi	1483	85	1.705	1.040	2.745
Goa	3702	1224	7.716	3.895	11.611
Gujarat	196022	18927	48.261	117.993	166.254
Haryana	44212	1559	4.893	15.268	20.161
Himachal Pradesh	55673	37033	321.314	21.146	342.460
ammu & Kashmir	222236	20230	227.388	147.745	375.133
harkhand	79714	23605	116.308	51.308	167.616
arnataka	191791	38284	315.156	101.733	416.889
erala	38863	11265	142.582	49.059	191.641
Madhya Pradesh	308245	94689	249.661	85.106	334.767
Maharashtra	307713	61939	293.669	147.029	440.698
Manipur	22327	17418	70.878	10.691	81.569
Meghalaya	22429	9596	45.411	20.964	66.375
zoram	21081	16717	68.042	9.392	77.434
Nagaland	16579	9222	40.955	12.681	53.636
Odisha	155707	58136	285.191	73.624	358.815
Punjab	50362	3084	15.710	19.305	35.015
Rajasthan	342239	32639	34.385	81.560	115.945
Sikkim	7096	5841	18.832	2.017	20.849
Tamil Nadu	130058	22877	144.404	70.328	214.732
Tripura	10486	6294	21.864	7.391	29.255
Uttar Pradesh	240928	16583	123.400	81.683	205.083
Uttarakhand	53483	34651	460.089	20.917	481.006
West Bengal	88752	11879	92.515	45.693	138.208
Andaman & Nicobar Islands	8249	7171	53.185	0.666	53.851
Chandigarh	114	34	0.288	0.084	0.372
Dadra & Nagar Haveli	491	204	3.994	0.833	4.827
Daman & Diu	112	8	0.010	0.108	0.118
Lakshadweep	32	0	0.000	0.049	0.049
Puducherry	480	13	0.093	0.320	0.413
Total	**32,87,263**	**7,69,538**	**4498.731**	**1548.427**	**6047.158**

Table 2.6. Annual Estimated Production of Wood and Fuelwood from Forests of India (Forest Survey of Ind Report, 2011)

State	Recorded Forest Area (km^2)	Production	
		Wood (million m^3)	Fuelwood (m. tonnes)
Andhra Pradesh	63814	0.138	0.002
Arunachal Pradesh	51540	0.064	0.007
Assam	26832	0.025	0.0004
Bihar	6473	0.007	Negligible
Chhattisgarh	59772	0.397	0.09
Gujarat	18927	0.099	0.008
Haryana	1559	0.108	0.002
Himachal Pradesh	37033	0.231	0.00005
Jammu & Kashmir	20230	0.055	0.02
Jharkhand	23605	0.013	0.004
Karnataka	38284	0.049	0.03
Kerala	11265	0.068	0.009
Madhya Pradesh	94689	0.397	0.017
Maharashtra	61939	0.203	0.07
Odisha	58136	0.029	Negligible
Punjab	3084	0.090	Negligible
Rajasthan	32639	0.057	0.05
Tamil Nadu	22877	0.013	0.005
Uttar Pradesh	16583	0.425	0.008
Uttarakhand	34651	0.250	0.05
West Bengal	11879	0.317	0.003
Other NE States	64988	0.124	0.02
Union Territories	8739	0.012	0.0005
Total	**769538**	**3.175**	**1.232**

In the past the abuse of high-quality teak wood for railway sleepers had ravaged the lush forests of the Tarai belt in Nepal, Uttar Pradesh, Uttarakhand, Himachal Pradesh, Jammu and of Karnataka a Kerala. The industrial wood extracted from the forests amounted to about 10 m m^3. The paper industry made a heavy demand on forest resource. Resin-tapping done in a very cruel manner was responsible for the untimely demise of pine trees, as witnessed in April 1981 when in one sweep a spring storm raz to the ground 6 to 7 lakhs of pine trees in the outer Himalayan belt between the Kali and Satluj valley Understandably, all the trees broke at a particular height of 2 feet above ground level, for the number blazes (cuts) were far in excess of the permissible limit and many times deeper, wider and longer than was good for the trees. Even the young trees were not spared (*The Times of India,* July 24, 1982).

Thirty years ago over 250 m m^3 of all kinds of wood were removed annually from the forests, which was 20 times the volume of annual timber growth (Bowander, 1982). The annual withdraw of the biomass from the Indian forests was 564.03 million t/yr (Singh and Misra, 1978). The result this alarming rate of extraction of wood and withdrawal of biomass (Table 2.6 Figure 2.4) is that mo of our forests have become savannas with scattered trees or shrubs on grasslands. The grasslands are emselves threatened by overgrazing, fire and disturbances.

The pressure of livestock population has extended the areas of grazing and deforestation. Overgrazing 'literally eating away forest vegetation without giving it a chance to regenerate" (D. N. Misra in *The Times of India,* January 29, 1984). In the Uttarakhand Himalaya, for example, the grazing pressure was .5 to 4.5 times greater than the carrying capacity of the forests in the early eighties of the 20th century ngh and Saxena, 1980).

More than 51% of the fuelwood requirement is met by biomass withdrawal from the forests, implying at nearly 86 million tonnes of wood mass is being annually removed from the forests — in excess of hat the forests are able to produce (K.D. Singh, 2008). Understandably the pressure on the forests is mendous; and is growing each day. The demand was 123 m m^3 in 1970; rose to 133 m m^3 in 1980, and to about 225 m m^3 by A.D. 2000, which was met from 34 m ha of forests (Khoshoo, 1982). Thirty-forty years o every year approximately 4.54 million m^3 of wood was extracted from the forests of the Kumaun and rhwal divisions (Fig. 2.13). And if this extraction rate of nearly 5.8% persists, the regeneration of these forests will cease by A.D. 2031 and that of grasslands by A.D. 2041 as prognosticated by S.L. Shah (1981) ulting in the incipient desertification of the Lesser Himalayan terrain (Valdiya, 1985).

In the Khimli valley in the midland of eastern Nepal, for example, 170,000 tonnes of wood was tracted against a regeneration of only 60,000 tonnes, thus leading to a catastrophic condition (Rieger, 77). **The global picture of the pressure on forests from demands of fuelwood was and continues to be equally gloomy. This explains the widening circle of deforestation around villages all over the developing world.**

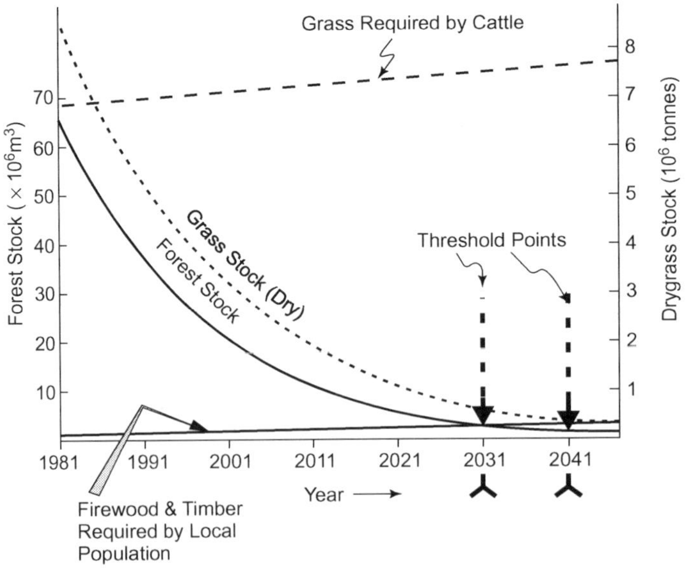

Fig. 2.14. If the present rate of extraction from forests in Kumaun and Garhwal (>4.54 million m^3/year) is not reduced, the regeneration of forests would cease by A.D. 2031 and of grasslands by A.D. 2041. (After S.L. Shah, 1981).

Fuelwood Consumption

Nearly 22% households in urban centres and 75% in the rural areas used firewood (along with chip as primary source of energy in the year 2004-2005. The average consumption of fuelwood is 312 k capita in the rural and 110 kg/capita in the urban areas (K.D. Singh, 2008). As already stated, 51% the fuelwood requirement is met by biomass withdrawal from forests and that **today more than 86 million tonnes of woodmass is being removed annually in excess of what the natural forests and tree plantations of the country produce on a sustainable basis** (K.D. Singh, 2008). The forests a bound to decline in consequence. The **natural forests** have thus declined at the rate of 1.5% to 2.7% p year i.e., from 514,137 km^2 in 1995 to 389,970 km^2 in 2005, implying forest clearing of 10,426 km^2 per year (Puyravaud et al, 2010), despite the fact the tree plantation expanded at the rate of 15,400 km^2/ye in that period. **The tree plantations have grown at the expense of natural forests.**

Rape of Forests

In the last six decades approximately 6 m ha of forestland has been converted into agricultural fields f rehabilitating displaced peoples, and lost as a result of water-resource development projects. This count in the early 1980s was losing forest cover at the rate of 0.16 million ha/yr or 441 ha/day or 18 ha/hour

The loss of forests is principally due to expansion of agricultural land, accounting for the loss of more than 71% (2.5 m ha) of the forest area between 1951 and 1973 (Table 2.7). In the year 1990 the loss w 63939 ha, in 2010 and 684,344, 692027 ha in 2011 (*FAO Forest Assessment Report*, 2011).

In Uttar Pradesh-Uttarakhand, vast tract of genetically rich Tarai forest between Dehradun a Gorakhpur was cleared for agriculture, and much larger areas (~ 16,000 km^2) on the fringes were left to be ravaged by the people. Large parts of the forests in Kerala, Karnataka, Maharashtra and the Himalay region were allowed to be submerged in reservoir waters behind dams, accounting for about 12% loss ju between 1951 and 1973. The forests of the Nilgiri massif have been ruthlessly destroyed for hydroelectr projects and by among others the labour force brought from outside. Much more forests have go under the reservoir waters behind high dams. To rehabilitate the displaced people of the Kali valley, t Ramnagar forest near Londa (Karnataka) was completely decimated (Gadgil, 1979). In Uttar Pradesh t river-valley projects took a heavy toll of 1200 km^2 of forests, many of the projects being of very doubtf value, and in some cases the land cleared for the river valley projects was leased out to agriculturists reservoirs could not be developed (D. N. Misra in *The Times of India,* January 29, 1984).

Table 2.7. Forests Lost Between 1951 and 1973 (Bowander, 1982)

Forest Lost to	Years 1951–1973		Upto 1979
	Area (ha)	% Loss	Area (ha)
Agriculture, horticulture, plantations	24,32,500	71.5	25,06,900
River valley projects	4,01,400	11.8	4,79,100
Establishment of industries	1,24,500	3.7	1,27,200
Road construction	54,700	1.6	57,110
Miscellaneous	3,87,900	11.4	9,56,400
Total	3.4 m ha		4.1 m ha

Road construction (Table 2.7) accounted for the loss of 1.6% and industries claimed 3.7% of t forests in the period between 1951 and 1973. Since 1973 the rape of forests has been rapid and disastrou virtually proceeding uncontrolled. The land was deprived of another 1.1 m ha of forest within a short riod of 30 years (1953–1981).

Elsewhere in the developing countries of the world, the situation is equally dismal. In a little more an a decade (1964-1979) the Tarai belt of Nepal lost 10,27,000 ha (> 31% of forest cover) through the ocess of rehabilitating farmers. The losses in countries of Asia and Africa are given in Table 2.8.

Table 2.8. Changes in Forest Cover from 1990 to 2010 in South Asian and African Countries (area in thousands of hectares) (Bawa et al., 2011)

Country	Forest 1990	Forest 2010	Total change 1990–2010	Annual % change
sia				
Bangladesh	1494	1442	–52	–0.2
Bhutan	3035	3249	214	0.4
India	63939	68434	4495	0.4
Indonesia	118545	94432	–24113	–1.0
Malaysia	22376	20456	–1920	–0.4
Myanmar	39218	31773	–7445	–0.9
Nepal	4817	3636	–1181	–1.2
Pakistan	2427	1687	–840	–1.7
Philippines	6570	7665	1095	0.8
Sri Lanka	2350	1860	–490	–1.0
Thailand	19549	18972	–577	–0.1
Vietnam	9363	13797	4434	2.4
frica				
Dem. Rep. of the Congo	22726	22411	–315	–0.1
Gambia	442	480	38	0.4
Ghana	7448	4940	–2508	–1.7
Kenya	3708	3467	–241	–0.3
Madagascar	13692	12553	–1139	–0.4
Rwanda	318	435	117	1.8
Zimbabwe	22164	15624	–6540	–1.5

Saving Genetically Rich Forests

e tropical forests that once covered a vast stretch of land in the Western Ghat, Kerala, Karnataka, Tamil Nadu, Sri Lanka and Bangladesh have drastically shrunk, and are in danger of being wiped out. The multistoried rainforests (Fig. 2.15) constitute biologically the most productive zone on earth, characterized by very rich genetic varieties and abundant biomass.

Mangroves along the coasts with their enormous gene pools are among the most sensitive ecosystems and need protection against reckless exploitation and marine erosion. The mangroves provide habitats r marine fish that live on the coastal belt, offer sites of sedimentation in which nutrients are trapped, d serve to moderate droughts and floods. India had nearly 2.5 m ha of mangroves, but largely in great distress. Today much more owing to fast expanding cities and growing tourist resorts. Those of

Kachchh in the early eighties of the last century, Khambhat and Mumbai on the west coast are now in highly degraded condition. The east-coast mangroves of the Kaveri, Krishna, Godavari, Mahanadi a Ganga deltas are in a process of deterioration as a result of over-exploitation, intrusion of saline waters and clearfelling for settlements (Thanikaimoni, 1980). The progressive degradation culminated in t encroachment of xerophytes and weeds, due mainly to the damming up of the rivers, thus reducing the suspended silt and dissolved nutrients in the river waters, — and also due to the influx of people w brought fungal infections (Caratini et al., 1980). The mangroves of Bengal, for example, shrank fro 6000 km^2 in the mid-fifties to 3500 km^2 in the mid-seventies (Blasco, 1977).

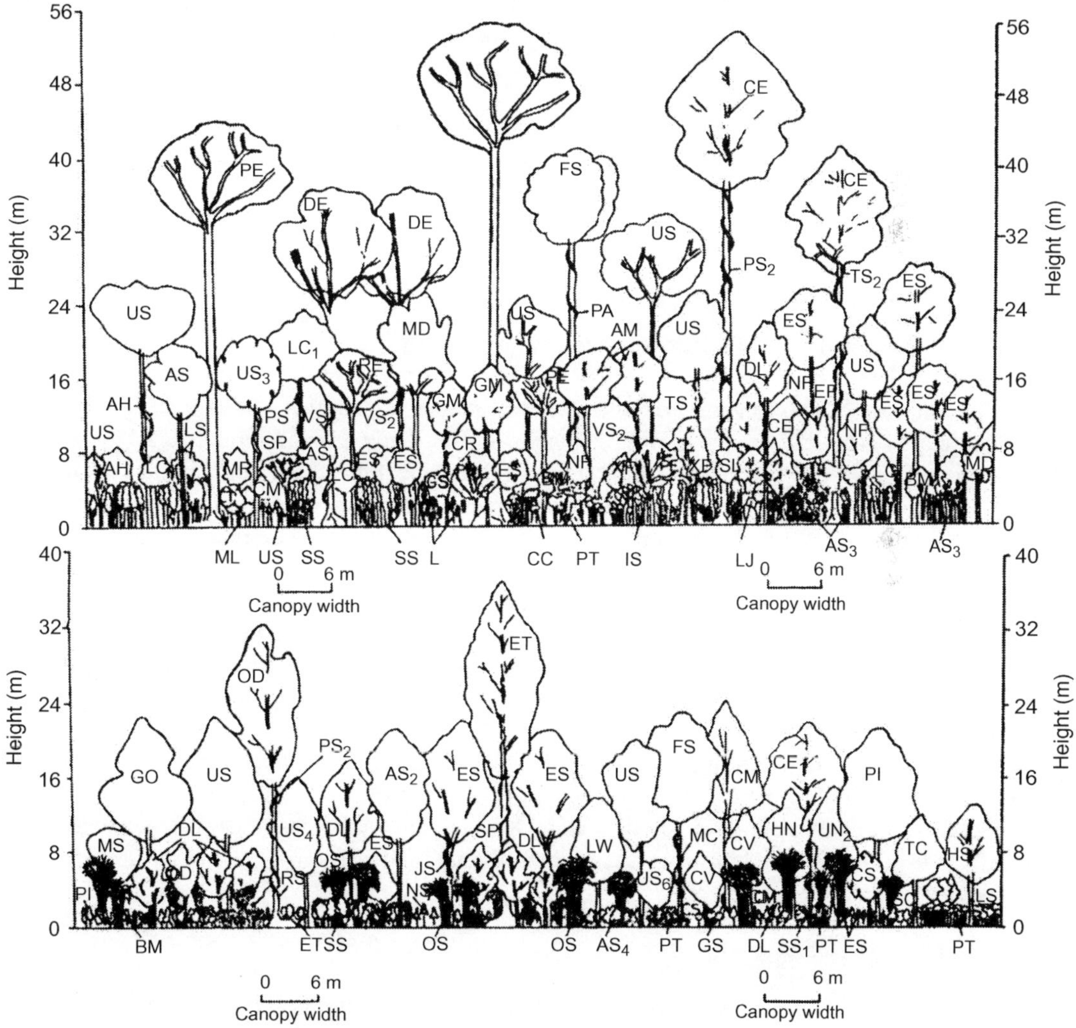

Fig. 2.15. Profile diagram of the multistoried rain forest in the Silent Valley, Kerala.

(a) *Mesic Upland Forest: AM =Agrostistachys meeboldii, AS1=Amoorasp., AS*$_3$*=Antistrophe serratifolia, CE=Cullenia exarillata*, CT = *Callicarpa tomentosa*, CR = *Calamus rotang*

CC=*Chasalia curviflora*, *V>E=Dry petes elata*, *EP=Erythropalum populifolium*, *GM=Gar-cinia morella*, IS=*Ipomaea silicosa*,*ECi=Litsea coriacea*, *hCi* = *Laportea crenulata*, LI= *Leea indica*, *LJ=Lasianthus jackianus*, *MD=Myristica dactyhides*, *MY=Macaranga pel-tata*, *ML=Microtropis latifolia*, NF = *Nothapodytes foetida*, *PE=Palaquim ellipticum*, *PSi* = *Pothos scandens*, *VA=Paranignya armata*, *SL=Syzygium laetum*, TSi = *Thunbergia* **sp.>** TS_2 = *Trichosanthes* sp., **US = Unidentified sp., B, VSi** = *Ventilago*, \S_2 = *Cissus* sp., XF= *Xanthophyllumflavescens*.

(b) *Riparian Forest:* AS_2=*Alseodaphne* sp., *ASA* = *Antidesma sp.*, BM = *Boehmeria malabarica*, CEi=*Calophyllum elatum*, CM=*Cinnamomum macrocarpum*, CV=*Clerodendrum viscosum*, *T>h—Dimocarpus longan*, ET=*Elaeocarpus tuberculatus*, ES=*Eugenia* sp., *FS=Ficus* sp., *GS* = *Glochidion* sp., GO = *Gordonia obtusa*, *HN^Holigarna nigra*, *HS* = *Helicia* sp., JS = *Jasminum* sp., *LS=*Litsea* sp., *LW=Litsea wightii*, *MC=Mexoneuron* sp., MS=Mewe-cy/tf/ zsp., *NS=Nothofagus* sp., *OD=Olea dioica*, OS = *Ochlandra scriptoria*, *PS2—Piper sp.*, *Vl=Poecilo neuron indicum*, PT=*Psychotria thwaitesii*, RSi = *Randia speciosa*, *SC= Syzy-gium caryophyllatum*, SS = *Strobilanthes* sp., SSi = *Stephania* sp., *S¥* = *Smilax prolifera*, *TC^Toona ciliata*. (After Singh et al., 1984b).

Grasslands and Grazing Grounds

pproximately 20% of the earth's land surface (30 million km^2) is grazing ground, supporting more an 3000 million heads of cattle (Alien, 1980). The 2003 census showed that the livestock population as then 4.91 million heads. In comparison to the well-managed pastures of Europe supporting 3 to 5 ads of cattle per hectare, the grazing grounds in North Africa and Saudi Arabia had 50 to 60 animals/ This explains the severe degradation and attendant desertification of the lands in these countries. In stern Rajasthan while the livestock population increased considerably, the area of grazing ground rank from 13 m ha in 1951 to 11 m ha in 1961. It is far, far less today. The intensity of grazing in the azing lands throughout the country is very high—19% of the world population of cattle Iive in India haudhury et al., 2011), making demands on the ever shrinking resources. In the Lesser Himalayan lt of the Uttarakhand the incidence of grazing used to be 2.4 times the carrying capacity of the forests Chamoli district, 4.5 times in Almora district and 4 times in the Chakrata tehsil of Dehradun district ngh and Saxena, 1978). This was comparable with the situation in Iraq (4 times) and Syria (3 times) West Asia.

Because of agricultural constraints, expansion or even reservation of land for grazing is no longer possible. The only solution lies in raising fodder crops in plantations, such as of 'subabul' (*Leucaena leucocephalia*), *Acacia ferruginea*, etc. and establishing and improving pastures in selected areas. Stall-feeding of genetically improved breeds and reducing the numbers of unproductive cattle would lessen the pressure on the pastures and forests.

Tree Cropping: Merits and Disadvantages

Taking into consideration the growing demand for woodstock and the low productivity of our natural forests, it is imperative to seek alternative sources of fuel and industrial wood (for pulp, plywood, sports goods, furniture, matches, packing cases, building-construction, etc.). One way in which this can be achieved is taking recourse to **plantations of fast-growing trees—not by replacing the natural forests, but by tree-cropping on marginal and degraded or unused lands. Tree plantations, captive**

of the concerned industries seem to be the only option left. The productivity of tree plantations is greater than that of the natural forests, for in the absence of interspecies competition in monocultural ecosystems, the trees grow faster and taller and also the selected trees are really more productive. Compared to the average productivity of 1 to 4 m^3/yr of trees in natural forests, the yield of the tre in plantations is 10 to 20 m^3/yr. In a well-irrigated farm of eucalyptus in Gujarat the production w 45 t/ha/yr compared to mere 4 t/ha/yr from rainfed plantations (K. S. Shah, in *The Times of India,* July 1983). With application of fertilizers and irrigation a yield of 8 t/ha/yr of fuelwood from a plantation a degraded field can be expected (Bachkheti, 1981).

The choice of the species of trees is of crucial importance; it should conform to the local needs of the people, industry and trade. Species of quick-maturing trees and shorter rotation are preferred, such as eucalyptus, poplar, 'subabul' (*Leucaena leucocephala* or Hawaiian giant), *Casuarina equisetifolia,* etc. The 'subabul', a leguminous plant and also a good protein-rich fodder and fuel, can grow even on rocky soils. Its fodder yield is 200 t/ha in 3 years and the firewood 2 to 25 kg per tree (Gangu 1983). A number of species of eucalyptus have been grown in a large number of forest-farms in Tam Nadu, Karnataka and Gujarat. In the drought-prone Bhavnagar district of Gujarat, the experiment wı high-density plantations of eucalyptus and 'subabul' yielded rich dividends, of course with the liberal application of fertilizers and irrigation. The spacing of trees in such plantations is 60 cm by 60 c so that 25,000 trees can grow in a hectare of land—6 to 10 times more than in traditional farms. T struggle for living space in densely populated areas makes trees grow straight, taller and faster. These can be harvested every three years, and the cut stumps regenerates 6 to 10 times faster and denser (Pat 1982).

In the outer Lesser Himalaya a number of species of poplar *(Populus ciliata, P. deltoides, P. yunnanensis, P. euramericana, P. nigra)* and in the Tarai belt *Populus deltoides* G-3, *PD-100,* PD-121, *P. cassale* have been grown successfully (Fig. 2.16) yielding rich harvest in 10 to 12 years (Chaturve (1983).

While a choice can be made in selecting the type of plant for cropping, there is none in the matt of choosing the land class. **The tree-crops must be grown in non-agricultural or unused marginal or submarginal and degraded lands**. There are 57.05 million hectares of wasteland in India—19.9 ha barren uncultivable, 13.93 m ha cultivable waste, and 23.22 m ha fallow land. In the >7.5 milli hectares of'usar' land (salinized and alkalinized soil) in 14 states can be grown the firewood plan such as *Prosopis juliflora, Acacia nilotica Albizzia ebbek, A, procera, Terminalia arjuna, Cass siamia, Cordia dichotoma, Syzygium cumini, Haplophragma adenophyllum, Pithecolobium dul* and *Eucalyptus* hybrid (Chaturvedi, 1985). The best time for planting these trees is immediate after the rainy season when the waterlogged condition just disappears. Not only do these trees ha tolerance for temporary waterlogging and are not good fodder in the juvenile stage, but are also able absorb salts and generate enough leaf litter to contribute nutrients to the nitrogen deficient 'usar' s• (Chaturvedi, 1983). For the degraded ravine land the following fuelwood trees are recommended *Prosopis juliflora* and *Acacia nilotica* on flat tops having clay soils; *Syzygium cumini, Haplophragm adenophyllum, Dalbergia sissoo* and *Cassia siamia* in ravine bottoms, *Cassia auriculata, Acacia nilotica, Pongamia pinnata, Pithecolobrium dulce* and *Holoptelea integrifolia* on gentle slopes; and *Eulaliopsis binata* and *Saccharum munja* on steeper slopes. In the brackish water districts of Mathura, Agra, Etah and Mainpuri in Uttar Pradesh, trials demonstrated the utility of these trees as well *Leucaena leucocephalia* and *Adhatoda vasica,* which apart from being valuable fuel trees, reduce the salt content of the groundwater.

Fig. 2.16. *Upper*: A natural forest in the Sahyadri. *Lower*: A man-made plantation in Karnataka. (Photos: Prof. K.N. Ganeshaiah, University of Agricultural Sciences, Bangalore) (See colour Fig. 2.16)

The efforts of some farmers (such as those in Gujarat and Karnataka) of growing tree crops in food farms have turned out to be counter-productive in the long run. **Agricultural, or for that matter, forest land, should never be diverted for tree cropping.** In Madhya Pradesh plantations of teak and 'bhair' re developed by clearing 'saal' forests; in southern Bastar the 'saal' trees (from 40,000 ha) were ar-felled for growing cashew plantations. In the Singhbhum-Chhotanagpur area in Jharkhand the 'saal' trees were removed to raise teak trees. In Nasik (Maharashtra) the natural forests were cleared at e rate of 600 ha/yr for tree-cropping (*The Indian Express,* July 11, 1983). In the mountainous tracts of Tamil Nadu and Kerala plantations of coffee, eucalyptus and teak replaced more than 14% of t natural forests in the later 1970s. **In the Central Himalayan forests the pine is fast replacing oaks and rhododendrons—71% of the newly afforested areas were under the pine and only 29% under broad-leaf oaks thirty years ago (S. L. Shah, 1982). Such unsound measures lead to eco-disast in the long run.**

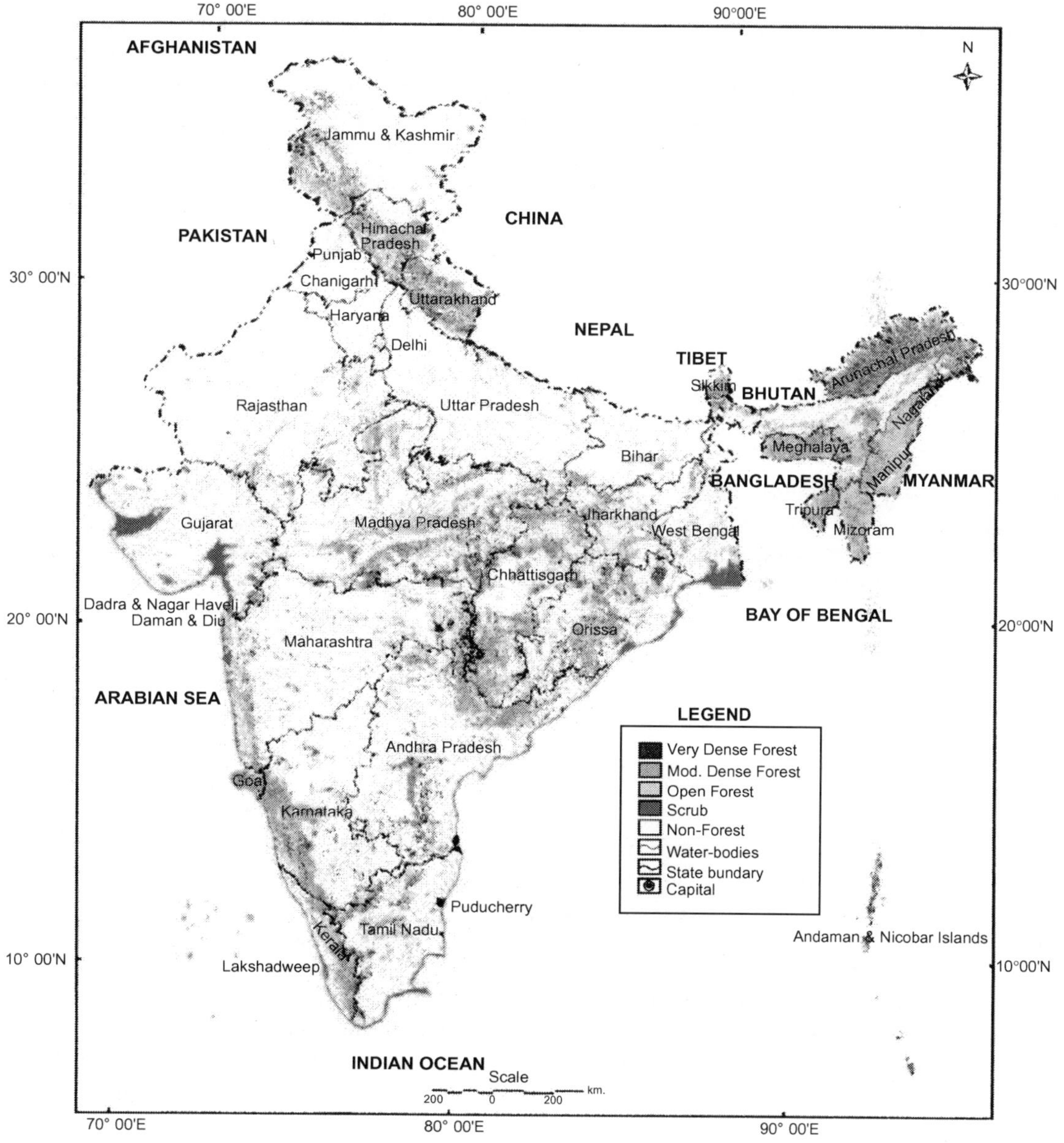

Fig. 2.17. Forest cover of India. (From: *Report Forest Survey of India-2011*)
(See colour Fig. 2.17)

In 2006, 3.57 m ha of land in India were under plantations for industrial purposes for fuelwood, and r soil conservation. Today nearly 90844 km^2 are under man-grown trees.

Admittedly, **there are quite a few disadvantages in the monocultural tree-farming also. Owing to reduced genetic diversity and more easy spreading of pests as compared to mixed plantation, the trees lose their capability to struggle against and resist invasion from pests, insects and weeds, and are thus vulnerable to the attacks of fungi and pathogens**. The Vazhachel eucalyptus plantation Kerala, for example, was wiped out by a fungal disease. Diseases are very common in humid regions, where weeds and insects overwhelm the vulnerable monocultures (drier climates control weeds, insects and diseases). Tree-cropping in deforested areas in humid regions thus face certain perils. For example, the obnoxious weed *Eupatorium* ('Assam lata') has taken deep hold in the deforested areas of Bengal and s spread rapidly into the Simlipal forest of Mayurbhanj, Gonasika forest in Keonjhar and Chandaka forest in Bhubaneshwar district (Odisha). This utterly useless weed climbs over plants, forms a close canopy and eventually strangulates plant growth. In this manner the valuable 'saal' forests of Bengal d the Darjeeling foothills have been ruined (*The Times of India,* September 24, 1984). The damage to monocultural tree-crops in this kind of a situation would be disastrous.

The tree-crops, particularly of eucalyptus and pines, are also vulnerable to fire occurring nearly ce every 1 to 2 years in Kumaun. The fire cause substantial nitrogen losses in pine forests occurring commonly on the driest slopes, leading to nitrogen shortage in soils and considerable decline of oaks (J. Singh et al., 1984). The eucalyptus-pine plantations do not provide shelter to birds and wildlife, and are incapable of meeting the needs of fodder for the stall-fed cattle. Also these trees generate recalcitrant er, which does not break up to produce humus. Consequently, the soil over the years becomes infertile, leading to the suppression of understorey plants. Another consequence of such litter is the diminished otection against erosion and reduced infiltration of rainwater so that groundwater is not recharged fficiently under the plantation. **Besides, the eucalyptus trees draw considerably more water from e soil than do other plants (Ruddle and Manshard, 1981) leading to lowering of the water-table and drying up of wells and springs**. The eucalyptus plantation to the east of the Yamuna River in the Delhi region lowered the water table so greatly that this once waterlogged area has now become a densely habitable place. The Israelis successfully used eucalyptus forests to reclaim their marshlands.

Social Forestry

farmer on an average annually needs 40 quintals of wood—24 quintals as firewood. This can be met by fast-growing trees within the limits or in the vicinity of his village. Social forestry was introduced to meet the needs of fuel, fodder, fruits, timber and biomass. This effort envisages afforesting the sides roads, rail-lines, canals, streams and rivers, the boundaries of fields and the degraded lands around lages. The objective is that each village should have its own bank of fuel, fodder and timber. In the rly 1980s the available land on the sides of roads, railways and canals amounted to 9,00,000 ha in hich approximately 1400 million trees could have been planted (Khoshoo, 1982). There are 19.09 m of barren lands together with 13.94 m ha of unculturable and culturable wastelands (Bhardwaj et , 2007). Even if 12 m ha can be covered with trees, the anticipated yield of fuelwood at the rate of /ha/yr would be quite sufficient to meet the requirements of the people. It may be emphasized that this production rate presupposes that the soil is not too degraded, there is no grazing and there is enough rainfall. In actuality none of these conditions are present.

The successful implementation of social forestry calls for active participation of village communities, including the needy and the poor. Financial assistance for developing and maintaining the social forests in terms of long-term easy loans (or salaries) and **sharing of benefits equitably wou provide an incentive and give a fillip to the programme**, as happened in Gujarat (Karamchanda 1981). **Provision will have to be made also against rich farmers and politically powerful people cornering loans and the benefits of this programme.**

Imperative of Forest Management

Whether it is a captive tree-farm or a social-forest, the aim must be to increase not only the producti capacity but also the genetic diversity. Tree-cropping will be ecologically unhealthy and socially unsound if it is not integrated with food-farming and animal-husbandry and horticulture (Douglas et. al., 198 Jackson, 1984). In other words, an ideal tree-farm would comprise large blocks or belts of econom trees interspersed with narrow grazing strips of grasses and herbage (Fig. 2.18). These varied stri

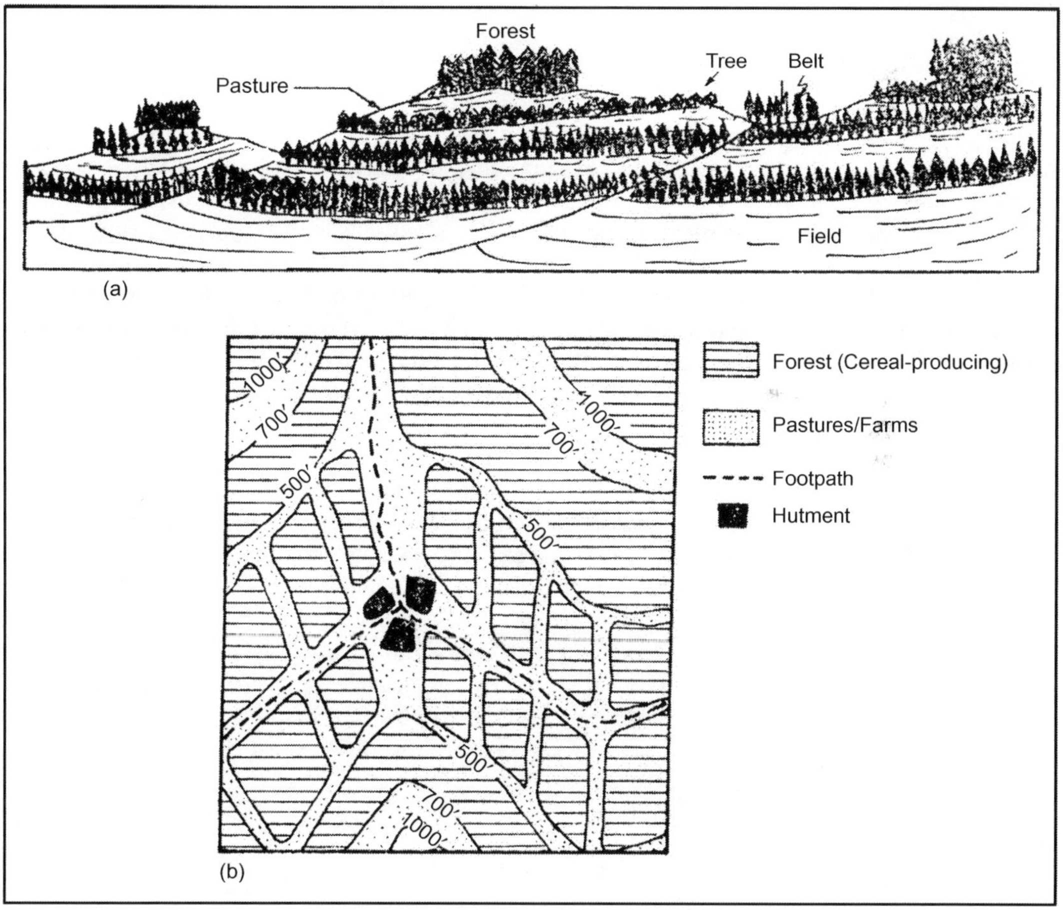

Fig. 2.18. An ideal design of the forest farm, involving integration (linkage) of agriculture, horticulture, animal husbandry and free cropping. (Based on Webar and Hoskins, 1983, and J.S. Douglas, R.A.J. Hart and G.S. Ranganathan, 1982, in: *Forest Farming*, Nataraj, Dehradun).

would provide fodder, fuelwood, fruits and seeds to the village community. **Priority should be given to the needs of the local people who depend on the forests for sustenance and even survival, thus eventing industries and contractors from becoming the prime beneficiaries (Moddie, 1977). he sharing of the benefits will ensure participation of the people in maintaining, protecting and expanding the tree-farms.**

In the matter of choice of the type of trees, the local needs must be the paramount factor. No exotic ecies should be introduced without adequate trials. Adoption of scientific methods of harvesting involving monitoring the growth rate, estimating total stock available, extracting only the annual harvests, etc. would ensure survival of the forests. These measures call for strengthening of forest management staff and investment of larger amounts of money. Forest exploitation in ecologically sensitive and geologically unstable zones must be discouraged if not banned, whatever the economic imperatives. **Subsidies on forest products should not be conceded to industries, and the wood-based industries must be linked with tree-farms developed on marginal or sub-marginal lands.**

SUMMARY

In order to understand the relationship between organisms and their environment, it is imperative that we know the functions of the living (biotic) communities in relation to soils, air and water. The self-nourishing producers (plants), the consumers dependent on them for food (animals) and the decomposers (bacteria) which rearrange the complex substances are inseparably and interdependently linked together rming the functional unit of the ecosystem. The ecosystems are characterized by stratification, the larger the number of stratum, the larger would be the habitats available, and lesser would be inter-ecific competition. Any act that reduces the stratification in an ecosystem, such as monoculture of tree plantation, horticulture, food-cropping (agriculture), etc. would cause shrinkage of habitat, leading to deterioration of the very ecosystem and its living community. It is unfortunate that the people all over the world are responsible for altering the terrestrial and aquatic environments. This has culminated in % reduction of the total net productivity of the earth.

The welfare and even survival of organisms (including man) depends on the stability of the otic community to which they belong. A modification of the habitat changes the character of the community and throws the organisms into distress. The species/genetic diversity imparts stability to the ecosystem as much as it ensures against disease, decline and death of the race. Any attempt to reduce species diversity, such as monoculture cropping and selective exploitation of forest resources, is bound to endanger the ecosystem. Gene pools are the irreplaceable capital for the future generations for drawing food, medicines, etc. In order to preserve gene pools and save the threatened species, conservation of all wildlife in sanctuaries and biosphere reserves will have to be undertaken on a larger scale. On the other hand, the numbers of domestic livestock will have to be drastically reduced to save the grasslands and forests from destruction.

The explosive growth population of humans — nearly one tenth of which is hungry and bereft of food, clothes, shelter—threatens to destroy the very base of the resources on which people depend for sustenance. This is leading the terrestrial ecosystems to collapse eventually.

Forests constitute the most precious and productive of all ecosystems—they are providers of practically everything that man requires for living. This resource base is being eroded at an alarming e in developing countries including India. Unless this rate of exploitation of woodstock is reduced, all rests in the tropical zone will vanish completely within a century; and in the central Himalayan region desertification would set in by the middle of this century. There is no option but to take to large-sca plantation of fast-growing trees, not by replacing the natural forests or food-farms, but by reclaiming degraded and marginal lands. Tree-farming can be made ecologically healthy and socially sound by integrating it with animal husbandry and horticulture, and by increasing not only their productive capacity but also genetic diversity.

3 Land – Its Use and Management

LAND — A RESOURCE

Land is not only a basic natural resource but also a cherished asset. It has to be utilized in accordance with its suitability and ecological capabilities, which depend on geological and biological factors. However, unbearable pressures such as of an explosively growing population, are brought to bear on the land to overuse it and over-exploit its limited capability. Therefore the ecological balance is upset. Forgetting the responsibility to the posterity and having learnt no lessons from the consequences of mismanagement of land by their forebears in ancient Phoenicia, Babylonia, Palestine and Sindhu–raswati valleys, the peoples have been abusing the finite resources, including the land.

The injury inflicted on the land may well result in an irreversible situation. It calls for appropriate land-use planning based on optimal use of land in accordance with its suitability and capability. Hill-lands, for example, if put largely under forests, would serve as a source of fuel, fodder and timber, besides recharging the groundwater and providing scope for cattle farming. Land should be so managed as to satisfy at least the basic needs of man — food, shelter, recreation, communication, transportation, etc. Food production from the agricultural lands should not only be sustained but also need to be appreciably augmented. **This calls for prevention of encroachment on fertile agricultural lands for non-agricultural purposes**. With productivity ratings of 2, 1 and 0.5 for three categories of land— good, medium and poor respectively — the use of 1000 ha of good land for non-agricultural purpose implies a loss of 2000 ha, while if poor category of land is taken, the actual utilization would be only 500 (M. Shafi, per. comm., 1985). Thus, **extreme care in land-use is called for, particularly in selecting es for industries, power plants, dams and reservoirs, and in expansion of towns and cities.**

More people means more houses and new sites for the housing, as also provision of land for recreation. Land-use planning should also encompass land for lines of communication, routes of transportation and sites for transport stations and airports. The needs of the defence forces, including strategic installations, sites for troop deployment, and their routine exercises, etc. cannot be ignored. Another essential of land-use planning is to recognize the need to keep the land fallow for some time after a particular activity is completed, and restoring degraded land. India is helplessly witnessing a repudiation of both these imperatives.

TRAGEDY OF COMMONS

Air, water, soil, animal and plant species, scenic beauty of the landscape are collectedly owned the society at large — 'owned by all and accessible to everybody' (Bawa et al, 2011). These constitute the ***common property resources***. The tragedy is that the governments, the industries and even the people over-use and misuse them recklessly, damaging the very base of the resources, leading what G. Hardin (1968, 1985) described as *tragedy of commons*. The resulting damage – due to appropriation of all or maximum benefits and gains by some individuals, and losses and pai shared by all other members of user groups – eventually culminates in the failure of the market, often economic collapse (Bawa et al, 2011), as we are witnessing today. In an individualistic socie such as exists in India, "a property belonging to everybody is a property belonging to nobody". The consequence of this misconception is that the public land which should have been utilized carefully has become degraded and almost derelict. Land which should have been covered with a canopy of forests (such as the watersheds in the mountains) have been converted into agricultural farms • horticultural plantations and for tourist resorts, the steeper unstable slopes have been brought under the plough or excavated for laying down roads, the valuable vegetations have been submerged as a result of impoundment of water. **The farmlands are being irrigated much beyond their abili to remain healthy, the productive fields are encroached upon to accommodate expandin cities and burgeoning industries, and so on. This mismanagement has continued inexorab unimpeded because we regard our land as a commodity belonging to all, none of whom feel t responsibility to take good care of it.**

SPHERES OF LAND MANAGEMENT

Land management embraces planning on rural, urban and regional scales. ***Rural land-use management*** involves recognition and demarcation of land primarily capable of supporting forests, pastures and agricultural crops, in addition to dwelling abodes or structures. Forest land management is tantamount to management of the upland watersheds, and cropland management means analysis of the capability of soils and availability of water for irrigation and drinking.

Urban land-use management encompasses habitat planning on the basis of the conditions of stability of the ground for foundation of structures, the freedom from natural hazards, the availability of water of requisite quality in sufficient quantity, the suitability of sites for disposal of wastes, the development sites for recreation, roads, water-supply, etc.

In ***regional land-use management*** broader aspects of development are taken into consideration— (i) lands for agriculture, tree-farming, conurbanization and industries, (ii) areas for water-resour development — dams, reservoirs, canals, etc., (iii) transport networks (roads and railways) (iv) e ploitation of natural resources — forests, minerals, aquatic life, (v) zones of bigger natural hazar (earthquakes, floods, cyclones) and (vi) defence requirements (airfields, dockyards, areas of tro deployment) — on the basis of terrain evaluation.

The role of the geologist is to evaluate the suitability of the land on the basis of geological conditions, stability of the ground, vulnerability to hazards, availability of resources, including water and construction-material and also point out the environmental stresses likely to be generated as a result of the use and abuse of the land.

The four major elements of a land management programme are:

1. Preparation of an inventory of land resources, making use of aerial and satellite photos complementary to comprehensive fieldwork, and then delineation on maps.

2. Identification of natural hazards likely to threaten the area, indicating the anticipated degree and frequency of the threats.
3. Investigation of various geomorphic, geological, hydrological, pedological and ecological properties of the lands, and formulating a policy for land-use commensurate with the needs of the society.
4. Monitoring changes that occur consequent on the use of the land.

As already stated, 377 million (31% of the total population) live in urban centres. The metropolitan ies have understandably high density of population — 26597 persons/km^2 in Chennai and 20057 rsons/km^2 in Mumbai (A. Srivathsan, *The Hindu*, March 9, 2012). **Despite India's city-centric and fluent-friendly policies and programmes, the urban centres are unable to cope with the "current vel of urbanization". It is therefore unthinkable that any programme of development of rural ntres would ever be taken up or succeed.**

In the **urban development programme**, the emphasis must be on three basic needs — transportation, energy and housing, in addition to coping with the problem of mounting pollution of all kinds. Unless an effective public transport system with dedicated lanes for public transportation is developed, and priority given in the right of the way to buses carrying more people compared to cars carrying an individual or two, there will be no lessening of the woes and hazards — and indeed no growth — of cities. The New York Transport Commissioner Janette Sadik-Kahn (*The Hindu*, March 9, 2012) mphasized that all cities and towns need "(1) an effective public transport system, (2) an effective cling network, and (3) an **effective ways for people to get around by walking. The motor car should not be the priority**".

The **rural development programme** should include developing all those basic facilities in the small towns and larger villages, the lack of which forces the rural population to migrate to cities and big town. ese basic needs are: (1) good schools, (2) better manned and equipped hospitals, (3) markets or stores oviding basic necessities, and (4) a medium or means of entertainment or facilities for enjoyment, including playing grounds.

LAND-CAPABILITY MAPPING

Objectives

A land-capability map shows ***carrying capacity of the land*** in terms of its productivity and its ability to pport various human activities in rural and/or urban areas (Fig. 3.1). These maps usually also provide information concerning engineering properties of the soils and rocks (including the stability of the ound), and the conditions of hydrologic regimes. The primary objective of preparing such maps is to obtain maximum amenities from the terrain at the cost of minimum ecological distress and reasonable onomic expenditure. One of this kind of maps relates to agricultural land classification.

The exercise of land classification is integrally linked with national planning. On the basis of quality, ee major categories of land may be recognized (M. Shafi, per. comm., 1985): the *good-quality land* characterized by a thick mantle of nutrient-rich soils and assigned the productivity rating of 2 is used for intensive cultivation; the *medium-quality land,* wearing thinner soil carpets, and having the productivity rating of 1 makes farmland; and the *poor-quality land* which includes mountainous terrain, moorlands, degraded lands, 'usar' lands, etc. have poor quality of soil covers.

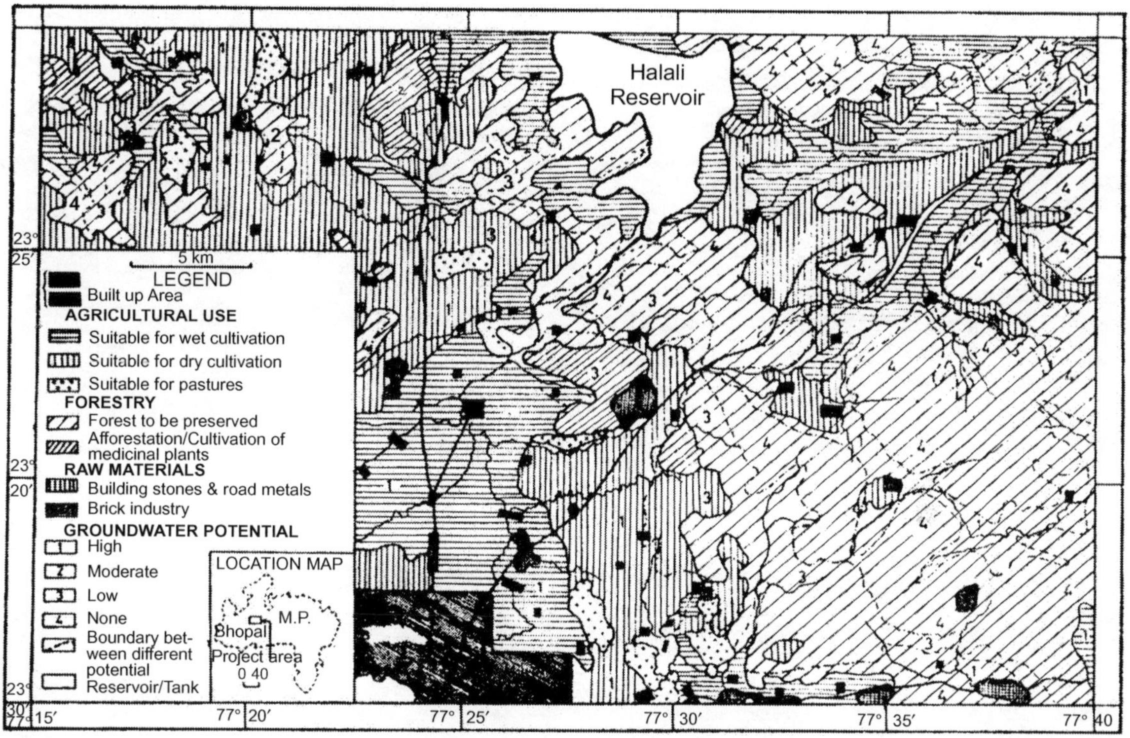

Fig. 3.1. The land-capacity map in the early 1980s of the Bhopal area in Madhya Pradesh. The subdivision has been done in terms of productivity of the land and ability to support various human activities in rural and urban areas. (After Ranga Rao et al., 1982).

Agricultural Land Classification

Agriculture is the mainstay of the human society, and the demand for bringing more and more land under the plough is becoming imperative. About 68.79% of the Indian population (i.e. 833 million Census, 2011) live in the rural areas (Table 3.1), the basis of their livelihood being agricultural lan Agricultural land classification is based on the properties of topsoils — on the features that promo productivity of the soils. The system that is widely followed is the one that was developed by the U.S. Agricultural Department. The eight classes of land, designated by Roman numerals, are based on the degree of limitations and the risk of soil damage (Anon., 1971; Olson, 1981):

Class I: Best quality of soil; suffering least from limitations and exposed to minimum risks of damage.

Class II: Soil suffering from some deficiencies and damage that can be corrected by modera conservation measures.

Class III: Soil suffering from severe maladies, which limit the choice of plant types, and require special treatment to restore it to its original health.

able 3.1. Pattern of Land Utilisation (in thousand ha) in 2007–2008 (Source: Directorate of Economics and Statistics, Ministry of Agriculture, New Delhi)

State/Union Territory	Forests	Not available for cultivation	Permanent pastures & other grazing lands	Land under misc. tree crops and groves (not included in the area sown	Culturable waste	Fallow lands	Current fallows	Net area sown
Andhra Pradesh	6210	4784	571	306	659	1500	2719	10756
Arunachal Pradesh	5154	64	19	37	67	69	41	210
Assam	1833	2626	160	196	77	59	126	2753
Bihar	622	2083	17	241	46	120	566	5665
Chhatisgarh	6350	999	857	1	344	258	255	4727
Goa	125	37	1	1	53	-	10	134
Gujarat	1854	3754	851	4	1977	12	670	9747
Haryana	40	561	26	12	28	8	104	3594
Himachal Pradesh	1101	1130	1501	69	127	14	60	543
Jammu & Kashmir	2023	592	126	66	147	26	67	734
Jharkhand	2239	1319	110	93	333	913	1428	1536
Karnataka	3072	2157	930	290	415	505	1262	10419
Kerala	1082	488	(a)	6	93	45	83	2089
Madhya Pradesh	8703	3392	1352	20	1170	643	790	14687
Maharashtra	5213	3146	1249	248	916	1188	1327	17473
Manipur	1693	27	1	6	1	(a)	(a)	235
Meghalaya	944	227	-	160	434	161	67	235
Mizoram	1594	134	5	67	5	166	45	93
Nagaland	863	78	-	111	64	87	100	316
Odisha	5813	2138	494	342	375	229	556	5624
Panjab	287	507	3	3	3	1	41	4187
Rajasthan	2727	4264	1703	16	4573	2167	1724	17096
Sikkim	319	250	4	5	2	30	5	108
Tamil Nadu	2106	2661	110	261	347	1449	981	5062
Tripura	606	134	-	27	1	1	1	280
Uttarakhand	3465	472	220	269	367	64	44	765
Uttar Pradesh	1658	3268	65	374	440	540	1408	16417
West Bengal	1174	1783	6	61	33	20	311	5296
Andaman & Nicobar Islands	717	24	6	18	12	3	1	13
Chandigarh	(a)	5	-	(a)	-	(a)	(a)	1
Dadra & Nagar Haveli	20	4	1	-	(a)	1	2	21
Daman & Diu	-	-	-	-	-	-	-	2
Delhi	1	92	(a)	1	10	8	12	23
Lakshadweep	-	-	-	-	-	-	-	3
Puducherry	-	18	(a)	1	4	2	3	20

Class IV: Very sick soil so that only very few types of plants can grow. The treatment has to be intensive and careful.

Class V: Excepting erosion, the soil suffers from other maladies so that the land cannot be used for agriculture.

Class VI: Land wholly unsuited for agriculture, but quite suited for pastures, forests and wildlife sanctuaries.

Class VII: Soil suffering from many limitations so that the land can be used only for grazing, commercial plantations and wildlife sanctuaries.

Class VIII: Severely sick soil, unsuited for growth of vegetation of any kind so that the land can be used only for developing recreation parks, water-resource development, etc.

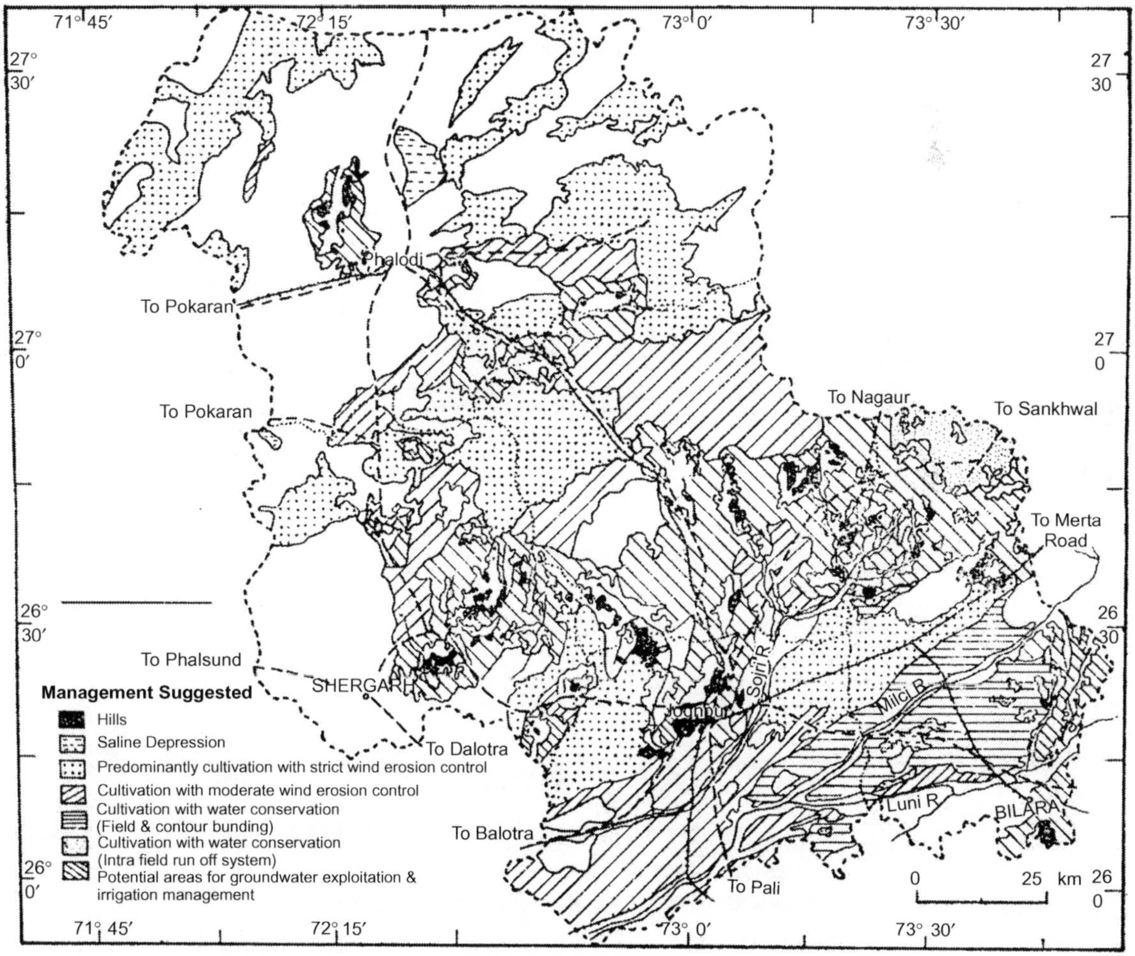

Fig. 3.2. As a result of measures to combat desertification and bring the land under canal irrigation, the fa of Jodhpur district in northwestern Rajasthan changed and the land-capability classification had be considerably modified. (CAZRI, 1983).

Table 3.2. Rural–Urban Distribution of Population and Proportion of Rural and Urban Population in India (National Census, 2011)

State/Union Territory	Population			Population (in %)	
	Total	Rural	Urban	Rural	Urban
India	**1,21,01,93,422**	**83,30,87,662**	**37,71,05,760**	**68.84**	**31.16**
Jammu & Kashmir	1,25,48,926	91, 34,820	34,14,106	72.79	27.21
Himachal Pradesh	68,56,509	61,67,805	6,88,704	89.96	10.04
Panjab	2,77,04,236	1,73,16,800	1,03,87,436	62.51	37.49
Chandigarh	10,54,686	29,004	10,25,682	2.75	97.25
Uttarakhand	1,01,16,752	70,25,583	30,91,169	69.45	30.55
Haryana	2,53,53,081	1,65,31,493	88,21,588	65.21	34.79
Delhi	1,67,53,235	4,19,319	1,63,33,916	2.50	97.50
Rajasthan	6,86,21,012	5,15,40,236	1,70,80,776	75.11	24.89
Uttar Pradesh	19,95,81,477	15,51,11,022	4,44,70,455	77.72	22.28
Bihar	10,38,04,637	9,20,75,028	1,17,29,609	88.70	11.30
Sikkim	6,07,688	4,55,962	1,51,726	75.03	24.97
Arunachal Pradesh	13,82,611	10,69,165	3,13,446	77.33	22.67
Nagaland	19,80,602	14,06,861	5,73,741	71.03	28.97
Manipur	27,21,756	18,99,624	8,22,132	69.79	30.21
Mizoram	10,91,014	5,29,037	5,61,977	48.49	51.51
Tripura	36,71,032	27,10,051	9,60,981	73.82	26.18
Meghalaya	29,64,007	23,68,971	5,95,036	79.92	20.08
Assam	3,11,69,272	2,67,80,516	43,88,756	85.92	14.08
West Bengal	9,13,47,736	6,22,13,676	2,91,34,060	68.11	31.89
Jharkhand	3,29,66,238	2,50,36,946	79,29,292	75.95	24.05
Odisha	4,19,47,358	3,49,51,234	69,96,124	83.32	16.68
Chhattisgarh	2,55,40,196	1,96,03,658	69,36,538	76.76	23.24
Madhya Pradesh	7,25,97,565	5,25,37,899	2,00,59,666	72.37	27.63
Gujarat	6,03,83,628	3,46,70,817	2,57,12,811	57.42	42.58
Daman & Diu	2,42,911	60,331	1,82,580	24.84	75.16
Dadar & Nagar Haveli	3,42,853	1,83,024	1,59,829	53.38	46.62
Maharashtra	11,23,72,972	6,15,45,441	5,08,27,531	54.77	45.23
Andhra Pradesh	8,46,65,533	5,63,11,788	2,83,53,745	66.51	33.49
Karnataka	6,11,30,704	3,75,52,529	2,35,78,175	61.43	38.57
Goa	14,57,723	5,51,414	9,06,309	37.83	62.17
Lakshadweep	64,429	14,121	50,308	21.92	78.08
Kerala	3,33,87,677	1,74,55,506	1,59,32,171	52.28	47.72
Tamil Nadu	7,21,38,958	3,71,89,229	3,49,49,729	51.55	48.45
Puducherry	12,44,464	3,94,341	8,50,123	31.69	68.31
Andaman & Nicobar Islands	3,79,944	2,44,411	1,35,533	64.33	35.67

Needless to state, the category (class) of the land changes with improvement through conservati measures. An example of this phenomenon was witnessed in Jodhpur district, Rajasthan. As a result of measures taken to combat desertification by the Central Arid Zone Research Institute, and t coming of a vast stretch of land in the northwestern part under the network of canal irrigation (Indira Gandhi Canal), the face of the region changed and the land classification had to be drastically modifi (Fig. 3.2).

HUMAN SETTLEMENTS AND LAND-USE

Growing Slums

Nearly 31.16 % of the Indian population, (i.e. 377.1 million in 2011) live in urban centres (Table 3.2 In 1901 there were only 25 cities with populations larger than one lakh people; in 1984 there were 21 Mumbai, Chennai, Kolkata, Delhi and Karachi were then among the world's 40 largest urban centres, each with over 10 million people. **These metropolis, like other cities of Asia "are growing too fast f their own good", because more than 40% of the people of the urban world in India are compell to live in slums and squatter settlements (jhuggi-jhoparis)**. If this growth goes unchecked, of every 10 city-dwelling Asians, six will be living in slums or on footpaths, and large cities would become graveyards of civilization. The bigger cities are growing faster than smaller ones, and the growth rate of the urban population is astoundingly high (Table 3.2).

The urban areas are facing more serious problems than the rural, for it is in the cities where the pressure of rapidly growing population is strongest, and the problems of shelter, water, sanitation and food are unsustainably acute. The urban population has less to eat than the rural population dominated by agriculturalists. While 15% of the rural population suffers from undernourishment and malnutrition, more- than 25% of the people in urban areas suffer from these deficiencies (Sukhatme, 1983). Thu there is greater need to look hard at the cities.

Geography of Cities

Cities, with their dense agglomerations of buildings and utility structures, have their own geography of land-use and areas of degradation. The inner part is always the most derelict socially, economically and environmentally—where the social structure is severely strained, homelessness very common, vandalism very rampant, and water, air and land in extremely deteriorated states. Tom Lehrer's comment echoes the lament on an Indian city, "If you visit a big Indian city, you will find it very pretty. Just two thin of which you must beware. Don't drink the water and don't breathe the air." It is this inner city which is the focus of all attention, including planned changes.

This kind of city is expanding at a very fast rate, nibbling and swallowing productive and healthier agricultural land at the rate of 30 ha/ minute all over the world (Coleman, 1982). Urbanization in Ind had swallowed 1.5 million ha of agricultural lands in the period 1950–1985, and another 0.8 million was lost by 2000 A.D. At present, cities and towns occupy 1.2% (22.5 m ha), primarily agricultural la (Bhardwaj et al, 2007). Suburbs after suburbs are coalescing with the inner cities, and neighbouring citı are merging to form ***conurban areas*****. These cities are constantly hungry for all kinds of resources, which flow from the rural hinterland into the cities, which in turn produce colossal volumes wastes and pollutants — fouling the atmosphere and water and making the landscape ugly.**

Requirements of Urban Planning

Land management for the urban sector requires planning not only for residential, commercial, institutional d public utility buildings, but also for purification of water, disposal of wastes, treatment of sewer ater, facilities for recreation, etc. (Fig. 3.3). There are, then problems relating to the stability of the ground-foundation conditions for construction of buildings, availability of construction material, and condition of soils, and rocks suitable for laying down of roads, sewerlines, pipelines, etc.

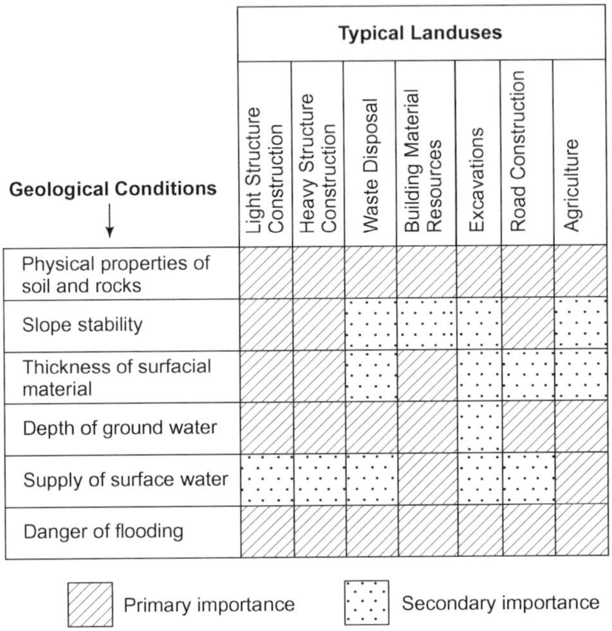

Fig. 3.3. Urban land-use planning requires consideration of various geological parameters relating to the stability of the ground, the groundwater condition, the availability of site for disposal of wastes, the proneness to hazards, etc. (Modified after E.A. Keller, 1982, *Environmental Geology*, Bell and Howell, London, 488 p.)

Another very important consideration is the freedom or safety from natural hazards like floods, ground subsidence, landslides, etc.

Rural Planning

The third vital problem relates to the unending tide of migration from the rural to the urban centres. If this trend is not reversed by A.D. 2025, more than half the population will be living in big cities and towns. **donesia has shown the way for reversing the trend by opening up new fields of employment peripheral areas of the country where tree-farms, oil refineries and fertilizer factories were tablished. The poor and the landless taken there from the densely populated districts, from eas prone to natural hazards, and from areas submerged under impounded water behind dams re given 0.25 hectare of land for housing and one ha for farming—*both of which are developed by the Government.* For eighteen months the food as well as agricultural inputs were provided by**

the Government. The scheme strengthened the population base in the border regions and reliev congestion in cities. India needs an ambitious programme of development of new towns and cities in non-agricultural areas. The programme would involve massive housing in which people should also come forward to invest their savings.

LAND-USE PATTERN IN THE COUNTRY

National Perspective

About 50.9% (142.82 m ha) of the geographic area in India is under agriculture, 22.9% (68.90 m h of the land is forested, 3.6% (11.04 m ha) used as grasslands and pastures, and 1.2% (22.45 m ha) utilized by cities and towns (Tables 3.1 and 3.3). **The land : man ratio in rural India** was 0.9 in 197 1972, ~ 0.8 in 1983-1984, **and only 0.45 in 2012 (Fig. 3.6), implying progressively shrinking spa for livelihood for the rural Indians**.

The per capita availability of land which was 0.5 ha in 1981 and reduced to 0.33 ha by 2000. So f as cultivable land is concerned, the per capita availability was 0.20 ha in 1981 compared to 0.48 ha 1951, and was down to 0.14 ha by 2000.

Table 3.3. Land-use Statistics in India (in m ha) (Bhardwaj et al 2007)

Type of land	Quantity
Forest area	68.90
Area covered by towns, cities	22.45
Permanent Pastures	11.04
Miscellaneous tree crops	3.57
Barren uncultivable land	19.09
Cultivable wastelands	13.94
Current fallows	13.33
Other fallows	9.89
Net sown area	142.82
Area for which no data available	22.7
Total (m ha)	334.80

Distressingly, more than 33 m ha (10.8%) of land in India is wasteland—19.1 m ha of which have lo their productivity forever. Additionally, 23.22 m ha of the farmland is fallow or currently kept fallow (Bhardwaj et al, 2007). A quarter of the land brought under irrigation has gone out of cultivation owi to mismanagement of soil and water resources (Chapters Four and Five). Indian planners still do n seem to realize the dimensions of the side-effects of irrigation and development programmes that are pursued without taking into consideration and without taking care of the environment.

Agricultural lands embrace the alluvial tracts made up of Quaternary to Recent sediments (Fig. 3.4 Even in the Himalayan domain river terraces support fertile fields. The greatest density of populati is seen in the valleys in the mountains and in the flood plains of the Ganga, Brahmaputra, Narmad Sabarmati, Godavari, Krishna and Kaveri systems. What is left of our once vast forests is confined the inner Himalayan regions, predominantly on the northern slopes of the ranges, the Indo-Myanmar ranges, the tops of the Satpura and Vindhya ranges in Central India and the higher tracts of the Sahyadri Range in western and south-western India.

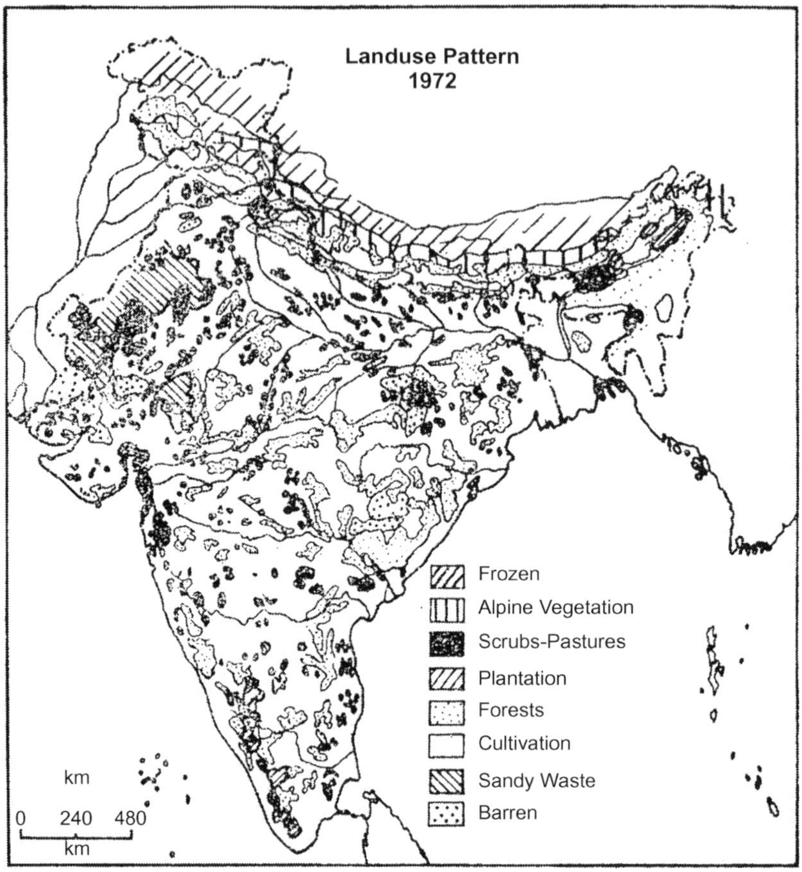

Fig. 3.4. Land-use pattern in India in 1972. (Based on *New Tajmahal Atlas*, 1974, Orient Longman, Mumbai, 9 p.)

Between the forests and the farms lie pastures, locally shrubby but commonly all grassy (Fig. 3.4). Wastelands have developed in the arid western Rajasthan and in those parts of the agricultural alluvial tracts of Gujarat, Rajasthan, Panjab, Haryana, Uttar Pradesh, Madhya Pradesh, Bihar and Andhra Pradesh where gross mismanagement in the past has damaged or killed the productivity of soils.

Land-use in the Himalayan Region

amples from the Himalayan region (Table 3.4) bring home the point of misuse of land resources. ore than 63% of the population lives in 9.7% land area of this region, the cultivation being confined to er terraces of extremely restricted area and on gentler slopes in lower altitudes (Fig. 3.5). In the late 70s the pressure on land (Table 3.4) was excessive— 17.6 people lived in one hectare of cultivable nd in the Kumaun division (i.e. 0.057 ha/capita) compared to 2.3 to 5.8 people per hectare in the Uttar

Pradesh plains (Moddie, 1981). The situation has not changed much in the last three decades (for near half the population live outside Uttarakhand, in different parts of India). This pressure is driving t people up the slopes, leading to encroachment of forests. In the last two decades the agricultural land has expanded. In the late 1960s the condition in Nepal was no better, where the per capita land available for agriculture was less than 0.1 ha (Rieger, 1977) more than thirty years ago.

Special attention is focused on the Himalayan region because "**what we do or do not do in the Himalayan region today will have considerable impact on the future of the agriculture in the Ind Gangetic Plains**" (Swaminathan, 1982). There is thus a greater need for appropriate land manageme in the Himalayan regions.

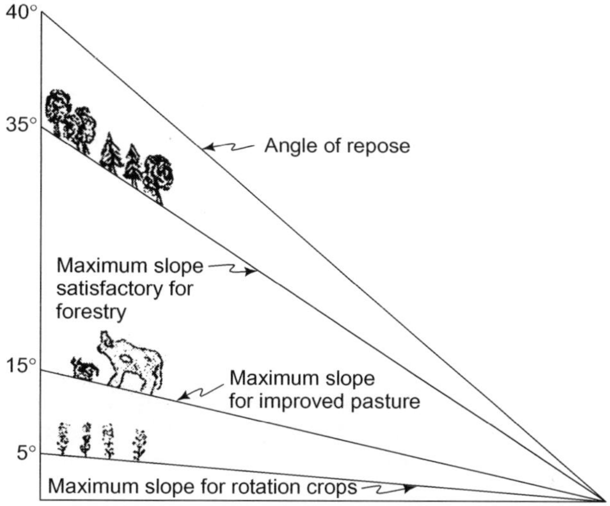

Fig. 3.5. In the mountain regions such as the Himalaya, the slope is the most important factor governing t use of land for dwelling, cropping, grazing and forestry. (Based on Bradshaw and Chadwick, 198

Table 3.4. Land-use Pattern in Indian Himalayan Region (After *Report of Task Force for Himalayan Region* Planning Commission, 2010)

Land use	Percentage in Indian Geographic Area	Whole Himalaya
Forests	65.4%	59.27%
Not available for cultivation	10.0%	3.62%
Cultural Wasteland	5.7%	4.3%
Permanent Pastures and Grazing Land	7.1%	5.5%
Miscellaneous tree crops and groves	2.8%	2.55%
Fallow lands and currently fallow	~ 5%	4.36%
Net Area Sown	13.6%	10.76%
Barren, non-cultural area		9.64%

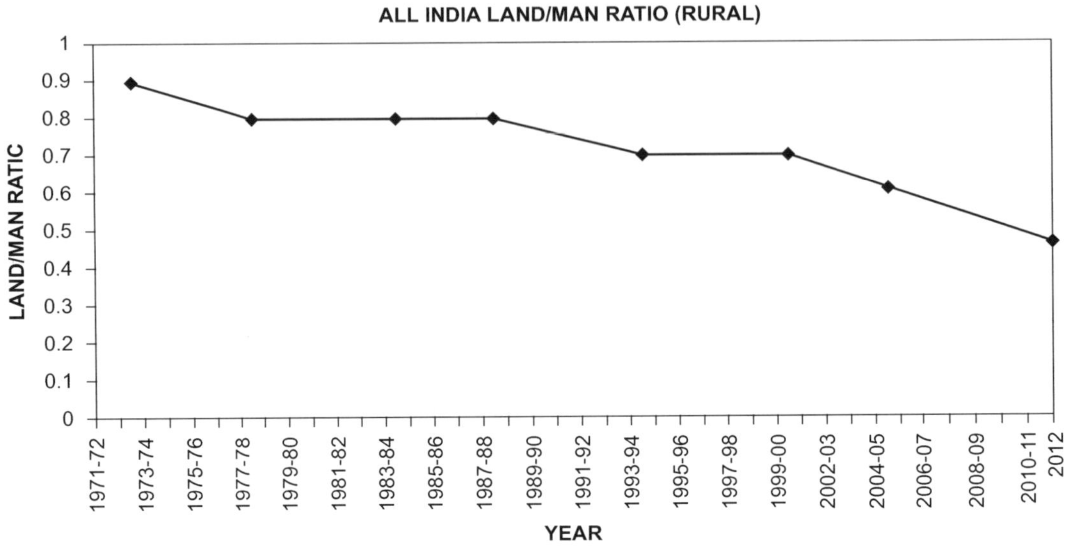

Fig. 3.6. All India land: man ratio in rural India.

An Imprudent Method of Land-use

An example of imprudent if not cruel land-use is the ***slash-burn-and-shift cultivation (jhuming)*** practiced by the tribals belonging to about 100 tribes of Arunachal Pradesh, Nagaland, Manipur, Meghalaya and zoram (~ 207,000 ha) and in the hilly tracts of Kalahandi, Koraput and Phulbani districts in Odisha 3,09,000 ha). The duration of the cycle of *jhuming* of 10 to 20 years in the past has come down to just 2 to 3 years in Tirap district in eastern Arunachal Pradesh, resulting in the decrease not only of the tility of the soil but also in the productivity of the crops grown (Wangpang and Tangjang, 2012). Quite more than half a million hectares of valuable rainforests in the country have been destroyed in this manner to grow paddy, maize and jowar. If jhumland is cropped continuously for more than two years, its fertility declines sharply, forcing the tribals to move on to other forests to repeat the slaughter d carnage of trees. Since soil replenishment takes place in about 8 to 10 years, the cultivators used to return to the original patch to till again only after a decade or so.

In addition to deforestation, the jhuming practice has brought in its trail an invasion of pest species such as the fast-growing blue-green alga and the lantana on the ash-laden scorched ground (Bordoloi, 76). These pests have made the jhumland very sick ecologically.

ASSESSMENT OF IMPACTS OF LAND-USE

The varied utilization of land carried out through a variety of agricultural and engineering activities s brought about considerable, locally drastic, modification of the morphology of the land, and caused profound ecological changes. There is thus a need for regular monitoring of the changes taking place and exhaustive appraisal of the impacts of land-use. This is an essential element of the land-management programme. Environmental impact assessment involves the following:

1. *Initial reference state*—embracing a description of the land-use pattern and existing state of the environment, the population density as well as the economic condition of the people.
2. *Environmental consequence*—involving discussion on the nature and magnitude of the changes likely to occur as a result of implementation of the proposed programme.
3. *Relevant human concern*—implying identification of social, emotional, psychological a economic stresses among the people affected.
4. *Alternatives*—or proposition of options after rigorous comparison of various alternatives.
5. *Remedial and preventive measures*—for recommendation to ensure maintenance, if not improvement, of the environment and the integrity of natural resources.

SUMMARY

Land is a finite resource, and its use must be governed by its stability and capabilities. The misconception that "land belonging to everybody is a property of nobody" lies at the root of the widespread and cruel abuse of this vulnerable resource. Land which should have been under forests, has been turned into agricultural fields and even the wildest of landscapes have been plundered. **The opening of roads have not only grievously hurt the stability of mountain slopes and damag hydrological systems, but also have encouraged vandals, poachers and plunderers to enter de inside the sanctuaries of forests.** Valuable productive lands and forests have been submerged under impounded water behind dams to generate power and to irrigate farms, the latter practice leading to soil-crippling salinization-alkalinization and water-logging of fertile farms. In rural, urban and regional land management exercises, recognition and demarcation of lands capable of supporting various land-utilization schemes should be undertaken. Identification of natural hazards that are likely to threat the land in use, and locating the various resources needed (e.g. water, construction material, timber, fuel-wood, etc.) are the imperatives of land-use planning. The earth-scientist prepares land-capabilı maps on the basis of the data collected through comprehensive field work. The third essential eleme of the planning is the appraisal of the likely impacts on the environment—social, psychological as well as natural.

About 50.9% of the geographic area in India is under cultivation, 22.6% under forest cover, 2.6 exposed to grazing cattle, and remaining land used for varying other purposes such as settlements, industrialization, tree-cropping, etc. More than 33 m ha of land is barren wasteland.

Nearly 31.16% of the Indian population resides in the urban areas. The cities are rapidly expandi all over the world — swallowing up productive agricultural land at the global rate of more than 30 h minute. Simultaneously, **the inner parts of cities are deteriorating socially and environmentally, n only because of the lack of space for various land-uses and human activities but also due to t absence of urge and responsibility for improvement. The resource-hungry cities have thus becom producers of pollution and have themselves deteriorated into slums as a result of failure of prop land-use planning.**

The only hope lies in reversing the trend of migration of villagers to urban centres by openi new avenues of employment and centres of production, and developing areas of habitation wi requisite amenities in peripheral areas and in remote rural sectors.

No project of resource-development or transportation can be embarked upon without first making comprehensive appraisal of the environmental, social and economic impacts of the implementation of the plan.

4 Soils – Types, Distresses and Conservation

NATURE OF SOIL

Soil is one of the most precious resources of the earth. Forming a carpet of variable thickness over e land, it has sustained a succession of life and civilizations, providing food, fodder, fuel and fiber to all kinds of life, storing life-giving water, and supporting shelters and dwellings, and the structures that man builds. Soil is integrally and intimately connected with the rocks beneath, the vegetation growing above, and the water percolating through it.

Soil consists of products of weathering of rocks, intermixed with living organisms and the products their decay, the moisture and air filling the interstitial space. It thus has a microclimate of its own, controlled by water, air and organisms. The silt particles constitute the framework (or skeleton) and the clays (such as montmorillonite, illite, kaolinite and organic matter (humus) hold together the framework, and also serving as the nutrient base. The colloidal clays and humus are endowed with the capacity for cation exchange, and are thus chemically active, attracting molecules of water which form a sort of mosphere around the clay particles. It is the water-attracting and water-holding capabilities of the colloidal humus and clays that keep the soil moist even during the drier seasons and permit microbial activities. **Undeniably, the soil is a living system. As it has the capacity to hold moisture or water, as it contains nutrients, and is endowed with an internal micro-atmosphere of air vapour, it is capable of supporting vegetation and promoting various biological and biochemical processes.**

Geologists are used to thinking of soil as the upper part of the weathered material (***regolith***) resting on the bedrock and containing inorganic and organic nutrients. Engineers look at it as a surface material that can be easily removed without requiring blasting opcration. But **to an environmentalist the soil is a dynamic living layer forming the foundation of all ecosystems**. To those who are concerned with the primary production systems, such as farmers, foresters, horticulturists, and livestock rearers, **the soil is a living and perpetuating base; as it not only sustains and nourishes the primary productions, but also serves as both a passage and store for groundwater which is so essential for survival after the cessation of rainfall** (Figs. 4.1 and 4.2).

There are two major categories of soils. *In situ* or *zonal soils*, which occur usually in relatively flat or undulating areas, are formed as a result of weathering. *Drift* or *azonal soils* are formed over the colluvial (emplaced by mass-movements) or alluvial (river-deposited) sediments, and are thus transported soils (Fig. 4.3).

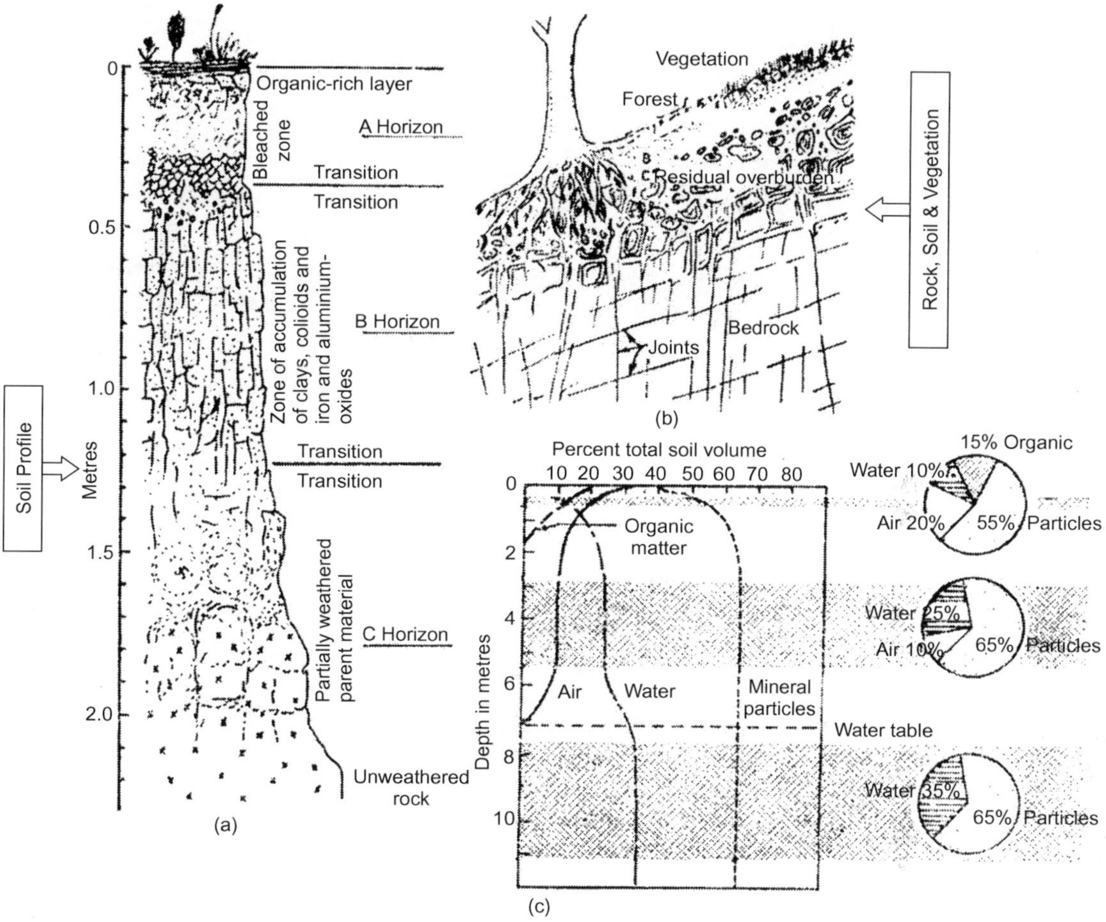

Fig. 4.1. Soil is integrally connected with the bedrock beneath, the vegetation growing above, and the water percolation through it. The compositions of the soils change with the kinds of vegetation that grows. (a) and (b) After Stahler, 1973), (c) After Marsh and Dozier, 1981)

SOIL PROFILE

The ***in situ soil*** grades downward into a zone of weathered rocks—"decayed" into an admixture of clays, silts and sand-sized particles. This horizon is called the ***regolith*** The uppermost part of the soil, known as the ***topsoil* or *A horizon*** (Fig 4.1), is the true soil supporting life through its contents—humus, mineral nutrients and moisture and air. The topsoil may further be subdivided into layers (A A_2.....A_n) **on the basis of progressive stages of humification of contained organic matter. The characte** and thickness of the topsoil varies from place to place, depending on the composition of the parental bedrocks, topography, climate conditions, groundwater regimes and the living environment (Fig. 4.2).

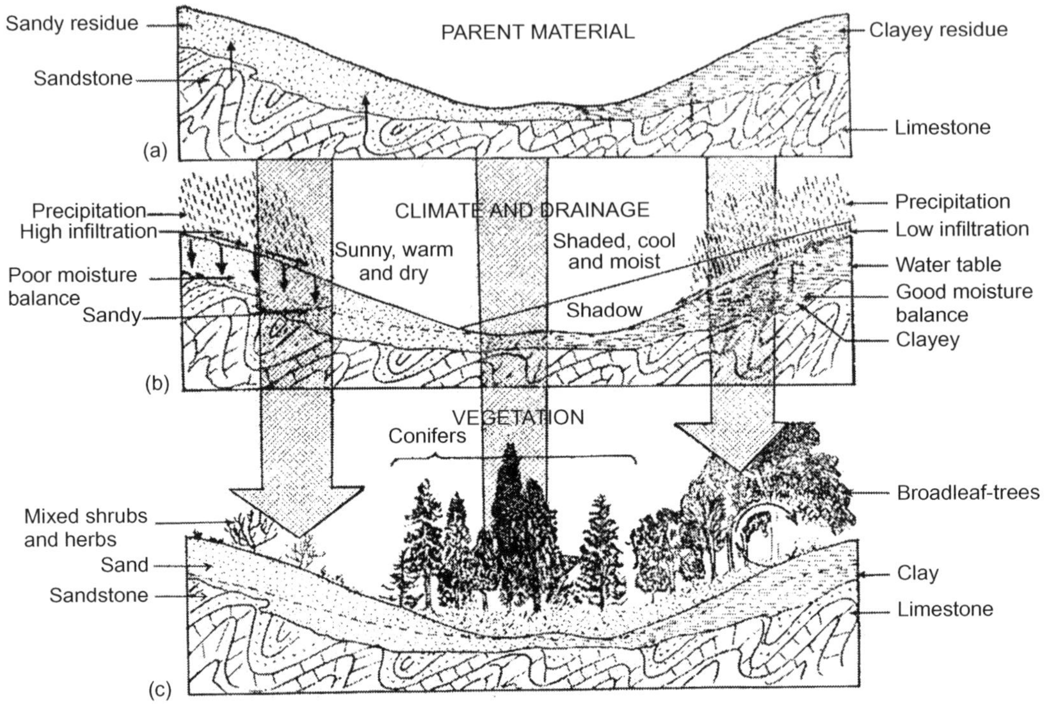

COMPOSITE SOIL TRAITS	Sunny slope	Valley bottom	Shaded slope
Composition	Sandy	Organic	Clayey
Organic content	Very low	Very high	High
Moisture	Low	Very high	Intermediate
pH	6-7	4.5 - 5.5	7 - 8

Fig. 4.2. The composition of the bedrock (lithology), the climate conditions, the drainage and the nature of vegetal cover govern the character and extent of soils. (Based on W. M. Marsh and J. Dozier, 1981, *Landscape: An Introduction to Physical Geography,* Addison-Wesley, Reading, 637 p., Fig. 16.6).

the forest soils the litter and roots decay slowly but mineralization takes place comparatively rapidly so that the resulting layers of humus are thinner. The grassland soils contain on an average 245 t/ha (600 tons/acre) of humus in comparison with 20 t/ha (50 t/acre) in the forest soil (Odum, 1959).

In the humid region the organic matter accumulates very rapidly, giving rise to a soil profile within a century or so but in drier regions the structural development of the soil may take a few thousand years.

The next lower layer, ***B horizon,*** is composed of mineral-rich soil in which organic matter occurs in small amounts owing to its decomposition but is secondarily enriched in solute material (leaching down from the ***A horizon)*** deposited within its framework. The ***C horizon*** may be tens of metres thick (Fig. 4.2) in comparison with the topsoil, which is commonly a metre or less thick. The ***C horizon*** is indeed e unmodified weathered rock material called the ***regolith.***

In the Himalayan region the steeper slopes (>30°) have usually a very thin mantle of soil (Fig. 4.4b) without evidence of maturity and commonly highly leached out and less productive (Ghildiyal, 1981).

ORIGIN OF SOILS

Controlling Factors

As already adumbrated, topography (relief), climate conditions, groundwater regimes and organic activity (primarily vegetal) govern the genesis of soils. The thickness is primarily controlled by relief and, of course, the length of time of soil formation. On the hillslopes the soil cover is thinner as a result of denudation, while on flat ground the mantle is thicker (Figs. 4.3 and 4.4). ***Pedogenesis*** **(formation of soil)** is a long-term process stretching over periods ranging from a few hundred to a few thousand years. In the tropical climatic zones it takes, as already pointed out, 200 to 1000 years to produce jus centimetre thick soil from the granitic parent rocks.

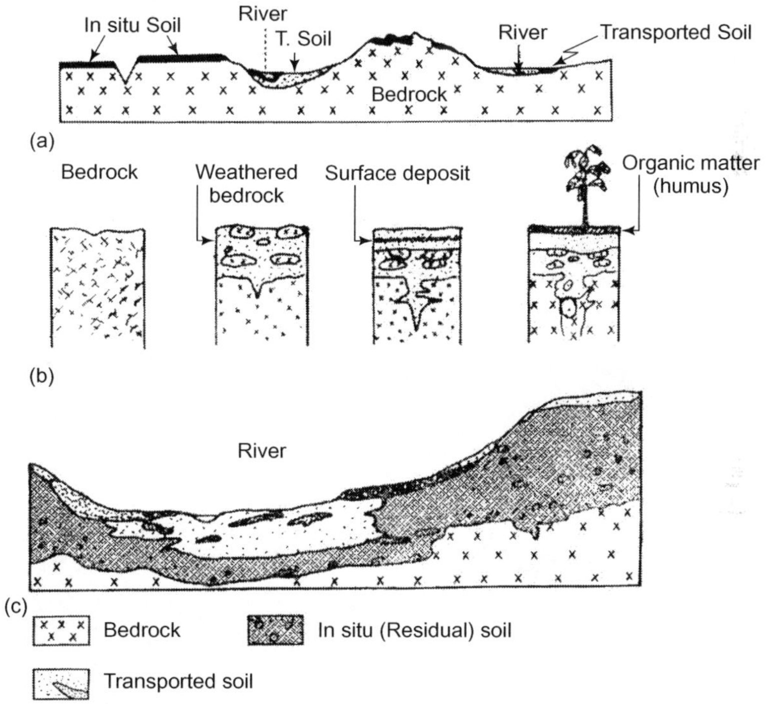

Fig. 4.3. The situation and development of *in situ* (residual) (a & b) and transported soils (c)—the latter in river valley. (Based on Marsh and Dozier, 1981, Fig. 14.5).

The role of climate, particularly higher temperature and rainfall, is very critical in the formation of soils. This is evident from the thickness and nature of the soil profile in relation to latitudes, that climate zones (Fig. 4.5). Water is the controlling factor in determining the chemical composition of soils, stimulating as it does the cation exchange on soil colloids (Fig. 4.7) and regulating the rate of production of plants (Fig. 4.6). In the ultimate analysis, the environment of a soil profile is regulated the quantum of moisture that exists at a point of time (Das, 1985). The stems, roots and leaves of plants yield humus. These vegetal matters are rich in Ca, Mg, K, Na, which the roots bring up from deeper layers of the soils. The presence of humus in the topsoil thus adds to its fertility and productivity.

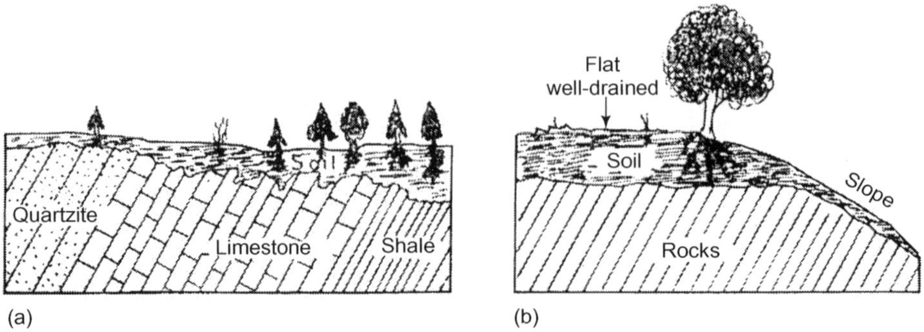

Fig. 4.4. On steeper slopes(>30°) the soil mantle is thinner and the composition deficient in nutrients, and therefore not productive. This is not so in flat terrains. (Based on S. N. Davis, P. H. Reiton and P. Pestrong, 1976, *Geology and Our Environment,* McGraw-Hill, New York, Figs. 8.8-8.9).

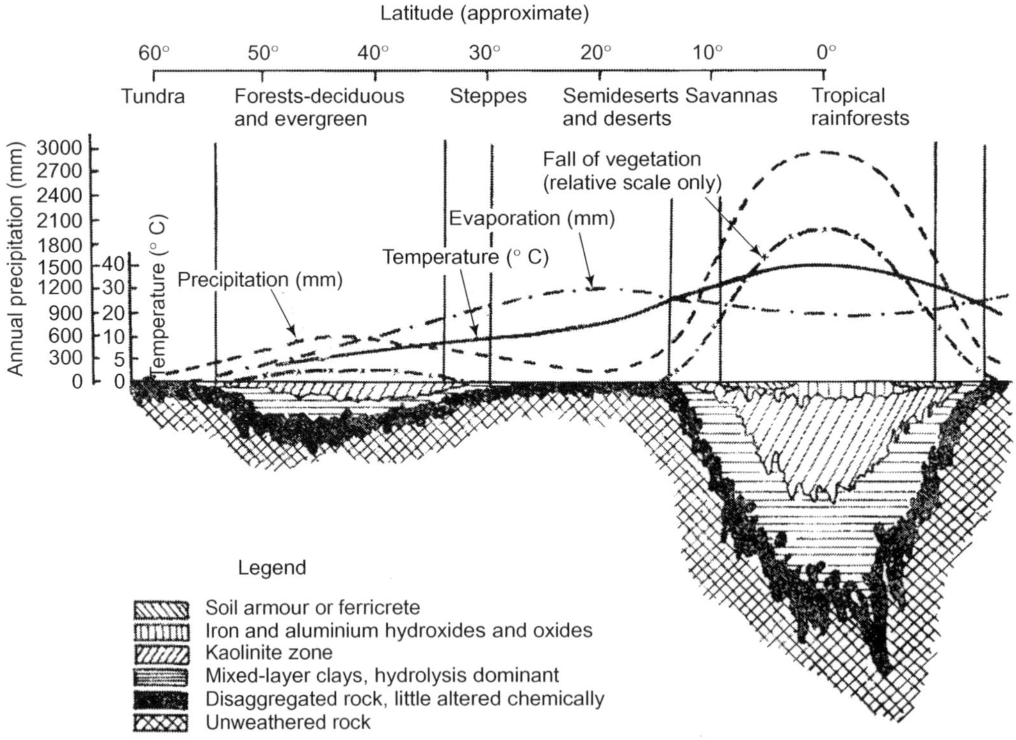

Fig. 4.5. The role of climate, particularly temperature and rainfall, is very critical in the formation of soils. This is evident from the thickness and character of the soils in relation to different climate zones. (After A. L. Bloom, 1978, *Geomorphology,* Prentice-Hall, New York, Fig. 10.6.)

Pedogenic Processes

A number of processes are involved in the genesis of soil (Fig. 4.7). Under cold humid climatic conditions, organic acids are produced during humification. These acids decompose the minerals a the hydrogen ions of the acid, and replace the basic cations K, Na, Ca, Mg. The resulting compounds are leached down by interstitial water from the upper layer, leaving it deficient in bases, colloids and oxid of Fe and Al, but richer in SiO_2. The base-deficient ashgrey acidic soil (pH 4.0 to 5.5) is called ***podsol*** It has low fertility, so that only larger green plants such as coniferous pines, spruces, cedars (devdaa and cypresses thrive. The process of the formation of ashgrey acidic soils is known as ***podsolization*** (Figs. 4.6. and 4.7).

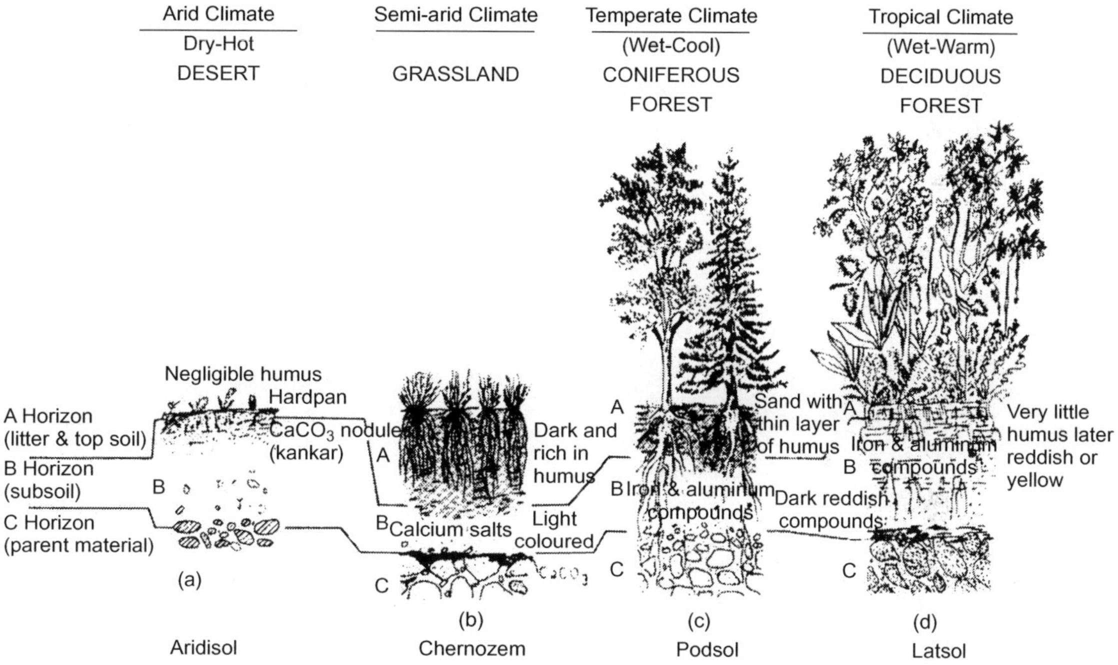

Fig. 4.6. Four major types of soils. The character and thickness of the topsoil varies from place to plac depending on the climate and the living environment (vegetal ecosystems). (Based on Marsh a Dozier, 1981).

In the Himalayan region the high altitude slopes above 2000 m are clothed with coniferous fores the soil being podsolic. The granite-mica schists ridge of Chaubatia-Ranikhet in Kumaun (Uttarakhand Himalaya), for example, supports forests of devdaar (*Cedrus deodara),* spruce *(Picia smithiana,* blue pine *(Pinus griffiths)* and cheer pine *(Pinus roxburghi)* on its northern slope (Ghildiyal, 1981). In t Chakrata Hills (western Uttarakhand) the soil mantle upon the dolomites and limestones interbedd with slates contains small amounts of Fe, Al and K, but greater amounts of Mg, Ca, P, Mn, organic matter and nitrogen (Ghildiyal, 1981). This soil supports forests of spruce, silver fir *(Abies pindrow)* and oaks (*Quercus semecarpifolia* and *Q. dilatata*). The shallow soils of the Siwalik ranges support forests of cheer pine in the higher altitudes and saal *(Shorea robusta)* in the lower heights.

In a very cold climate in poorly drained areas such as swamps and marshes, where the pH is 5 to 5.5 and anaerobic conditions prevail, thick layers of black structureless compact clay called ***glei*** are formed (Fig. 4.7). The humus, in the absence of bacterial activity, accumulates on the soil to produce layers of peat.

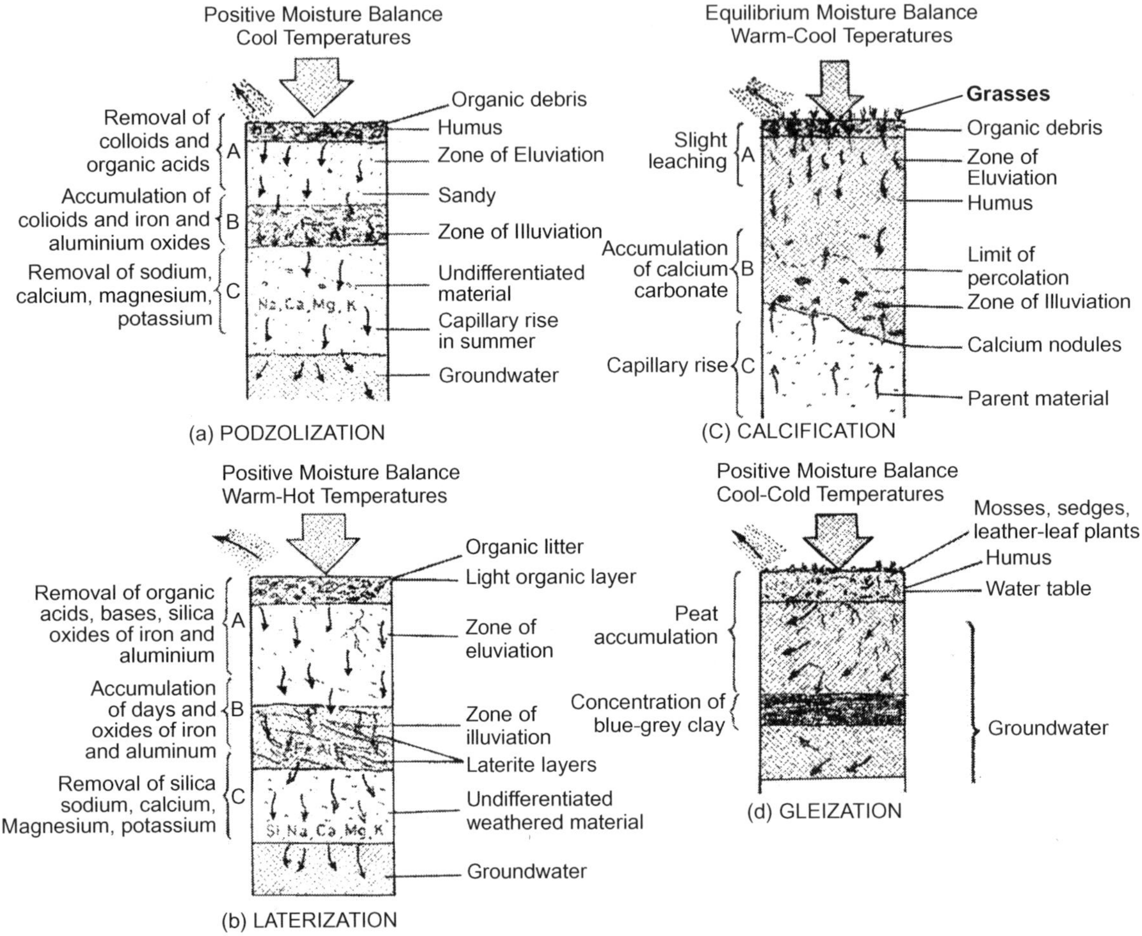

Fig. 4.7. Processes involved in the four types of pedogenesis. (Based on Marsh and Dozier, 1981.)

In the hot, wet climate belts characterized by abundance of water, the bacteria are very vigorous d destroy humus considerably so that the soil becomes deficient in organic acids (Fig. 4.8). In the ficiency of organic acids (pH 6.0 to 6.7), sesquioxides of Fe and Al are not converted, dissolved and removed by leaching. Consequently, the sesquioxides $A1_2O_3$ and Fe_2O_3 (with FeO) accumulate in the soil as red and yellow clays and pisolitic nodules (Fig. 4.7). Bases accumulate in larger proportion in relation to silica, which is leached out. The result is the formation of ***laterosol*** of low fertility. The process is known as ***laterlzation,*** witnessed in larger parts of the Peninsular India including Jharkhand, Madhya Pradesh, Odisha, Andhra Pradesh, Maharashtra, Karnataka and Kerala.

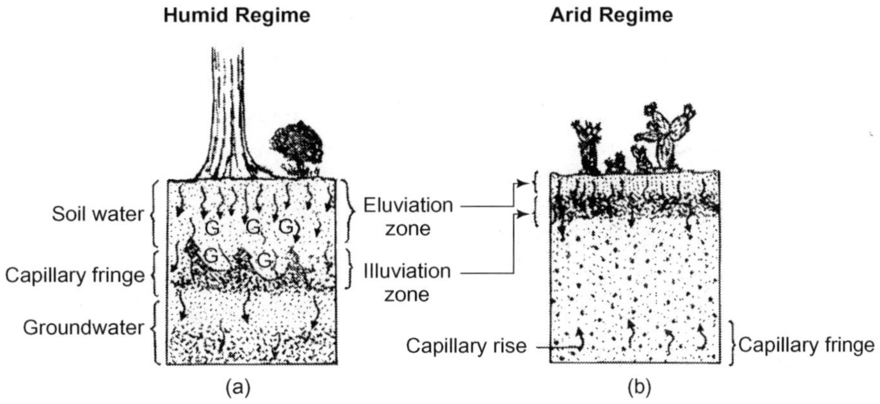

Fig. 4.8. Under humid conditions, the rate of production and destruction of organic matter is faster. At temperature of >20°C the destruction is not high enough to eliminate all organic matter. Above 24°C, the rate of destruction is faster, eliminating all organic matter so that soils in arid regions are deficient in the although rich in down-leached minerals and colloids. (Based on March and Dozier, 1981 Fig. 16.2).

Post-Pedogenic Processes

One of the crucially important functions of bacteria such as *Rhizobium* living in the roots and nodules of leguminous plants (Fig 4.9) is the fixation of atmospheric nitrogen. The molecular nitrogen of t atmosphere is converted into nitrogeneous compounds serving as plant nutrients. Earthworms and other burrowing animals like rodents mechanically rework the soil, making it more porous and permeable.

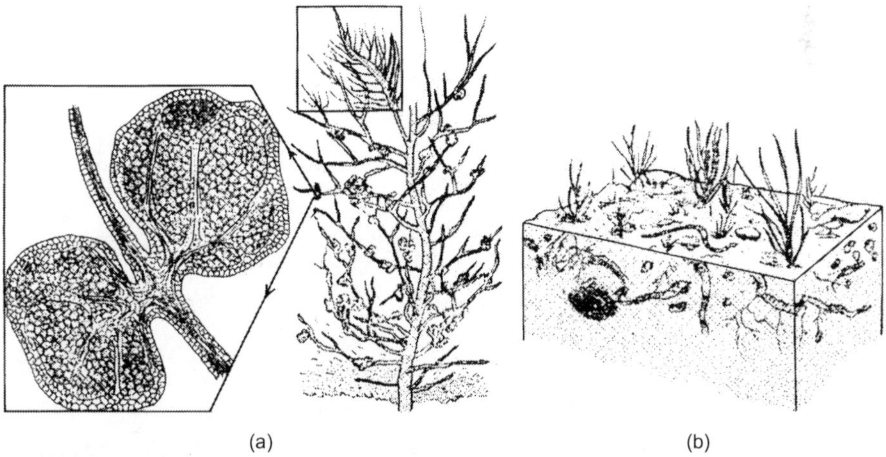

Fig. 4.9. (a) *Rhizobium* bacteria living in the roots and nodules of legumes fix nitrogen in the soil. (After Richter, 1978, *Plant Metabolism,* Croom Helm, London, 475 p.) (b) Earthworms and other burrowi animals mechanically rework the soil, making it permeable to water and air. (W.M. Marsh and J. Dozier, 1981, *Landscape,* Addison Wesley, Reading, 637 p., Fig. 2.18.)

This promotes movement as well as storage of water and air within the soil (Fig. 4.9). By admixing the soil with their excrements, they add to the organic nutrient content. Moreover, moving through the soil, they tend to distribute bacteria, fungi and protozoa, and thus maintain the biological equilibrium within the soil horizon. ***Bioturbation*** is thus an important process which takes place within the soil.

Not all the post-pedogenic progresses are beneficial. In the arid and semi-arid regions such as the ains of Gujarat, Rajasthan, Punjab, Haryana, Uttar Pradesh, Madhya Pradesh and Bihar, where apotranspiration is higher than precipitation, the soil becomes progressively deficient in moisture, rticularly during summer. In the absence of sufficient water, the bases (Ca, Mg) and alkalies (K, Na) dispersed thinly through the soil are not leached out, leading to their enrichment. There is also enrichment in colloids which occur in a flocculated state and hold the soil in aggregate form. To

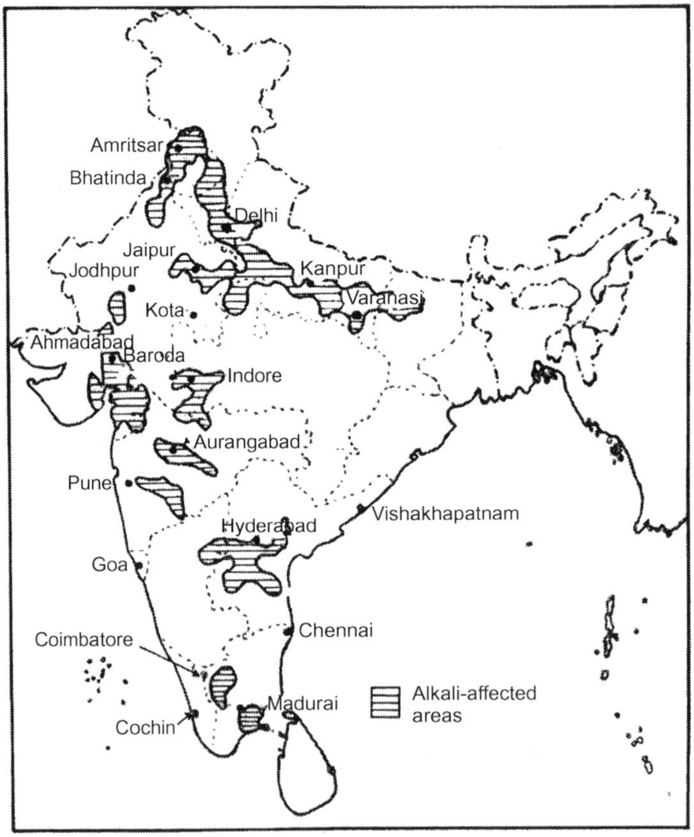

Based upon Survey of India map with the permission of the Surveyor General of India. The territorial waters of India extend into the sea to a distance of twelve nautical miles measured from the appropriate base line. Responsibility for the correctness of internal details shown on the p rests with the publisher The boundary of Meghalaya shown on this map is as interpreted from the North Eastern Areas (Reorganisation) 1971, but has yet to be verified. Government of India Copyright 1986

Fig. 4.10. In semi-arid regions where the soils are deficient in water, the capillary action brings up carbonates, sulphates, nitrates and chlorides of bases and alkalis dissolved in groundwater. This has caused extensive alkalinization and salinization of the irrigated farmlands in India. (Based on K. K. Metha, 1983, Fig. 1.)

aggravate the problem, capillary action brings up additional quantities of, say $CaCO_3$, from the deeper layers and deposits in the topsoil in the form of concretionary nodules called ***calcrete*** or '***kankar***'. The process of calcification in the middle-latitude grasslands has produced a type of soil called ***chernozem*** or black earth (Fig. 4.6). In place of $CaCO_3$ the carbonate and/or chlorides, sulphates, nitrates of sodium may also be brought up from the deeper layers. This represents alkalinization of the soil, entailing concentration of Na_2CO_3 and $NaHCO_3$. Likewise, the nitrates, sulphates and chlorides of Na, Ca and rarely Mg— dominantly Na_2SO_4 and NaCl and less commonly Na_2NO_3, $MgSO_4$ and $MgCl_2$—come up through capillary action, and get deposited in the topsoil, causing sickening or deterioration of the soil leading to ***salinization*** and ***alkalinization***. These are widespread in the poorly drained but overly irrigated fields in Uttar Pradesh, Punjab, Haryana, Rajasthan, Gujarat and Andhra Pradesh. In the coastal areas Tamil Nadu and Kerala, the salinization is due to inundation and salt deposition during high tides. Thus about 6.73 m ha and 9.30 m ha of our irrigated lands are alkalinized and salinized respectively (Anon, 2010). Far more than 2.5 m ha in the Indo-Gangetic Plains were affected by alkalinization (Mehta, 198 in the early 1980s. Needless to state, these processes render the soil infertile practically forever.

On top of this 1603 m ha land is degraded due to ***acidification*** (Anon, 2010).

CLASSIFICATION OF SOILS

The classification of soils is based on determination of observable mineral and measurable chemic compositions and textures of the soil, and on the climate conditions that control their genesis. The classification adopted by the U.S. Soil Survey Division (Anon., 1975) seems to offer the best model it is applicable to both cultivated and virgin lands (Fig. 4.11).

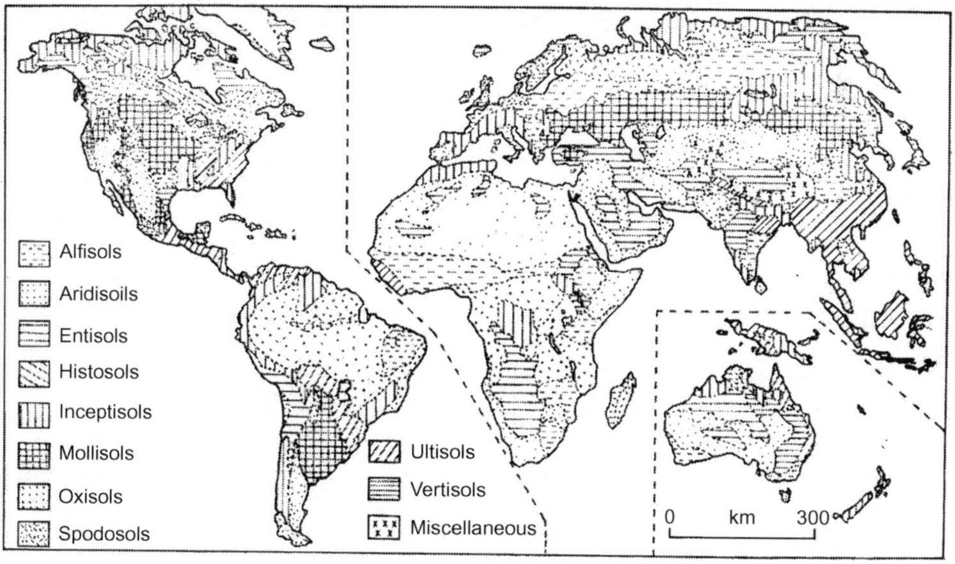

Fig. 4.11. Global distribution of ten categories of soil types (as classified by the U.S. Soil Survey Divisio based on gross morphology, nutrient matter and colour. (After G.W. Olson, 1981, *Soils and the Environment,* Chapman & Hall, New York, p. 178, Fig. 27).

On the basis of gross morphology, nutrient status, quantity of organic matter, colour and general climate consideration, the soils are placed in ten broad categories or orders; and these orders are subdivided on the basis of formative elements into suborders (Olson, 1981, Keller, 1982). The orders are *fisol, aridisol, entisol, histosol, inceptisol, mollisol, oxisol, spodosol, ultisol*, and vertisol (Fig. 4.11).

SOIL TYPES OF INDIA

e larger part of the Peninsular India (Fig 4.12) is covered with in situ soils formed under monsoon climate conditions characterized by alternating hot-dry and hot-wet seasons, which favour dissolution and leaching out of mineral nutrients and concentration of the sesquioxides of aluminium and iron. Only cally this soil may contain organic matter generated by the vegetation that was able to find a foothold. ing very old in terms of time-span, the soils are very mature, as reflected in the profound modification in the clay fraction of the red and black soils.

The ***oxisols*** produced under hot-humid conditions are indeed residual accumulation of inactive clays edominantly (50 to 80%) kaolinite with oxides of Si, Fe and Al. The soil profile is deep, but exhibits little differentiation, barring slight clay enrichment at depths and that the surface horizon of granular to subgranular material is dark red, brown or yellowish red (Fig. 4.13). The deeply ferruginous (Fe_2O_3 + FeO) oxisol, known as ***laterosol*** (Table 4.1; Fig. 4.13), is usually 60 to 90 cm in thickness and caps plateaus, low hills and undulating terrains where the rainfall is 100 to 300 cm/yr, the mean temperature gher than 22°C and the dry season lasts less than four months in a year. The type of oxisol characterized by 50 to 80% dominantly kaolinite clay and abundant sesquioxides (Al_2O_3 and Fe_2O_3) is also called ***nitosol***. The laterosol (pH 4.3 to 6.5) supports rain forests and savannas, despite deficiency in such tural nutrients as magnesia, lime, phosphates and nitrogen compounds. Its good physical property (granularity and permeability) and resistance to erosion have made it a useful soil. With suitable inputs of organic manure and fertilizers, it supports cultivations of potatoes, bananas, pineapples and rubber, among others.

A vast part of the Peninsular India is constituted of orecambrian metamorphic rocks (charnockites, khondalites, granulites, gneisses, amphibolites and schists), with huge igneous intrusive bodies (granites, gabbros, dolerites, pyroxinites, carbonatites) and sedimentary covers (sandstones, quartzites, shales, limestones, dolomites) with a variably thick carpet of lateritic soil. The soil composition varies from place to place, depending on the parent rock. The sandstones of the geological groups or formations called 'Cuddapah' of Andhra Pradesh, the 'Upper Vindhyan' of Madhya Pradesh, Andhra Pradesh and arkhand and the 'Delhi' of Rajasthan have yielded thin infertile siliceous soil, while the shales, marls, limestones and dolomites of the groups known as the 'Kurnool' and the 'Bhima' of Andhra Pradesh, the 'Raipur' of Chhattisgarh, the 'Semri' and the 'Bhander' in southeastern Uttar Pradesh, northwestern Madhya Pradesh and adjoining Rajasthan, and the 'Ajabgarh' of the Aravali range have produced argillocalcareous soils.

The Deccan and Malwa plateaus of western and central India (Fig. 4.12) built up of basaltic lavas d pyroclastics, wear mantles of highly argillaceous (>62%) black and loamy soil (pH 6.0 to 7.8 at the surface) characterized by carbonates of Ca and Mg (6 to 8%), Al_2O_3 (<10%), iron oxides (9 to 10%), small amounts of potash (0.5%) and soda, and traces of phosphates which make the soil very fertile (Wadia, 1966) supporting mixed grasslands-forests, savannas, and crops of sugarcane, groundnut, soyabean, rice, etc. This ***'regur'*** or **black cotton soil or v*ertisol*** (verto=turn) is characterized by high mounts (40 to 48%) of swelling clay montmorillonite (Fig. 4.14). In dry seasons the 'regur' shrinks, so

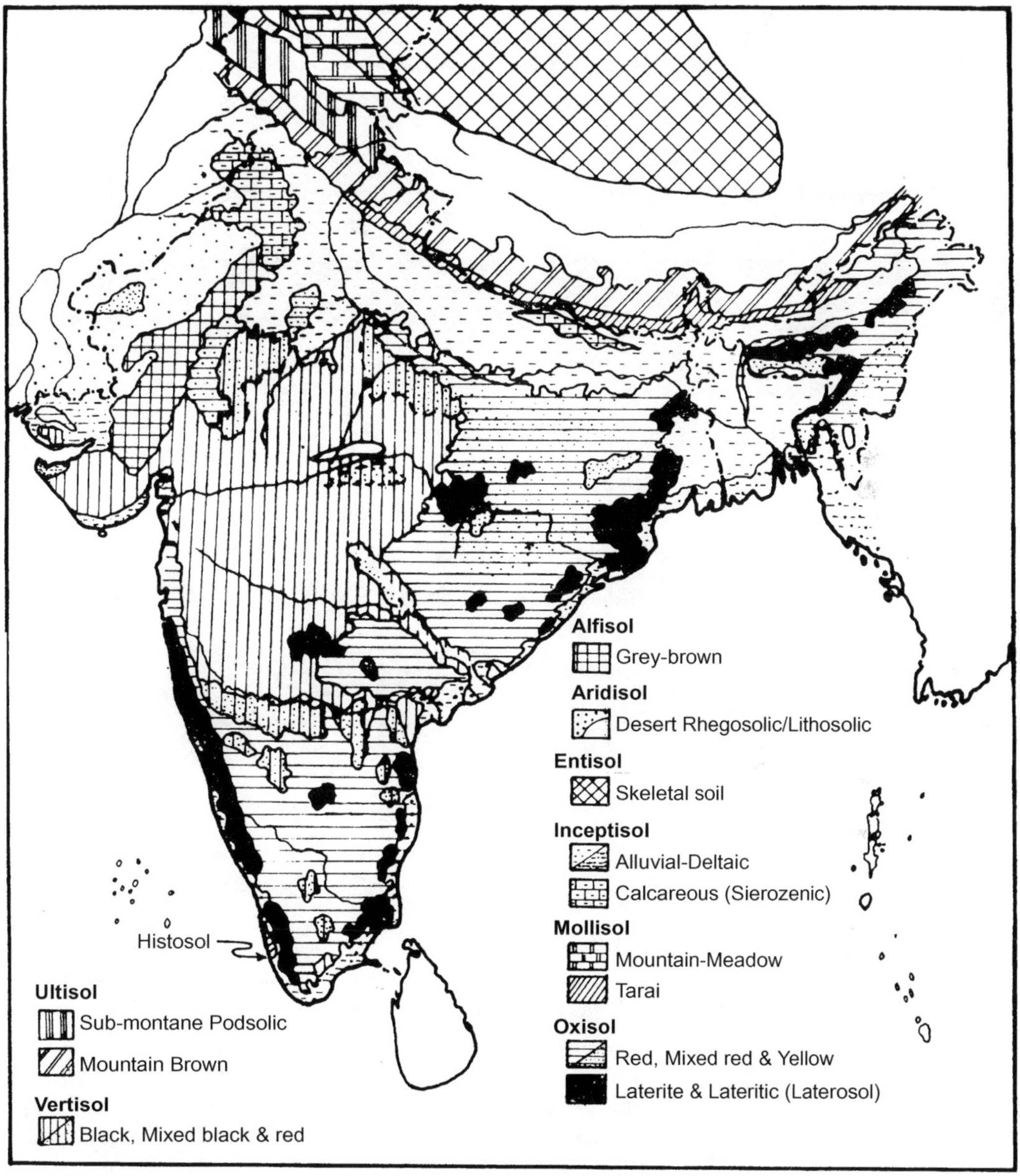

The territorial waters of India extend into the sea to a distance of twelve nautical miles measured from the appropriate base line. Responsibility for the correctness of internal details shown on the map rests with the publisher. Based upon Survey of India map with the permission of the Surveyor General of India Government of India Copyright 1986.

Fig. 4.12. Indian soil types. (Based on plate No. 10 of the *National Atlas of India,* National Atlas Organizatio Kolkata).

at its surface becomes riddled with a network of 5 to 10 cm deep cracks. In the rainy season the soil is converted into a plastics sticky mass. This swelling and shrinking characteristics of the vertisol makes it a bad foundation material for buildings and civil structures. Marginally, the 'regur' soil grades into grey-red soils ***alfisol*** with medium to high base contents and a subsurface clay horizon, usually moist but dry during warmer seasons.

Table 4.1. Subtropical Red Soil from Zimbabwe in Africa (After H. Jenny, 1980). (The composition of the soils of the Dharwar Craton, South India, is somewhat similar to the Zimbabwe soils).

Depth (cm)	pH	Clay %	Exchangeable bases per 100 g	SiO_2/ Al_2O_3	SiO_2/ R_2O_3	Description
0-15	6.1	48	10.0	2.06	1.46	Red and reddish-brown friable
15-38	5.8	62	6.7	2.00	1.44	Soils, many roots, no stains
76-107	6.0	53	5.4	1.99	1.45	
137-145	5.3	48	5.0	2.00	1.45	Hard, partly cemented (Fe, Mn)
145-183	5.6	49	5.3	2.01	1.44	Stains
213-244	5.6	47	5.6	2.05	1.47	Brown, loamy friable, pebbly
244-274	5.8	47	6.1	—	—	Concretions (Mn, Fe)
274-305	5.7	45	6.6	2.15	1.50	Red-yellow mottling
305-335	5.7	39	8.6	—	—	
335-366	6.4	32	11.7	2.34	1.59	Reddish-brown clay
396-427	5.9	29	11.7	2.32	1.64	Light brown clay
427-457	6.2	16	11.9	2.60	1.79	Gritty, last stages of decomposition of lock
457-488	6.4	10	11.3	3.35	2.26	
518-549	7.2	10	13.6	3.27	2.20	Brown-green decomposed rock
579-610	6.8	9	14.1	4.02	2.60	Grey-green decomposed rock
610+						Fresh dolerite rock

In Dakshin Kannada district in coastal Karnataka (Fig. 4.14) on the Paduvari plateau the soil profile consists of a lower zone of lithomarge clay, the middle zone of laterite-aluminous and laterite-bauxite sequence, and the upper zone of enriched laterite cap. During the process of laterization the laterites and bauxites have suffered excessive depletion of Ca, Na, K, Sr, Li and Rb, low mobility of Mg and Be and retention of Na:K and Li-Rb rations of the parental rocks. The laterite contain dominant gibbsite and kaolinite clays (Khanadali and Devaraju, 1987).

The vast Indo-Gangetic Plains extending from Pakistan to Assam and embracing the riparine and deltaic gion of Bengal and Bangladesh, and the flood plains of Mahanadi, Godavari, Krishna, Pennar, Kaveri, the many rivers of Kerala, Tapi, Narmada, Sabarmati, etc. (Fig. 4.12) are covered with drift soil—the ***inceptisol*** (inceptum=beginning). The pedogenic horizon is made up of the considerably altered material transported by rivers. Being very youthful (Subrecent to Recent in age), the alluvial soils exhibit but little pedologic variations. They are monotonously light in colour, loamy in texture, drier in composition and extremely variable in thickness from place to place. These are highly productive soils, constituting the life-pport system of the Indian people. The soil is acidic (pH 5.5) or calcareous (where the content of bases ries from 40 to 70%) or even alkaline and thus unfit for cultivation. This is a result of primarily human interference with the natural balance. In the marshy tracts of the deltas of the Ganga (Sundarban in Benga Godavari, Krishna, Pennar, Kaveri and rivers of northern Kerala between Kochi and Alappuzha, the soils contain abundant amounts of such organic matters as woody peat and decomposed muck forming the ***histosol*** *(histos*=tissue). In the northern border of the Indo-Gangetic Plains in the foothills of Kumaun a the Nepal Himalaya the wet or waterlogged ***Tarai*** belt is mantled with ***mollisol*** (mollis=soft) characterized by a high to moderate content of organic matter and bases. The black surface grades downwards into a grey or olive-grey mottled horizon of fine to very fine sand or loam.

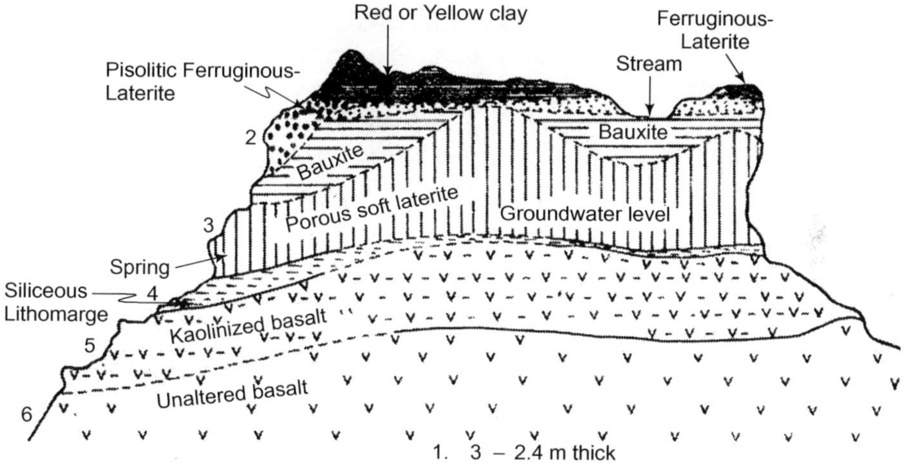

Fig. 4.13. In hot humid climate belts characterized by abundance of water and bacterial activity in the biomass, the humus is destroyed and there is rich production of sesquioxides of Al and Fe (Soil profile of Deccan volcanics, Maharashtra).

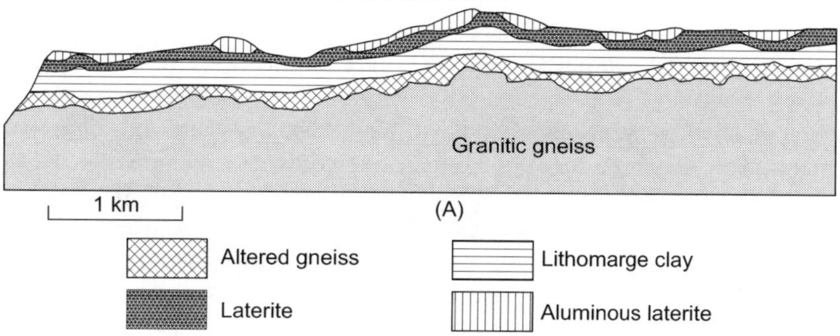

Fig, 4.14. Laterization on the Paduvari Plateau, made up of gneissic granites, in Dakshin Kannada distri showing different zones of the laterite profile. (After Khandali and Devaraju, 1987).

In the mountainous Himalayan region the soil cover is commonly thinner and occurs only on the gentler very commonly northern dip-slopes, the southern scarp slopes being characterized by aprons and cones of screes and colluvial (mass-movement) deposits. The northern slopes thus support lush forests, hile the southern faces are generally bare. At least four types of soils are recognizable in the Himalayan region. The submontane ashgrey to pale yellowish brown podsolic mountain soil, belonging to the order ***ultisol*** (ultimus= lost) (Fig. 4.12) is characterized by deficiency in bases, colloids (including organic matter) and oxides of Fe and Al (due to leaching out by interstitial waters) supporting coniferous forests rich in SiO_2 (pH 4.0-5.5) and oak-rhododendron forests where bases are available. The underlying clay rizon is moderately thick, and brown or yellowish red-brown in colour. Having low fertility, it can support only larger coniferous plants such as pines *(Pinus roxburghi,* P. *griffithi),* devdaars *(Cedrus deodara),* spruce *(Picia smithiana),* cypress (*Cypressus torulosa*), etc. seen covering the middle altitude the entire Himalayan region where the products of podsolization on the usually steeper slopes have suffered a considerable leaching of soluble components (Ghildiyal, 1981). The ***podsols*** developed on the rbonate rocks limestones and dolomites of the geological groups in Vaishnodevi of the 'Jammu', the hali' of Himachal Pradesh, the 'Deoban' of Garhwal and Kumaun, the 'Krol' of Himachal, Garhwal and Kumaun, the 'Dhading' of Nepal, the 'Buxa' of Bhutan containing larger amounts of Ca, Mg, P, n, organic and nitrogenous matters besides Fe, Al and K, support forests of spruce, silver fir *(Abies pindrow),* and oaks *(Quercus semecarpifolia, Q. dilatata, Q. incanda),* etc.

The higher altitudes (>3000 m) with youthful landscape have brown mountain meadow or ***bugyal soil*** belonging to the order ***mollisol*** (Fig. 4.12). The thick topsoil layer is dark, rich in humus and dominated by bivalent cations and the underlying horizon is represented by either altered rock material or clays, which are calcic or sodic depending on the parental rocks underneath. The cold deserts of northern Ladakh and adjacent Tibet have thin covers of skeletal soil or ***entisol,*** characterized by absence of pedogenic horizons, and great variability of texture, colour and composition.

OIL AND VEGETATION

Microclimate Modification

The character of the vegetation depends on the kind and quantity of nutrients present in the soil. nifers, for example, grow prolifically in acidic soils devoid of or deficient in bases and alkalis (Ca, Mg, K, Na). On the other hand, grasses including grain-producing wheat, oat and barley thrive only if e soil contains an adequate amount of Ca and Mg and is a little deficient in moisture.

The vegetation in turn determines the microclimate within the soil. In the absence of a cover of vegetation, the soil gets rapidly heated up, and is beaten by wind and water leading to the impoverishment in its moisture content. Later, in the absence of humus, the soil becomes hard and compact so that rain-water and air fail to enter into the matrix of the soil. This inhibits growth of vegetation (Fig. 4.15). The production humus stops at 40°C or above, resulting in total breakdown of the nutrient cycle and microclimatic equilibrium (Kollmannsperger, 1977). **Since the soil and vegetation are inseparably linked in a close cycle, the destruction of the one brings about degeneration or demise of the other (Fig. 4.15).**

The adverse affects of drastic microclimate change due to deprivation of vegetal cover are quite parent in the once lush terrain of the Himalayan domain. Drought-resistant (xerophytic) plants erant to dryness have taken hold of the deforested slopes, rapidly replacing the local flora on the •ded moistureless soil in many parts of the Lesser Himalayan terrain in Uttarakhand (such as in the Kosi valley). According to Kollmannsperger (1977), out of the 53 plant species in Nepal only 30% are

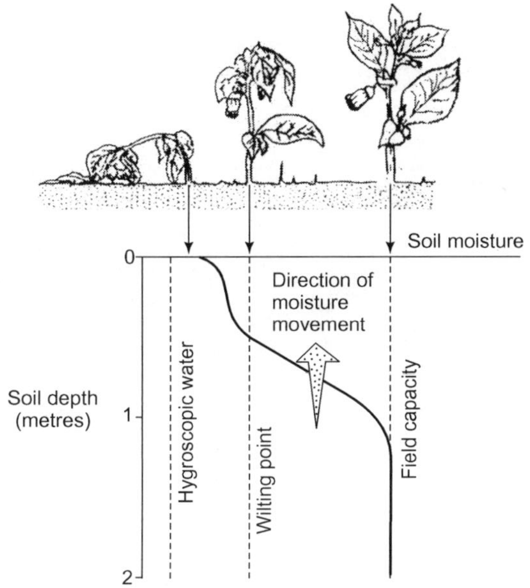

Fig. 4.15. During summer due to high evapotranspiration the content of moisture in the soil diminishes considerably. There is wilting of plants if the drought is prolonged. (W.M. Marsh and J. Dozier, 1981, *Landscape,* Addison-Wesley, Reading, 631 p., Fig. 15.2.)

indigenous, the rest being exotic, including 6 of the 16 species of trees. The signs of desertification the Himalayan terrain are thus fairly apparent.

Change of Shear Strength

The roots of the trees and shrubs provide additional shear strength to the soil (Fig. 4.1b).

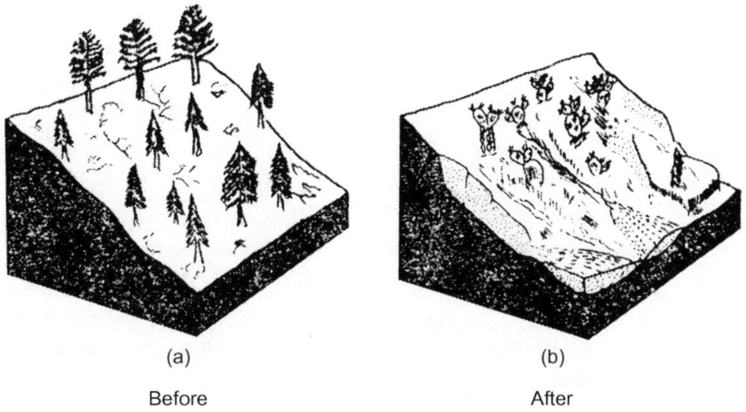

Fig. 4.16. Clear-felling in forests greatly accelerates the rate of erosion as manifest in gully-formation. (After W. M. Marsh and J. Dozier, 1981, Landscape, Addison-Wesley, Reading, 637p., Fig. 26.1).

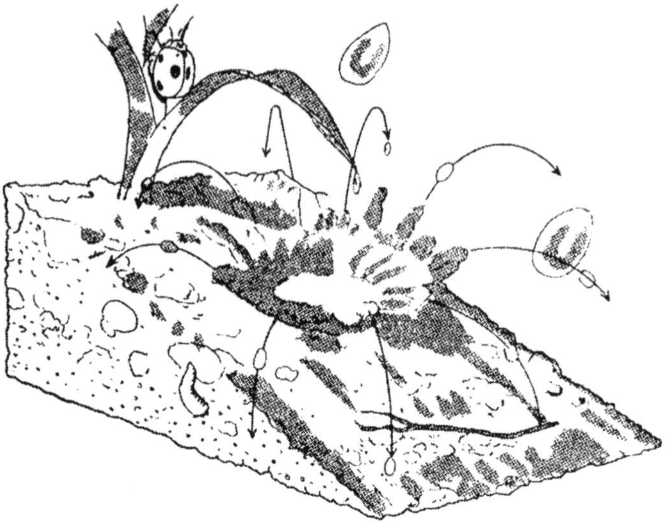

g. 4.17. Deprived of the protection of vegetation, the soil is exposed to uninhibited beating of raindrops. Note the relative lengths of the downhill and uphill trajectories. (W.M. Marsh and J. Dozier, 1981, *Landscape,* Addison-Wesley, Reading, 637 p., Fig. 26.1).

The destruction of vegetation causes weakening of the anchorage the roots provide, and contributes to the loss of shear strength. The soil then becomes vulnerable to mass movements and erosion (Fig. 4.16). Similarly, the degeneration of vegetal cover resulting from the soil infertility mentioned above exposes the bare unprotected soil to the uninhibited beating of falling raindrops (Fig. 4.17), which accentuates the pace and extent of erosion (Chapter Seven). A few examples may prove the point. **In the state of Oregon in U.S.A., clear-felling in forests increased the erosion of soil 100 mes and the incidence of landslides more than five times** (Coates, 1983). In a nearly burnt area der slash-and-burn cultivation in north-eastern India, the loss of soil is 7.24 t/ha/yr compared to less than 0.76-1.13 t/ha/yr in the grass-covered land (Ramakrishan and Tokay, 1978).

The rate of loss of soil, primarily due to erosion, is 15.93 t/km^2/yr in the forested Himachal Pradesh d Jammu-Kashmir Himalaya, 50 t/km^2/yr in the forest-covered northeastern Himalaya, but 330 t/km^2/ in the Punjab-Haryana plains, 1410 t/km^2/yr in the Upper Ganga plains, 940 t/km^2/yr in the Lower Ganga plains, and 4095 t/km^2/yr in the *jhum* land and adjoining alluvial plains in the northeastern India hruva Narayana, 1993). In Sikkim the conversion of a dense mixed forest to open degraded forest increased the soil loss from 4.18 t/hr/yr in 1988 to 8.82 t/ha/yr in 1992 (Rai and Sharma, 1998). Similarly a stream catchment in central Kumaun (Uttarakhand Himalaya) the rate of loss recorded on the slopes covered by oak and pine trees was 0.02 to 0.04 mm/yr and the deforested tract and the farmlands were denuded at the rate four times higher — 0.09 to 0.18 mm/yr (Rawat and Rawat, 1994).

In Nepal the largely deforested terrains registered erosion rate of >6 mm/yr (Arita and Ganzawa, 97). In Tripura, the erosion-caused loss is 14.6 t/ha/yr (Bhattacharyya et al, 2007). The latest estimation non, 2010) shows the following overall rates of erosion in the Himalaya:

Western Himalaya	362 to 3596 tonnes/km^2/yr
Central Himalaya	110 to 3703 tonnes/km^2/yr

Eastern Himalaya   67 to 616 tonnes/km^2/yr

The very higher rate of erosion in Eastern Himalaya is due to jhuming of the forest land.

Another reason why the soil becomes vulnerable to escalated erosion is that the hardened and humus-deficient soil is incapable of absorbing and percolating rainwater. The litter that covers the soil under forest not only protects it from the impacts of falling raindrops but also absorbs a considerable part of the precipitation (Fig. 4.18) which as runoff water would cause erosion and contribute to the floodi of streams (Chapter Seven).

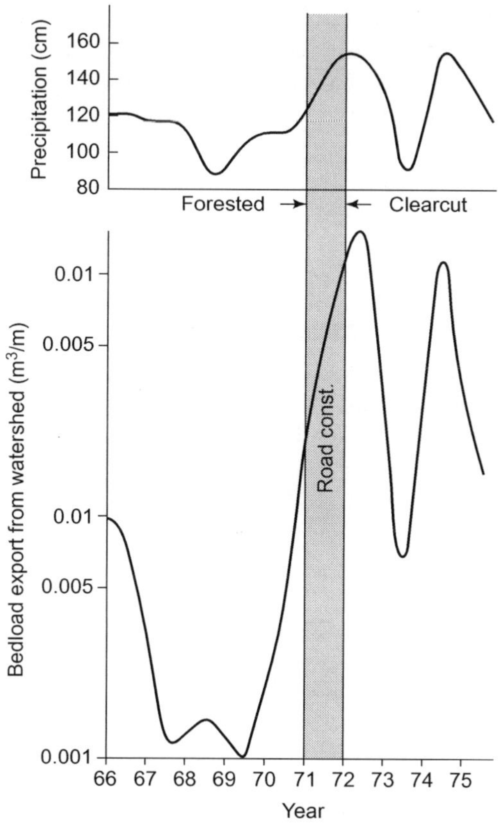

Fig. 4.18. Due to road construction and clear-felling, runoff water would cause erosion, and greatly increase erosion on hillslopes. (Based on Swanson and Swanston, 1977).

Agricultural and Engineering Practices and Soil Deterioration

Unscientific agricultural practices that cause reduction in permeability and roughness of the soil invariab lead to greater erosion (Fig. 4.18). The shallow topsoil becomes saturated with water in a shorter period, so that surface runoff develops sooner than it would if the soil were thicker. Hence the shallow soils are mo prone to erosion, particularly if the degree of slope is higher (>30°) and their strength lower. Since the s loss is commonly higher on convex slopes (compared with that on concave slopes) **any agricultural or engineering activity that breaks the slopes and results in building up of convex slopes (such as the dumping of debris) would escalate the pace of erosion**. Unsystematic terracing for farming, quarrying and mining, road building, excavation for foundations of buildings, etc. always increase the rate of soil s. In the mountains, making plough furrows in the direction of the slopes, **tilling on steeper slopes (>30°) without contour benches and without provision of drainage ditches, and over-cropping without giving rest to the overtaxed soils have greatly contributed to the deterioration of the soils as scernible in large parts of Uttarakhand and Nepal in the Himalaya. Unscientific mining activities and improperly planned road construction have not only marred the landscape but also dislodged the soil layers from the vulnerable slopes in many parts of the country.**

Over-irrigation of the very productive agricultural land has created a serious problem of waterlogging and attendant salinization. In parts of southwestern Punjab the water table rose at the rate of 20 to 50 cm/yr in the period 1975 to 1995, raising the extent of waterlogging from 60,090 ha in 78 to 104,000 ha in 1997 (K.P. Singh, 2000). In areas of Ferozpur, Muktsar, Bathinda and Sangrur, where water level went up at the rate of 15 to 20 cm/yr, about 85,000 ha of agricultural land is seriously waterlogged (Shakya and Singh, 2010). Over 5% agricultural land in the adjoining northwestern jasthan is afflicted with the problem of waterlogging and salinization by the Indira Gandhi Canal. is can be seen in and around Jodhpur (Paliwal and Paliwal, 2010).

In western Haryana the water table has risen by 3.0 to 9.0 m in seven districts, causing waterlogging over 2346 km^2 of irrigated land (Lunkad and Sharma, 2010).

Today more than 6.41 m ha of agricultural land of India is waterlogged (Anon, 2010).

More-than-necessary irrigation also causes soil sickness resulting from its impregnation with salts and alkalies. Over 6.73 million hectares of land in India is salt-affected and another 9.30 m ha bedeviled with problem of alkalies (Anon, 2010).

OIL CONSERVATION

trategy

By adopting a package of biological and engineering techniques, the deteriorated soils can be restored to their original health. As an environmentalist, the geologist is concerned primarily with erosion control, while the agricultural scientist endeavours to restore the productivity of degraded soil. The objective of all efforts is not to allow the wind and the water to move unchecked and sweep away the soil. **The best strategy therefore is (i) to put a protective cover on the vulnerable soil, (ii) to drain away the runoff water before it starts destroying, and (iii) not to permit dislocation of the material forming the slopes**. It must be emphasized that conservation measures would vary from situation to situation, as each soil terrain is different from the other in respect to climate condition, texture and composition and factors causing injury.

The various methods of minimizing and halting soil erosion are dealt with at length in Chapters ven and Eight. Here only those measures are mentioned which relate to agricultural soils.

proved Agronomic Practices in Hills

the hillslopes the intensity of erosion by runoff water can be reduced and even controlled (Table 4.2) by building terraces or benches parallel to or conformable with the contour lines—***contour terracing***

(Fig. 4.19)—and providing reinforcement through ***wattling*** with trees and shrubs planted at suitable intervals of 5 to 7 m (Fig. 4.20). Each of the crop-beds is made slightly sloping with levees ('bunds') on the outer margin, which would serve as a barrier to runoff flow. A furrow on the inner side would he drain out excess water and prevent erosion (Fig. 4.21). Partial levelling or cultivation along contours not only minimizes erosion but also promotes retention of moisture and increased crop yield. The benches, with longitudinal gradient to drain out runoff, should not be longer than 100 m (Das, 1977). These benches can best be developed behind graded barriers of grasses, shrubs or stone/earthern walls. As a matter of fact all 'bunds' and benches are stabilized with vegetation, such as *Tripsacum laxum* in the Nilgiri Hil *Eulaliopsis binata, Ricinus cummunis,* etc. The vetiver grass (*Vetiveria zizanoides*), called "khas khas" in Hindi, has proved a powerful protector of soils on slopes. Forming dense hedge within a short period of ju 3 to 4 months of planting, its roots penetrate 2 to 3 metres deep underground, thus greatly strengthening the stability of thc slopes. In the black soil region of central and western India and the red soil belt of southe India, graded 'bunding' (graded narrow-base terracing) with small ponds for storage and reuse of t runoff water has proved very effective (Das, 1977). In Bashkir (erstwhile U.S.S.R.), erosion was reduc 6.2 fold by ploughing along the contours, and 4.6 fold by resorting to terracing (Zacher, 1982).

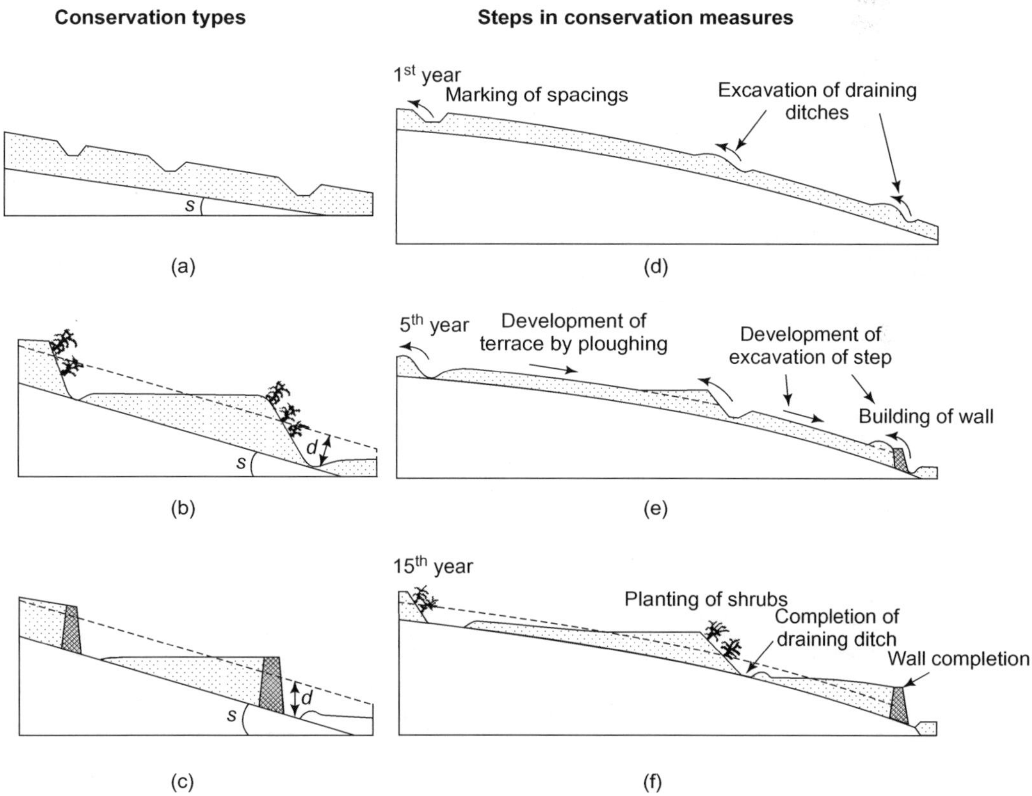

Fig. 4.19. On the hillsides, the terraces are developed parallel to the contour lines. The terraces are then strengthened by retaining walls or by shrubs. (A.D. Bradshaw and M.J. Chadwick, 1980. *The Restoration of Land,* Blackwell, Oxford, 317 p.)

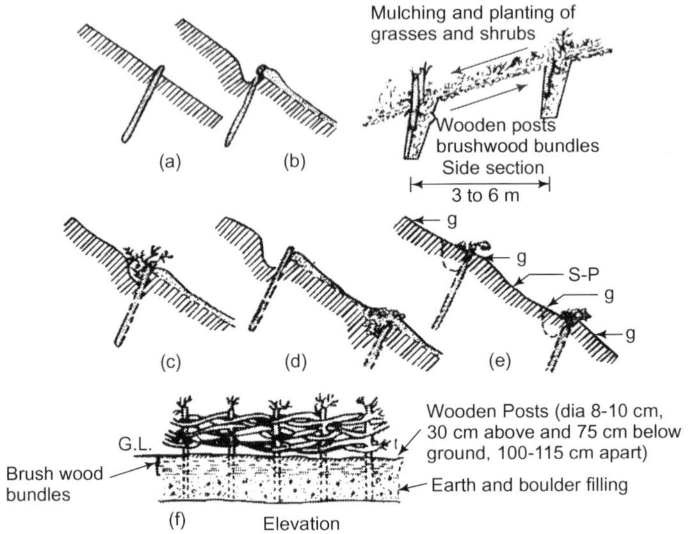

Fig. 4.20(A) The contour-terraces are reinforced by wattling of trees and shrubs that sprout easily. (After R.K. Gupta, 1979. *Plants for Environmental Conservation,* Bishan Singh Mahendra Pal Singh, Dehradun, p. 247).

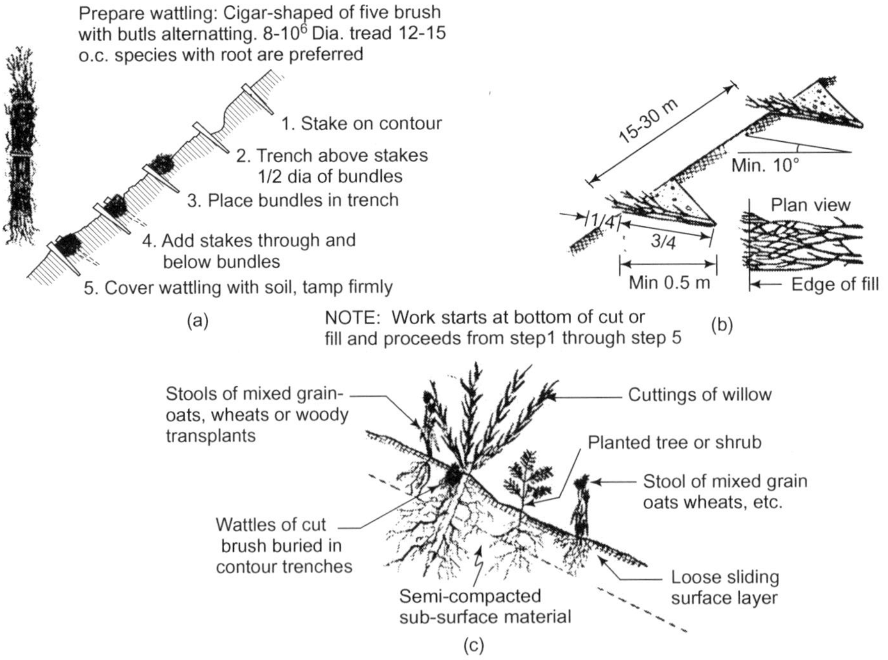

Fig. 4.20(B) The biotechnical method of stabilizing slopes (a) wattling and (b) contour brush-layering, using willow cuttings. In course of time, the cuttings grow and form a live willow stake. (From D.H. Gray and A.T. Leiser, 1982, Biotechnical Slope Protection and Erosion Control, Van Nostrand-Reinhold, New York, P.271 Fig. 7.10, 7.18 and 7.30)

Table 4.2. Reduction of Soil Loss from Up-Down-Slope Tilling (After Coates, 1981)

Slope	Contour Tilling	Strip Tilling	Terracing
1-12°	50-60%	25-30%	10-12%
12-18°	80%	40%	16%
18-24°	90%	45%	18%

Table 4.3A. Statewise Ranking of Saline Soil (Source: After Anon, 2010.)

State	Areal Extent (thousand ha)	Area (%)
Andaman and Nicobar Islands	71	9
Gujarat	1,559	8
West Bengal	408	5
Rajasthan	192	1
Maharashtra	171	1
Odisha	137	1
Kerala	21	1
Haryana	46	1
Andhra Pradesh	77	1
Bihar	45	0
Uttar Pradesh	22	0
Tamil Nadu	13	0
Karnataka	1	0
Arunachal Pradesh	0	0
Assam	0	0
Chhattisgarh	0	0
Delhi	0	0
Goa	0	0
Himachal Pradesh	0	0
Jammu and Kashmir	0	0
Jharkhand	0	0
Madhya Pradesh	0	0
Manipur	0	0
Mizoram	0	0
Nagaland	0	0
Punjab	0	0
Sikkim	0	0
Tripura	0	0
Uttarakhand	0	0
Others	124	4
Total	2,887	

In the Nilgiri Hills, where potato cultivation on red lateritic soil caused a soil loss of 39 t/ha/yr, resort to simple contour farming brought down the erosion to 15 t/ha/yr, and the cultivation on benches reduced the soil loss to 1 t/ha/yr (Das, 1977). Contour bunding or narrow base terracing on the contour at table spacing has been practiced in the Indo-Gangetic Plains to reduce excessive runoff and resultant erosion.

Table 4.3B. Statewise Ranking of Acid Soil (Source: After Anon, 2010)

State	Areal Extent (thousand ha)	Area (%)
agaland	1,516	91
Manipur	1,597	72
Tripura	709	67
Kerala	2,426	62
Mizoram	1,163	55
Meghalaya	1,023	46
Goa	103	28
Assam	1,995	25
unachal Pradesh	1,769	21
Chhattisgarh	2,342	17
Jharkhand	735	9
Sikkim	38	8
Uttarakhand	401	7
West Bengal	418	5
mil Nadu	427	3
adhya Pradesh	482	2
Odisha	203	1
machal Pradesh	76	1
Maharashtra	269	1
Karnataka	93	1
Bihar	41	0
Jammu and Kashmir	78	0
aryana	2	0
ndhra Pradesh	1	0
unjab	0	0
ndaman and Nicobar Islands	0	0
Delhi	0	0
Gujarat	0	0
Rajasthan	0	0
tar Pradesh	0	0
Others		
Total	17,930	

To overcome nutrient deficiency in the soil due to overuse, ***rotation of crops*** has been advocated different types of cover crops and legumes are grown (Table 4.3). The legumes add to the nitrogeno and other nutrients and thus increases the soil-holding power.

Inter-seeding of legumes is one way of accomplishing this. Alternatively, ***minimum tillage,*** entaili leaving roots and stubbles and the residue in the field, helps recuperate from nutrient deficiency a ensures minimum damage from the impact of heavy rain drops. This practice requires elimination weeds and insects which would thrive on the organic compost of the stubbles and roots.

In ***strip cropping*** the cultivated beds of crops alternate with strips of erosion-resistant vegetation includi grasses, maize, sugarcane, cotton, tobacco, onions, etc. (Fig. 2.18). In the erstwhile U.S.S.R. the growing maize combined with application of fertilizers reduced erosion by a factor of 12.37 (Zachar, 1982).

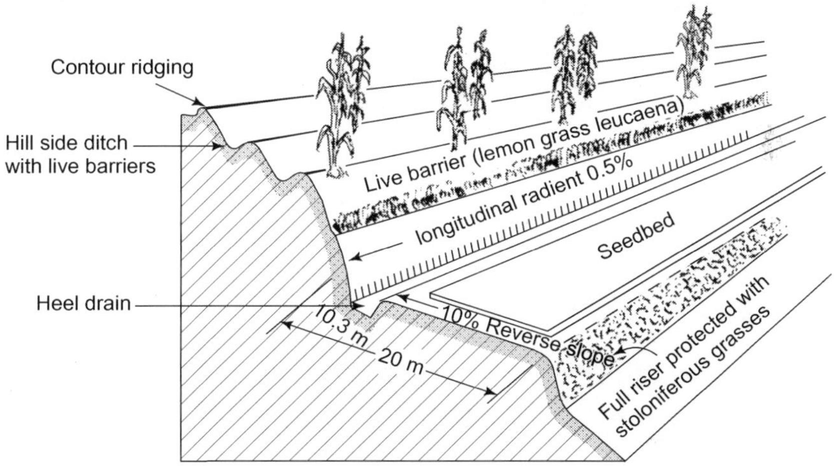

Fig. 4.21. The intensity of erosion can be minimized by providing furrows or drainage ditches in the inner side of the terraces and building levees on the outer margin. (Wiggins, 1981, In: R.P.C. Morgan (Ed.), *Soil Conservation, Problems and Prospects,* John Wiley, New York, Fig. 3.)

Controlling Gully Erosion in Ravine Lands

A very effective method of controlling gully erosion is to build small dams—earthen, masonry or wood (using sproutable branches or stems)—to impede the flow of running water and check widening the gullies (Figs. 4.22 and 7.19). The Sukhomajri experiment near Chandigarh achieved considerab success in halting expansion of badland created by severe erosion. Within a decade, the sediment load a major stream declined from 50 t/ha/yr to 5 t/ha/yr as a result of various conservation measures. Effo made in the ravinelands of Yamuna (south-western Uttar Pradesh), Chambal (Madhya Pradesh), a Sabarmati-Saraswati (Gujarat), and Mayurakshi (Bengal) likewise paid dividends (Tejwani et al., 197 In the catchment of the Damodar in Jharkhand, the sediment yield went down by 50.23% and in t Mahanadi (Odisha) at Hirakud and Machikud by 13.11% and 31.95%, respectively (Kaul and Das, 198 1983). The engineering measures are best supplemented by planting trees and shrubs in the ravines. T slopes can be stabilized by sodding with such grasses as *Cenchrus citiaris, Dichanthium annulatum* a *Eremopogon foveolalus.* The shallow ravines (< 2 m deep) were successfully reclaimed for agricultur d horticultural purposes. In the east Indian badlands, resorting to engineering and biological measures had resulted in increased rainwater retention by 53% and in the central and western regions the ravine-lands were brought under fodder plantation and treated with such engineering measures as construction of peripheral 'bunds' and drop-structures to reduce peak discharge and consequent soil-loss (Das, 1977).

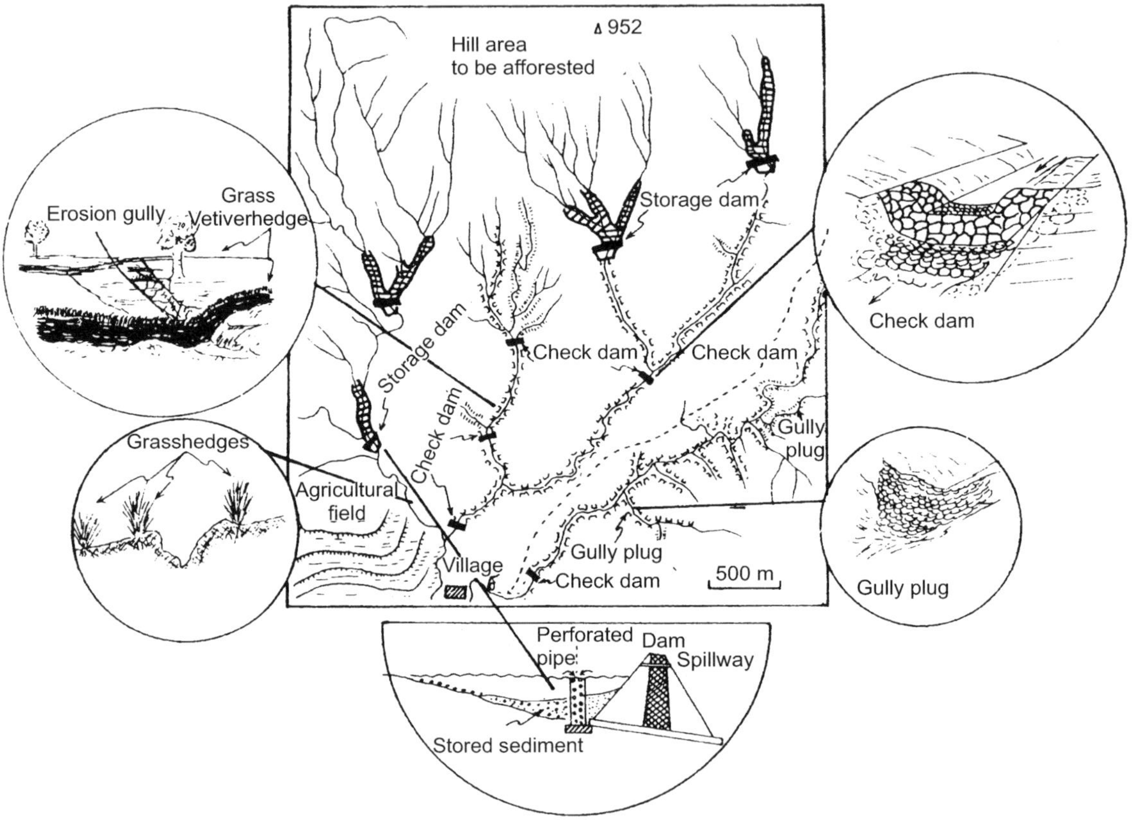

Fig. 4.22. Watershed Management Programmes embrace a package of biological and engineering measures. These include: (i) covering the land with vegetation, (ii) plugging the gullies and ravines, and (iii) draining out the erosion-causing water. (From Valdiya, *Geology, Environment and Society*, 2004, Universities Press, Hyderabad)

Shelter Belts in Deserts

areas where wind is the principal agent of denudation the deterioration of the farmland can be checked by raising impediments in the way of the wind. The ***windbreaks*** or ***shelter belts*** consist of rows of trees and/or shrubs (Figs. 4.23 and 8.8) including *Acacia Senegal, A. tortilis, A. nilotica indica, Balanites gyptiaca, Prosopis cineraria, P. juliflora, Parkinsonia, Dalbergia sissoo, Albizzia lebbek, Tecomella undulata, Azadirachta indica, Ricinus communis* (R. K. Gupta, per. comm., 1983) and grasses like *Spinifex littoreus*.

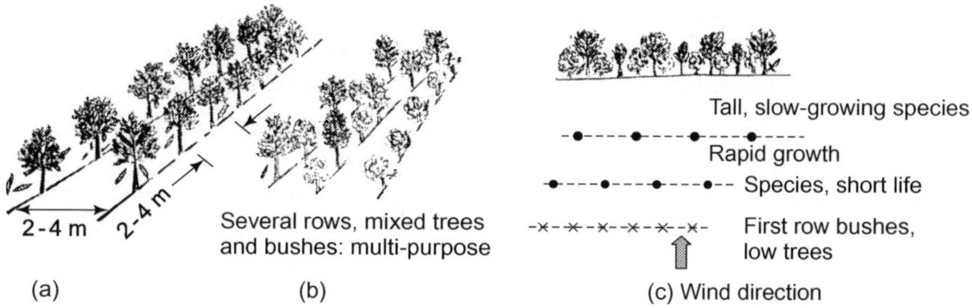

Fig. 4.23. The windbreaks or shelter belts consist of rows of trees and/or shrubs capable of halting the march of sand dunes. (After F. Weber and M. W. Hoskins, 1983. *Soil Conservation,* University of Idaho, Moscow, 94 p.).

These measures indeed stopped the advance of the Rajasthan desert into southern Haryana and sout western Uttar Pradesh.

In Saudi Arabia plants of tamarisk, acacia and eucalyptus prevented the march of the desert towar the Hofuf oasis. Similar attempts were made in the coastal belts of India by planting *Palmyra, Acacia auriculae, Acacia planiforms,* and *Casuarina equisetifolia* that survive salinity in the soils, moist breeze and the onslaught of tidal winds and waves. In addition to the trees, shrubs like *Calotropis gigantea, Pandanus tectorius,* and grasses like *Spinifex littoreus* were found useful (R. K. Gupta, per. comm., 1983).

Restoration of Sick Soil in Irrigated Fields

Waterlogging and salinization-alkalinization—the two maladies of soils resulting from over-irrigation in arid and semi-arid plains—can be treated by resorting to a five-phase programme

1. Construction of network of drainage to drain out excess water and lower the water-table—realizing that a rise of the water table within 1.5 m of the surface may become instrumental in development of salinity/alkalinity. There should be complementary efforts to seal all points and strips of leakage from canals, tanks or ponds by lining them.
2. A recent technique developed to combat the problem of waterlogging (Fig. 5.20) is to install multiple-pump pumping or siphoning system combined with cyclic use of canal water and bio-drainage through underground perforated lateral pipes buried below the root-zone depth (Shakya and Singh, 2010). The pipes are surrounded by filtering material.
3. Application of only that much of water to the fields as is sufficient to meet the need of the cro In Israel incredible success was achieved by resorting to **trickle (drip) irrigation through perforated pipes**. Water is carried to the field in rows of thin perforated pipes so that plants a irrigated by drops of water dripping out through the holes. In some areas in India, application this method has proved rewarding.
4. Planting salt-tolerant vegetation such as barley, pearlmillet, raya *(Brassica juncea),* cotton, spinach, date, palm and certain grasses used for fodder (R. K. Gupta, 1979, 1983) may help abate the salinity of soils (Table 4.4). Crop-rotation involving 'dhaincha' (grass)-cotton in the Deccan plateau, dhaincha-rice in Uttar Pradesh and Bihar, dhaincha-rice-berseem in Punjab and Harya have been advocated for solving the salinity problem (Mehta, 1983). Trials in Uttar Pradesh

demonstrated that such fuel plants as ***Prosopis juliflora, Acacia nilotica, Alblzzia lebbek, A.*** *procera, Terminalia arjuna, Cassia siamia, Cordia dichotoma, Syzygium cumini, Haplophragma adenophyllum* and ***Pithecolobium duke*** not only grow successfully in these 'usar' soils but also by absorbing salts and also increase percolation through the addition of leaf litter and by reducing degraded soil (Chaturvedi, 1985).

5. Treatment of alkaline soil requires replacement of sodium salts (which are 2% at the surface and increase to 20% at the depth of one metre) by calcium salts. Liberal application of gypsum (which contains 70 to 80% $CaSO_42H_2O$) in the upper 15 cm thick soil at the rate of 2 to 15 t/ha would convert the alkalis into soluble compounds (Fig. 4.24). Organic residue such as rice husks and rice straw may be added to promote formation of mild acid as a result of their decomposition. Other amendments include phosphogypsum (containing 80 to 90% $CaSO_4$), press mud (an organic byproduct from sugar factories), zinc sulphate (@ 45 kg/ha), farmyard manure and green manure (Mehta, 1983).

Table 4.4. Alkali-Tolerant Crops (After K.K. Mehta, 1983)

Tolerant	Semi-Tolerant	Sensitive
Bermuda grass	Wheat	Cowpea
Para grass	Barley	Gram
Rice	Oats	Groundnut
Sugarbeet	Raya	Lentil
	Senji	'Mash'
	Berseem	'Mung'
	Sugarcane	Pea
	Millet	Maize
	Cotton	Cotton

Table 4.5. Gypsum Requirement (t/ha) to Treat Alkalinized soil (After K.K. Mehta, 1983)

pH of Soil	For soil having texture		
	Light	Medium	Heavy
9.2	1.7	2.5	3.4
9.4	3.4	5.0	6.8
9.6	5.0	7.5	10.0
9.8	7.5	10.0	12.5
10.0	8.5	12.5	15.0
10.2	10.0	15.0	15.0

Such a package of measures as this will yield the desired results in any reclamation programme. Mehta (1983) suggested the following action programme: the ploughed land is levelled and smoothed, ving a slight (0.1%) slope from the side of the water channel. The field is enclosed by levees to prevent try of saline/alkaline water from adjoining fields. The alkalis/salts are then leached down upto a depth 15 cm by using tubewell water, primarily to lower down the water table and secondarily to flush out alkalis/salts from the soil. The excess water is necessarily drained out through ditches. The soil is then tested to determine the requisite dose of gypsum and/or other amendments, the quantity depending on the extent of sickness and texture of the soil and also the crop to be grown (Table 4.5).

The common amendments suggested for Haryana and Punjab are gypsum and phosphogypsum; a for Uttar Pradesh pyrite in powdered state. Highly deteriorated soil needs gypsum at the rate of 12 15 t/ha spread (broadcast) uniformly with hand when the soil is dry, that is, before the onset of rains. It is then mixed up with soil in the upper 10 cm layer by ploughing or tractoring. Water upto 4 to cm depth is put and kept standing continuously for about 10 to 12 days in order to leach out the salts/ alkalis. Farmyard manure is added to provide the plants with nutrients and improve the condition of the soil. Nitrogen sulphate (150 kg/ha) and zinc sulphate (50 kg/ha) are applied in the first year. ***Crop rotation*** (rice-wheat, rice-wheat-dhaincha and rice-berseem) is recommended for the first five yea About 100,000 ha of land in Haryana and Punjab were restored to health in this manner by the ear eighties of the last century.

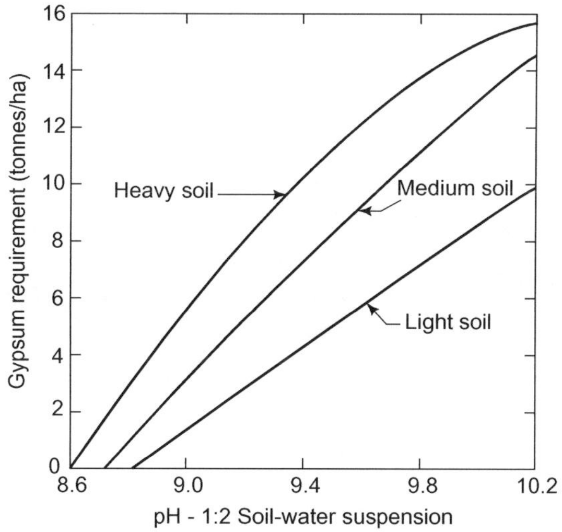

Fig. 4.24. The treatment and restoration of alkalinized soil of a farmland. (After K. K. Mehta, 1983. *Reclamation of Alkali Soils in India,* Oxford & IBH, New Delhi, 280 p).

Reclamation of Derelict Lands

If the soil is suffering from hardening of the surface due to excessive consolidation, the only way restore it to health is to excavate and dig upto a depth of about 50 cm, so that water is able to infiltra through openings in the dug up material. Weather permitting, the soil is then seeded with application of requisite amounts of fertilizer, including organic manure and leguminous plants. In some cases t seeded soil may have to be covered with a mulch of fibrous material such as chopped straw, shredd bark and wood pulp to protect the surface from rains and wind. In only extreme situations, a plastic latex stabilizer (Table 4.6) is sprayed after seeding. **A simple and pragmatic method is to mix the grass seeds with a small proportion of rapidly growing stemmy plants, such as sorghum which acts as a nurse crop (Bradshaw and Chadwick, 1980).**

These efforts must be supplemented by construction of a very efficient system of drainage to prevent cess of runoff flow into the seeded ground. Since the derelict lands are devoid of nutrients, ample doses of fertilizers with proportionate amounts of leguminous plants are added to the soil. Legumes with *Rhizobium* bacteria attached to them (Fig. 4.9a) would provide nitrogen to the recuperating soil. noculation of seeds with *Rhizobium* culture would expedite the process of soil recovery. Soil suffering from toxicity requires special treatment. Acidity can be countered by application of lime. And the salinity can be leached out by excess irrigation or by planting species that ensure eradication of this malaise.

Table 4.6. Mulches and Stabilizers Suitable for Restoring Derelict Lands (After Bradshaw and Chadwick, 1980).

Material	Rate (t/ha)	Resistance	Stabilization	Soil water	Nutrients	Toxicity
Mulches						
xcelsior	4	xx	xx	xx	-	-
Wood shavings	4	xx	xx	xx	-	-
Wood chips	10	xx	x	xx	-	-
Shredded bark	4	xxx	x	xx	-	-
eatmoss	2	x	x	x	-	-
Jute netting		xx	xxx	x	-	-
Corncobs	10	xxx	x	x	-	-
ay	3	xx	x	x		-
Straw	3	xx	x	x		-
Fiberglass	1	xxx	xx	x	-	-
Stabilizer Mulches *(as Slurry)*						
ood cellulose fibre	1 to 2	xx	xxx	x	-	-
Sewage sludge	2 to 4	x	xx	-	-	-
Stabilizers *(as emulsion)*						
Asphalt (1:1)	0.75	x	xx	x	-	x
Latex	0.2	x	xx	-	-	xx
Alginate/Colloidal Carbohydrate	0.2	xx	xx	-	-	-
olyvinyl acetate (1:5)	1	xx	xx	-	-	xx
Styrene butiadene	0.5	xx	xx	-	-	xx

xxx high; xx moderate; x low; — nil

Retiring Severely Degraded Lands

Severely degraded lands are best retired permanently for plantation of fuel and fodder-trees, or species that are suited to the local climate and soil conditions. The usual method is to transplant 1 to 3 year old edlings raised in nurseries (or less than a year-old grown in small fibre tubes or polythene containers). case of certain plants, chipped branches or seed-bearing shoots can be spread on the soil and allowed to sprout. Needless to state, the application of fertilizers to supply nutrients and also irrigation would be necessary to ensure easy or rapid growth of seeds, seedlings and sproutable chips (Table 4.7).

Table 4.7. Types and Proportion of Fertilizers used for Regenerating Derelict Lands (After Bradshaw a Chadwick, 1980)

Fertilizer	Composition (%)				Special Characteristics
	N	P	K	Other	
Fast Release					
N ammonium nitrate	35	0	0		Nitrogen balanced NH4-fNO*
Nitrochalk	250	0			Contains lime
P. superphosphate	08	0			
K Potassium sulphate	0	0	41		
Balanced 17:17:17	17	8	14		
High N 25:10:10	25	4	8		Many other formulations available
High P 10:25:25	10	10	12		
High K 15:15:35	15	7	20		
Slow Release					
Sulphur-coated urea	32	0	0		Rather fast release
Magnesium ammonium sulphate	5	10	8	Mg 10	Satisfactory release
Urea formaldehyde	38	0	0		Release slow and dependent on bacterial action
Isobutylidene diurea	31	0	0		Expensive
Osomocote	18	4	7		Other formulations available – expensive but effective
Rock phosphate	0	13	0		Phosphorus only release in acid soils — cheap
Basic slag	0	7	0		Variable
Hoof and horn	14	0	0		
Bone meal	3	10	0		

SUMMARY

Soil is a precious natural resource that the nature has gifted to us all

Soil is a living resource. It consists of finely divided rock material and organic matter with som air and water filling the interstitial space. The upper layers rich in humus and other nutrients, call topsoil, is capable of supporting vegetation and animals. Since the genesis of the soils is governed by the factors of climate conditions, topography or relief of the terrain, groundwater regime, and the living environment, the characters and thickness of the soil vary from place to place. In a cold humid clima the acidic soils (podsol) bereft of bases and colloids are formed. The podsol supports a luxuriant growth of conifers as in the Himalayan domain in the 1500 to 2000 m altitudinal belt. In a hot humid clima on the other hand, red and yellow soils rich in sesquioxides of Al and Fe but deficient in SiO_2 (laterosol) are formed, which have low fertility.

After their formation the soils are subjected to various changes. Certain bacteria living in the roots d nodules of leguminous plants fix atmospheric nitrogen and contribute to the nutrient stock in the soil, while burrowing animals make it more permeable and porous, that is, more receptive to accommodation of water and air. Capillary action in dry and semi-arid regions causes very serious damage to soil by destroying its productivity.

The soil and the vegetation are integrally linked in a close cycle, the damage or destruction of the one brings about the demise of the other. The vegetation on the soil determines its microclimate and also provides an umbrella against attack of denuding winds and water. The roots of the trees and shrubs impart shear strength to the soil mantle.

The most significant conservation measure involves application of techniques for putting covers of grasses, shrubs and trees on damaged or derelict soil. Other means of controlling or minimizing osion and consequent loss of soil is the provision of scientifically planned drainage to prevent ater flowing down freely from slopes, and building check dams in the gullies and ravines to pede the flow and reduce the violence of water. Shelter belts formed of trees and/or shurbs control the drift of sand dunes in deserts.

Waterlogging and alkalinization-salinization of the soil in arid and semi-arid irrigated tracts can be checked by lowering the water table through intensive pumping and developing a network of deep drainage, and by applying appropriate amendments such as gypsum, pyrite, phosphogypsum etc.

Reclamation of severely damaged or derelict soil can be achieved through plantation of desired species of plants, and growing or cropping turfs or grasses in specially prepared well-manured (or fertilizer-treated) beds and then spraying the beds with emulsion or mulch to ensure protection from wind and rain. Severely damaged lands are best retired for permanent plantation of trees.

5 Water Resources – Hydrogeological Considerations, Problems and Management

आपोहिष्ठा मयोभुवस्ता न अर्जे दधातन महे रणाय चक्षसे ।

—यजुर्वेद ११:५०

PRECIOUS RESOURCE

In the '*Yajurved*' (11:50) water has been described as the elixir of life, the source of energy that sustains life on earth, and the factor that governs the evolution and functioning of the universe. The invocation to rivers Ganga, Yamuna, Sindhu, Saraswati, Godavari, Narmada and Kaveri in all Hindu religious ceremonies indicates the exalted position the rivers occupy in the scheme of things. The Ganga is also known as 'Jahnavi', meaning a life-giving system, and the Saraswati has been extolled in the '*Rigveda*' as a purifier endowed with riches and the treasure of intellect and enlightenment.

महोअर्णना सरस्वती प्रचेतयति केतुना, धियो विश्वाविराजति ।

—ऋग्वेद १:३:१२

Even though apparently abundant, **water is a finite resource and quite scarce in many parts of the rth, including a very large part of India**. In the global perspective (Table 5.1), only 2.7% of this uid resource is freshwater, although water covers nearly 71% surface of the earth. The water that flows down all rivers and streams makes merely 0.0001% of the total water on our planet, and the underground water amounts to only 0.62%—two-thirds of which lies at a depth of more than 750 m. Freshwater lakes store another 0.009%; and about 2% of water is locked as snow-ice and in glaciers and snowpacks. Thus, ly about 1% of the water on earth is usable. Indeed water as a resource is extremely finite, but happily replenishable. **This situation calls for not only a comprehensive hydrogeological investigation of e country but also formulation of an effective water management policy involving judicious use**. As Deming (1975) points out, "**The wealth of a nation does not reside in mineral deposits, fossil fuels, or abundant freshwater resource, but in the ability to use these resources intelligently**."

Table 5.1A. Water Resources of the Earth (After A.N. Strahler, 1973)

Resource	Volume (km^3)	Percentage
Atmosphere	13,000	0.001
Oceans	1,32,20,00,000	97.20
Icecaps and glaciers	2,92,00,000	2.15
Saline lakes and inland seas	1,04,000	0.008
Soil moisture	67,000	0.005
Underground	41,70,000	0.625
Rivers	1,250	0.0001
Freshwater lakes	1,25,000	0.009
Total	1,35,56,80,250 km^3	100%

Table 5.1B. Water Resources of India (After Mall et al., 2007)

Resource	Volume (billion m^3)
Annual precipitation	4000
Available water resources	1869
Utilizable	1122
Surface water (Storage and diversion)	690
Groundwater (replenishable)	438
Present utilization	
(Surface water 63%, groundwater 37%)	605
Irrigation	501
Domestic	30
Industry, energy and other uses	74

WATER RESOURCES OF INDIA

Surface Flow and Groundwater

Indian rivers carry to the ocean about 18,69,000 million cubic metres of water annually (K.L. Rao, 1975; Sukhatme, 2011). The volume of water that flows down our rivers at any moment and fills the lakes just 0.03% of the earth's total water in transport held back in lakes (Table 5.2). Between 80% and 90% of the water flowing in the Indian rivers comes from rains during the monsoon period, and 20 to 30% of t water of the Himalayan rivers is contributed by the melting of snow-ice. More than 18,69,000 million m^3 of water is available for being tapped, but only one-third of this quantity is gainfully utilized.

Much of the stored groundwater is restricted to the Sindhu-Ganga-Brahmaputra plains and the coastal sedimentary belts in the Peninsular India. The total annual replenishable groundwater resource of the country has been estimated at 4,38,800 million m^3, out of which presently only 37% per year is being used, leaving thus a balance available for future development (Mall et al, 2007).

In addition to liquid form, water is locked in the bulk of ice caps and glaciers in the Himalayan domain, covering over 49,200 km^3 areas (C. P. Vohra, 1978). Between 20 and 30% of the waters of the Himalayan rivers is the product of melting of ice and snow during summers and the rainy seasons. More than 60% (varying from 30% in winter to 76 to 70% in summer) of the Satluj water is meltwater (Vohra and Sahgal, 1979). The Gangotri glacier alone had 20 km^3 of frozen resource (till the late 1970s), capable of yielding three times the water held in the Govind Sagar behind the Bhakra Dam. If the entire cover of the world were to melt slowly, all the rivers of the world would keep flowing for 830 years (Deming, 1975).

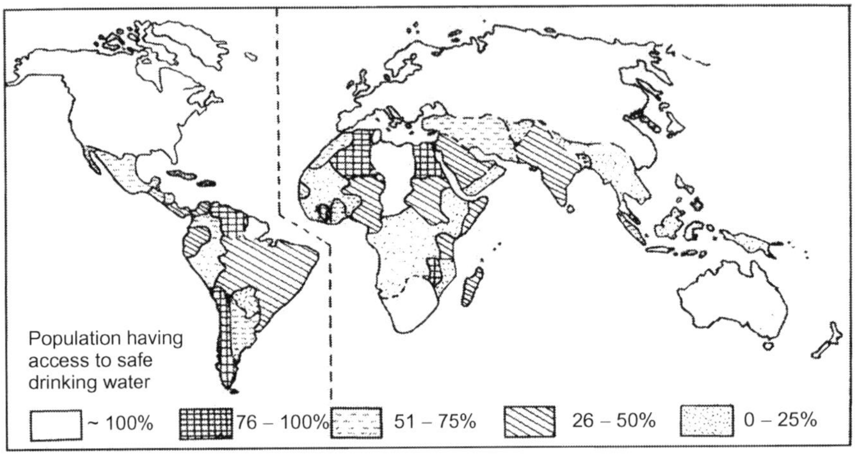

Fig. 5.1. Peoples of a large part of the world in developing countries do not have access to safe drinking water. (After M.M. Moulin, 1982).

Water Availability

According to Sharad K. Jain (2011), against the total surface water availability of 1614 billion cubic metres for 1210 million people, the utilizable surface water is 570 billion cubic metres.

From the 2011 census it is more than apparent that only 17.9% of rural and 62% of the urban (total mere 32%) households in India have access to tap water from treated sources (*Census of India's Houselisting and Housing: Data Highlights*-2011) (*The Hindu*, March 14, 2012) the remaining 82.1% rural and 38% of urban households depend on sources such as wells, tanks, ponds, rivers or streams — all untreated water. **It is galling to realize that 32% of India's population do not have access to safe ater supply**. Everybody knows how safe is the tap water supplied from "treated sources".

The *FAO Aquastat Report* 2010 shows that the per capita use of water is 627 m^3/per day in India, 341 /person/day in Nepal, 253 m^3/p/d in Bangladesh, 993 m^3/p/day in Pakistan, 618 m^3/p/d in Sri Lanka, and 323 m^3/p/d in Malaysia and 693 m^3/p/d in Japan, compared to 1518 m^3/per/day in the U.S.A.

The situation is much worse in villages where against the need of 40 litres per day per head, the availability is merely 11 to 20 l/day/head (Kayastha, 2000). A decade ago in the towns, the consumption was less than 90/l/day/head, but in big cities like Kolkata the consumption far exceeded 214 l/day/head (Kayastha, 2000). By 2025 when the Indian population will be 1459 million, very large part of the untry will suffer from chronic deficiency of water. There is no denying that **the current rate of water nsumption is unsustainable and the overall demand will exceed availability in another 35 to 40 ars (T.N. Narasimhan, 2011)**.

In spite of the abundance of water resources in the country, ironically about 152,000 out of 570,000 villages have no supply of reliably pure water. In other words, **only 162 million people of our rur population have access to drinking water through supply systems**.

Table 5.2. Average Annual Flow of Indian Rivers (Based on K.L. Rao, 1975 and Sukhatme, 2011)

Rivers (Measuring Point)	Flow (million m^3/year)
Sindhu System	2,07,800
Sindhu	73,330
Jhelam	27,890
Chenab	29,000
Ravi	8,000
Satluj	16,660
Beas	15,800
Ganga System (Farakka)	5,25,020
Ganga (Hardwar)	23,900
Yamuna (Saharanpur)	12,300
Gandak (Patna)	53,200
Tista (Anderson Bridge)	19,400
Other rivers	4,16,220
Brahmaputra System (Barak)	5,85,600
Brahmaputra (Guwahati)	4,55,000
Manas (Mathanguri)	47,100
Barak, Meghna, etc.	83,500
Peninsular India Rivers	
Mahanadi	66,880
Brahmani	28,480
Subanrekha	7,940
Godavari	1,10,540
Krishna	78,120
Pennar	6,320
Kaveri	21,360
Tapi	14,800
Narmada	45,640
Mahi	11,020
Sabarmati	3,810
Banas	8,674
Luni	8,000
Tupan	4,000
Madha	2,515
Saraswati	2,935
Ghaggar	1,300
Others	1,17,088
Total	18,69,000

Surface Water

ater flowing in rivers and streams, and locked in lakes, ponds and glaciers amounts to only 3% of the total available water on earth, but to humans and animals, this is the most important source. The lakes and ponds contain 100 times more water than all rivers carry at any given moment. The rivers are fed by countless springs and seepages that drain out from underground reserves. The spring water is supplemented by melt waters emanating from the ice caps and glaciers. Seasonally, the rivers carry cess runoff water flowing often in floods.

Table 5.3. Water Requirement for Different Uses (billion cm^3) (After Mall et al, 2006)

Use	Year 1997-98	Year 2010		Percent age	Year 2025		Percent age	Year 2050		Percent age
		Low	High		Low	High		Low	High	
Surface water										
rrigation	318	330	339	48	325	366	43	375	463	39
Domestic	17	23	24	3	30	36	5	48	65	6
ndustry	21	26	26	4	47	47	6	57	57	5
Power	7	14	15	2	25	26	3	50	56	5
nland navigation	-	7	7	1	10	10	1	15	15	1
Environment ecology	-	5	5	1	10	10	1	20	20	2
Evaporation loss	36	42	42	6	50	50	6	76	76	6
Total (A)	399	447	458	65	497	545	65	641	752	64
Groundwater										
rrigation	206	213	218	31	236	245	29	253	344	29
Domestic and municipal	13	19	19	2	25	26	3	42	46	4
ndustry	9	11	11	1	20	20	2	24	24	2
Power	2	4	4	1	6	7	1	13	14	1
Total (B)	230	247	252	35	287	298	35	332	428	36
rand total (A + B)	629	694	710	100	784	843	100	973	1128	100
Total water use										
rrigation	524	543	557	78	561	611	72	628	817	68
Domestic	30	42	43	6	55	62	7	90	111	9
ndustry	30	37	37	5	67	67	8	81	81	7
Power	9	18	19	3	31	33	4	63	70	6
nland navigation	0	7	7	1	10	10	1	15	15	1
Environment ecology	0	5	5	0	10	10	1	20	20	2
Evaporation loss	36	42	42	1	50	50	6	76	76	7
Total	629	694	710	100	784	843	100	973	1180	100

When it rains, a part of the precipitation evaporates into the atmosphere, a proportion of it flows on the surface down slopes to stream channels as surfacial ***storm runoff***, and the balance infiltrates into the ground percolating slowly as subsurface groundwater.

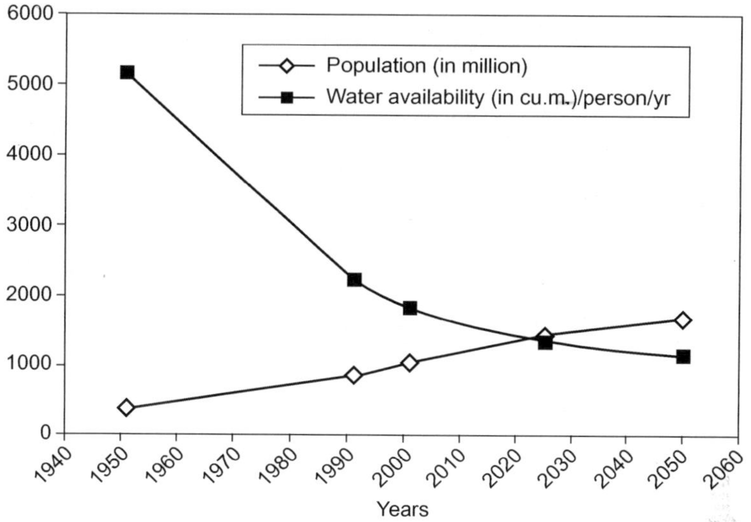

Fig. 5.2. The lines showing the observed and projected decline in per capita average annual availability of freshwater and of the growth of population intersect at A.D. 2025, implying the time of the beginning of acute crisis. (From Mall et al, 2006).

A part of the groundwater reappears on the surface down the slope in hills, after the rainfall, to give rise to ***base flow*** that sustains streams during rainless periods (Fig. 5.3). The surface runoff fi depressions on the ground and then spilling over runs down in irregular sheets, called ***overland flo*** increasing in volume and velocity as more and more rainwater is added. The highest rate attained by the overland flow ranges between 0.1 and 1.0 km/hour. As already pointed out, the annual discharge all the Indian rivers is 18,69,000 million m^3—85% of this volume being carried by major rivers, 7% by minor ones, and 8% by ephemeral streams in dry lands.

NATURE OF GROUNDWATER

Infiltration of Rainwater

A part of the precipitation is absorbed by the receptive ground, the amount of infiltration depending three factors, viz. texture of the soil, nature and extent of vegetal cover, and degree of ground slope (Fig. 5.4). The rate of infiltration declines one or two hours after rainfall (Fig. 5.3) and attains a consta value. This is because the fine pores get filled up, capillary forces driving up water diminish, swelli clays clog the pores, and raindrops demolish the framework of the soil texture. The coarse texture of the soil or the fractured and/or weathered nature of rocks of the ground favour greater and easy infiltratio Fine-textured soil or massive structureless rocks, on the other hand, hinder or inhibit percolation of groundwater. The covering of vegetation and litter not only protects the soil from the pounding of raindrops but also provides organic matter (humus) for binding soil particles in open aggregates, and thus promotes infiltration. The elimination of the vegetal cover, including that for cultivation, considerab reduces infiltration and proportionately augments runoff. Different landuses, likewise, have substant impact on the hydrologic cycle (Fig. 5.4). In short, the soil is able to absorb greater amounts of rainwater

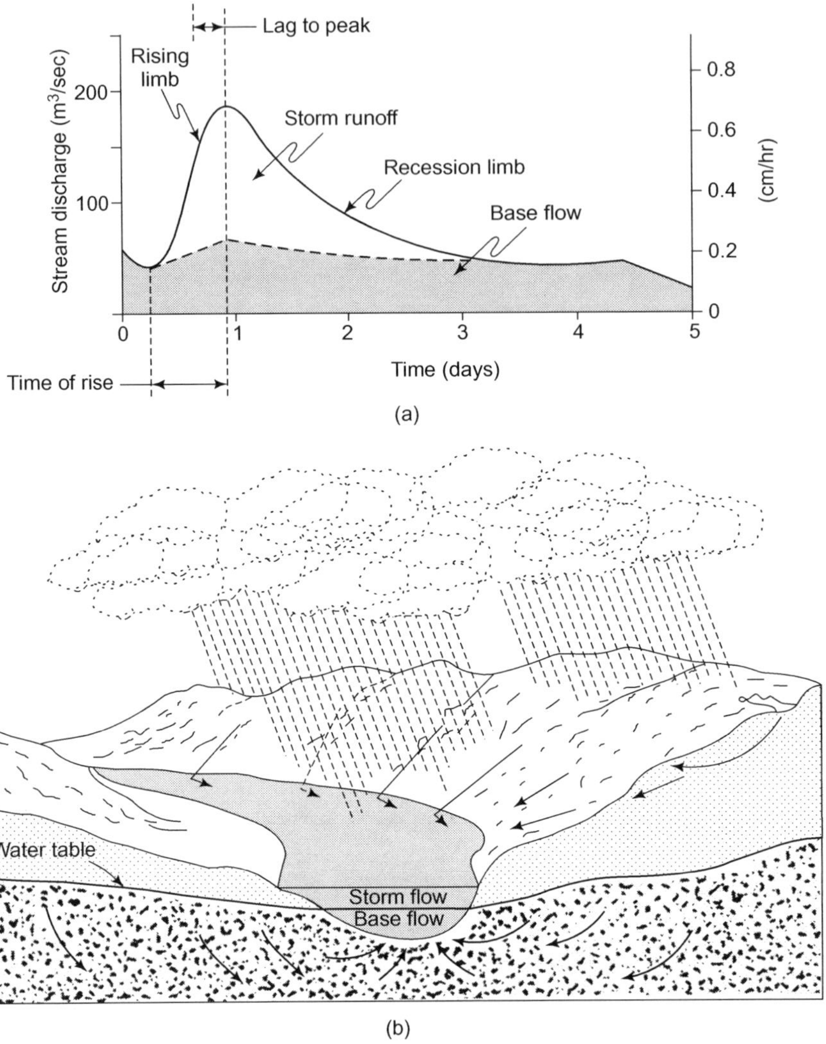

Fig. 5.3. (a) Graph showing the time-lag of peak flow after rainfall. (After Dunne and Leopold, 1978, Fig. 9.3.) (b) Storm flow derived from overland flow (runoff), rainfall in the channel, and the base flow contribute to the flood in a river. (From W. M. Marsh and J. Dozier, 1981, *Landscape*, Addison-Wesley, Reading, 637 p. Fig. 13.8.)

the rainfall is gentle and prolonged, if the ground is flatter, if the soil is open textured (porous or fractured), and if the surface is covered with a carpet of litter and/or canopy of vegetation. On the other hand, intense rain loosens the soil particles and causes the pores to get clogged, thus inhibiting filtration.

Percolating through the permeable soil or rocks, the absorbed rainwater enters deeper levels, to be stored or temporarily held as ***meteoric water*** (meaning water which comes from the sky). Some sedimentary rocks, particularly the open-textured sandstones and carbonate rocks may additionally

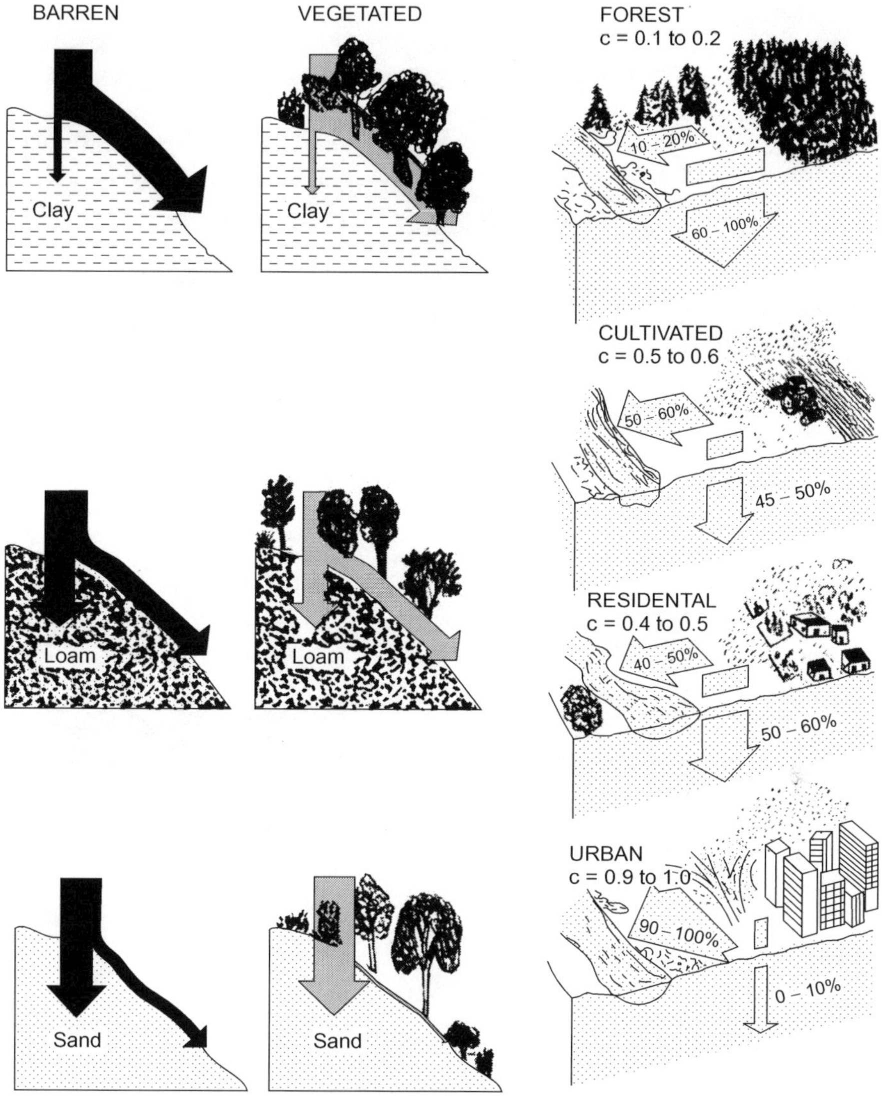

Fig. 5.4. The amounts of runoff and infiltration depend, among other factors, on the landuses and the textu of soils and the nature and extent of vegetal cover on the soils. (c=coefficient of runoff). (From W. Marsh and J. Dozier, 1981, *Landscape,* Addison-Wesley, Reading, 637 p., Figs. 11.4 and 13.20.)

contain indigenous or ***connate water***—water that was trapped in the pores at the time of sedimentation in the basin. Some rocks may also contain water derived from deep-seated ***plutonic*** sources. The quantity of meteoric water is overwhelmingly large in comparison to the other two types of water.

The infiltrating water fills the pores and openings in the deeper layers, forming a ***zone of saturation* or *phreatic zone*** (Fig. 5.5). Above the saturation zone the pores are filled with air and moisture in t zone of aeration. The growth of vegetation depends on the amount of moisture present in this zone.

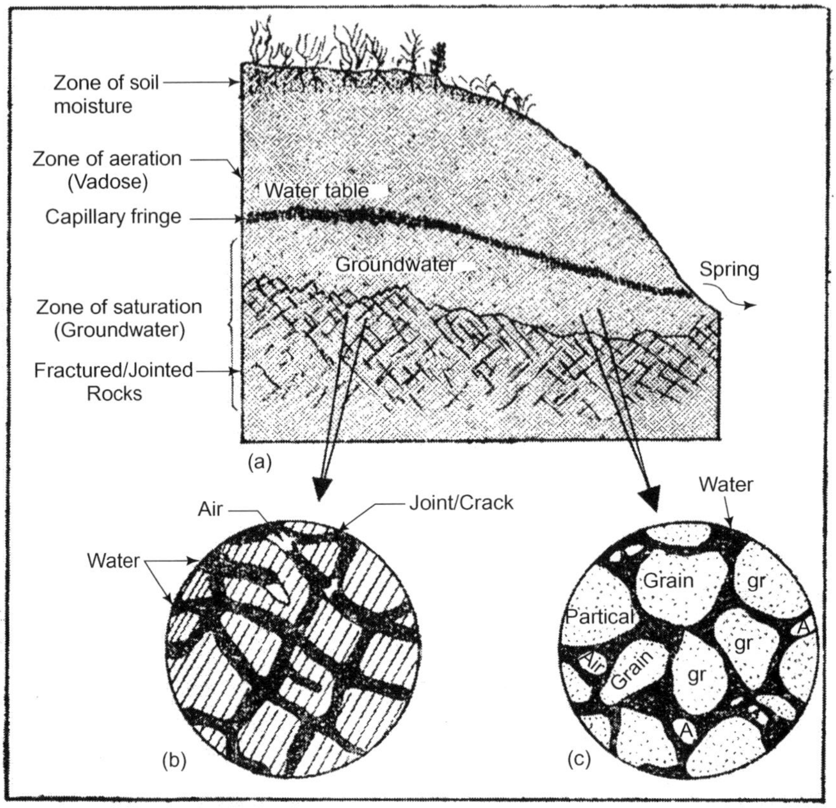

Fig. 5.5. The infiltrating water fills pores and openings within the soil or rock to form a zone of saturation (phreatic zone). Above it, the pores are filled with air and moisture (zone of aeration), on which depends the vegetation. Between these two zones lies the capillary fringe (1.3 cm thick).

tween these two zones, occurs a very thin layer, varying in thickness from 1 to 3 cm, in which films of water are attached to the soil particles. The water here is drawn up through capillary action. This is called the ***capillary fringe*** (Fig. 5.5).

Water Table

The upper surface of the groundwater is called the ***water table***. Its position conforms roughly to the nfiguration of the topography—it is close to the surface in depressions and valleys, and deeper on hilltops and prominences. The depth of the water table depends on many factors. In areas with moderate rainfall and gentle slope, it may be within a few metres of the surface; in rainless regions as deep as m; in the forested coarse-textured soil quite near the surface, and deep if the soil texture is fine and thout vegetal cover. The water table fluctuates conspicuously from season to season, and situation to situation. It rises after abundant rainfall, sinks during dry summers or spells of drought, and when there heavy withdrawal of groundwater through pumping. It may be emphasized that the rate of fluctuation is gradual owing to the slow movement of interstitial water.

Movement of Groundwater

Below the water table, water flows under the force of gravity from a higher level to lower levels, that down ***the slope of the water table***. The rate of movement varies widely from a few centimetres to tens of metres per day, depending on a number of factors, including the elevation difference of the areas of **recharge, infiltration and discharge (springs). The ability of the soil or rocks to allow passage of wat** without impairment of structure— ***permeability***—is the most important factor. The permeability in turn depends on the size of the pores and openings and their interlinkages. In coarser permeable sands or **soils, the movement is a hundred times faster than in fine-textured less permeable material. In soils ri** in clays, there is practically no movement of the groundwater, leading to water-logging.

In larger artesian basins it takes as much as 200 years for the rainwater to move from the area of recharge to the location of withdrawal. Radiometric dating of the artesian water in an area in the Sahara desert indicated that the recharge took place more than 5000 years ago. **Even in the case of alluvial basins, it takes several months for the groundwater to travel the distance. The flow of the groundwater in the rivers in the Ganga Basin, for example, lags five to six months behind the rainfall.**

The direction of the flow is of crucial importance in locating sites for wells and in monitoring polluti trends.

FAVOURABLE GEOLOGICAL CONDITIONS

Fracture and Fault Zones

Fracturing of rocks increases their permeability, commonly a hundred to a thousand times, depending on the degree of fracturing and the depth of the rocks (Fig. 5.6). The fractures tend to close up below the depth of 30 m. The shallow fractured zones are therefore the favourable locations of groundwater. Fault zones, particularly upstream, serve as excellent repositories of groundwater, for the impermeable fault gouge (material crushed to powdery state) prevents downstream movement and thus causes accumulation of water in crushed rocks. Fault zones store large volumes of groundwater.

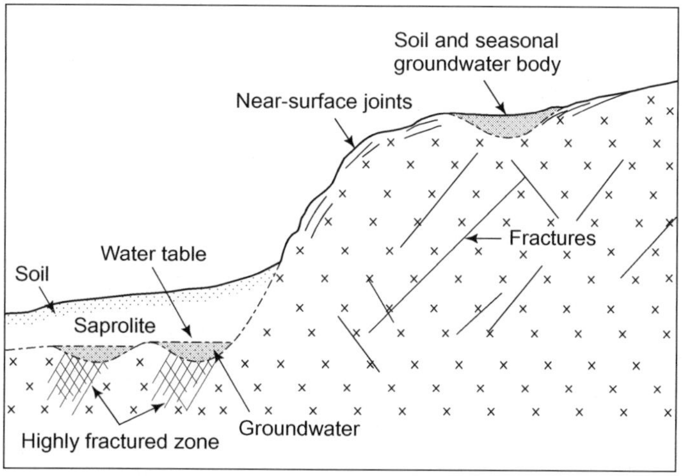

Fig. 5.6. Shallow fractured rocks and zones of deep weathering are favourable sites of groundwater, the water collecting in the depressions. (After T. Dunne and L. B. Leopold, 1978.)

Subsurface Barriers

ykes and sills cut the lavas and other rocks to form effective subsurface barriers to groundwater flow d are instrumental in the confinement of underground water. For example, in the Barata basin north-east of Jhansi in the Bundelkhand Granite terrain a NNE-SSW trending quartz reef serves as a barrier subsurface flow in the meandering channel of the Betwa River in Uttar Pradesh (Chaturvedi et al., 1983). Not only the upstream channel of the present Betwa but also its many abandoned channels now ed with sands and gravels, have been recharged on the convex side of the meanders (Fig. 5.7). About 30 tubewell sites with high potential discharges were located in the Betwa valley as far back as 1982.

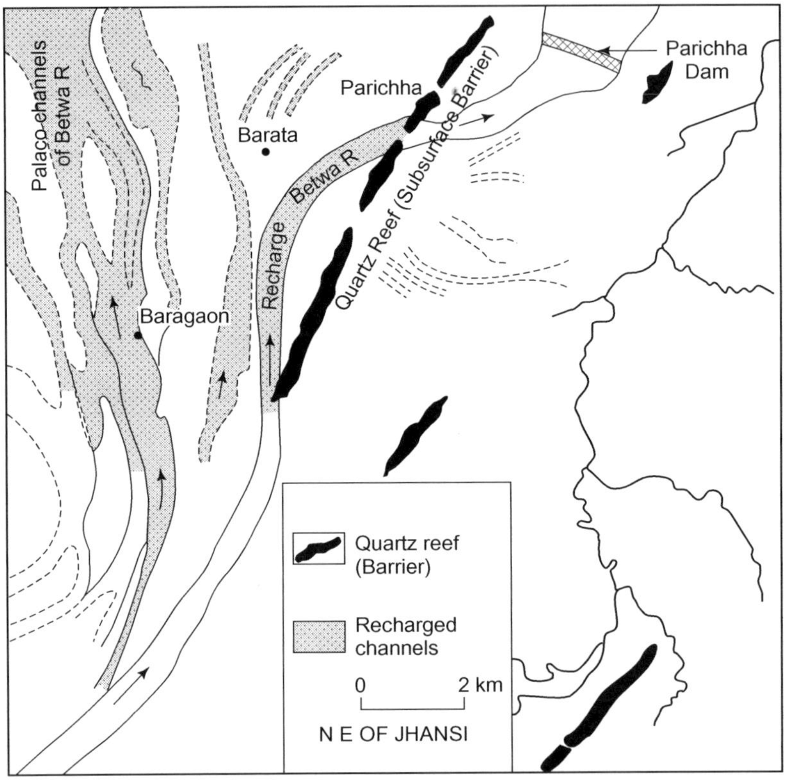

Fig. 5.7. The NNE-SSW trending dyke-like quartz reef serves as a barrier to the subsurface water flowing in the present channel of the Betwa River and in its many abandoned past channels in the Bundelkhand area, Jhansi District, Uttar Pradesh. (After R. S. Chaturvedi, D. C. Bhattacharya, P. Kamal, J. Krishnamurthy, and N. Sunder Raman, 1983.)

Colluvial-Alluvial Deposits

In the foothills of the Himalaya the piedmont belt (Bhabhar) is constituted of a prism shaped apron of lluvial-fluvial deposits of gravels and sands. The belt is perennially recharged by mountain streams d rivers. The downstream valleys are filled with permeable sediments serving as good aquifers, particularly close to the mountain front, and nearer to the mouths of the gorges (Figs. 5.13 and 5.12).

Weathered Zones and Vesicular Lavas

Under hot and humid conditions, igneous and metamorphic rocks are deeply weathered, giving rise to a 5 to 15 m thick soil mantle. The soil mantle is capped by a few metres thick, very permeable lateritic layers, the larger middle part is constituted of deeply weathered rock material, and the bedrock is usually highly fractured and weathered. The latter two layers are good repositories of groundwater (Figs. 5.6 and 5.8) as seen in large parts of Kerala, Tamil Nadu, Karnataka, Andhra Pradesh, Madhya Pradesh, southern Uttar Pradesh and western Jharkhand. The deeply weathered tip of the highly jointed and fractured Bundelkhand Granite with intruding dolerite dykes in southern Banda district (Uttar Pradesh), for example, stores abundant reserves of water, the average yield of wells used to be of the order of 1000 to 100,600 1/minute (Ahmad, 1983).

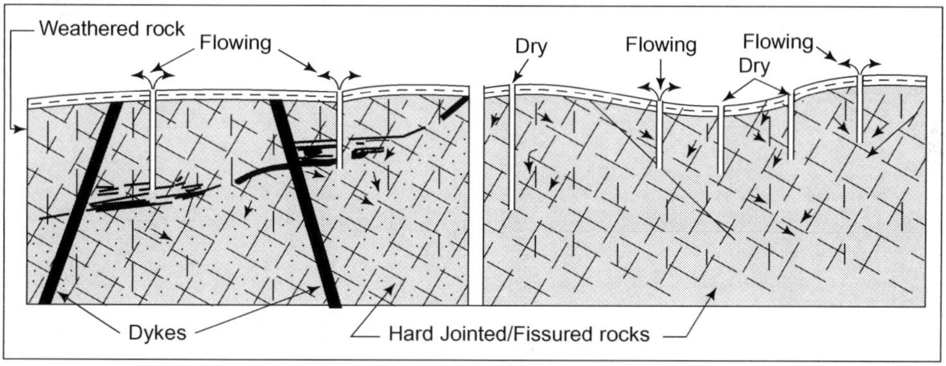

Fig. 5.8. Recharge of unconfined aquifer (permeable), development of perched water above an aquiclu (impervious), and artesian well and spring.

In the Deccan volcanic terrain of Maharashtra and Malwa (Madhya Pradesh), the jointed (including columnar-jointed) and vesicular-cavernous lavas and associated beds of volcanic ash and buried soils developed between eruptions have been found to be good aquifers.

WATER-HOLDING CAPACITY OF SOILS AND ROCKS

Aquifer

The porous or fractured and permeable zone of a soil or a rock saturated with freely flowing interstit water capable of supplying adequate quantities of water to wells and springs is called an ***aquifer***. The capability and capacity of an aquifer for holding water depends on its porosity—the volume of empty space between the grains constituting a soil or sediment, or the cracks, fissures, fractures, joints a faults in igneous or metamorphic rocks. It may be emphasized that porosity governs the total volume of water a given aquifer can hold, but it has no bearing on the amount of water which is available for withdrawal or supply. Stored water, in soil or rock, is available only if the pores are interconnected, and when the interstitial water is free to move from one point to another.

The porosity depends on the degree of sorting—uniformity of size and shape of particles (grains) of the soil or sediment (Table 5.5). Well-sorted sediments characterized by uniformity of grain size are generally more porous than sediments having grains of heterogeneous size (poorly sorted). Structural homogeneity gives rise to or increases the porosity as testified by cavernous carbonate rocks, scoriaceous and vesicular lavas and jointed, fractured or sheared granites. Weathered parts of granites and gneisses have higher porosity near the surface and lower at the bedrock level.

Clays and humus have a higher water-holding capacity, and hold water more tenaciously than do the sands. But this water in the clays is not amenable to recovery because of their lower permeability. On the contrary, gravels and gravelly sands hold water so loosely that gravity drains it out very rapidly leaving little in the sediment to hold.

Table 5.4. Porosity and Permeability of Sediments and Rocks (Data from T. Dunne and L.B. Leopold, 1978)

Material	Porosity	Permeability (m^3/day)
Unconsolidated Sediments		
Soils	0.30 to 0.50	
Weathered rocks	0.01 to 0.50	
Glacial tills	0.25 to 0.45	0.001 to 10.0
Dune sands	0.35 to 0.40	2 to 20 (8–10)
Loess	0.40 to 0.55	0.0001 to 1.0
Clays	0.45 to 0.55	<0.01
Silts	0.40 to 0.50	0.0001 to 1.0
Fine sands	0.30 to 0.52	0.01 to 10.0
Medium sands	0.30 to 0.35	10 to 3,000
Coarse sands	0.30 to 0.35	10 to 3,000
Sandy gravels	0.20 to 0.30	0.3 to 10.0
Gravels	0.25 to 0.40	1000 to 10,000
Sedimentary Rocks		
Conglomerates	0.50 to 0.25	0.3 to 3.0
Sandstones	0.50 to 0.30	0.3 to 3.0
Siltstones	0.05 to 0.20	
Shales	0.05 to 0.15	
Crystalline limestone	0.01 to 0.10	0.00003 to 0.10
Chalk and oolitic limestone	0.05 to 0.20	
Marl	0.05 to 0.15	
Metamorphic Rocks		
Slates	<0.01 to 0.50	
Schists	<01 to 0.05	
Gneisses	<0.01 to 0.05	
Igneous Rocks		
Tuffs	0.10 to 0.80	0.0003 to 3.0
Lavas	0.01 to 0.30	0.0003 to 3.0
Granitic rocks	<0.01 to 0.05	0.0003 to 0.003
Weathered rocks	0.01 to 0.10	>0.0003

Aquiclude and Artesian Condition

Shales, mudstones, and dense impervious rocks which do not possess permeability and do not allow passage of groundwater in sufficient amount are called ***aquiclude***. An aquiclude *within a body* of a permeable horizon impedes or even effectively stops downward percolation of groundwater to the water table. This gives rise to what has been called ***perched water***, quite separate from the main zone of saturation (Figs. 5.9 and 5.14 a). Accumulation of iron oxides, salts and calcrete 'kankar' in layers or hard pans within the soil often results in development of a shallow perched water table (Fig. 5.12), or even marshes as discernible in some parts of the Indo-Gangetic Plains.

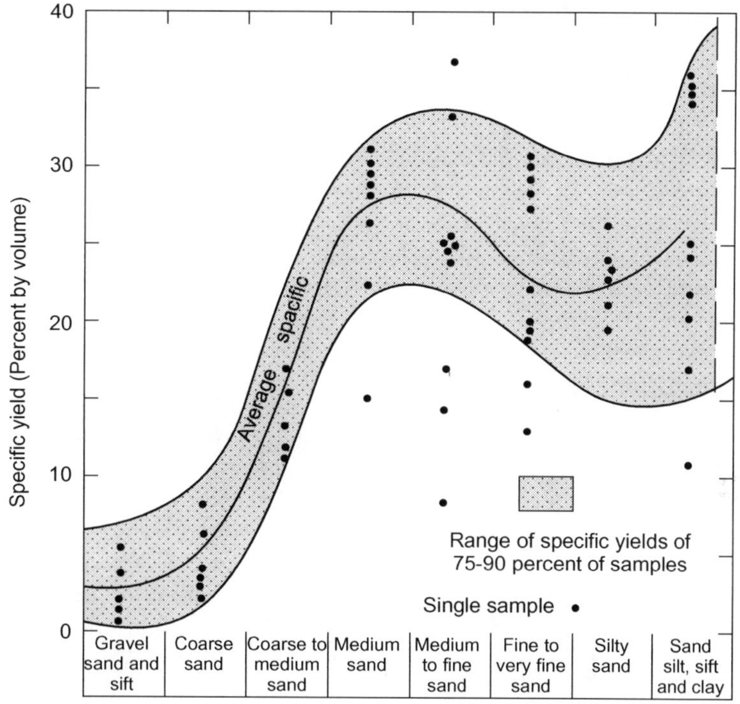

Fig. 5.9. **Specific yield (quantity of water that an aquifer parts with) of different textural types of sedimen expressed in terms of the ratio of water draining out of the aquifer to the total volume of the aquif** (From Dunne and Leopold, 1978).

Commonly, groundwater is connected, indirectly of course, with the atmosphere through the network of pores and fractures, joints, other openings etc. so that the aquifer remains ***unconfined***. In other situations, a horizon of aquiclude may prevent this kind of connection and the groundwater becomes ***confined*** in the aquifer. Water pressure may build up within this confined aquifer, giving rise to what called the ***artesian condition*** (Fig. 5.24). Water under artesian conditions flows out or is prompted o by built-in pressure when an opening such as a fault or well is provided. The voluntarily flowing sprin and wells are called ***artesian springs*** **or** ***artesian wells***.

The development of the artesian condition in the Brahmaputra Basin (Assam) is related to the existence of porous permeable lenses within the clayey succession, and to synclinal cores enclosed by impermeable clays (aquiclude). The artesian aquifer extending from the western border of the Darrang district to the frontier of Kamrup yields water at the rate of 250 to 450 l/minute (Biswas and Saha, 1983). An artesian condition exists in the 'Tarai' belt in Nepal, Uttar Pradesh, Uttarakhand and Panjab (Sharma, 1977; Pathak, 1975) and in the Banaskantha-Mehsana basin in northern Gujarat at a depth of 2 to 20 m (Choubey and Tanwar, 1982).

WATER-YIELDING CAPACITY OF AQUIFERS

Permeability

If the porosity determines the amount of water held by an aquifer, it is the permeability that governs the volume of water that can move out and be available for recovery. The permeability depends, as already stated, on the interconnection of the pores and cavities and cracks within the rocks and soils. The interconnection ensures passage or easy movement of water from place to place. The ability of a rock or soil to transmit water is described as ***transmissibility***, which depends on the product of the thickness of e aquifer and its coefficient of permeability. The permeability, expressed in terms of volume of water wing per day, depends on the difference of head of the water table and the dimension of the medium ough which the water is flowing.

Specific Yield and Specific Capacity

Not all the water in an aquifer is available for recovery through wells or springs. The quantity of water that the aquifer readily parts with is called the ***specific yield*** (Fig. 5.9). It is expressed in terms of the ratio of the volume of water draining out of the aquifer to the total volume of the aquifer that is being drained (Dunne and Leopold, 1978). The proportion of water left behind in the aquifer, i. e. the part not available for recovery, is expressed in terms of ***specific retention***. It is the specific retention that is conducive to the growth of vegetation which draws out deep-seated water through roots. It is noticed at coarse-grained sediments have higher specific yield, **fine-grained sediments are better retainers d therefore provide better support to vegetation**. The specific yields of gravels is usually 25 to 35% and of coarse sand 25 to 30% (B.B.S. Singhal, per.com. 1985). The performance of an aquifer is also expressed in terms of the rate of discharge of wells during pumping divided by the consequent lowering of the water table (drawdown)—m^3/d for *n* metres of drawdown.

Some thirty years ago in the Dehradun valley of Uttarakhand, the wells intersected the water table at depths of 2 to 14 m. The wells of the Tarai belt fringing Bhabar in the Nainital district had a yield of 3000 litres/minute from the thick aquifer at the depth of 150 m. In western Rajasthan and the adjoining plains of northern Gujarat the yield of deep tubewells was upto 3000 l/day/ well. From the aquifer of semi-consolidated sediments in western Rajasthan some of the deep (upto 250 m) tube wells drew water at the rate of 1000 to 2500 l/minute per well, while those from the cavernous limestones of Barunda in Jodhpur district had a yield varying from 100 to 10,000 l/minute at the drawdown of 10 m (Choubey and Tanwar, 1982).

In the upland alluvial tracts in the Narmada valley the water table lies at a depth of 2 to 30 m. The yield of an average well was 35 l/minute (Adyalkar, 1975, 1983), while in the Tapi valley the wells yielded at the rate of 35 l/minute from the aquifer at a depth of 2.5 to 44.5 m.

RECHARGE OF GROUNDWATER

The aquifers in the flood plains are naturally recharged by rivers and streams, both seasonal and perenn (Figs. 5.12, 5.24). The flood plains become inundated during spates, and if the ground surface is n covered with an impenetrable layer such as a salt pan, asphalt (Fig. 5.4), cement or any civil structure, the floodwater infiltrates and recharges the groundwater. The recharge can be induced or accelerated pumping in water under pressure or through numerous wells sunk along or in the river bed (Figs. 5.25 and 5.27).

In the desert terrain such as Rajasthan (Jodhpur-Jaisalmer), the main sources of sustenance are the fossil groundwaters, confined as they are within aquifers under high pressure and higher temperatu (20°C). The fluvial aquifers are recharged in the distant hills of the Aravali (Fig. 5.24) bordering t desert. There are lenses of freshwater in the regional saline groundwater regime. In the heart of the Thar Desert in northwestern Rajasthan large deposits of fresh water have been discovered in the winding-curving aquifers representing palaeochannels of the now vanished Saraswati river. Despite being intensively used for 40 to 45 years there was no indication of any decline in the tubewell discharges until about 1998 (Soni et al, 1999). Interestingly, the freshwater in the aquifer at the depths of 30 to 50 m dated 1800 to 5000 years, and at 60 to 250 m about 6000 to 22000 years (Nair et al, 1999). Much older fossil water has been recently discovered in the Jaisalmer area.

Similar examples of underground reserves of water in the heart of the desert are Nafud (4000 km Riyadh (1500 km^3) and Ral-el-Khali (14,500 km^3) at depths of 1800 m in the great basin of west Asia, Chad (14,000 km^3), Fezza (4800 km^3), Kufra (3400 km^3) and the Western Desert of Egypt (1,88,000 km^3) in northern Africa (Burden, 1982).

EFFECTS OF WITHDRAWAL OF GROUNDWATER

Depression of Water Table

The withdrawal of groundwater causes depletion of its reserves, which manifests itself in the lowering of the water table. The pumping out of water creates a ***cone of depression*** in the water table around the well, if the rate of inflow is less than the quantity withdrawn (Fig. 5.10). The degree of depressi depends on the rate, quantity and duration of the withdrawal. The larger the number of wells working at faster rates, the greater would be the depth of depression. The result is the widening of the base of the inverted cone of depression.

A large number of wells discharging large quantities of water cause general lowering of the water table. This is called ***overdraft*** of groundwater. **The culmination of the overdraft is complete depleti of the groundwater reserve. As a result of lowering of the water table, shallow wells go dry and the springs and streams loose their waters. In the coastal belts saline seawater encroaches upon t depleted reservoirs, rendering the groundwater reserves unfit for consumption** (Fig. 5.23).

The replenishable reserve of groundwater in India is of the order of 433 billion cubic metres, out of which 399 bm^3 is available annually and 231 billion m^3 is utilized annually — 92% for irrigation 18 billion cm3 for domestic and industrial uses (K.D. Sharma, 2009). The extent of irrigation from groundwater increased from 6.5 million ha in 1951 to 54.75 mha in 2002 (Das, 2006) — the states which critically overexploit groundwater are Panjab (>100%), Haryana (>100%), Rajasthan (>100%), Delhi

(>100%), Gujarat (>76%), Tamil Nadu (>85%), Karnataka (>70%), and Uttar Pradesh (>70%) (Das, 2006). In the country as a whole, in some areas in Panjab, Haryana and Rajasthan, the overexploitation is to the tune of 109 to 145% (K.D. Sharma, 2000, 2001). The groundwater constitutes more than 50% of total irrigation.

Table 5.5. Total Water Withdrawal Per Capita (m^3/person/yr) (FAO: *Aquastat Survey*, 2011)

Country	Withdrawal (m^3/person/yr)	Source year
Afghanistan	1055.00	2002
Argentina	8645.00	2002
Australia	1149.00	2002
Bangladesh	224.20	2008
Bhutan	492.00	2008
Brazil	305.40	2007
Canada	1468.00	2002
China	414.60	2007
Egypt	937.00	2002
France	512.40	2007
Germany	392.30	2007
India	644.10	2010
Israel	281.90	2007
Italy	788.60	2002
Japan	708.40	2002
Malaysia	511.30	2002
Myanmar	701.60	2002
Nepal	339.50	2002
Norway	622.70	2007
Pakistan	1037.00	2008
Russian Federation	455.50	2002
Saudi Arabia	959.10	2007
Singapore	81.97	1997
South Africa	270.60	2002
Sweden	285.60	2007
Switzerland	359.00	2002
Thailand	855.60	2007
United Kingdom	212.50	2007
United States of America	1550.00	2007

Note: Withdrawal indicates the volume of water used, **not necessarily equitably**. These figures imply the pressure on the water-resources (this author's comment).

Adverse Consequences of Overdraft

Over-irrigation in rural region and misuse due to over-consumptive lifestyles in cities have created serious water stresses. A few examples of the consequences of over-use or overdraft will suffice.

(1) In northern Gujarat (Saurashtra and Kachchh) the annual rate of decline of water table was 1 m/yr upto 1970, and 2 m/yr thereafter. In the Mehsana district the rate of decline was 5 to 8 m/yr (Das, 2006). In parts of Ropar, Hoshiarpur, Gurdaspur, Amritsar, Kapurthala, Jalandhar, Patiala and Sangrur, the water table went down at the rate of 23 cm/yr — the extent of decline varying from 5 to 10 m (Das, 2006). In parts of Bikaner and Churu districts, the water level is going down at the rate of 0.5 m/yr, while in parts of Jodhpur and Nagaur at the rate of 1 m/yr (Paliwal and Paliwal, 2010).

The consequence of the declining water table was that millions of shallow wells — t mainstay of marginal farmers and the poor — became dry. This brought about a situation which the rich farmers owning deep wells dictated terms to the poor, including selling wat at high costs.

(2) The lowering of water table resulted in the decrease in the base flow of streams and frequent turning them into dry *nalas*.

(3) The content of material such as salts, fluorides, nitrates and arsenic dissolved in groundwat increased in all the states where groundwater has been overexploited. In Haryana, for example (Figure 21.4) the groundwater shows high concentration of nitrate and fluorine (Thussu, 1999 The situation in Bengal and Bangladesh has assumed alarming proportion — the water drawn from 50 to 80 m deep wells are highly enriched in arsenic (Figure 21.4), and causing serious debilitating sickness (Mallick and Rajgopal, 1998).

(4) Along the coasts, the lowering of water table, such as in Tamil Nadu and Saurashtra in Gujarat, has resulted in the ingress of saline seawater. The saltwater-freshwater interface advanced landward from 3 km in 1969 to 9 km in 1986 north of Chennai and 6 km in Puducherry and Karaikal (Das, 2006). In the Chorwad–Verawal tract (Gujarat), the saltwater incursion had affected 100,000 ha area by 1977 (compared to 35,000 ha in 1971) affecting 280,000 people (Ghassemi et al, 1995). Similar development took place in the 24-Parganas of West Bengal and in the Visakhapatnam sector in Andhra Pradesh.

Subsidence of Ground

An overdraft of groundwater causes subsidence or sagging of the ground surface over the depleted aquifer constituted of unconsolidated or poorly consolidated sediments. This happens because of the reduction in volume of the aquifer. Following this loss of strength in grain-to-grain surface and attendant readjustment in packing there is a reduction in the aquifer volume. Subsidence of the ground in this manner was discerned in Tokyo and Osaka (3 to 4 m) in Japan, Mexico City in Mexico (7.5 m), Santa Clara Valley in U.S.A. (1.52 m) (Figure 5.10 Coates, 1981) and so on. Disruption of drains, pipelines, rail-lines, canals and rivers, and damage to buildings are some of the consequences of the ground sagging. The general recession of the piezometric surface to the tune of 6 m in the period 1956 to 1981 under the city of Kolkata was caused groundwater subsidence of 0.3 m in 25 years (Biswas and Saha, 1985).

In order to obviate the adverse consequences of overdraft, it is desirable to know before han the limit to which water can be withdrawn without depleting the reserve, also called the *safe yield* factor. The rates of withdrawal must take into consideration the rate of recharge.

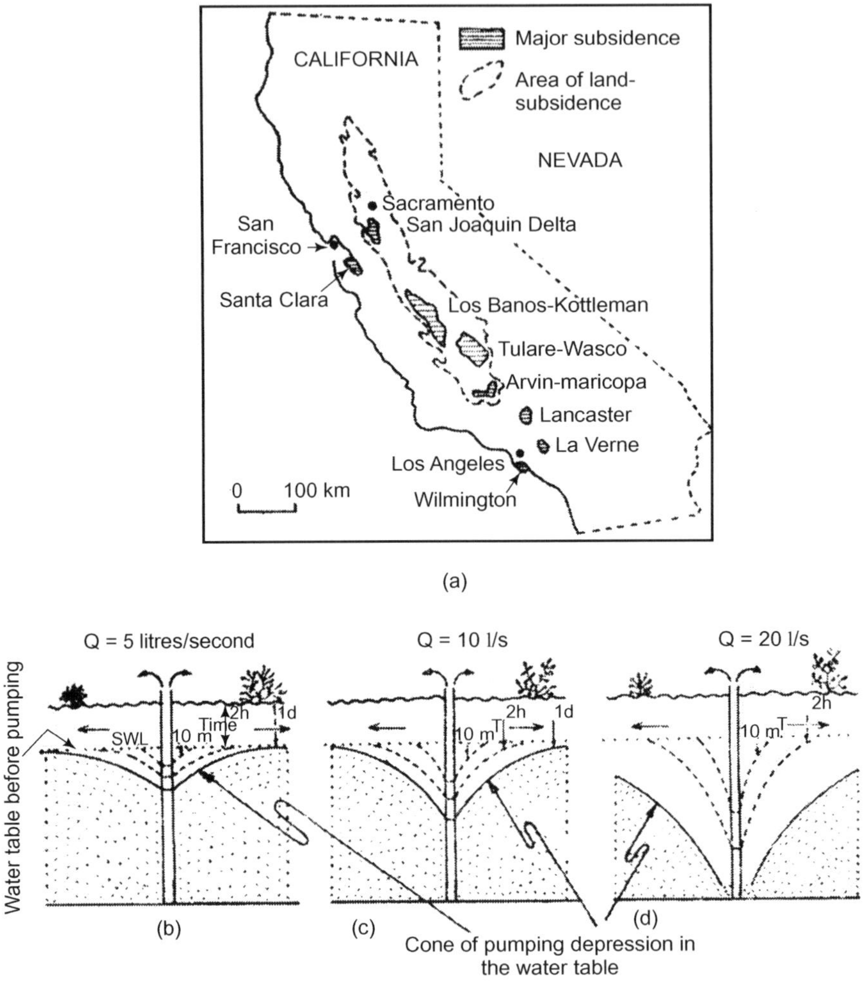

Fig. 5.10. (a) Map showing subsidence of the ground in California (U.S.A.) as a result of heavy and proloned withdrawal of water. (After R.W. Tank, 1973). (b) As a result of heavy and sustained withdrawal the water table sinks to form a cone of depression around the well, the degree of depression depending on the rate and amount of withdrawal as well as aquifer parameters. (After J.E. Poland, P.H. Reiton and P. Pestrong. In: R.W. Tank, 1976. *Focus on Environmental Geology*, **Oxford, New** Yourk, 354 p.)

GROUNDWATER PROVINCES OF INDIA

ree major groundwater provinces are clearly identifiable in the Peninsular India—the thick accumulation of porous unconsolidated sediments in the river systems, the thin covering of porous sedimentary rocks exposed in patches in the coastal belts from Odisha through Kerala to Saurashtra, and the fissured and weathered hard rocks of southern, south-western and central India (Fig. 5.11). T Himalayan region and the mountain belt of Nagaland, Manipur, Mizoram and Tripura constitute the fourth province.

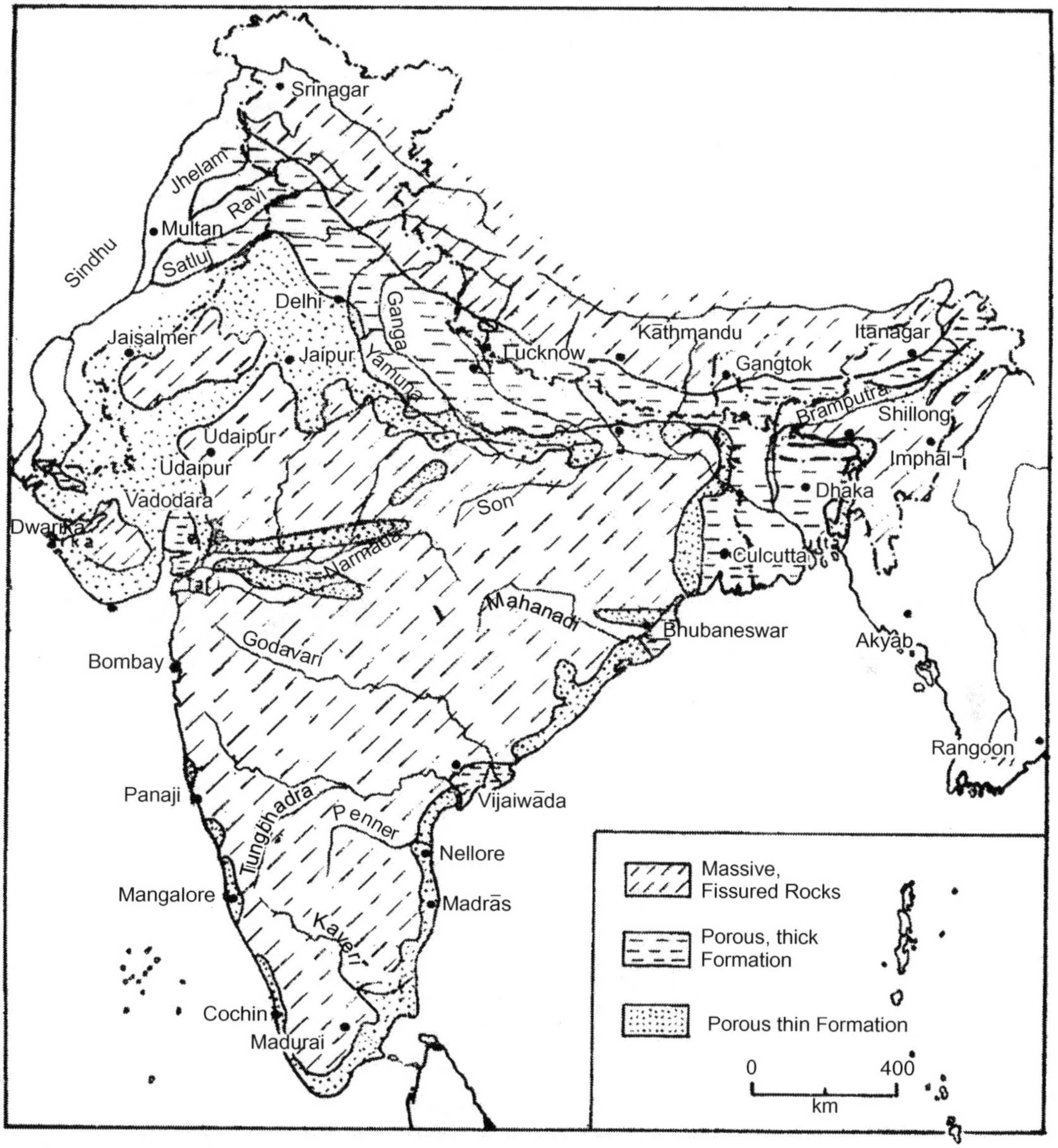

Fig. 5.11. Groundwater provinces of India. The fourth province includes the mountain domain. (After B. Pathak, 1975).

Alluvial Belts

Covering 9,05,000 km^2 area in the alluvial plains of the riverine systems of Sindhu, Ganga, Brahmaputra, Mahanadi, Godavari, Krishna, Penner, Kaveri, Tapi, Narmada, Mahi and Sabarmati the considerably though variably thick accumulations of unconsolidated recent sediments constitute the richest groundwater province of the country (Fig. 5.11). This regime is characterized by large lensoid and cally interfingering aquifers. The thickness of the aquifers varies from less than 20 m to as much as 330 m in the Ganga Basin (Fig. 5.12). The prominent and persistent zone constituted of silt and sand with

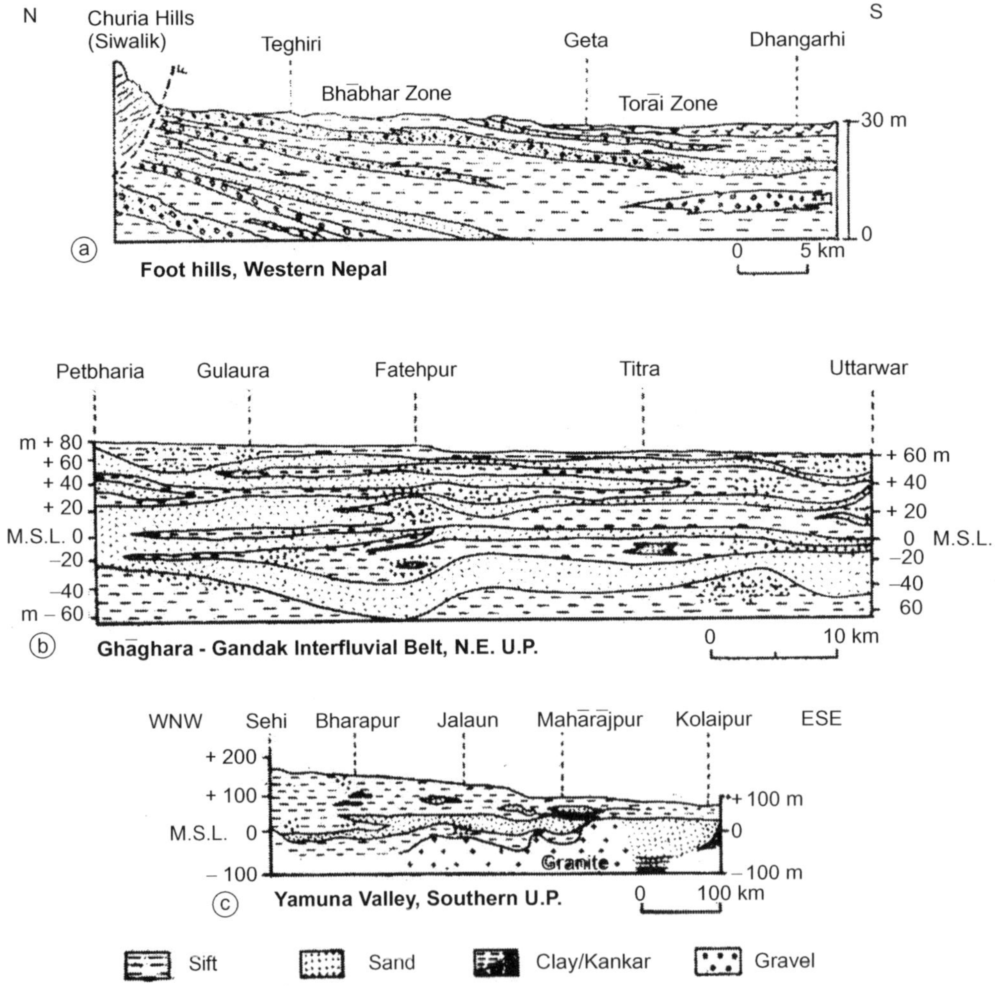

Fig. 5.12. The richest groundwater province—the Ganga Basin—is characterized by large lensoid and locally interfringering aquifers constituted of sands, local gravels and kankars. [(a) after C.K, Sharma, 1974; (c) after B.D. Pathak, 1982; and (b) after S. Mukherjee, 1983.]

Table 5.6. Groundwater Resource of India (billion m^3/yr) as on 31 March 2033 (Mall et al, 2006)

State and Union Territory	Total replenishable groundwater resource	Provision for domestic, industrial and other uses	Available groundwater resources for irrigation	Net draft	Balance ground water	Level of ground water development
Andhra Pradesh	35.29	5.29	30.00	8.57	21.43	28.56
Arunachal Pradesh	1.44	0.22	1.22	-	1.22	Neg.
Assam	24.72	3.71	21.01	1.84	19.17	8.75
Bihar	26.99	4.05	22.94	10.63	12.31	46.33
Chhattisgarh	16.07	2.41	13.66	0.81	12.85	5.93
Delhi	0.29	0.18	-	0.12	-	-
Goa	0.22	0.03	0.19	0.02	0.17	8.30
Gujarat	20.38	3.06	17.32	9.55	7.77	55.16
Haryana	8.53	1.28	7.25	8.13	0.00	112,18
Himachal Pradesh	0.37	0.07	0.29	0.03	0.26	10.72
Jammu & Kashmir	4.43	0.66	3.76	0.03	3.73	0.81
Jharkhand	6.53	0.98	5.55	1.84	3.71	33.13
Karnataka	16.19	2.43	13.76	4.76	9.00	34.60
Kerala	7.90	1.31	6.59	1.46	5.13	22.17
Madhya Pradesh	34.82	5.22	29.60	8.02	21.58	27.09
Maharashtra	37.87	12.40	25.47	9.44	16.04	37.04
Manipur	3.15	0.47	2.68	Neg.	2.68	Neg.
Meghalaya	0.54	0.08	0.46	0.02	0.44	3.97
Mizoram	1.40	0.21	1.19	Neg.	1.19	Neg.
Nagaland	0.72	0.11	0.62	Neg.	0.62	Neg.
Odisha	20.00	3.00	17.00	3.61	13.39	21.23
Panjab	18.66	1.87	16.79	16.40	0.00	97.66
Rajasthan	12.71	1.99	10.71	9.26	1.45	86.42
Sikkim	0.07	0.01	0.06	Neg.	0.06	Neg.
Tamil Nadu	26.39	3.96	22.43	14.45	7.98	64.43
Tripura	0.66	0.10	0.56	Neg.	0.06	33.43
Uttar Pradesh	81.12	12.17	68.99	32.33	36.62	46.89
Uttarakhand	2.70	0.41	2.29	0.82	1.47	35.78
West Bengal	23.09	3.46	19.63	7.50	12.13	38.19
Andaman and Nicobar Islands	0.326	0.013	0.313	Neg.	0.313	Neg.
Chandigarh	0.030	-	-	0.025	-	-
Dadra and Nagar Haveli	0.043	0.006	0.04	0.005	0.031	12.81
Daman and Diu	0.013	0.002	0.01	0.008	0.003	70.00
Lakshadweep	0.002	-	-	0.007	-	-
Puducherry	0.029	0.004	0.02	0.116	0.000	-
Total	**433.882**	**71.165**	**362.364**	**149.97**	**213.128**	**41.57**

varying proportions of gravel and kankar occurs at a depth of 30 to 170 m in the northern part and 45 to 110 m in the south-western part in the Agra division (Pathak, 1982). Additionally, there are lenses sand and gravel of limited extent in which water occurs both in confined and unconfined conditions. In the vast Indo-Gangetic domain there are large reserves of water — 52.52 m ha-m in the Ganga sub-basin, 53.73 m ha-m in the Brahmaputra sector, and 5.98 m ha-m in the Barak sub-basin alone 1982). The sustained yields of the tubewells varied from 25,000 to 4170 l/minute at a drawdown of 3 to 4 m. the eastern part of the Ghaghara-Gandak interfluvial area ('doab') where the regional water table lay at the depth of 3 to 5 m, the two major aquifers extending upto a depth of 130 m had transmissivity of 1200 to 2850 m^3/day about thirty years ago (Mukerjee, 1983).

In the Midnapur district of Bengal, aquifers at depth of 30 to 50 and 200 to 350 m have been encountered. In the Balasore and Cuttack districts in Odisha, the water table lay at a depth of 130 m in the early 1980s.

The old abandoned channels of rivers in the geomorphologically mature terrain of crystalline rocks in central and southern India have been found to be rich aquifers. As already stated, from the abandoned channels of the Betwa River in Jhansi district Uttar Pradesh (Fig. 5.7) the wells yielded water at the rate of 30,000 to 40,000 gallons per hour (2500 l/minute) in the early 1980s (*The Times of India,* 14.3.83). number of buried channels have been recognized in the wide flood plains of the Narmada, Son, Mahanadi and Indravati rivers in central India at depths varying from 10 to 22 m (Ghosh, 1978). And in Haryana and northwestern Rajasthan, in the Yamuna–Ganga basin in Uttar Pradesh, and in the Kaveri lta in Tamil Nadu.

The shallow (<20 m) unconfined aquifers of this kind in the alluvial deposits of Maharashtra (Nanded, Wardha Yeotmal), Madhya Pradesh (Satna, Rewa, Raipur), Rajasthan (Jhunjhunu, Sikar), Jharkhand (Ranchi, Santhal Pargana) cover more than 100,000 km^2 area, the specific capacity being 25 to 250 minute per metre of drawdown, the specific yield 8 to 20% and individual dug wells yielding 100 to 500 m^3/day at the drawdown of 3 to 10 m (Adyalkar, 1984). The ephemeral Shivaganga, Mula and drayani rivers of the Pune region in Maharashtra, for example, contain ample water in unconfined ndition in the colluvial and alluvial deposits—the specific capacity being 380 to 455 l/minute per metre drawdown, compared to 80 to 200 l/minute of the lava aquifers (Thigale, 1983; Thigale and Zambre, 1982). The spring discharges of in this region varied from 70 to 122 l/minute and the well discharges from 53 to 95 l/minute in the early 1980s.

In the upland Narmada alluvium (Fig. 5.11) the water table lay at a depth of 2 to 3 m, the hydraulic gradient is 1.6 to 7.75 m/km towards north-west and the total potential was 57,000 to 25,100 million olitres in the shallow aquifers and 32,000 million kiloletres in the deeper ones in the early 1970s (Adyalkar, 1975). In the adjoining Tapi alluvium, the water table was encountered at a depth of 2.5 to 44.5 m, the hydraulic gradient being towards south-west (3.1 m/km) and southeast (5.5 m/km), and the total potential of the shallow and deeper aquifers respectively 8,000 and 3,000 million kilolitres (Adyalkar, 1975). The withdrawal rate in the two alluvial belts was then, on an average, 50,000 l/minute per well.

Continent-Margin Sedimentary Belts

The Indian Peninsular shield has a narrow to wide marginal mantle of sedimentary rocks, Mesozoic (Gondwana) to Cenozoic in age. The consolidated to semi-consolidated porous sandstones and imestones, including the miliolite-limestone of Saurashtra, cover about 55,000 km^2 area. There are maller discontinuous aquifers. About thirty years ago the specific capacity was in the range of 100 250 l/minute per metre drawdown, the specific yield 8 to 15% and the individual well potential of the order of 400 to 1000 m^3/day for the drawdown of 5 to 10 m (Adyalkar, 1984). From the semi-consolidated aquifers the rate of water yield per dug well was 100 m^3/h compared to less than 50 m^3/h per well from the indurated sedimentary rocks, such as the Cuddalore Sandstone in Tamil Nadu. Three interconnected aquifers under the lignite belt in Neyvelli in the Cuddalore terrain had discharged 218 m^3/minute through 15 to 30 pumps. From the Umia Sandstone (Jurassic) in Gujarat, where the water table lay at a depth of 150 to 300 m, the yield per tubewell, was 0.6 to 500 l/minute at the drawdown of 15 to 45 m (Biswas and Saha, 1983). In western Rajasthan where the water table had by then sunk to a depth of more than 120 m, there lies under the vast cover of dry desert sand, large usable groundwater reserves amounting to 1200 million m^3—sufficient for 200 years (Dasgupta, 1982 In the Lathi Sandstone (Jurassic) the aquifer occurs at a depth of 150 m, the tubewell yielded on the average was 3,333 l/minute per well at the drawdown of 5.5 m (Biswas and Saha, 1983). The deeper tubewells yielded from the cavernous limestones, at the rate of 1,000 to 2,500 l/minute per well, and in the Barunda area in Jodhpur district the yield varied from 10 to 10,000 l/minute per well at the draw-down of 10 m (Choubey and Tanwar, 1982).

The aquifers in the Gondwana sandstones such as the Barakar, Pachmarhi, Kamthi, Bogra and Jabalpur formations in central India are being tapped by wells having a specific capacity of 50 to 2 l/minute per metre of drawdown and the specific yield of 5 to 10%. The yield potential of larger diamet wells was considerably higher—250 to 750 m^3/day for drawdown of 5 to 15 m, such as at Durgapur and Mahakali coalfields in Bengal (Adyalkar, 1984).

Fractured and Weathered Hard Rocks

The jointed, fractured and deeply weathered hard rocks (igneous and metamorphic) of southern, centra and eastern India (Figs. 5.6 and 5.8) cover about 17,00,000 km^2 area. The thick mantle (6 to 20 m) of weathered material makes good aquifers, with the water table at a shallow depths of 2 to 10 m. The weathered mantle is also associated with the buried erosion surfaces (unconformities) where porosity is high (18 to 20%) and permeability appreciable. The yield of the wells, however, is generally poor—less than 5 to 25 m^3/h (Pathak, 1975). The specific capacity of coastal lateritic soil ranged from 100 300 l/minute per metre drawdown, the specific yield being 3 to 8%, the larger-diameter wells had yield potential of 400 to 1000 m^3/day for a drawdown of 2 to 5 m. With the lateritic cappings in Goa, Kolhapur district in Maharashtra, Ranchi in Jharkhand, and Medak in Andhra Pradesh and Bidar in Karnataka are related the many oozes, seepages and springs which used to discharge water at a rate of 36 to 1362 m^3/day (Adyalkar, 1984). The lateritic mantles not only gets rapidly depleted, but are al quickly recharged during rains. The wells have the specific capacity of 100 to 300 l/minute per met drawdown and the specific yield of 3 to 8%. Larger dug wells had a discharge of 10,000 m^3/d for a drawdown of 2 to 5 m.

In the Deccan volcanic terrain of Maharashtra and Malwa in Madhya Pradesh the lavas are characterized by a high degree of jointing (including columnar), interconnecting vesicles and pipes or tunnels, and are interbedded with porous-permeable volcanic ash and sediments, and locally 0.3 to 1.0 m thick red boles, such as seen in the Solapur-Osmanabad area. There are thus multiple-layered aquifers upto the depth of 15 to 20 m (Singhal, 1973). More commonly, there are continuous water table aquifers. Intersecting dykes, such as discernible in the Dhulia areas, Thana–Kolaba belt and Mumbai–Pune region, serve to impound subsurface flowing waters in buried channels and the present sediment-filled rivers, and in t paths of groundwater. The specific capacity of vesicular volcanics (aquifer) is in the range of 100

0 l/minute per metre of drawdown, the specific yield was 3 to 10% and the yield potential of larger-diameter wells was 400 to 1000 m^3/day (Adyalkar, 1984) in the early eighties of the last century.

Bhabhar-Tarai Belt in the Himalayan Foothills

The Siwalik ranges are fringed on the southern side by an apron of coalescing fans of colluvial and vial debris. This piedmont belt is known as 'Bhabhar'. The mountain rivers loose their water in the gravelly accumulation (Fig. 5.13), thus recharging the unconfined aquifers, which extend down to a depth of 90 to 150 m. The depth of the water table is extremely variable feeding the tubewells with a discharge as much as (thirty years ago) of 3000 l/minute (Biswas and Saha, 1982). In the Pathankot area in the northwest, the hydraulic gradient of the water table, which was at a depth of 4 to 10 m, was quite steep—9 m/km in the north and 3 m/km in the south. The total available discharge was 103 million m^3 in the southern fringe (Verma and Dogra, 1983). In the southwestern extremity of Nepal, the water table ctuated between 6 and 9 m. The aquifers have been penetrated upto the depth of 240 m, there being three horizons of 9 to 15 m thickness (Sharma, 1974). The wells reaching the depth of 90 m yielded water at the rate of 1320 litres/minute at 3.6 m drawdown (Pathak, 1975).

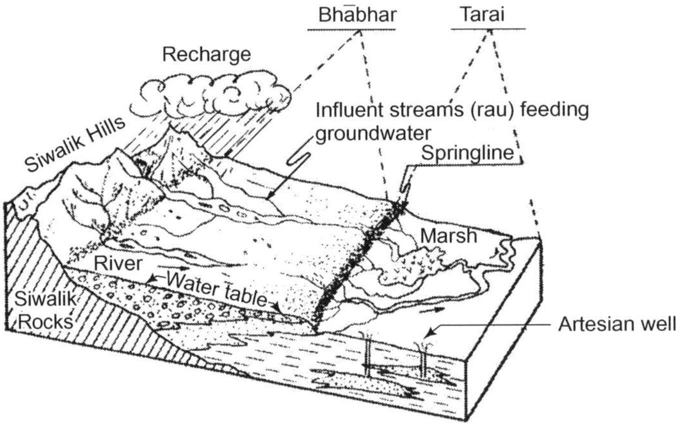

Fig. 5.13. The Bhabhar comprises coalescing fans of colluvial and fluvial gravelly deposits into which the waters of the myriad mountain rivers and streams get lost, thus giving rise to deep (90 to 150 m) unconfined aquifers. The water emanates at the margin in a series of springs and seepages, causing development of wet land or Tarai.

The 'Bhabhar' is fringed to the south by a line of countless numbers of perennial springs and seepage (Fig. 5.13), making the whole belt very wet and locally even marshy. This is the 'Tarai' belt (tarai = tness). In the vertical column of 300 m, there can be as many as three to four different confined aquifers under varying artesian conditions. Artesian conditions are found at Malangwa, Parsa, Janakpur, rthern Siraha and southern Jaleswar in southcentral Nepal. Nearly five confined layers of sands and ocal gravels were recognized in southwestern Nepal at depths of 3, 11 and 19 m (Sharma, 1974).

Intermontane Valleys

The Kashmir valley in the Tethys Himalayan domain is filled with lake deposits capped with layers wind-borne loess—the combined succession known as the Karewa Group. The wells reaching the depth of 150 m had discharged water at the rate of 5000 l/minute and the springs at the base of the succession discharged at the rate of 7 to 420 l/minute (Biswas and Saha, 1984).

In another very similar geomorphic-hydrogeological setting in the Kathmandu valley (Nepal), the water table in the lacustrine deposits lay at the depth of 4 to 12 m, the water occurring in confin (artesian) conditions upto the depth of 450 m below the 50 m thick black clay horizon. More than thirty years ago the total reserve of water was of the order of 28 million m^3 (Sharma 1974).

In the Dehradun valley in the Siwalik sub-province a big synclinal valley is filled with gravels. T water table is at a depth of 2 to 14 m and the aquifer intersected at a depth of 75 m, the tubewells yielding at the rate of 3,000 l/minute for a drawdown of 1 to 7 m.

SPRINGS IN THE HILLS

Types of Springs

Where the surface of the slope of the hill meets or intersects the water table, groundwater moves out in the form of a spring or seepage (Fig. 5.14). Marshes or ponds are formed when the water tab rises above the floor of the topographic depression and the groundwater issues out through myriads openings. Water may come out at the contact point of the permeable and impervious layers or where a dyke cuts the permeable layers (***contact spring***), it may move along the fractures or faults before surfacing (***fracture spring***), or the pressure build-up within the aquifer may force the water out as ***artesian spring***. Thermal springs emit heated water coming from deeper levels, the surface or rainwater having earlier penetrated great depths and got heated. In some places, the vaporization of water throws out heated water as forceful fountains called ***geysers***.

Recharge of Springs

Rainwater infiltrates the ground through intergranular pore spaces, openings, fissures, fractures, join etc. and reappears down the slope as springs and seepages. The yield of a spring depends not only on the rate of flow of the groundwater but also on the rate at which the soil or rock of the recharge ar absorbs the rainwater. It is the watershed where maximum recharge occurs (Fig. 5.24). It varies from place to place and season to season. During the rainy season, immediately after the rainfall, the recharge is higher, and following prolonged droughts the discharge of springs drops, sometimes even to the zero level in the absence of recharge in the watershed. The slope, the nature of soil and the character and extent of vegetation on the watershed govern the extent of recharge, as already emphasized (Fig. 5.4). To recapitulate, **for desirable recharge the watershed has to be treated appropriately by coveri it with multistoried vegetation comprising trees, shrubs and grasses having an adequate canop of foliage and the floor carpeted with moss and/or a litter of leaves**. One of the reasons why the springs in parts of Kumaun and Garhwal in Uttarakhand are drying up or becoming seasonal (40 to 45% in the catchment of the Gaula River in Nainital district—Valdiya, 1985) is that the vegetal cover had drastically dwindled to just 28.7% of the surface area, and that forests of trees with more than 60% crown density were restricted to bare 4.4% (J. S. Singh et al., 1984) thus considerably diminishing the replenishment of sub-surface water.

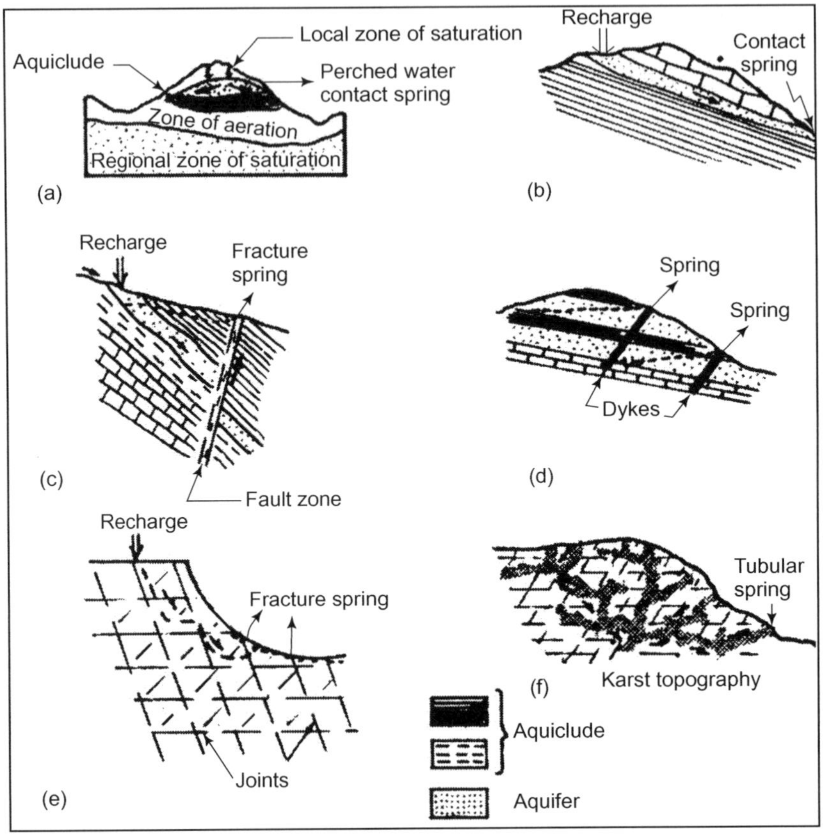

Fig. 5.14. Types of springs and their origin.

Despite the official claim that the forest cover has increased, the springs continue to be in stress — much more than in yesteryears.

On an average, about 60% of the rainwater will infiltrate if the forest ground is covered with moss and/or litter as can be deduced from the data (Table 5.7) collected in an experimental forest near Chandigarh (Samar Singh, 1980) or in an eucalyptus plantation, where the peak runoff was reduced by 55% (Patnaik, 1980). **The greater the infiltration, the greater will be recharge of water and the eater the discharge in springs, and vice versa.**

Table 5.7. Extent of Infiltration of Rainwater i.e. Degree of Recharge in a Watershed under Different Forest Covers (Based on data from N. Patnaik, 1980)

Vegetal cover	Slope 5°	Slope 10°	Slope 20°	Slope 30°
1. Pasture of grazing land	22%	18%	10%	5%
2. Spruce forest with thin carpet of litter and moss	72%	66%	47%	23%
3. Spruce forest with thick carpet of moss and little	98%	97%	95%	92%

Too-Little-Too-Much Water Syndrome

Consequent on widespread deforestation, degradation of land and unscientific and indiscrimina excavations for roads and mineral/rock mines, the discharge of hill springs have declined considerab and drastically in some areas. A study in the catchment of the Gaula River in southcentral Himalaya showed that the spring decline in spring discharge varied from 25% to 75% in 5 to 50 years, culminating in the reduction of stream flow from 29.2% in the period 1951–1970 to 38.5% in the interval of 19 to 1981 (Bartarya and Valdiya, 1989). In the Dugar Valley in Pauri Garhwal (Uttarakhand), the spring discharge decreased by 36.4% in a very short period of 1994 to 1998 (Negi et al, 1998).

Since spring discharges and seepages feed large streams and rivers, lessening of spring discharges culminated in what is called ***two-little-too-much water syndrome***. The ratio of outflow of Himalay rivers between rainy and the dry seasons is 120:1 for the Karnali, 15:1 for the Saptakosi and 180:1 for Kali Gandaki in Nepal (Uprety, 1988).. Significantly, **the flow in the dry season in the deforest Lesser Himalaya rivers in Nepal is 60,000 to 30,000 times less than the flow in the rainy seas•** (Kattleman, 1987).

The implication is clear and loud — the mountain springs are unable to contribute enou water to rivers.

Inducing Surface Water

Infiltration (percolation) tanks, trenches and pits dug on large scale in the Yawat taluk in Jalga district (Maharashtra) resulted in additional recharge of 410,000 cubic metres annually of water to the groundwater, bringing 81.9 ha land under cultivation following 5 m rise of the water table (Karmarkar et al, 1995). Similar development was witnessed in Aurangabad, Osmanabad, Beed, Ahmadnagar, Solapur and Sangli.

In the Alwar district, Rajasthan, a large number of check dams built across the Anwari river and its five rivulets to hold rainwater raised the water-table from the depth of 90 m to 9 m, thus providi benefits to 650 villages (Agrawal et al, 1999). In Hunsur taluk in Karnataka, the weathered gneiss permitted recharge to the extent of 10,238 m^3 to 17,776 m^3.

The chronically drought-prone Ralegan Siddhi area in Maharashtra produced in 1994 food-grains at the rate of 138 tonnes/year compared to 29 t/yr in 1976, due to the rise of water table in all the wells following construction of infiltration tanks at appropriate sites in *nalas* and streams (Deshmukh, 1995).

QUALITY OF WATER

Drinking Water

In the early eighties of the 20th century not more than 162 million people living in the villages in India had access to safe drinking water. In cities and towns the situation was only slightly better. **Today, according to official claim, only 17.9% households in the villages get the right quality of water f drinking** (*The Hindu*, March 14, 2012). The rich have taken to bottled mineral water, and to have water-purifying gadgets installed at homes. Most Indians have developed immunity, and do not bother to set any standard on the quality of water, except ensuring reduction or elimination of turbidness. The Wor

Health Organization (WHO) suggested certain limits of qualities for drinking water, particularly with regard to toxic elements (Table 5.8). The qualities which are of common concern are the hardness of water, the pH value and the biological oxygen demand (BOD). The presence of salts and toxic elements poses problems locally. The hardness is a result of the presence of dissolved carbonates of calcium and/or magnesium. If the amount exceeds 200 ppm the water becomes hard, if less than 50 ppm it is soft. The acidity-alkalinity (pH) of the water is expressed in terms of concentration of hydrogen ions which is measurable in exceedingly low fractions of unity and thus given in the logarithmic scale. Water having greater than 7.8 pH is alkaline and less than 7 acidic, and between 7.8 and 7.0 it is neutral.

Bacteria need oxygen for the reduction of carbonaceous-organic matter present in the water. The higher the amount of these materials the greater the utilization or depletion of oxygen dissolved in the water. In other words, there would be a greater ***biological demand for oxygen*** **(BOD)** if there were larger amounts of organic mater. It is the dissolved oxygen that sustains life within the water, and its depletion s adverse effects on the health and survival of aquatic life. The discharge from affluents contributes greatly to the organic content of the river or lake waters, and thus increase the BOD, which makes the ater unfit for all kinds of organisms. In drinking water the value of BOD should ideally be nil.

Turbidity results from the presence of suspended particles of silts and clays, which can be eliminated by allowing the water to stand still for a period of time (depending on the size of the particles) when e particles settle down. But it is difficult to get rid of the dissolved material such as lime, salts and trace elements. The total salt content of the groundwater can be determined by measuring its electrical conductivity (expressed in micro-mhos/cm at 25°C).

Water for Irrigation

The permissible limit of dissolved salts in water for irrigation is 1,000 ppm under ordinary conditions. In sandy and well-drained soils, a little higher content of salts can be endured and the bicarbonates are mparatively less deleterious. Vegetable gardens can be irrigated by effluents after proper treatment from municipal or some industrial drains. The following limits should be maintained in such cases—pH 5.5 to 9.0, BOD 500, total inorganic solid<2100 mg/1, boron<2 ppm and total dissolved salts of K, Na, Ca, Mg less than 60%.

The waters of the Ganga, Gomati (Uttar Pradesh) and Mahi (Gujarat), for example, are by and large hard. The Ganga drains the Lesser Himalayan terrain (and also from the Tethyan domain across the snowy ranges), where thick formations of dolomites and limestones (e.g. the Buxa, Dhading, Deoban, Shali, Jammu, Mandhali and Krol formations) are exposed to erosion and dissolution. The Gomati, though originating in the Tarai belt, gets the carbonates from the carbonate grains, cement and calcareous pans in the alluvial deposits derived in the past from the Himalaya. The Mahi drains a part of the Aravali made up of the phosphatized dolomitic limestones. The Sindhu and the rivers of southern India on the whole carry softer water, for there are very few carbonate formations in their paths. The waters of the Ramganga, Jhelam, Ravi, Betwa and Chambal are by and large acidic, for they drain the regions of iceous rock-formations such as granites, gneisses, quartzites, sandstones and shales. The Godavari (at Huvenhedgi) and the Tungbhadra (at Harlahatti) are generally alkaline with low to moderate salinity due to chlorides, sulphates and bicarbonates present (Mitra, 1982).

Table 5.8. Drinking Water Standards (Potability) Prescribed by WHO in 1971 (Data from Coates, 1981)

Constituents	Highest Desirable Level (ppm)	Maximum Permissible Level (ppm)
Total Dissolved Solids	500	1500
Anionic detergents	0.2	1.0
Arsenic	0.01	0.05
Barium	-	1.0
Cadmium	-	0.01
Calcium	75	200
Chloride	200	600
Chromium	-	0.05
Copper	0.05	1.5
Cyanide	0.01	0.05
Fluorine	0.8–1.7*	1.6–3.4*
Iron	0.1	1.0
Lead	-	0.1
Magnesium	150@	150@
Manganese	0.05	0.5
Nitrate	40	45
Silver	-	0.01
Sulphate	200	400
Zinc	5.0	15.0
BOD	9.51	9.20
pH	7.0–8.5	6.5–7.0

* Varies inversely with mean annual temperature.*
@ If 250 mg/1 of sulphate is present, Mg should not exceed 30 mg/1.@

WATERLOGGING—THE BANE OF EXCESSIVE IRRIGATION

Causes of Rise of Water Table

If the soil is given water more than the vegetation or crop needs the water table rises. This happens particularly in the wetter periods and in places where there is inadequate drainage such as in topographic depressions. Not only excessive irrigation, but seepage from canals, channels, reservoirs, tanks, etc. is also responsible for waterlogging (Fig. 5.15). It is estimated that 15 to 40% of the water flowing throu the canals in India is lost due to primarily seepage and secondarily to evaporation (Dakshinamurthy et al., in Kayastha, 1981). In the Ganga Basin the loss due to seepage is as high as 50%, and in the Sindhu Basin in Pakistan about 33%. Even as early as the mid-sixties of the last century, 13 to 20% of the water transported in the upper Bari Doab Canal was lost through seepage (Usha Rai in *The Times of India,* January 11, 1985). The efforts expended and money spent to bring river waters to agricultura farms have thus become counterproductive and infructuous to a some extent. The lack of knowledge of e ground conditions (geomorphology), texture of soil and inefficient system of supply are principally responsible for the deplorable state of affairs.

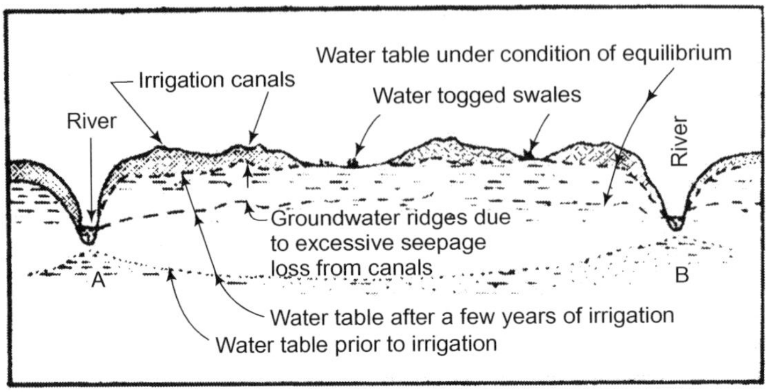

Fig. 5.15. Seepage from canals and over-irrigation is responsible for the rise of the water table and consequent waterlogging.

In the Saryu canal system in eastern Uttar Pradesh—remote sensing has revealed that — the low-ng flood-prone Ghaghara valley in the Fakharpur, Tejwara, Hazurpur, Bisheshwarganj, Taraliganj and Karailganj blocks in the Gonda and Bahraich districts is waterlogged, pronouncedly so during the monsoon period (Chaturvedi et al., 1983). This is evident from the comparison of the satellite photos ken on March 16, 1975 and November 9, 1981 (Figs. 5.16 and 5.17). Significantly, even the higher surface in the Sharada Sahayak Canal command area has developed problems of waterlogging (Fig. 5.17) since the commissioning of the Daryabad and Barabanki branches (Chaturvedi, et al., 1983).

Adverse Effects on Vegetation

In the waterlogged soil, water rises to the root zone of the plants, retarding their growth. This happens due to the expulsion of oxygen from the zone of aeration, and consequent, cessation of all processes of ygen so necessary for plant growth. In the oxygen-deficient environment, moss and aquatic weeds get the upper hand, and in the severe competition that ensures, the desirable crop-plants are eliminated. Bad irrigation had thus reduced the crop yields in half the world's irrigated land, rendering nearly 25 m ha of land out of production at the rate of 100 ha/day over the world till the 1980s (*The Times of India,* June 10,1984).

The productivity of the waterlogged fields has declined considerably. Against the anticipated oduction of 4 to 5 t/ha/yr, the actual was merely 1.7 t/ha/yr (Bhumbla, 1983). Thus there was not even a recovery of the money invested in developing a canal irrigation system at the rate of Rs. 4270 llion/yr (Rs. 55,000 million in the period 1951-1978) to irrigate 5.5 m ha as Bhumbla points out. It may be stated that **the benefits of canal irrigation last only the first 10 to 20 years, and then follow e backlashes**.

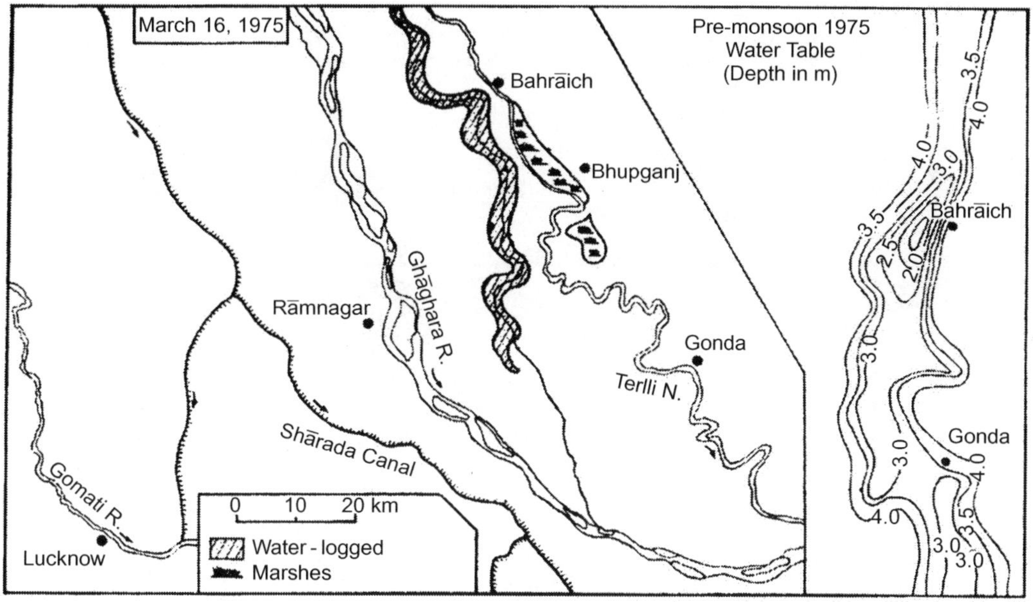

Fig. 5.16. Groundwater condition and extent of waterlogging in the Ghaghara valley (eastern Uttar Pradesh) scanned on March 16, 1975 by satellite. (After R.S. Chaturvedi, R.S. Sharma and A.K. Tangri, 1983).

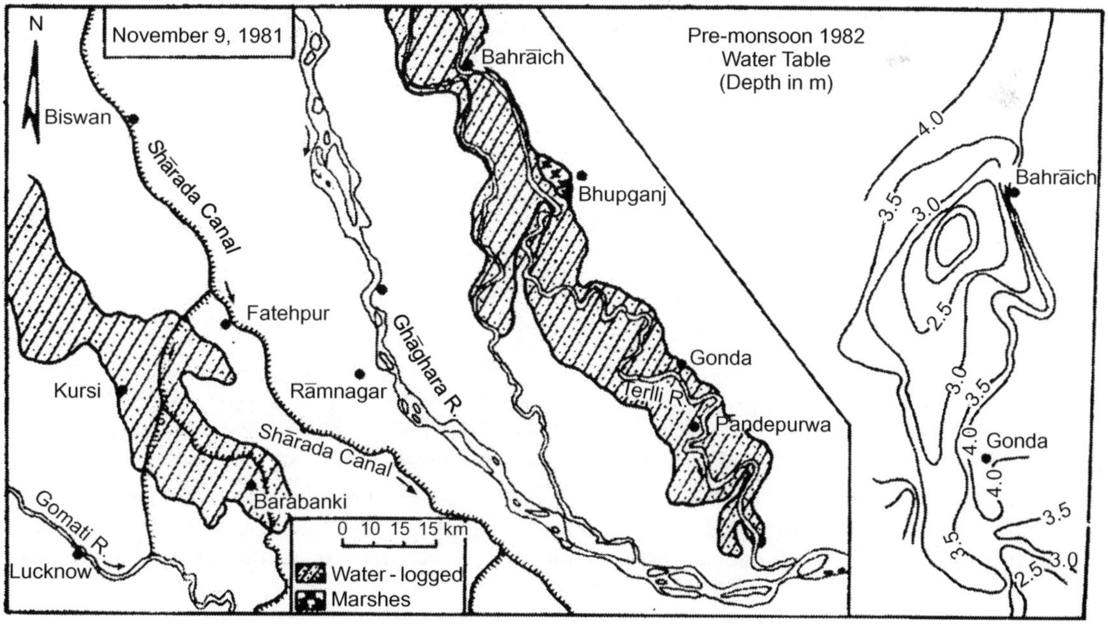

Fig. 5.17. Extension of waterlogged ground and rise of the water table as on November 9,1981 in t Ghaghara valley resulting from irrigation through the Saryu Canal System. (After R. S. Chaturvedi, R. S. Sharma and A. K. Tangri, 1983).

Foundation Problems

Another manifestation is the turning of topographic depressions, ditches, trenches and pits into pools marshes as the water table rises to a level higher than the floors of the depressions. The foundations of buildings and civil structures are greatly weakened as a result of loss of as much as 50% of bearing strength of the ground. In the town of Muktsar in south-western Panjab, where the water table had risen from 10 to 12 m in 1973 to 1 to 2 m by 1983 (Fig. 5.18), a large number of buildings had developed cracks (*The Times of India,* September 1, 1983).

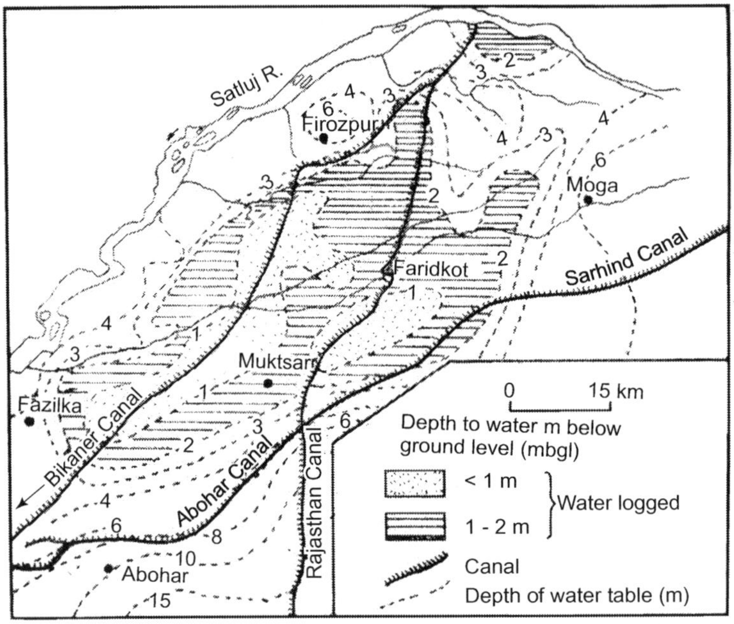

Fig. 5.18. Over-irrigation has waterlogged a large part of extremely rich agricultural fields in south-western Panjab. In the township of Muktsar the water table rose from 10 to 12 m below the surface in 1973 to 1 to 2 m in 1983. (Based on D. S. Saini, S. B. Singh, V. M. Sikka and N. C. Bhatnagar, 1983).

Extent of Waterlogging in India

A very large tract of the well irrigated south-western Panjab (Fig. 5.18) has developed serious waterlogging problems. In the cotton belt of Hissar and Sirsa districts in southern Haryana (Pandit and Chadha, 1983), the subsoil water had risen from 20 m in 1964 to 1.5 m by 1983. Even the basement of the Agricultural University building at Hissar was then waterlogged (*The Times of India,* October 15, 1983). About 350-odd villages in the Rohtak, Bahadurgarh and Jhajjar subdivisions had become aterlogged. In the very first decade of the functioning of the unlined canals in Panjab, the water table had risen by 7 to 8 m at the rate of 0.5 to 0.75 m/yr, the total rise in the 70 to 75 year period, being 20 25 m in the fields of Ferozpur, Faridkot, Bathinda, Fazilka, Abohar, Ganganagar and Hanumangarh districts (Mann and Prakash, 1983). The problem has assumed serious proportion in many areas of

Ferozpur, Muktar, Bathinda, and Sangrur in Panjab where the water table rose at the rate of 15 to 20 cm/year over the 85,000 ha of agricultural land (Shakya and Sinha, 2010). In western Haryana the water table has risen by 3.0 to 9.0 m in seven districts, causing waterlogging over 2346 km^2 of irrigated land (Lunkad and Sharma, 2010). The irrigation through the Sharada Sahayak Canal in the north-eastern part of the Ghaghara-Ganga 'doab' (Fig. 5.18) had raised the water level to 0.5 to 1.6 m below the surface within the one kilometre belt from the canal (Singh and Singh, 1983).

Likewise, in the command area of the Mahi canal (Gujarat), unlimited irrigation coupled with inadequate drainage had raised the water table, which rose at the rate of 0.375 m/yr in clayey soil and 0.76 m/yr in the sandy loam soil (Kalubarme and Sahai, 1983). Part of the Khambhat and Matar talukas are thus waterlogged; so are parts of Thasra and Nadiad talukas.

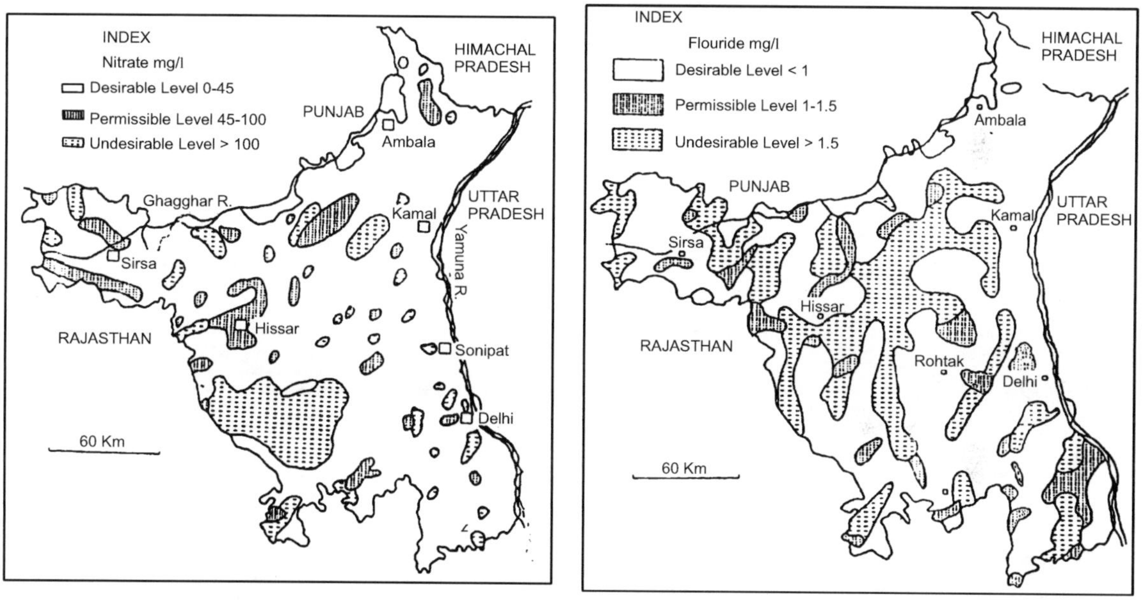

Fig. 5.19. The nitrate and fluoride contents in the groundwater of Haryana. (From T.N. Thussu, 1999).

In the well-irrigated region of Rajasthan while the recharge from rainfall and irrigation causes the water table to rise by 2.05 m, the withdrawal through all kinds of wells accounts for the lowering of not more than 1.05 m, so that the net annual rise is 1.4 m. This explains why in the Anupgarh area (Ganganagar district) in the Ghaggar valley, the water table rose at the rate of 1.52 m/yr (locally 3 m/yr) in just 15 years and creating very serious problems then in about 15.4% of the irrigated area (Mann and Prakash, 1983). The problem has now assumed very serious dimensions in the command area of the Rajasthan Canal. Over 5% of agricultural land in northwestern Rajasthan is waterlogged (Fig. 5.21) with attendant salinization (Paliwal and Paliwal, 2011). There is also a hard pan of gypsum in the subsurface at shallow depth which helps create waterlogging. There are now 42 marshy lakes along the canals.

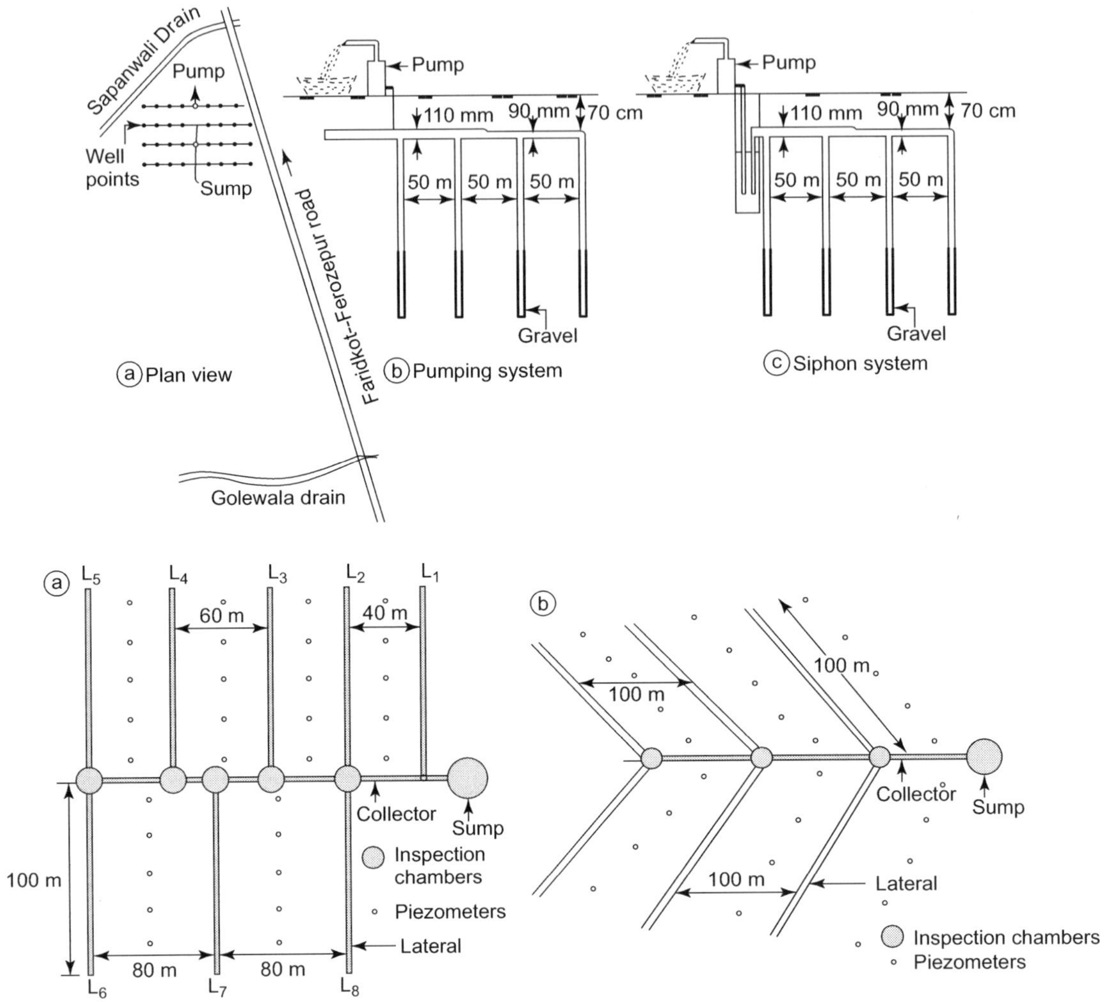

Fig. 5.20. Overcoming problem of waterlogging. *Upper*: A system of multiple pumps or tubewells are installed systematically to withdraw groundwater intensively and channelise them out. *Lower*: Layout for subsurface perforated PVC pipes (a) Gridiron pattern and (b) Herringbone pattern. (From: Shakya and Singh, 2010).

In the Chambal valley (Madhya Pradesh) nearly 25% of the command area had become waterlogged thin 10 years of introduction of canal irrigation. Similar affliction of farmland was witnessed in the Mahi-Kadana and Kakrapara-Dantiwada canal-irrigated belts in Gujarat, the Tungbhadra command area in Karnataka, the Nagarjunasagar area in Andhra Pradesh, the Purna, Pravara, Meera and Jaililawadi valleys in Maharashtra, the Son valley in Bihar, the Hirakud region in Odisha, and so on. **It is now alized that vast tracts of the well-irrigated regions in the country are exposed to the threat of waterlogging.**

In the Sindhu valley in Pakistan the problem was far more serious—the water table had risen from 30 m to just a few centimetres below the surface in about 2.4 million ha of irrigated land until about the mid 1970s (Jackson, 1977), waterlogging even the great archaeological sites of Mohenjodaro.

As already stated, more than 6.41 m ha of agricultural land in India is waterlogged (Anon, 2010).

Fig. 5.21. Waterlogged tract of land due to overuse and misuse of water of the Indira Gandhi Canal in parts of western Rajasthan (From Paliwal et al., 2010). (See colour Fig. 5.21)

DEVELOPMENT OF ALKALINITY-SALINITY IN SOILS

Capillary Action

In the arid and semi-arid regions another problem far much vexatious than waterlogging, is created by imprudent irrigation through canals—the introduction of salts and alkalies in the soil (Figs. 5.19 and 5.22). The interstitial water containing dissolved salts and carbonates rises to the surface through minute tubelets by capillary action and deposits in the pore spaces the dissolved salts in the form of encrustation on the topsoil such as 'kankar' and 'reh' (Chapter 3). The alkalies and salts are injurious to plants and impede infiltration of water. In the absence of rainwater and deprivation of nutrients, t soil deteriorates eventually.

Extent of Damage

As already stated, about 9.30 m ha of our cultivated lands have gone out of production as a result of alkalinization and 6.73 m ha due to salinization of the soil (Figs. 5.19 and 5.22) (Anon, 2010), resulting from over-irrigation chiefly in the Indo-Gangetic Plains. Between 30 and 80% of all the lan under irrigation throughout the world had suffered grievously from waterlogging and alkalinization-salinization—the rate of loss of the world's best land is 2,00,000 to 3,00,000 ha per year in the mid 1970s (Worthington, 1977). In the erstwhile U.S.S.R. there were then 140 m ha of saline soils. Nearly 50% the irrigated soils in Iraq and Syria, 30% in Egypt and more than 15% in Iran were afflicted wı the malady born of over-irrigation (Worthington, 1977) some forty years ago.

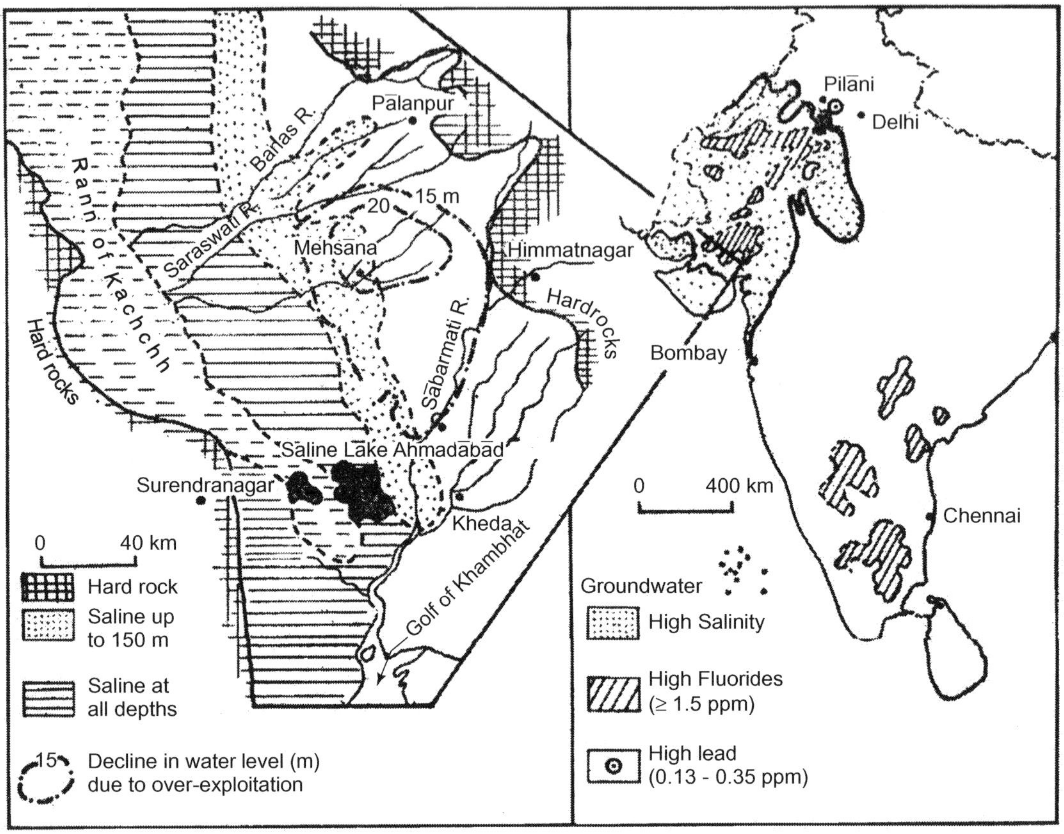

Fig. 5.22. The groundwater in northern Gujarat and adjoining Kachchh was already naturally saline. Heavy withdrawal for intensive irrigation lowered the water table, aggravating the condition of salinity. (Based on P. N. Phadtare and E. Bradley, 1983.)

Natural Salinity

Groundwater dissolves soluble salts occurring in the soil and subsoil and thus becomes saline, as witnessed in parts of Mathura, Agra, Banda and Hamirpur districts in southern Uttar Pradesh. In a large tract of Haryana (Hissar, Mahendragarh, Gurgaon) and adjoining Rajasthan (Suratgarh, Ganganagar) in the Quaternary basin of the now vanished Saraswati River (the Ghaggar of today) the groundwater is very saline (Figs 5.19 and 5.22), the total dissolved salts (TDS) exceeding 8,000 ppm, and the ratio of chloride to the total solids approximating 0.5.

Very large extent of the land of Haryana is underlain by saline waters in unconfined aquifers. Admittedly, the salinity is due to the original sediments containing salts and existence of the barrier of the Aravali range. But irrigation undeniably aggravated the condition (Prakash and Chadha, 1983). In e lower Ghaggar basin, saline water had been identified over 4400 km^2 area underlying the fresh to partly fresh water in the early eighties of the 20th century, the freshwater being restricted to abandoned channels and to flood plains (Pandit and Chadha, 1983). In the adjoining area in Rajasthan about 60 of the area has saline groundwater. In about 1,05,000 km^2 area the groundwater has chloride content higher than 1500 ppm, particularly in the Luni basin embracing Barmer, Jodhpur and Jalore districts (Dasgupta, 1982). In the marine Tertiary and the Jurassic sedimentaries of the Jaisalmer, Barmer and Jalor districts and Kachchh and northern Gujarat (Fig. 5.22) the salinity is moderately high, the TDS concentration being 3,000 to 8,000 ppm. However, in the Eocene of Bikaner, the Jurassic Lathi Sandstone of Rajasthan, and the Jurassic of Upper Bhuj Sandstone of Kachchh, the salinity is low—500 to 3,000 ppm (Table 5.9).

Table 5.9. Natural Salinity of Groundwater in Some Parts of Rajasthan (After Dasgupta, 1982)

District	Area (km^2)	Percentage Area having Saline Waters		
		>2500 mg/l	Upto 2500 mg/l	Upto 1500 mg/l
Barmer	29,030	68%	32%	10%
Jaisalmer	38,470	71%	29%	9%
Bikaner	27,100	78%	22%	5%
Churu	16,820	82%	18%	9%
Pali	11,810	56%	44%	25%
Nagaur	17,290	62%	33%	20%
Jodhpur	22,210	55%	45%	30%
Jalor	10,430	65%	35%	20%
Average		52%	32%	16%

Intrusion of Saline Seawater

In the coastal area of Saurashtra in Gujarat the groundwater is very saline (Fig. 5.22.) This is partly due to tidal flooding, but principally due to the intrusion of saline seawater following heavy withdraw through pumps and wells (Nair at at., 1983) (Fig. 5.23). The invasion of the seawater rendered waste a large tract of the land in the 5- to 8-km wide belt between Una and Madhavpur, embracing the coastal stretch of Junagadh and Amreli. The extent of the saline area increased from 3,500 ha in 1971 to 1,00,000 ha in 1977 at the rate of 0.5 to 1 km/year (Ahluwalia, 1983).

A similar situation developed in the Vishakhapatnam area in the Andhra coast where the saltwater interface advanced through structural lineaments and fissures (Hari Ram and Rama Rao, 1983). coastal Tamil Nadu, overpumping caused lowering of the water table by more than 25 to 30 m during the seventies of the last century, resulting in the intrusion of oceanic saltwater into the aquifer at the rate of 120 to 150 m/yr (*The Hindu,* August 1985).

The **remedial measures** to combat saltwater intrusion include: (i) reduction of pumping draft; (ii) artificial recharge of the aquifer by spreading freshwater on the surface or through injection wel (iii) developing a trough of depression in the level of the saltwater table through a series of pumping wells dug in a line, which will prevent the advance of the salt water; (iv) creating subsurface barriers of impermeable material; and (v) developing pressure ridges by recharge wells. In the coastal stretch of Bengal the salinity is of a slightly different kind, regarded to be connate in origin (Singhal, 1963).

Between Digha and Ramnagar (Midnapur district) the freshwater aquifers (150 to 200 m) lie under the brackish water aquifers (upto 130 m), the two being separated by clay aquiclude of 10 to 20 m thickness (Singhal, 1963); and between Pinskura and Dabra there is alternation of brackish saline and freshwater aquifers. The salinity is attributed to tidal waves. But in the Haldia area in the 24-Pargana district, groundwater withdrawal had caused serious invasion of saltwater in the early 1980s (Sinharoy, 1983).

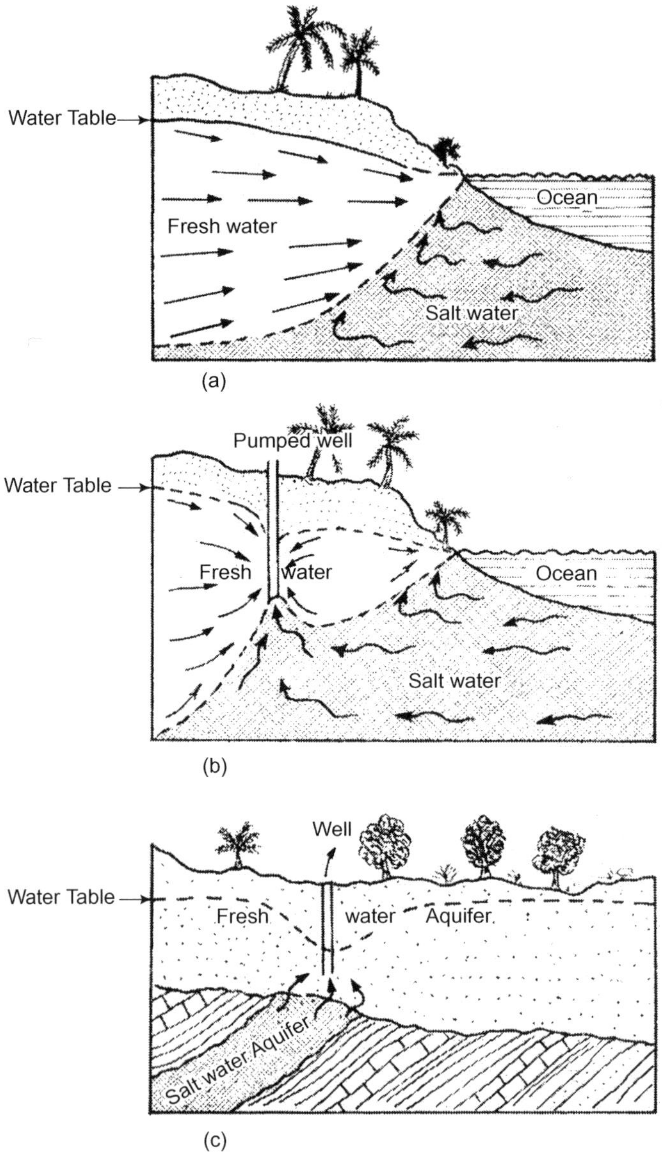

Fig. 5.23. Intensive groundwater withdrawal causing depletion of the reservoir of freshwater and stimulating intrusion of saline seawater along the coasts.

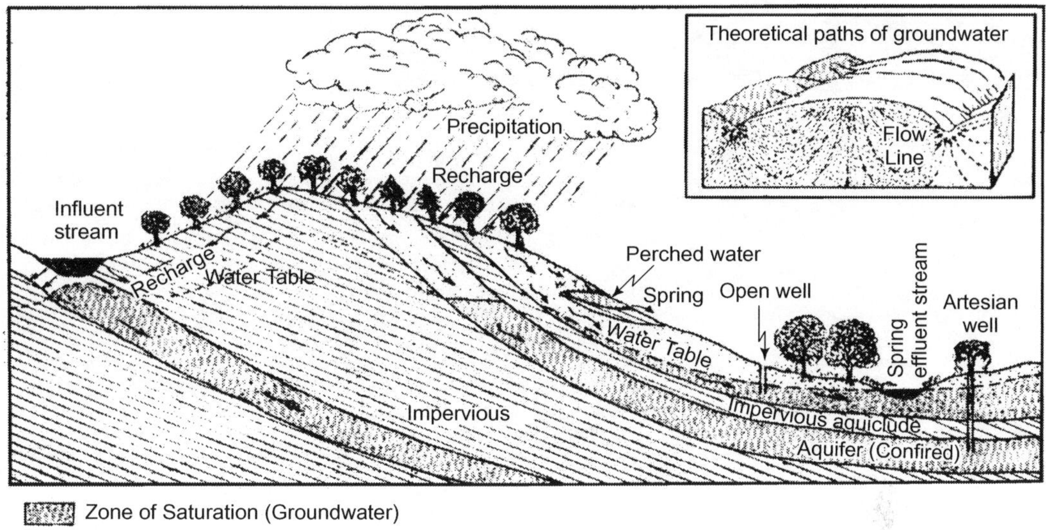

Fig. 5.24. Afforesting the watershed is the most effective way of recharging groundwater.

WATER MANAGEMENT

Approach

The management of water resources means a programme for ensuring a continual and adequa supply of water of requisite quality for varied uses *without endangering the life of the reserve or t source*. It implies efforts directed at maintaining or preserving better quality water, controlling the use and checking the abuse of this precious resource, and ensuring the recharge of aquifers by appropriate management of watersheds or the recharge area. Diverting surplus water from an area of abundance to a region of deficiency is yet another important component of an efficient water management programm (Table 5.10).

Table 5.10. Use and Conservation of Water in Arid Lands (Based on: *Renewable Natural Resource and Environment*, UNEP, Nairobi, 1983).

	Technology	Methods
A.	*USE*	
1.	***Harvesting Rainwater***	Ponds constructed in small depressions and modification of catchment b making soil surface more impermeable in order to increase the amount of runoff.
		The principal methods are:
		(i) *Land Alternation*—constructing ditches or rock walls along contours clearing rocks and vegetation and compacting soil surface.
		(ii) *Chemical Treatment*—applying sodium salts, silicones, latexes, aspha or wax to fill pores and make the soil water repellant.

Contd

	Technology	Methods
		(iii) *Soil Covers*—placing the water-shedding covers of plastic, rubber or metal foil over the soil surface.
	Runoff Agriculture	Practices and principles depend on rainwater harvesting.
		(i) *Runoff Farming*—watersheds are divided into small catchments of 1-3 ha to permit collection of water in specially constructed channels that direct water to terraced fields connected by water spillways.
		(ii) *Water Spreading*—diversion of waters from natural channels to spread over adjacent floodplains or valleys. Diversion also by ditches, dykes, small dams or brush fences,
		(iii) *Microcatchment Farming* — where precipitation from a large area is forced to irrigate individual plants.
		(iv) *Desert Strip/Contour Catchment Farming*—a series of terraces are used to shed water into all adjacent strips of productive soil.
	Saline-Water Irrigation	Salt-tolerant crops can be irrigated with water containing 3,000–5,000 mg/litre of dissolved salts; 500-1500 mg/litre can be widely used on all but the most salt-sensitive crops. Various techniques of irrigation may be used but proper irrigation and drainage management is essential to prevent excessive salt build-up in soil. Frequent irrigation (**such as with the trickle method**) minimize salt stress on plants.
	Reuse of Waste Water	(i) *Agriculture*—Use of waste water to irrigate farmland may prove particularly important in arid regions, especially with some **prior biological treatment of the sewage**.
		(ii) *Industry*—Municipal waste water from secondary treatment plants can be used for cooling, ore-separation and other pro-cesses without high water-quality requirements.
		(iii) *Municipal Use*—**Requires secondary and tertiary treatments** to make waste water potable.
B.	*WATER CONSEVATION*	
	Reducing Evaporation from Water Sources	Covering the water surface with a barrier to inhibit vapourization. Covers on roofs can be placed over small tanks, but the problem is more difficult for larger water bodies. Methods tested include: (i) *Liquid Chemicals* — such as aliphatic alcohols that cover the water surface with a film one-molecule thick.
		(ii) *Wax*—floating blocks of wax soften to form a continuous, flexible film on the water surface,
		(iii) *Solid Blocks*—such as light-weight concrete, polystyrene, wax, rubber and plastic.
		(iv) *Sand-filled Reservoirs*—in which water is stored in the pores between the particles of sand and loose rock.
	Reducing Seepage	Methods include soil compaction, chemical treatment of soil and soil covers such as butyl rubber, sheet plastic, asphalt reinforced with plastic or fibreglass and ferrocement.
		Bank stabilization can be accomplished by constructing a revetment of used automobile tyres, ferrocement or soil cement packed into long thin plastic bags.
	Reduction of Evaporation from Soil Surfaces	(i) *Plant Residues*—*new* crops are planted directly into the standing residue of the previous crop to both retard evaporation and reduce soil erosion, because residues and hard soil provide a better moisture-barrier than loose-tilled soil,

Contd.

	Technology	Methods
		(ii) *Gravel Mulches*— water infiltration and conservation of soil moistur enhanced by spreading of a thin layer of gravel or volcanic cinders,
		(iii) *Paper and Plastic Mulches*—widely used to retard weeds, increase germination and growth,
		(iv) *Latex Asphalt, Oil Mulches*—used to conserve soil moisture, create runoff to concentrate moisture and suppress wind-blown sand. Used commercially to establish vegetation on water-bearing sand dunes Libya, India and Australia.
4.	***Trickle (Drip) Irrigation***	At frequent intervals small quantities of water are screened and delivered by pipe to a specific site near each plant. To conserve energy and minimiz flow rates, trickle irrigation systems normally operate at low pressures.
5.	***Reducing Cropland Percolation Losses***	In areas of sandy soil, underground moisture barriers of continuous film of water-resistance material are placed some 60 cm below the soil surface with drainage gaps left every 150 m. In theory, any impervious material can be used, plastic sheets have been employed in East Africa.
6.	***Reducing Transpiration***	Destruction of unwanted phreatophytes.
		Breeding varieties that transpire less.
		Enclosing crops within a structure to recycle transpired water.
		Reducing air movement over a crops.
		Removing unproductive leaves.
7.	***Controlled Environments***	Controlled environments consist of cultivating crops in transparent fibreglass structures, air-inflated bubbles or low-tunnels over growing plants,
		(i) *Close Environments*—plants are grown in inflated plastic green house in which there is little or no connection with the outside atmosphere,
		(ii) *Partially Open Environments.*
		(iii) *Plastic Tunnels*—placed over plants to reduce evaporation and transpiration losses.

Recharging Aquifers

In the mountainous terrain the most effective and appropriate method of recharge of groundwat is to clothe the watershed with vegetation—multistoried forests with trees, shrubs and grasses an having thick carpet of absorbent moss and litter (Fig. 5.24) as already explained. The litter-covered watersheds allow greater infiltration of rainwater (Table 5.7).

Finding that there is appreciable decline in discharge of springs of the catchment of the Khulgad, near Almora in Uttarakhand, a mode of development for augmenting their discharge by **developme of spring sanctuaries** was suggested and successfully tested (Rai et al, 1998). The method proved effective in Dugargad basin in Pauri Garhwal, where the discharge increased four fold in just four years — from 595 l/day in June 1995 to 2170 l/day in June 1998 (Negi et al, 1998). The subcatchment-scale watersheds are enclosed by constructing stonewalls or fences to ward off fuel-fodder gatherers, poachers and grazing animals. This measure allows unimpeded growth of all kinds of vegetation. The litter generated by the leafy trees, shrubs and grasses that are grown within the enclosure absorb moisture and add to the groundwater. Percolation tanks, trenches and ditches are dug across the joints, fissure fractures, shear zones and faults in the rocks and the permeable beds. This allows the runoff water to filtrate into the underground (Rai et al, 1998).

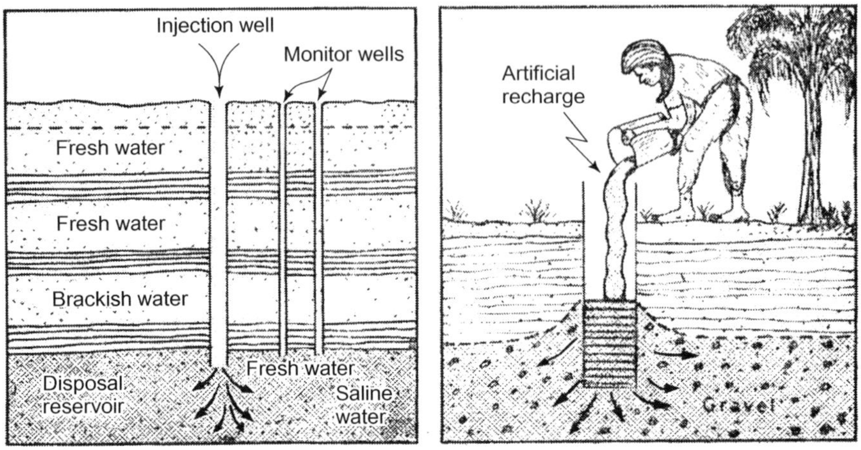

(a) (b)

Fig 5.25. (a) Recharging groundwater by injecting surface water into underground aquifers through wells or pumps. (b) It is like filling an empty tank.

In flatter terrains there are many artificial ways of recharging the groundwater in addition to putting the ground under forest cover. **Stormwater or used water flowing down the municipal and domestic ains can be fed into wells, pits and trenches or depressions and allowed to infiltrate** (Fig. 5.25). This will lead to a perceptible rise in the water table as was evidenced in the city of New York. Untreated fluents, however, cause pollution.

A much better way of recharge is to allow the floodwater to spread on the fields and fill the cavated trenches, tanks, ditches and furrows on the sides of the roads and railway lines. g. 5.26) The stored water eventually finds its way to the underground reserve.

Spreading of floodwaters on crop fields, such as of paddy, would greatly augment the groundwater reserve (Figs. 5.26 and 13.9). In the Chhattisgarh division the farmers used to harness the rainwater wing over slopes of 'bhatas' by digging channels leading to tanks. **Impoundment of river water hind the dam accomplishes much the same thing. In the arid and semi-arid regions earthen mbankments (*levees*) or 'bunds' at suitable places across seasonal streams would impound e water to recharge the groundwater. The accumulated sediments in such reservoirs need be removed periodically**. The construction of percolation tanks since 1972 in the Pune district (Maharashtra) proved very useful (Limaye, 1975).

The third method of artificial recharge is injecting rain and flood waters through wells in areas where excessive water has been withdrawn, such as the southern and central parts of Mehsana district Gujarat. In the Kasuliwara area, for example, the rainwater flowing in channels of the ephemeral Saraswati is injected through 120 m deep wells sunk in the river bed (Fig. 5.27). This is done after the m deep wells located 130 m away from the injection wells have first pumped out the groundwater and thus depleted the aquifer to make room for the input. **It is tantamount to filling an underground tank.**

In the Mehsana Basin, there are five major confined sand zones, 5 to 35 m thick, down to a depth of 6 m (Tiwari and Srivastava, 1983). Over-exploitation for intensive irrigation at the rate of 145 million m yr from the 50 to 250 m deep aquifers caused, some thirty years ago, appreciable difference in water level of the phreatic aquifers and the first confined aquifer provided a natural pressure head of 15 to m, thus stimulating effective injection (Phadtare and Bradely, 1983). Similar efforts made in Ghaggar basin at Kurukshetra (Haryana) were successful in recharging depleted groundwater.

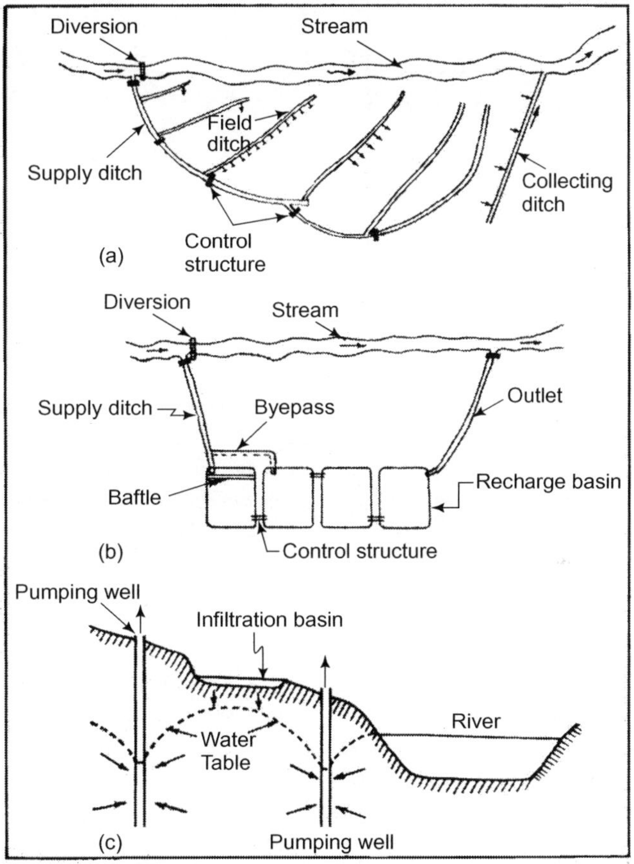

Fig. 5.26. Spreading floodwater on absorbent ground of farms along rivers through networks of ditches a controlling the spread through levees cause a considerable recharge of groundwater. (After H. M. Raghunath, 1982).

Near Ahmadabad the experiments to inject water of the Sabarmati River through siphon pumps (Fig. 5.28a) met with success (Desai and Gupta, 1979). Resort to heavy pumping out before the onset of monsoon rains, through appropriately spaced pumps installed close to the channels of the ephemeral streams/rivers and canals on both sides, would appreciably lower the water table. During the rainy season the floodwater is induced underground by pumping in through a chain of injection wells (F 5.27) sunk along the channels (Rama, 1977; Chaturvedi et al., 1975).

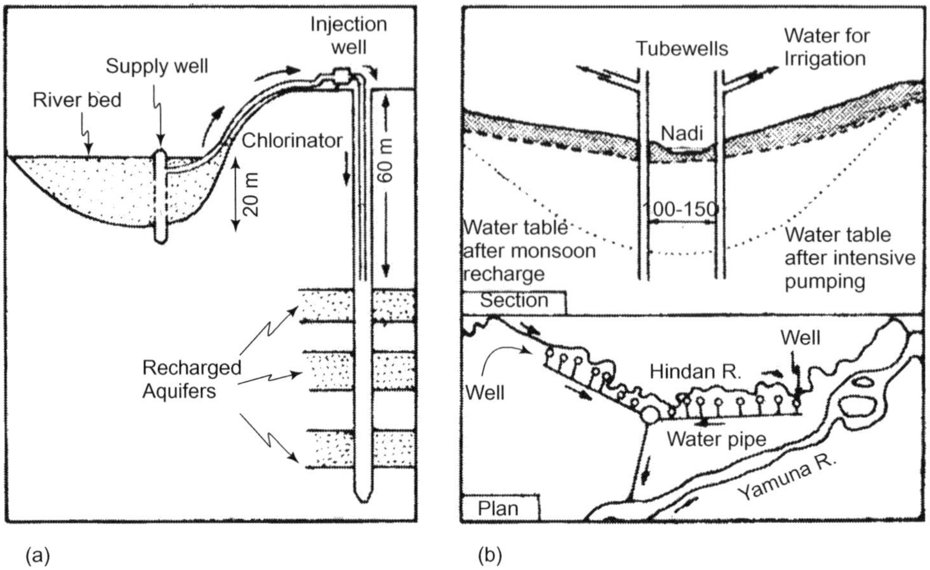

Fig. 5.27. Groundwater can be recharged artificially by siphoning water from riverbeds (B.T. Desai and S.K. Gupta, 1979) and by installing a series of pumps close to dry riverbeds or canals in order to withdraw considerably during summer and then allowing floodwater to enter underground through well sunk in the riverbeds (After Rama, *The Hindustan Times*, August 5, 1973).

Conserving Water

he most effective and extremely beneficial way of conserving water is to let the flood-flows and rface runoff fill the underground palaeochannels in the meandering belts of rivers. The three ajor palaeochannel (Fig. 5.28) identified in the Ganga basin, covering an area of 396 km^2, have the cumulative water storage capacity of 4.65 billion cubic metres (R.P. Gupta et al, 2011). Interconnected with the aquifers of the adjacent alluvial succession, these palaeochannels have higher rate of recharge — 17.0 to 28.7% compared to 6.3 to 11.0% of the plains.

Hundred of rivers all over India are intimately associated and many connected with palaeochannels. These hold the promise of storing very large volumes of water.

Diversion and Storage of Excess Water

A large part of India — nearly one-third — is prone to drought, the recurrence interval of which is three five years. In western Rajasthan, where the interval is two years, the scarcity of water is extreme. Diverting water from a region of surplus to an area of scarcity will considerably abate the distress. model is the transbasinal transfer of water from the west-flowing Periyar River to the east-flowing Perambikulam River to irrigate land 77,000 ha and 7,900 ha respectively, in Tamil Nadu. The attempt to take the Krishna water in Andhra Pradesh to meet the needs of the Chennai metropolis and the villages, wns and fields on the way including the ever thirsty Rayalseema region, is another commendable

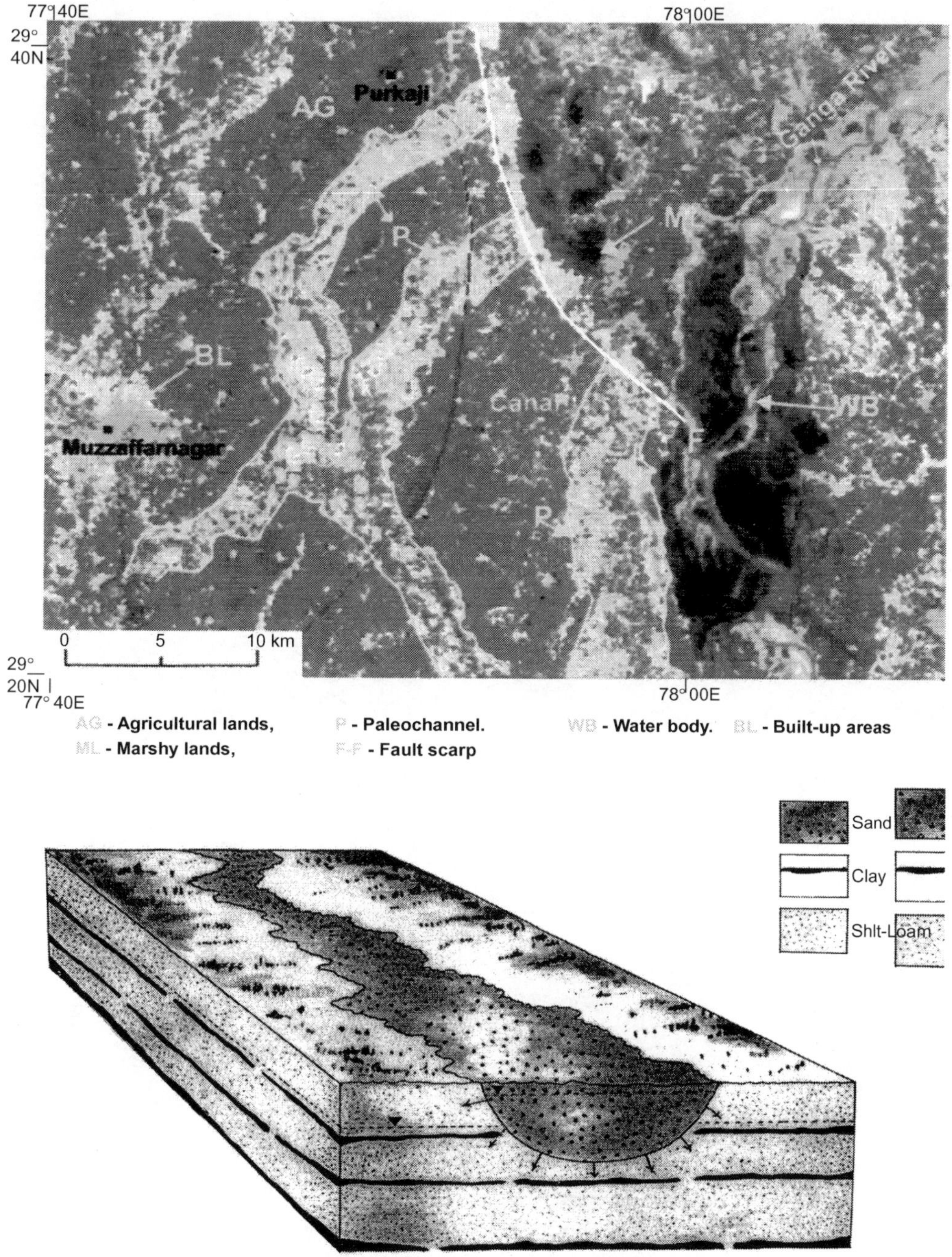

Fig. 5.28. The palaeochannels, abandoned by shifting rivers, have proved to be great storage of floodwate They exhibit higher rate of recharge than the adjoining alluvial aquifers. (From R.P. Gupta et 2011). (See colour Fig. 5.28)

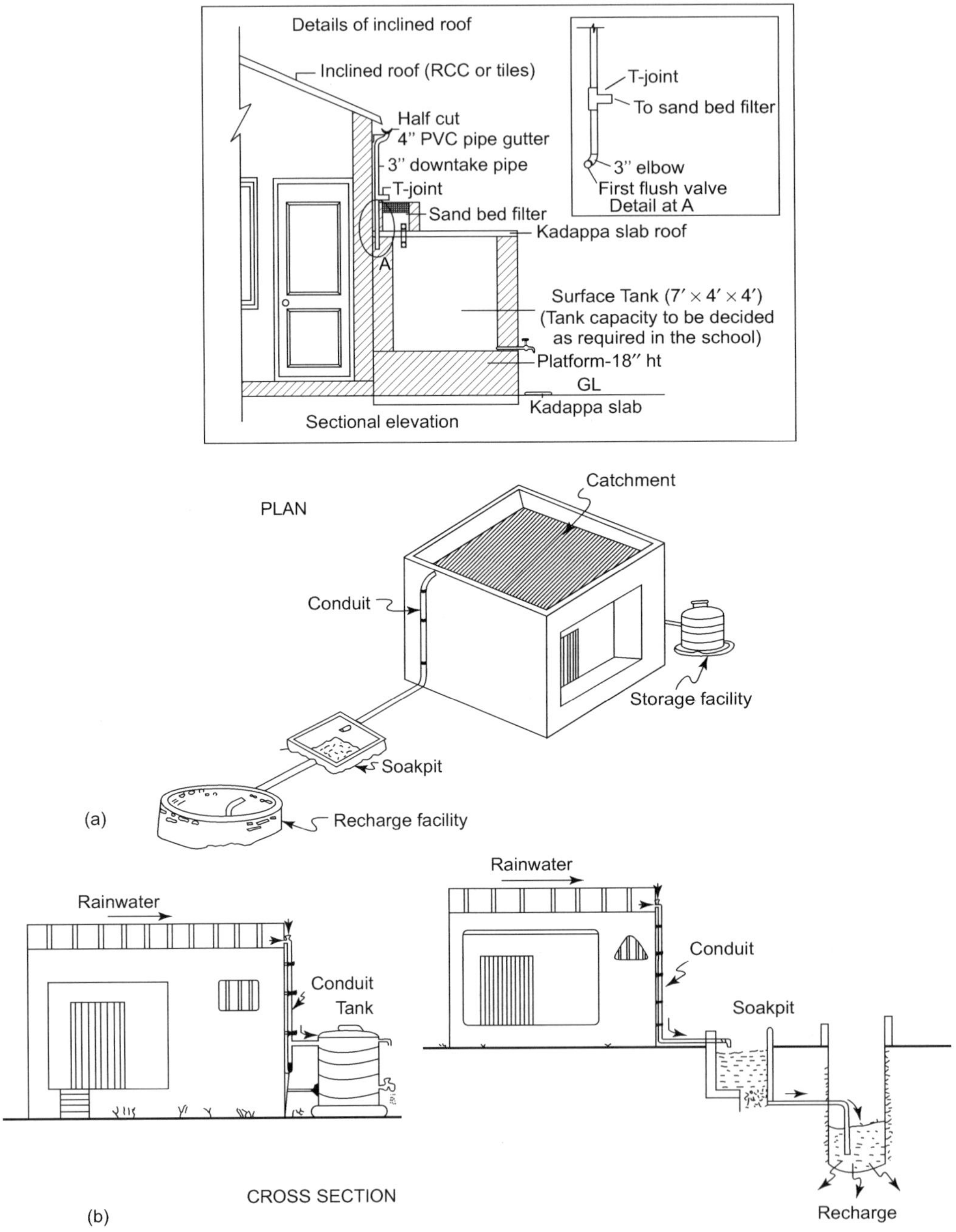

Fig. 5.29. (a) In Karnataka schools the water collected from roofs passing through filter of sand in a pit is collected in completely enclosed tanks. (From: Shivakumar, 2007). (b) Roof water can be collected in storage tanks as well as used to recharge the groundwater, that is, water in the wells. (From Valdiya, 2004).

experiment in this direction. The Indira Gandhi Canal in western Rajasthan brought the water of the Himalayan rivers—Satluj and Beas—to the chronically thirsty land of the Jaisalmer–Bikaner divisions in the desert. (Fig. 5.18) The Yamuna Canal transfers water from the Ganga Basin to irrigate a vast tract of Ghaggar basin in Haryana and adjoining Rajasthan. To prevent 30 to 35% loss of water through seepage, it will be necessary to line the canals; otherwise the entire venture would prove counterproductive.

However, the transfer of water from one river to the basin of another river is just an idea—a grandiose but non-pragmatic idea—in the Indian context. When more than two decades after the construction of a canal, the water of the Satluj cannot be brought to the Yamuna, and when sharing of water of the same river by different states is a matter of unending acrimonious disputes, one cannot think of pursuing the grandiose plan of "**Interlinking of Rivers Programme**". There are other more compelling reasons for untenability of this idea or programme.

(i) The drier-season discharges of almost all rivers in India have considerably diminished for a variety of reasons already explained. Therefore, no river has 'surplus' water to spare (only flood water c be shared !). On the contrary, most rivers are in great need of influx of additional water in order remain in a healthy state or to just survive.

(ii) The transfer of water from any river would have far-reaching impact on the aquatic flora and fau of the river channel — downstream of the point of transfer. **The downstream human populati would suffer even more**. The aquatic flora and fauna of the receiving rivers would take long readjust to the increased drier-season discharges.

(iii) Topographic conditions of India being what they are it would require enormous quantity of electricity to lift water across the barriers of hills and plateaus. When the country does not have enough energy to illuminate homes even in high class urban centres, where would the energy come from to operate the very large number of heavy-duty lift pumps! The programme, if implemented would be prohibitively costly economically and environmentally unsound.

(iv) The lithological constitution of the country varying from terrane to terrane, region to region, zone to zone, it would not be easy to excavate canals without endangering the very integrity, stability, and ecology of the lands they would cross, and without uprooting the people from their homes and forcing them to resettle elsewhere—which is nowhere available today.

The better option is to store floodwater as well as rainwater. Rainwater and floodwater can stored behind dams and in depressions with impervious floors of clays or bitumen or films of silicone butyl rubber or plastic. In order to minimize loss from evaporation, very thin coatings of paraffin w or polythene foam can be thrown upon the surface of water. Very thin films—one molecule thick— octadecanol or hexadecanol provide the best cover. This allows oxygen and carbon dioxide to pass through, but prevents water vapour from escaping. **If the film can be maintained unbroken over t entire surface, there will be a reduction in evaporation by 15 to 80% as has been demonstrated Israel and the U.S.A.**

It may be emphasized that the dams provide only a temporary and short-term storage. The amount of impounded water is equal to the total river flow during the time it takes to fill the reservoir behind t dam, i.e. a few year's supply. Then the accumulation of sediments reduces its capacity drastically, to make the proposition not very gainful.

Roof Rain-Water Harvesting

If the rainfall is 250 mm, a 1500 square metre area of roof top is capable of yielding 300 cub metres of water—enough to meet the drinking water need of 30,000 persons per day. In Karnataka,

23683 schools were taken up for this purpose. The roof water is collected through PVC pipes or gutters d stored in fully enclosed tanks (Shivakumar, 2007). The rain water is first passed through filtering sand beds built over the surface of the tanks before storing. No light is allowed to enter the tanks in order to prevent the growth of bacteria, algae, etc., and no manholes provided to stop withdrawal of water by dipping vessels and to fall inside of animals (Fig. 5.29).

It may be mentioned that storing rainwater through a variety of methods has been an age-d tradition in the drier parts of Rajasthan, Madhya Pradesh and Maharashtra. Collected in nks, the runoff water is transferred through tunnels to the wells of settlements, or conserved *in situ (Fig. 5.30)*. In the **high** Sinhagarh hill top, the rainwater was stored in underground tank dug in a volcanic ash bed, enough to support a large army of Chhatrapati Shivaji for long period of encirclement by his enemy.

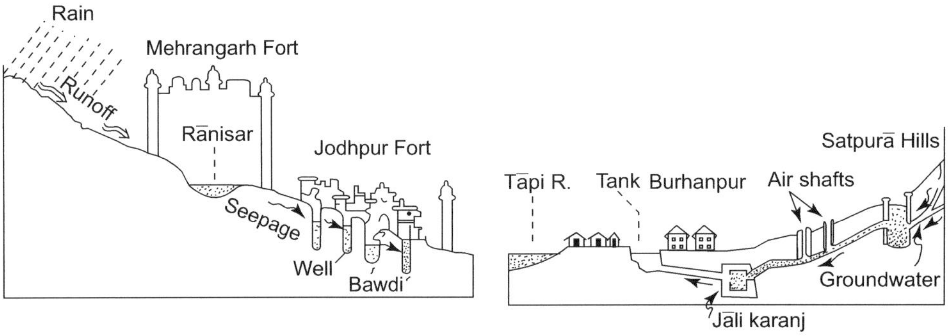

Fig. 5.30. *Left*: In western Rajasthan the people of Jodhpur used to collect rainwater runoff on the hill slope through channels which converged into a large tank within the fort, and then transfer through subterranean conduits to the wells in the city situated at the foot of the hill. *Right*: The people of Burhanpur in Madhya Pradesh used to get the runoff collected in large tanks up the slope, and then tunnelled it down to the township on the bank of the Tapi River. (Both are modified after Agrawal et al, 1999 in Valdiya 2004).

euse and Recycling

e effluents from domestic and municipal sewers and drains, **after proper treatment**, can be used for industrial and agricultural purposes, which do not require high-quality water. Water once used can again be used for the same or other purposes after recycling, **i.e. alter eliminating pollutants, pathogens and desirable elements**. Recycled water is not desirable for drinking because the pathogens and organic elements are hard to get rid of (Table 5.8).

esalinization of Sea Water

Desalinization of seawater can provide adequate quantities for the coastal cities. Solar-powered distillation plants can be made use of. Undoubtedly, the technology is very costly. Isreal desalinized 3 million m^3 of water each year and Saudi Arabia got 2.3 billion litres of freshwater everyday from the sea in 1980. The thirsty coastal cities of India can afford to tap seawater in this manner. The Salt and Marine Chemical Research Institute at Bhavnagar developed a technology, using solar energy to distill saline water through flash distillation, electrodialysis and reverse osmosis. The solar stills ha been found ideal for small-community needs, for they have the capacity of recycling 5,000 l/day of water as successfully demonstrated at Avania village in Bhavnagar district, and Bhaleri in Churu district (Rajasthan). In the flash distillation technique the treated saline water is allowed to flow though a seri of chambers maintained at different pressures below the atmospheric pressure. The saline water gets evaporated in each stage corresponding to its temperature, and the vapours are collected in tubes and condensed.

Drip Irrigation

Drip-irrigation leads to saving of 50 to 70% of water, besides the elimination of problem generated by over-irrigation, reduction of labour, check on growth of weeds, lower consumpti of fertilizers, faster maturing of crops, and higher yield (20 to 25%). It is this method of irrigati that is advocated for our farmlands.

SUMMARY

Only 2.7% of the total water on earth is freshwater, the groundwater accounting for merely 0.62%, and the water flowing in streams and held in lakes making about 0.0091%. The volume of water th the rivers of India annually discharge into the seas is 18,69,00 million m^3—80 to 90% flowing dow during the rainy season. The amount of annual recharge to underground water is 4,33,882 million m Approximately 2,13,128 million m^3 of this underground reserve remains untapped at present. In spite of the abundance of water, **far more than 17.9% rural household in India have no reliable and sa supply of drinking water**.

A proportion of rainwater infiltrates the ground and moves through the spaces between grain openings, cracks, fractures and joints to be either stored in the porous-permeable zones called aquifers or come out on the surface in the form of springs and seepages. The quantities of water held in the aquifers and discharging through the springs depend on the proportion of rainwater absorbed by the ground, that is, on the magnitude of recharge. The ability of the ground to absorb and permit percolation depends on the slope of the ground, the texture of the soil or the structure of the rocks, and the nature and extent of vegetation covering the recharge area. The lesser the slope, the higher the openness of texture and greater the jointing or fracturing, and the greater the vegetal cover, the greater will be the amount of rainwater infiltration into the ground contributing to augmentation of the underground reserve.

The amount of water held in the aquifer depends on the proportion of the empty space available in the rocks/soils (i.e. porosity), but the capacity to yield this water, and the rate of outflow depends permeability which is controlled by the interconnection of pore spaces, fissures and openings. Fast and heavy withdrawal of water causes lowering of the water table, leading to drying up of wells and springs in the hills and to intrusion of saline water in the coastal belts as witnessed in Gujarat, Andhra Pradesh and West Bengal. **Excessive irrigation causes waterlogging (moist areas) and alkalinizati or salinization and acidification (dry areas) of the soil following a rise in the water table. The phenomena have destroyed the productivity of our farms and diminished the bearing capacity the ground.** Thus about 6.41 m ha of farmland in India are waterlogged and another 9.30 m ha seriously alkalinized and salinized, mostly in parts of Panjab, Haryana, northern Rajasthan and Uttar Pradesh.

Today more than 80 m ha of our cultivated lands are being irrigated, out of which the groundwater is contributing to irrigating at the rate of 34.362 million m^3/year. As of now the groundwater reserves are fficient to provide assured irrigation to nearly 40 m ha.

The quality of water is crucial to sustain life. Quality control implies elimination or reduction of toxic inorganic elements, soluble carbonates that render the water hard, salts that increase the salinity-alkalinity, and organic material that makes heavy demands on the dissolved oxygen so essential for the survival of aquatic life.

able 5.11. Annual Water Requirement (BCM) for Different Uses (S.K. Jain, 2011)

Uses	Year 2025			Year 2050		
	Low	High	Percentage	Low	High	Percentage
Irrigation	561	611	72	628	807	68
Domestic	55	62	7	90	111	9
Industries	67	67	8	81	81	7
Power	31	33	4	63	70	6
Inland navigation	10	10	1	15	15	1
Environment – ecology	10	10	1	20	20	2
Evaporation losses	50	50	6	76	76	7
otal	**784**	**843**	**100**	**973**	**1180**	**100**
Surface water	447	558		641	752	
Groundwater	247	252		332	428	
Population (million)	1286	1333		1346	1581	

Realizing the limited quantity of water of requisite quality, water must be distributed among ers on the basis of genuine needs in terms of quality and quantity. This must also take into account the impact on the environment of the utilization of water-resources. Industries and agriculture, which can do with less than the best quality of water, must not be allocated high-quality water. **Cities should ve two systems of distribution—one for drinking and bathing water, the other for washing and irrigating lawns, gardens, etc. *And water must be priced suitably and distributed equitably.***

Measures to combat the menaces of waterlogging and alkalinization-salinization include reduction of seepage from canals and reservoirs by lining them, construction of a network of deep drainage to lower the water table, **resort to the practice of irrigation that would enjoin application of just e requisite quantity of water (such as trickle or drip-irrigation),** and application of chemicals to counteract the salts and alkalies.

Effective measures for replenishing groundwater reserves include putting the watershed or recharge area under multistoried forests or thick vegetation which produces abundant litter, spreading e stormwater or floodwater over low-lying areas, depressions, tanks, trenches and wells to ensure filtration of water into the aquifers, and artificially induced recharge through injection wells on dry er beds. Storage of floodwater in **reservoirs behind dams in degraded lands** and diversion of surplus water to areas of water scarcity through transbasinal canals are among the desirable steps of a water management programme.

Those of us who plan for the development of India, and those who want to live in environmentally healthy India must not forget the great amount of water that would be annually required for a variety of purposes in the years to come as the Table 5.11 by S.K. Jain (2011) shows.

6 Conservation of Mineral Resources and Impacts of Mining

यं सर्वशैलाः परिकल्प्य वत्सं
मोरौ स्थिते दोग्धरि दोहदक्षे
भास्वन्ति रत्नानि महौषधीश्च
पृथ्विदष्टां दुदुहुर्धरित्रीम्
कालिदास (कुमार सम्भवम्)

alidās in his *Kumār Sambhavam* describes how the Earth came to be called "prithvi" after King Prithu, ho transformed the Earth into a cow and used the mountains as efficient milkmen. He milked the rth for its mineral and plant resources but did not exploit it (P. R. Pisharoty, per. comm., 1983). India day is faced with the problem of rapidly depleting mineral resources due to over-exploitation and controlled, locally reckless, plunder.

MINERAL CONSUMPTION AND RESERVES

Growing Consumption

Minerals form the base of all industrial, economic and commercial activities. In one form or another, they are used for the generation of energy that touches all spheres of life of the common man. Coal, iron, copper, aluminium, petroleum and mineral fertilizers have become indispensable. The e of consumption of some of these minerals in India is doubling every 9 years. The U.S.A. with about % of the world population consumed about 33% of the world production of petroleum and 30% of etals forty years ago; the per capita consumption in 1979 being iron ores 510 kg, base metals 25 kg, d clays 250 kg (Keller, 1982). Compared to the American, an Indian was then a very poor consumer minerals and fossil fuel. The per capita consumption of minerals in India (based on the 1980 data of e Indian Bureau of Mines) was as follows: limestones 54.79 kg, iron ores 26.7 kg, dolomite 2.99 kg, uxite 2.66 kg, clays 2.42 kg, gypsum 2.24 kg, mineral-fertilizers 1.93 kg, manganese 1.74 kg, silica-nds 1.31 kg, sulphur 1 kg, base metals 0.46 kg, magnesite 0.39 kg, coal and lignite 0.17 kg (Figs 6.1 d 6.2).

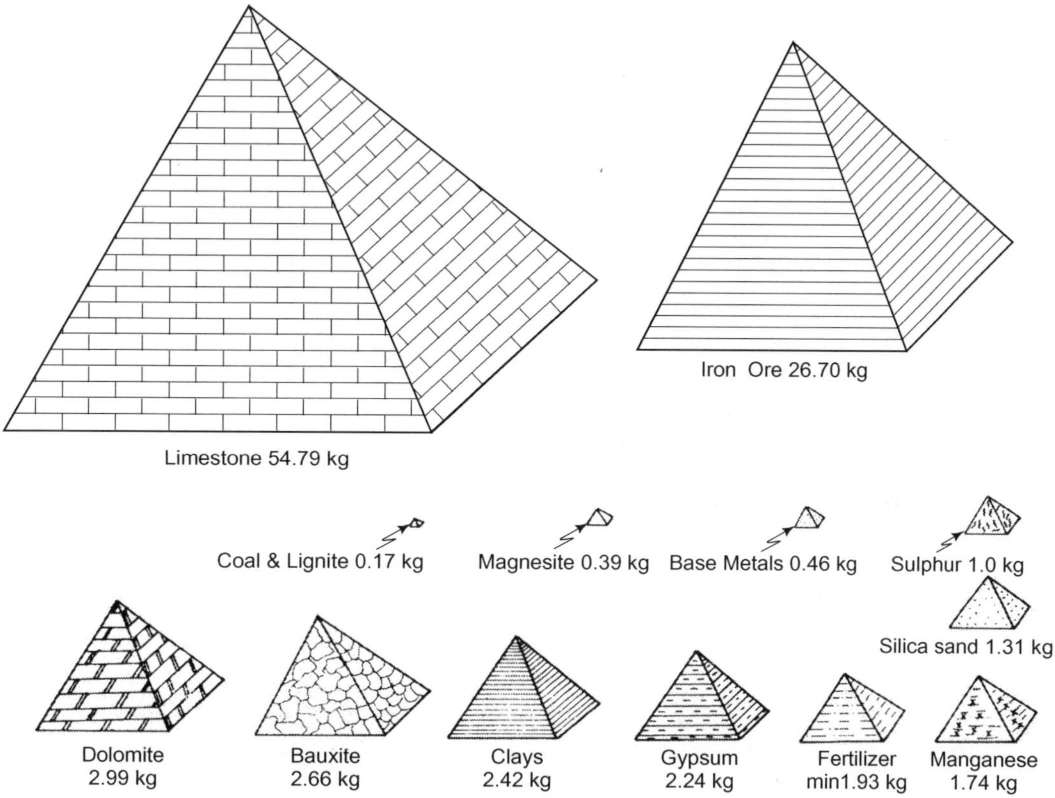

Fig. 6.1. Per capita use of minerals in India in the late 1970s.(Based on data from the Indian Minerals Year Book, 1980, Indian Bureau of Mines, Nagpur, 1983.)

The situation today is quite different—the Indians are as voracious consumers as other developed countries, but are considerably less concerned about the consequences of mineral utilization. Today the per capita consumption in India of copper is 0.5 kg, of aluminium 1.3 kg, of ste 51.7 kg, of cement 170 kg, of fertilizers 335 kg (Annual Report 2010-2011 of Ministry of Mines), give just a few examples.

Geological Constraints on Reserves

Each deposit is a product of a complex set of interactions of temperature, pressure and chemic environment. The formation of the deposits takes place very slowly over a geologically long perio of time, measurable in millions of years. As the process of growth of a mineral deposit is extreme slow, there is never a second crop—once the deposit is used, it becomes exhausted. **The minerals thus constitute a non-renewable, non-replenishable one-crop resource. Secondly, the mineral reserves, however large in size, are finite and would be exhausted if continually extracted.** Thirdly, th occurrence is restricted to certain geographic areas for all times, and the distribution is quite unequ from country to country. **Over-consumption and waste of the limited and depleting resource for the benefit of the present generation of people at the expense of future generations cannot be allowe**

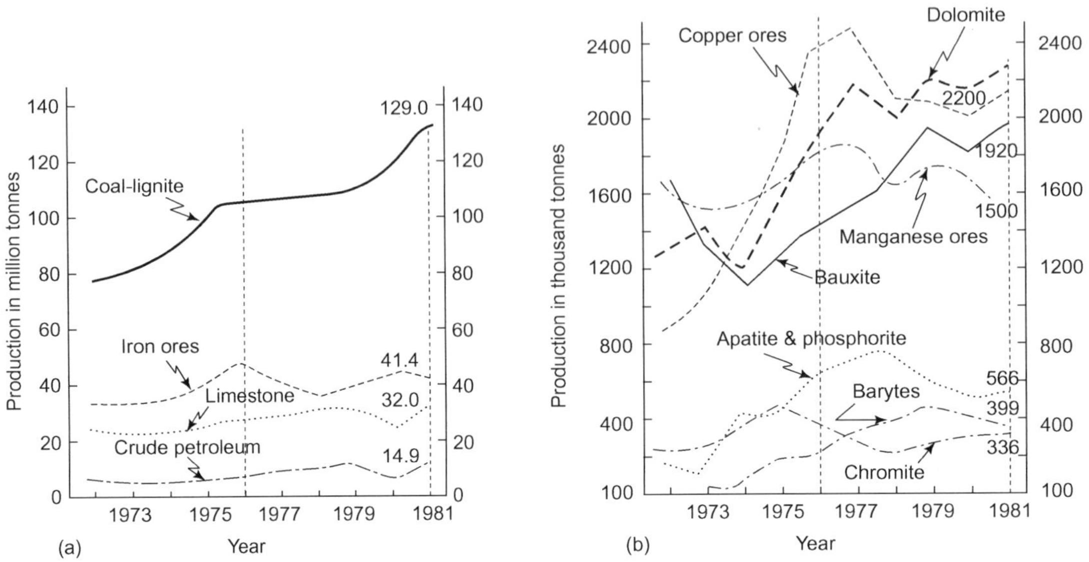

Fig. 6.2. The annual production in the seventies of the 20th century of some important minerals in India in the last decade. (Based on data from *Indian Mineral Industry,* 1983, Indian Bureau of Mines, Nagpur.)

To meet the exponentially growing demand of society, and appreciating the even greater needs of the ure generations new deposits must be found through application of new technology of exploration, conservation of existing reserves should be taken up. The third option is to reuse or recycle the used nerals and metals. Admittedly, the geologists continue to make discoveries of new deposits, but the e of discovery has slowed down considerably. Thus **minimum waste, judicious and efficient use, d extensive reuse of minerals should be the central points of the national mineral policy.**

APPRAISAL OF MINERAL RESOURCES OF INDIA

e Indian Bureau of Mines periodically makes an appraisal of mineral reserves, production and use Indian minerals. These data and their analysis are needed to formulate the plan of development d to adopt means of conserving reserves that are fast dwindling. As matters stand, **India is wholly largely self-sufficient in 35 minerals that constitute the raw materials of basic industries. In this category are ores of iron, steel and ferroalloys, aluminium, cement, refractories, glass and ceramics, abrasives, paints and a wide variety of inorganic chemicals.** This is not to state that ecial-grade minerals for very specific requirements of these industries are not important. Among such nerals are bentonite, fluorite, graphite, spathic iron ore, dead-burnt and seawater magnesia, billing-ade manganese oxide, etc. India is a leading producer of mica, kyanite and sillimanite and holds a nking position among the world producers. **India is a globally important producer in the matter iron ores, manganese, barytes and talc-steatite. The productions of bauxite, coal, chromite and agnesite are also quite significant in the global context. But then going by the magnitude of ploitation of various mineral reserves, their life is quite less than what was estimated at the ginning of the twenty-first century (Tables 6.3 and 6.4).**

Table 6.1. Degree of Self-Sufficiency in Important Minerals and Metals in 2007-2008 (Source: ***Indian Minerals Year Book-2008***).

Commodity	Order of self-sufficiency (%)
Minerals	
Asbestos	1
Barytes	100
Bauxite	100
Chromite	100
Dolomite	100
Felspar	100
Fireclay	86
Fluorite	10
Gypsum	50
Iron ore	100
Ilmenite	100
Kyanite	31
Limestone and other calcareous minerals	100
Magnesite	98
Rock phosphate (including apatite)	48
Rutile	100
Sillimanite	100
Silica minerals	100
Sulphur	28
Talc/Steatite/Pyrophyllite	100
Metals	
Aluminium	94
Copper (refined)	100
Lead (primary)	30
Zinc	95
Ferroalloys	
Ferro-chrome	100
Ferro-manganese	100
Ferro-silicon	100

According to Narayana (2004), the anticipated crude production of oil should be in the range of 25 70 million tonnes per year in 2016–2017, which means that India would be left with the balance reser of only 507 million tonnes by 2016, unless new discoveries are made or alternative sources of gas a tapped.

However, **the production of rock phosphates, crude-petroleum, oil and gas and nonferrous metals including copper, zinc and lead, the Indian deposits are not large to meet the demands of**

our industries. A huge amount of foreign exchange is thus invested for their import, despite the country ıng happily placed so far as their reserves are concerned (Fig. 6.3 and Table 6.2).

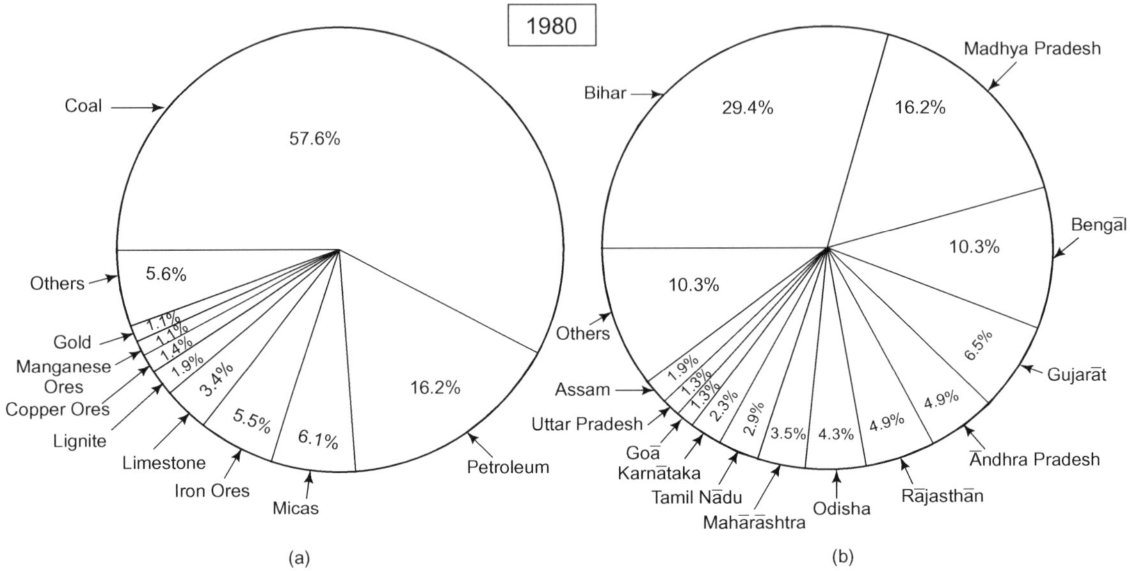

Fig. 6.3. (a) Values of various minerals produced in 1980.(b) Shares of different states in the total value of minerals produced in 1980. (Data from *Indian Minerals Year Book,* 1980, Indian Bureau of Mines, Nagpur, 1983.)

Table 6.2. Production of Selected Minerals in India in 2006–2007 and 2010–2011 (Source: *Indian Minerals ear Book,* 2010, I.B.M., Nagpur)

Mineral	Unit	2006-2007		2010-2011 (Estimated)	
		Quantity	Value (in Rs. Crores)	Quantity	Value (in Rs. Crores)
All Minerals			104490.44		200609.98
uel			66158.29		135243.81
Coal	M.Tonnes	431	34836.79	537	49011.64
Lignite	M.Tonnes	31	2626.03	38	4240.19
Natural Gas (Utilised)	M.C.M.	31747	9764.16	58982	18488.83
Petroleum (crude)	M.Tonnes	34	18931.32	38	4240.19
Metallic Minerals			18285.81		41828.44
Bauxite	Th.tonnes	15733	384.78	13363	502.66
Chromite	Th.tonnes	5296	1450.17	3865	2089.76
Copper Concentrate	Th.tonnes	150	311.71	161	486.34
Gold	Kg	2488	229.09	2214	400.97

Contd.

Mineral	Unit	2006-2007		2010-2011 (Estimated)	
		Quantity	Value (in Rs. Crores)	Quantity	Value (in Rs. Crores
Iron Ore	Th.tonnes	187696	14204.31	212613	34852.37
Lead Concentrate	Th.tonnes	107	133.14	136	181.70
Manganese Ore	Th.tonnes	2116	557.37	2901	1428.49
Zinc Concentrate	Th.tonnes	947	971.47	1342	1641.02
Other met. Minerals			43.76		245.12
Non-metallic Minerals			3351.46		4802.69
Ball Clay	Th.tonnes	627	13.87	1021	20.13
Barytes	Th.tonnes	1681	94.71	1774	196.45
Diamond	Carats	2180	1.47	71381	54.45
Dolomite	Th.tonnes	5172	112.57	4724	154.22
Fire clay	Th.tonnes	497	6.59	285	5.02
Garnet Abrasive	Th.tonnes	859	25.92	2724	144.78
Gypsum	Th.tonnes	3006	49.42	4557	137.38
Kaolin	Th.tonnes	1460	162.22	2488	56.64
Laterite	Th.tonnes	1373	15.76	915	10.35
Lime shell	Th.tonnes	104	7.29	23	1.88
Limestone	M. tonnes	197	2405.01	240	3220.21
Magnesite	Th.tonnes	239	34.17	291	42.17
Phosphorite	Th.tonnes	1587	218.46	1651	423.21
Pyroxenite	Th.tonnes	302	8.27	303	14.61
Sand (others)	Th.tonnes	1770	6.71	2457	11.86
Silica Sand	Th.tonnes	2663	37.79	3861	41.15
Sillimanite	Th.tonnes	26	10.13	32	33.03
Steatite	Th.tonnes	740	40.12	1006	66.16
Wollastonite	Th.tonnes	132	11.41	200	16.74
Other Non-metallic					
Minerals			89.59		152.23
Minor Minerals			16694.93		18734.45

M.Tonnes – Million tonnes Th.tonnes – Thousand tonnes
M.C.M. – Million cubic metre Kg – Kilogram

Note: (1) The value figures pertain to pit-head value
(2) Value of Petroleum (crude) is revised for the year 2008-09 and 2009-10 on the basis of the data received from the Ministry Petroleum & Natural Gas

Source: Coal and Lignite : Coal Controller, Kolkata
Petroleum (crude) and Natural Gas : Ministry of Petroleum & Natural Gas
Minor Minerals : State Governments.

The minerals in which India is very deficient—and therefore woefully dependent on import include cryolite, spinning-grade asbestos, sulphur, borax, acidic fillers (diatomite and perlit petroleum oil and gas, industrial diamond, and alloy metals. The production of steel, aluminiu copper, lead, zinc, ferrochrome, and ferrosilicon are very small.

It may not be out of place to state that in the year 2007–2008, various minerals (unprocessed and miprocessed) brought 14.5% of India's export earning, while manufactured mineral-based commodities ch as alloys contributed 26.8% of the India's trade (*Indian Minerals Year Book-2008*). **The total rning from the export of India's minerals is thus of the order of 41.3% of the total export trade. his is not a small contribution, if some believe that India's natural wealth does not count in the atter of India's economic growth and health.**

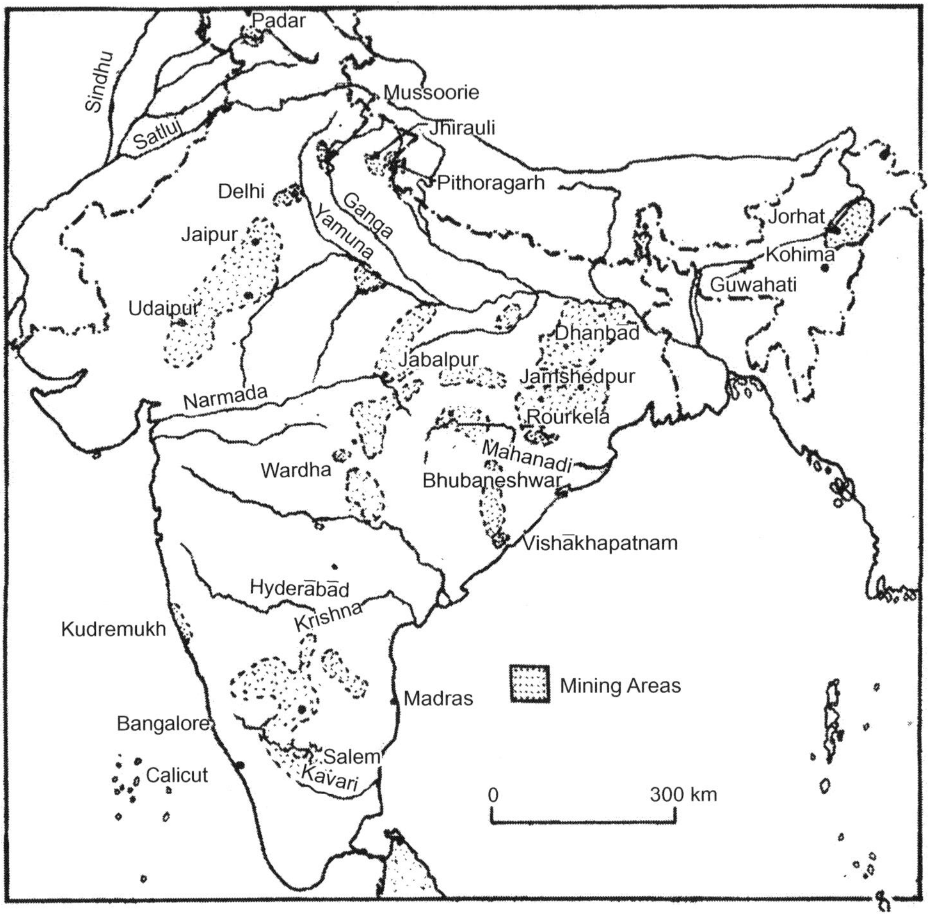

Based upon Survey of India map with the permission of the Surveyor General of India. The territorial waters of India extend into the sea to a distance of twelve nautical miles measured from the appropriate base line. Responsibility for the correctness of internal details shown on the map rests with the publisher © Government of India Copyright 1986

Fig. 6.4. More than 3,10,000 ha of the land in India is directly affected by mining activities and a much larger area by associated activities in A.D. 2011.

The proven reserve of coal at the beginning of this century was 21 billion tonnes in the Gondwana sin and 254 billion tonnes in the Tertiary basins (D.D. Sharma, 2003). The prognosticated reserve of al-bed methane (described as clean fuel gas) in all the coal and the lignite fields is 94,873 billion m^3 Kumar, 2003). It may be mentioned that the Indian coals have gas content values ranging from 1 to 23 m^3/tonne. The proven reserve was 820 billion m^3 in just the 16 fields in central India (Chand 2004). The reserve of total petroleum oil is 725 million tonnes and that of natural gas 1055 billion m (T. Kumar, 2009). In the submarine Palakudi–Pasarlpudi horizon offshore of the Krishna-Kaveri de the established reserve of gas is 24 million metric tonnes although the potential reserve is estimated be of the order of 3400 million tonnes (Majumdar et al, 1995).

Table 6.3. Reserves and Remaining Resources of Some Minerals of India (Annual Report 2010-2011, Ministry of Mines; National Mineral Inventory)

Mineral	Reserves		Remaining Resources (measured)
	Unit	Proved	
Apatite	Tonne	61,25,946	2,28,521
Asbestos	Tonne	29,73,825	90,687
Barytes	Tonne	3,16,39,934	2,44,159
Bauxite	000 Tonnes	5,38,945	5,79,561
Calcite	Tonne	32,18,256	90,97,635
China clay	000 Tonnes	1,01,520	2,87,435
Chromite	000 Tonnes	30,892	39,673
Cobalt ore	Million tons		30.63
Copper ore	000 Tonnes	1,35,461	1,12,019
Corundum	Tonne	317	13
Diamond	Carat	6,05,577	2,98,638
Dolomite	000 Tonnes	4,07,794	1,91,266
Dunite	000 Tonnes	12,714	23,909
Felspar	Tonne	1,92,20,596	22,35,339
Fireday	000 Tonnes	26,897	49,149
Fluorite	Tonne	85,85,124	17,13,833
Garnet	Tonne	67,19,626	1,19,831
Gold Ore (primary)	Tonne	1,55,54,089	2,70,62,651
Granite	000 cu m	23,010	8,41,284
Graphite	000 Tonnes	5,164	51
Gypsum	000 Tonnes	40,803	9,033
Iron ore	000 Tonnes	49,59,668	
Kyanite	Tonne	9,21,762	6,05,209
Lead & Zinc ore	000 Tonnes	62,860	12,193
Limestone	Million tons	7,492	7,883
Magnesite	000 Tonnes	20,863	59,010
Marble	000 Tonnes	4,700	
Mica	Kg	12,70,873	10,94,94,690
Manganese ore	000 Tonnes	76,844	4,175
Molybdenum ore	Tonne		36,000

Contd

Mineral	Reserves		Remaining Resources (measured)
	Unit	Proved	
Nickel ore	000 Tonnes	28,620	30,700
Perlite	000 Tonnes	188	334
Rock phosphate	Tonne	3,30,90,127	3,03,690
Pyrite	000 Tonnes	27,129	9,590
Pyrophyllite	Tonne	95,84,547	8,67,742
Quartz	000 Tonnes	2,71,614	55,566
Rock salt	000 Tonnes	8,470	13,530
Ruby	Kg	143.1	5270.74
Sillimanite	Tonne	4,57,305	53,58,481
Silver ore	Tonne	5,57,52,053	19,00,000
Talc/Steatite/Soapstone	000 Tonnes	65,012	143
Tin ore	Tonne	2,00,828	1,68,326
Ilmenite @	Tonne	1,17,27,540	10,90,326
Rutile	Tonne	8,26,121	4,460
Tungsten ore	Tonne	22,30,000	1,97,84,073
Vandium ore	Tonne	4,90,663	13,82,875
Zircon	Tonne	24,84,687	42,75,660

Figures rounded off
@ The resource of ilemanite are 461.37 million tonnes as per Department of Atomic Energy.

Table 6.4. Life Indices of Various Mineral Deposits (Anupendu Gupta, 2003).

Mineral	Total Recoverable Reserve as on April 1, 2000 (in million tonnes)		Life Index after 2002 (years)	
	Proved	Probable	Proved	Probable
Bauxite	2464	1354	300	166
Chromite	97	56	45	25
Copper	537	332	150	92
Lead-Zinc ores	176	118	60	40
Iron ores	10,052	7475	122	90
Manganese ores	167	89	35	20
Limestone	75,678	28,766	580	218
Dolomite	4,386	1301	1459	430
Rock phosphate	145	105	112	80
China Clay	1042	347	1298	430
Gold	17.8	14.17	32	24
Diamond (in carats)	9,81,515	8,51,156	14	12

In the 160-km long Tummalapalli belt in the Cuddapah Basin (Andhra Pradesh) 150,000 tonnes medium-grade uranium ore deposits have been proved (Maithani, 2011). Very rich deposits of uraniu ores occur at Domiasiat and Phamgdiloin in southwestern Meghalaya (Maithani et al, 1995).

CONSERVATION OF MINERAL RESOURCES

Keynote of Mineral Policy

Conservation of mineral resources implies scientific and rational utilization with minimum was in extraction, judicious blending of sub-marginal and higher-grade ores and 'beneficiation' of low grade ores so that maximum use is made of the mineral material. Also implicit in the philosop of conservation is the need for restriction if not a ban on the export of minerals having small reserves and those which are of strategic importance. Discovery of new deposits, and finding subs tutes for the present minerals are the other elements of the national mineral policy aimed at conservi the nonrenewable assets.

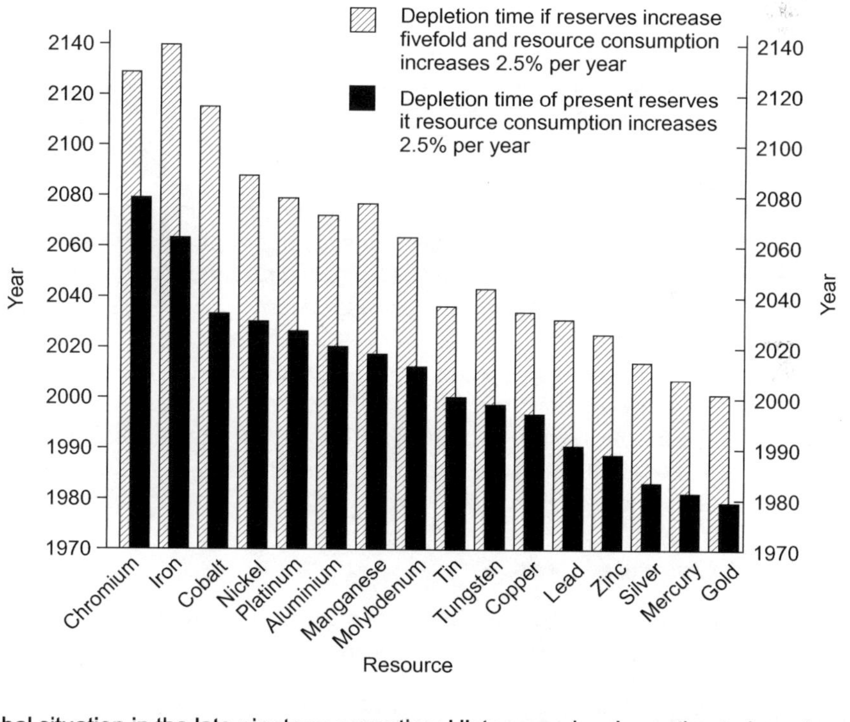

Fig. 6.5. Global situation in the late nineteen seventies: Histogram showing estimated number of years to per cent depletion of key mineral reserves under two different sets of assumptions. (After G. T. Miller and P. Armstrong, 1982, *Living in the Environment, Wads-worth,* Belamont, 491 p., Fig. 13.4.)

It is unfortunate, however, that the Indian mining industry, with few exceptions, continues to follo by and large the policy of exploitation—the unsound practice of producing a maximum quantity minerals with a minimum application of scientific and systematic methods of extraction and lea nsideration for the integrity of the environment and its ecological balance. The result is not only rapid pletion of the resources (Fig. 6.5 and Tables 6.2, 6.3 and 6.4), but severe injury to the environment. e following points are suggested for mineral conservation.

Scientific Mining Operations

cept for the mining of gold, coal, copper and manganese, all operations conducted for the extraction of e majority of minerals are by and large far from scientifically satisfactory and environmentally sound. e prevailing methods of extraction entail colossal waste and loss of the valuable ores and degradation land and water resources. **The aim of mining operations is to recover all material which can be worked out and utilized without allowing wastage. The marginal grades of the material should her be left in the mines in a condition that would permit future extraction, or properly stacked art to ensure utilization later. And in the case of strategic minerals, such as radioactive ores, gold, diamond, etc., the recovery must be total, regardless of the grade and cost.** Not only are ese strategic minerals stockpiled for specific purposes, and kept as raw material or in partly or fully ocessed forms to meet emergencies, their mines are kept in a state of readiness so that their production n be stepped up at a moment's notice. (A. K. Dey, per. comm., 1985).

To ensure minimum waste and greater recovery, it is necessary to resort to modern and advanced hnology. Sandstowing in underground coal mines, for example, has helped almost complete covery of the coal and lignite and elimination of the hazard of accidental collapse. The gasification of derground lignite, likewise, in Palana in Bikaner district (Rajasthan) led to the production of coal gas sed as fuel and for the making of fertilizers). Applying the Fischer-Tropisch Synthesis technology, the derground coal can be gasified to produce diesel, as already being done in China, Qatar and South frica. **Four such plants gasifying 60 to 76 million tonnes of coal per year can generate 12 million tonnes of petroleum liquid** (Bhardhwaj et al, 2007). A vast coal reserve of the order of 63 billion tonnes ng at a depth of 700 to 1700 m in the Mehsana area of Gujarat was exploited by gasifying the coal *in u,* producing 15,000 billion m^3 gas.

Constructing terraces (benches) of appropriate width and wall height in open cast mines linked with rvice roads and deposition of debris in proper dumps (stacks) have ensured maximal recovery of nerals as was seen in the magnesite mine at Jhirauli, Almora district in the Kumaun (Uttarakhand).

Similarly, improved methods of rock-cutting and tunnelling, automation, greater efficiency in crushing d grinding of raw material, etc. minimize waste and contribute to environmental preservation.

Blending Different Grades of Ores

The practice of utilizing only the best, and leaving the rest as waste needs to be curbed. Instead using only the higher grade of material it is more prudent to blend, in appropriate proportions, the arginal and sub-marginal grades of ores with the higher-grade material. This will not only considerably nimize the waste but also ensure longevity of the reserve.

Beneficiation of Marginal Grades

e cut-off grade of mineable copper ores in the U.S.A. was reduced by a factor of 250 since mining rted in the 19th century (Tyler and Armstrong, 1982); and in Papua, New Guinea, the grade containing low a content of copper as 0.64% was used profitably. This was achieved through processes known **'beneficiation' or ore-dressing that removes undesirable material (gangue), leading to the**

concentration of the desired material. The separation of gangue and the resultant concentration desirable material is done through gravity, electromagnetic or electrostatic, or high tension separation a by floatation or flocculation methods. Mainly physical methods are adopted to concentrate the valuab mineral, and in some cases a combination of both physical and chemical methods are used. In the ca of coal washing, the undesirable ash-forming mineral matter is removed by specific gravity separati and also by floatation. Floatation is a method by which the surface property of the mineral is modifi to a hydrophilic nature and made to attach itself to air bubbles and floated selectively. Sulphide minera like Cu, Pb and Zn, graphite, kyanite and phosphates are concentrated by the floatation technique. Mi is separated by crushing, grinding and sieving. Though floatation can be used to separate mica it is n employed in India for this purpose. Gold is recovered by the gravity, cynidization or amalgamati process. Bauxite is mainly concentrated by chemical methods such as leaching. Barytes, chromi zircon, beryl, wolfram, fluorspar and glass sand are concentrated by gravity processes using the Wilfl Table, Humphrey's Spiral and centrifugal concentration using hydrocyclones (N. G. K. Nair, per. comm 1985). In most of the mining industries a processing plant is attached so that the valuable minerals c be concentrated at the mine site in order to reduce the transportation cost.

Recycling Used Metals and Scraps

In the U.S.A. a little less than one-third of copper consumed and 14% of the lead used (Fig. 6.6) w produced by recycling the scrap and used material (Keller, 1982) more than thirty years ago. In 20 out of the total metal supply of 132 m t in the USA, nearly 66.9 m t was the recycled products, the ir and aluminium being the dominant ores. Recycled lead was 80%, iron and steel 52%, tin 19% and zi 26% of the apparent metal supply (U.S. Geological Survey Report). The scrap is generated in lar quantities in mills and foundries. **Old used metals of abandoned machines, vehicles, aircrafts, ships, buildings, etc. can be properly recycled to meet a part of the growing demand for scarce metals like copper, zinc, lead, platinum, molybdenum, etc.** This requires developing better automated methods collecting scraps, wastes and abandoned articles. India can meet a considerable portion of its needs f a variety of metals, if a machinery for waste collection is strengthened and *seriously managed* togeth with the establishment of plants to process the wastes. This venture would prove not only very profitab but also promote preservation of the environment.

As far back as 1985 about 3.6 million tonnes of iron and steel scraps (worth Rs. 4000 million) we dumped in various plants during 1983-1984 alone *(The Times of India,* March 20, 1985). These scra should have been recovered and processed.

Flyash produced when coal is burnt is not only used for manufacturing bricks and light-weig aggregate but is also known to contain germanium and gallium—the Garo Hills "ash" (Meghalay containing 0.095 to 0.122% germanium, the Singreni coal mines 0.055%, and the Tandur coal mines (Andhra Pradesh) 0.017% (Dey, 1978). In China they successfully recovered germanium from flyas Likewise, quartz from the old dumps of the Kangudi gold mine in Chittoor (Andhra Pradesh) has be found to contain appreciable amounts of tungsten and gallium (Dey, 1978). These rare elements can recovered if new technology is applied.

Developing Metallurgy for Substitutes

One way of getting over the problem of scarcity of certain minerals is to find substitutes, i.e. repla the scarce minerals with the ones that occur in abundance and are easily available. The new alloys uminium, magnesium and beryllium and recently recognized graphene for example, have been found be more useful than the older alloys in the making of aircrafts, cars and the like. Alloys of magnesium e fast replacing steel, thus lessening the demand on copper, lead and tin which are in very short supply. ere is a need for finding substitutes for mercury, asbestos, chromium, gold, silver, palladium, tin, timony, barytes, vanadium, zinc and petroleum.

Shale Gas

e Americans have perfected techniques of cracking the impermeable (but porous) shale by injecting nd and water under high pressure, resulting in the release and free flow of natural gas that fills the re spaces and is also adsorbed onto kerogen. It is the **shale gas. One single deposit—the Barnett ale in Texas—produced gas at the rate of 1.1 trillion cubic feet per year, and the cost of the gas us produced was half the cost of imported gas (Sah, 2010). Sizeable estimates of shale deposits in e Indo-Gangetic Plains extending from Panjab to Assam, in the floodplains of the Sabarmati in ujarat, the Kaveri in Tamil Nadu, the Krishna-Godavari in Andhra Pradesh, and the Mahanadi Odisha need to be explored to tap the reserve of natural gas.** The Oil & National Gas Corporation s a pilot project to drill wells at Ichapur near Durgapur (West Bengal) in the Damodar River Basin ah, 2010). This would be over and above the attempt to mine the coal-bed methane by injecting CO_2 ough pumps. In 7,800 km^2 area, the proven reserve of 820 billion m^3 of gas has the potential of oducing at the rate of 23 million m^3 per day in Central India, just in the identified 16 fields (Sibal et 2006).

Discovering New Deposits on the Seafloor

e seabed has been a source for many sedimentary minerals in the geological past. It is on the seafloor at some minerals continue to be deposited today, albeit at an exceedingly slow rate. With the resources land fast dwindling, the focus of attention must now shift to the seabed. While glass sands, gravels, ell deposits and heavy minerals (zircon, rutile, ilmenite, monazite) are mined in the beaches on the astal belts, and the offshore continental shelf has been found to be the repository of very rich deposits petroleum fluids and phosphates, the abyssal floor is the locus of deposition of polymetallic manganese dules at a depth of 4000 to 5000 m. The ocean beds can therefore be mined for oxides of manganese d sulphides of cobalt, nickel and copper, which invariably occur in these nodules.

Phosphorite deposits occur on the continental margin at the depth of 380 to 390 m off Goa and 186 293 m off Chennai (Purnachandra Rao et al, 2000).

The 100,000 to 150,000 km^2 bed of the deep Indian Ocean studied revealed promising deposits of manganese nodules characterized by presence of copper, nickle and cobalt in addition to iron. Sea-bed mining although prohibitively costly today, promises to provide much needed cobalt, nickle, copper and manganese (Srinivastava, 2009). The coastal belts in Kerala, Goa, Okha (Gujarat), ahabalipuram contain deposits of phosphorite in the form of nodules.

The extraction technology for the polymetallic nodules is in the developing stage (Fig. 6.7). How eful this venture will prove can be judged from the fact that **one cobalt extraction plant alone is pable of increasing world production of this mineral by 33% (Tyler and Armstrong, 1982).** About 5 trillion tonnes of manganese nodules occur in clusters on the ocean beds, including the 10 million m^2 area of the Indian Ocean. On the floor of the Indian Ocean the carpet of one-nodule thickness is ade up of manganese nodules containing 15.25% Mn, 13.35% Fe, 0.534% Ni, 0.241% Co and 0.29%

Cu (Qasim, 1981). The estimated capital cost of hydraulic mining (Fig. 6.6) at the rate of 1 m t/yr thir years ago was Rs. 250 to 800 million and the total investment in the order of Rs. 2,000 to Rs. 8,0 million. In the long run, the ocean-bed minerals would be less expensive than importing minerals.

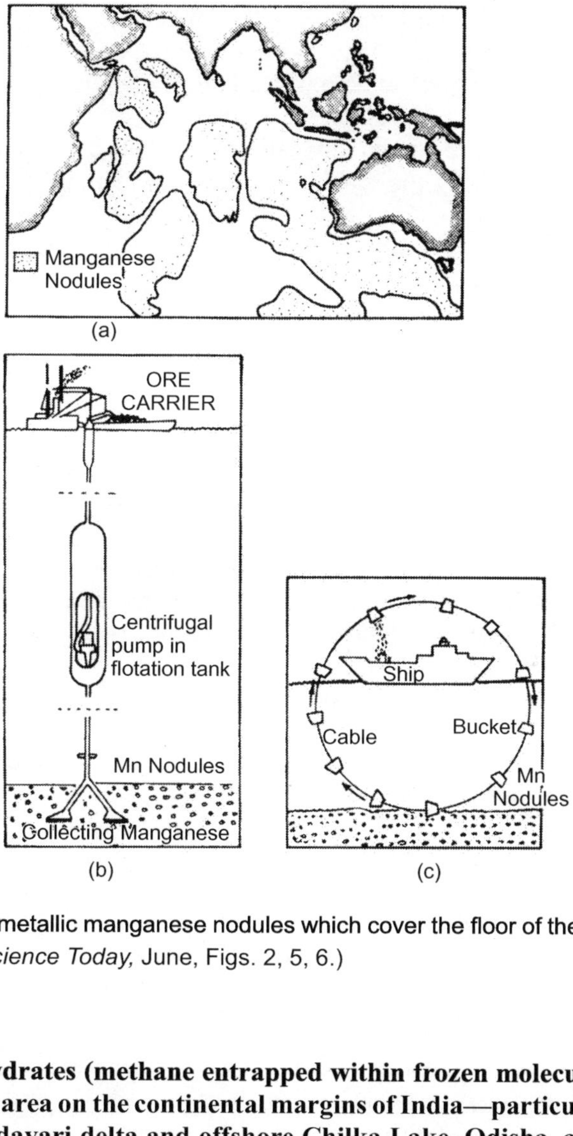

Fig. 6.6. Recovering polymetallic manganese nodules which cover the floor of the Indian Ocean. (After S. Qasim, 1981, *Science Today,* June, Figs. 2, 5, 6.)

Gas Hydrates

Large deposits of gas hydrates (methane entrapped within frozen molecules of water) have be identified over 1400 km^2 area on the continental margins of India—particularly offshore Narsapur part of the Krishna–Godavari delta and offshore Chilka Lake, Odisha, as well as in the western flank of the Andaman Islands. One cubic metre of gas hydrate formed in deep sea can produce 1 m^3 of methane (T. Kumar, 2009). The estimated amount of gas deposit is 2000 billion cubic metres sufficient to provide twice the energy resource compared to the conventional form of energy (Rama et al, 2007). The difficulty is that we have as yet not developed a technology to tap this large reser under the seawater.

IMPACTS OF MINING ACTIVITIES ON THE ENVIRONMENT

Problem

ore than 310,000 ha of land in India are presently under strong stress of mining activities of various nds (Fig. 6.4). A much larger area is disturbed by other activities associated with mining. The recovery minerals and construction material requires removal of vegetal cover with underlying soil mantle and cavating overlying rock masses (overburden) which more commonly exceed the volume of material ught. The result is the reshaping of topography, the generation of great volumes of debris (waste) and e disruption of surface- and ground-water circulations. The network of roads constructed to service ning operations further aggravates the problem of land degradation. So severe is the problem that en when well-managed as in the U.S.A., more than 16 m ha of land had been damaged by surface ning till 1980 (Coates, 1981). The damage is manifest in sheet and gully erosion, landslides and ckfalls, pollution of surface and groundwaters, etc.

The magnitude of the environmental problem created by mining can be judged from the example of al mining in India.

Assuming the depth of overburden in the opencast mines to be about 100 m, the amount of debris m about 100 km^2 area removed was about 8,800 million m^3, produced at the rate of 4 m^3 per tonne of al mined amounting to 200 mt (Ghosh, 1979) in the late 1970s. The disposal of this colossal volume debris must have been a serious problem. Furthermore, another 100 km^2 area was disturbed adversely activities ancillary to mining. By A.D. 2000, the 270 million tonnes of coal produced from the aria coalfields in Jharkhand resulted in the degradation of land at the rate of 604 m^2/yr and generated 0 million km^3 of wastes (Rai, 2000). The waste material covered approximately 11,000 ha of land •und the mined areas. For the extraction of 2700 million tonnes of coal in the Singrauli Coalfieds in uthwestern Uttar Pradesh, more than 7800 million m^3 of debris had to be disposed of, which was done 6560 ha of forested area (Rai, 2000).

Likewise, in A.D. 2000 India mined annually 7 m t of copper ores. The yearly production of tailing mounted to 36,50,000 m^3. The problem of disposal of the tailings was quite acute.

In addition to destruction of forests and laying waste of the otherwise productive lands, the dumping the overburden material has created very serious problem practically in all areas (>800,000 ha) der mining in Haryana, Rajasthan, Jharkhand, Odisha, Karnataka and Goa. Far much larger areas are turbed due to activities associated with mining.

As the demand for minerals grows, the area of mining will also expand at a faster rate, threatening creasingly larger areas of landscape with scarification, debris dumps, soil degradation, and a widening circle of deforestation and distress to the affected population.

Types of Mining

pending on the conditions of the terrain and the depth of ore deposits, two main types of mining erations are carried out—*opencast mining* or *quarrying* to extract surficial and shallow deposits, d *underground mining* to recover deeper ores. The Singrauli coalfields in Mirzapur district (Uttar adesh), the Kudremukh iron mines in the Western Ghat region of Karnataka, the Noamundi iron mines Jharkhand and the Jhamarkotra phosphorite mine in Udaipur (Rajasthan) are examples of opencast ning. The opencast mines account for more than 50% of mines in India, This type of mining is done strips (strip mining)or along contour lines in hill and mountainous terrain (contour-strip mining)

(Fig. 6.7). Cave-ins of quarry walls are common accidents, such as in the stone quarries. In order prevent this hazard, quarry walls are propped up by providing supports. If the quarries are deeper th a metre; fencing is installed to check walking on steep banks, and rainwater or other surface water diverted. The stacking of overburden and building of spoil banks create problems of landslides a severe erosion whereby roads and streams below may be adversely affected. In spite of these problem opencast mining has many merits—there is greater recovery of raw material, the practice is quite sa and free from risks that are inherent in underground mining, and there is less problem of air pollutio The acid water generated by oxidation of pyrite present in the coal mines can pose quite a serio problem of water pollution. Selective stripping, removal and safe disposal of high-sulphur beari materials would obviate this problem.

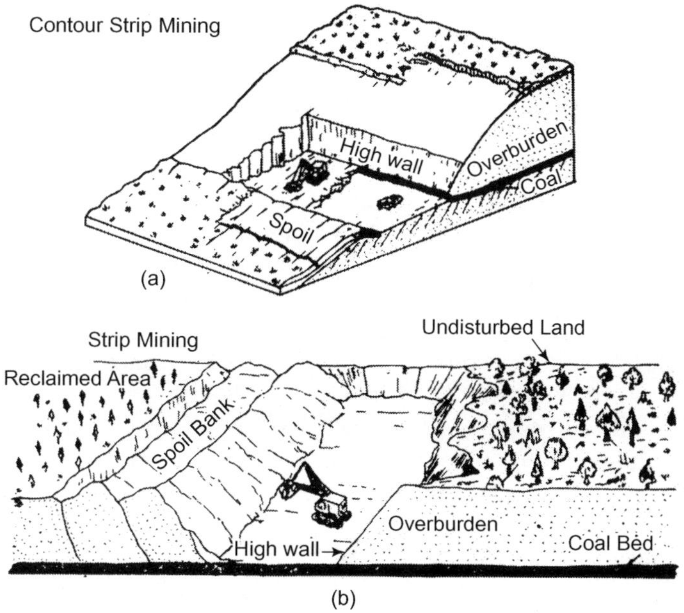

Fig. 6.7. Contour-strip mining in hill terrain and strip mining in flatter country. The top soil is scraped off a stacked away, and the overburden with debris systematically heaped in continuously excavated parts and then levelled, covered with soil and afforested eventually.(From E.K. Keller, 1976. *Environmental Geology,* Bell and Howell, London, 488 p., Figs. 13.16 and 13.17.)

In ***strip-mining*** the overburden is stripped away to recover the minerals by use of bulldozers, scrape or by manual operations. ***Contour-strip mining*** (Figs. 6.7 and 6.12) involves removal of the overburde This is done by starting from the outcrop and proceeding along the hillside so that one cut after anoth is made and a series of benches are formed. The inside walls range in height from a few metres to mo than 30 m, but generally within 10 m. On the outer side, debris is dumped on the slopes. The net resu is disfigurement and degradation of the landscape.

Pieces of ore that have been broken and transported from their source by weathering and erosio and normally occurring near their source are called *float ores,* as recovered in the Sandur mangane nes in Karnataka (A. K. Dey, per. comm., 1985). Their treatment before use prevents pollution of air d water.

If the ore deposits lie deep ***underground mining*** is resorted to, as practiced in Jharia (Jharkhand) and niganj (Bengal) for coal, in Kolar (Karnataka) for gold, in Zawar (Rajasthan) for lead and zinc, and on. Networks of vertical shafts or deep wells and/or horizontal tunnels and adits are developed to get the ore deposits.

To recover ores from the seabeds or minerals and petroleum fluids in the water-covered coastal belts, edgers, giant augers and hydraulic drills are used. Hydraulic mining and surface scraping of beach nds are in use to recover ilmenite, monazite, zircon and gold from the inundated placer deposits along e South Indian coasts. The tin ores in offshore Indonesia and Thailand and diamonds off the coast of amibia are won in this manner.

Problems in Contour-Strip Mining

the opencast contour-strip mining in the mountains, rich soil together with overburden is scraped off d pushed downslope. The sliding of great volume of debris causes great damage to the vegetation and springs and streams (Fig. 6.8).

Indiscriminate mining induces landslides and aggravates erosion, as was witnessed on the Mussoorie lls in Uttarakhand when extensively mined for limestones and dolomites, and in Sikkim where (in latter se) foliated and jointed phyllites are being mined for uraniferous polymetallic sulphides. In northern rts of Bageshwar district and adjoining belt in Pithoragarh district of Uttarakhand the mountain slopes e recklessly and extensively damaged by the diggers of talc and soapstone.

A study carried out in the mining district in Kentucky (U.S.A.) demonstrated that the rate of erosion the strip mined area was 27,000 t/km^2 compared to just 27 t/km^2 in the normal hillslope—1,000 fold gher (Tank, 1973). In La Utica county, Illinois the rate of ground loss due to mining was 29 to 33 mm/ (Haigh, 1982). Reclamation of the mined area on a slope more than 14° is rarely successful, according Coates (1981). One can understand the seriousness of the problem in the Himalayan region where the ning is carried out on slopes as steep as 30 to 50°.

Added to the so-called wastes generated by mining is the debris produced in the construction of e network of roads servicing mining operations. This aspect has been dealt with at length in Chapter ven.

About 14 ha of forested hillslopes above the village Khirakot in the Kosi valley (Almora district) d several hundred hectares of the Hiunpani forest in the Chandak area and elsewhere in Pithoragarh trict in Kumaun in Uttarakhand were haphazardly and crudely mined for soapstone and magnesite, pectively. More than 4,820 ha of land area in Kumaun Himalaya, 11,471 ha in the Darjiling Hills, 438 in Himachal Pradesh, and 886 ha in Jammu-Kashmir (Negi, 1982) had been ravaged or very seriously fected by mining until about the beginning of the 1980s.

Impacts of Area-Strip Mining

relatively flat terrain such as the Jharia, Raniganj and Singrauli areas strip mining involving cutting trenches (Figs. 6.7 and 6.8) resulted in extensive deforestation, removal of soil, lowering of the water ble and pollution of surface water. In the Singrauli Coalfield, Mirzapur (Uttar Pradesh), 36 million m^3 overburden had to be removed and dumped in the forests, killing all vegetation. More than 184 km^3 area in parts of Bina, Jayant, Gorbi and Jhingardih blocks was rendered derelict. For, to extract 2,700 m t of coal (at the rate of 76 m t/yr) about 7,800 million m^3 of overburden was removed and dump somewhere in that forested area (Sinha et al., 1982).

The debris disrupt the circulation of the surface as well as groundwater reducing channel capacity some streams, promoting erosion in others, and causing water pollution in the region.

Fig. 6.8. Excavation for extracting mineral talc and soapstone in part of Bageshwar district in Kumaun Himalaya generates large volumes of wastes (debris) which cover productive agricultural fields a bury springs.(Courtesy: Dr S.K. Bartarya, Wadia Institute of Himalayan Geology, Dehradun; Fro Valdiya 2004). (See colour Fig. 6.8)

Adverse Effects of Open-Cast Mining

In addition to destruction of forests and laying waste of the otherwise productive lands, the dumpi of the overburden material has created very serious problem practically in all areas (>800,000 h under mining in Haryana, Rajasthan, Jharkhand, Odisha, Karnataka and Goa. Far much larger are e disturbed due to activities associated with mining. A couple of examples will suffice to show the mension of damage and distress.

By A.D. 2000, the 270 million tonnes of coal produced from the Jharia coalfields in Jharkhand ulted in the degradation of land at the rate of 604 m^2/yr and generation of 700 million km^3 of wastes ai, 2000). The waste material covered approximately 11,000 ha of land around the mined areas. In e Singrauli Coalfieds in southwestern Uttar Pradesh, for the extraction of 2700 million tonnes of coal, •re than 7800 million m^3 of debris had to be disposed of which was done on 6560 ha of forested area ai, 2000). The Malanjkhand mine, which accounts for 70% of India's copper reserve, covers more an 115 ha of land while the waste dump is spread over 120 ha (M.P. Singh, *The Hindu,* June 29, 12).

Decimation of Plant Life

the Gandhmadan Reserve Forest in Odisha, the mining of bauxite has endangered 2700 species of wering plants and 150 species of medicinal plants. The sandstone quarrying in the Bijolia belt in jasthan resulted in the loss of 47% of forest cover, 14% decline in agriculture and 67.5% increase wasteland (Anon, 2000b). In the Jharia mining operations 36,385 ha of forest land was adversely fected (Rai, 2000).

The destruction of forests and agricultural land in the Bellary district in northeastern Karnataka is • well-known to be described. In the Chitradurga and Tumkur districts in Karnataka, the deposition iron dust on crops affected 3312 ha of agricultural land, causing dramatic decline in the productivity *ragi* crop in areas within 2 km radius of the mining. Likewise, the production of arecanut and banana l down by 40% according to the report of the Indian Council of Forestry Research and Education (*The ndu,* December 24, 2011).

Indiscriminate mining since 1961 had destroyed 50,000 ha of forests in Goa (J. D. Souza in *The mes of India* February 13, 1984) and 2500 ha between 1988 and 1997 (*The Hindu,* December 29, 11). Goa lost 2500 ha of forest area due to mining between 1988 and 1997 (*The Hindu,* December , 2011). Against the export of 12 m t/yr of high grade iron ores, 30 m t of iron ore rejects are scattered discriminately over 10,000 ha of paddy fields and coconut groves, thus diminishing the fertility of the l. The washing of mineral ores for beneficiation has caused serious water pollution in the rivers as ll as in wells and springs.

Underground Mining Hazards

rious underground mining hazards in India had taken a toll of 3,400 lives in the period 1900 to 1980. llowing withdrawal of coal at depths of 80 to 200 m from the underground mines in Jharia, the strata erlying the coal seam subsided 10.6 cm, 5.9 cm and 25 cm in different places (Kumar et al., 1975). e subsidence, at a slow rate in an area of 32 km^2 by the end of 1966 had caused damage to buildings ading to the collapse of many (Fig. 6.9). In North Kujama colliery, the sudden collapse on March 30, 79 of a 200 x 150 m block resting on 14-m high coal pillars caused subsidence of about 2.4 m of the ound-surface over 2,000 m^2, and the Jharia–Baliapur road sank 1.2 to 1.6 m (Banerjee, 1981).

The subsidence of the ground over underground mines is a result of unscientific practices. A relatively w process of sinking extending over decades, eventually leads to drastic accidents such as floods, lure of dams and levees or embankments on the rivers that flow on the ground over the underground nes, collapse of buildings, and destruction of sewage, drains and canals. The water of the Barakar

River often fills the depressed area with flood waters in the Jharia coalfields. The Chasnala tragedy 1978, that took a toll of 376 lives, is an example of this kind of hazard.

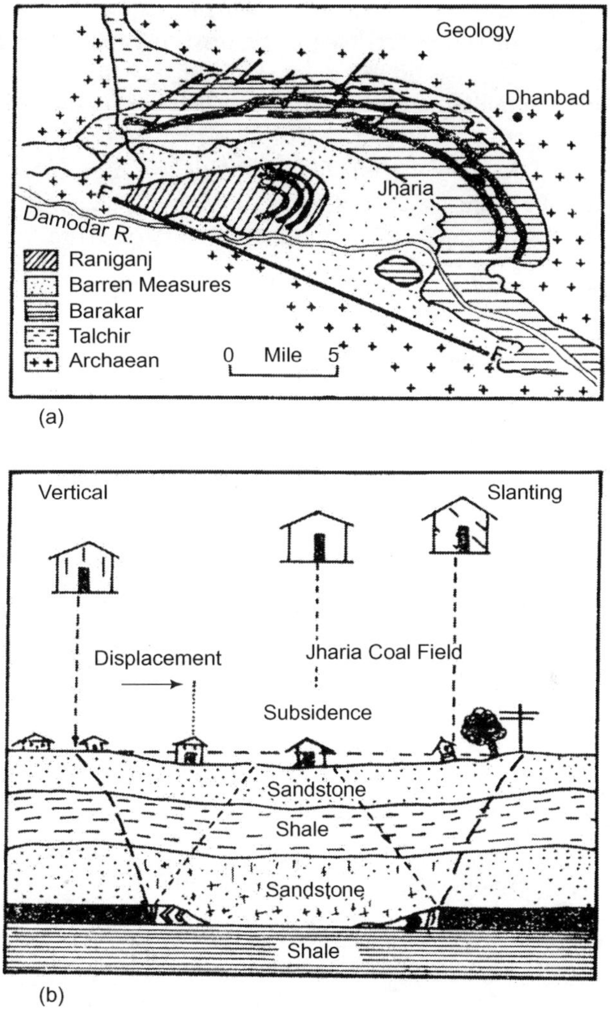

Fig. 6.9. (a) The geological map of the Jharia coalfields in Jharkhand delimited to the north by a fau (J. Coggin Brown and A. K. Dey, 1955, Fig. V).(b) Removal of coal from an underground seam has caused subsidence of the ground, as evident from the damage to buildings, utility structures and the flooding of the sunken area.

Withdrawal of oil and gas from deeper reservoirs results in the reduction of fluid pressure, culminati in the subsidence of the ground as witnessed in the Inglewood oilfield Los Angeles, where the subsiden in the period 1937 to 1967 was of the order of 9 m at the centre (J. F. Poland in Tank, 1973). Similar in the Lake Maracaibo oilfield, Venezuela, the cumulative subsidence was about 3.3 m between 19 d 1957; in the Nago Gasfield, Japan, the ground by 1954 sank at the rate of 48.8 cm/yr, and in the sfield in the Po delta (Italy) the subsidence rate was 25 cm/yr. Minor subsidence has been reported in e media in the southern Gujarat plain where many oilfields are located.

Underground fire in coal mines is yet another hazard which is difficult to control. Till 1978 about 110 ajor fires were reported from Jharia coalfield alone, destroying 34 m t forever, and 79 mines are still mouldering in underground fires. It is said that a fire was lit in a shallow incline in 1940 (or according another version in 1919) and it continues without stop. The conflagration has gravely threatened the fety of the township of Jharia. The burning of coal means production of such gases as sulphur dioxide, rbon dioxide, carbon monoxide, methane, etc. together with coal dust. It goes without saying that the derground fires in coal mines cause great pollution of the atmosphere. In Jharia the carbonization, quetting and beneficiation plants generate fly ash at the rate of 10 m t/yr in addition to colossal lumes of smoke and gases.

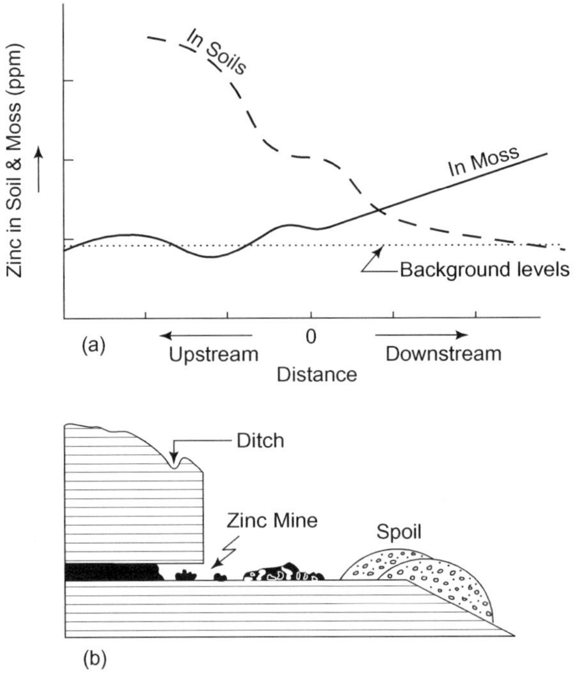

Fig. 6.10. (a) Distribution of zinc in water and soil around a lead-zinc mine. The zinc in water has moved down the valley, but the zinc held in the soil has been used up by the plants. (b) Stacking of spoils from zinc mine. The ditch atop the overburden ensures diversion of rainwaters. (After A. D. Bradshaw and M. J. Chadwick, 1980. *The Restoration of Derelict Land,* Blackwell, Oxford, 317 p., Fig. p.3.)

Adverse Impacts on Hydrologic System

eper excavation on the surface or underground digging causes the water table to sink locally, en drastically, resulting in the drying up of wells and springs of the neighbourhood area, or their

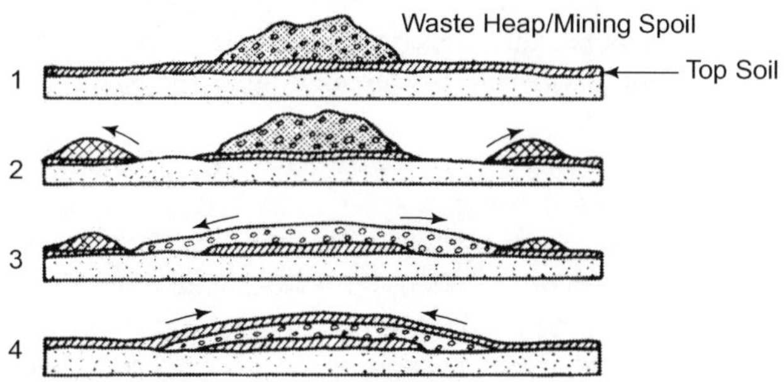

Fig. 6.11. Method of covering an infertile mining waste or spoil with a mantle of nutrient-rich topsoil—that was earlier scraped off and stacked away. This measure would not only prevent pollution of water by the mining wastes, but ensure restoration of the ground through afforestation. (After A. D. Bradshaw and M. J. Chadwick, 1980, *The Restoration of Derelict Land, Blackwell,* Oxford, 317 p., Fig. 6.6).

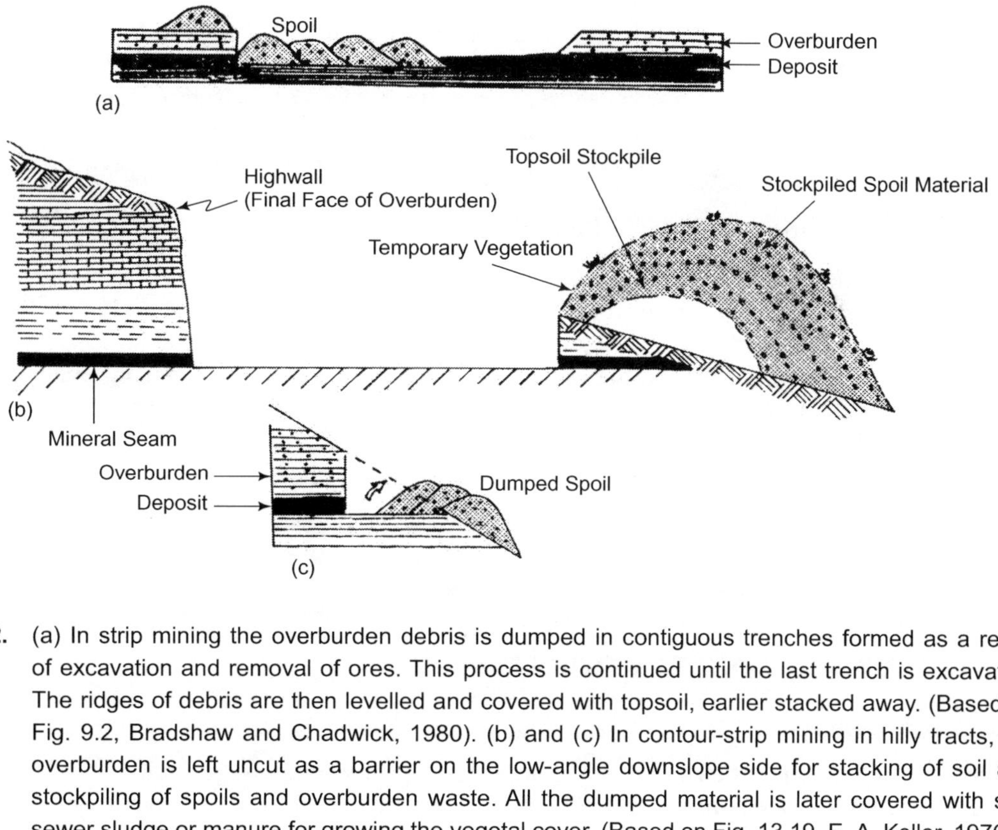

Fig. 6.12. (a) In strip mining the overburden debris is dumped in contiguous trenches formed as a result of excavation and removal of ores. This process is continued until the last trench is excavated. The ridges of debris are then levelled and covered with topsoil, earlier stacked away. (Based on Fig. 9.2, Bradshaw and Chadwick, 1980). (b) and (c) In contour-strip mining in hilly tracts, the overburden is left uncut as a barrier on the low-angle downslope side for stacking of soil and stockpiling of spoils and overburden waste. All the dumped material is later covered with soil, sewer sludge or manure for growing the vegetal cover. (Based on Fig. 13.19, E. A. Keller, 1976).

coming seasonal. In the hilly terrains, the process of excavating and attendant landslides frequently pose passages of underground water, thus depriving the springs of their content. This had happened the Mussoorie Hills, where the discharge of the celebrated Sahasradhara seepage had gone down nsiderably due to mining in the area.

Exposure of fresh rocks due to mining initiates weathering with inevitable generation of substances hich cause water pollution. The spoil banks, likewise, contain minerals that weather easily so that e production of pollutants takes place at a much faster rate. Among the minerals involved are pyrite, arcasite, ankerite and siderite, which produce sulphuric acids and other soluble salts such as sulphates d oxides. All these substances adversely affect flora and aquatic fauna. In the U.S.A. the acid waters m mines had contaminated 10,000 km of streams and rivers, as well as large number of lakes, ervoirs and marshes (Coates, 1981).

The Tiri stream emanating from the lead mines at Zawar in Rajasthan was polluted for 15 km wnstream, resulting in a decline in the growth of many species of timber plants, and elimination of *ringtonia tinctonia* and *W. tomentosa* (Fig. 6.12). A similar situation was discernible around Jhamarkotra osphorite mines near Udaipur, and in the Khetri copper mines in northeastern Rajasthan.

Effects of Blasting of Rocks

quarries and mines explosives are used in increasingly large quantities. In stone quarries close to urban tlements, the blasting of rocks not only annoys the local population but results in hazards from flyrocks, and mage due to ground vibrations. The vibrations generated by detonations also contribute to the destabilization hillslopes by opening out joints, fractures, fissures and cracks. This phenomenon causes diversion of oundwater, leading to drying up of springs in some places and increased discharges elsewhere.

Effects of Dredging and Hydraulic Mining

ean-floor mining on an extensive scale to recover polymetallic nodules containing sulphides Zn, Pb and Ni will have a considerable effect on the benthic community, owing to stirring up of diments and attendant formation of plumes of suspended material. Disposal of deep-sea sediments ought out with the ores will be a big problem—1,000 t/day of dry nodules will entail disposal of [illegible]00 t of sediment brought to the land everyday. This sediment and the tailing produced—if allowed be washed back to the sea—will affect the life on the shallow sunlit coastal belts. The effects on awning of fish and other fauna will be particularly serious.

Problem of Pollution

is aspect has been dealt with in Chapter Twentyone.

ENVIRONMENTAL MANAGEMENT IN MINING

Basic Strategy

There are four important elements in the management of mining operations.

First, **a running programme of overall assessment of reserves, embracing not only the benefits to derived from the extraction of deposits, but also the magnitude and cost of adverse impacts on her natural resources, such as water, soil, forest, agriculture, grazing land, etc. (Table 6.4).**

Table 6.5. Analysis of Environmental Impacts of Mining Activities (From Munn, R.E., 1975, *Environmental Impact Assessment*, John Wiley, New York)

A. ***General Information***
1. Location of the mine(s)
2. General topography and climate
3. Nature of mining operation
(a) Opencast
(b) Underground
4. Date of initiation of mining operation
5. Owners of the mines
B. ***Size of the Mineral Deposit***
1. Length, width and thickness
2. Mineral reserve
(a) Proved
(b) Estimated
3. Average daily production
C. **(i) *Open cast Mining***
1. Overburden to ore ratio
2. Volume of overburden to be removed (excavated)
(ii) *Underground Mining*
1. Depth of the working
2. Mode of entry (shaft, adit, incline)
3. Methods of mining
D. ***Disposal of Waste***
1. Nature of solid waste
2. Quantity of solid waste generated
(a) From mining operations
(b) From ore processing operations
E. ***Problem of Pollution***
1. Of surface and groundwater
2. Of air
(a) From mining operations
(b) From ore-processing
3. Noise pollution
4. Health hazards
(a) Existing maladies
(b) Potential hazards
F. ***Proposed Programme of Reclamation of Excavated / Mined Area***
1. Restoration of excavated areas

	2. Stabilization/disposal of debris dumps
	3. Treatment of spoil banks
	4. Pollution and noise control
Socio-economic Considerations	
	1. Landuse pattern (within 10 km of site)
	2. Extent of forest/pasture/farm/settlement affected
	3. Number of persons employed
	(a) Daily-wage workers
	(b) Salaried employees
	(c) Ratio of locals/outsiders

Second, **legislative provision for placing full responsibility of treatment and environmental habilitation of excavated areas squarely on the mining agencies or mine owners. There should be legal binding on the mine owners to restore the land they have damaged by resorting to systematic and appropriate methods of disposal of wastes or debris and regeneration of vegetal cover. Stacking and preserving of topsoil must be made compulsory in order to cover back the cavated parts with debris produced and then put a mantle of this soil over it** (Figs 6.11 and 6.13) order to grow grasses, shrubs and trees. Incentives should be given to mine-owners for successful ogrammes of land-restoration, soil conservation and afforestation.

Third, serious attempts should be made to stabilize the slopes not only during the excavations but o afterwards to ensure that the stability is maintained in future also. This is done by constructing a ainage network (Fig 6.13) and retaining wall.

Fourth, **In the context of the Himalayan region, it would be prudent to restrict or even ban mining in the geodynamically sensitive and ecologically fragile belts, such as the zones of the Main undary Thrust and Main Central Thrust where the rocks are in a greatly weakened condition and landslides are recurrent and severe** (Fig. 11.19). Where mining is unavoidable in such belts, ecial care should be taken to keep the slopes under stable condition, to avoid blasting operations and take steps to install ropeways in place of roads for hauling material, and resort to underground mining preference to opencast mining.

Fifth, the provision of safe water to the workers and the inhabitants around the mines and the treatment water pollution from mining and ancillary activities should be made the essential responsibility of ne owners.

Extensive quarrying for limestone on the southern slope of the Mussoorie Hills had completely scarred d seriously damaged the forest cover as already mentioned. Resort to biological and engineering medial measures successfully ensured reduction in the rainfall runoff from 57% to 37% and the diment outflow from 550 t/ha/yr to 8 t/ha/yr (Sastry et al, 1995; Juyal et al, 1998). There was also nsiderable improvement in the outflow of dry-weather streams. The scars have vanished and the rests are regenerating.

Rehabilitation of Strip-Mined Areas

e topsoil is scraped off and stacked away safely for use later (Fig. 6.11). The overburden from the ntiguous cut is dumped in the trench formed as a result of removal of ore from the first cut (Fig. 2). This process is continued until the last trench is excavated. Then the ridges of dumped debris are levelled and covered with the topsoil that had been safely stacked apart (Fig. 6.13). Soon aft appropriate species of grasses, shrubs and trees are planted, making liberal use of fertilizers to accelera plant growth. In the quarries for construction material, little waste is left so that the depressions form can be converted into lakes (as was done at Timlia in Gujarat), or used for dumping municipal a industrial wastes. These waste dumps are subsequently covered with topsoil and afforested.

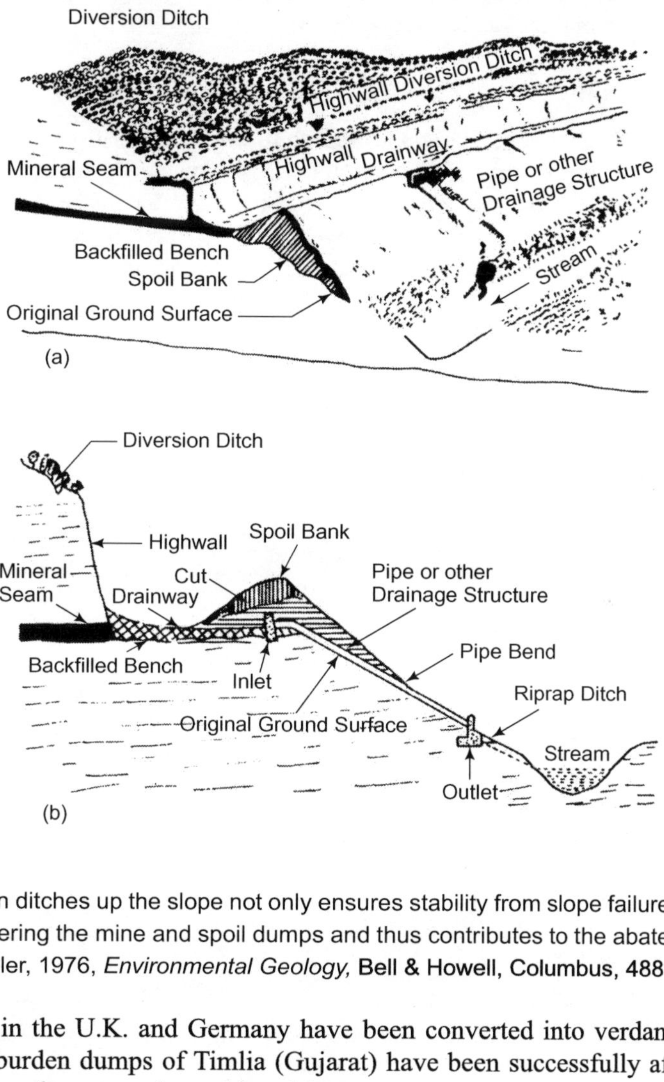

Fig. 6.13. Diversion ditches up the slope not only ensures stability from slope failure but also prevents water from entering the mine and spoil dumps and thus contributes to the abatement of pollution. (From E. A. Keller, 1976, *Environmental Geology,* Bell & Howell, Columbus, 488 p., Fig. 13.15.)

The coal mines in the U.K. and Germany have been converted into verdant land by this metho Likewise, the overburden dumps of Timlia (Gujarat) have been successfully afforested, while deep (>30 m) trenches have been transformed into lakes.

Stabilizing Debris Dumps in Contour-Strip Mining

The stabilization of debris dumps involves construction of diversion ditches above the area of mini to save the debris from the effect of running water. Equally important is the appropriate disposal bris by stacking it at a low angle (Fig. 6.13). In order to ensure stability, a part of the ore-bearing d with its overburden is left intact to serve as a barrier. Then the debris of the second excavation is read in such a manner as to form a terrace. These accumulations are then covered with the topsoil, plied with appropriate fertilizers and finally seeded with grasses, shrubs and trees having deep roots. r example, planting of such flora as *Woodfordia fructicosa, Mimosa himalayanensis, Spermadictyon aveolens, Adhatoda vasica, Mallotus philippinensis,* etc. have been suggested for the carbonate mines Uttarakhand–Himachal Himalaya (R. K. Gupta, 1978).

Treatment of Spoil Banks

the case of coal mines of Raniganj, Jharia and Singrauli, the amount of wastes (spoils) generated is ry great in comparison to the quantity of coal extracted. Usually, the wastes also contain lower-grade es, which in the coal-mine debris are very acidic due to oxidation of sulphides present. The result is e destruction of plant nutrients, and eventual infertility of the soil. To neutralize the acidity, lime can applied at the rate of 30 to 50 t/ha to a depth of 20 to 50 cm by ripping the dumps deeply.

In the gold mines (such as of Kolar and Hutti in Karnataka), vast quantities of quartzose (siliceous) bris bereft of plant nutrients are generated. The application of heavy doses of lime and addition of anure or organic waste may help improve the condition.

Table 6.6. Use of Waste Products in Restoration of Derelict Soil (A. C. Bradshaw and M. J. Chadwick, 1980)

Material	Composition %				Special Problems
	N	P	K	Organic matter	
armyard manure	0.6	0.1	0.5	24	Can be toxic when used direct on plants
Pig slurry	0.2	0.1	0.2	3	High water content
Poultry manure (broiler)	2.3	0.9	1.6	68	High levels of ammonia
(battery)	1.5	0.5	0.6	34	High levels of ammonia
Sewage sludge (air dried)	2.0	0.3	0.2	45	Possible toxic metals, higher moisture contents
Peat (partly dried)	0.1	0.005	0.002	50	Variable, especially calcium content
Mushroom compost (dried)	2.8	0.2	0.8	95	None, except high lime content
Domestic refuse (municipal)	0.5	0.2	0.3	65	Miscellaneous objects
Straw	0.5	0.1	0.8	95	Adverse C/N ratio
Building rubble	0.05	1.8	2.0	0.5	Brick and pieces of masonry
Colliery spoil	0.03	0.04	0.4	0	Possible high levels of pyrite causing acidity
Pulverized fuelash	0	0.05	2.2	0	High boron

In the well-managed Kudremukh mines in the Bhadra basin in the Western Ghat, the production in e early 1980s of iron ores at the rate of 22.6 m t/yr each year generated about 13.1 m t of dominantly iceous tailings. Appropriate measures of control of erosion and pollution of the Bhadra River required equate treatment of the spoils. In the Zawar lead-zinc mines located in the Aravali hills, the waste ck produced was used for the construction of roads, buildings and levelling of valleys. The inflow groundwater in the mines is small and is being utilized in the mining operation. The beneficiation ocess comprises crushing of ore to 12 mm size and thus generation of dust and tailings, and the use a number of toxic chemicals. The dust is collected and the tailings are deposited behind dams in the valleys, where water is recycled for reuse after proper treatment for removal of toxic elements. T tailings are covered with vegetation (Paliwal, 1985).

In the case of alkaline debris (pH>10), in which hydroxides of Ca, Mg, Na, K are present, acid nutrients may be applied to neutralize the material. Such nutrients are contained in sewage sludg domestic or municipal refuse, colliery spoils, farmyard manure, peat, etc. (Table 6.6). The nutrients a the plant seeds can be mixed together and sprayed over the debris dumps so that vegetation acquir a foothold in the quickest possible time. Addition of chopped straw, wood pulps and shredded plas ensures not only shade against sun and wind for the seeds, but it also augments the moisture-retaini capacity of the waste. Latex or some oil-based emulsions, which form a porous but firm crust, may added to impart a higher viscosity (Bradshaw and Chadwick, 1980).

The spoils produced in the mining of heavy metals such as zinc and lead are very toxic, particularly the metal content is more than 1,000 ppm (0.1 per cent). Such dumps can be covered with crushed roc and/or granulated slag and then with the mantle of topsoil. Covering the spoil heaps with sewage slud or domestic refuse before putting the mantle of soil is another very effective and inexpensive techniq of tacking the problem of pollution. The mantled heaps are then treated with phosphorus fertilizers the rate of 400 to 800 kg/ha), legumes and grassroots in order to grow turfs (Table 6.7).

Table 6.7. Hydroseeding Specifications for Establishing Grass/Legume Cover in Temperate Climates (A. C. Bradshaw and M. J. Chadwick, 1980)

Material	Rate of Application
1. Mixture (appropriate) of grass-seeds (including nursecrop in exposed situations)	70 kg/ha
2. Wild white clover or other legume (innoculated and pre-germinated, if necessary)	10 kg/ha
3. *Mulch*—woodfibre, chopped straw or glasswool	1 to 2 t/ha
4. *Stabilizer*—woodfibre, alginate, PVA or latex (if likelihood of severe erosion)	Depending on stabilizer
5. *Fertilizer*—Complete 15:15:15 (slow release N-fertilizer value if legume is not	200 kg/ha followed by
included) OR	300 kg/ha after 8 weeks
Dry organic manure	500 kg/ha
6. *Lime* (if pH requires this and may have to be spread separately)	0 to 5000 kg/ha

Pollution and Noise Control

The most effective way of controlling pollution from mines, spoils and tailing would be to divert wat away from the heaps and prevent it entering into mines. As already stated, the application of lim neutralizes the acidity of debris-generated pollution.

In order to abate noise and vibration generated by detonation of explosives used to blast rocks, would be advisable to restrict the charge weight per delay by keeping delay intervals of 17 millisecon or more, to reduce the number of charges per hole, increase the number of stemming columns, minimi the blast size by using multiple delays, to deck and limit the number and size of blast holes and t frequency of blasting, maximize charge depth, and cover detonating cords with at least 7.6 cm thi layer (A. K. Ghosh, 1982).

SUMMARY

inerals are wasting assets. They constitute non-renewable, non-replenishable finite resources, occurring only in geographically restricted areas. Over-consumption and waste of the rapidly dwindling reserves must be curbed and effective measures taken to conserve them if the society sires to preserve the reserves of such crucial minerals as petroleum, sulphur, phosphates, base metals, . At the present rate of consumption, the reserves of the ores of managanese, chromite, copper, zinc, ad, etc. will not last beyond a decade or two. **"Efficiency in use and avoidance of waste" should be the keynote for policy in mineral conservation.**

The aim of the mining operation is to recover all material which can be worked out and utilized thout allowing wastage, and adoption of new advanced techniques that would ensure greater recovery d less waste in extraction. **Blending of appropriate proportions of marginal-submarginal grades of ores with higher-grade ores will not only eliminate wastage, but increase the life of the reserve. Stockpiling of yet unworkable lower-grade ores for future utilization is a very important measure of conservation.** By removing undesirable rock material from the ores through a variety of beneficiation ocesses, the concentration of the desired minerals can be considerably increased. The supply of scarce nerals can be appreciably augmented by recovery of metals from spoils, tailings and wastes such fumes and sludges, and by recycling old used materials and scrap. Efforts to find new deposits the offshore belts (with their petroleum and heavy-mineral resources) and in the seafloor which e carpeted with polymetallic manganese nodules, would prove considerably rewarding. Finding bstitutes for minerals that are scarce is yet another profitable venture, which could eliminate the oblem of deficiency.

Planning for long-term development of mining areas should ensure reinvestment of a part of e profit of mining so that the life of the deposit is extended. The development programme should also involve measures of protection of the natural environment which mining tends to despoil and mage. Social cost and social security and commitment must not be overlooked in the exercise of cost-benefit analysis. The benefit of mining must also go to the people adversely affected by the mining operations.

The various methods of extraction of minerals involve not only drastic reshaping of the landscape d scarification of the land, but also destruction of forest cover, removal of the soil mantle, generation enormous volumes of polluting debris and serious disruption of water circulation, both surficial and derground. Presently, nearly 3,10,000 ha of land in India is under various degrees of stress due to ning operations and a larger area is disturbed in the hilly regions which have become prone—as much 1,000 times in some places—to erosion and mass-movements. Strip mining in flatter terrain has not ly caused deforestation but also polluted surface waters and drastically altered the underground water lance, rendering springs and wells dry or seasonal. Underground mining is responsible for subsidence the ground, leading to hazards such as floods (as in Jharia, Jharkhand). Deep ocean-dredging and draulic mining stir up sediments, damage the habitat of benthonic life and the deep-seated sediments ought to the land may cause pollution in the coastal belts.

The restoration of land damaged by mining should be the essential item in environmental management enforced through legislation. The rehabilitation of damaged ground involves stabilization the slopes not only during the excavations, but also after mining operations are over. Systematic and propriate disposal of debris without harming the slopes, their vegetation, water resources and human tlements is another very essential measure. Prevention of entry of water into the sites of excavation and spoil dumps through drainage networks would ensure stability and minimize pollution. **Stacking and preserving topsoils should be made compulsory in order to make use of it later to reclaim the excavated parts and debris dumps and then seed the soil mantle with grasses, shrubs and trees an attempt to restore the land through afforestation.**

7 Erosion—Causes and Control

DIMENSION OF EROSION

Indian Subcontinent

More than 45.37 m ha of the Indian land (Fig. 7.1 and Table 7.1) suffer from various types of erosion—8.89 m ha by wind alone (Anon, 2010). About 4.36 m ha of forested land under *jhum* cultivation has become seriously degraded (Anon, 2010). **The erosion is particularly pronounced where man has tampered with the ecological balance by reshaping or defacing the topography.** The extent of erosion by water can be judged from the volume of sediments the rivers are carrying to the oceans. In the late 1970s all Indian rivers together annually transported about 1490 million tonnes of material in suspended and dissolved states, eroding the catchments at a furious rate of 4.5 m t/ha/yr (Subramanian and Dalavi, 1978). According to another estimate (Bali and Kanwar, 1977) in the mid 1970s about 6,000 m t of sediments containing 5.37 to 8.0 m t of nutrients were eroded from about 8,00,000 km^2 area and deposited in the seas. Ravines and gullies had eaten 10.37 m ha of land (Anon, 2010) at the rapid rate of 8,000 ha/yr (D. C. Das, per. comm., 1985).

Table 7.1. Extent of Erosion of Different Types (After Anon, 2010)

Type of Erosion	Area affected
Loss of topsoil by water erosion	13.25 m ha
Gully erosion and ravine formation	10.37 m ha
Loss of topsoil by wind erosion	5.65 m ha
Terrain deformation by wind actions	3.24 m ha
Forest area under accelerated erosion primarily due to jhuming	4.36 m ha
Total	45.37 m ha

In the mountainous realm of the Himalaya, the rivers are eroding the catchment areas at much faster rate—more than 6 mm/yr in the middle populated Nepal (Arita and Ganzawa, 1997). In the valley of the Marsyandi River in Central Nepal, the rate of erosion is 5.1±1.2 mm/year and the sediment yield 14,000±3400 tonnes/km^2/year in the tectonically active Main Central Thrust zone of high relief (Garzanti et al., 2007). The rate of erosion, however, varies—it is 1.6±0.6 mm/year in the low-relief Lesser Himalaya and 2.49±0.9 mm/year the high Manaslu Granite area in the Great Himalaya. The

Ganga alone at Kolkata carried annually, in the early 1980s, nearly 411 m t of material (328 m t in suspended state and 83 m t as dissolved chemical load), implying erosion in the catchment at a rate of 549 t/km^2/yr (Abbas and Subramaniam, 1984). Compared to the Himalayan domain, the rate of erosion in the stabler terrain of the Peninsular India is a much slower—the laterite-capped Deccan plateau was worn out at the rate of 40 t/ha /yr (Kaul and Das, 1982) in the early 1980s.

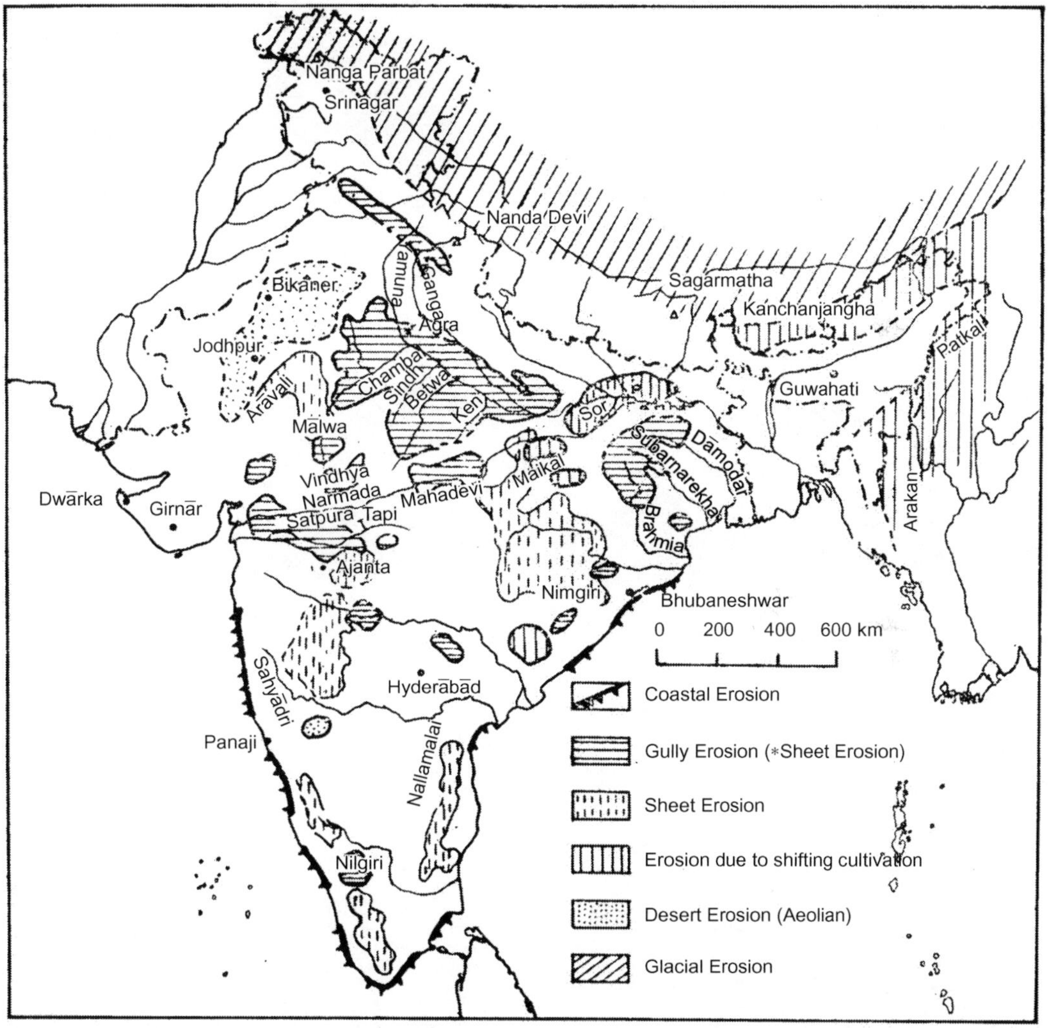

The territorial waters of India extend into the sea to a distance of twelve nautical miles measured from the appropriate base line. Responsibility **for** the correctness of internal details shown on the map rests with the publisher. Based upon Survey of India map with the permission of Surveyor General of India Government of India Copyright 1986.

Fig. 7.1. Types and extent of erosion in the Indian subcontinent. (Based on D. C. Das, 1977; and R. K. Kaul and D. C. Das, 1982).

Table 7.2. India's Land Degraded by Water Erosion (After Anon, 2010) [Source: National Bureau of Soil Survey & Landuse Planning]

State	Areal extent (000 ha)	Area percentage
Uttar Pradesh	12,884	54
Madhya Pradesh	13,465	44
Karnataka	7,799	41
Jharkhand	3,181	40
Andhra Pradesh	8,864	32
Meghalaya	706	31
Assam	2,366	30
Maharashtra	8,822	29
Rajasthan	8,632	25
Odisha	3,328	21
Delhi	28	19
Uttarakhand	1,009	18
Chhattisgarh	2,422	18
Himachal Pradesh	984	18
Tamil Nadu	2,134	16
West Bengal	1,264	14
Bihar	1,049	11
Jammu & Kashmir	2,001	9
Tripura	74	7
Manipur	150	7
Haryana	303	7
Panjab	302	6
Gujarat	1,011	5
Arunachal Pradesh	380	5
Kerala	117	3
Nagaland	31	2
Sikkim	2	0
Goa	1	0
Mizoram	0	0
Andaman & Nicobar Islands	0	0
Total	**83,310**	

Global Perspective

By the late seventies of the last century about 75% of the total soil surface of Pakistan, 80% of Iran and 37% of Israel had been damaged by erosion (Fig. 7.3). In Iraq 2.4 m ha had been adversely affected by wind erosion and 1.6 m ha by water erosion; the Euphrates and Tigris eroded their lands at the rate of 690 and 100 t/km^2/yr, respectively (Zachar, 1982). The Irrawady in the neighbouring Myanmar was eroding the country at the rate of 850 t/km^2 per year. The situation in China was no better (Fig. 7.3). About 15% of the Chinese land was then seriously eroded—the loss of top-soil being 5 billion tonnes per annum (Snil, 1984). The Huangho River used to wash away 920 to 2000 tonnes of soil from each square kilometre of its catchment and cultivated flood-plain. Half a century ago the extent of erod land was 160 m ha (16.67%)—58 m ha in the cultivated part.

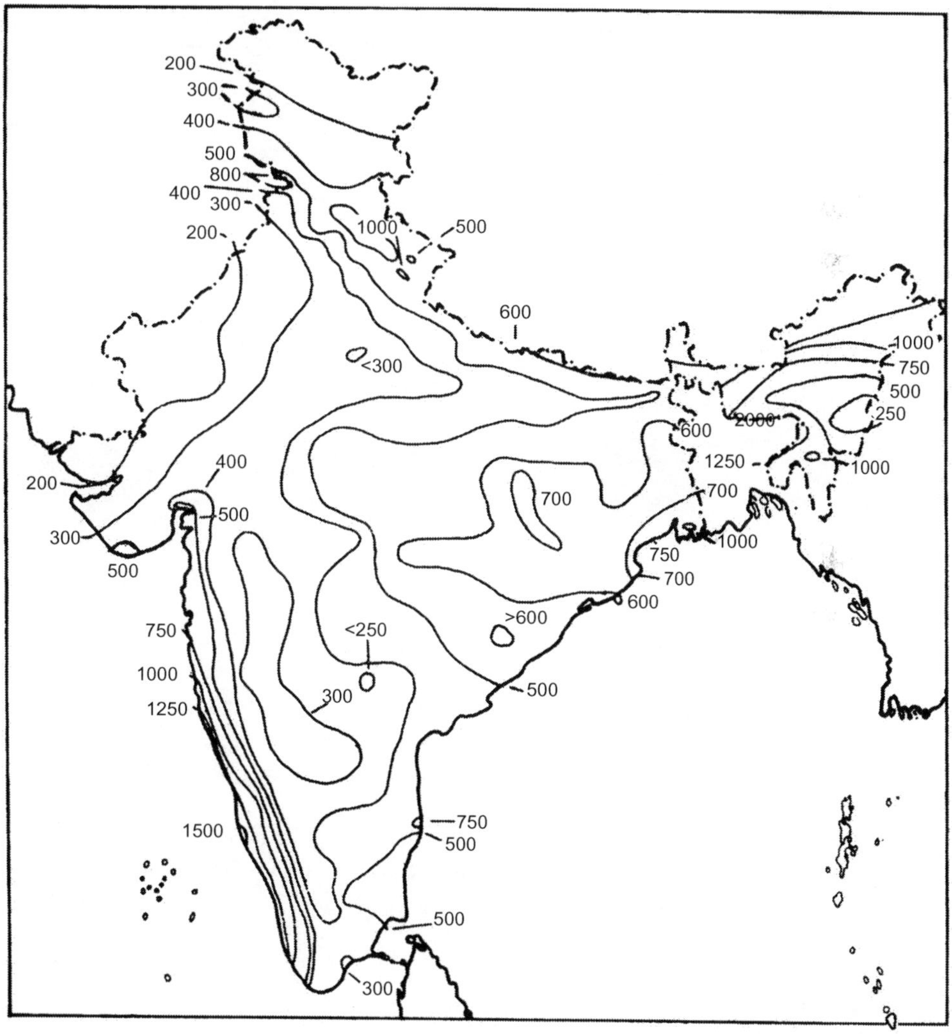

The territorial waters of India extend into the Sea to a distance of twelve nautical miles measured from the appropriate base line. Responsibility for the correctness of internal details shown on the map rests with the publisher

Fig. 7.2. Rainfall erosivity index is a measure of the intensity of average erosion. (The map is after Babu et al, 1979).

In the later seventies of the 20th century, the global rate of water erosion (Fig. 7.3) was between 0.342 mm/yr and 0.407 mm/yr (Zachar, 1982), being highest in the developing countries (0.50 to 0.43 mm/yr) and lowest in Europe (0.06 mm/yr). **The use and abuse of land by humans considerably augmented the pace of erosion. The amount of sediments the rivers used to carry to the seas before man colonized the land was 9,300m t/yr and it rose to 24,000 m t/yr after he started reshaping landscape and exploiting natural resources** (Judson, 1968). In the arid and semi-arid regions, such as western Rajasthan, the rate of wind erosion and attendant deterioration of the land is very serious, primarily because of the indifference of man to his fragile environment.

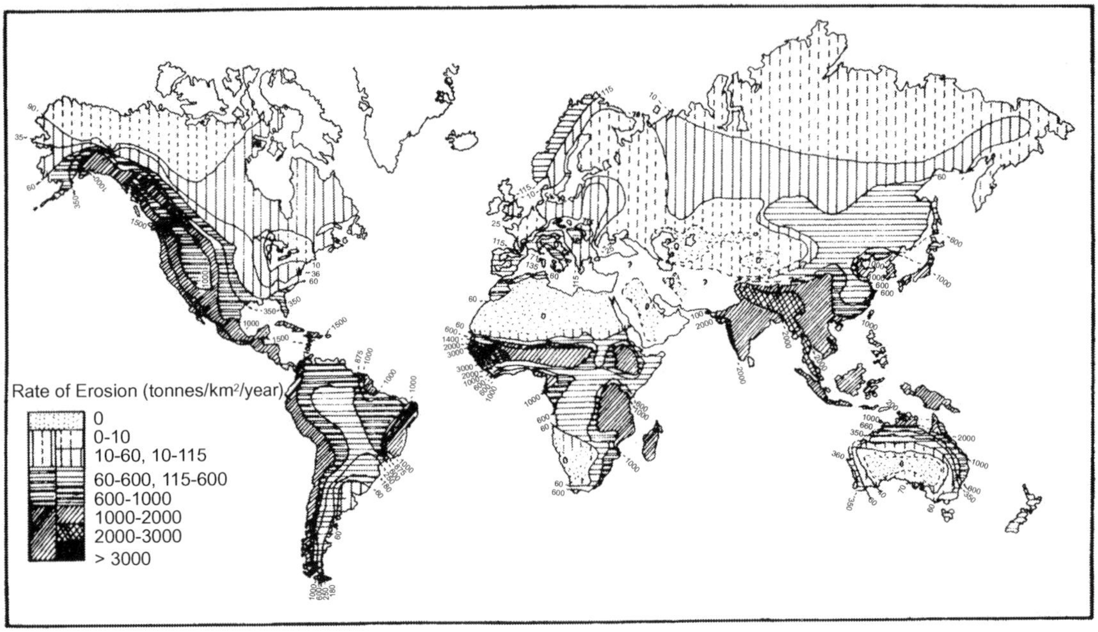

Fig. 7.3. Intensity of water erosion—global perspective. About 80 per cent of the sediments the oceans receive are delivered by the rivers of Asia—the Indian rivers contributing at the rate of 1500 m t/yr in the late 1970s. (From Dussan Zachar, 1982, *Soil Erosion*. Elsevier, Amsterdam, 547 p., Fig. 19.3).

For India one estimate shows that the annual water erosion rate value is less than 5 t/ha/yr from dense forests and from arid regions of western Rajasthan, more than 80 t/ha/yr in the Siwalik Hills and is quite more than 40 t/ha/yr in the ravine lands in the Yamuna, Chambal, Mahi, Tapi and Krishna, and in the *jhum* land in Odisha (Singh et al, 1992).

Impacts of Erosion

The seriousness of the problem engendered by the increasing pace of erosion in the country can be uged from the extent of damage done to our agricultural fields—more than 60% of our farmlands need soil conservation measures more or less, the damage by erosion taking a toll at the rate of 7 to 9 m ha each year. This is on top of 22.44 m ha of land rendered unproductive due to waterlogging

(6.41 m ha), alkalinization (9.30 m ha) and salinization (6.73 m ha). **The country is thus losing land at an alarming rate. Admittedly, the erosion is not noticeable within a shorter span of time, i.e. not until the ground is eroded to the depth of the bedrock or infertile subsoil—when productivity abruptly declines. This may take several years, keeping the farmers blissfully unaware of the lurking peril. This has been demonstrated in Chapter 4.**

The second major impact of land erosion is the passage of great volumes of sediments through the channels of rivers and streams. Not all the material eroded is carried away to the oceans, rather quite a part of the sediment load is deposited in the channels and on the floodplains—nearly 70% sediment load in the case of the Ganga–Brahmaputra (Goodbread and Kuehl, 1998). Locally the rate was quite high—98.20 ha m/100 km^2/yr in Tista valley (Sikkim) and Sukhna valley (Chandigarh), 31.40 ha m/100 km^2/yr in Pagladiya (Bhutan-Assam) and 20.09 ha m/100 km^2/yr in Mayurakshi valley in Bengal (Das et al., 1981). The reservoirs that have been formed by damming up rivers have become repositories or traps of the sediments. In the early 1970s the rate of accumulation of sediments in different man-made reservoirs (Chapter Seventeen) was 2.5 to 18 ha-m/100 km^2/yr area of the catchment (G. P. Gupta, 1975), which was three to five times greater than stipulated at the time of project planning. the reservoir behind the Tehri Dam on the Bhagirathi Ganga, for example, 146 ha m/100 km^2 (110 acre feet/100 mi^2) of sediment is annually depositing (Pant et al., 1968). The major portion of the sediments encroach upon the live storage space, thus drastically reducing the lifespan of the reservoir. About 62 per cent of the live storage space of the Nagarjunasagar had been filled up with sediment by 1974 (G. Gupta, 1975). Due to silting up of reservoirs, the country was then losing irrigation potential at the rate of 60,000 ha/yr (Agrawal et al., 1982). **As a result of filling of reservoir with sediment the effect the potential generation of electricity would be considerable.**

Table 7.3. Sediments Carried by Some Selected Rivers (Data from Abbas and Subramaniam, 1983, Subramaniam, 1987, 1993)

Drainage Basin	Annual Load (m t/yr)			Rate of Erosion in Catchments t/km^2/yr
	Dissolved	Suspended	Total	
Ganga	84	329	413	549
Yamuna	17	125	142	392
Brahmaputra	51	597	648	953
Son	6	50	56	805
Gandak	9	24	33	523
Godavari	17	170	187	610
Krishna	10.4	1.74	12.14	48
Kaveri	3.5	0.04	3.54	40.5
Huangho	22	1080	1102	1432
Irrawady	90	285	375	873
Mississipi	131	210	341	104
Mekong	59	160	219	274
Amazon	287	900	1187	192

In the floodplains of northern India, the rate of alluvial sedimentation is 0.25 to 0.33 mm/yr in the 'doab' of Ludhiana in Panjab (Naruse, 1977) and much higher in the northern parts of Uttar Pradesh and Bihar. Although some of the sediments brought down by rivers do contain organic nutrients derived from the denuded forestlands and croplands, the material derived from landslides, ep gully erosion and debris flows from road and mine excavations are deficient or even devoid of ything that would contribute to the fertility of the floodplain soil. Their deposition indeed adversely fects the productivity of the farms on the floodplains.

The accumulation of the sediments within the channels have diminished their carrying capacity. For example, the bed of the Puthimari River, a tributary of the Brahmaputra coming from Bhutan, was rising at a rate varying from 0.5 to 1.35 cm/yr, the bed of the Brahmaputra had itself risen by 1 metre in some parts by early 1980s (Samar Singh, 1982). The Brahmaputra channel aggregated 16 to 21 cm in just eight years from 1971 to 1979 (Goswami, 1985). The Rispana stream in Uttarakhand is a wide rennial stream flowing past Dehradun in a rock-filled, boulder-strewn channel, its surface discharge declining as its channel widened and vanishing in its own deposits after a few tens of metres (Haigh, 1984). The beds of the Kosi and other rivers in the south-eastern foothills of Nepal were rising at the e of 15 to 30 cm/yr. The channel of the Bagmati River, for instance, is being filled at the rate of 9.3 cm/yr (Sinha and Friend, 1994). **The rise of the river bed is responsible for reduction in the carrying pacity of the channels, and consequent recurrence of floods, greater extent of damage, and ayward shift of the river courses during floods,** as exemplified by the Kosi which shifted 167 km westward between 1838 and 1964. The bed of the Huangho River of China had risen 12 to 15 m over tretch of 640 km resulting in floodwaters breaking the intricate systems of dykes and embankments (Zachar, 1982). There were 973 major breaks in 2100 years upto 1950. In the nineteen seventies when the sediment load was 1.6 billion tonnes/yr, the bed rose at the rate of 12 cm/yr (Snil, 1984).

The shallowing of channels make the rivers difficult or unfit for inland navigation. The Hooghly ver flowing past Kolkata is so silted up that vessels cannot negotiate upto the port without release of additional water from the Farakka barrage.

PROCESSES OF EROSION

Water Erosion

In the absence of vegetal cover, rainfall initiates a set of processes—rain-splash erosion, sheet erosion and gully erosion. The processes are accelerated if the topography is steeply sloping, the soil has been loosened by activities of humans, and precipitation is heavy and prolonged. The falling rain-drops possess considerable kinetic energy and loosen the soil and initiate erosion by dispersing its aggregates and splashing them around. In the absence of intercepting trees, shrubs and grasses, the ***rain-drop erosion*** (rain-splash erosion) assumes serious proportions as billions of drops beat the soil surface and ash the particles downslope (Fig. 4.17). This kind of erosion is pronounced in agricultural fields without crops, in clear-felled forests, on road-cuttings, and naked slopes (see Chapter 4).

When the rainfall becomes very heavy or when the soil has lost the capacity to absorb water, the nwater accumulates on the ground and then flows down the slope in irregular sheets rather than as threads of water. As a consequence, the depth and the velocity of water progressively increases downslope as more and more waters mingle together. The intensity of ***sheet erosion*** depends on the epness of the slope (gradient) and on the volume of water flowing. The sheets denude the soil layer more or less uniformly, making it thinner and thinner. In contrast, the threads or tenors of water may get separated and carve out tiny ephemeral channels or rills down which flow gathering volumes water, giving rise to gullies. The gullies vary in width and depth from tens of centimetres to a couple of metres. They may be continuous, evolving on the ground-surface as a result of concentration of overland flow, discontinuous and particularly associated with subterranean concentration of flow in soil pip caused by local lowering of the water table. The ***trenched gullies*** are the product of rejuvenation as a result of changes in the hydrological balance (Haigh, 1984b) or due to uplift of the ground. Vigorous and continued ***gully erosion*** gives rise to gorges and canyons. The ramification and merging togeth of gullies in unmanaged terrain lead to development of ***badlands*** or ***ravinelands*** (Fig. 7.4), particularly where the rocks are softer and easily erodible. Examples of this type of erosion are seen in the Chambal ravineland in north-western Madhya Pradesh and adjoining Rajasthan, where the ravines had spread over 2.4 million ha area, embracing the districts of Bhind, Morena and Gwalior of Madhya Pradesh (Kaushal et al., 1975) and 0.125 million ha in Kota district in Rajasthan (Khybri, 1983). Until about the early 1980s the rate of expansion of the badlands was 809 ha/yr. The ravines of the Yamuna valley in south-western Uttar Pradesh then covered 0.39 million ha (Prajapati et al., 1975). The valleys of Sabarmati, Mahi and Watrak in Gujarat are, likewise, characterized by development of ravinelands. In eastern India the catchments of the Mayurakshi. Kangasabati and Gumti have been subject to serious gully erosion (Table 7.4).

Table 7.4. Extent of Gully-Eroded Land in India in the early 1960s (Data from S. P. Chaudhuri, 1966)

State	River Valleys and/or District	Areal Extent (ha)
Uttar Pradesh	Yamuna, Chambal, Betwa in Agra, Etawa and Jalaun	1,23,000
Madhya Pradesh	Chambal, Asan, Betwa, Kali Sind in Gwalior, Bhind and Morena	2,43,000
Gujarat	Sabarmati, Mahi, Watrak, Meshon, Banas, Narmada in Bharuch and Panchmahal	4,00,000
Rajasthan	Western districts	2,80,00–3,20,000
Jharkhand	Damodar in Palamau, Ranchi, Hazaribagh, Santhal Pargana	6,00,000
Bengal	Purulia	4,000
Panjab-Haryana		1,20,000

In the unconsolidated soil constituted of rather loose aggregate of particles, the rainwater infiltrat easily and uninhibitedly through small openings and intergranular pores, washing down silts and clays. The result of continued activity of interstitial flow is the widening of these openings and the developme of a network of tunnels which eventually collapse so that the ground breaks up chaotically into badland. This process, witnessed in dryland alluvium and finer deposits is called ***piping*** or ***tunnel*** erosion. Local lowering of the water table increases the hydraulic gradient and thus initiates piping in soils that contain planes of structural weakness such as shrinkage cracks (Haigh, 1984 b). The solution of calcareous layers in the sediment in the Chambal valley is stated to have promoted piping (Sharma, 1980).

The development of ***karst topography*** in the terrain of carbonate rocks, such as the Shali belt in Mandi district (Himachal Pradesh), the Krol belt in Mussoorie-Nainital Hills (Uttarakhand), the Raipur (Vindhyan) region in Chhattisgarh and so on, is a phenomenon resulting from piping and internal solution of carbonate rocks. As a result of underground dissolution of carbonate rocks such features

as tunnels, caves and caverns are formed. The collapse of roofs of the caves and cavities gives rise *swallow* ***holes***— depressions that may or may not be filled with water. In the extreme case, the entire terrain breaks up into an inpassably rugged and desolate terrain.

(a)

(b)

Fig. 7.4. (a) The Mahi valley in Gujarat is characterized by badland, formed due to ravines and gullies. (Courtesy: L.S. Chamyal, M.S. University, Vadodara). (b) Gully erosion at the foot of the Nandidurga Hill, north of Bangaluru (From K.S. Valdiya, 2001). (See colour Fig. 7.4)

Intensity of Water Erosion

It is well-known to geologists that the intensity of erosion depends on both volume and velocity of water. **There is fourfold increase in the eroding ability when the velocity doubles (while the carrying capacity increases eight times).** The kinetic energy in channels may rise upto 2500 times greater than the kinetic energy of the sheet runoff. This explains why the intensity of erosion is greater in gullies and ravines.

The Universal Soil Loss Equation of Wischmeir and Smith (Dunne and Leopold, 1978), with suitable refinements, provides the measure of intensity of erosion—the amount of soil removed from the origin position:

$$A = RK\ LSCP$$

where A = Soil loss (t/ha), R = Rainfall erosivity index, K = Soil-erodibility index, L = Factor of the hillslope, C = Cropping management factor, P = Erosion control practice factor.

The erosivity of rainfall is best expressed as the product of the kinetic energy of the rain-storm and its maximum thirty-minute intensity divided by one hundred (Dunne and Leopold, 1978).

While there is a good correlation of this quotient with measured soil losses at Dehradun, the relationship is weak in the case of Udagamandalam in the Nilgiri Hills. Nonetheless, the index is the best available tool for predicting the geographical variation in climatic erosivity as demonstrated by Babu et al. (1978) (Fig. 7.2).

According to Zachar (1982), the rate of removal at 4 m^3/ha/yr is very critical, for at this intensity of erosion in cultivated fields the seeds are washed away and the nutrients of the soil deplete Erosion at the rate of 15 to 50 m^3/ha/yr means complete removal of the topsoil (Table 7.5).

Wind Erosion in Drylands

The process of erosion by wind has been dealt with at length in Chapter 8. Wind erosion involves ***deflation*** or carrying away of finer particles of soils and abrasion of coarser particles (particles bigg than 0.1 to 0.15 mm being not removable by wind). The wind erosion depends on the wind velocity, the cohesion of the soil aggregates and the moisture in the soil. It thus works selectively, resulting in segregation of coarse skeletal drift sediments deprived of finer particles, the latter carried far and wi (Fig. 7.5). Further, if in one place the soil surface gets dessicated and its organic substance destroyed as a result of the passage of dry wind, at another the soil with its crops could be buried under the heap of blowing sediments. **Wind erosion is thus doubly damaging.**

The rates of erosion during dust storms exceed 1000 t/ha. The March 1901 dust storm over the Sahara carried 19,60,420 tonnes of dust far and wide over Europe, forming a 0.25 mm thick layer of dust (Zachar, 1982). Today nearly 20% of the continents of the earth are threatened with wind erosion of one kind or another. About 8.89 million ha of the Indian land is affected by wind erosion (Anon, 2010), particularly the drier tracts of Rajasthan and adjoining Gujarat and Kachchh. In the early fifties of t 20th century it was felt that the desert was marching towards Haryana and Uttar Pradesh at the rate of 0.8 km/yr. However, the advance of the desert sands has been checked even though dust storms and resultant erosion have seriously affected the fertile lands of the Indo-Gangetic Plains (see Chapter 8).

Table 7.5. Annual Sediment Yields in Indian Himalayan Catchments at Different Locations (From: Omvir ingh, 2012)

Basin	State	Station	Sediment yield (1 km^{-2} yr^{-1})
Spiti	Himachal Pradesh	Khab	1334
Satluj	Himachal Pradesh	Khab	192
Kinnaur	Himachal Pradesh	Wangtu	1597
Giri	Himachal Pradesh	Dadahu	957
Chenab	Himachal Pradesh	Tandi	371
Chenab	Jammu and Kashmir	Benzwar	1597
Chenab	Jammu and Kashmir	Premnagar	1363
Chenab	Jammu and Kashmir	Dhamkund	1900
Chenab	Jammu and Kashmir	Ghousal	513
Chenab	Jammu and Kashmir	Tilar	373
Chenab	Jammu and Kashmir	Sirshi	939
Chenab	Jammu and Kashmir	Kuriya	878
Nubra	Jammu and Kashmir	Siachin Glacier Snout	707
Bhagirathi	Uttarakhand	Tehri	1946
Bhagirathi	Uttarakhand	Dokami Glacier Snout	2800
Gaula	Uttarakhand	Kathgodam	3703
Nana Kosi	Uttarakhand	Nana Kosi	200
Khulgad	Uttarakhand	Khulgad	186
Dhauliganga	Uttarakhand	Markura	461
Chauliganga	Uttarakhand	Reni	418
Dhauliganga	Uttarakhand	Tapoban	670
Dhauliganga	Uttarakhand	Rishiganga	419
Mamlav	Sikkim	Riniikhola	616
Ganga	Uttarakhand	Haridwar	178
Ganga	Uttar Pradesh	Kannaur	91
Ganga	Uttar Pradesh	Allahabad	718
Ganga	West Bengal	Farakka	1235
Ramganga	Uttarakhand	Kalagarh	424
Yamuna	Haryana	Tajewala	2178
Yamuna	Delhi	Delhi	1075
Yamuna	Uttar Pradesh	Allahabad	308

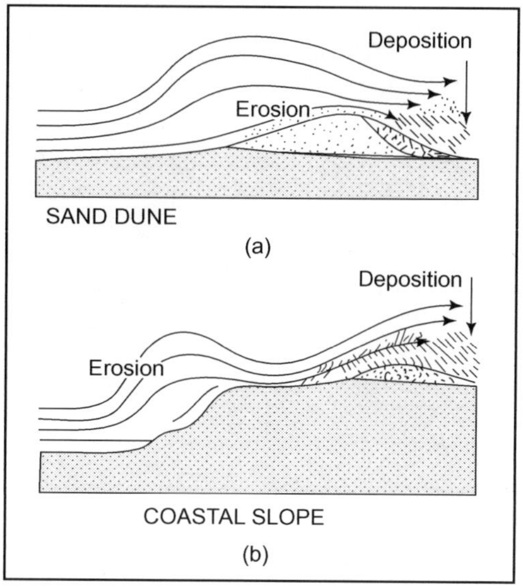

Fig. 7.5. The mode of operation of wind erosion in deserts and along coasts. Wind works selectively, carrying only the finer particles and leaving behind a skeleton of coarse drift sediments deprived of nutrien (After Fig. 30.11 W. M. Marsh and J. Dozier, 1981.)

Coastal Erosion

The sea coasts are ceaselessly and heavily pounded by waves, and exposed to the rigorous action of coastal current. The erosion is pronounced during storms and cyclones, which possess enormous amount of energy (Fig. 7.6). This is evident from an incidence that occurred in a harbour in 1972 when a block weighing 1371 tonnes was lifted up and placed intact high up inland. Together with the force exerted by waves, there are jet impulses, viscous drags and hydrostatic pressures. Along the straight coasts the breakers strike obliquely, giving rise to longshore currents which flow parallel to t beach. **When high tides combine with high waves, the erosion becomes very severe especially on headlands and promontories.** The translation waves reflected from the cliffs exert six times as mu pressure on the walls as oscillation waves of the same size (Nayak, 1980). And when the breakers strike the coast, jets of water are thrown up. Being two times faster than the waves, these jets cause great damage.

On the other hand, there is simultaneous deposition of sediments on the downdrift side of the headland promontories. Jetties and other artificial impediments to the flow of drift currents (Nayak, 1980) likewı cause erosion on one side and deposition on the other (Fig. 7.6).

In the mid 1970s while the north-easterly littoral drift carried about a million tonnes of sediments, the south-westerly accounted for only 0.25 m t so that the net drift in the north-eastern direction along the coasts was 0.75 m t (Das, 1977). The entire coast of the Western Ghat and some parts of the Eastern Ghat (50,000 to 80,000 km^2) are subject to varying degrees of erosion as evident from the development

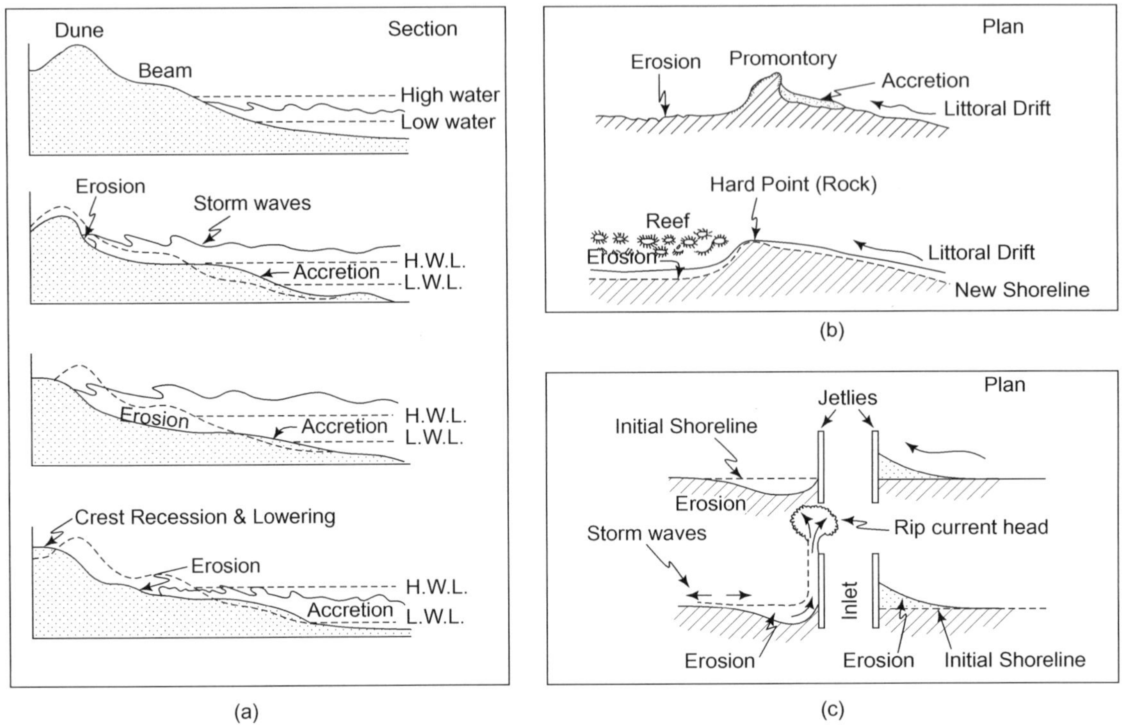

Fig. 7.6. The breakers strike the coasts obliquely, giving rise to longshore currents flowing parallel to the coasts, which are particularly active on promontories, depositing sediment on the downdrift side. Similar phenomena occur on the two sides of the jetties. (Based on B. U. Nayak, 1980.)

of arches, stacks, blowholes, caves, notches, etc. resulting from the destruction of coastlands. An earlier estimate showed that the coastline of the Western Coast was being eroded at the rate of 20 to 30 t/ha/hr. •wever, it is not possible to give specific rates, due to the fact that the human intervention has altered the natural balance (K.V. Thomas, per. com., 2012). **The problem is acute in the coasts of Odisha-Bengal and Kerala** (Figs. 7.7 and 7.8).

The global average rate of coastal recession is 1 to 2 mm/yr. The rate of erosion in the 1970s er and above this normal recession was 0.5 m/yr along the Western Ghat (Brunn and Nayak, 1980). In the 70 years period of the last century, from 1910 to 1980 about 100 to 400 m of coastal nd in Kerala was devoured by the sea at the rate of 3 to 5 m/yr (Figs. 7.8 and 7.9). If the time span is stretched to 120 years, the extent of loss of land would be upward of 600 m (Raju and Raju, 1982). **The principal cause of erosion is possibly the shortage of sediment supply (drift) in the longshore current as a result of trapping of riverine sediments in reservoirs behind dams (Baba, 1979).** The spent energy of the waves deficient in sediments is utilized in eroding the coasts.

The Kannur and Moplay bays in Kerala, Kanyakumari and Pudimadakam water fronts in Tamil Nadu, and Waltaire and Machhlipatanam **coasts in Andhra Pradesh are experiencing erosion primarily**

due to the excavation and reshaping done, or due to the removal of beach sands as construction material. Dredging for and construction of jetties in the harbours of Kochi and Vishakhapatnam have likewise triggered erosion on the coasts (Brunn and Nayak, 1982). *The Coastal Vulnerability Index Atlas* prepared by the Indian National Centre for Ocean Information Service, Hyderabad (2012) shows that a large extent (~ 64 km) of Kancheepuram district and parts of the Chennai district in Tamil Nadu and 37 km stretch of the Andhra Pradesh coast between East Godavari district and Vishakhapatnam district are very high-risk areas (*The Hindu,* May 18, 2012).

The Digha coast (Fig. 7.10) in Bengal had been retreating since it was first investigated in 19 (Niyogi, 1970). Coastal Zone Management Project report shows that 67.9% of the entire coastline of Bengal is being eroded (*The Hindu,* December 25, 2011).

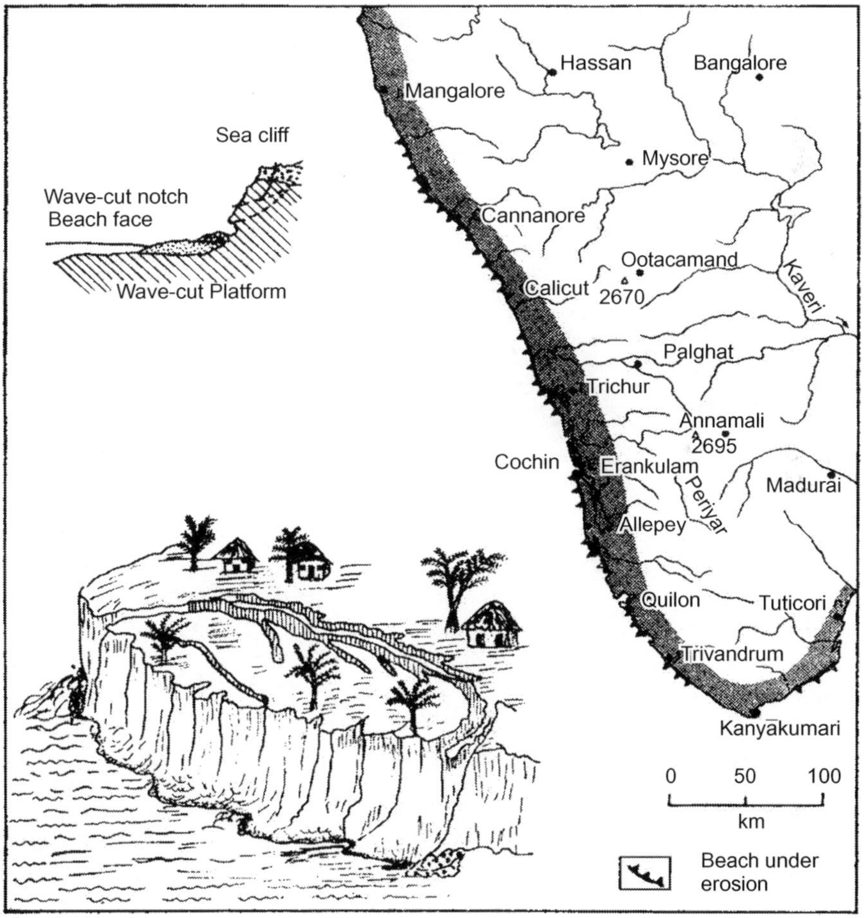

Fig. 7.7. Practically the entire Kerala coast is prone to marine erosion. Undercutting by waves and currents and collapse of coastal cliffs and beaches is responsible for the inexorable retreat of the coastline. (Map after D. V. Raju and K. C. C. Raju, 1982.)

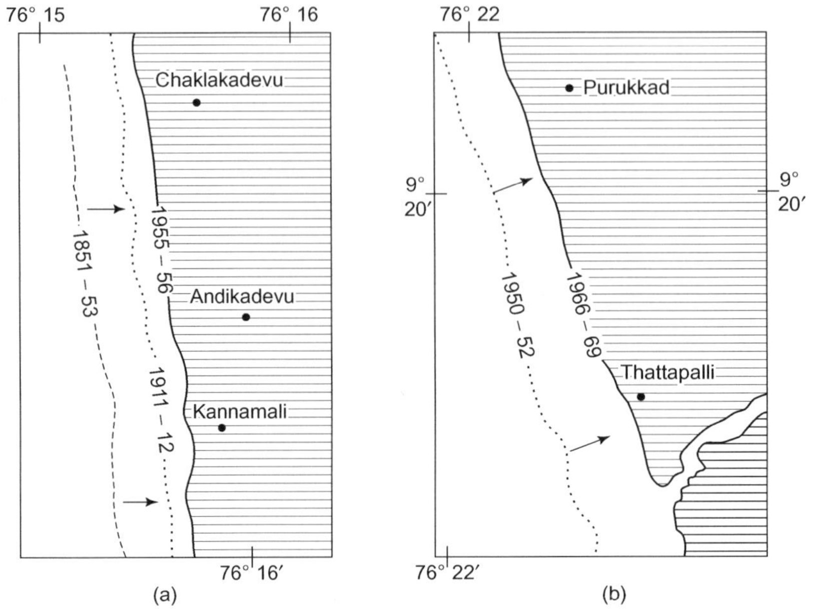

Fig. 7.8. In the period of 70 years from 1910 to 1980 between 100 and 400 m strips of the Kerala coast disappeared into the sea. (After D. V. Raju and K. C. C. Raju, 1982.)

CAUSES OF ACCELERATED EROSION

Mass movements like the slow downslope movement of soil material (creep) and the rapid slide of debris and rock masses (landslides) cause destruction of the slope and ground surface, initiating a chain of uncontrolled erosion in the mountain terrain. Heavy rainfalls or cloudbursts and rapid melting of snow in higher altitudes contribute considerably to destabilization of the slopes. **Excavations for mining, ads, canals, buildings, etc. have added greatly to the mass movements and intensification of erosion in the hilly terrains.** As a result of various processes of erosion more than 45.37 m ha of the land of India has degraded to varying orders (Table 7.1). Some of the major causes of erosion, for which people are responsible, are discussed.

Overgrazing in Forested Areas

The pounding and trodding by hoofs of grazing cattle not only destroy the vegetal carpet but also disaggregate the soil mantle. Cattle tracks (Fig. 7.11) later grow into gullies. **The greater the grazing stock, the more intense would be the erosion.** As already stated, the population of livestock in this country is growing as explosively as the human population, and the phenomenally mounting pressure of grazing is responsible for severe damage not only of the forests, but also of the soil. The trees are defoliated, the nutritious elements of the pastures devoured, and the once lush hillsides become extensively furrowed. Moreover, the soils of the overgrazed area are left with only 8 to 36% clays (compared to 50 to 69% in the unaffected soils of the area where entry of cattle is restricted) so that e stability is impaired owing to the deficiency of binding material (Haigh, 1982). The clays also hold trients. Their deficiency thus diminishes the productivity of the soils.

Fig. 7.9. Rough sea waves lashing against the beach and eroding away road and coconut plantations Kerala coast. (Courtesy: Dr C.P. Rajendran, Centre for Earth Science Studies, Bangalore) (See colour Fig. 7.9)

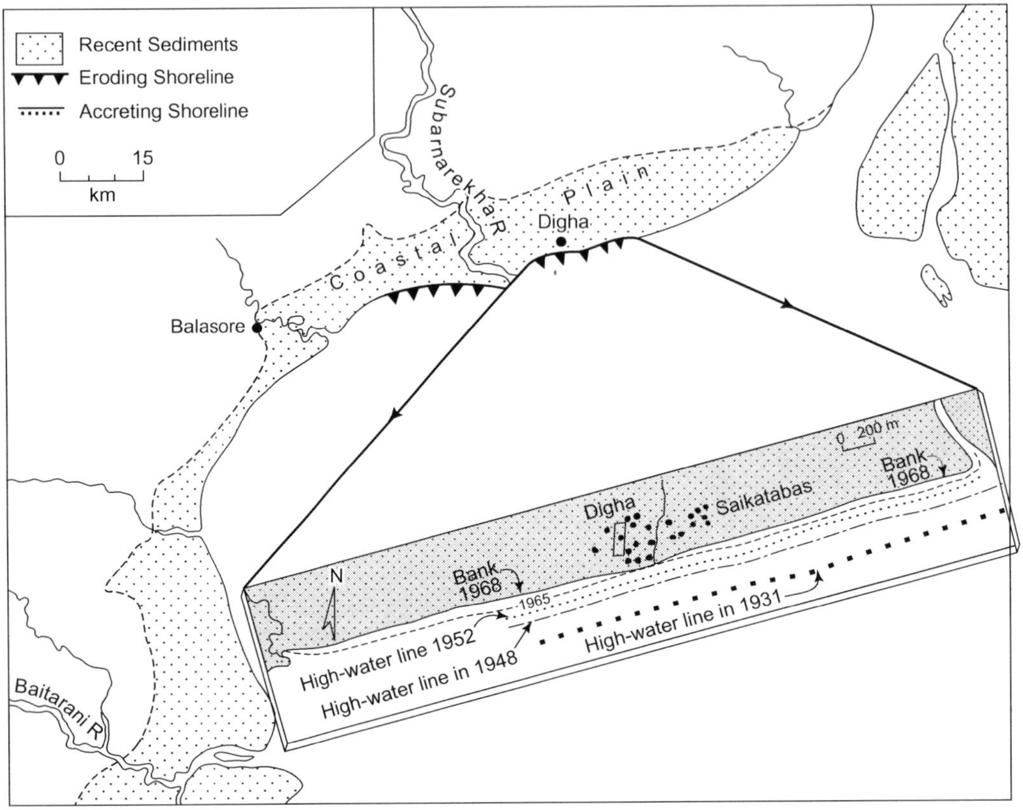

Fig. 7.10. Since 1931 until about 1968 the Digha coast of Bengal was retreating as a result of beach impoverishment. (Based on D. Niyogi, 1970).

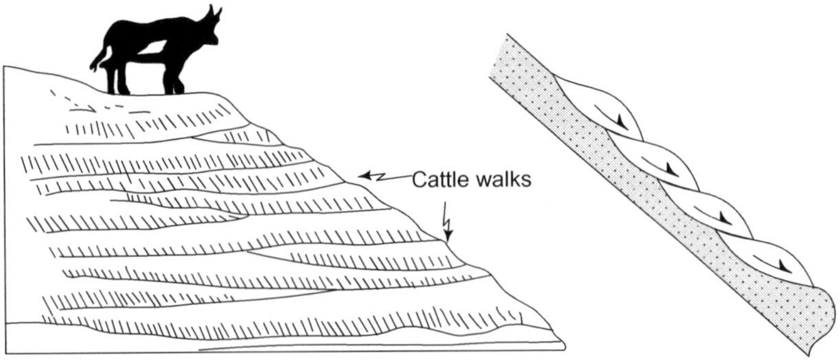

Fig. 7.11. Pounding by hooves of cattle is responsible for the formation of cattle tracks which eventually develop into gullies and initiate a cycle of erosion. The inset to the right shows the mini landslides caused by pounding of cattle hooves. (From J. P. B. Mears, 1970, *Essentials of Geology*, Van Nostrand, New York, 420 p.)

In an experimental farm at Chander in Jaisalmer district (Rajasthan) after the removal of grass cover **from a 60 m by 40 m patch, the wind blew away soil upto a depth of 50 to 60 cm in the first three yea** (Mann and Prakash, 1983). The runoff and consequently the erosion in overgrazed grassland in Western India was very high in the early 1980s—respectively 27% and 1050 kg/acre (Mann and Prakash, 1983). **The arid land of Rajasthan (Thar) has expanded primarily due to the pressure of grazing. The livestock population here is very high—there were 25.5 units of cattle trampling and defoliati every hectare of pasture land (R. K. Gupta, 1980).** The livestock population in 2003 was 49.14 million (Chaudhry et c., 2011). Today it is much more. This is one of the reasons why the once verdant land fed by the ancient Saraswati River became a vast growing bowl of sands. A similar fate had overtaken the once green belts of Palestine, Israel, Libya and Algeria which today are deserts.

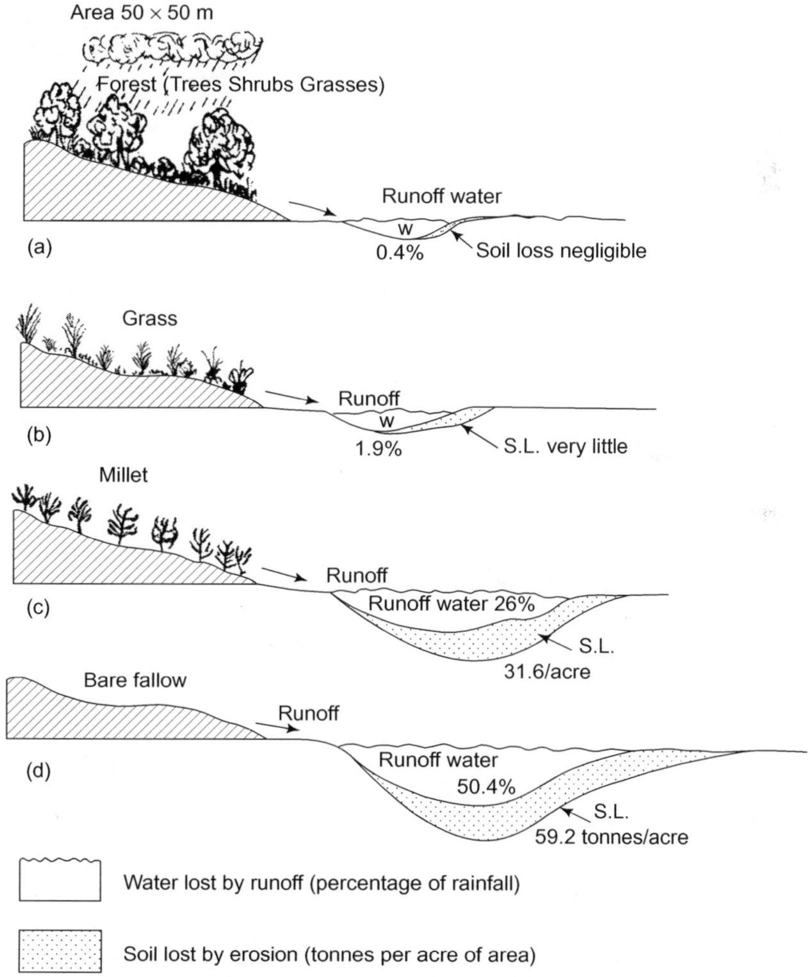

Fig. 7.12. The accentuation of erosion intensity as a result of agricultural practices, measurable in terms of the volumes of sediments carried by streams and runoff water.

The incidence of grazing in the Himalayan state of Uttarakhand was 2.5 to 4.5 times higher than the carrying capacity of the forests (J.S. Singh and Saxena, 1978)— a situation in the 1970s then comparable with that in Iraq (four times) and Syria (three times). The result of this excessively high grazing pressure is an alarming reduction (in that time) in the protective vegetal cover-dwindling to a mere 16.8 % at an altitude of 763 m, 24% up to a height of 1080 m, 25.8% upto the altitude of 1650 m in Himachal Pradesh and 38% upto an elevation of 1800 m in Jammu (Singh and Saxena, 1980).

Deforestation—Deprivation of Protective Cover

Migrating from the desert that was expanding and becoming inhospitable, the Phoenicians settled down along the eastern coast of the Mediterranean where they built the celebrated townships of Tyre and Sidon. Then followed the rape of the lush cedar forests that covered the hills of Lebanon and Syria. The plunder of the forests by the conquering Romans later converted the whole region into a wasteland.

This disturbing consequence of the destruction of the plant cover was discernible in the Siwalik Hills in Hoshiarpur district (Panjab). As a consequence of deforestation the extent of degraded (bad) land had increased from 195.2 km^2 in 1852 to 2000 km^2 by 1980. One deforested hill alone in the catchment of the stream feeding the Sukhna Lake near Chandigarh used to bring—before conservation measures were taken up—sediments at the rate of 50 t/ha/yr, thus reducing 60% of the lake capacity just in 15 years of its existence. In the Mussoorie Hills, the volume of sediments generated by erosion in the areas under forests was 5.7 times less than that in the area covered with grass and shrubs. Likewise, the gentle slopes covered with 'saal' trees lost 39% *less* **sediments than the adjoining agricultural fields of Dehradun** (Haigh, 1984). Investigations carried out during 1981–1982 in a part of Kumaun (Uttarakhand) indicated that the soil-loss from eight mixed oak-pine forests varied between 14 and 32 kg/ha (mean 24 kg/ha) compared to 37 to 81 kg/ha (mean 57 kg/ha) from nonforested areas including croplands (J.S. Singh et al., 1983).

In an experimental area the burning of forests and clear-felling increased the surface runoff 69% and 34%, respectively (R.K. Gupta et al., 1975). More runoff means greater erosion.

Australia bushfires are known to increase sheet erosion fourfold, and in western California (U.S.A.) the logging of trees and conversion of forests into grasslands in India has increased the annual sediment production from 4 m^3/ha to 24 m^3/ha (Samar Singh, 1982). As stated in Chapter Two arly 53% of the forest of India periodically suffer from forest fires (Singh and Kushwaha, 2008).

The accentuation of erosion intensity due to deforestation (Figs. 7.12 and 7.13, Table 7.6) is tributed chiefly to three causes—destruction of foliage that intercepts raindrops, decay of roots that hold together and stabilize the soil on rocky foundations, and deprivation of humus that binds together soil aggregates. Broad-leaf oaks and multi-storied forests accomplish maximum interception. Coniferous trees generally intercept more water than the hardwoods in the snowy zones at Himalayan heights, and grasses are capable of intercepting 10 to 20% of the precipitation—as much as the trees (Dunne and Leopold, 1978). The pine trees in a Dehradun experimental plantation intercepted 22 to 27% of rainfall while broad-leaf 'saals' as much as 34% (Ghosh et al., 1980). The leaf litter in the Himalayan pine forests are capable of intercepting over 7% of rainfall, while under the broadleaf 'saal' forest over 9%, (Dabral et al., 1963). It is thus obvious that the vegetal litter in the forests provide very effective protection to the ground and prevent more than a third of the total rainfall from reaching the ground. **It is the extent of the canopy of foliage in the forests that is significant, not the number of trees. The larger the canopy the better would be the protection.**

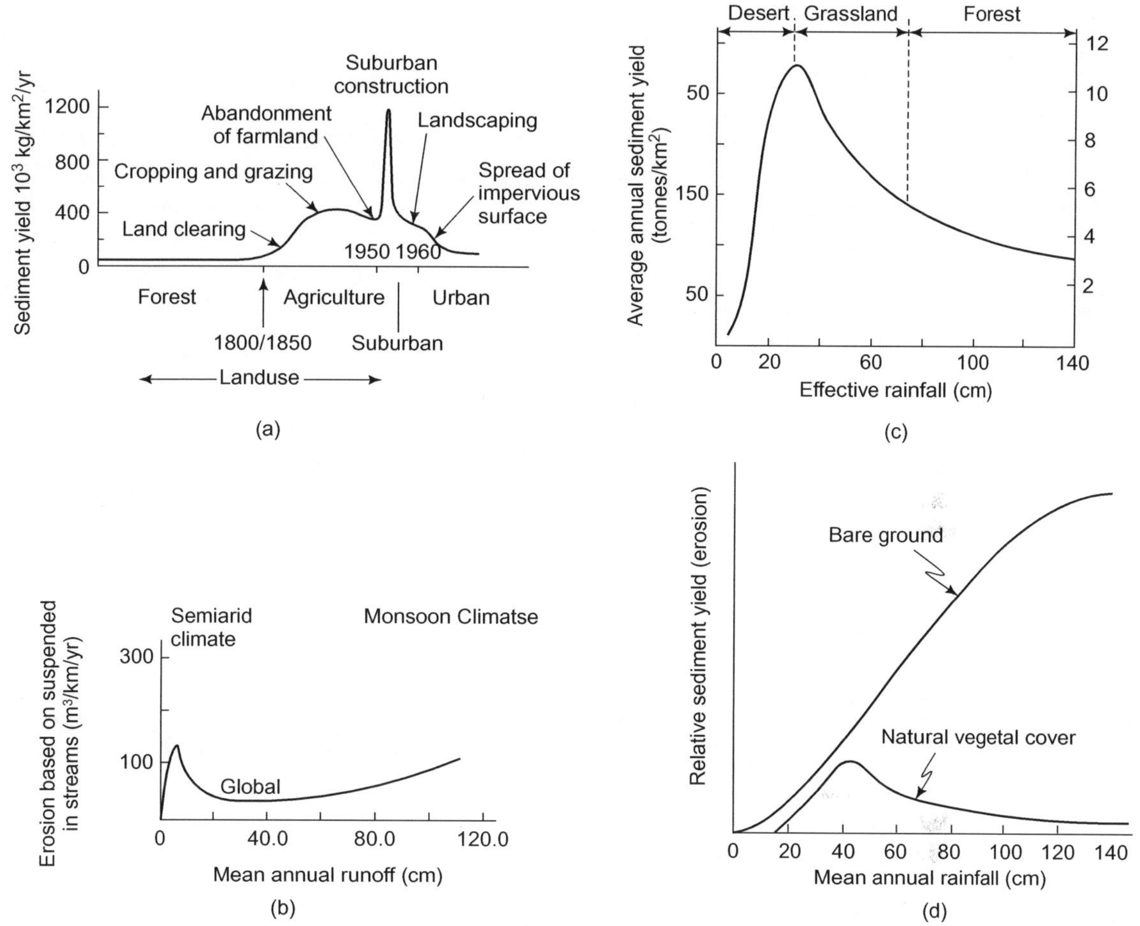

Fig. 7.13. The degree of erosion measurable in terms of the amount of sediment eroded away from t catchment annually under different landuses. (a) From E.A. Keller (1976) and after I. Douglas (1967) and W.B. Langhein & S.A. Schumm (1967). (b) After M.G. Wolman (1967) from W.M. Marsh and J. Dozier (1981), Fig. 26.5. (c) After S. Judson (1968. (d) Anon. (1977), U.N.E.P., 115

In this perspective has one to view the deforestation that is taking place in the Himalaya. The forest cover is reduced to less than a third of the surface area on an average. In the early 1980s it was merely 28.7% in Kumaun (Uttarakhand) where the forests having canopy more than 60% was less than 4.4% (J.S. Singh et al., 1984). Less than 11% of the land surface of our country then had this vegetal protection against erosion by rainwater. As a result more than 45.73 m ha of the land of India is today suffering from various types of erosion (Fig. 7.1, Table 7.1).

The 2011 official report (by Forest Survey of India) shows that in the year 2009 the forest cover (all kinds) was 39.74% in India and 45.80% in Uttarakhand. **However, the very dense forests covered only 5.9% of the Indian land and 8.9% of Uttarakhand respectively. Obviously the land of Ind is scarcely getting protection by its forests against erosion.**

Table 7.6. Erosion in Forests, Clearcut and Roaded Areas in Western U.S.A. (After Swanston and Swanson, 78)

Record	Period of Record (yr)	Area (%)	Number of Sliders	Erosion Rate ($m^3/km^2/yr$)	Erosion Rate Relative to Forested area
Stequaleho Creek, Olympla Peninsula					
Forest	84	79	25	71.8	1
Clearcut	6	48	0	0	0
Road R/W	6	3	83	11,825	X165
Alder Creek, Western Cascade Range, Oregen					
Forest	25	70.5	1	45.3	1
Clearcut	15	26.0	18	117.1	x2.6
Road R/W	15	3.5	75	15,565	X344
Selected Drainage, Coast Mountain, SW British Columbia					
Forest	32	88.9	29	11.2	1
Clearcut	32	9.5	18	24.5	X2.2
Road R/W	32	1.5	11	282.5	X25.2
H.J. Andrews Experimental Forest, W. Coscade Range, Oregon					
Forest	25	7.5	31	35.9	1
Clearcut	25	19.3	30	132.2	x3.7
Road R/W	25	3.2	69	1772.0	z49

Unscientific Agricultural Practices

If the rows of grooves made by ploughing on agricultural lands are not kept parallel to the contour of the hillslope, they become channels for rainwater. The furrows become gullies, and the gullies enlarge to become streams and rivulets. During heavy rainfall the furrows deepen and widen, and a network of gullies develops, eventually giving rise to badlands. The mechanism is simple. The rainwater enters into cracks and crevices, forming tunnels as the water moves, which later collapse to form rills that entually enlarge into gullies. The longer the rains, and less protected with vegetation a field is, the greater is the erosion. Farms are really tilled ground with soft loose soil without protection of natural vegetation. Even if they have vegetation, it is of monocultural crops which do not have enough resilience d capacity for protection as the natural and generally multistratal ecosystems. In the agricultural fields it is therefore natural to have erosion as much as 100 times greater than in grass-covered lands (Figs. 7.12 and 7.13a and Table 7.7).

How indiscriminate ploughing destroyed the once fertile Palestine is well-known in history. The promised land of springs and streams, of wheat and barley described by Moses some 3000 years ago, lestine is now a barren land primarily due to erosion initiated by unscientific ploughing in the direction of the slope without contour-terraces and drainage ditches.

In Nepal and Uttarakhand today farmers, indifferent to the needs of having marginal levees on rraces and ignorant of the imperative need for drainage ditches, are extending their fields upslope, occupying steeper and less stable parts and depriving the watersheds of their vegetal umbrella.

Table 7.7. Soil-loss Related to Different Landuses in India (Data from Vasudevaiah et al., 1965; and Batlaw and Rao, 1962).

Landuse	Soil-loss (t/acre/yr)
Cropland	
Fallow, bare	14.6 to 15.4
Maize	4.7 to 10.54
Hill paddy	1.0 to 9.4
'Arhar dal'	5.5
Cowpeas	1.9
'Urad dal'	21.0
Peanuts	6.2
Grassland	
Hay grass	1.05 to 8.30
Tropical pasture grass	0.20
'Doob grass'	0.47
Common grasses	1.4 to 1.8

The second **extremely unsound and insensitive agricultural practice that has greatly increased erosion in large parts of Arunachal Pradesh, Nagaland, Manipur, Tripura, Mizoram and the hil tracts of Odisha and Chhattisgarh is the slash-burn-and-shift *(jhum)* cultivation.** In a nearly burnt down area, the loss of soil material was found to be 7.24 t/ha/yr compared to less than 0.76 to 1.13 t/ha/yr in the area covered with grass (Ramakrishnan and Tokay, 1974), thus registering a 6.2 to 9.5 times increase in erosion. The soil-loss on the slopes 10 to 85 per cent varied from 3.3 to 204 t/ha/yr. As of now 4.36 m ha of forest land is under accelerated erosion primarily due to *jhum* cultivation (Anon, 2010).

Indiscriminate Road Construction

The amount of sediments eroded away from sites of road construction is ten times greater than the amount derived from cultivated land, and about 200 times greater than from grassy area, and 2000 times greater than from the forested area (Dendy, 1968). This is quite understandable, for the construction of roads disturbs the stability of the hillside, inflicts serious damage to t hydrologic system, and removes the protective vegetal cover from the vulnerable slopes. The disturbances—usually very drastic in nature and tragic in consequence—are the results of clumsily planned and improperly executed excavations—cutting unstable or marginally stable slopes (Figs. 11.5 and 11.6) using high-intensity explosives to blast rocks, felling trees with robust roots that provide a grip to the soil and rocks, dumping loose unprotected debris on steep slopes, not caring for drainage ditches, and so on.

In the Himalayan region states, including those of eastern India, there are more than 2,10,000 km of highways and asphalted roads (*Road Transport Year Book, 2009*, Ministry of Transport & Highways). **More roads are being constructed without regard to geological conditions and sensitivity to t ecological balance. They cut through cones and fans of fossil landslide-debris or scree-fans, thus activating them periodically.** This is seen along the pilgrim routes, particularly between Pakhi and

Gulabkoti on the Badarinath road, and between Bhatwari and Gangnani on the Gangotri route, to give just a few examples. These roads span the geodynamically very sensitive zones of sheared, shattered and crushed rocks of the boundary thrusts and tear faults, where incidence of landslides has increased and intensity of erosion heightened considerably.

The excavation of each kilometre stretch of the road requires removal of 40,000 to 80,000 m^3 (average 60,000 m^3) of debris. Thus the construction of 44,000 km of roads until 1983 in the Himalaya generated about 2640 million m^3 of debris which slid down the slopes, destroyed vegetation, killed springs, and clogged watercourses downslope (Valdiya, 1985). One can estimates of the volume of debris generated and pushed down the slopes by the 210, 000 km long roads in the Indian Himalayan region. After the roads are constructed, they are ravaged recurringly during rains by landslides and rockfalls. In the late 1970s the Border Road Organization was clearing debris from its roads at the rate of 729 m^3/km/yr on the Jammu-Srinagar highway, 411 m^3/km/yr on the Tanakpur-Tawaghat road in eastern Kumaun, and 691 m^3/km/yr from a highway in Arunachal Pradesh (Patnaik, 1980). In other words, at the average rate of 550 m^3/km/yr, the total landslide debris on Himalayan roads would be of the order of 24 million annually. One can estimate the amount of debris generated and is being produced annually by the 2,10,000 km aggregate length of the roads.

Road construction activities are known to increase the occurrence of landslides greatly— 25 to 5 times (Table 7.6). In the early 1970s, along the Dehradun–Tihri road there were 10 medium and small landslides in every kilometre stretch, each of which generated 500 m^3 of debris, and there was one chronic landslide in every 10 km stretch (Bansal and Mathur, 1976). The situation is much the same today.

Accounts of road-related landslides are numerous (Krishnaswamy and Jain, 1975). The Nashiri Slide between Batote and Ramban on the Jammu-Srinagar highway once occured in a highly jointed sandstone-mudstone sequence of the Murree Formation. Over a stretch of 100 m about 4 million m^3 mass of debris moved down following a downpour and attendant erosion by the Nachar stream. The Khuninala slide between Ramban and Banihal developed on multiple-jointed gneisses and schists of the Salkhala Formation. In Uttarakhand, the Kaliasaur Slide, east of Shrinagar on the Rishikesh-Badarinath road, is an endemic phenomenon of incessant movement of exceedingly jointed and shattered quartzites ernating with slates. The toe erosion by the swift-flowing currents of the Alaknanda has considerably aggravated the instability of the hill slope. Further north between Pakhi and Gulabkoti the pilgrim road has reactivated old landslide cones on a massive scale. The situation is worse in the Tista valley in the arjiling-Sikkim region where ten important landslides were identified along the highway in the early 1980s. In 1983 the whole mountain slope came down into the valley killing 125 people on the road in northern Sikkim. In 2010 a stretch of over 20 km road along the bank of the Kosi River in Kumaun was most wholly ravaged by landslides, debris flows and undercutting by the river, following a cloudburst in the catchment.

Migration of Stream Channels in Flood Plains

hen a river is in flood, it often changes its course, or gradually migrates by cutting its bank. Excess sediment load causing silting-up of channels forces the river to spread and erode its banks. In the mountains, increased sediment yield from one bank (e.g., by landslide) can push the channel to undercut the opposite side and thus cause ***toe-erosion*** and more landslides. As a result of erosion and resultant bank recession, the currents cut more directly against the downstream side of the eroded bank so that erosion and bank recession continue uninhibitedly (Fig. 7.14). Until the 1960s the Brahmaputra River, for example, had moved 90 km west in the 200 years period and the bank width had increased from 2.5 km to 10 km in the 144 km-long stretch between Pasighat and Dihingmukh, mainly during falling flo stage when the excess sediment is deposited on bars within the channel which cause change in local flow direction (Coleman, 1969). The 1950 earthquakes caused the riverbed to rise by 3 m at Dibrugar forcing the river to change its path. This resulted in damage of 1000 ha of land covered by tea gardens. In Bangladesh the Brahmaputra changed its course drastically many times—between 1720 and 1830. It changed its original course past Mymansingh through Sylhat area and after joining the Meghana River now goes straight south to join the Ganga. In Bangladesh the banks of Jamuna (the present course of Brahmaputra) river is eroded at the rate of 50 m/yr (Barua, 1991).

The Tista used to be a tributary of the Ganga, but it suddenly changed its path in 1781 and join the Brahmaputra. The notoriously wayward Kosi in northern Bihar migrated 167 km westward in 126 years (1838 to 1964), eroding and damaging vast areas of farmland in the districts of Purnea and Saharsa. In 2008 during very high floods the then westward shifting Kosi River suddenly returned eastward to flo through its abandoned channel in the middle of the fan, bringing wholesale disaster. A recent report (*The Times of India,* March 3, 2011) shows that the river is now drifting eastward, nibbling its bank. Both the Yamuna and the Ramganga in western Uttar Pradesh have been shifting eastward, and the Sindhu westwards, the Yamuna had migrated 10 to 40 km and the Sindhu 160 km in the last 3000 years (Valdiya, 1984, 2010).

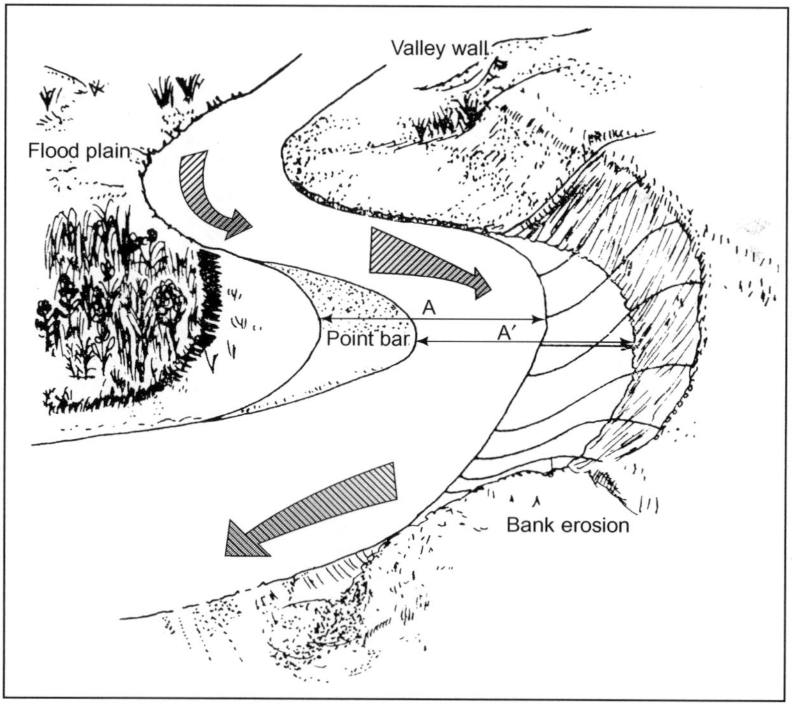

Fig. 7.14. The slow, gradual migration or shift of the course of the river is responsible for streamback erosion in the floodplains.
(After W. M. Marsh and J. Dozier, 1981, *Landscape,* Addison-Wesley, Reading, 637 p., Fig. 27.17).

REMEDIAL MEASURES

Strategy

The strategy for controlling erosion and reducing degradation of land embraces measures that protect it from the actions of wind and running water. Prior to embarking on these measures it would be appropriate to make an assessment of erosion risk—an investigation involving evaluation of what can be called the erodibility of the land or soil on local as well as regional scale. The subcatchment and then the catchment of the river form the unit of erodibility investigation. Scanning the surface by satellite sensors and aircraft-borne cameras is the most effective and reliable way of preparing the erodibility maps. Space Applications Centre of Indian Space Research Organization has done commendable work on the assessment of the land degraded by erosion.

Watershed Treatment

Reduction of loss of soil can best be achieved by covering the watershed with vegetation—multistratal forests with trees, shrubs and grasses. The deteriorated or threatened area is first put under cover of quick-growing plant species and then deep-root long maturing appropriate trees to restore the original condition. In the Chambal ravines in Madhya Pradesh, planting of 'sheesham' (*Dalbergia sissoo*), 'khair' (*Acacia catechu*) and *A. arabica* together with the fodder grass *Dicanthim annulatum* reduced the peak discharge of runoff from 0.27 m^3/second to 0.12 m^3/sec within seven years (Kaushal et al., 1975). This considerably reduced gully erosion, a chronic problem in the Chambal valley.

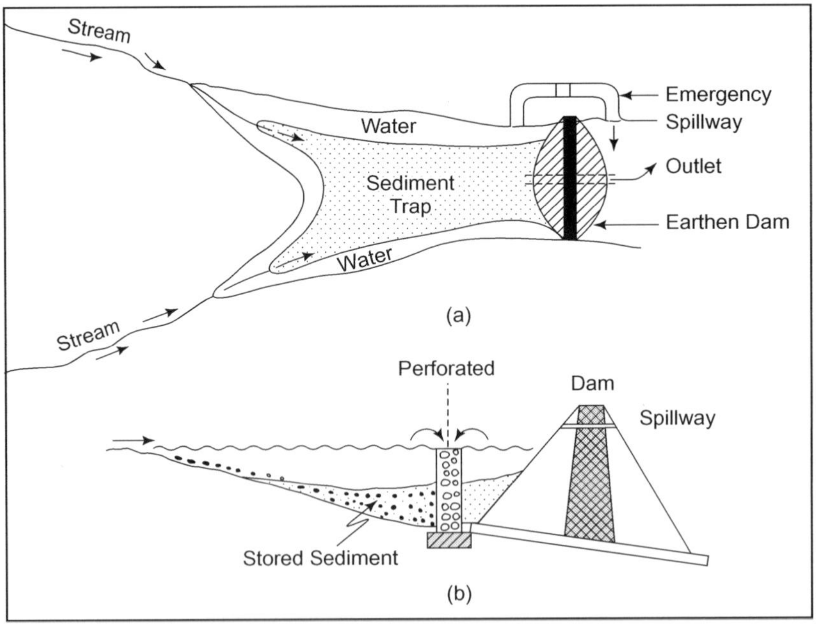

Fig. 7.15. In larger gullies or ravines bigger check dams are built to trap eroded sediments. Perforated pipes or wells serve to drain out water. (Based on Fig. 9.30, and Fig. 9.31, E. A. Keller, 1976).

The land conservation measure in the ravine land involves plugging the gullies by constructing a series of check dams at intervals of one to two metres (Figs. 4.22, 7.15 and 7.16), then levelling the ground with bulldozers, and finally covering it with grass. In the construction of check-dams, use stakes of bamboo, and/or self-regenerating trees like poplar, willow (*Salix tetrasperma*), *Ficus cunia*, 'pāngara' or 'mandār' (*Erythrina suberosa*), *Phragmites kaka*, *Lannea grandis* 'beshram' (*Ipomea carnea*), 'nam' (*Arundo donax*), 'gainthi' (*Boehmeria rugulosa*), 'semal' (*Bombax ceiba*), *Sapium insigne etc.* are made (R.K. Gupta, 1979; Natrajan and Gupta, 1980). The stakes are driven in staggered rows across the gullies which are then partly filled with straw and bush to trap sediments (Fig. 7.16). O bigger gullies larger check dams are built (Fig. 7.15).

In the Ramganga catchment in the Uttarakhand Himalaya these conservation measures have reduced the sediment discharge rate from 0.18 ha-m/km^2/yr in the interval 1958–1962 to 0.14 ha m/km^2/yr in the period 1967–1971 (Pathak, 1974). Planting of eucalyptus in place of shrubs in a Dehradun farm reduced the peak discharge by 77% and erosion proportionately, while narrow-base terracing (contour bunding) reduced the runoff peak by 40% (Das, 1977). Measures adopted in the Damodar valley in Jharkhand were successful in reducing the sediment yield by 50.23%, and in the Hirakud and Machikud valleys in the Mahanadi basin by 13.11% and 31.95%, respectively (Das et al., 1981; Kaul and Das, 1982).

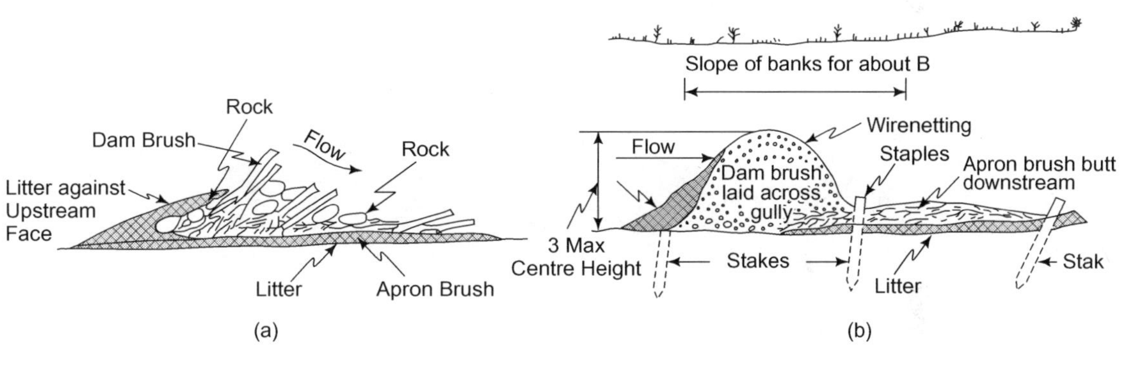

Fig. 7.16. The biotechnical method of strengthening rock-and earth-check dams using (a) brush and (b) brush and wire netting.(From D.H. Gray and A.T. Leiser, 1982, *Biotechnical Slope Protection and Erosion Control,* Van Nostrand Reinhold, New York, 271 p., Figs. 8.16 and 8.17).

Protection of Agricultural Fields

The subject of conservation of agricultural fields has been dealt with at length in Chapter Four. It m be emphasized that in addition to building contour-terraces, the contour wattling (Fig. 7.17A) would considerably strengthen the stability of the ground. The length of the slope is broken into benches by planting wattles at intervals of 5 to 7 m on the slope (upto 33°). The trenches for wattles are commonly less than a third of a metre wide and more than one quarter metre deep. They are then filled with bundl of brushwood 5.6 cm in diameter lying horizontally (Fig. 4.20 A, B). Supplementary treatment of the soil involves putting 5 to 7 cm thick layers of locally growing grass such as 'gorda' *{Chrysopogon fulvus),* lemon grass, 'kikiyu' grass, etc. in the inter-wattle areas. The 'khas khas' (*Vertiveria*) has proved to be very effective (Fig. 7.17B). To add to the stability of the soil, jute netting can be placed on the grass bed.

(a)

(b)

Fig. 7.17. (a) Contour terraces are strengthened by wattling at the terrace edges. Self-generating and sproutable plants make good wattles. (b) The "khas khas" or *Vertiveria* grass is a very effective binder of soil and loose material. (From Valdiya, 2004). (See colour Fig. 7.17)

Rotation of crops, minimum tillage, strip-cropping (Fig. 2.17), etc. would prove beneficial in arresting the soil degradation. In the Siwalik hills near Chandigarh a package of measures involving construction of earthern debris basins, pondage banks, staggered contour trenches and afforestation in watershed had reduced the runoff by 54.2% and the rate of sediment production from 80 t/ha/yr to 5 t/ha/yr (Das, 1977).

Controlling Landslides

Chapter Eleven devoted to landslides deals in detail with measures that are taken **to control mass movements and minimize hazards. The strategy, it may be recapitulated, is to provide efficie and effective surface drainage, and prevent seepage of water from drains, canals, pipelines, tanks, etc.** The routine measures of building retaining walls, breast walls, cribwalls, buttresses, piles, etc. are quite effective in cases of minor hazards, but will not succeed without first draining out the age causing mass movements. On the landslide scars and landslide debris, protective nets together with asphalt mulch help the regeneration of soil (Fig. 7.18).

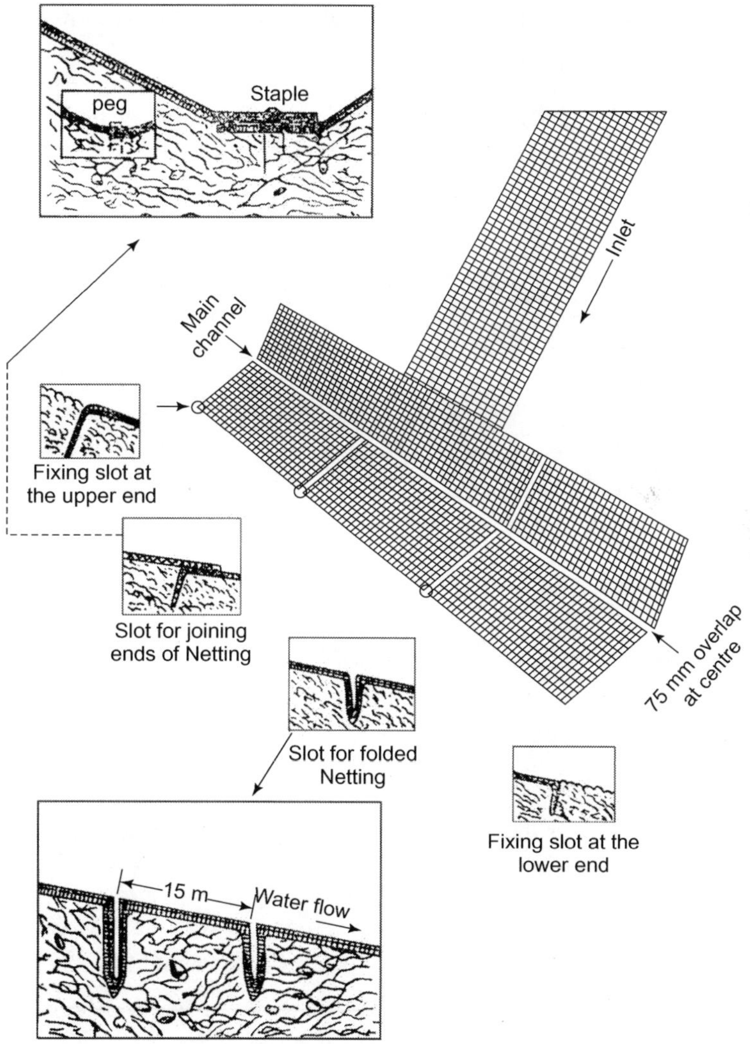

Fig. 7.18. Method of laying coir netting to protect the ground prone to erosion or mass-movements. (Courtes Dr. R. K. Bhandan 1985.)

Stream-bank Stabilization

Channel beds and sides can be protected by introducing vegetation—trees and bushes on levees or embankments. Among the trees may be mentioned willows (*Salix tetrasperma*), napiers (*Pennisetum purpureum*), 'nara' (*Arundo donax*), 'beshram' (*Ipomoea carnea*), 'shamalu' (*Vitex negundo*), 'jalmala' poplars, *Jatropha curcas,* etc. (R. K. Gupta, 1979). These are planted during the rainy season at spacings of 15 to 25 cm. The 'nara' grows luxuriantly in moist conditions and the willows thrust roots deep even in running water.

In order to keep erosion at a minimal level, the channel grade is made flat by inserting massive intervention such as dams and weirs. The energy of water is reduced by making **stilling basins (which serve as shock absorbers), or making the beds irregular by putting a series of drop structures with adequate aprons and spillways.** To keep away the currents from the valley slopes and ensure stability the banks and embankments, spurs are built. These structures deflect the currents and reduce their velocity (Fig. 7.19). **Ripraps and gabions (Fig. 7.20) prevent not only washout but also landslips and related mass movements.**

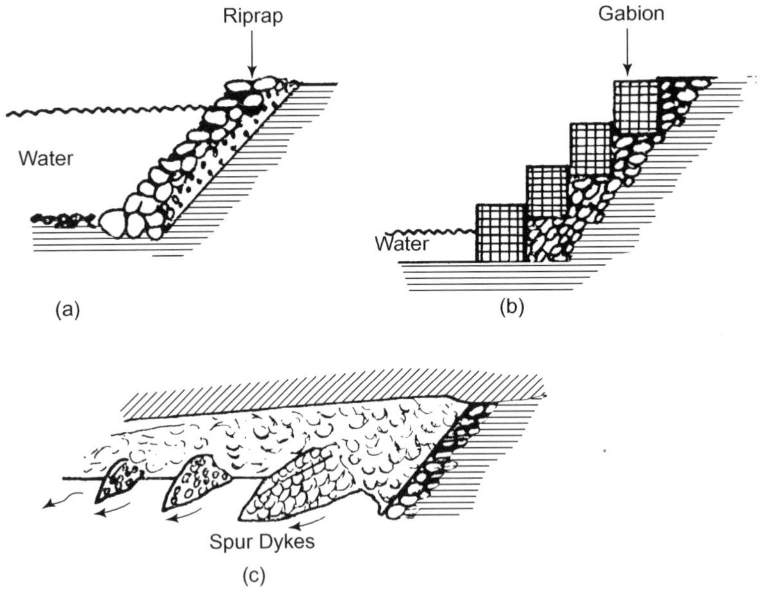

Fig. 7.19. Ripraps and gabions protect the embankment against lateral erosion and the spurs deflect and reduce the velocity of currents.

Saving Coasts and Nourishing Beaches

There are three major types of engineering structures that are built to save the coasts from erosion by waves, currents and tides. Offshore ***breakwaters*** intercept waves and provide protected areas for mooring (Figs. 7.20e and 7.21d). Those of Kannur and Vizhingjam in Kerala have not proved very successful. The ***seawalls*** (Fig. 7.21c) and ***bulkheads*** are located on or near the upper beach and seaward face of the strand scarps. Along the Indian coasts similar structures built of laterites and granite have proved only partially beneficial, for in many places (such as the Kerala coast) fishing and navigati have become difficult or risky (Baba,. 1979). In other places they have caused depletion of beach san as a result of high wave action, which pulls sands away from the base of these walls (Brunn and Nayak, 1980). However, the less heavy bulkheads are effective where sediments occur in low-energy areas.

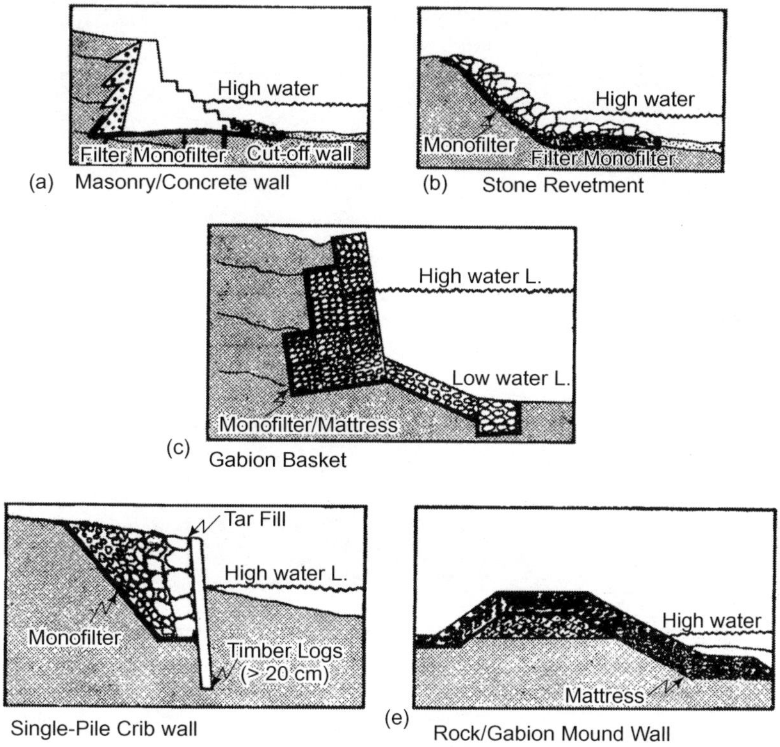

Fig. 7.20. Various types of engineering structures for protecting coasts from the pounding and erosive action of waves and tides. Note the use of monofilters, fabric and mattress for stabilizing the ea material. [(a), (b) after Fig. 8.19, E. A. Keller (1976); (c) after Fig. 13.37, D. R. Coates (1981); and (d) and (e) after B. U. Nayak (1980).]

Groynes **have proved successful in arresting erosion by building beaches in a majority of cases** (Fig. 7.21 a and b). These are linear structures built in groups and placed perpendicular to the coastline at intervals. The coastal longshore currents that carry sediments deposit their loads in front of the series of groynes although there is always downdrift erosion.

This erosion can be countered by filling sands artificially through pumping of sands from the offsho belts. The groyne at Veltoor near Varkala has been successfully protecting the coast for more than a century, but those of Kochi harbour and Moplah Bay have not yielded the desired result.

Jetties are massive structures, generally in pairs, constructed along the channels of estuaries and lagoons and in harbours, primarily to improve navigation by halting or minimizing accumulation of sediments and for retarding erosion (Fig. 7.6c). They check littoral transportation of beach sediments, resulting in updrift development of the beach, but downdrift erosion.

Beach nourishment is accomplished by building groynes and jetties, and by pumping watery diments from the offshore belts and filling them on the downdrift sides of the groynes. These measures have been successful in building the Parakkad beach in Kerala and could be adopted for those parts of the Indian coasts where the waves are not very strong. A combination of the methods described above would not only prove effective but also less expensive.

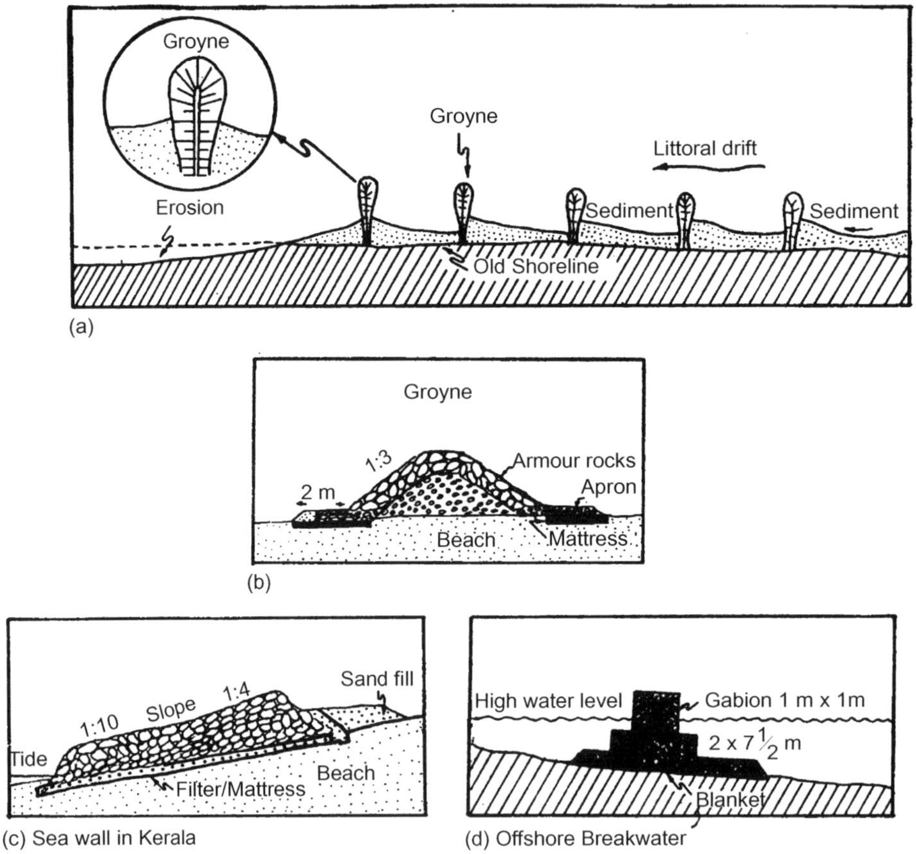

Fig. 7.21. The structures that protect the sea coasts from wave erosion. ['a' 'b' 'c' after B. U. Nayak (1980). 'd' after E. A. Keller (1976)]

Combating Wind Erosion

As already stated in Chapter Four, **live fences of trees and shrubs and windbreaks help check the advance of sands blowing with the wind. Application of chemicals, asphalt and rubber solutions stabilize the surface of the dunes.** ***Windbreaks*** **(Fig. 4.23) consist of several rows of trees and/ or shrubs placed at right angles to the direction of the wind.** Plants are grown close enough to provide a continuous barrier against the wind. On the windward side, shrubs and short trees, in the middle fast-growing short-lived trees and in the leeward side tall and long-living trees are planted. Quick growing species of tall trees such as *Acacia albida, A. tortolis, A. nilotica, Dalbergia sissoo,* and tamarind are preferred. The interspaces are planted with short living but quick-growing plants such as 'neem' and eucalyptus. In between the rows of shrubs of *Euphorbia* and *Salvadora persica,* some fodder plants can be grown to strengthen the windbreaks. Grasses like *Cenchrus ciliaris, C. setigerus, Lasiurus sindicus* and *Dichanthium annulatum* have been found useful. Plantations of *Casuarina equisetifolia* and *Jatropha curcass* on the sandy foreshores backed by inland tree-farms of coconut, cashewnut and eucalyptus have successfully checked the onslaught of tidal erosion.

SUMMARY

More than 45.37 m ha of land in India is seriously affected by erosion of one kind or another. The Indian rivers together carry to the seas annually about 1600 m t of eroded material produced at the rate of >4.5 t/ha/yr, compared to the global amount of 24,000 million t removed from the vulnerable parts of the earth at the rate of 3.6 cm/1000 years. Another 1800 m t of silt is blown off by wind from the drylands. **Erosion has become critical and pronounced in areas where man has indiscriminately used the land and tampered with the natural balance by reshaping, modifying and defacing the topography—such as faulty agricultural practices, excavation for roads, canals, buildings and mining on hillsides and deforestation, particularly where the geodynamic conditions do not allow such disturbance.** On the plains, gully erosion and ravines have damaged 10.37 m ha of land, including agricultural and pasturelands at a rate of > 8000 ha/yr. In the Himalayan region denudation is taking place at the rate of more than 2 mm/yr, and locally as much as 6 mm/yr locally.

As a result of such accelerated erosion the country is losing valuable land at the rate of 7 to 9 million ha/yr. Reservoirs created for the purpose of electricity generation and irrigation are being filled rapidly that their lives have been reduced four to five times. The deposition of sediments in the riv beds has reduced their carrying capacity, leading to overflow and recurrent floods that now spread ov vaster and vaster areas.

The principal causes of this accelerated erosion are the (i) overgrazing of our forests and pastures by an explosively growing population of livestock; (ii) deprivation of protective vegetal cover due to exploitation particularly in areas where this protection is critically needed; (iii) excavations on slopes for laying of roads and canals and for mining of minerals; (iv) natural migration and induced shifting in the courses of rivers in the floodplains; (v) incessant pounding by waves and tidal currents along t shores; and (vi) human interference with the dynamics of coastal currents.

Among the measures taken to lessen the intensity of erosion are treatment of the watersheds by bringing them under cover of multistratal forests of trees, shrubs and grasses, building a series of check dams across and drop-structures in the gullies in ravine lands, and leveling the ground to put vegetal cover upon it, building spurs and embankments covered with trees, shrubs and grasses. In the hilly and mountainous terrain, contour terraces must be provided with levees and drainage ditches, and terrace walls reinforced with wattling of sproutable plants. Steeper slopes should be retired for bringing them under forest plantations, or the beds of crops should alternate with strips of grasses and/or forests. In the dryland or deserts, the march of the sands can be effectively arrested by building windbreaks consisting of rows of quick-growing shrubs and trees. Along the coasts, groynes have proved very effective in controlling erosion and building up beaches.

8

Desertification and Degradation of Land

MEANING OF DESERTIFICATION

The loss of productivity of soil as a consequence of degradation or pervasive dryness is called sertification. There are many dimensions of this manifestation of environmental deterioration in semi-arid and arid regions—loss of regenerative capacity of the soil resulting from extreme depletion or absence of moisture or water, alkalinization-salinization, waterlogging and deep erosion that deprives the soils of their nutrients. Although **climate is one of the most important factors in the development of desertic conditions, the abuse of land and over-exploitation of its vegetal and water resources e primarily responsible for the expansion of desert or aggravation of desertic conditions. In the words of Tolba (1982), "man himself is both the culprit and victim, for the degradation of land ading to desertification is invariably accompanied by deterioration in socio-economic conditions d quality of life."**

Too much or too little rainfall and recurrence of drought have greatly aggravated the deterioration of land. Rains cause widespread and deep erosion and resultant destruction of soil in the over-used lands, and droughts periodically generate severe stresses on the soil productivity ultimately leading to sertification of the drought-prone areas.

EXTENT OF DESERTIFICATION

More than 800 m ha or 33% to 36% of the earth's drylands are deserts (Figs. 8.1 and 8.2). More than 16,500 km^2 area of Asia is afflicted with arid to semi-arid conditions. **About 5.7% (18.8 million ha) of e landmass of India is desert** (Samar Singh, 1982). And, 88% of Pakistan, 55% of Central Asia, 95% of Arabia, 40% of North Africa, 8% of South America, 10% of North America, and 49% of Australia is under varied conditions of desert (Petrov, 1976). In the early eighties of the last century, more than 135 m people (20% of the population) living in 63 countries of Asia (Fig. 8.2) and Africa were threatened with the consequences of the inexorable march of the desert at a rate of 600 million ha/yr, and irretrievable loss at the rate of 6 million ha/yr and productivity loss at the rate of 20 million ha/yr (Mpinga, 1984). In other words, **the land lost annually to desertification was then about equal to the land brought der irrigation each year (Anon, 1977).**

The Thar desert of Rajasthan (Fig. 8.1) advanced 1500 km eastwards in the last 8000 years. In the period 1930 to 1980 the desert encroached upon 13000 ha of land in Haryana and Uttar Pradesh—the rate of extension being 0.8 ha/yr (Das, 1977).

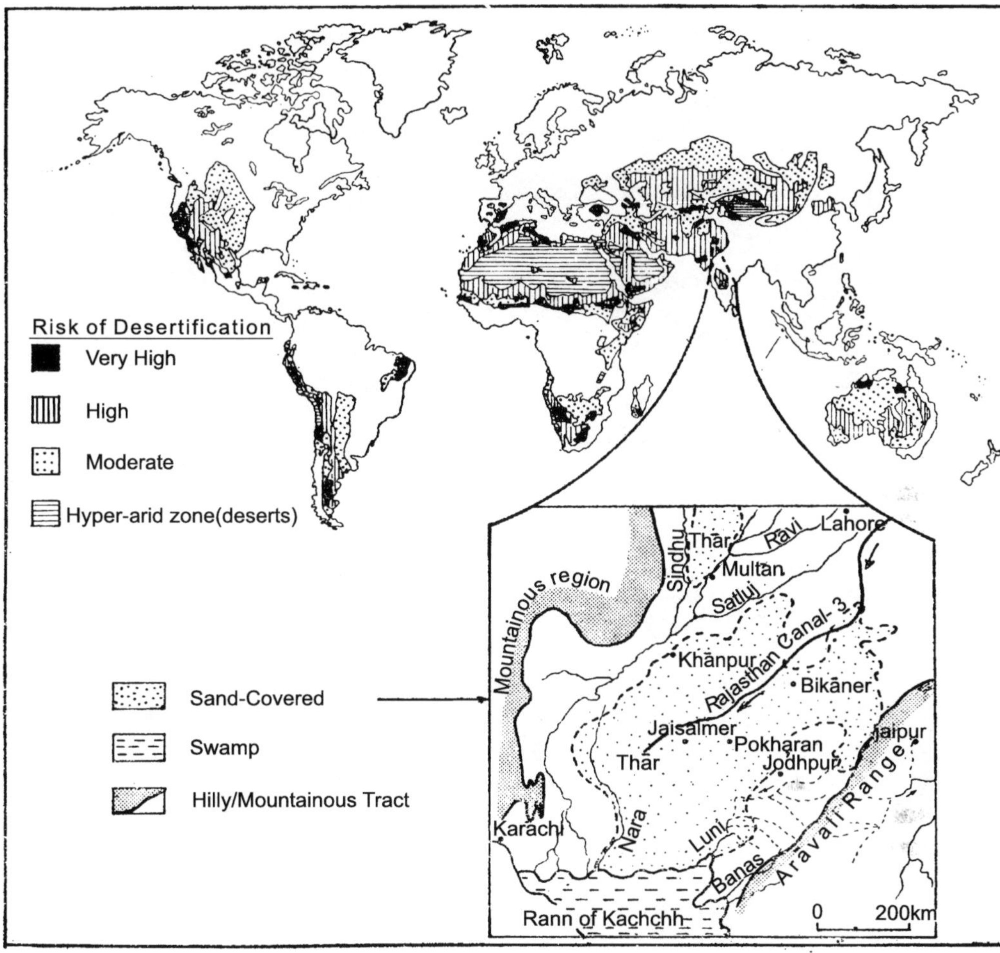

Fig. 8.1. Various degrees of global desertification. (After Holgate et al., 1982.) The inset shows the expan of the Thar desert in western Rajasthan and the attempt to regenerate parts of it through can irrigation.

If this deterioration of land continues unabated or unchecked, the generation that follows will be left with a dismal heritage (Fig. 8.3), less productive, less diverse and with fewer optio and more people (Allen, 1980).

CAUSES OF DESERTIFICATION

As already mentioned, there are various manifestations of desertification (Table 8.1) and a variety factors responsible for the deterioration of land leading to the development and expansion of deserts (Table 8.2). A few of these factors are discussed here.

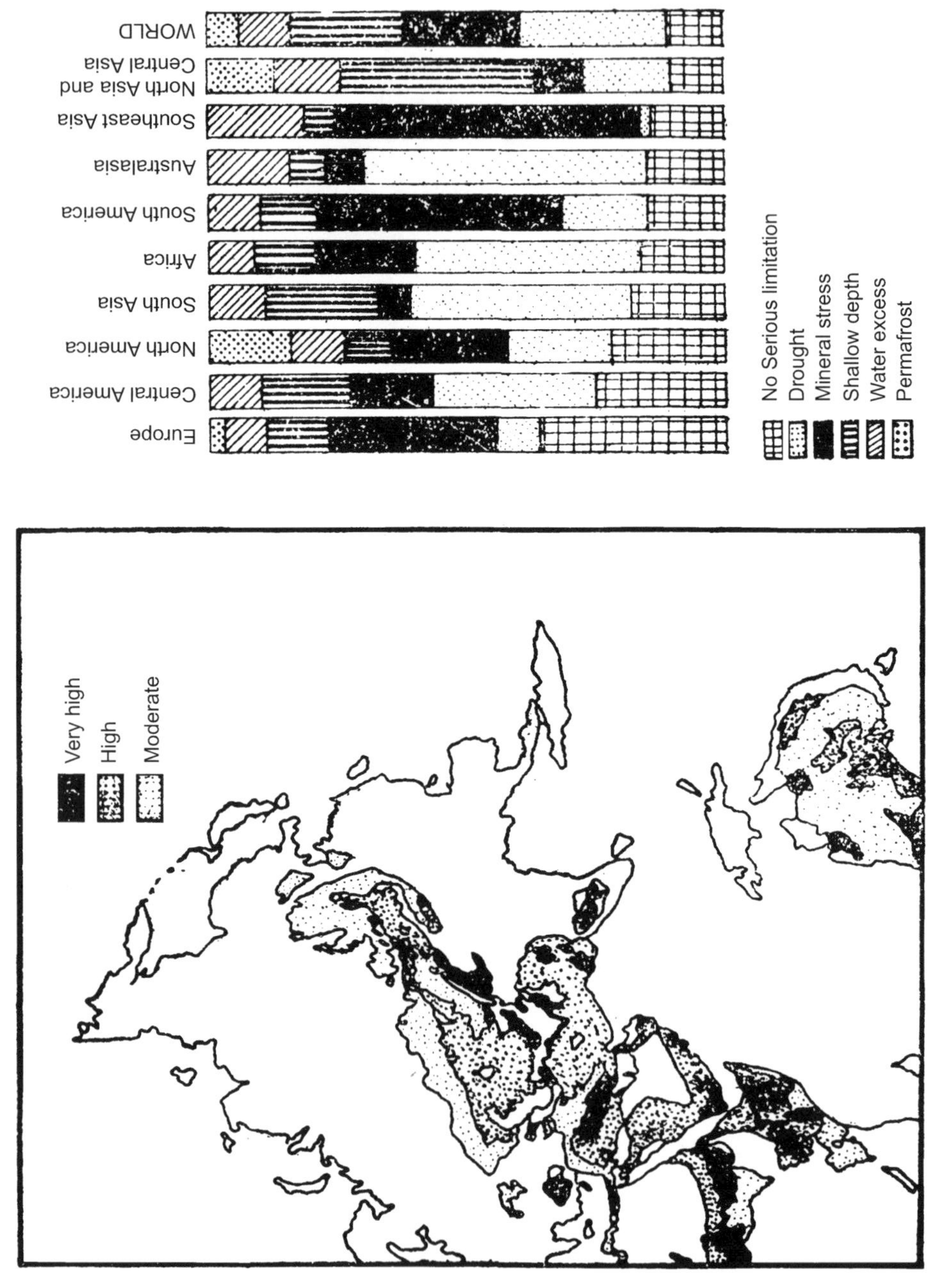

Fig. 8.2. Various kinds of stresses that the soils of Asia are facing. (After Mpinga, 1984)

Table 8.1. Signs of Increasing Desertification (After G. Singh et al., 1979)

Desertification Process	Slight	Moderate	Severe	Excessive
Water erosion	Rills	Soil hummocks	Piping	Rapid siltation in reservoir
	Shallow tunnels	Silt accumulation	Gullying Coarse washout deposition	Extensive gullying
				Landslides
Water erosion	Fluting	Wind mounds	Desert pavements	Extensive active dunes
Water and wind	Small-scale			
combined in	erosion	Wind sheeting	Scalding	Extensive scalding
irrigated lands		Minor saline patches	Extensive saline patches	Abandonment of lan due to excessive salinization

Climate Changes

High atmospheric pressures prevail over the desert domain that stretches eastward from Algeria to Oman and southward from Syria to Aden between 15° to 30° north and south latitudes. The high pressure prevents entry of moisture-bearing winds into this region. Since, the sky over the dryland remains clear of clouds, solar radiation heats the ground surface excessively and causes very high potential evaporation at the rate of 6 mm/day. The little moist air that is able to get into this high pressure zone, subsides and gets warmed up and absorbs moisture from the surroundings, including the ground. This development inhibits the formation of rains. Unable to rise to higher levels of the atmosphere, the small amount of moisture fails to condense. This explains the deficiency or absence of rains in the low latitude subtropical belts.

The drought over the Sahel in North Africa is also connected in some ways with the anomalies of surface-water temperature over the Atlantic. The very small amounts of rainfall over the Thar are likewise connected with the surface-temperature anomalies over the Arabian Sea (Anon., 1977).

Mountain barriers, such as the Himalaya and the Andes, induce localized subsidence of moist air across the barriers and inhibit condensation of moisture (Fig. 13.6). This is the cause of the development of deserts in Central Asia, Chile, Peru and Atacama. The Himalaya does not let the moist humid air-streams reach the Tibetan plateau, with the consequence that "the roof of the world" has become a cold desert.

The modern phase of climate began with the rapid warming up of the earth about 11,000 BP (Before Present), leading to melting of Pleistocene ice and snow that had covered a large part of the earth's surface. At that time, the Sahara and Thar deserts must have had a moist climate, and must have been inhabited by people and animals as testified by rock paintings portraying abundant wildlife and the remnants of ancient oak and cedar trees in the Saharan soil. The growth and development of the Mohenjo Daro and Harappa Civilization in western Rajasthan, Sindh and Haryana, Panjab took place during that period— more than 5500 years before the present (Valdiya, 2002). That was the period when rhinos and elephants roamed in the lush green forests of the vast Indo-Gangetic Plains. Subsequently,

Table 8.2. Causes of Desertification and Methods of Combating the Problem (Anon, 1977, *Desertification: Its uses and Consequence* (Pergamon, London.)

Factor	Problem	Causes	Solutions
eather	Scarcity	Low preciptation erratic distribution of rainfall. Mismanagement of irrigation water. Over-exploitation of groundwater and surface reservoirs. Uncomtrolled water losses by eveporation	Improved water supplies, water conservation
	Mismanagemet in drgland agricIture	Poor and erratic rainfall distribution Uncontrolled runoff.	Runoff management, plant management, soil conservation
	Irrigation mismanagement	Unkonwn consumptive use. Drainage system deficiency.	Improved irrigation methods, drainage systems, slinity control, dependable water supplies
		Deficient land levelling Inadequate wather distribution. Imprecise wather measurements. Poor irrigation methods.	
	Floods	Poor and erratic rainfall distribution. Uncontrolled runoff	Flood control
Soil	Erosion (by water and wind)	Reduction of vegetation cover. Uncontrolled runoff. Sedimentation and siltling. Soil structure degradation. Improper tillage practves. Strong winds. Solil profile depath depletion. Loss of fertile surface soil. Reduction of water-holding capacity.	Soll conservation, soil-moisture conservaton, plant estabils- ment, plant conservation.
	Soil salinity and waterlogging	Canal seepage. Waterlogging. Poor-water quality. Deficient leaching practives irrigation-whater mismanagement. Deficient drainage systems. Floods.	Irrigation water salinity control, dependable water supplies, soil salintiy control, drainage plant establshment, plant conserveation, runoff management, canal lining.
ants	Reduction in vegetal productivity	Land clerance. Plant mismanagement. Over- cultivation. Invasion of undesirable plant species. Uncontrolled wood gathering. Forest felling. Fire (uncontrolled burrning).	Irrigation, runoff management, water supplies, soil conservation, plant establishment, plant management.
nimals	Reduction in animals	Water scarcity. Deficient fodder crops and food reserves, Health and nutrtion. Overpopulation.	Water supplie, water conservation, range management, livestock management, soil conservation, plant productivity, improved breeds, plant conservation, control of pests, game ranching.
nergy	Scarcity and misuse of fuel	Uncontrolled wood gathering. Misuse of available energy	Afforestation, solar energy, wind energy, energy conservation biogas.

the land was overwhelmed by the processes of desertification. The desiccation seems to have beg around 3500 BP.

There is no denying that the aridity is increasing, and the atmospheric and surface temperatures are rising in some pockets as a result of increase in dust, aerosols and hydrocarbons and carb dioxide in the lower atmosphere, and the reflection of solar heat from the severely deforest drylands. Definite changes in local climate are taking place.

The fragile ecosystems of drylands with their delicate equilibrium and reduced water supply and sparse life are very vulnerable to the stress of fluctuating climate—the **irregular and unpredictable alternations of drought and excessive rainfall as witnessed in many parts of western India (F 8.6, Table 8.7)). The threat of desertification is very real in the dry arid and semi-arid tracts wi deeply ploughed and pulverized farmlands, or trampled and defoliated forestlands, or deeply excavated or mined areas, or excessively irrigated agricultural fields (Fig. 8.3).**

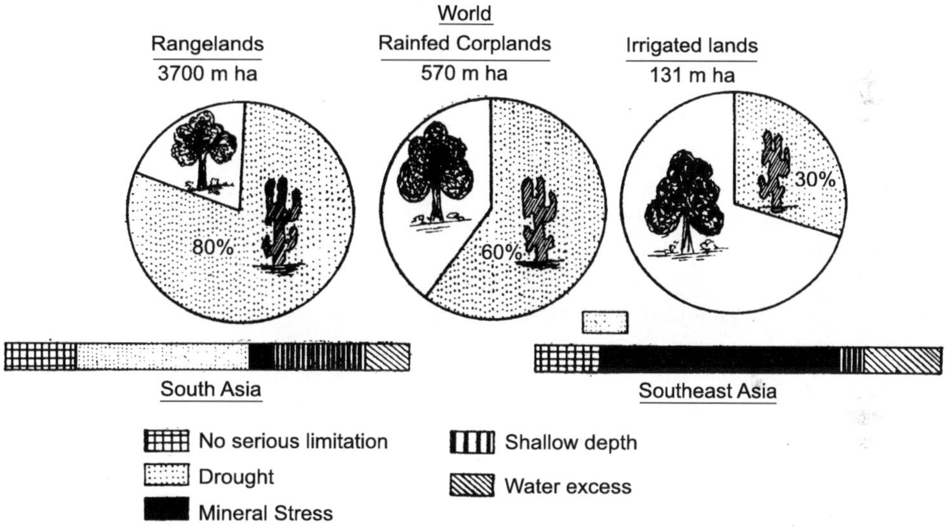

Fig. 8.3. Extent of the land the world over affected by moderate to severe desertification as a result of vari landuse in the early 1980s. The two horizontal bars show distribution of soils in Asia with or witho limitations. (Based on a map in the World Environment Day pamphlet, 1984, UNEP, Nairobi).

Dissection and Erosion of Land

The formation of furrows on the slopes due to plouging intitiated gully erosion in the first half the twentieth century in the south-central plains of what constitutes Palestine and Israel today. The furrows served as channels for rain storms, the rills became gullies, and the gullies grew into ravines. The attendant piping and collapse enlarged the area and depth of deterioration. The railway line on the major 'wadi' courses only accelerated the distress. The consequence was the expansion of the Negev desert (J. Aghassay in Coates, 1973). The ravines in south-western Uttar Pradesh and adjoining Madhya Pradesh, and those of the Mahi, Sabarmati and Saraswati valleys in Gujarat are examples of serious land degradation of this kind. In Maharashtra the districts of Satara, Solapur, Bijapur and Ahmadnagar are exposed to severe erosion and soil deterioration resulting in the loss of productivity.

In Rajasthan, owing to the unpredictability of the rains, farmers are constrained to plough their fields during the dry summer months, thus promoting wind erosion., In the early 1980s the wind blew away dust and silt at the rate of 0.58 to 5.11 quintal per hectare per day (0.5 to 4.2 in Jodhpur and 5.11 in Jaisalmer, scraping the soil at Chandan (Jaisalmer) to a depth of 50 to 60 cm (Mann and Prakash, 1983).

n the average, five to eight major dust storms sweep the desert domain each year during May and June, and the dust clouds rise to a height of 6,000 to 10,000 m. **One dust storm, according to Strahler 973), is capable of lifting on an average 875 tons of dust per cubic kilometre of air, and a major st storm of 500 km diameter would sweep away as much as 90 million tonnes of fine sedi-ents. Thus, western Rajasthan is one of the dustiest places in the world raising great amounts of sediments, which are blown far and wide as evidenced by the dust storms that sweep northern dian plains and adjoining Rajasthan year after year for hundreds of years.**

The Sahara is a major supplier of dust—every year 25 million tonnes are carried away to the Mediterranean basin and the Arabian Peninsula, 25 to 35 m t to the west as far as Barbados, 60 m t accumulates in the bed of the Atlantic Ocean, and quite appreciable quantities reach France, England, Sweden and even Greenland (Mainguet, 1983).

Deprived of the fine sediments with nutrients, the soil becomes incapable of retaining the water it receives from rains, because the water either rapidly evaporates or runs off down the rills and gullies. e stormwaters become downstream flash floods of destructive potential. In the extreme deficiency of water or moisture, the land becomes hostile to all forms of life. Under this condition the animals migrate to adjoining areas in search of greener pastures, which soon get depleted due to extra pressure. **he desert thus expands "feeding on itself". This is what one witnesses in western India;** During droughts the Rajasthan desert people with their livestock migrate to adjoining Malwa (Madhya Pradesh) and Banaskantha (Gujarat), putting these regions in turn to stresses of varied kind.

If the wind scrapes off finer sediments from one place, it deposits them elsewhere, covering the lds, grasslands, forests and water bodies. The area of distress thus becomes very much wide.

Salinization and Alkalinization of Soil

e demise of the flourishing Sumerian and Babylonian civilizations in Mesopotamia (West Asia) some 6000 years ago, is attributed to the destruction of their agricultural land due to concentration of salts and kalies in the top soil resulting from excessive irrigation and flooding caused by the rivers Tigris and phrates (Jacobson and Adams, 1971). The silt-choked canals caused floods, and the salts accumulated in the soil and poisoned their crops. In the early 1980s, nearly half the irrigated soils in Iraq and Syria, nearly 50% in Egypt and more than 15% in Iran were in a state of severe distress from salinization-alkalinization.

More than 16.03 m ha of India's best farms in 14 states (Fig. 8.4) including Rajasthan, Haryana, njab, Uttar Pradesh, Maharashtra and Gujarat are afflicted with this "cancer on land"—9.30 m ha due to alkalinization and 6.73 m ha owing to salinization and salinity (Anon 2010)—mostly in Panjab, Haryana, Rajasthan and Uttar Pradesh (Figs. 4.10, 8.4, and 8.5; Tables 8.3 and 8.4). In Uttar Pradesh the 'usar' (alkali-affected) soils cover 1.15 m ha of land (D.C. Das, per. comm., 1985).

As a result of over-irrigation, the water table rises and with it the capillary fringe of the saline or alkaline water rises, invading the root-zone. Consequently restricting or inhibiting plant growth. The ts cannot be easily flushed out or removed, not without very expensive and elaborate treatment of the soil. Therefore, once the intergranular or interstitial space within the soil is filled with alkalies or sal they remain there for ever, rendering the soil infertile. The salts include chlorides and sulphates, and less commonly nitrates; and the alkalies comprise carbonates and bicarbonates of sodium, calcium, magnesium and rarely potassium. Even as little as 0.2% of sodium salts including Na_2CO_3 is enough to make the soil unproductive. The alkali soils are known as 'usar', 'kallar', 'chopan', 'kshar' and 'chandu', and the saline soils as 'thur', 'reh', 'kallar', 'kari' and 'khar' in different parts of the country. Higher ESP and pH, low organic matter (<0.05%), low nitrogen (<100 ppm) and low zinc (70 to 285 ppm in Uttar Pradesh and 25 to 30 ppm in Maharashtra), but higher content of phosphorus (68 to 81 ppm) and a significant proportion of potassium are some of the features of the Indian alkaline soils (Mehta, 1983 Salinization occurs more commonly and readily in bare soils without vegetation, in coarse-textured soils, and during warmer spells of weather when the water table is high.

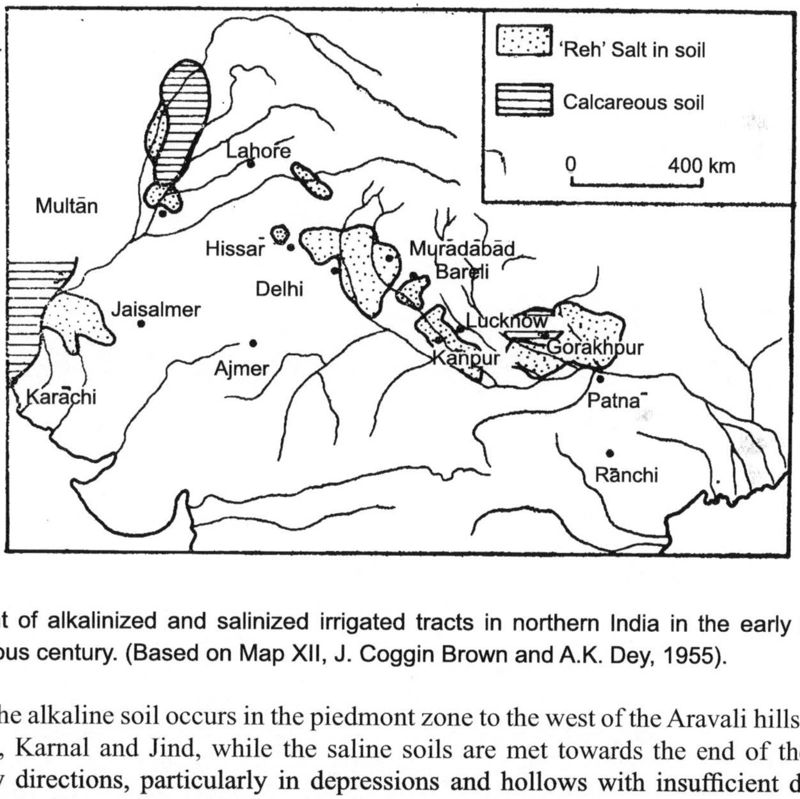

Fig. 8.4. Extent of alkalinized and salinized irrigated tracts in northern India in the early twenties of t previous century. (Based on Map XII, J. Coggin Brown and A.K. Dey, 1955).

In Haryana the alkaline soil occurs in the piedmont zone to the west of the Aravali hills in the districts of Kurukshetra, Karnal and Jind, while the saline soils are met towards the end of the surface- and subsurface-flow directions, particularly in depressions and hollows with insufficient drainage in t districts of Sonipat, Rohtak, Jind, Hissar, Bhiwani and Gurgaon.

The formation of a brittle encrustation of 'reh' (KNO_3) is taking place at a brisk rate in the floodplai of Hissar, Karnal, Gurgaon (Fig. 5.22), Rohtak and in the adjoining districts of Uttar Pradesh. The salt-affected soil is scraped off upto a depth of 3 to 4 mm within an interval of 8 to 10 days during summer (Awasthi, 1966, 1977). **The canal-irrigated areas in Panjab, Bihar, Bengal, Andhra Prades Tamil Nadu (Table 8.4) are likewise affected by alkalies, the upper 15 cm being acutely sick. T introduction of the Western Ganga Canal** (which was first constructed in the 14th century and rela and realigned in 1839) increased the alkali patches in 10 out of 12 fields until about the end of 197

Table 8.3. Statewise Ranking of Sodic Soils (After Anon, 2010)

State	Degraded and wastelands classes (000 ha)	Area (%)
Uttar Pradesh	1,320	6
Haryana	184	4
Panjab	152	3
Gujarat	545	3
Tamil Nadu	352	3
Maharashtra	421	1
Andhra Pradesh	194	1
Rajasthan	181	1
Karnataka	145	1
Bihar	145	1
Madhya Pradesh	106	1
Chhattisgarh	124	0
Andaman and Nicobar Islands	13	0
Arunachal Pradesh	0	0
Assam	0	0
Delhi	0	0
Goa	0	0
Himachal Pradesh	0	0
Jammu and Kashmir	0	0
Jharkhand	0	0
Kerala	0	0
Manipur	0	0
Meghalaya	0	0
Mizoram	0	0
Nagaland	0	0
Odisha	0	0
Sikkim	0	0
Tripura	0	0
Uttarakhand	0	0
West Bengal	0	0
Others	0	0
Total	**3,737**	**0**

(Mehta, 1983). **Until about that time the Lower Ganga Canal had reduced 2000 ha of fertile land 12 villages between Narora and Ramghat in Bulandshahar district (Uttar Pradesh) into 'usar' wasteland, forcing 5000-odd families to give up agriculture for good** *(The Times of India, December 7, 1981).* **In central India 15,000 tonnes of salts accumulated in the topsoil in the Betwa basin tween 1977 and 1979 (Das, et al., 1983), following intensive irrigation.**

Fig. 8.5. Salinized tract of land in the command area of the Indira Gandhi Canal, western Rajasthan. (Paliw and Paliwal 2010). (See colour Fig. 8.5)

Since the commissioning of the Rajasthan Canal (Figs 8.1 and 8.5) in the desert, some parts of Jodhpur and Pali district (western RajasJhan), have developed salinity conditions (Paliwal and Paliwal 2010). In the Luni block 15 km^2 of once productive area was salinized beyond use. The extent of salinized area was 8.3 km^2 in 1953 (Mann and Prakash, 1983).

The well-irrigated farmlands in the Sindhu valley in Pakistan experienced an acute problem of salinity—about 11 m ha out of the 15 m ha of the irrigated land was either waterlogged or salt-affected by the late 1970s (B.B. Vohra, 1981).

Human Abuse of Land

Surveying the prevailing present-day situation and analyzing the causative factors, it becomes obvious that man is the biggest desert maker. By over-exploiting the resources of dryland, he h greatly contributed to the expansion of the domain of desert. Industries pour out fumes and carbon dioxide that have warmed up the atmosphere and brought about climate changes locally. Forests that control and moderate the climate are being decimated at an alarmingly fast rate, marginal lands have been cultivated to such an extent that they have lost their fertility, grasslands have been "grazed to death" and mismanagement of farmlands have converted them into either dust bowls or saline wastes, and so on. But then, **for one-third of humanity—90% of which lives in the developing countries—livi means surviving. They cannot think about the future and of the posterity.**

As already emphasized, bringing semi-arid lands with sparse vegetal cover under the plough has been the biggest contributing factor for the degradation and eventual desertification of that land. Storm and winds that blow in these regions—characterized as they are by quick and steep fluctuations temperature—lift dust from the dry soil into the atmosphere and spread it on the adjoining areas, damaging crops and water bodies.

Table 8.4. Alkali-Affected Areas in India in the Early 1980s (After K. K. Mehta, 1983)

State	Districts Affected
Uttar Pradesh	Aligarh, Etah, Mainrpuri, Hardoi, Kanpur, Unnao, Lueknow, Fatehpur, Raibareli,
	Sultanpur, Azamgarh, Ghazipur, Pratapgarh and Meerut.
Haryana	Karnal, Kurukshetra, Jind and parts of Rohtak, Sonepat and Gurgaon.
Panjab	Amritsar, Kapurthala, Sangrur, Patiala, Gurdaspur and Ropar.
Rajasthan	Parts of Jaipur, Kota, Tonk, Bharatpur, Sawai-Madhopur and Bhilwara.
Madhya Pradesh	Gwalior, Morena, Bhind, pockets in East Nimar and West Nimar along the Narmada.
Bihar	Muzzaffarpur and Shahabad.
Tamil Nadu	South Arcot, Salem, Thanjavur, North Arcot, Coimbatore, Madurai and Ramnathpuram
Karnataka	Bellary
Maharashtra	Sholapur
Gujarat	Navasari

Table 8.5. Human Activities and Desertification in Drylands (From Holgate et al 1982)

Human Activity	Causative Factors	Resultant Problem	Remedial Measures
ommercial Ranching stems	(a) Animal specialization	(a) Heightened environmental and commercial vulnera-bility. Inefficient use of pasture complex.	(a) Government assisted mar keting and price-stabilization schemes; measures to diversify stock composition.
	(b) Poor management of grazing practices.	(b) Overgrazing of certain plant communities.	(b) Development and imple-mentation of ecologically and socially sound range-land management techni-ques; stricter government regulation.
	(c) Dependence on external markets.	(c) Maximization of profits leading to poor ecological management and short-term overstocking.	(c) Rangeland management techniques; stricter govern ment regulation.
	(d) Ignoring subsistence elemets.	(d) Heightened vulnerability of system	(d) Efforts to maximize locall employment, where desirable. increase in on-ranch processing of products.
	(e) Large units which Labour economics	(e) Little employment contributing to outmigration by limiting access to productive land	(e) Development of better monitoring techniques; Improved extension and technical assistance services.
		(f) Slow response to changing environmental conditions (delated adjustment to overgrazing can lead to irreversible deterioration and to economic collapse).	(f) Careful design of tracks and roads, and channeling of runoff; use of stabilizing surfaces; revegetation and soil treatment.

Contd.

Human Activity	Causative Factors	Resultant Problem	Remedial Measures
(g) Large-scale use of heavy machinery and fixed installations.	(g) Severe environmental disturbance and locally irreversible deterioration.		
(h) High capacity for ecological manipulalation	(h) Bushfire control, for example, can lead to invesion of undesirable plant species.		
Rainfed cropping systems	(a) Extension of rainfed cropping systems into crologicaliy marginal areas in which traditional systems become either too intensive or maladjusted to environamental constaints.	(a) Relatively drastic modification of the natural ecosystem which exposes and disturbs the soil, heightening risk of erosion espe daily during seedbed preparation, fine-tilling and barefallwoing, sheet and gully erosion; duststorms and increased sedimentation; hydrological altera tions.	(a) Further agroclimatologica study to define suitable areas for rainfed aficulture areas for rainfed agriculture as weel as ecologically based government incentives and restorations; increased use of drought-resistant plants and research into local crop potential; use of antierosion techniques,
	(b) intensive land clearance, particularly in plain areas, and under systems of shifting systems of shifting cultivation.	(b) Removal of trees and shrugs accelerating wind erosion and strippin soils of nutrients an humus.	(b) Prevention of foreshortening of fallowing slash and-burn systems; introduction of economically valuble perennials into fallow phase.
	(c) Specialization both in crops and techniques an greater mechanization.	(c) Undue removal of plant residues leading to loss of soil fertility and gradual breakdown of soil struture which increases suseptibility to erosion; little incorpation of livestock or legumes into system.	(c) Crop and animal diversl-fication; improved techniques better technological assessment.
	(d) Systems supportiong relatively high population densities, many having operated for several millenia.	(d) Increased pressure on resources in regions already highly degraded through centuries of mismanagement an area exploitation; impoverishment of environ ment and resource base leading to out-migration and lack of manpower to maintain agricultural systerm.	(d) Integrated resource use planning and resettlement schemes.
Hunting and Gathering	(a) Over-exploitation by hunters.	(a) Some species near extinction.	(a) Strict enforcement of regulations and quotas for hunting (very difficult to polic effectively).
	(d) Use of offroad vehicles.	(d) Local destruction of vegetation and heightened risk of soil erosion.	(d) Limitations on use of vehicles.

Contd

Human Activity	Causative Factors	Resultant Problem	Remedial Measures
	(c) Destriuction of animal habitats by other economic activites and competition for grazing	(c) Decline in faunal populations.	(c) Strciter control of main economic activites,
	(d) Over-exploition of wild vegetation to provide complemaentary foodstuffs durring droughts.	(d) Degradaion of plant communitles.	(d) Integrated resource use planning and better organization of natural hazard relief measures;
mber and Fuelwood xtraction	(a) Growth of local and external demand for timber, fuelwood and charcoal, leading to severe over-exploitation of trees and shrubs.	(a) Vast areas, especially around urban centres denuded of trees and shrubs, heightened soil erosion: destruction of animal ands plant habitats; loss of complementary and famine recouse foodstuffs.	(a) Establishmient of woodlots and energy foreste; Regulation to prevent indiscriminate cutting and foster ustained yield,
			(b) Deveopment of inexpensive alternative local energy systems.
ning	(a) Mining and mineralbased industries which cause direct diturbance to soil vegetation and terrain in mining operations, processing facilities and transport.	(a) increased soil erosion, sand drift an duststorms, and hydrological disturbaces.	(a) Design and enrorcement of regulations to protect environment and other local resource users from exploitation by the mineral extraction sector,
	(b) Air-borne or waterborne waster production.	(b) Airborne noxious effluents which kill vegetaion which in turn leads to soil erosion and hydrological disruption; water-borne wastes which pollute scarce water supplies and distupt other users.	(b) Measures to control impact of mining communtiy on vulnerable local societies,
	(c) Competition for local fulewood, water, energy and labour supplies.	(c) Over-exploitation of local resources with detrimental environmental consequences; removal of labour leading to breakdown of other local resource ststems.	(c) Social forestry to meet the need of fuelwood snd fodder.

The second biggest factor is overgrazing the grasslands and forests. The continued trampling of fine-textured soil and defoliation have combined to produce deserts in many parts of Asia. Palestine from 200 B.C. to A.D. 200, used to feed 3 million people with its bountiful harvests; today it is very hard to support even 300,000 people, thanks to the activities of Bedouin farmers. There were only 14,000 people in 1948 when the nation of Israel was born. The expansion of desertic conditions over wide parts of western India also bear eloquent testimony to the deeds of resource-hungry marauders and the uncontrollable growth of population of that terrain. A large part of what is today a dryland in Rajasthan was under cover of dense forests during the Harappa (5500 to 3900 yr BP), Mahabharat times (~3000 to 4000 years BP). Even at the time of Alexander's invasion the Sindhu valley was a forested tract (though Balochistan had already become a desert by then). The climate must have been fairly humid and hospitable (Valdiya, 2002; 2010) as evident from the remains of the elephants that roamed the forests of Rajasthan and the testimony of pores and pollens buried in lake sediments. **If man is responsible f causing desertification (Table 8.5), he is also the greatest sufferer.**

DROUGHT

Extent of Drought-Affected Land

Drought means not only deficient (less than 75% of the normal) rainfall (Fig. 8.6) but also acute scarcity of moisture in the soil. The deficiency of moisture leads to loss of soil productivi deprivation of vegetation, deterioration of land and the initiation of desertification. T disappearance of vegetation causes an exodus or even the demise of wildlife and the nomadism among people. Migrant people (>3 million) and livestock then prey on the adjoining areas regions, bringing distress to both the natural environment and society. The 1967 to 1971 droug in western Rajasthan brought total damage to the land and the little vegetation it supports. Flocks of

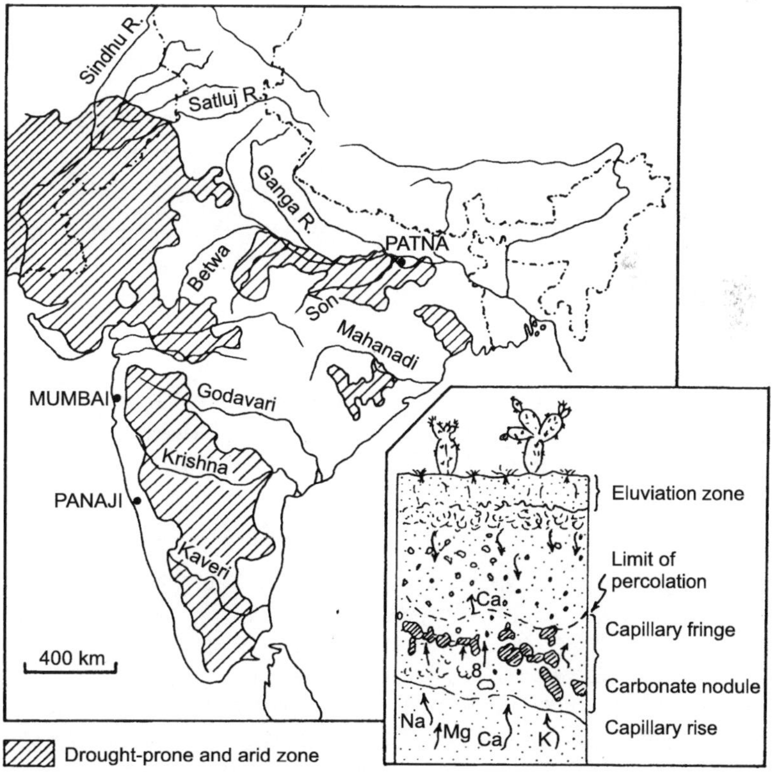

Based upon Survey of India map with the permission of the Surveyor General of India. The territorial waters of India extend into the sea to a distance of twelve nau miles measured from the appropriate base line. Responsibility for the correctness of internal details shown on the map rests with the publisher. **Government** India Copyright 1986.

Fig. 8.6. Pattern of annual average rainfall and extent of drought-prone area in the Indian subcontine Notice the expanses of cold and hot deserts.(Based on Singh and Ghosh, 1975).

cattle and people spread out as far as the Malwa plateau in Madhya Pradesh, Banaskantha in Gujarat and southern Haryana, putting severe strain on their resources. The 1968 to 1973 drought of Sahel (North Africa) damaged pastures so severely that it took more than two decades to restore them to their former conditions. If not treated carefully in time, drought-hit land will never recover completely, owing to the irreversible biological changes that take place during droughts.

Thus nearly 94 million ha (31%) of the 305 million ha geographic area of India is dry, suffering •m various degrees of water-related stresses, affecting quite more than 300 million people. Of ıs, dry region, more than 50% areas are affected by drought once every year (A.K. Gupta et al, 2011). On an average 28% of the geographical area is vulnerable to drought. The dry region encompasses parts of Rajasthan, Bundekhand, Odisha and Karnataka, where water-stresses are strong.

Recurrence Interval

e Thar Desert experiences very severe droughts at regular intervals of two to five years. The years 1895 to 1905 and 1962 to 1971 were periods of extremely severe conditions and pronounced drought (Table 8.7). Even in the normal years between 1875 and 1970, the total rainfall during the south-west •nsoon was 28% less than the normal, and one year in every two to five years faced drought (Mann and Prakash, 1982). Some parts of Gujarat and Tamil Nadu also experience drought conditions at intervals of two to three years. In other dryland regions of the country, the recurrence interval is 10 to 20 years (or even 30 years). But there are intervening wet periods in these parts (Table 8.7).

Since two-thirds of the India's crops, including 42% foodgrains and industrial raw-materials like tton and groundnuts, depend on rains, the national economy is profoundly affected by droughts. When the country was hit by a severe drought in 1979-1980, the production of foodgrains, pulses and oilseeds fell by 28 million tonnes.

DRYLANDS OF THE INDIAN SUBCONTINENT

Extent

bout 5.7% of the land in India and 88% in Pakistan is desert. Eight states in India have vast panses of arid and semi-arid tracts, the hot desert being localized in western Rajasthan and joining Sindh. This is known as the Thar desert (Figs. 8.1 and 8.6).

Thar Desert

The Thar desert is 640 km long and 160 km wide, embracing 4.35% of the area in Rajasthan. The desertic conditions, however, prevail over a much larger area embracing 61% of the land of Rajasthan, parts of Sindh, Gujarat and Haryana. About 76% of western Rajasthan is thus vulnerable to moderate desertic conditions and 19% to highly desertic conditions (Fig. 8.1). The annual rainfall varies between 15 and cm, the coefficient of variation being 60 to 70%. The heart of the desert—Thar—registers the highest temperatures in the subcontinent, the average being 41°C during summer and 23 to 29°C in winter. The mean diurnal temperature variation ranges from 13°C to 19°C. The relative humidity during summer varies from 36 to 50% in the morning and 20 to 35% in the afternoon, but during winter it is 66 to 78% in the morning and 98 to 60% in the afternoon. The mean evaporation during summer exceeds 10 mm/day. The mean wind velocity during summer and monsoon is 8 to 20 km/h, the maximum being 70 km/h.

Table 8.7. Drought Years with Percentage Area of the Country Affected by Drought (Mall et al, 2006)

Year	Area affected by drought (as percentage area)		
	Moderate drought	Severe drought	Total
1877	30.6	28.9	59.5
1891	22.4	0.3	22.7
1899	44.1	24.3	68.4
1901	19.3	10.7	30.0
1904	17.5	16.9	34.4
1905	25.2	12.0	37.2
1907	27.9	1.2	29.1
1911	13.0	15.4	28.4
1913	24.5	0.0	24.5
1915	18.8	3.4	22.2
1918	44.3	25.7	70.0
1920	35.7	2.3	38.0
1925	21.1	0.0	21.1
1939	17.8	10.7	28.5
1941	35.5	0.0	35.5
1951	35.1	0.0	35.1
1965	38.3	0.0	38.3
1966	35.4	0.0	35.4
1968	21.9	0.0	21.9
1972	36.6	3.8	40.4
1974	27.1	6.9	34.0
1979	33.0	1.8	34.8
1982	29.1	0.0	29.1
1985	25.6	16.7	42.3
1987	29.8	17.9	47.7

This climate setting has developed over the ground made up of Quaternary alluvial sands covering a thick sequence of sedimentary rocks, ranging in age from very early Palaeozoic to middle Eocene. Two cycles of erosion are responsible for shaping the present geomorphology characterized by dunes, whalebacks, flat domes, inselbergs, koppies, piedmonts and pediment passes. Sand dunes of all varietı cover 58% of the area, particularly in Jodhpur and Barmer districts (Fig. 8.7).

Groundwater is abundant, especially in the northern region, fed by subterranean flows from t Aravali, and partly by infiltration from the Sindhu River and its system of canals and within the deep buried palaeochannels of now vanished rived Saraswati (for references see Valdiya 2002; 2010). The average depth of the water table is 100 to 120 m in the west, 20 to 30 m in the northeast and 140 to 200 m in the south-west. A perched watertable is encountered in the area of hardpans of evaporites at shallow depths in many sandy flats and interdune depressions, where the water is usually saline. The salinity water increases from 0.5 g/l in the east to 625 g/l in the west. However, in the basin of the Ghaggar River in the north-east, there is no problem of salinity because of the existence of buried channels through hich flows fresher water.

Fig. 8.7. Sand dunes in Jaisalmer. (Courtesy: Prof. M. S. Sisodia, JN Vyas University, Jodhpur) The *inset* shows grazing cattle defoliating the trees and shrubs. (From: Chaudhry et al, 2011) (See colour Fig. 8.7)

Aggravation of Desertic Conditions

The high pressure of grazing by a very large population of cattle, which is growing very pidly, is the principal cause of aggravation of desertic conditions. The density of cattle (Fig. 8.7) rose from 72 cattle per hectare in 1951 to 175 cattle/ha in 1971. This had inevitably led to destruction of vegetation, especially of the palatable and nutritious varieties (Gupta and Prakash, 1975). The runoff in the overgrazed land was 27 %, and the maximum soil loss was 2614 kg/ha/yr (Gupta and Prakash, 1975). **The population of rodents had likewise increased at a terrific rate—there were 200 rodents per hectare, devouring grasses and crops and digging deep**

inside vulnerable soils at the rate of 100,000 kg/ km^2/day. The human population densi is highest anywhere in the deserts of the world—57 persons/km^2. The Thar can thus described as the world's most crowded desert. Astoundingly, the population was growing at the rate of 158% (in the 1909 to 1971 period), compared to 132% in the rest of the country during the comparable time (Gupta and Prakash, 1975). **The livestock population in Rajasthan in 20 was 49.14 million compared to 25.52 million 1951 (Chaudhry et al, 2011).** Understandably, the resultant degradation of natural resources as well as agricultural and pasture land reduced the average size of the landholding to just 6 ha in 2010, compared to 17.8 ha in 1957 (Chaudhry et al, 2011). **It is quite obvious that there is an unbearable pressure on all woody biomas including the seeds which are collected for consumption, resulting in severe inhibition of the regeneration processes.**

Furthermore, excessive irrigation of farmlands in the districts of Suratgarh and Ganganagar have created not only waterlogging, but introduced weeds that have been able to secure foot-holds, and in the absence of competition from the indigenous vegetation, proliferated rapidly over wide stretches on the sides of canals.

SIGNS OF DESERTIFICATION IN THE HIMALAYA

The accelerated pace of erosion in this geodynamically sensitive region, coupled with the growth of human and livestock populations have brought about certain irreversible changes, endangering the life-support systems for the flora and fauna. This is particularly noticeable in parts of Nepal (Rieg 1977) and western Uttarakhand (Valdiya, 1985). Over-exploitation of forests for commercial purposes, destruction of slope stability by indiscriminate and ruthless excavations for mining and roads, and overgrazing in forests are responsible for this state of affairs. **In Uttarakhand region there were million people growing at the exponential rate of 2.28 % in the period 1970-1980 (Shah, 1981) an living on extremely restricted cultivable stretches. The density of population in these stretch was four times higher than in the plains, while the productivity of the land was very much less (Moddie, 1980), so that the per capita income was nearly half the subsistence consumption level (Shah, 1981). The pressure of grazing was 2.5 to 4.5 times higher than the supporting capacity the forests (Singh and Saxena, 1980)** with the consequence that the forest cover was reduced to less than 30 % of the surface area. Landslides have become endemic, particularly in the belts of the Main Boundary Thrust and Main Central Thrust and the many faults that tear apart the mountain ranges. The rate of erosion—quite more than one millimetre per year—is understandably five times higher than was in the geological past—producing sediments at the rate of 16.5 ha-m/100 km^2/yr in the mid nineteen seventies (G.P. Gupta, 1979). The sediments are filling reservoirs and choking streams. Springs a drying up or becoming seasonal—in the areas east of Nainital 40 to 45 % springs have met with this fate until about mid nineteen-eighties. Since then the dismal development has become quite alarming throughout Uttarakhand.

In the tributaries of Seti River in western Nepal the difference in volume of water during d and rainy seasons was as much as 1000 times (Kollmannsperger, 1977) resulting in the too-littl too much water syndrome—a common feature of the desert country. With greater runoff on t forestless slopes, the incidence of floods greatly increased in the mountain streams and rivers.

The evaporation of moisture in the soil on the treeless slopes in the Kathmandu Basin was as high as 190 % (Kollmannsperger, 1977) in the mid 1970s leading to progressive drying up of the ground surface.

MEASURES TO COMBAT DESERTIFICATION

Strategy

Since man himself is primarily responsible for aggravating environmental stresses and acce- ating processes leading to desertification, the measures to combat this hazardous development should incorporate a programme of sustaining and improving the livelihood of the people, in dition to conserving natural resources and treating the environment. ***A willing participation of the local communities is essential, for their experience and capabilities will go a long way in mbating the problem of desertification.***

The first step in this programme is the preparation of a desertification hazard map. Amelioration of surface microclimate through proper and **controlled landuse offers an effective strategy for generation of the degraded environment.** ***The history of Israel is a story of magnificent success in nning and reclaiming the land from the desolate Negev desert. Roses that bloom in the desert areas e flown to Amsterdam markets, the fruits and vegetables they grow supply many vegetable market in European cities, and 80% of the food they require is met from cultivating the dry soils covered th plastic mulch to slow down evaporation. To conserve water and prevent waterlogging, perforated pipes are used to deliver to each plant (grown in rows) just the amount of water they need, that too ly at the roots.***

Desertification Hazard Map

e essential steps in hazard mapping embrace firstly, the identification of areas vulnerable to sertification on the basis of prevalent environmental stresses, and secondly, the regular and repeated monitoring of changes occurring in the desert-prone ecosystems. The map would depict degrees of lnerability (very high, high, moderate, low and slight), and the extent of desertification (very severe, severe, moderate, low and slight) in terms of injury or destruction of vegetal cover, erosion of ground, crease in surface flow or discharge of subsurface waters, decline or loss of productivity of soils, etc. Simultaneously, the monitoring of ecosystems provides early warning of the trend of changes taking place and helps in identifying areas of maximum distress. The monitoring is best done through remote-sensing techniques, using false-colour satellite imageries or in black and white. The false colour mosaics of space imageries of Rajasthan indicate that the bulk of sand dunes and sand sheets occur in the extreme west of the Aravali upto the Sindhu River and discontinuous sheets cover the belt stretching north-east from the Rann of Kachchh to Haryana and Delhi. There is no piling up of sand in the Rann, but there are concentrated patches around Jaipur, Kuchamara-Daira, south of Dholpur and Agra.

Desert hazard mapping should be supplemented by surveys related to socio-economic conditions the people, embracing growth of populations of humans and livestock, cropping pattern, farm oductivity, malnutrition and health, jobs and professions and earnings, degree of literacy, etc.

Changes in Grazing Practices

Livestock cannot be allowed to range and graze freely if the soil is to regenerate and nurtu vegetation. Realizing that the cattle constitute the mainstay of the desert communities, they have to be protected but permitted to graze only in the specially fenced reserves. Planting of fodder trees inside and around the fencing of the grazing land—in contour furrows 60 cm wide and 20 cm deep and at intervals of 8 to 10 cm across the slope—would save the pastures and allow growth of protective and benefic vegetal cover on the loose soil.

Stabilizing Active Dunes

To prevent wind and storms from sweeping away finer sediments from the exposed and loosened so of the dryland, a cover of grasses, shrubs and trees having deep roots may be grown (Fig. 8.8). The ***Silvipastoral plantation*** attempted by the Central Arid Zone Research Institute at Jodhpur successfully stemmed the advance in parts of the Rajasthan desert. The soil stabilization was done primarily by planting of *Saccharum* and by mulching with brushwood. Grass species such as *Crotalaria burhia, Calligonum polygonoides, Laptadenia pyrotechnia, Zizyphus nummularia* (R. K. Gupta, 1979), *Cenchrus ciliaris, C. setigerus, Lasiurus sindicus, and Dichanthium annulatum* (Das, 1977) have been found useful. The trees and shrubs recommended by R. K. Gupta (per. comm., 1982) are *Acacia* ***senegal, A. nilotica, A. indica, A. tortolis, Prosopis cineraria, P. juliflora, Balanites aegyptica, Jatroph*** *Eucalyptus, Parkinsonia, Dalbergia sissoo, Albizzia lebbek, Tecomella undulata, Azdirachta melia, A. indica and Ricinus communis.* The "khejri" *(Prosopis cineraria)* thrives even in the deserts, its roots penetrating deep down and thus stabilizing the soil. Its leaves are not only excellent fodder but also give nutrient-rich litter to the soil. **To ensure safety of seedlings and quicker growth of plants, it desirable to spray asphalt mulch on the dunes and cover the seed bed in the initial stage with ju nettings.**

Construction of *windbreaks or shelter-belts* in the windward side of the settlements and agricultural farms reduces the velocity of winds and checks the advance of the dunes (Fig. 4.23) as seen in Rajasthan. Paving the roads and restricting traffic within the tree-fenced tracks have help control physical degradation of the ground in Jodhpur, where the Central Arid Zone Research Institute planted three staggered rows of trees on either sides of the roads.

Water Harvesting

The practice of water-harvesting adopted by Israel has proved very rewarding. The rainwat runoff is gathered through converging channels on the hillslopes and diverted to settling basin The gathered water is then led through underground cisterns hewn in bedrocks (Hillel, 1982 The cistern tops are surrounded by boulders to form protective circles, and the sides and bottoms lined with plastic sheets to prevent seepage. The surface of the basin is covered with a mono-molecular film of hexadecanol, which though easily breakable by wind, does not allow evaporation of wat **The simplest technique is the filling of reservoirs with coarse gravels or stones which effective minimize evaporation even though the storage capacity is considerably reduced.** Floodwaters are likewise diverted and stored in reservoirs. The stored water is covered with an appropriate film prevent evaporation.

The Chhattisgarh people have long been conserving rainwater in tanks led through channels dug on hill slopes.

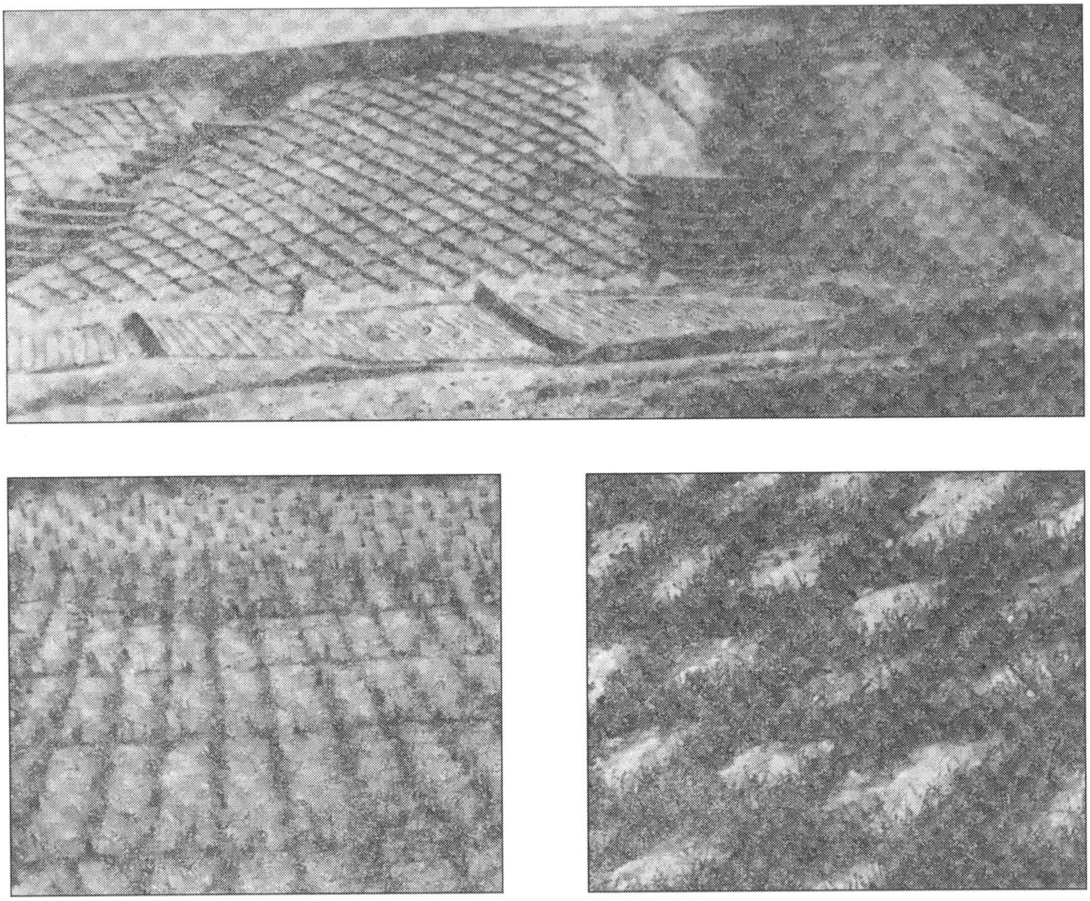

Fig. 8.8. The most effective way of stabilizing active sand dunes is to cover them with grasses and shrubs in the first instance, followed by resort to jute or choir netting and application of asphalt mulch and fertilizers. (Photos courtesy: Prof. Dussan Zachar, Forest Research Institute, Zvolen, Slovak Republic.)

Treatment of Salinized-Alkalinized Soils

is soil sickness can be treated painstakingly in five ways as already explained in Chapter 4. To recapitulate; (1) prevention of seepage of water from the irrigation channels by lining them—a measure at will have to be rigorously implemented in the canals flowing through the sandy-silty terrains; **application of water to fields in accordance with the need of the crops—just the quantity**

needed—such as the Israeli practice of trickle (drip) irrigation through perforated plastic pipe (3) building networks of drains to remove excess water introduced into the ground ensuring that the drainage channels are kept clear of silts and weeds; (4) planting salt-resistant species of vegetation such as barley, pearl millet, raya, *(Brassia juncea),* cotton, sugarcane, sugarbeet, spinach, date palm, certain fodder grasses, rice, berseem, etc. The promising species for planting—which not only absorb salts, increase percolation by contributing leaf litter, but also enrich harvests of fuelwood—are *Prosopis juliflora, Acacia nilotica, Albizzia lebbek, A. procera, Terminalia arjuna, Cassia siamia, Cord dichotoma, Syzyguim cumini, Pithecolubium dulce* (Chaturvedi, 1985). Crop rotation of 'dhaincha'-sugarcane-cotton in the Deccan, 'dhaincha'-rice-berseem in Panjab-Haryana, and 'dhaincha'-rice in Uttar Pradesh has proved very effective in curbing the menace of alkalinity. (5) The alkalinized ground in the upper 15 cm is treated with a liberal application of powdered gypsum at the rate of 2 to 15 t/ha. This treatment removes exchangeable sodium and replaces it by calcium. Addition of organic residue such as rice husks and straws, which leads to the formation of mild acid on decomposition, helps quicker regeneration of the soil. Other amendments include sulphuric acid; phosphogypsum, press mud (a byproduct of sugar factories), zinc sulphates, farmyard manures, etc.

SUMMARY

Not only drier climate conditions but also the activities of man are responsible for the creation and expansion of deserts. By letting water and wind erode away and damage the land in mountains, river valleys and plains, by causing waterlogging and alkalinity-salinity in the soil and attendant destruction of the productivity of irrigated fields, and by wiping out the canopy of vegetation that protects vulnerab lands, man has contributed considerably to the expansion of the desert domain.

Although the livestock he rears is greatly responsible for the deterioration of the land, m himself is the biggest desert maker.

About 5.7% of our landmass is a desert or desert-like and quite a large part (one-third of the geographic area of India) chronically drought-prone, of which nearly 13% area is on the brink of total desertification.

The strategy for combating the menace of desertification involves changing grazing practices restricting grazing to fenced areas surrounded by fodder plantations, stabilizing sand dunes and sheets by covering them with vegetal cover, harvesting rainwater to make use of it for varied needs, and treating soil sickness due to salinity-alkalinity by appropriate use of technology including construction of a drainage network.

9 Natural Hazards, Their Zoning and Risk Assessment

NATURE OF NATURAL HAZARDS

Sudden and Swift Events

Natural events occurring suddenly and swiftly, and consequently causing harm to the people and their institutions are called natural hazards. To the common man such a geological event may be due to the wrathful power of Nature or God, over which there is no control. **But many such events result from man's own acts, particularly his abuse of the environment triggering off or stimulating geological phenomena resulting in tragic consequences. Whether naturally occurring or man-induced, geological phenomena are hazardous to people who build their structures in or adjacent to areas susceptible to destabilizing processes.** A landslide, for example, occurring in a rural or urban area is a hazard, while in a remote unpopulated terrain it would be just another geological event regardless of its intensity and dimension. Avalanches striking the pilgrim or strategic routes are hazards, but not in the snowy wilderness unvisited by man.

During the period 1960 to 2010 more than three and half million people were killed all over the world as a result of various natural hazards. Of this number, more than 75% were of under-developed or developing countries.

Usually it is the poorest sections of the population (>80%) in developing countries of the world that suffer most from hazards as they have no option but to remain where they are constrained to live. Poorer people have been forced into hazardous areas by the pressure of population, particularly the higher Himalayan slopes and the upper reaches of valleys and in the floodways in alluvial plains. Larger numbers of poor people tend to be clustered together and are thus more vulnerable to hazards. This explains why the number of the people affected by various natural disasters is increasing at the rate of more than 6% per year. For example, in the Kathmandu-Kakani area in Nepal in the later 1970s, the poorest subsistent farmers sustained landslide damages more easily than the richer ones, because ey were confined to unproductive land (Ives and Messerli, 1981, 1984). If a disaster in Japan kills 63 rsons on an average, the disaster in Peru takes a toll of more than 3000 people (*News Alert*, 1-2 : 4, 84). It must be confessed, this comparison is odious.

Various Hazards

rthquakes are perhaps the most disastrous of all the natural phenomena afflicting repeatedly and destructively very large parts of the earth in geodynamically active zones. Approximately 2000 tremors strike the earth's surface each day, of which two are of sufficient magnitude to cause damage to ma made structures. About 15 to 20 earthquakes world-wide each year produce widespread damage and destruction. Volcanic activity is a violent and hot manifestation of natural forces, inflicting destructi in volcanic belts, and polluting the atmosphere on a large scale. There are 516 active volcanoes in the world which erupt approximately once every 15 days. Floods are a recurrent phenomenon in the floodplains of rivers—the agricultural tracts—where the population is densest. Thus nearly 900 milli people on earth live in some of the most vulnerable areas. The coastal belts are pounded and ravaged by cyclones, hurricanes and seismic sea waves, causing destruction year after year. On land, more than 1800 thunderstorms can be seen at any given time, and in the late summer at any given time 60 or mo hurricanes and 700 to 1000 tornadoes ravage the drier lands somewhere on earth.

The loss from the various geological hazards each year on the average was 250,000 deaths, mostly in developing countries. In just one decade 1965–1975 about 35,00,000 lives were lost (Davies, 1984).

Anthropogenic Stimulation

As already pointed out, **the abuse of land by man in many cases stimulates occurrence of catastrophic events or at least aggravates their consequences.** Landslides in the mountains, for example, are triggered when the natural drainage of rainwater, along with vegetal cover, is disrupted or damaged by the inhabitants or visitors such as road makers and or dam builders. The people of the Indo-Gange Plains in Uttar Pradesh and Bihar suffer grievously and recurrently from the flood hazards, because th live in old abandoned stream-channels or in the present low-lying floodways. In Bangladesh more th 22 million people inhabit the land that is less than 3 m above sea level and are therefore vulnerable annually occurring floods and cyclonic surge waves. More than 3,00,000 people of Bangladesh perish in the cyclonic storm of November 1970. In the May 1985 cyclone the human toll was 60,000. But the is no option for the majority of the people other than living out their traditional way for life in the places where they have lived all along.

All geological hazards are not necessarily sudden and swift. The slower manifestation of natural hazards are the direct consequence of human activities. Landslides are aggravated by man, because the forests are cut down thereby exposing the soil to direct rainfall, colluvial creep and landslides. A related effect is soil erosion—removing topsoil needed for agriculture. The third effect is increased du because the windbreak effect of trees is lost. Likewise, "the dams aggravate coastal erosion, because they hold back sediments which earlier reached the coasts and moved along the shores by long shore drift and built beaches. Free of sediments, the waves tend to cut back into the coast. Reduced water flo causes salt-water intrusion and pollution of coastal wells. Overpumping of wells near the coast produces land subsidence (R. S. Yeats, per. comm., 1985).

ENVIRONMENTAL SECURITY AND HAZARD ZONING

Coping with Hazards

Natural events cannot be prevented from occurring but their impacts can be reduced if effective measures are taken to reduce their severity, frequency and, possibly, size. The hazard-coping measures include planning for disaster preparedness based on appraisal of the magnitude of risks in the past and those anticipated in the future. The preparation of hazard-zoning maps is the first pre-requis of these measures. These maps require not only identification of vulnerable zones and sites of pa occurrences and of predicted future events, but also an estimation of the actual danger to the settlements, communication systems, and essential supply lines. On the basis of these maps, the geologists investigate ways in which the intensity of the anticipated event could be moderated. And if man's acts themselves are found to aggravate the ecological imbalance leading to hazardous events, a way must be found to identify and stop or modify such activities.

Principles of Hazard-Zoning Maps

anning for environmental security requires thorough knowledge of (i) the geological conditions,) topographic layout of the region, (iii) demographic data and landuse pattern, and (iv) quantitative praisal of the magnitude of risks anticipated. In other words, the geologists cartographically portray their answers to two major questions:

1. Which geological event would produce the greatest damage.
2. How the impact of the event will change with the passage of time.

Hazard-Preparedness Plan

Once the answers to above two questions are known, **the whole spectrum of the community—people, government agencies and voluntary organizations—prepares the plan to cope with the anticipated hazard.** There are three ways of doing it. In a *short-term response* the community prepares to adjust itself to the changed situation either bearing the consequences stoically or sharing and distributing the distressful effects among themselves, but simultaneously taking measures to reduce the intensity and extent of hazards by modifying the trend of the event. The *long-term response* is to adapt to the changed hazardous conditions as one is forced to do in volcanic areas. **The third softer option is to evacuate the threatened or affected area**. In the face of mounting population pressure and extreme paucity of land for resettlement, this **softer option is not available in India. In fact, it is the very pressure of population that is responsible for encroachment or occupation of areas prone to hazards.** In the Himalayan region the people have been constrained to move up the steeper slopes and remove the forests for settlement and farming, thus inviting occurrence of landslides. The ancient hillfolk cultivating the t terraces on banks of rivers and streams first extended their cultivation to gentler slopes (1500 to 00 m), but today they are forced to encroach upon the higher regions (> 1900 m) at the cost of rests. In the Indo-Gangetic Plains the floodways have been occupied for habitation and cultivation. The concentration of people in cities, likewise, exposes them to hazards of pollution and pestilence.

RISK ASSESSMENT ANALYSIS

has been explained that a ***hazard*** is a natural event taking place or likely to take place in an area or locality during a given time. The hazard involves ***risk*** when things that people value very much are stake (Dibble et al., 1984). An area which is susceptible to damage or destruction by a hazard is vulnerable area. The analysis of risk thus involves (i) identification of the nature and dimension of zard(s), (ii) estimation of the economic cost of the damage likely to be incurred by the people and eir institutions, and (iii) assessment of the degree of probability of the hazard in the identified area and sociated consequences (Petak and Atkisson, 1982).

STRATEGIES FOR HAZARD MITIGATION

Priorities

The strategy for hazard mitigation embraces measures to reduce if not eliminate the probability or frequency of occurrence and intensity of impacts. It may be emphasized that individually or severally any one measure taken can never be effective; only the combination of measures can yield desired results. To ensure participation of the people of the affected/threatened area—which is extremely desirable—hazard-coping measures must be compatible with the local traditions and culture. What is effective in the U.S.A. or Philippines, for instance, may not be appropriate Uttarakhand. It is also imperative that the cost of the measures should be within the economic capabiliti of the local people or the government. To stabilize a slope, an Indian geologist would not necessari recommend the same technology as a Swiss geologist would do.

As the vast majority of hazard victims belong to the poorest section of the population, deaths are primarily due to the vulnerability of the locations of settlements and to the constructional weakness of their dwellings. Hazard-coping measures should therefore give priority to risks to the poorest sections of the community for whom the investment today is less than 5%, even though their fatality exceeds 80• **Recognizing the priorities of the local people and appreciating their sensitivity to tradition and culture, appropriate measures of rehabilitation should incorporate providing the affected people with key materials at prices which they can afford, and also offering them expertise for building houses and engineering structures for the prevention of hazard and protection of settlements.**

Prediction, Warning and Public Information

By tapping the reserve of accumulated experience of generations, by reading various signs of instability in the natural systems, by analyzing various geophysical changes that take place on the ground, and by studying the erratic behaviours of insects, birds and animals who have an extraordinary perception of impending disasters, it maybe possible to forewarn the people of a potentially hazardous event. Capability of predicting to some degree the occurrence of earthquakes has been developed in China, Japan and the U.S.A. At Haicheng in China a major seismic event (Februa 4, 1975) was predicted which otherwise would have taken a toll of lakhs of people. However, aft Haicheng, the Tangshan earthquake occurred on July 28, 1976 without warning, killing 240,000 peop In the U.S.A. the success in prediction has been only partial—the place and magnitude have be successfully predicted, but not the time.

The confident prediction of likely event is to be followed by warning the people through vario media and by enforcement of a hazard-preparedness programme including evacuation if necessary. The public must be educated simultaneously about the nature of the hazard, the risk involved, and the likely impacts on the natural environment. This education should spur the people to voluntary action.

Brochures on the hazard-preparedness plan should be distributed beforehand among various organizations and agencies identified for mobilization of relief work, including fire-fightin medical care and sanitation, food supply, rehabilitation and shelter and law and order. The plan must ensure smooth and coordinated functioning of the various agencies.

Minimizing the Probability of Hazards

Measures should be taken to reduce the frequency of occurrence of hazards, such as developing effecti drainage systems which prevent triggering of landslides, or constructing engineering structures to protect the people and their houses from the hazard. Table 9.1 offers various methods for the abatement hazards of different kinds, as discussed in depth by Petak and Atkisson (1982).

Table 9.1. Measures for Abatement of Natural Hazards and Protection of Settlements (Modified after W. J. tak and A, A. Atkisson, 1982).

Hazard	Specific Action or Example
Riverine Flood	1. Reforest or reseed denuded areas.
	2. Mechanical land treatment of slopes, such as contour farming, to minimize runoff.
	3. Protect watersheds from forest and grass fires.
	4. Control or eliminate large-scale clear-cutting of forest land, as in oak forests.
	5. Manage watersheds to minimize erosion (e.g. by preventing overgrazing).
	6. Design and construct new urban development so as to provide temporary on-site water retention.
	7. Construct catchment basins for debris and sediment
	8. Clear debris and sediment from upper streams.
	9. Create water-holding areas (i.e. small lakes).
	10. Restore or preserve natural detention areas such as sloughs, swamps, small lakes.
	11. Construct and maintain levees and flood walls as physical barriers.
	12. Construct and maintain channel improvements to increase flow capacity.
	13. Emergency flood control.
	14. Construct dams and reservoirs to store water.
	15. Minimize failures of engineering works with inspection, rehabilitation, and removal of substandard structures.
eadwater Flood	1. Local stream and storm drainage management with drains, street gutters, culverts, etc.
orm Surge	1. Seed hurricanes to reduce the pressure differential (the technique is still quite impractical).
	2. Maintain and preserve barrier islands and coastal wetlands.
	3. Preserve, maintain, and replace coastal sand dunes.
	4. Construct and maintain break-waters, levees, and flood walls as physical barriers.
arthquakes	1. Relieve earth strain resulting from tectonic movement (the technique is still quite impractical).
xpansive Soils	1. Cover sites with a blanket of non-swelling soil.
	2. Ensure proper drainage
andslides	1. Reforest and reseed denuded areas.
	2. Protect hillside areas from forest and grass fires.
	3. Manage hillside areas to minimize vegetation removal and erosion.

Public Policy for Hazard Management

blic policy for hazard management should include landuse classification, building codes, capital investments by governments, public information, education and warning. **The Policy plans should include** ***formulating and enforcing laws and regulations for*** **preventing or restricting development**

and use of the lands prone to hazards as indicated in the hazard-zoning maps. Since preventive and restrictive measures have proved ineffective in India, the most effective way of curbing the tendency to occupy hazardous tracts would be to impose a series of disincentives such as (i) den of government assistance for development of roads, hospitals, schools, etc. in the identified hazar prone areas, (ii) non-availability of loans from the banks, (iii) denial of essential supplies such as wat and electricity, and public services, such as communication lines, sewer system, bus service, etc., and (iv) denial of insurance against hazards, natural or man-made. Alternatively, the government can acqui the areas for alternative landuse such as recreation parks, wildlife sanctuaries, afforestation or allowing the original owners to pursue agriculture but forbidding construction of buildings, etc. In this w productive use is made of land that is in short supply while the degree of risk is reduced.

The following policy actions involving legislative and/or administrative measures are suggested.

1. **Formation of a fully empowered national commission —which might be called the Natural Hazards Management Commission (NHMC)**—to provide the community with all information relating to vulnerability of the areas to hazards, extent and magnitude of risks and likely impacts, and the mitigation measures to be taken up in time. This can be conveyed through periodic bulletins on hazards and media containing all the information, including the hazard-zoning maps. The Commission would plan and coordinate the efforts of different government agencies and voluntary organizations mobilized to cope with hazard relief and rehabilitation. It may emphasized that each government agency through its normal activities is expected to provide leadership and take action to reduce the risk and minimize impacts. **When disasters happen the chief of the NHMC will act as the fully empowered coordinator (in the manner of the Chief Election Commissioner during elections) and help mobilize or commandeer government personnel and civil defence organization, and strive to provide financial assistance for measures, including relief, medicare, food, sanitation, shelter, unemployment allowances, loans for economic recovery, etc.**

   The NHMC would develop and promote an integrated programme of hazard-zone mapping a landuse classification. This programme can be pursued through academic institutions, if need be a contractual basis. The various maps and comprehensive reports may be put in the public domain, and published and distributed among all agencies, organizations and the interested public.

   Local governments should also be required to identify areas prone to any hazard and publish information regarding frequency of the hazard and anticipated magnitude of risks. Simu taneously, they should identify buildings and structures that are too dangerous for continued or future use. **The local governments would have their own hazard-management plan to be approved and coordinated by the NHMC.** However, other kinds of problems may arise when such an agency starts functioning. The owner of the hazardous land, for instance, may claim that he is affected by loss of value of his property because of the Commission's assertion that the land is dangerous and seek legal aid to counter. This is also true in the U.S.A. (R. S. Yeats, per. comm 1985). Also of interest is **the Swiss case where all property is enlisted and the existence and degree of risk are recorded in the legal offices of the authorities so that any possible buyer land will see immediately if there is any natural hazard risk** (Ives, 1985).
2. As already pointed out, the critical parts of the identified hazard zones should be compulsorı acquired by the governments and converted to parks, forests, fuelwood farms, fodder farms, horticultural gardens, etc. **There is a strong need for legislation enforcing disincentives for use and occupation of the identified hazard-prone areas.**

3. A distinction should be made between a critical structure and ordinary structure. **A *critical structure,* such as a nuclear power plant or a high dam is one whose destruction or severe damage by a natural disaster would cause such extensive damage that it should not be built even if the chance of hazard is relatively small** (R. S. Yeats, per. comm., 1985). An *ordinary structure* (e.g. a building or a bridge) might fail or cause property damage or loss of life but the destruction would not be catastrophic.
4. **Governments should lay down building codes in all populated areas identified as hazard-prone. Building standards should be enforced even by lending and mortgaging agencies, such as banks,** which should not support or finance any construction without first ascertaining that adequate safeguards against hazards have been incorporated into the construction plan. Implied in this regulation is the full declaration by the applicants of reports on the state-of-the-environment of the area, including its vulnerability to hazards.

There is also a need for legislation for mobilizing financial resources for monitoring, research and management of hazards.

Hazard Zone Mapping

e Building Material Technology Promotion Council, Ministry of Urban Development, has published *Disaster Vulnerability Atlas* on the scale 1:2.5 millions (see Gupta et al, 2001). Embodying multi-hazard maps, the *Atlas* shows that the Indian subcontinent is among the world's most hazard-prone area—nearly % of the land is vulnerable to earthquakes, 8% to cyclones, and 5% to floods.

Remote sensing images provide vital information—comprehensive synoptic multi-temporal coverage large areas in real time and at frequent intervals (Rao, 2000). An example is the INSAT picture taken May 9, 1990, that helped timely evacuation of 1,70,000 people from the coastal belt of Andhra Pradesh, before it was ravaged by a mega cyclone.

The various methods of identifying hazard-prone or threatened areas are given in Table 9.2, taken m Petak and Atkisson (1982). These methods form the basis for preparing hazard-zoning maps of different lands.

The mountainous regions, such as the Himalaya, are particularly vulnerable to hazards of avalanches, landslides and earthquakes. The alarming expansion of agricultural farms upslope at the cost of forest covers, and various engineering activities related to development of resources and augmentation of communication-transportation have greatly exacerbated occurrence of hazardous events. This calls for an effective combined hazards map for the mountains.

SUMMARY

ddenness, severity and shortness of duration distinguish hazards from normal natural phenomena. ddenness is not always the rule. Slow moving landslides and swelling clays have caused doors and ndows of houses to stick and floors to crack. **While man has no control over their occurrence, his acts of commission and omission certainly aggravate the consequences of hazards, and sometimes even trigger their occurrence.** The hazard occurrences cannot be prevented in all cases, but the impacts can be made less severe and less risky by taking effective measures of coping with them and by planning for disaster preparedness.

Table 9.2. Methods for Identifying and Avoiding Hazard-Prone Areas through Geophysical Studies a Landuse (W. J. Petak and A. A. Atkisson, 1982)

Hazard	Specific Action or Example
Riverine Flood	1. Delineate floodway, floodway fringe, flash flood, or other Hazard areas
Storm Surge	1. Restrict usage so as to prevent increased water depth or Flow velocities
Tsunami (Giant Sea Waves)	1. Restrict usage so as to minimize and control runoff
	2. Alert people as to the true degree of hazard of their property.
	3. Public purchase of flood-prone lands to restrict land to open-space use
	4. Public acquisition of easements used to prevent or control development
	5. Tax adjustments to encourage open-space use and control developmen
	6. Flood-conscious governmental policies in extension of public services to flood-hazard areas.
	7. Regulations to prevent obstruction of floodways
	8. Encroachment regulations. Regulation of dykes, levees, dams, sea-wall or channel modifications
	9. Require sellers and brokers to disclose flood hazards in real estate transactions.
	10. Relocation of existing structures.
Wind, Hurricane and Tornado	1. Microzonation to establish sever wind areas
Earthquake	1. Regional investigation of earthquake history of the area.
	2. Geological map of regional and/or local faults, epicenters map(s), fault strain and/or creep map.
	3. Local investigation of the geological conditions at the site that might indicate recent fault or seismic activity.
	4. Surface investigation: geological mapping, study of aerial photographs, review of local groundwater data.
	5. Subsurface investigation: trenching, soil borings, geophysical surveys.
	6. Land-use management: limit types of use in high-risk areas (e.g. no hospitals in an active fault zone).
Expansive Soils	1. Determine swelling potential from soil index properties, colloid content, plasticity index, shrinkage limit, liquid limit, grain-size composition
	2. Microzoning to establish swelling-prone areas.
	3. Alert people as to the hazard to their property.
Landslides	1. Regional mapping and investigations: topography, precipitation, geology
	2. Local investigations: soil borings, geophysical on-site inspections.
	3. Determine slide potential from soil investigations: geometry, soil or rock strength, water table, forces analysis.
	4. Microzonation to establish slide-prone areas
	5. Limit use in high-risk areas subject to catastrophic sliding
	6. Alert people as to the hazard to their property.

The first step in this direction is the preparation of hazard-zoning maps on the basis of geological d geomorphological information and the records of past occurrence of hazards—their frequency, magnitude and extent. These maps specify areas of future possible occurrence or potential threats and ovide estimates of the degree of risks involved—in terms of the probability of frequency of occurrence, relative severity in comparison with other situations, and impacts and cost of relief and rehabilitation of the population affected. On the basis of analysis of previous records and tell-tale changes indicating incidence of natural events, it is possible to warn the public beforehand and prepare for coping with the hazard.

The hazard-management plan requires setting up a fully empowered national Commission for monitoring geophysical changes indicative of occurrence of hazards, preparation of hazard-zoning maps and their publication, and dissemination of information to educate the public and mobilize resources and personnel from different government agencies and voluntary organizations for coordinated and smooth hazard-coping operations including evacuation, relief, sanitation, medicare, shelter, resettlement, etc. **There is an urgent need for legislation to provide for strong disincentives for use and occupation identified hazard zone.**

10 Earthquakes and Seismic Hazards

TOLL OF HUMAN LIFE

Occurring restrictively on the margins of crustal plates (Fig. 10.1) and in the zones of faults and thrusts that cut the crustal plates into blocks and sheets, the earthquakes cause far greater hazards than any other natural phenomena. This is evident from the death of more than 100 million people since the beginning of the recorded history of man (Table 10.1). In India, more than 70, 000 lives were lost during that period, the hazard being restricted to the northern Indian segments, including Kachchh, Pakistan–Afghanistan.

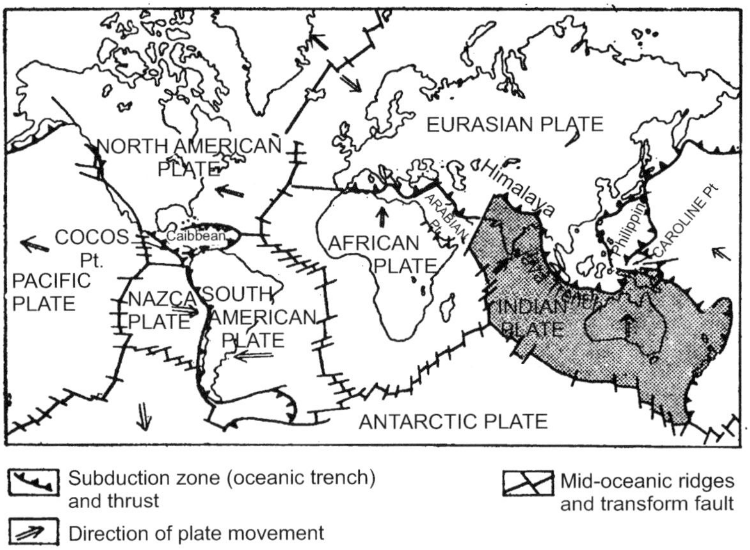

Fig. 10.1. Major crustal boundaries and intracrustal deeper faults where earthquakes are not only frequent but commonly very severe. Thicker lines indicate zones where the crustal plates split up and move apart, while the barbed lines demarcate zones where one plate is sliding under another. Thinner lines in the oceans indicate strike-slip (transform) faults. All these are seismic source zones. (After W. H. Hays, 1981, *Facing Geologic Hazards: Earthscience Considerations,* U.S. Govt. Printing Office, Washington, pp. 36-37.).

Table 10.1. Toll of Human Life from Earthquakes in Asia (Data from D.R. Coates, 1981; H.K. Gupta, 1984 and Valdiya 2004)

Year	Locality	Death Toll
1038	Shensi, China	23,000
1857	Chihlih, China	25,000
1268	Silicia, Asia Minor	60,000
1290	Chiklih, China	1,00,000
1493	Kamakura, Japan	30,000
1556	Shenshi, China	8,30,000
1667	Shemaka, Caucasia	80,000
1737	Kolkata, India	3,00,000
1755	Northern Iran	40,000
1759	Baalbek, Lebanon	30,000
1819	Kachchh, India	15,000
1822	Aleppo, Asia Minor	22,000
1828	Echigo, Honshu, Japan	30,000
1855	Badgam, Kashmir, India	3,000
1897	Meghalaya, India	1,600
1905	Kangra, Himachal Pradesh, India	19,000
1920	Kansu, China	1,80,000
1923	Tokyo, Japan	1,43,000
1932	Kansu, Japan	70,000
1934	Northern Bihar, India	11,000
1935	Quetta, Pakistan	30,000
1939	Erzincan, Turkey	23,000
1946	Honshu, Japan	2,000
1948	Fukui, Japan	5,000
1950	Northeastern Arunachal Pradesh, India	>1,500
1956	Northern Afghanistan	2,000
1957	Northern Iran	2,500
1957	Outer Mongolia	1,200
1957	Western Iran	2,000
1962	Northwestern Iran	10,000
1968	Northeastern Iran	11,600
1975	Patan, northern Pakistan	700
1976	Tangshan, China	2,40,000
1991	Uttarkashi, Uttarakhand	>1,500
1993	Latur, Maharashtra	>9,000
2001	Bhuj, Gujarat	>20,000

Ranges, Himalaya and Assam. In the year 1897 the Meghalaya earthquake—one of the most disastrous natural phenomena—laid waste nearly 380,000 km^2 area and killed in less than a minute more than 1600 people of then sparsely populated region.

The annual toll of human life all over the world is over 16,000. More than a million earthquakes occur annually, but only a small proportion are strong enough to be recorded, i.e. only less than 10 earthquakes have magnitudes of 7.0 to 7.9.

ORIGIN AND SEVERITY OF EARTHQUAKES

Faulting and Earthquakes

Rock masses subjected to pressures or stresses are strained, leading eventually to their breaking along weak zones or faults. Not only is there a sudden stress drop but also rebound of the strained rocks along the fault so that the stresses are suddenly released fully or partially. The breaking of rocks is accompanied by their displacement. **The rupture spreads along the fault surface at a rate of 2 to 3 km per second in all directions in a series of uneven movements. This uneven spreading of displacement leads to bursts of high-frequency waves which travel in all directions producing seismic shocks.** The high frequency content of radiated waves is determined by the level of stress working in the rock failure process.

Table 10.2 Approximate Relationship between the Richter Scale Magnitude *(M)* and Modified Mercalli Scale Intensity *(MM)*(E. A. Keller, 1982)

Richter scale (M)	Area over which seismicity felt (km^2)	Distance upto which felt	Mercalli Scale (MM)	Ground Acceleration (g)
3.0 to 3.9	1,950	29	II–III	<0.15
4.0 to 4.9	7,800	50	IV–V	0.15 to 0.04
5.0 to 5.9	39,000	110	VI–VII	0.06 to 0.15
6.0 to 6.9	1,30,000	200	VII–VIII	0.15 to 0.30
7.0 to 7.9	5,20,000	400	IX–X	0.50 to 0.60
8.0 to 8.9	2,80,000	720	XI–XII	>0.60

The influence of the vibrations causes horizontal as well as vertical accelerations which in the most severe earthquakes may be comparable with the acceleration due to gravity 'g'. The shocks consist of (i) ***surface waves,*** which cause maximum damage by shaking and displacing the ground and thus, resulting in augmentation of the vertical and horizontal acceleration of ***g*** (measurable in terms of cm/s/s), (ii) ***primary or compressive*** *waves* which cause particles to vibrate back and forth in the longitudinal direction of propagation of waves, and (iii) ***secondary or shear waves*** which are responsible for vibration or particles left and right as well as up and down in planes perpendicular or transverse to the direction of propagation.

The locations of stress buildup are the junctions of crustal plates—such as delineated by the oceanic trenches of Oman, off coast Balochistan and of Andaman-Nicobar-Java—by the boundary thrust planes such as the Main Central Thrust and Main Boundary Thrust in the Himalayan region, and by the transcurrent faults in the continents, such as the Palghat-Cauvery Shear Zone in southern India, the Son-Narmada Rift in central India, the Chaman Fault in Pakistan, the Dauki Fault in Meghalaya, the Altyntagh Fault in Tibet. In these belts high pressure builds up due to plate motion, and when it (pressure) exceeds the strength of the rocks, they fail, giving rise to earthquakes.

The length of the fault associated with destructive seismicity (Fig. 10.2a) ranges from about 20 km for earthquakes of *M* 6 to over 1000 km of *M* 8.25 (Press and Brace, 1966). The 1905 Kangra earthquake (*M* 8.6) originated in a 250-km long rupture of the crust, the 1934 Bihar earthquake originated from a 300-km long rupture (Khattri, 1983), the 1897 Meghalaya event (*M* 8.7) is associated with a 400-km long faulting (Chen and Molnar, 1977), and the 1950 Arunachal earthquake was related to a 250-km long rupturing (Singh and Gupta, 1980).

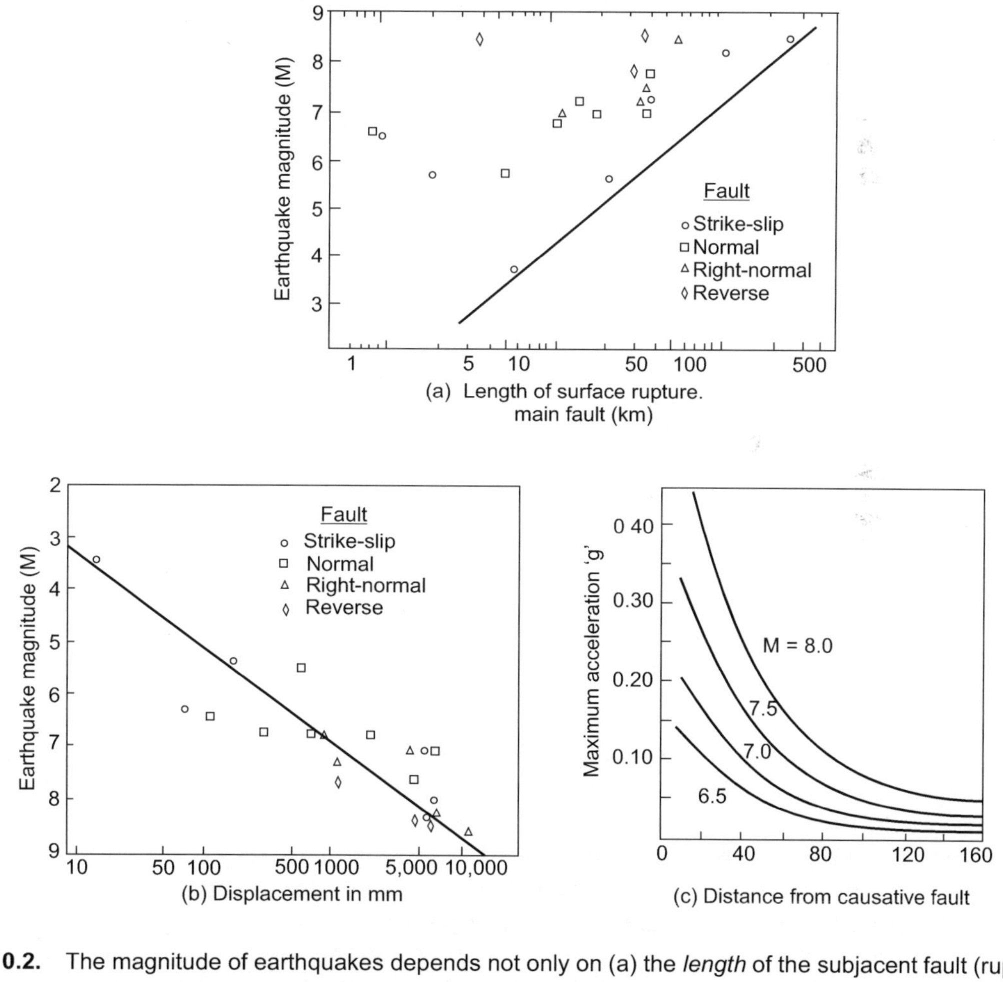

Fig. 10.2. The magnitude of earthquakes depends not only on (a) the *length* of the subjacent fault (rupture), but also on (b) the *amount of movement* taking place on the active fault, (c) the ground acceleration 'g' and thus the degree of hazard increases towards the causal faults. (Based on F. G. Bells (1980), Figs. 11.2, 11.3 and 11.4).

Earthquake Intensity

The severity of earthquakes is expressed in terms of the energy released during faulting. The energy released by the 1950 Arunachal earthquake was 3×10^{27} ergs—several million times more than the energy released by the 1945 atomic explosion in Nagasaki and Hiroshima. The energy is measured on the basis of the amplitude of the largest wave recorded in the seismograms. The amplitude is measured on a linear scale, and the magnitude M in the Richter Scale, the logarithm of the amplitude (Fig. 10.3).

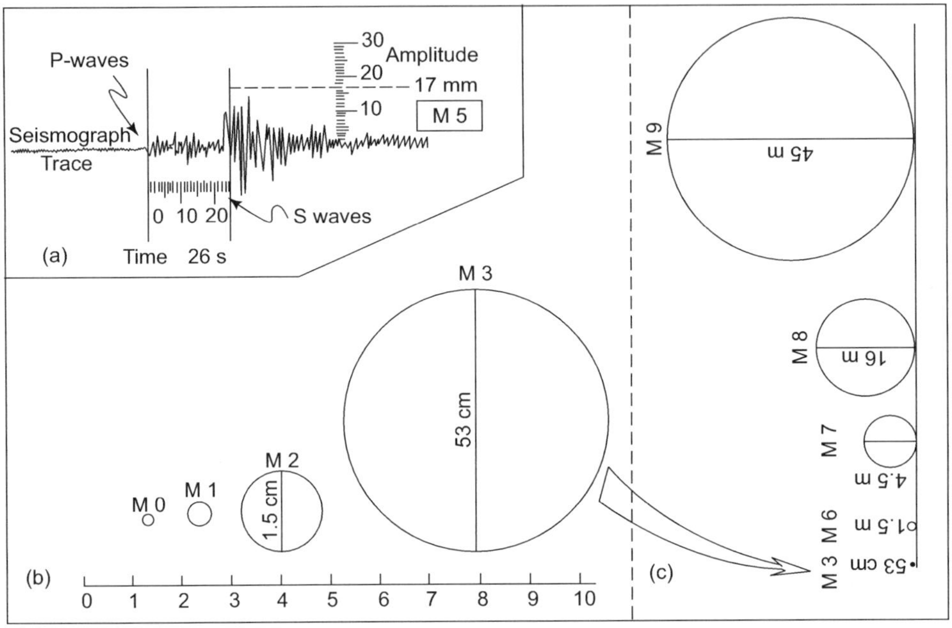

Fig. 10.3. The circles express the measure of comparative severity of earthquakes expressed in terms of magnitude M in the Richter Scale. Each number of this scale indicates 31.5 times greater energy release than the preceding number. The magnitude M measured on a logarithmic scale is expressed in terms of the largest amplitude of the earthquake recorded on a seismograph (inset upper left).

Developed by Charles H. Richter in 1935. The ***Richter Scale*** has a range upto 10. The severest earthquake recorded so far has been M 9.3 (in the Sumatra–Java Trench) in 2004 and the mildest *M* 2.5. **Compared to the preceding number any next higher whole number of this scale represents about 31.5 times more energy release and a tenfold change in vibrational amplitude** (Table 10.3). **Thus an earthquake of *M* 5 has ten times greater amplitude and 31.5 times greater energy release than a seismic event of *M* 4.** The 1934 Bihar earthquake of *M* 8.4 released 2.04×10^{25} ergs of energy compared to 1.65 10^{25} ergs released by the 1935 Quetta earthquake of *M* 7.5, and 3.0×10^{27} ergs released by the 1950 Arunachal Pradesh earthquake of *M* 8.7 (Singh and Gupta, 1980). It may be mentioned that during the Bihar earthquake, within three minutes the towns of Munger in Bihar and Bhatgaon (in Nepal) were in ruins—most of the houses in Purnea and Sitamarhi collapsed, and about 28,500 km^2 area riddled with fissures through which sand and mud erupted in vast quantities.

Moment Magnitude

Now-a-days a new scale called **Moment Magnitude** (M_w) is used. It is a physical quantity related to total energy released in an earthquake, which is estimated by examining the geometry of the causative fault in the field and/or by analyzing the seismogram in the laboratory. The advantage of this scale that it is possible to determine magnitude of all kinds of earthquakes—big and small, near and distant, of past and recent. It is obvious that it is a very reliable scale.

Table 10.3. Energy of Earthquakes (Based on W. J. Petak and A. A. Atkisson, 1982)

Richter-Scale Earthquake Magnitude	Approximate Earthquake Energy (equivalent of explosion of T.N.T.)
1.0	170 gm
1.5	1 gm
2.0	6 gm
2.5	28 gm
3.0	179 gm
3.5	895 gm
4.0	5 t
4.5	29 t
5.0	180 t
5.5	907 t
6.0	5,687 t
6.5	28,616 t
7.0	90,700 t
7.5	9,07,000 t
8.0	56,86,890 t
8.5	2,86,15,850 t
9.0	18,04,93,000 t

The severity of the earthquakes is also measured and described in terms of ***observed*** impacts on the environment, extent of damage caused to the structures and human sensitivities. This meaningful measure is the Modified Mercalli Scale with twelve divisions of intensity, each denoted by a Rom numeral (Table 10.4). The relationship between Richter and Mercalli scales and the range of impacts of earthquakes of various magnitude are given in Table 10.3.

Foreshocks and Aftershocks

Foreshocks and aftershocks occur in the geographic proximity of the main shock. The larger earthquakes are generally followed by aftershocks within the first hour (Jones, 1995). **On the average the magnitude of the largest aftershock is 1.2 unit less than that of the main shock, the difference in intensity**

Table 10.4. Measure of Intensity of Earthquakes According to Modified Mercalli (MM) Scale (E.A. Keller, 1982)

MM Scale	Observation
I.	Not felt except by a very few under especially favorable circumstances.
II.	Felt only by a few persons at rest, especially on upper floors of buildings. Delicately suspended objects may swing.
III.	Felt quite noticeably indoors, especially on upper floors of buildings, but many may not recognize it as an earthquake. Standing motor cars may rock slightly. Similar to the vibration of a passing truck.
IV.	During the day felt indoors by many, outdoors by few. At night some are awakened. Utensils, windows, doors disturbed. Walls make cracking sound. Sensation like heavy truck striking building. Standing motor cars rocked noticeably.
V.	Felt by nearly everyone, many are awakened. Some utensils, windows, etc. are broken. A few instances of cracked plaster. Unstable objects overturned. Disturbance of trees, poles and other tall objects sometimes noticed. Pendulum clocks may stop.
VI.	Felt by all, many are frightened and run outdoors. Heavy furniture may move. A few instances of fallen plaster damaged chimneys. Damage slight.
VII.	Everybody runs outdoors. Damage negligible in buildings of good design and construction, slight to moderate in well-built ordinary structures, but considerable in poorly built or badly designed structures. Some chimneys broken. Noticed by persons driving motor cars.
VIII.	Damage slight in specially designed structures, considerable in ordinary substantial buildings with partial collapse, and great in poorly built structures. Panel walls thrown out of frame structures. Fall of chimneys, factory stacks, columns, monuments, walls. Heavy furniture overturned. Sand and mud ejected in small amounts from underground. Changes in well water. Disturbs persons driving motor cars.
IX	Damage considerable in specially designed structures. Well designed frame structures thrown out of plumb. Great damage in substantial buildings, with partial collapse. Buildings shifted off foundations. Ground cracked conspicuously. Underground pipes broken.
X.	Some well-built wooden structures destroyed, and most masonry and frame structures with foundations destroyed. Ground badly cracked. Rails bent. Landslides considerable from river banks and steep slopes. Shifted sand and mud. Water splashed over banks.
XI.	Few, if any, masonry structures remain standing. Bridges destroyed. Broad fissures in ground. Underground pipe-lines go completely out of service. Earth-slumps and landslips in soft ground. Rail lines bent greatly.
XII.	Damage total. Waves seen on ground surfaces. Lines of sight and level distorted. Objects thrown upward into the air.

varying between 1.0 and 3.0. The intensity and number of aftershocks decrease progressively with the passage of time. The aftershocks can last a few days, a few weeks, a few months or even a few years.

EFFECTS OF EARTHQUAKES

Geological Effects

Surface waves entailing violent ground motion in earthquakes of greater magnitude cause greater damage; for these waves rupture the ground and displace rock masses, sometimes of the order of several metres. In the June 1819 earthquake of Kachchh, the great east–west trending 90 km long Allahband

Fault was formed, resulting in 5 to 8 m subsidence of the ground (with Fort Sindri) and the evolution of the Allahband Ridge. Likewise, the 1897 Meghalaya earthquake caused vertical displacement of 11 m to 30 m along a 20-km long N-S trending fault parallel to the Chidrang River, giving rise to a number of waterfalls and as many as 30 lakes in the course of the river (Wadia, 1975). During the 1892 earthquake a railway line across the Chaman Fault in Pakistan was offset by 75 cm. (The cumulative displacement since the Cretaceous times along this fault is of the order of 200 km). **Uplift and subsidence of the ground is a common phenomenon related to an earthquake**. The movement may be of regional scale affecting large segments of the crust. The great Alaska Earthquake of 1964, for example, uplifted 25,000 km^2 area of Alaska (U.S.A.) giving rise to two major upwarps 1,000 km long and 200 km wide. The inner part subsided by 2.5 m (Keller, 1982). The 1762 earthquake in the Arakan coastal belt uplifted the coast by 3 to 6 m, and brought about the tsunami that ravaged the coastal belt, including Chittagong and Kolkata (Rajendran et al., 2011). During the 1905 Kangra earthquake the Dehradun area is stated to have been raised by 13.4 cm (Middlemiss, 1910).

The ground vibrations cause landslides on unstable hill slopes and trigger avalanches on snow-covered mountains—both of great intensity and wide dimension. The 15th August, 1950 seismic event of *M* 8.6 originating in eastern Arunachal Pradesh ravaged the mountain ranges over a 300 km by 100 km area, the debris amounting to 45,900 million m^3 (M. B. Rao, 1953). East of the Subansiri valley the outer ranges, made mainly of the Tertiary rocks, were scarred with landslides, and the higher ranges of the overthrust metamorphic rocks pronouncedly stripped bare from crest to base in unbroken stretches between Sessini and Dihang valleys in the Abor Hills. About 130 km upstream of Sadia, a 6.5 km by 0.5 km block slid down into the valley of the Subansiri, forming a natural dam. A number of such dams were formed in the valleys all over the ravaged region. The bursting of these dams brought disastrous floods in Assam. The bed of the Brahmaputra was silted up, the floor raised by 1.5 m on average (M. B. Rao, 1953).

Similar widespread landslides occurred in eastern Nepal during the 1934 earthquake and in western Himachal Pradesh following the April 1905 Kangra event. A massive landslide near Larji blocked the passage of the Tirthan and Sainj streams, the tributaries of the Beas. The Maling, Spiti and Parachu valleys in Kinnaur were blocked by huge landslides during the January 1975 earthquake. The 1999 Chamoli earthquake and the 2011 Sikkim earthquake caused excessive landslides.

The valleys in snowy realms witness widespread avalanches, as happened during the 1975 Kinnaur earthquake. In Peru, the 1970 earthquake brought down large avalanches killing more than 70,000 people in the valleys.

Seismites

In the alluvial succession of unconsolidated sediments the earthquake shaking causes the pore-pressure to build up within sediments. If this pressure exceeds the pressure of the overburden , the sediments loose their shear strength and behave like a liquid (Bhandari, 2002), and begin to flow and fill cracks the overlying material, thus giving rise to sills, dykes and what is known as ***seismites.*** It could also come out on to the surface in the form of ***sandboils, sandblows*** and ***mud volcanoes.*** This phenomenon was witnessed in the Rann of Kachchh when the M_W 7.7 Bhuj earthquake hit the region in 2001.

Earlier this phenomenon was witnessed in northern Bihar struck by the 1934 seismic shock of *M* 8.4. Millions of mud volcanoes were later submerged in the vast pool of water that was also squeezed out from the mobilized aquifers in Munger, Darbhanga, Motihari, Muzzafarpur, Sitamarhi, Madhubani and

Purnea districts. The 1897 Meghalaya earthquake likewise caused inundation of low-lying plains with ater with liquefied sediments sprouting out through fountains. Another impact of the earthquake on the groundwater is the dismemberment and choking of artesian wells and tubewells due to earth movements as witnessed in upper Assam during the August 1950 earthquake.

Tsunamis

In the oceanic realm ***giant sea waves or tsunamis*** are generated which travel long distances with a terrific speed of 800 to 900 km/h and strike coast with a speed of 50 to 180 km/h, bringing utter destruction in the seaboards. The November 1945 earthquake that originated in the Makran (Balochistan) or Oman subduction zone was accompanied by a violent tsunami—the tidal waves rising 12 m high in the high seas and about 2 m on the distant Mumbai coast, sweeping away people to death. The 26th December 2004 tsunami generated by an earthquake of M 9.3 along the Sumatra coast in the Java Trench ravaged the eastern coastal belt from Odisha to Tamil Nadu and further south to Sri Lanka. The tidal waves rose as high as 10 to 30 m, and the salt water invaded upto 2 km inland, damaging water bodies and ruining ricultural fields.

Secondary Effects

The destruction of buildings, utility structures, lines of communication, water supply, gas and electricity lines, etc. constitute the secondary effects of earthquakes.

STABILITY OF STRUCTURES AND RISK EVALUATION

Factors Governing Stability

The stability of structures depends on (1) their design and construction, on (2) the strength of the ground, and on (3) the intensity and duration of earthquakes. Ordinarily, structures are built bearing in mind only vertical loads and therefore when they are subjected to horizontal stresses produced by earthquake waves, they are damaged particularly when constructed inhomogeneously or dissimilarly. On the other hand, reinforced concrete (combining the strength of concrete with the tensile strength of steel), withstands greater stresses *when integrated with the foundation.* In the unconsolidated sediments seismic motion is transmitted with higher magnitude as compared to that in harder rocks. As already mentioned, **the extent of damage depends primarily on the amplitude as well *duration of the ground shaking* and the failure of the soil or slope.** The 1897 Meghalaya earthquake lasting 2.5 minutes affected a 200,000 km^2 area, while the 1950 Arunachal earthquake which shook the ground for 4 minutes, completely ravaged 45,000 km^2 area and affected another 30,00,000 km^2 region. **The duration of the earthquake depends on the magnitude, which in turn, depends on the length of the rupture (faulting)** (Fig. 10.2 B).

The second factor is **the *peak acceleration* of the strong ground-shaking. In the construction of buildings, utility structures, dams and bridges, quantitative estimation has to be done of the ground-shaking—peak acceleration—which the particular structure can withstand.** The peak vertical acceleration of ground is about half the mean horizontal ground acceleration (Bolt et al., 1975). The acceleration, understandably, is maximum at the epicentre and diminishes away from it.

Therefore, the closer an area is to the active fault, the greater would be the damage to the environment and structures.

The geological (structural) conditions, particularly active faults, joints and fractures, have great bearing on the stability of structures.

Seismic Risk Evaluation

Seismic risk evaluation is a three-point programme: *First*, the pattern of distribution and the frequency and severity of earthquake-induced phenomena such as landslides, mudflows, subsidence, ground upwarpin liquefaction of underground sediments, tsunamis, etc. are studied and analyzed. The information on the rate of slip along active faults is significant and also the data pertaining to post-faulting even *Second*, the fluctuations of the water table, and variations in discharges of wells and springs are studi and inferences drawn. *Third*, the seismicity history of the region in terms of location, magnitude and intensity is compiled, and attempt made to construct recurrence curves of the frequency of all magnitudes of events in order to estimate the frequency of occurrence of the damaging earthquakes.

The geological and seismological data provide the basis for predicting the magnitude and intensity of damaging earthquakes likely to occur in the region.

The next step is to investigate the shear- and bearing-strength of the soils or foundation material on the basis of their physical properties.

SEISMICITY CONDITIONS IN INDIA

Regional Pattern

About 50 to 60% of the Indian landmass is prone to seismic activity of varying intensity (Figures 10.4 and 10.7). This is particularly strong along the margins of crustal plates or blocks where they go down or slide under the adjoining plate or block and along the zones where moving plates were faulted and sideway movements have taken place in the past. In the first category is the Main Boundary Thru

Table 10.5. Region-wise Earthquake (M>5.0) Occurrence in India (1897–1937) (Gupta et al, 2001)

Seismic region	Number of Earthquakes of Magnitude				Return period
	5.0–5.9	6.0–6.9	7.0–7.9	8.0+	
Kashmir and Western Himalaya	25	7	2	1	2.5 to 3 yrs
Central Himalaya	68	28	4	1	1 yr
Northeast India	200	128	15	4	< 4 months
Indo-Gangetic basin and Rajasthan	14	6	–	–	5 yrs
Gulf of Khambhat and Rann of Kachchh	4	4	1	1	20 yrs
Peninsular India	31	10	–	–	2.5 to 3 yrs
Andaman & Nicobar Islands	80	68	1	1	< 8 months

with its associated Himalayan Frontal Fault in the frontal belt of the Himalaya (Figure 10.4 and Table 10.5). In the second category is the Narmada–Son Rift constituting the great geological divide between northern and southern India.

Seismicity of varied degree is exhibited by the fault-ridden south-central part of the Peninsular India, the Koyna-Bharuch-Kachchh belt in western India, and the Himalayan-Indo Myanmar ranges in the north (Figs 10.4 and 10.7). The largest number of epicentres are located in the Himalayan belt where the seismicity is (Table 10.5) strongest.

Himalayan Belt

The tectonic stress in this mountain world is primarily caused by the Peninsular Indian plate sliding under the Himalaya at the rate of 17.7±2 mm/yr (Molnar, 1990) to 20±3 mm/yr (Freymueller et al, 1996), even as the Indian subcontinent converge towards mainland Asia at the rate of 58±4 mm/yr (Bilham et al, 1998). The tectonic compression, produced by the converging continents is responsible for strain accumulation in the Himalaya as well as in the framework of the northward-moving Peninsular Indian shield (Valdiya, 2001, 2010). The abrupt and fast relaxation of the accumulated strain causes rock bodies to break and to reactivate old faults and consequently the Indian plate to shake violently.

The zone of the Main Boundary Thrust (MBT) has been the locus of great earthquake. Three great earthquakes occurred in the last one hundred years: 1905 (M 8.6) of Kangra in northwest Himalaya, 1934 (M 8.4) in northern Bihar and adjoining Nepal and the 1950 (M 8.7) in northeastern Arunachal. **e MBT extends northwards, flattening with depth as a subsurface detachment plane, between the** very thick Himalayan sedimentary pile and its foundation or basement made up of the northerly prolongation of the Indian Shield. Movements caused by the sliding of the Indian plate along this detachment thrust plane are responsible for frequent earthquakes of moderate to low magnitudes in the belt of the inner Lesser Himalaya — along the middle longitudinal terrain that rises in elevation between 900 and 2500 m. This nearly 50 km wide belt of high seismicity (lying south of the zone of welding of Indian subcontinent with the Asia continent) is among the most populated regions of the mountain domain.

A few of the significant earthquakes that struck this region may be mentioned—1803 (M>7.5<7.9) western Uttarakhand, 1833 (M>7.5<7.9) of west Nepal, 1916 (M 7.5) of far-western Nepal, 1975 (M 6.8) of Kinnaur in eastern Himalachal, 1980 (M 6.3) of Dharchula in northeastern Kumaun, 1991 (M 6.6) of Uttarkashi, and 1999 (M 6.9) of Chamoli in Uttarakhand, September 18, 2011, M 6.9 of Sikkim. The northwestern and northeastern ends of the Himalayan arc exhibit extraordinarily high seismicity (Figure 10.4). This is because the projecting promontories of the northward-moving Indian Peninsula are consistently pushing these parts of the mountain arc (Fig. 10.5).

The higher seismicity of the northern part of the Tibetan plateau and adjacent regions is ascribed to dominant strike-slip movements along the many E-W to ENE-WSW trending transcurrent faults (Molnar and Tapponier, 1975). These movements represent accommodation of the crust consequent upon the northerly drift of the Indian plate. The seismicity in southern Tibet adjacent to the Himalaya is mild, and the distribution of epicentres is sparsely dispersed indicating that the tectonic junction of the Indian and Asian plates is now practically inactive.

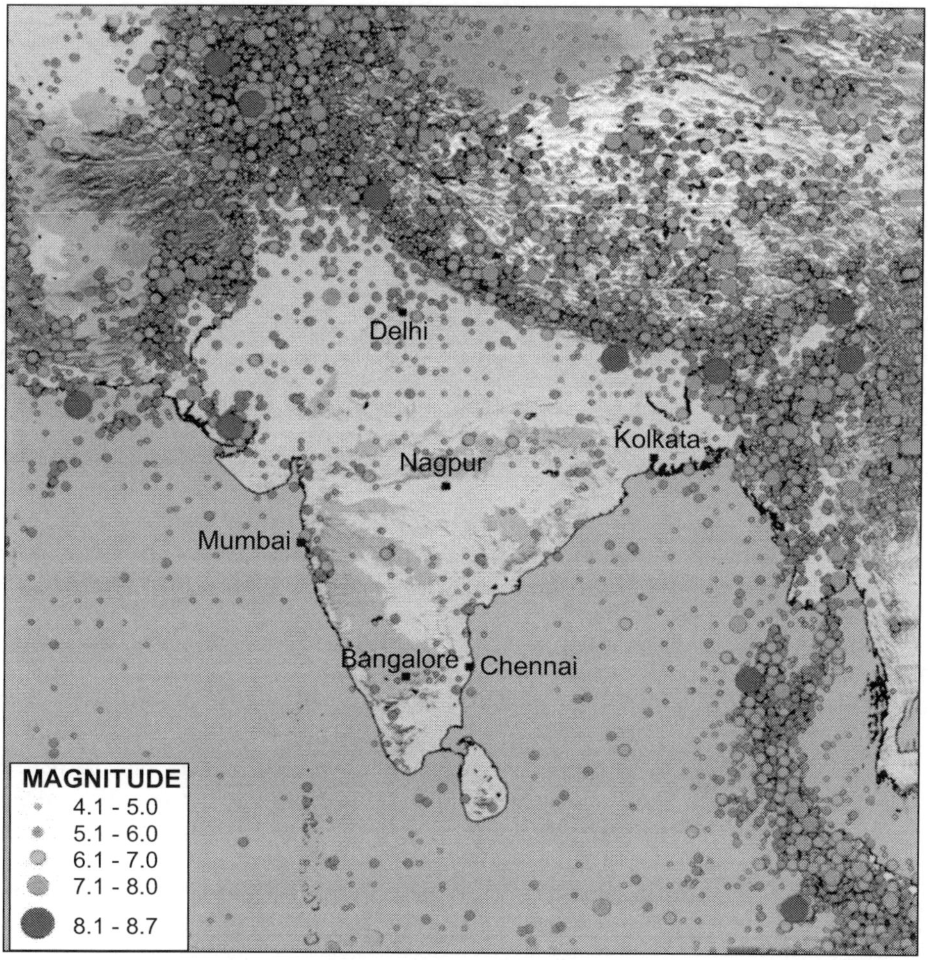

Fig. 10.4. Distribution of earthquake epicenters of magnitude M>4. (From an Annual Report of the Centre for Mathematical Modelling and Computer Simulation, Bangalore; Courtesy: Dr Gangan Prathap). (See colour Fig. 10.4)

Stable Continental Region

The seismicity in the Peninsular India (Figures 10.4 and 10.7) is mild and restricted principally to NNW-SSE, N-S and NE-SW trending faults and fractures. The northward-advancing Indian Shield itself has experienced about a dozen earthquakes of M>6 since the eighteenth century (Ramalingeswa Rao, 2000). Of the many earthquakes that have struck this stable continental region since the fourteenth century, the significant ones are listed as: 1341 of the Coromandal coast in Tamilnadu, 1964 (M 6.0) Mahabaleswar in Maharashtra, 1803 (MM IX) of Mathura in Uttar Pradesh, 1819 (M 6.0) of Kachchh in Gujarat, 1827 (MM IX) of Lahore in Pakistan, 1846 (M 6.5) of Jabalpur in Madhya Pradesh, 1848 (MM IX) of Mount Abu in Rajasthan, 1897 (M 8.7) of Shillong in Meghalaya, 1900 (M 6.0) of Coimbatore

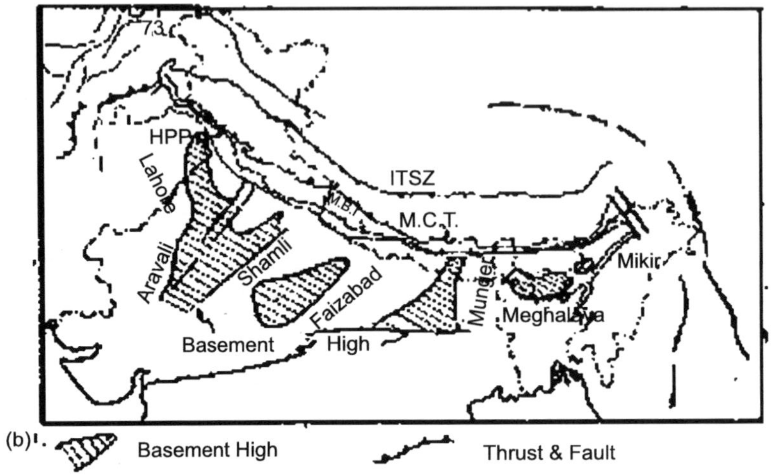

Fig. 10.5. The Himalayan arc is proded by the projecting promontories of the Peninsular shield. The promontories are of the hidden basement highs under the thick piles of sands of the Sindhu, Ganga and Brahmaputra. (After K. N. Khattri, 1983 and Khattri and Tyagi, 1983.

in Tamilnadu, 1918 (M 7.6) of Shrimangal in Bangladesh, 1919 (M 6.3) Bhavnagar in Gujarat, 1927 (M 6.5) of the Son valley in Madhya Pradesh, 1930 (M_s 7.1/M_w 6.4) of Dhubri in Assam, 1938 (M 6.2) of Khandwa in Madhya Pradesh, 1956 (M_s 6.1) of Anjar in Gujarat (Kachchh), 1967 (M 6.4) of Koyna in Maharashtra, 1993 (M_b 6.3) of Latur in Maharashtra, 1997 (M_b 6.0) of Jabalpur in Madhya Pradesh, and 2001 (M_w 7.7) of Bhuj in Gujarat (Kachchh) (Valdiya, 2004, 2010). It is noticed that seismicity largely confined to the linear belts representing faults, shear zones of rift valleys of great antiquity. Obviously the older faults and shear zones were reactivated giving rise to these earthquakes.

The earthquake that shook Coimbatore in Tamil Nadu on February 8, 1900 (M 6) originated in the E-W trending Palghat-Cauvery Shear Zone (Krishna Brahman and Kanungo, 1976). The small-magnitude seismicity of the Kerala coast is attributed to vertical movements on the NNW-SSE striking lineament extending from the Periyar Reservoir to Kozhikode on the coast. Primarily the intersections of the NNW–SSE and E–W trending faults are the locations of the earthquakes. The Bhadrachalam event of April 13, 1969 (*M* 5.7) on the bank of the Godavari River in Andhra Pradesh, which originated at the depth of 10 to 15 km, is related to the movement along the graben fault (H. K. Gupta et al., 1970).

able 10.6. Some Larger Earthquakes in the Indian Subcontinent

Year	Richter Scale (M)	Focal Depth (km)	Locality	MM Scale
5 July 1720			Delhi	VIII–IX
Oct. 1737			Kolkata	VIII–IX
5 July 1803			Mathura, Badrinath	VIII–IX
7 June 1819	8.0		Kachchh	IX–X
4 Sept. 1827			Lahore	VIII–IX

Contd.

Year	Richter Scale (M)	Focal Depth (km)	Locality	MM Scale
6 June 1828			Srinagar	IX–X
19 June 1845			Lakhpat, Kachchh	VII–VIII
30 May 1885			Badgam, Kashmir	IX–X
12 June 1897	8.7	18–25	Meghalaya	XII
8 Feb. 1900	6.0	30–40	Coimbatore, Tamil Nadu	VII
4 April 1905	8.6	44–100	Kangra, Himachal Pradesh	X
12 Dec. 1908	7.5		Northern Myanmar	
28 Aug. 1916	7.5		Bajang, Northwestern Nepal	
8 July 1918	7.6	21–25	Srimangal, Bangladesh	X
4 July 1930	7.1	25–32	Dhubri, Assam	IX
15 Jan. 1934	8.4	39–79	North Bihar	
30 May, 1935	7.6		Quetta, Pakistan	X
20 Oct. 1937			Dehradun, Uttarakhand	VII–VIII
7 Nov. 1937			Srinagar, Kashmir	VII–VIII
26 June 1941	8.1	92.5	Andaman Island	VII–VIII
23 Oct. 1943	7.2		Northeastern Assam	
27 Nov. 1945			Makran, Pakistan	>X
10 July 1947			Bhadarwah, Jammu	VII–VIII
29 July 1947	7.9		Northeastern Assam	
15 Aug. 1950	8.7	28–50	Eastern Arunachal Pradesh	XII
18 Feb. 1955			Quetta, Pakistan	VII–VIII
21 July 1956	7.0	23–25	Anjar, Kachchh	IX
29 Sept. 1960	6.0	30–40	Delhi	VII
15 April 1964	5.5	4–9	Kolkata	VIII
12 Jan. 1965	6.1		Central Nepal	
27 June 1966	6.0		Dharchula–Bajand (Indo-Nepal Border)	
10 Dec. 1967	6.4	11–14	Koyna, Maharashtra	VIII
13 April 1969	5.7	16–18	Bhadrachalam, Andhra Pradesh	VII
23 March 1970	5.8	25	Bharuch, Gujarat	VIII
29 Sept. 1973	7.1		Northeastern Assam	
19 Jan. 1975	6.8	18–22	Kinnaur, Himachal Pradesh	IX
31 Dec. 1984	5.4	5	Kachar, Assam	
25 Aug. 1988	>7		Udaipur, Rajasthan	
20 Oct. 1991	6.6		Uttarkashi, Uttarakhand	
30 Sept. 1993	6.3		Latur, Maharashtra	
26 Jan. 2001	7.7		Bhuj, Gujarat	
18 Sept. 2011	6.8		Sikkim	

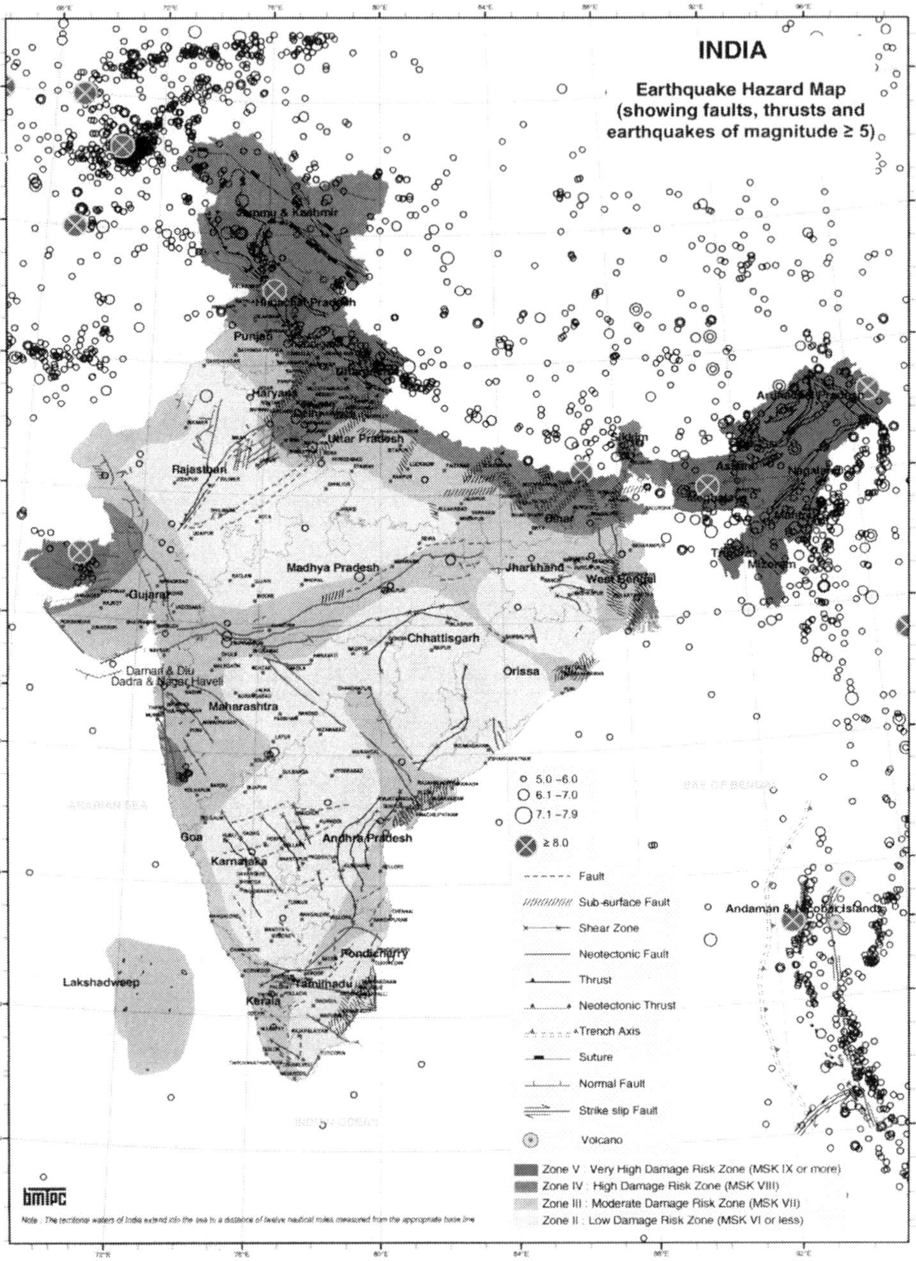

BMTPC : Vulnerability Atlas –2nd Edition; Peer Group, MoH&UPA; Map is Based on digitised data of SOI, GOI; Seismic Zones of India Map IS:1893 -2002, BIS, GOI, Seismotectonic Atlas of India and its Environs. GSI. GOI

Fig. 10.7. Earthquake hazard map showing causative faults, thrusts and also the earthquakes of magnitude ≥ 5. (Courtesy: Dr C.P. Rajendran, IISc., Bangalore)
(See colour Fig. 10.7)

In western Maharashtra the Koyna earthquake, particularly of December 10, 1967 (M6.4) resulted from movement at the depth of 10 km on a N 10°E-S 10°W trending fault beneath the pile of Deccan lavas (H. K. Gupta et al., 1979, 1980). This area is repeatedly shaken by earthquakes, six of which were of M5 and more than a hundred of $M \geq 4$ (1967-1984). This region is under tensional stress, causing normal faulting associated with grabens. The epicentre of the Bharuch earthquake of March 23, 1970 was located on the intersection of the boundary faults of the Narmada and the Sabarmati grabens, both concealed under thick cover of alluvial sediments. The earthquakes that frequently rock the Delhi area spring from movements along the NNE-SSW trending faults including the Sohna-Sonipat faults delimiting the Aravali ridge.

SPATIAL AND TEMPORAL GAPS IN SEISMICITY

Locked Segments

Investigations on the active San Andreas Fault in the U.S.A. have revealed very significant seism behaviour along its different segments—some parts are seismically very active and characterized by displacements and other parts quite inactive or locked (Fig. 10.8). In these locked segments, where no movement is taking place today, the strain energy is progressively stored, as manifest in the occurrence of very small-magnitude micro-earthquakes. Eventually, the concentrated stored energy would be released suddenly and violently in the form of major earthquakes. **The locked parts of the seismic belts are therefore likely candidates for greater earthquakes in future** (Larkiser et al., 1969). Thus the seismic gaps (Figs. 10.8 and 10.9) are the segments of faults where no large earthquakes occurred in the duration comparable with the recurrence time (Khattri and Wyss, 1978). **These have been identifi as potential sites of future greater earthquakes and thus areas of increased seismic hazard.**

Sites of Potential Seismic Hazard

The seismic hazard map of the northern part of the Indian subcontinent (Khattri and Wyss, 1978; Khattri and Tyagi, 1983; Khattri, 1983) shows three major spatial seismic gaps—the 500-km long Jammu gap west of the 1905 Kangra epicenter, the 900-km long Uttarakhand–Nepal gap between the Kangra and Bihar (1934) epicenters, and the 240-km wide Arunachal gap between the epicentres of the 1934 and 1950 earthquakes (Figs. 10.5 and 10.9). These are the sectors which have not ruptured in the last 200 years or more while the intervening segments show high seismicity. Since the repeat period of rupture in some sections for great earthquakes was 150 years (Khattri and Tyagi, 1978), these locked segments were recognized as sites of potential larger earthquakes by some seismologists. The 1905 Kangra, 1934 Bihar and 1897 Meghalaya earthquakes are related to the ruptures in the basement highs, respectively the Lahore-Sargodha, the Munger-Saharsa and the Shillong forming locales asperities and thus developing high stresses. It can therefore be surmised that the Faizabad basement high in the Ganga Basin (Fig. 10.5) could be the future site of the rupture and attendant seismicity, as Khattri has pointed out. The 18 September 2011 M_w 6.9 earthquake in Sikkim is a pointer to this sector being a geodynamic hotspot to be watched (Valdiya, 2011). Likewise the Mikir Hills high can serve as the location of a future high-magnitude earthquake in the seismic gap in Assam plain (Fig. 10.9). The north-eastern extension of the Delhi ridge (Aravali), which is geophysically recognizable upto Shamli in Muzaffarnagar district in northwestern Uttar Pradesh (Fig. 10.5), provides yet another asperity for developing high stresses in the Uttarakhand gap. As pointed out by Gaur (1984), the earthquakes of magnitude less than M 8 do not release sufficient strain, and the last major earthquake of M>7 having occurred way back in 1828, the is a probability of a major seismic event in Garhwal.

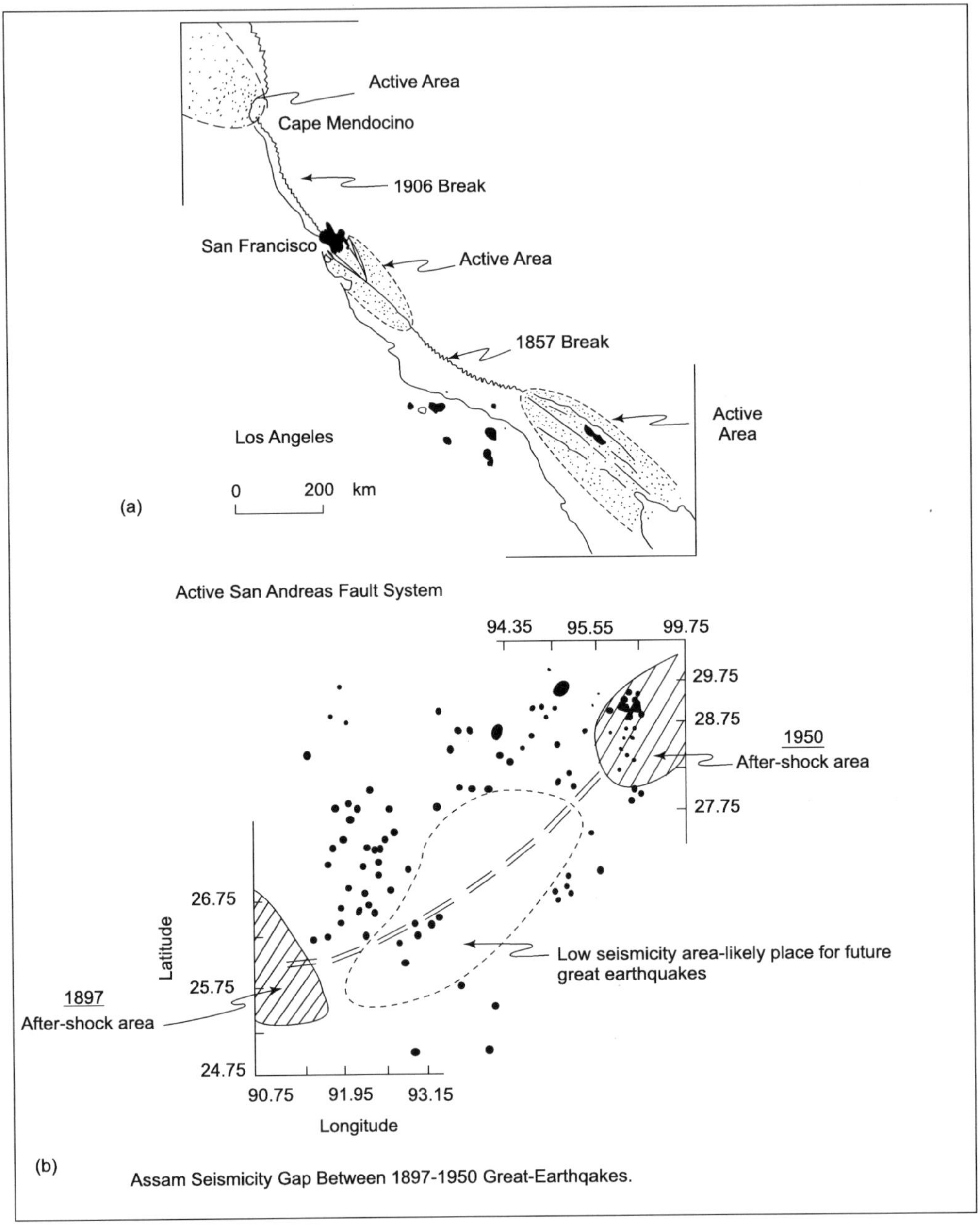

Fig. 10.8. (a) The seismotectonically locked sections of the San Andreas Fault (California) which have remained inactive since 1906 and 1857. (After R. W. Tank, 1923, *Focus on Environmental Geology*, Oxford University Press, New York, 534 p.)
(b) The Mikir seismic gap in Assam between the segments that ruptured in 1897 (Meghalaya) and 1950 (Arunachal Pradesh). (After K.N. Khattri and Wyss, 1978)

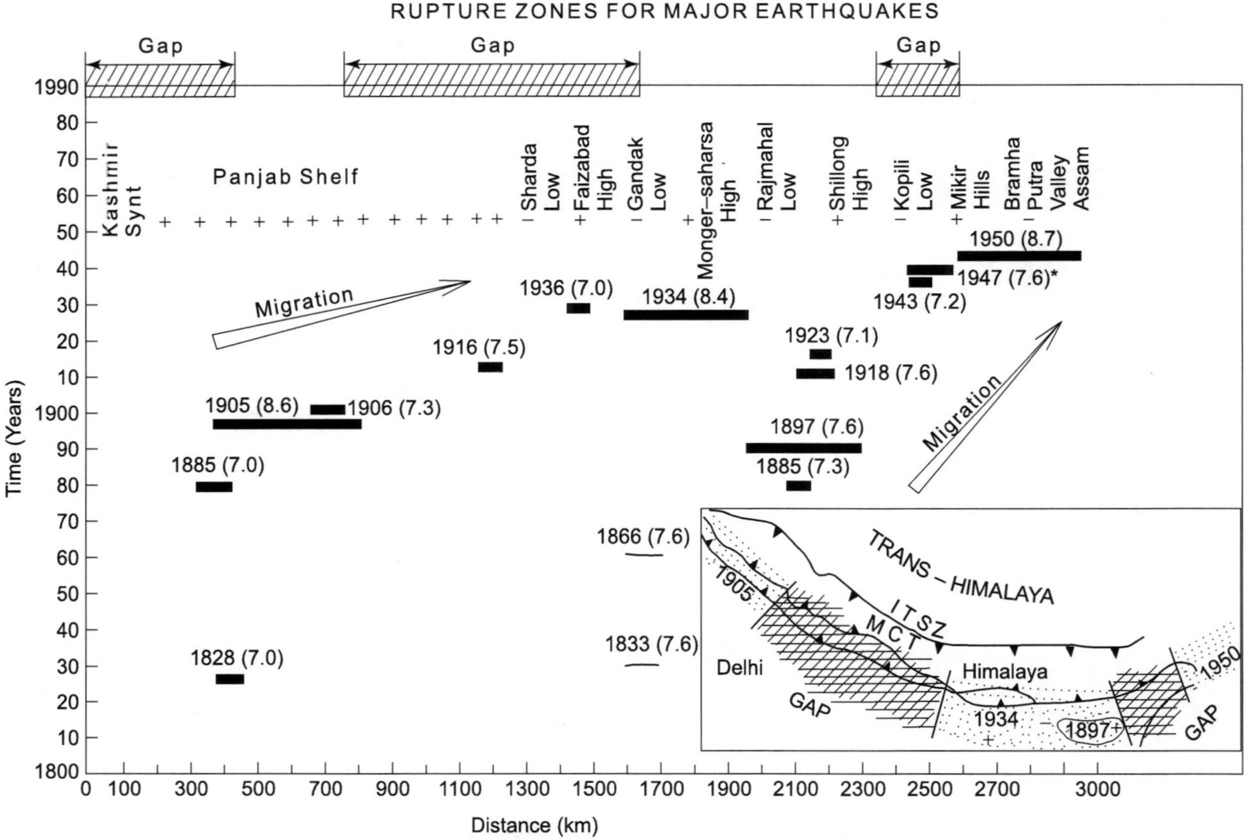

Fig. 10.9. Three possible seismic gaps and the pre-existing rupture zones of major earthquakes (M≥7) in the Himalayan seismic source zone. The plus (+) sign indicates basement high and the minus (–) denotes depressions under the Indo- Gangetic Plains. (After K.N. Khattri and A.K. Tyagi, 1983).

Taking into consideration the data from 1800 to 1976, and noting that the seismicity has been migrating eastward in the Himalayan front and north-eastward in the Assam region (Fig. 10.9) Khattri and Tyagi (1983) concluded that all great earthquake of $M \geq 8.0$ were preceded by seismically quiescent periods—19 years in the case of the Kangra event and 28 years in the case of the 1897 Meghalaya earthquake. **There is in fact a general reduction in seismicity following the great earthquakes** (Table 10.7). A remarkable decrease in seismicity following the year 1970 was observed along the 400-km section of the Himalaya east of the Kangra epicentre. Since 1967 the seismicity decreased in the region between the 1897 and 1950 epicentres in Assam (Table 10.7).

able 10.7. Duration of Seismic Quiescence before the Assam Earthquakes (K. N. Khattri and Wyss, 78).

Zone	Year	M	Duration of Quiescence (Yr)
I	1869	7.8	23
	1897	8.7	28
	1918	7.6	19
	1922	7.1	5
II	1947	7.7	17
III	1950	8.7	>30

COPING WITH SEISMIC HAZARDS

The prerequisites for the measures to cope with seismic hazards include preparation of seismic hazard maps and collection of information leading to forecasting. **The forecasting is related to the anticipated severity of earthquakes, the likely places of risks, and the bearing capacity and shear strength of earth material and its sensitivity to ground-shaking.** On the basis of hazard-zoning maps and information gathered, necessary steps are taken to cope with the hazards, including application of technology to reduce underground stress buildup. Another **important item of the programme of hazard management is the continual monitoring of the geophysical properties of the ground, changes in the water table and the physiography and the microseismicity.**

Hazard-Zoning Map

The seismic zoning maps (Fig. 10.7) delineate various source zones of earthquakes on the basis of location of epicentres, predicted level of shaking and acceleration of ground-motion that can be expected in future and the estimated risks. Seismic zoning maps provide records of the time and magnitude of occurrence and the location of epicentres. **The seismic zoning map is then superimposed on a detailed geological map showing in particular active faults and thrusts and geomorphic features bearing evidence of neotectonic activities, including landslides. These maps provide the basis for designing and building dams, nuclear power plants and other heavy structures that would remain safe during the occurrence of earthquakes.**

Earthquake Prediction

To anticipate an event is tantamount to preparing for coping with the consequences. From a study of the seismic gaps and locked sections in the source areas, and the cyclicity of seismic activity, the seismologists are in a position to indicate the possibility of recurrence of earthquakes in potential areas. **It must be confessed, however, the precise prediction remains an elusive, perhaps as yet unattainable goal.** The modern technique of mapping of seismic hazards is based on studies related to accumulation and transmission of elastic energy in geological structures. A few successful predictions of occurrences of events on this basis do not imply that a generally applicable and reasonably reliable scheme of prediction is available.

(i) **Perhaps the most dependable method of forecasting is the analysis of the patterns of microseismicity (swarm of small earthquakes) that precedes major events.** More than 40% of the larger earthquakes have foreshocks that occur days or hours before the main events (Jones and Molnar, 1976). Using this experience gathered by the Japanese,—together with the interpretation of a number of parameters of the phenomenon—the Chinese were able to predict accurately the Haicheng earthquake of February 4, 1975 (*M* 7.3) which ravaged 90% of the structures but without loss of life among the population of 9,00,000. The 15th August, 1950 earthquake (*M* 8.7) in Arunachal Pradesh was preceded by foreshocks on August 12—three days ahead, but no notice was taken of this incidence. On the contrary, the 28th July, 1976 earthquake (M 7.8) of Tangshan, 150 km east of Beijing, was not preceded by any foreshock. The human toll exceeded 240,000. In the U.S.A. seismologists had noticed significant decrease from 1.75 to 1.6 in the velocity rat of P and S waves before the major earthquakes but increasing to normal after the main event. Yash Aggarwal et al. (1973) had then successfully predicted the earthquake of August 3, 1973 (*M* 2.5) in the Appalachians on this basis.

(ii) By monitoring the variation of geoelectric parameters of rocks in space and time followed by careful analysis of the profile of earth resistivities and natural indicators, it is possible to identı epicenters, estimate the magnitudes and times of earthquakes as is being done in Japan, China, Russia and Greece (Nayak, 2004).
There is "surface manifestation" of regional thermal upwarp of the electromagnetic spectrum of thermal regime which is detectable and mappable by satellite and computer assisted systems with periodic monitoring" (Ramamurthy, 2011). It was noticed that 10 to 40 hours before the occurrence of an earthquake, changes in geomagnetic field cause shifting of radiofrequency, disturbances television reception, and difficulties in wireless and telephone communication as was notic prior to the 1993 Latur earthquake (Ramamurthy, 2011).

(iii) The earthquake originates as a result of faulting following buildup of stresses in brittle rocks. The stress buildup entails volume expansion (dilatancy) and development of tiny fractures until the rocks break up along fault planes. The penetration of groundwater—a good conductor—into these fractures lowers the electrical resistivity and geomagnetic field of the rocks (Figs 10.10 a 10.11). **The continual monitoring of electrical resistivity thus provides a precursor indication of the event.**

(iv) The amount of radioactive radon gas dissolved in deep groundwater also increases significant before the occurrence of an earthquake as was first noticed in the case of the 1966 earthquake th rocked Tashkent in Uzbekistan (Figs. 10.10 and 10.11). There is thus a strong correlation between precursor fluctuation in randon concentration and the earthquake to occur as noticed in China and Japan (Majumdar, 2004).

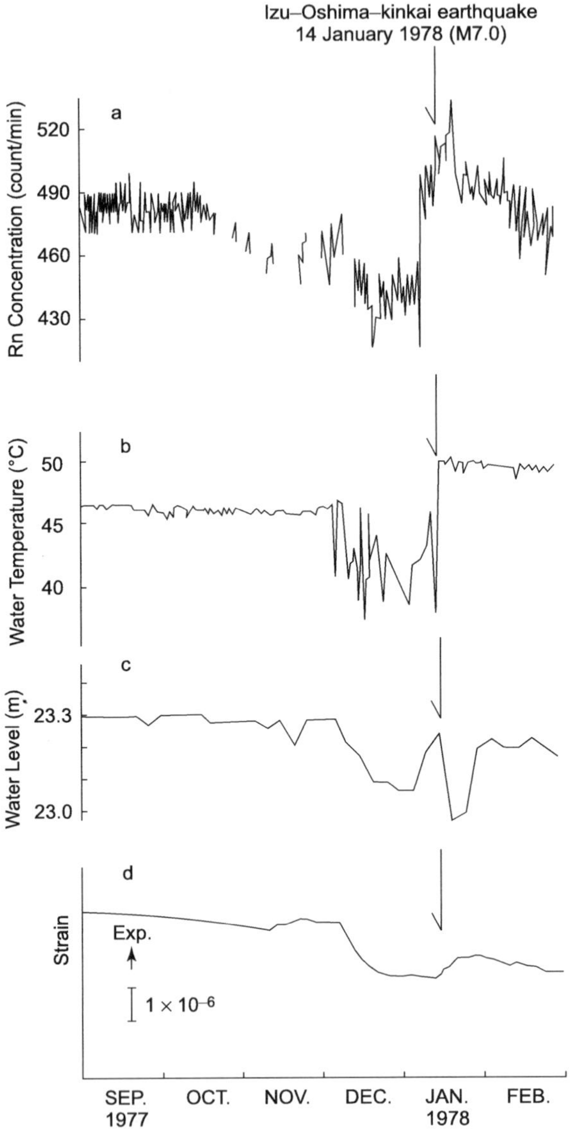

Fig. 10.10. Before the occurrence of the 1978 Izu–Oshimakinkai earthquake of M 7.0 in Japan, the precursory changes in water level, water temperature and radon concentration in well water 30 km away from the epicenter (From Majumdar, 2004).

v) Closely related to the development of fractures in the rock masses under stress is the fluctuation of water level in wells and the variations in discharges of springs. In some cases the water level in wells changed as much as four metres, as noticed before the 15th August 1950 earthquake in Upper Assam, and the February 1975 Haicheng earthquake in China (Figs. 10.10, 10.11 and 10.12). There was perceptible to drastic changes in the spring discharges prior to the 1999 Chamoli earthquake in Uttarakhand.

(vi) **Prior to the main shock, the physiography also exhibits perceptible progressive changes—the ground bulges, as witnessed in California** over the 84,000 km^2 area which rose as much as 35 cm during the period 1959-1968 and by 45 cm between 1972 and 1974 (Keller, 1982). The uplift of the Californian land was followed by subsidence by 17 cm between 1974 and 1977. The present author had noticed similar phenomena—bulging up and sinking of ground—in Kachchh before the 2002 Bhuj earthquake. The uplift of the ground, measurable through tilt-meters, is taken as a long-term precursor of a large earthquake.

(vii) The Chinese—who have been keenly interested in seismology from the ancient times — have successfully monitored abnormal animal behaviours to forecast earthquakes. Cattle, sheep, horses, dogs, cats, rats, snakes, etc. have an uncanny perception of the coming events and start behaving very abnormally and queerly—dogs bark untiringly, agitated cattle and horses break away from their leashes and run in circles, rats come out of their holes and appear dazed or stunned, hens refuse to lay eggs, rabbits jump around with raised ears, terrified pigeons keep on flying, toads st laying eggs and may run out of their places and so on. This happens presumably because there is pronounced changes in atmosphere and the ionosphere over ground or possibly because there is sharp rise in the number of charged particles in the atmosphere. The 4th February, 1975 earthquake of Haicheng was successfully predicted, partly on the basis of the anomalous behaviour of animals and partly on erratic fluctuation in the water table (Fig. 10.12) over a large area extending 150 k from the epicentre (Deng Qidon et al., 1981). Another successful prediction on the basis of animal behaviours coupled with geophysical changes was made for the August 1976 Songpan-Pingwa (Szechuan) earthquake in Yunnan, southern China.

A day before the Kangra earthquake on April 4, 1905, animals in the Lahore Zoo were extremely agitated, just as one day ahead of the December 2004 earthquake that generated a destructive tsunami, the animals of the Andaman Island were highly disturbed (Ramamurthy, 2011). A day before the January 17, 1995 Kobe earthquake in Japan the sea lions and hippopotamus in the city zoo showed strange behaviour—they sank within the pools and would not surface. Likewise in a garden at Iza-Atagawana the crocodiles stopped eating and the silkworms and fish aligned themselves in a common directi before the earthquake (Ikeya et al, 1997). Obviously, the animals had strongly felt the perturbation in the ionosphere or the effects of piezoelectric changes in the ground.

Among the animals, the common toads *(Bufu bufo)* have proved to be most reliable indicators of the occurrence of earthquakes. At L'Aquila in Italy prior to the occurrence of an earthquake of M>6.5 in 2009—which killed over 300 people, and rendered 50,000 homeless—the breeding activity the toads in the areas 74 km away from the epicenter declined by 96% a week before, and 100% on the day of the earthquake (Grant and Halliday, 2010). Similar phenomena was noticed in the Wenchuang area in China hit by an earthquake of M 8.0 on May 12, 2008.

Induced Seismicity and Earthquake Control

One of the ways of moderating the intensity of earthquakes would be to precipitate small harmless events *(M < 5)* in order to prevent the stress buildup to the levels leading to larger, dangerous earthquakes.

is can be achieved by artificially causing rupture of tectonically stressed rocks by detonation including nuclear explosion, which induces creeps in the fault zone. This is achieved also by injecting ids through wells and holes. Properly spaced (20 to 50 km apart) nuclear detonations (each of 1 to 10 megatons) placed deep underground (3000 to 5000 m) and exploded every 10 to 25 years have successfully released accumulated stresses in the erstwhile U.S.S.R. and in the U.S.A. Alternatively, by locking the fault zone by pumping out water from deeper levels through series of drill holes in an active zone would accomplish much the same thing, i.e. temporarily control or delay the occurrence of earthquakes.

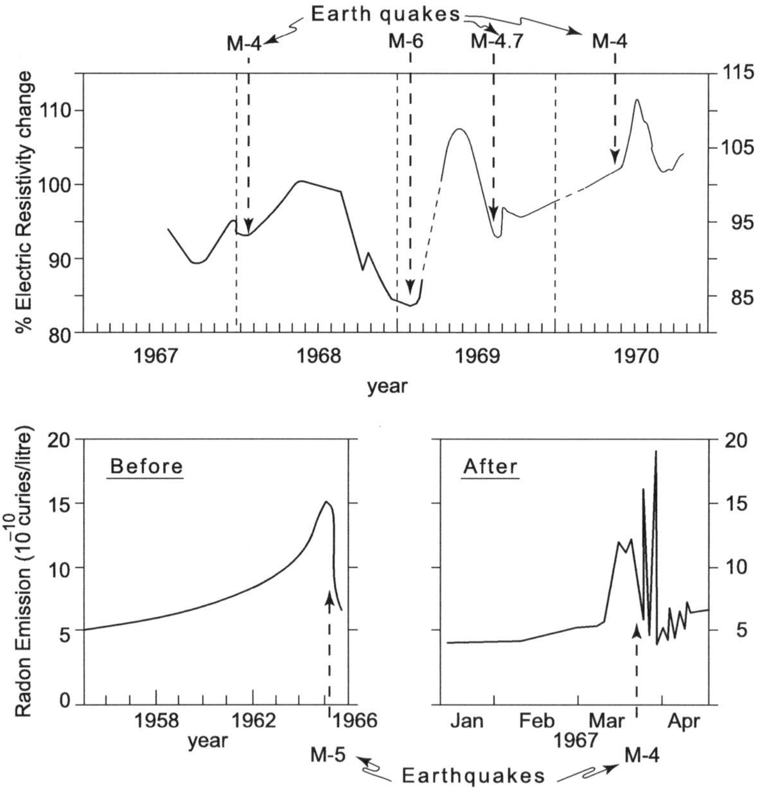

Fig. 10.11. Before the earthquake the electrical resistivity of the earth's crust dramatically drops. Likewise, the emission of radon gas (measured in deep wells) suddenly increases just before the earthquake and then drops precipitately after the event.(After E. A. Keller, 1982, *Environmental Geology,* Bell & Howell, London, 488 p., Figs. 621 and 630.)

Seismicity can also be induced by placing load, such as of impounded water in reservoirs on the ground. Induced earthquakes due to reservoirs formed behind dams more than 100 m high have been discussed in Chapter Seventeen on the dams. The Koyna earthquake of 1967 and subsequent events lesser magnitude in Maharashtra provides an eloquent testimony. The injection of fluids in deep lls can increase seismicity, as first noticed in Denver (Colorado, U.S.A.) where the waste water from chemical industries fed through deep (3600 m) wells in the highly fractured gneisses caused increase in seismicity—to 700 earthquakes of M 4.3 in the period April 1962 to November 1965. There was direct correlation (Figure 17.22) between the rate of injection and the frequency of earthquakes (Keller, 1982).

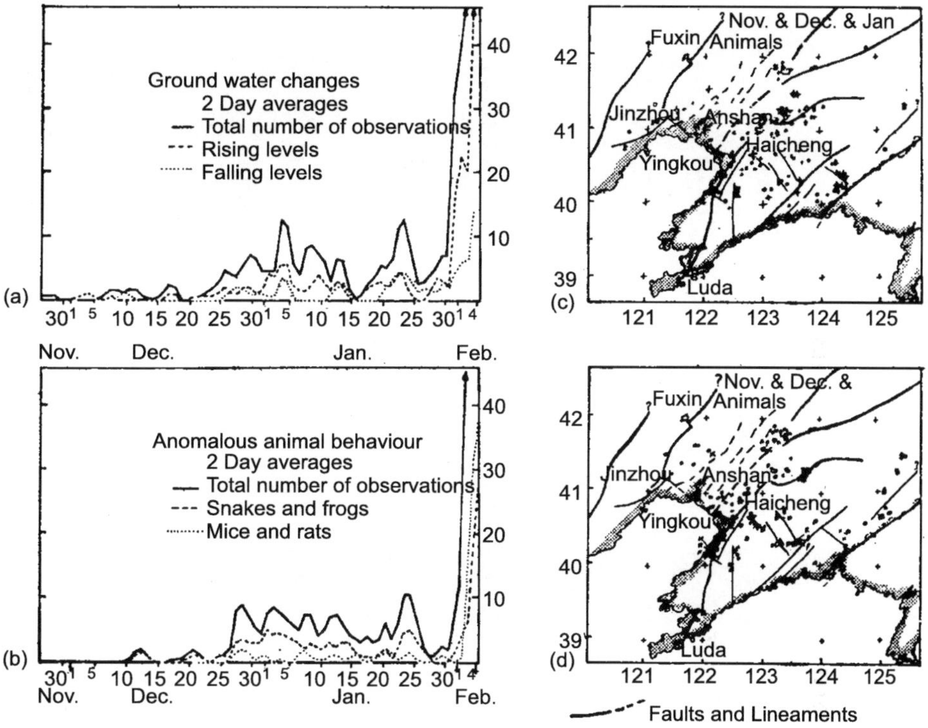

Fig. 10.12. The tectonic map of Haicheng region in the northeastern corner of China, where groundwater fluctuation and anomalous animal behaviour for the months November to January 1974-19 were precisely recorded. The graphs show 2-day averages made daily of the data on changes recorded. (After Deng Qideng, Pu Jiang, L. M. Jones and Peter Molnar, 1981).

EARTHQUAKE-RESISTANT HOUSES

The majority of people falling victim to natural disasters belong to the poorest sections of the poor countries of the world. **The mortality is largely due to house collapse. It is therefore imperative that greater attention be paid to designing and constructing earthquake-resistant dwellings for the poor communities.** There are, however, many constraints—the low income of the poor, their inability to obtain requisite building material, and the continued decline in the supply of timber, so greatly needed for lighter structures suitable in seismicity-prone areas. Then there is the crucial factor of the sensitivity of the people to local traditions of housing, which allows little modification in the traditional designs buildings.

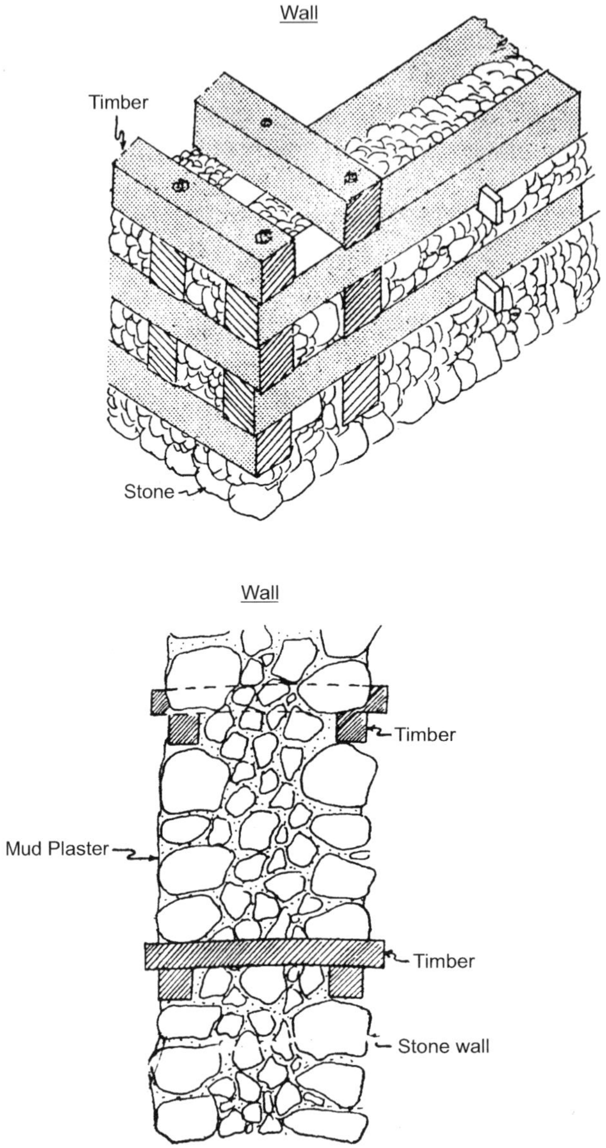

Fig. 10.13. The earthquake-resistant timber-laced houses, as seen in the Patan area, Kohistan (Pakistan). After A.W. Coburn, R.E. Hughes, D. Illi, D.F.T. Nash and R.F.S. Spence, 1984; and Davis, 1984).

The primary requirements for preventing house collapse are (i) **monolithic roof with sufficient in-plane rigidity**, (ii) a strong and durable vertical load-carrying system, and (iii) a lateral load-resisting system of enough residual capacity to safely resist lateral force (R. Razani in Coburn et al., 1984).

For the Poor

Keeping in view the constraints and primary requirements, the following modifications in buildin (Davis, 1984) can be introduced: (1) The foundations should be deeper than 0.3 m in order to prevent subsidence. (2) The height: width ratio of the walls should be small to ensure stability. (3) The structure can be built on a grid of wooden or reinforced concrete beams or some form of horizontal timber coursing. The close layers of timber (when and where available aplenty) are interlocked with notches (Fig. 10.13). (4) The wall plate of timber or reinforced concrete tie the walls together and serve to distribute the load of roof beams (Fig. 10.14); and (5) The masonry or brick or mud blocks are inserted between the wooden components in the timber frame. The mortar or mud should be of sufficient adhesive streng If the load of the roof can be transmitted independent of the wall to the ground through columns sunk into the ground and are resting on stone pads, the building will be able to withstand earthquake shock (Coburn et al., 1984). The stone pads are surrounded with stone to provide stability against displacement. The roof structure can be strengthened by addition of a second transverse set of beams and by connecting together all beams and wall plates. The lateral stability can be improved by providing column-caps or metal straps or bracelets at the top of the column (Fig. 10.14).

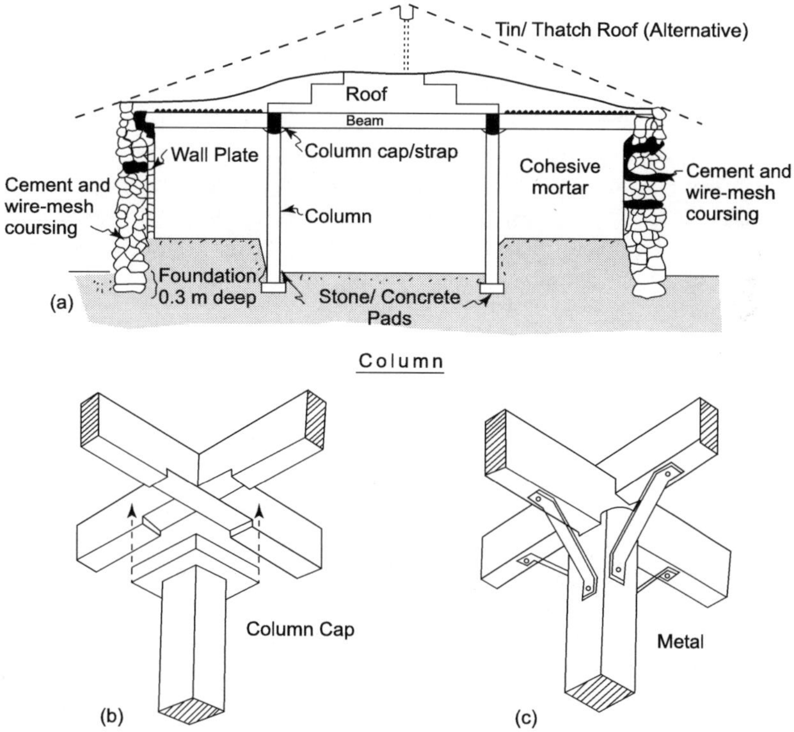

Fig. 10.14. Structural modification in the dwellings (of the Yasin-Gilgit area in northwestern extremity Kashmir Himalaya) to increase earthquake resistance. These can be used with suitable chang in the Indo-Gangetic Plains and elsewhere also. The improvisation of column caps or metal brackets increases lateral stability of the column heads. (After A. W. Coburn, R. E. Hughes, D. Illi, D. F. T. Nash and R. J. S. Spence, 1984).

For All Sections of Society

In order to ensure safety of buildings in earthquake-prone areas, it is necessary that the (i) tensile strength and shearing strength of walls be high, (ii) the bends between walls and corners and T-junctions of walls e unbreakable and (iii) floors and roofs are stiff and connected to the walls, as they serve as a medium to distribute the load (Arya, 2001).

The **building foundation** should not only be solid and go deeper into the ground but also be continuous — without break all along the perimeter (Fig. 10.15A). **The buildings having continuous solid foundations are stronger than the ones that stand on pillars or piers.** This is because when pressed, pushed or shaken, a continuous foundation does not permit shearing or displacement or shifting of component layers of the structure in relation to each other.

The loosely aggregated sediments and soils of flood plains, coastal belts and river-bank terraces have very low capacity of bearing the loads of buildings. In such places, **the foundations should have wider footing so that the load of the structure is distributed over larger area.** The walls of the superstructure need to be bolted to the foundations, and also linked together so that the entire building acts like one unit. Ordinarily, the width of the foundation for one-storey building should be one half of a metre and for double-storey building about three fourth of a metre.

The building walls should be mutually bonded by steel rods, wall plates or wire nets at a few vels (Figs. 10.15A, B). This would impart sufficient rigidity to the structure. In the earthquake-prone areas, houses are being made with reinforced concrete pillars. **It should be ensured that the lars are firmly linked together with horizontal concrete beams or bands. The horizontal steel rods in the bands/beams must be held together by lateral ties.** Where the earthquakes are frequent or strong, the buildings need to be provided with two reinforced concrete bands — one at the lintel level d another at the foundation level (Figs. 10.16B, C). Mere tying the steel rods with finer wires would not work — they would snap when shaken. The steel rods that join the walls with the foundations or a wall with another wall should be either riveted together at the extremity or twisted in a manner that there is no loosening when the structure is shaken (Figs. 10.15B, C).

Since windows and doors usually cause problems when the buildings are shaken, it would be desirable to have reinforced concrete beams above their frames (Figs. 10.16C). Column caps or metal brackets increase the lateral stability of the columns.

It is always **better to have light roofs,** more so in an earthquake-prone areas. If that is not possible, **the load of the roof should be transmitted to the ground by constructing additional pillars independent of the walls** (Fig.10.16A). These pillars should either penetrate deeper into the ground or t on stone pads or firm foundation.

The gable band in the building (Fig. 10.16B) can be strengthened by rafter held down to the walls by bolt.

It is better to have a building with square plan. Rectangular buildings are vulnerable to shaking. The buildings with T shape, L shape and H shape plans (Fig. 10.17) are quite unsuitable in the hazard-prone areas, particularly if shaken repeatedly or strongly owing to their proneness to displacement of parts in relation to each other. Partitioning of the rectangular building into square-shaped rooms (Fig. 10.17B) strengthens the frame of the structure. Construction of pillars at the edges of walls and in the middle provides a degree of rigidity to the rectangular building.

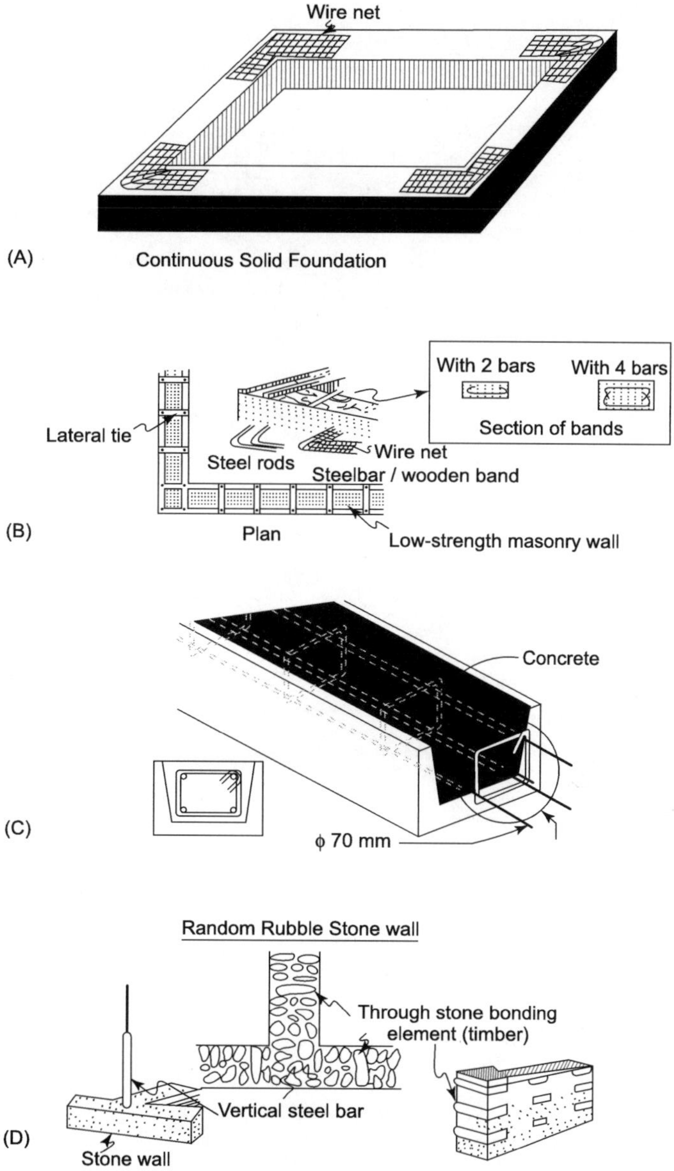

Fig. 10.15. (A) Building foundation must be made continuous all along the perimeter; and its sides link together by wire nets or steel rods so that the whole frame behaves as one unit.(B) Stonewalls are strengthened by placing steel rods that are laterally tied up. In the inset is shown how t rods are twisted firmly at their ends to prevent slipping.(C) Horizontal steel rods in a beam band are held together by lateral ties by twisting their ends.(D) Foundations and walls made of rounded stones are strengthened by fixing vertical steel rods, and/or horizontal lacing—even by putting elongate rough stones across the length.(From Valdiya, 2008).

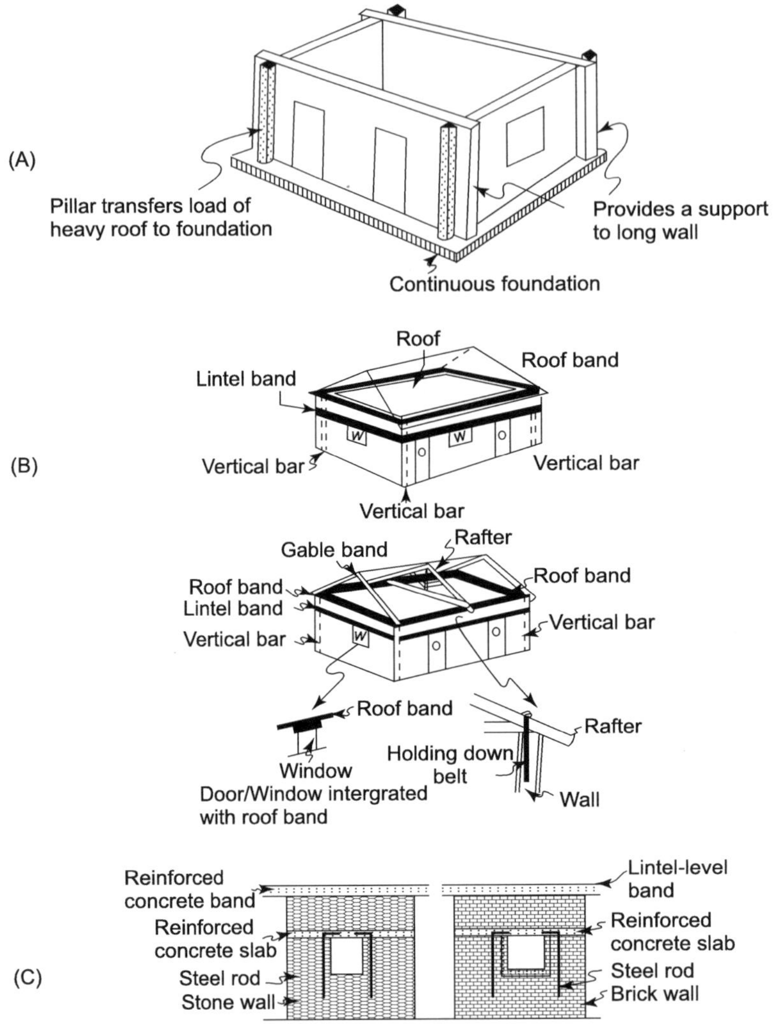

Fig. 10.16. (A) The pillars on the corners of the building transfer the load of the heavy roof to the ground and make the building stable. (B) Concrete bands or beams, reinforced appropriately by steel rods, are constructed at the levels of the foundation and the lintel to impart strength to the buildings. (C) Over the frames of doors and windows reinforced concrete slabs ensure stability of the walls. (From Valdiya, 2008).

If there is no option but to make a rectangular building, it would be advisable to reinforce it by constructing strong pillars at the four edges as well as in the middle (Fig. 10.17B). Appropriate partitioning into square rooms would also make the structure resilient to the hazards. **The T-shaped, L-shaped and H-shaped buildings can, likewise, be split up into two or three units, each independent of the other** (Fig. 10.17A). In other words, each unit should have its own foundation, walls and roof, but can

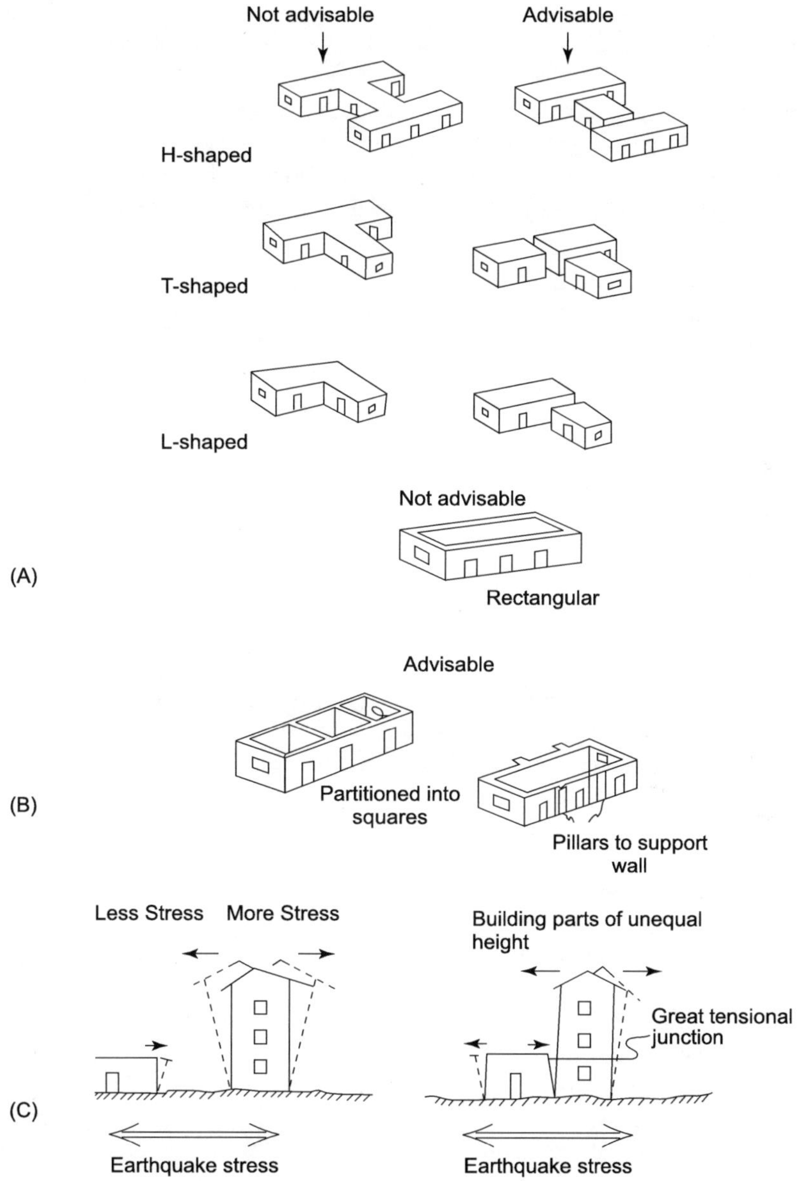

Fig. 10.17. The shape of the building is an important factor in its stability against ground shaking.(A) There is possibility of displacement or shifting of one part of the structure in a T-shaped, L-shap or H-shaped building (left side). However, if the buildings are split into two or three units, ea with its own foundation, walls and roof, they become safer.(B) Rectangular buildings should partitioned or provided with extra pillars to support the walls to ensure their stability.(C) Taller buildings suffer more from shaking than the smaller ones. And if the different parts of the buildin have different height, the damage would be relatively great.(From Valdiya, 2008).

stand together snugly forming a T, an L or an H. In the event of earthquakes, each unit will respond individually, independent of the other units. There will be no shearing when the units are independent of each other.

Another important fact to be borne in mind is that **the buildings should have uniform height**. Unequal height of different parts results in shearing and collapse when shaken (Fig. 10.17C).

Older houses need to be retrofitted, by fitting with a component or accessory that would increase the rigidity of the building frame. This is a job of the architects or engineers, and is an expensive exercise.

UMMARY

Although the occurrence of earthquakes is limited to the belts riven with intracrustal or deeper faults, the hazards are very severe and widespread, commonly along the boundaries of crustal plates. The annual toll the world over is in the order of 16,000 people. In the Indian subcontinent more than 70,000 people were killed since the beginning of recorded history.

When the strain on a part of the crust exceeds the limit of endurance of rocks, it ruptures along weaker zones or faults, and the strained body rebounds along the plane of rupture until the elastic stress is wholly or partly relieved. This and the uneven dislocation generates bursts of high-frequency seismic waves spreading out as ground-shaking in horizontal and vertical directions. It is the acceleration of ground movement that is most hazardous. The severity of an earthquake depends on the amount of energy released, and is expressed in terms of amplitude of the largest wave recorded on the seismograph—measured on a logarithmic scale and denoted by magnitude *M*. Compared to the preceding number, each whole number of this Richter Scale represents about 31.5 times more energy release. The severity of an earthquake is also indicated in terms of observed impacts on the environment, extent of damage caused various structures, and human perception. This is the Modified Mercalli *(MM)* Scale.

Among the primary effects of earthquake-waves are rupturing and displacement of the ground, widespread and severe mass-movements or landslides on hillsides (such as those produced by the 34 Bihar and 1950 Arunachal earthquakes), changes in the groundwater circulation and fluctuation the water table, squeezing out of liquified sediments with water from the aquifers in the form of mud-volcanoes (e.g. 1934 Bihar, 1897 Meghalaya and 2001 Bhuj earthquakes), bulging or uplift of the ground as happened in the Dehradun valley during the 1905 Kangra earthquake. The secondary effects resulting from ground shaking are destruction of or damage to buildings, dams, bridges, cultural features, lines of communication and transportation, essential supply lines etc. The extent of damage depends on the duration of ground shaking, which depends on the magnitude of the earthquake (which turn is influenced by the length of the crustal rupture). The second factor having significant bearing on the stability of civil structures is the peak acceleration of ground shaking. The structures have to be designed to withstand ground shaking, taking into consideration the shear strength and bearing capacity of the earth material.

Seismicity of various degrees is exhibited by fault ridden Karnataka shields, the Koyna-Bharuch-Kachchh belt in western India, the Pakistan-Afghanistan ranges, the Himalayan orographic arc and the Indo-Myanmar mountain ranges. The severest and highest seismicity is shown by the triangles between the Himalayan and Indo-Myanmar ranges in the north-eastern extremity and the Himalayan and Hindukush ranges in the north-western corner.

In order to cope with seismic hazards, it is necessary to identify potentially hazardous belts, and if possible, to forecast the time, magnitude and place of future earthquakes, and also make a guess-estimation of the risks likely to be generated in the event of occurrence of earthquakes. Short-range prediction is achieved through monitoring ground-tilt, underground strain-buildup, microseismicity pattern, changes in electrical resistivity and magnetism of the rocks, increase in radon emission, and fluctuation of the water table underground.

The larger earthquakes are commonly (in 40% cases) heralded by foreshocks and phenomenal increase in microseismicity and dramatic drop in electrical resistivity. Strange and erratic behaviours of birds and animals, particularly domesticated animals and the common toad who become extremely nervous and agitated, are reliable indicators of impending larger earthquakes. It must be conceded, however, that the value of anomalous animal behaviour in predicting is as yet not fully established.

In order to reduce the dimension of hazards, stress buildup to a dangerous level must be prevented by initiating or precipitating smaller events more frequently. This can be done by artificially induci small-magnitude ($M < 4$) earthquakes in seismically locked belts (gaps) of major fault zones. Th has been achieved by detonating nuclear devices placed deep underground at appropriate intervals of time and space, and by injecting fluids through deep wells which causes increase in fluid pressure a thus facilitates movements along faults and microfractures. By putting the weight of water impounded behind high dams, earthquakes can be induced.

The seismic gaps are the sections of fault planes where no large earthquakes have occurred for the duration comparable with the recurrence time period. These locked sectors are potential sites for great earthquakes in future. The crust in these segments is progressively storing strain energy, which when exceeding the limit of endurance, is released suddenly and violently. In the north Indian seismic zone, the sectors west and east of the 1905 Kangra earthquake, particularly the mountain arc in front of the Aravali Ridge, and subsurface Bundelkhand–Faizabad Basement Ridge, and the sector of the Brahmaputra valley between the 1897 Meghalaya and 1950 Arunachal earthquakes are the likely candidates for a future earthquake of $M \geq 8$.

Those living in the regions frequently rocked by earthquakes, should design earthquake-resistant houses and utility structures.

11 Instability of Hill-slopes and Landslides

LANDSLIDE-PRONE AREAS

In mountain belts such as the Himalaya and the Shyadri, the problem of hillside instability is very serious. The occurrence of landslides is particularly common in geodynamically sensitive belts, i.e. in the zones and areas repeatedly rocked by earthquakes and affected by other neotectonic activities (Fig. 11.1), particularly where human activities are strong. The Darjiling Himalaya once recorded more than 20,000 landslides in one day (Ives & Messereli, 1981) in the vulnerable belts. Landslides in Nepal take a human toll of 350 per year on an average (*ICIMOD Newsletter* No. 39, 2001, p. 26). Between 1983 and 2000, more than 2300 heads of livestock were lost, over 6700 houses destroyed and 4800 ha of cultivated land damaged by landslides debris. The situation in Uttarakhand is no better. The rockfall companied by debris flow killed 221 pilgrims to Mount Kailas on August 1998 at Malpa in the Kali lley (Sharma and Sanwal, 1998; Paul et al, 2000). Disastrous landslides killed 300 people in Assam ls in July 1991, 500 people with 200 houses in Kohima (Nagaland) in August 1993, 107 people and 3 animals with 525 houses in Okhimath area (Uttarakhand) in August 1998 (Bisht and Sah, 1999), d 60 persons with 14 houses at Dhaila, Himachal Pradesh in August 2007 (Anon, 2010). In September to 21, 2010 following 33.6% higher rainfall all over Uttarakhand and cloudbursts in the Kosi basin umaun (Uttarakhand) witnessed large-scale mass movements—landslides, debris flows, rock falls, etc., killing 220 people, more than 84 heads of cattle, 2672 houses wholly destroyed or substantially damaged (Sati et al., 2011).

The principal factors that initiate or trigger mass-movements are: (i) heavy and prolonged rainfall, (ii) cutting and deep excavations on slopes for buildings, roads, canals and mining without appropriate posal of debris, and (iii) earthquake shocks and tremors. All these factors operate in the Himalayan d Sahyadri belts, the latter including the Nilgiri massif. In every slope, gravity-induced shear stresses ıst which increase with the slope height, slope inclination and unit weight of slope-forming materials. earing stresses are also added—particularly in the surface zone—because of thermal expansion and contraction, freezing and thawing, shrinkage and swelling and by action of other agents of weathering. When the shearing stresses build up to a level that exceeds the shearing resistance of the slope along the most vulnerable or critical or the weakest surface, a landslide takes place, inducing downward and outward movement. When supplemented by seismic activity, they could be very serious.

The Himalaya is seismically the most active segment of the Indian subcontinent, its north-eastern and rth-western ends being repeatedly and violently rocked by earthquakes of considerable magnitudes. The rainfall over wide areas is in excess of 2000 mm/yr in the outer ranges. The network of roads (>

225,000 km in aggregate length) constructed—without any consideration for the vulnerability of t region—entailled generation of colossal volumes of debris that intensified erosion. There is widespre deforestation. The population pressure is forcing people to move up the steeper forested slopes with their ploughs and livestock. A similar situation prevails on the equally vulnerable slopes of the Nilgi Annamudi massif in South India.

Fig. 11.1. A massive landslide in the Guptkashi area in Uttarakhand Himalaya blocked the passage of the river. (Courtesy: Dr M.P. Sah, Wadia Institute of Himalayan Geology, Dehradun). (See colour Fig. 11.1)

DESTABLIZING FORCES AND MASS-MOVEMENTS

Forces Involved

Factors governing stability of slopes and causing mass-movements in any area are: (i) slope angle a altitude, (ii) lithology, (iii) structure, (iv) hydrological conditions, and (v) seismicity in that area. Forces operating on slopes are (a) weight of the material that build the slope, i.e. rocks, soils, vegetation and man-made structures, which act downwards, and (b) the shear strength of these materials acting on t potential slip planes which resist sliding or dislocation.

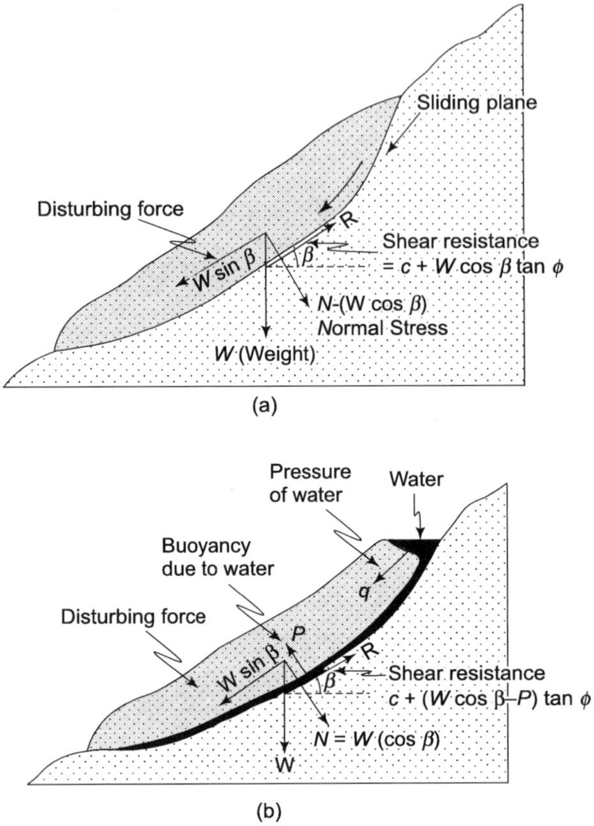

$$\text{Safety Factor} = \frac{c + (W \cos \beta - p) \tan \phi}{W \sin \beta + q}$$

c = cohesion, ϕ = angle of internal friction

Fig. 11.2. (a) Relationship of shear strength (the force resisting movements) with shearing stresses (the driving forces) in a sliding mass.(b) Water is the principal agent of destabilization.(Based on F. G. Bell, 1980; B. A. Bolt, H. L. Horn, G. A. Macdonald and R. F. Scott, 1975; and S. N. Davis, P. H. Reiton and P, Pestrong, 1976).

Water is not only the main agent of weathering of rocks producing soft cohesionless and yielding aterial, but also the principal cause of mass-movements. It reduces the shear strength (force of resistance). The relationship (Fig. 11.2) of shear strength*s* to the causal forces is given by the following rmula (Dunne and Leopold, 1978):

$$s = c + (N - P) \tan \phi$$

where c = cohesion of the components,

N = component of weight of the saturated mass acting perpendicular to the plane of failure or slippage, i.e. $W \cos \beta$,

P = pressure of water within the voids of sliding mass, and

ϕ = angle of internal friction, reflecting degree of interlocking and surface friction.

The safety factor (F) which is taken into consideration to ensure the stability of the slopes (Fig. 11.2) depends upon the ratio of forces of resistance (shear strength) to the driving force.

$$\text{Safety Factor } (F) = \frac{\text{Resisting Moment}}{\text{Disturbing Moment}}$$

$$F = \frac{cA + \cos\beta\tan\phi}{W\sin\beta} \quad \text{(for translation-slides)}$$

$$= \frac{cA + (W\cos\beta - \mu)\tan\phi}{W\sin\beta} \quad \text{(if water is trapped behind the upper face)}$$

$$= \frac{cL + W\cos\beta\tan\phi}{M\sin\beta} \quad \text{(for rotation slides)}$$

Causes of Mass-Movements

The following causes of mass-movements are recognized:

(1) Absence of surface drainage and the existence of channels or openings for seepage from intern sources greatly promote slope failure (Figs. 11.2 and 11.3). This explains the frequent occurrence of landslides during or after heavy, prolonged spells of rainfall.

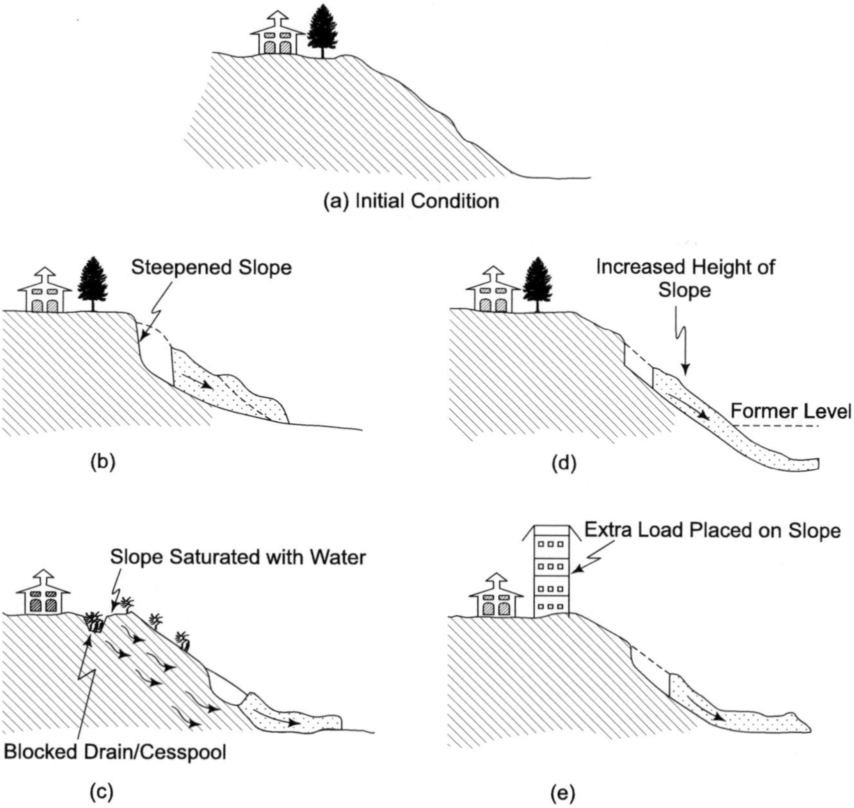

Fig. 11.3. Four ways in which the stability of a slope is destroyed and landslides initiated. (After Fig. 5.19 in E. A. Keller, 1976, *Environmental Geology,* Bell & Howell, Columbus, 488 p).

However, the rainfall duration and intensity alone are not a major factor. In Nepal heavy rainfall in summer often does not produce landsliding. Rather later in July-September when the groundwater table comes within one metre of the surface, the heavy or continuous rains produce landslides (J. D. Ives, per. comm., 1985). An example of rain induced mass-movement is the landslide which wholly ravaged the upper part of the township of Nainital in 1880. More than a hundred people perished. In the alternation of highly jointed shale and limestone there was a buildup of pore pressure due to impediments to drainage following heavy and continuous rain. **People often contribute to failure by altering the hydrology of the slopes by diverting rainwater and roofwater, and not doing anything to stop leakage from septic tanks, swimming pools, canals, sewer drains, reservoirs, etc.**

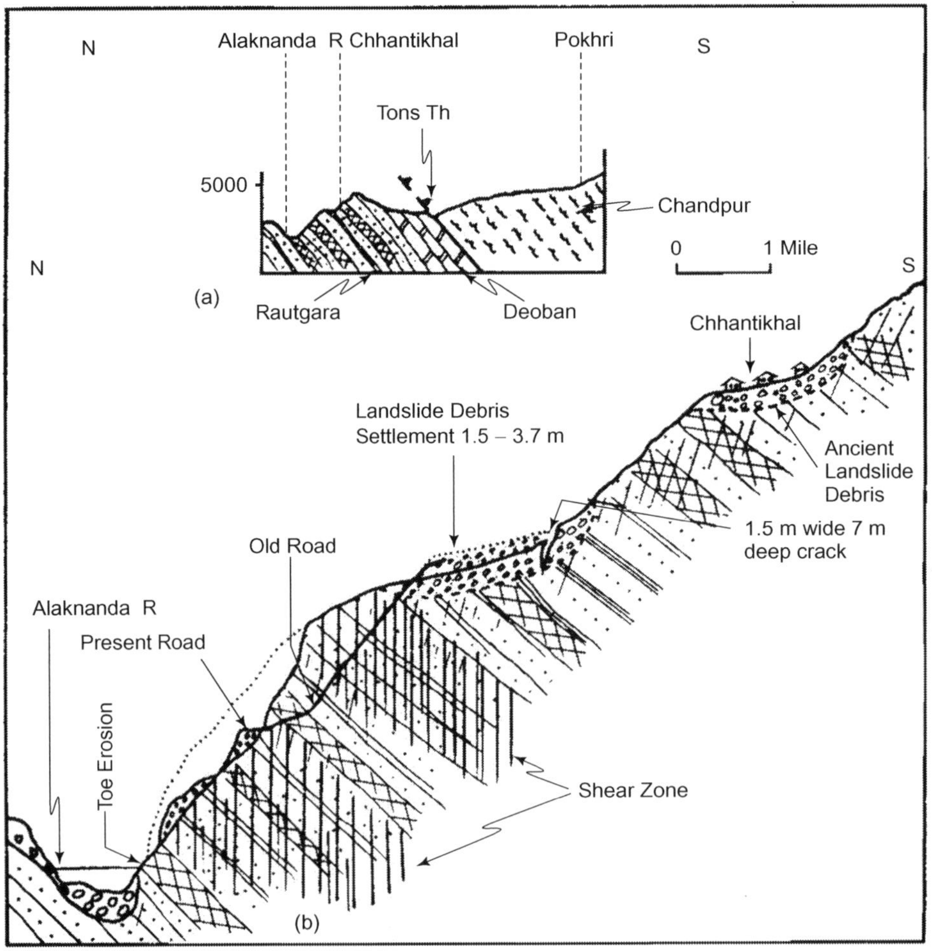

Fig. 11.4 The result of the combined action of toe erosion by the swirling water of the Alaknanda and steeper slope due to road cutting is the continual creep with attendant rockfalls in the Kaliasaur slide, east of Shrinagar in Garhwal (Uttarakhand). (After V. S. Krishnaswamy and M. S. Jain, 1975).

If the rocks are sheared and shattered, they will naturally become saturated with water. Their shear strength will consequently be less, so that the driving force will overwhelm it, leading to perceptible sliding. Thus the degree of fracturing or jointing has a strong influence upon the she strength of rocks.

(2) Earthquake shocks, particularly those of shorter-duration, acceleration of ground motion a resultant tilt of the slope modify the system of forces in a manner that the driving forces get the upper hand. **Thus the seismic shocks are the biggest triggering factor.** Extensive landslides caused by the 1950 earthquake in eastern Arunachal Pradesh, and by the 1934 North Bihar earthqua bear eloquent testimony to this fact. **The vibrations generated by the passage of heavy vehicles (trucks, tanks, trains, etc.) create oscillations of different frequency in the rocks and thus change the stress pattern, reducing shear strength and inducing mass-movements.**

(3) The slope angle greatly influences the strength of the driving force — the steeper the slope t greater the chances of its failure (Figs. 11.3 and 11.6). On the Haridwar-Badarinath highway, t 30-m wide and chronically active Kaliasaur landslide is a result of combined action of rushi water of the Alaknanda (toe erosion) and excessive road cutting entailing formation of steep a very high wall of the cut (Fig. 11.4). The 45 to 50° S-dipping quartzite alternating with min slates are ridden with close and multiple joints and fissures (2 to 4 m wide and 200 m lon which allow free passage to rainwater coming down in torrents (1500 mm/yr). The shattered and cohesionless rocks creep down continually.

(4) The addition of weight on the slopes like the dumping of debris or wastes and the construction of heavy structures (dams, reservoirs, debris dumps etc.) increase the intensity of the driving force and reduces the slope stability (Fig. 11.3). In the townships of Shimla in Himachal Pradesh and Nainital in Uttarakhand the deposition on the slopes of debris produced from the excavation for building foundations and for roads, and construction of multistoried concrete structures (Fig. 11.5) have appreciably changed the balance in favour of the driving force (Fig. 11.2), leading to creep movements as evident on the slopes of Lakkar Bazar (in Shimla) and Sher-ka-danda (in Nainita The Lakkar Bazar area of Shimla had sunk by more than 30 cm in 1978, and parts of Kaint subsided by 15 to 30 cm in 1985.

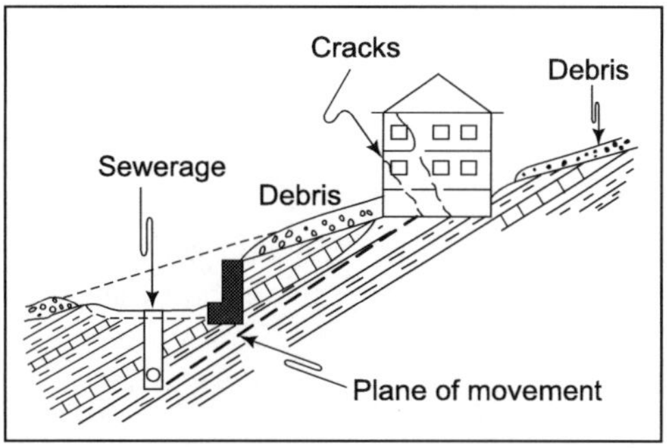

Fig. 11.5. Additional load of tall and heavy buildings on slopes coupled with removal of lateral support (such as the cutting for sewer lines) changes the balance in favour of the driving force and cause instability.(After Q. Zaruba and V. Mencl, 1969).

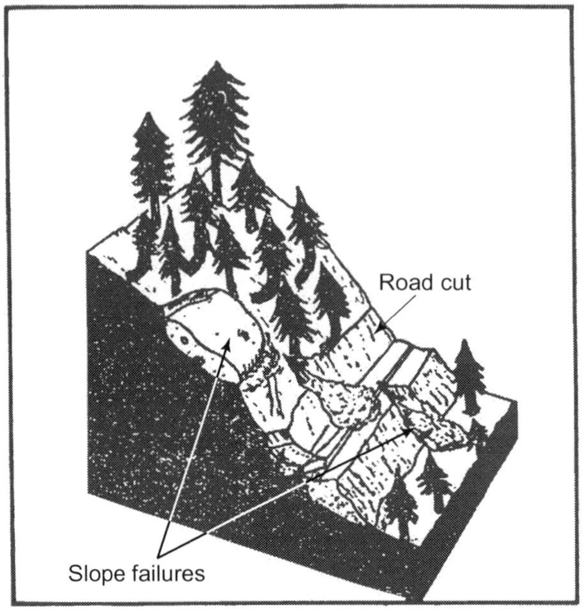

Fig. 11.6. Road-cutting is responsible for manifold increase in the incidence of landslides. In the Rockies (U.S.A.) the increase has been by a factor of 25 to 345. (After W. M. Marsh and J. Dozier, 1981).

5) As already demonstrated in Chapter 7, **road-building activities greatly contribute to the instability of hillsides and manifold increase in the incidence of landslides** (Figs. 11.6 and 11.7b). This is evident from studies carried out in western U.S.A., where road building increased landslides by a factor, varying from 25 to 345 (Table 7.7).

The excavation and removal of material reduces lateral support and promotes down-slope movement particularly where bedding planes or cleavage planes dip downslope (Fig. 11.7).

Various factors responsible for causing or stimulating mass-movements are listed in Table 11.1. It also brings out the mechanism of landslides.

Geological Deformation and Mode of Failure

slope failure or ground subsidence can be best understood with reference to associated geological formation. For example, 'back-tilting' can almost always be associated with rotational slips, whereas pearance of a 'graben' gives an indication of 'composite movement'. Similarly, 'karst topography' may be construed as a proof of subsurface solution of limestone and dolomite rocks, leading to underground collapse. This can also happen in chalk, gypsum, anhydrite and halite terrain.

In the case of marine erosion, presence of 'blow holes' may provide a clue to 'roof collapse' in caves rmed by the sea. In fact, falls in the inner parts of such caves in Cornwall, England, provide a striking ample of such a phenomenon. Curious steepwall depressions, called "volcanic sinks", are known to cur as a consequence of outflow of magma in some volcanoes. If the slide-prone formation consists of relatively softer material capped by a stiffer one, multiple rotational slips may develop. **Accumulation of argillaceous debris along with a plentiful supply of water often generates mudflows.**

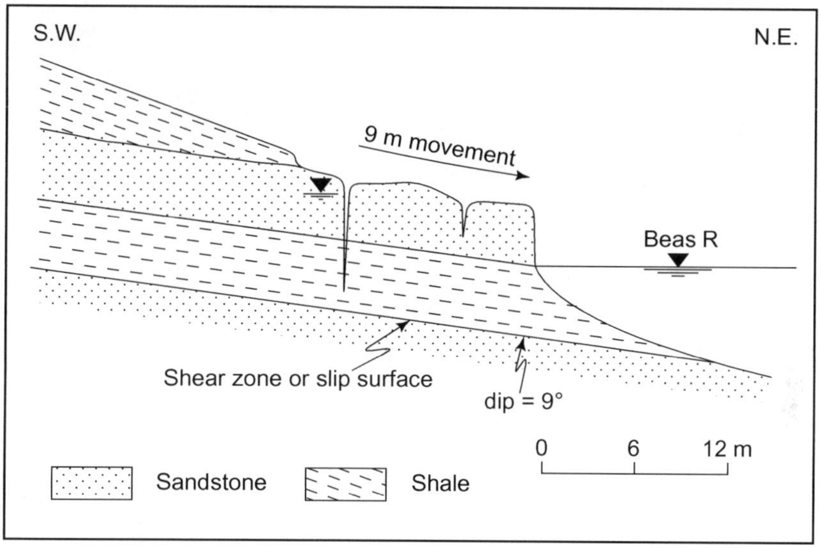

(a)

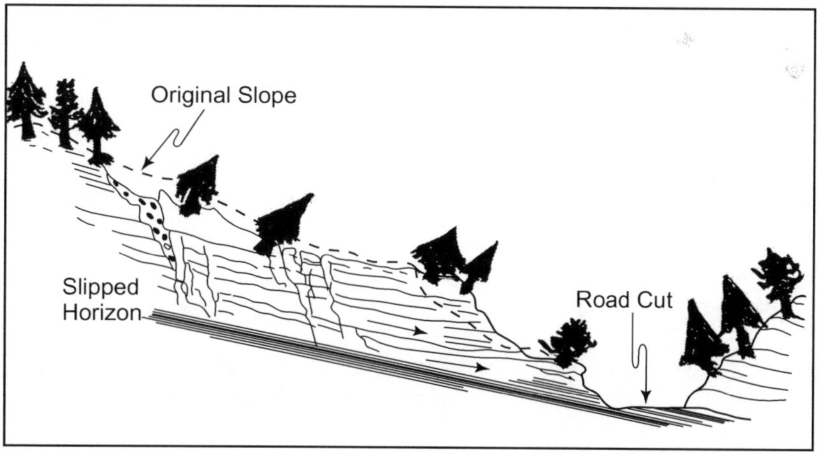

(b)

Fig. 11.7 Withdrawal of lateral support causes a mass of rocks to slide down on the bedding planes.(Low section after Q. Zaruba and V. Mencl, 1969.)

Table 11.1. Agents Influencing the Occurrence of Landslides (From Petak and Atkisson, 1982, *Natural zards, Risk Assessment and Public Policy,* Springer, New York, 489 p).

Agent	Event or process that brings the agent into action	Mode of height or rise of slope
ransporting agent	Construction operations or erosion	1. Increase of height or rise of slope
ectonic stresses	Tectonic movements	2. Large-scale deformation of the earth crust
ectonic stresses or explosives	Earthquakes or blasting	3. High-frequency vibrations
Weight of slope-forming material	Process that created the slope	4. Creep on slope
		5. Creep in weak stratum below foot of slope
Water	Rains or melting snow	6. Displacement of air in voids
		7. Displacement of air in open joints
		8. Reduction of capillary pressure associated with swelling
		9. Chemical weathering
	Frost	10. Expansion of water due to freezing
		11. Formation and subsequent melting of ice layers
	Dry spell	12. Shrinkage
	Rapid drawdown	13. Seepage towards foot
	Rapid change of elevation of water table	14. Rearrangement of grains
	Rise of water table in distant aquifer	15. A rise of piezometric surface in slope Forming
	Seepage from artificial source of water (reservoir of canal)	16. Seepage towards slope
		17. Displacement of air in the voids
		18. Removal of soluble binder
		19. Subsurface erosion

TYPES OF MASS-MOVEMENTS

ndslides and other mass-movements depict extremely wide band of hazard, ranging from creeping pes with movements of the order of 1 mm/yr on the one hand to high-speed mudflows with movements well over 50 metres per second on the other.

Creep

ost persistent of all mass-movements, **the creep is an imperceptibly slow movement of partially viscous soil and rock material. The movement is of the order of a few millimetre to a couple of centimetre per year.** It brings about permanent deformation without developing discrete failure planes (Fig. 11.8). In the shallower part of the slipping ground exceedingly slow movements take place on minute slip planes, but the real deformation occurs on grain boundaries and within the structure of the clay minerals. In these open spaces respectively, interstitial and adsorbed water play a contributing role by reducing the friction. In this manner the clay fraction is converted into a slimy material, which serves as a lubricant.

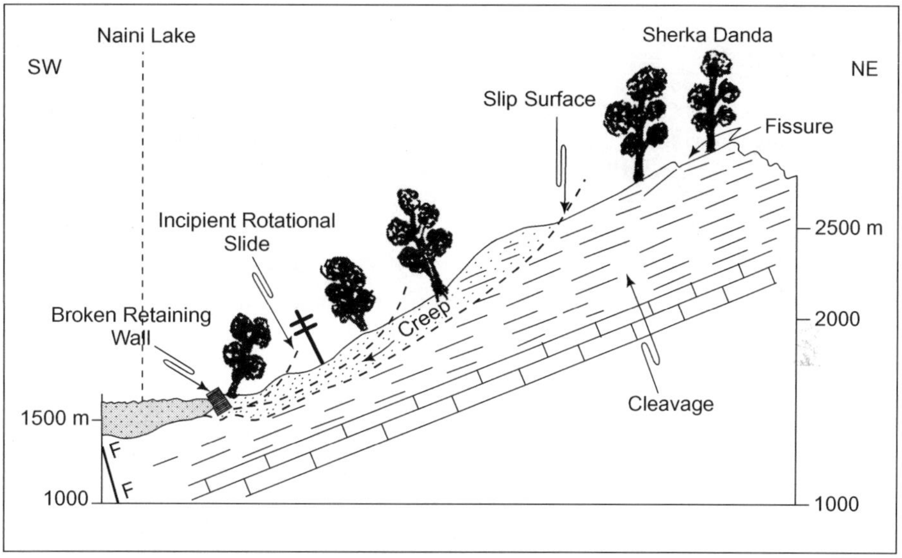

Fig. 11.8. Imperceptibly slow movements of a partially viscous mass called creep manifest in tilting of trees and poles and cracking of ground, which takes place without developing discrete failure planes on a hillslope as seen in the Naini lake basin in Kumaun (Uttarakhand).

Bending of rock beds, their disjointing, disruption and drag-folding are characteristic features of creeping mass. **Tilted trees and poles (Figs. 11.7b and 11.8) are indicative of creep movement as seen on the hillslope overlooking the Naini Lake in the Uttarakhand Himalaya. The curvature of the tree trunks bear record of the rate of creep during the period of growth of the tree.** It hardly needs stating that the rate of creep movement is quicker during the rainy season. There may be long periods stretching over years when no movement at all takes place. The temporal span of this phenomenon stretches over thousands of years. **The rate of creep is accelerated by clear-felling (of trees) in forests and by road construction and mining on steeper slopes.** The depth of the creep movement is variable and depends largely on the nature and degree of weathering of rocks, the subsurface structure and the amount of water present within.

The creep movements are discernible on practically every mountain slope in the Himalayan region, more pronouncedly in the zones of the Main Boundary Thrust, Main Central Thrust and the transver faults (Fig. 11.19). As already mentioned, the southwest-facing slope of the Sher-ka-danda Ran overlooking Nainital demonstrates this phenomenon as evidenced by tilted trees (5 to 10°) and pillars, bending and crumbling of retaining walls and formation of cracks on roads and buildings. (Fig. 11.20). The measurements done since 1895 on the Sher-ka-danda ridge and the Kailakhan spur indicate th there were regular creep with regular periods of halts (Hukku et al 1974}. The movement was of the order of 2.5 metre by a pillar near Ravenswood to 15 centemetre by the pillar near Ramsay Hospital er a period of 60 years, and 6 cm in the period 1952 to 1965 in the St. Loe-Edge Hill segment. In the ailakhan spur the magnitude was very high—the maximum value being 35 to 36 cm in the period 99–1966. And across the Balia stream on the right bank the movement was rather slow (6.1 cm) during the same period.

Slump and Sheet Slide

In the event of shearing stresses—the driving force— becoming strong, failure takes place along discrete planes called slip or shear surface (Fig. 11.9). If the surface of failure is planar, the resulting movement is called ***sheet slide,*** and when broadly concave the phenomenon is described as ***slump.*** (Fig. 11.9) The movement of the slumped mass is rotational without causing much impairment of the solid block even •ugh the modification of the landscape may be considerable (Fig. 11.10a) as witnessed along roads and in mines. Usually 3 m deep sheet slides affect extensive areas extending as they do even upslope. e result of recurrent sliding is a piling up of slid blocks one on top of another (Fig. 11.10b). Cohesive soils and claystones and mudstones are especially prone to slumping and sheet sliding particularly where loose materials rest upon hard competent rocks.

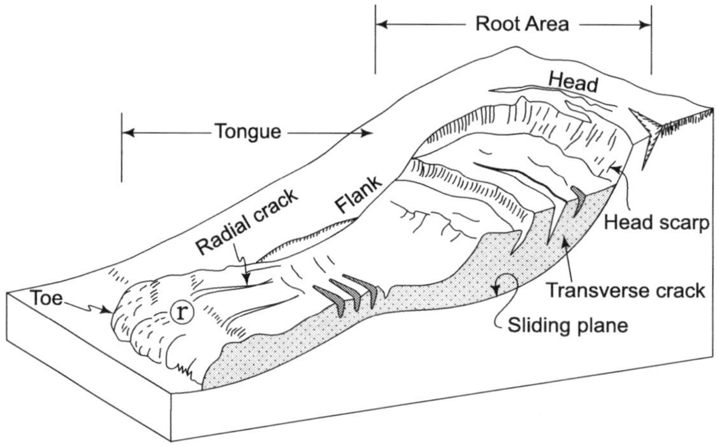

Fig. 11.9. If the failure takes place along discrete planes at depth, the mass-movement is called slumping or sliding. (After Q. Zaruba and C. Mencl, 1969.)

The rate of slumping-sliding varies from imperceptible to as much as one metre per day. **The major cause of this kind of mass-movement is undercutting of the slope due to excavation for buildings, ads and mining, and the provocation provided by the water that finds access to the unstable areas. Slumping is therefore very common in poorly drained areas and during rainy times.** A catastrophic slump, involving greatly weakened black carbonaceous slates of the Kailakhan spur (Nainital) slumped wn on August 17, 1898 into the Balia valley, following heavy rains. The 60 m high slumped mass blocked the stream, buried the settlement of a brewery (Fig. 11.11) and later spread several kilometres downstream (Valdiya et al., 1984).

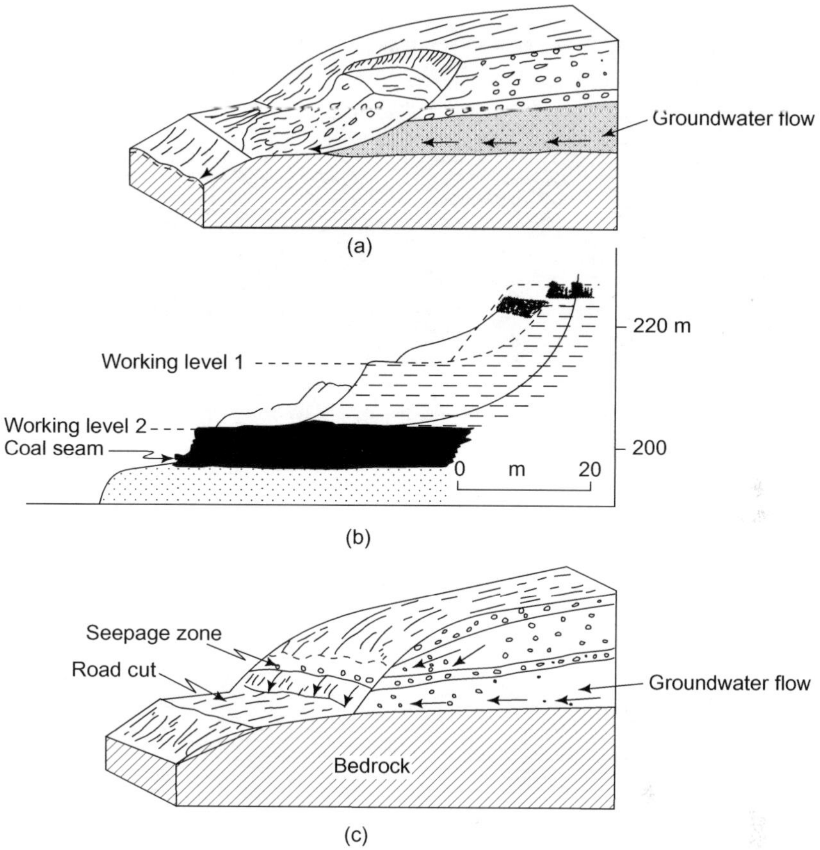

Fig. 11.10. In the case of slumps, the plane of movement is concave, and the movement rotational. Slumping due to road cutting, (b) mining, and (c) widening of roads (Based on (a) Fig. 13.10D Coates 1981; and (c) R. C. Parizek in Zaruba and Mencl, 1969.)

The northern slope of the ridge on which the town of Shimla in Himachal Pradesh is situated slumped considerably in two decades of 1960s and 1970s. The 3.5 km stretch of the slope east of Joshimath on the way to Badarinath slumped down into the Dhauli River about 15 to 20 m in some spots. The sliding necessitated realignment of the highway. Slumping continues to bedevil this strategic highway.

In the landslide-prone Kolpukhola catchment in the Kathmandu-Kakani belt (Central Nep landslides over 1% areas were initiated in two ways: first by rotational and planar sliding of surfac soil mantle (upto 3 m thick), commonly in topographic hollows and valley heads, and secondly, by fluvial erosion (Caine and Mool, 1982). Eventually, the two processes converged on a common for as linear gullies extended headward over the original landslide scar. Significantly, in the southern p of the basin where the phyllites wear thick mantles of weathering products,— piping or tunnel erosi is quite common, giving rise to steep walled trenches and hollows of badland, according to Caine a Mool (1982). The footpaths between Tinpipale and Pacmane had deepened into gullies, as much as 4 m deep, as far back as the late 1970s.

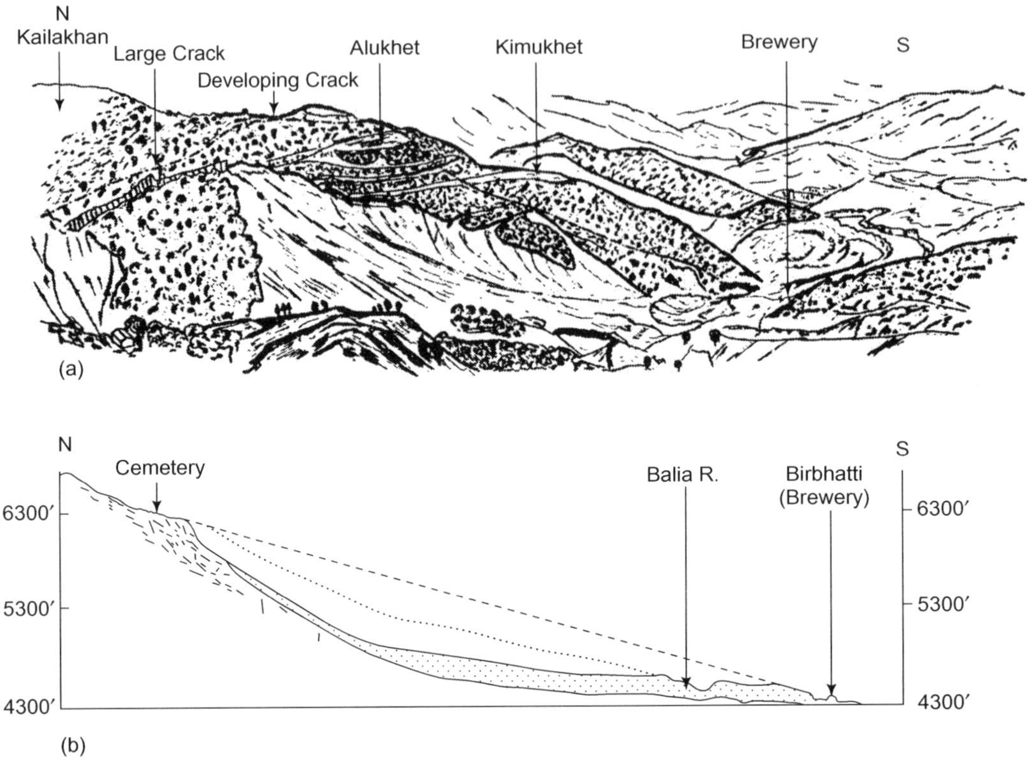

Fig. 11.11. Following incessant rain, an enormous mass of black-grey slates slumped down on August 17, 1898 into the Balia stream, burying the settlement of a brewery ("Birbhatti") in Nainital, Uttarakhand Himalaya. (Based on C. S. Middlemiss, 1898.)

Debris Avalanche

e rapid movement of shallow non-cohesive or loose material down a steep slope following heavy rainfall is called ***debris avalanche***. The debris masses form fans and cones at the foot of the slope, hile the scars left behind are often spoon-shaped depressions (Fig. 11.12b). If a large amount of water able to soak the moving material, the viscosity is considerably lowered, resulting in the debris flow reading far and wide. In the lower part of the slope, the water-soaked debris flows as a torrent, eroding and scouring the channel deeply. Later the scar of the debris avalanche is highly susceptible to erosion rill wash and gullying and also recurrence of debris flow.

The elimination of anchoring effects of tree-roots as a result of clear-felling (of trees) in forests greatly accelerates the occurrence of debris avalanches. In certain experimental forests in Oregon (U.S.A.) clear-felling increased the incidence 4.4 to 8.8 times (Swanston and Swanson, 1976). Road cutting had a far greater impact on slope stability, the incidence of debris avalanches having dramatically increased 25 to 345 times.

(a)

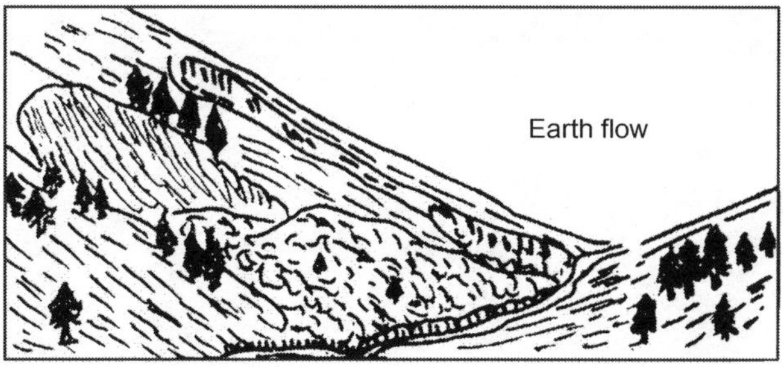

(b)

Fig. 11.12. Rapid descent down the slope of loose aggregate of debris is called debris avalanche, a the flow of muddy material is called earth flow. (Based on B. Mears, 1970, and A. N. Strahl 1973.)

The September, 18, 1880 event of less than half-a-minute duration that took a toll of 143 lives in t upper part of the township of Nainital in Uttarakhand was largely a debris avalanche. It swept away Victoria Hotel, some buildings and the ancient temple of Naina Devi before plunging into the lake. Following several hours of incessant heavy rain (2600 to 2900 mm) this disastrous event was preceded in 1867 by development of a 1.25 km long crack on the highly jointed slate-marl succession of the She ka-danda Ridge (Fig. 11.20) where until then there was no provision for the drainage of storm runoff. ropeway now spans the scar of this landslide.

In the early 1980s a 400-m stretch of the road in the Tista valley in Sikkim was destroyed by debris avalanches coming down 185 m from highly faulted and jointed and sheared phyllites and slates dipping to 50°N. The dimension of the mass-movement progressively increased as block after block detached and broke down into debris mass.

Debris avalanches together with slumping and sliding are quite common in the belts of the Main undary Thrust, Main Central Thrust and many transverse tear faults and thrusts in the Lesser Himalaya (Fig. 11.19). One of the most spectacular debris avalanches occurred in the small hours of ugust 6, 1978 in the Kanodia stream—a tributary of the River Bhagirathi. Following a cloudburst, a large mass of unconsolidated moraine moved down from the 3400 m high ridge and blocked the channel the small Kanodia stream, forming a 30-m deep 35-m wide and 3-km long lake. Subsequently, the natural dam burst and the material that was washed down, formed a second huge dam in the Bhagirathi g. 11.13). Some 14 hours later this dam also gave way, bringing incalculable destruction by flood wnstream upto Uttarkashi (Gupta and Dave, 1982). The remnants of the short-lived lake was soon mpletely filled up with sediment.

A small tributary of the Satluj River, the Parachu, was blocked in the summer of 2004 by a massive ndslide, resulting in a formation of >20 m high debris dam and nearly 2-kilometre square large lake. e bursting of the dam in September 2004 generated a devastating flood which had the discharge of 2000 m^3/second—reaching 40 m high in the early stage.

Fig. 11.13. A stupendous debris avalanche of 1978 blocked the passage of the Bhagirathi River and formed a big dam which burst 14 hours later. The remnant of the newly formed lake was filled up very rapidly—burying the bridge near Joti. (From Valdiya, 2004). (See colour Fig. 11.13)

Debris avalanches are frequently witnessed in the Western Ghat belt (Thigale, 1984), particularly where lateritic debris masses with underlying layers of lithomarge disintegrate into avalanches during rainy season (Figure 11.21). In the Ratnagiri district in Maharashtra, high rainfall, extensive deforestation and construction of roads and railway line are responsible for very damaging landslides, particularly on slopes steeper than 35% (Thigale and Khandge, 2000). About 45 km north of Trivandrum, a mas debris flow on November 9, 2001 swept away a village killing 36 people (in Valdiya, 2004).

Earth Flow

As a result of heavy rainfall, water-soaked finer loose material lying along the upper part of a slo flows down towards the foot of the slope and forms fans. The fluctuation of water content within the s is primarily responsible for this kind of mass-movement of mud clays and silts (Fig. 11.14b). ***Mudflow*** are commonly witnessed in the Siwalik terrain and along the Western Ghat.

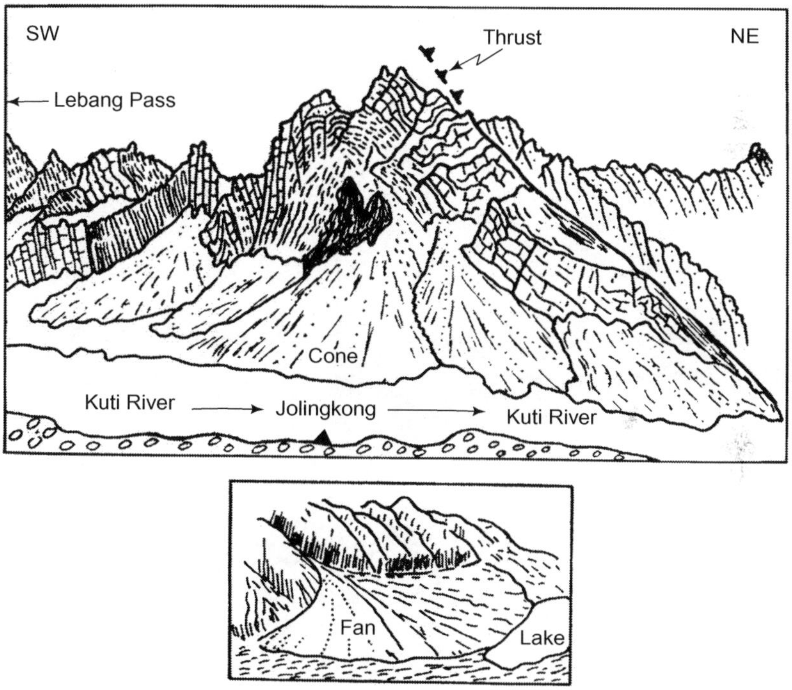

Fig. 11.14. Rockfalls are very common on the higher peaks of the barren Tethys Himalaya, (a) A scene north-west of Kuti in north-eastern Uttarakhand. (After A. Heim and A. Gansser, f939). (b) Debris avalanche/mudflow.

A special type of earthflow seen in high glacial areas results from thawing of ice that fills t intergranular space within clays. As the ice melts, the whole mass becomes waterlogged and unstab leading eventually to downslope flow called ***solifluction***. The solifluction is in fact downhill creep flow of any mass (usually regolith), involving creep, frost heave and collapse (J. D. Ives, per. comm 1985).

Rockfall

brupt and extremely fast movement of loosened blocks and pieces of jointed or fractured rocks from cliffs and steep slopes is called ***rockfall*** (Figs. 11.14a and 11.15b). Hydrostatic pressure of the water that ds its way into joints and fissures, the widening effect of freezing of this water and of the tree roots are all responsible for dislodging blocks and pieces which hurtle down the precipitous slope and bounce ricochet (Figs 11.5b and 11.16). The 2610-m high Naina Peak, overlooking the lake of Nainital in Uttarakhand, has an enormous apron of debris generated by recurrent rockfalls (Fig. 11.15a).

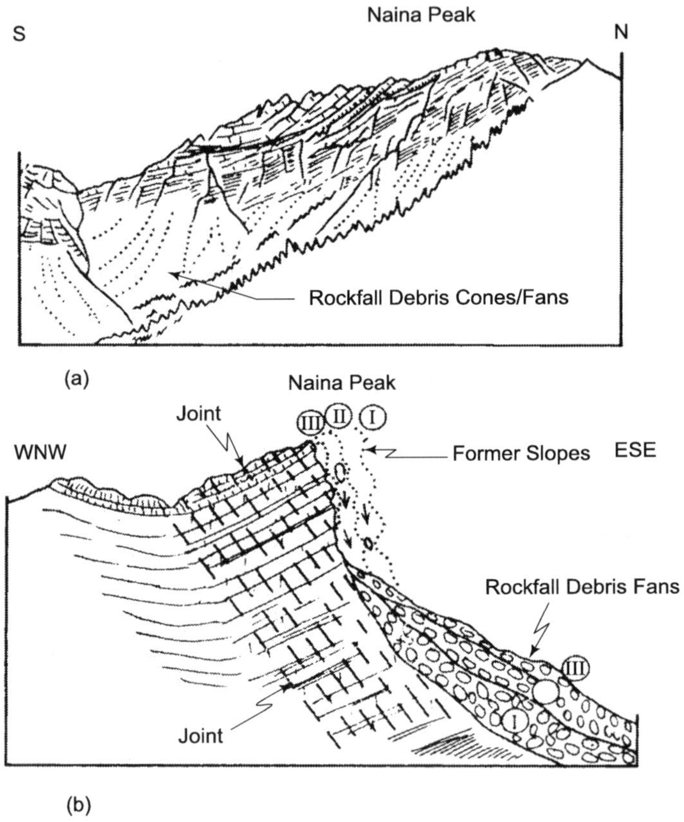

Fig. 11.15. The northwestern part of the township of Nainital is located on the fans and cones built by recurrent rockfalls coming down from the 2600 m high Naina Peak. (Upper section (a) after A. Gansser, 1964)

Rockfalls are very common along the Khuni Nala section on the Jammu-Srinagar highway in Kashmir. both the cases several sets of joints (including master joints) have greatly weakened the rocks—mestones and dolomites in Nashri Nala and gneisses and schists in the Khuni Nala (Fig. 11.17).

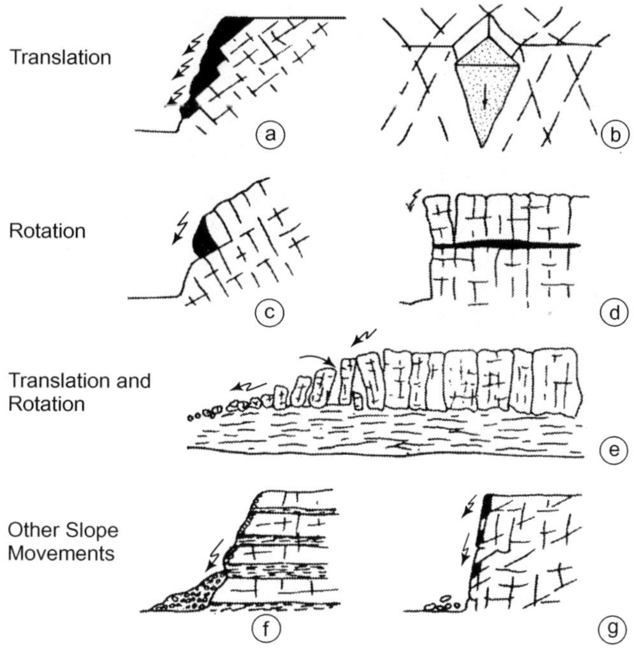

Fig. 11.16. Rockfalls result from a number of causes (a) sliding of blocks along joint planes, (b) sliding wedges of compact rock, (c) tipping of unstable joint blocks, (d) tipping of blocks due to removal of underlying layer, (c) block movement on plastic layer, (f) falling off of fine slope material, a (g) rock fall. (After K.J. Klengel, 1973).

IDENTIFICATION OF LANDSLIDE ZONES

Hazard-Zoning Maps

The first step in a programme for landslide hazard mitigation is the identification of zones prone mass-movements. This is done on the basis of geomorphic features revealing past history and indicati existing stresses, and recognition of the destabilizing effects that some developmental works commonly have or are likely to produce.

A hazard-zoning map involves comprehensive investigation (Table 11.2) embracing: (1) details structural and lithological settings, (2) geomorphic features with special emphasis on those relating to instability of slopes, (3) seismicity pattern of the region, including epicentral distributions and ground acceleration due to earthquakes (4) signs of instability such as cracks on the ground and walls, distortion of structural frames, tilting of trees and poles, leakage of underground pipes, tautness in pipes, wires and cable wires etc., (5) instrumental measurements of movements of residual horizontal stresses electrical potential in the soil, and (6) determination of cohesion and strength of the soil. **On the basis of combined evidence, the hazards or real threats can be evaluated. It must be emphasized that one or a few of the features mentioned above would not provide an absolute proof of the occurrence of movement.**

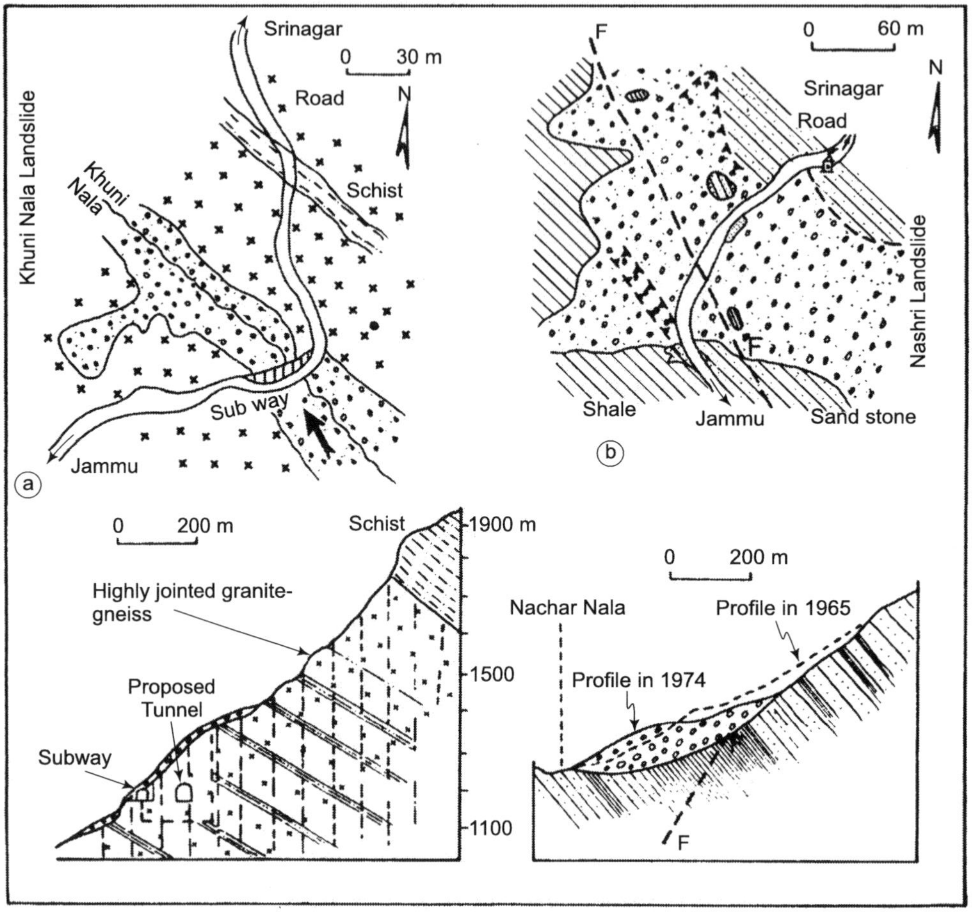

Fig. 11.17. Two sites of endemic landslides, (a) Khuni Nala and (b) Nashri Nala on the Jammu-Srinagar highway in the Kashmir Himalaya. (Based on B. M. Hukku and P. L. Narula, 1975).

The hazardzoning maps (Figs. 11.18, 11.19 and 11.20) thus prepared show a number of units picting different degrees of instability—and each unit is described in terms of slope stability (very high, moderately high, moderate, moderately low, low), landslide potential (very high, high, moderate, w), solifluction potential, slope (in terms of % or in degrees) and geological conditions. On the basis of verity of landslides and other manifestations of mass-movements, the vulnerable belt can be divided into three or four zones of instability. The ***Very Unstable Zone*** includes areas where the mass-movement likely to take place presently in view of prominent signs resulting from favourable lithological-structural and geomorphic conditions. In the ***Moderately Unstable Zone,*** the geological setting is the me as in the first zone and landslides have occurred here only in the past, but climate and slope conditions today do not favour instability. In this setting if the natural drainage is disrupted, or the slope is excavated for road buildings or mining, or if its toe is allowed to be eroded, there would be triggering

Table 11.2. Information Needed for Landslide-Hazard Mapping and Risk Evaluation (After Petak and Atkisson, 1982)

Information Needed	Uses
Geological Maps	For understanding causes of landslides for predicting density, severity and frequency of landslides in large areas. Aerial photographs and satellite imegery enable recognition and mapping of landslides in larger areas.
Topographic Maps	Provide a base for plotting landslide information. Geomorphic (topographic) forms are used to identify landslides. Topomaps provide basic data for slope analysis.
Slope Maps	Slope-stability maps combine steepness, landsliding and physical characteristics of different materials
Climatic Data	Climatic data based on specific events and annual climate are needed for assessments • influence for case histories and probabilistic predictions.
Subsurface Water	Buildup of pore-water pressure in subsurface materials and changes in water-levels needed for predicting initiation of specific landslides and predicting probability of failure.
Seismic Data	Enable preparation of maps of susceptibility to landsliding by earthquakes.
Vegetation Maps	Vegetation tending to support and stabilize slopes need to be depicted in terms of types, density and canopy cover.
Weathering of Rocks	Information on thickness and kinds of weathered rock products is used for estimating hillslope strength and landslide susceptibility.
Landuse Maps and Socioeconomic Data	Learning present and planned development for assessment of risks.
Chronology Data	Determining past intensities of landsliding for comparison to climatic and human activity data through use of historical records.
Computerization	Data collected are stored and manipulated to check for sensitivity of different map parameters and to superimpose large sets in different ways.

of mass-movement of one kind or another. The ***Moderately Stable Zone*** embraces dissected slopes of a hill characterized by stable geological structures and lithological setting that can be destabilized by erosion and engineering activities. In the ***Stable Zone,*** both the rock-formation and slopes are und very stable conditions and no action of man can threaten the balance.

Landslide Belts in Uttarakhand Himalaya

In the Himalayan region the zones of the Main Boundary Thrust (MBT) in the southern front and Main Central Thrust (MCT) in the interior are two extremely vulnerable landslide-prone zones. The MC separating the Great Himalayan ranges from the populated Lesser Himalayan terrane, had served as the plane along which the Lesser Himalayan domain had slid under the Great Himalaya and continues to slip time and again — causing tremendous uplift of the latter and wholesale splitting, shattering a crushing of the Lesser Himalayan rocks. Along the MBT the Indian plate along with its Siwalik fro is slipping under the Lesser Himalayan pile of rocks. The shattered and highly deformed rocks have been tightly folded and repeatedly split by thrust planes into multitudes of rock-slabs some of which a highly mylonitized or are in a friable condition. In the MCT and MBT zones, the weak rocks easily gi way to onslaughts of rains, shocks of explosions and earthquakes, vibrations generated by movements of heavy vehicles, and human interference. **As a result, these 5 to 20-km wide MCT and MBT belts in the Uttarakhand Himalaya are marked by almost continuous fringes or aprons of gigantic fans**

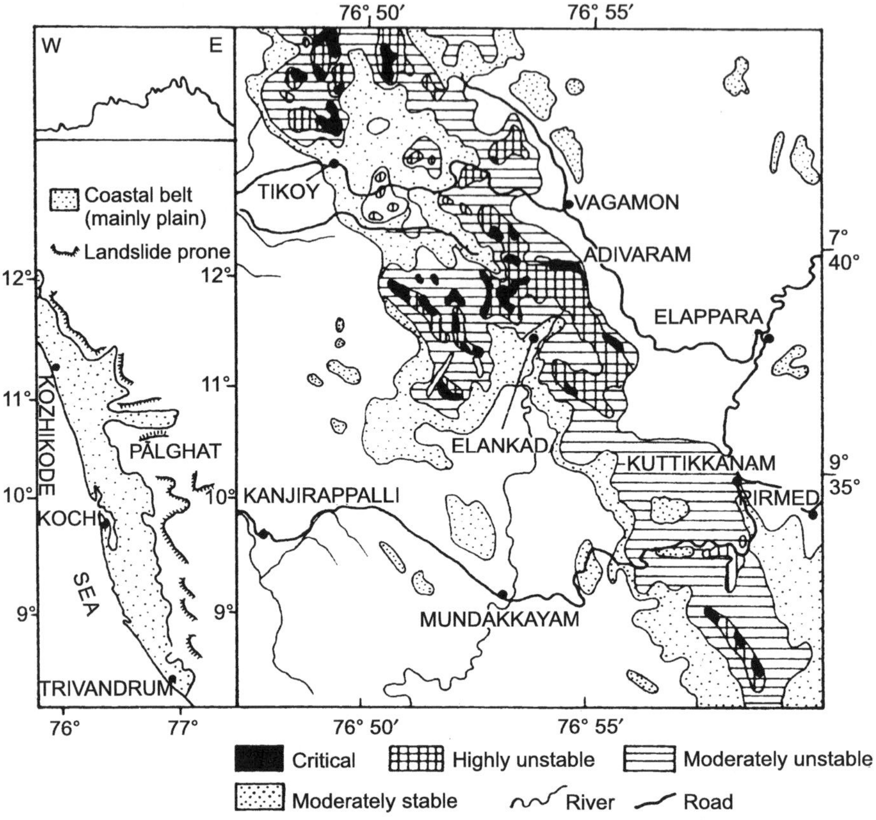

Fig. 11.18. Hazard-zoning map of an area in Kerala based on anticipated degree of severity of landslides and debris flows. (After Tampi et al., 1997).

and cones of landslide debris, both ancient and recent (Fig. 11.19). Most of the ancient landslide nes are camouflaged by thick vegetation. The situation is particularly severe along the MBT, registering neotectonic movements (Valdiya, 1985).

The Kumaun region of Uttarakhand, particularly its north-eastern area, is frequently rocked by moderate earthquakes ($M \geq 6$). The Kapkot–Dharchula–Bajang area in this region registers highest smicity quantity-wise (in terms of frequency) anywhere in the Himalaya. The larger quakes have caused considerable damage to the life and property due to destructive landslides over extensive areas. During the 1928 earthquake, for example, a 2500 km^2 area between Bajang (Nepal) and Kapkot lmora) was ravaged by landslides. The energy released by the 28th October, 1916 event was about 5000 times more than the energy released by the 1958 earthquake. The Uttarkashi earthquake of 1991, d the Chamoli earthquake of 1999 broke the southern face of the Great Himalaya extensively in the Garhwal region.

There are many tear faults trending transverse to the direction of the mountain ranges (i.e. NNE–W and NNW–SSE) along which neotectonic strike-slip movements are evident. **These NNE–SSW**

to NW–SE trending fault zones are characterized by extreme shattering and shearing of rocks and are very prone to landslides and erosion. These narrow belts also fall into the hazard zone map (Fig. 11.19). Thus the MCT and MBT belts are geodynamically very sensitive and the ecosystems they support are in a very fragile condition (Valdiya, 1985). Due to the precarious balance even small disturbances can trigger changes that rapidly assume alarming proportions. Any activity that affects the balance such as clear-felling (deforestation), uprooting shrubs, quarrying and mining and road and dam construction is bound to destabilize the hillsides with undesirable consequences as is happening for the last forty years, with progressively increasing frequency, severity and attendant disasters. Figure 11.19 was a preliminary attempt to identify broad landslide hazard zones—it shows areas that require special and expeditious attention, as well as detailed work f preparing hazard-prone areas.

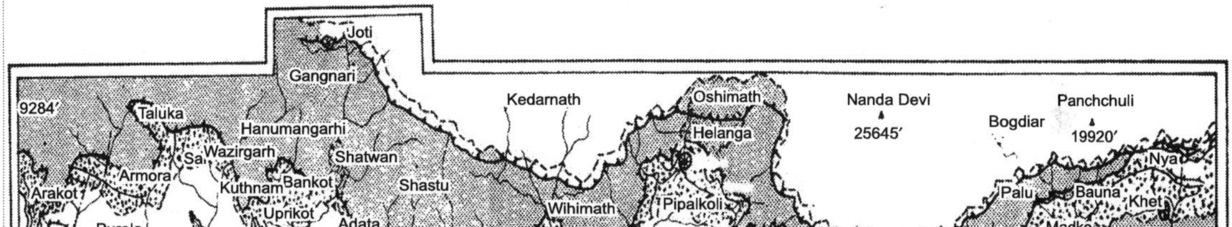

Fig. 11.19. Zone of instability and landslide-prone areas in the Uttarkhand Himalaya — a preliminary attempt made in the early 1980s. (K.S. Valdiya, 1985)

Landslide Zoning in Central Nepal

Exploring in some detail the dynamism of the man-land relationship in carefully selected areas in centr Nepal, the research team of J. D. Ives and B. Messerli (Keinholz et al., 1983, 1984; Johnson et al., 198 ine and Mool, 1982; Ives, 1982, 1985; Ives and Messerli, 1981, 1984) developed prototype maps of ndslide hazard of the Kathmandu-Kakani belt.

Not only did they take note of the progressive shift from potential instability to massive instability a consequence of heavy pressure on resources in the face of widespread mountain desertification, t also discerned the beneficial impacts of substantial energy directed towards reterracing of landslide scars and reduction of slope instability in culturable tracts.

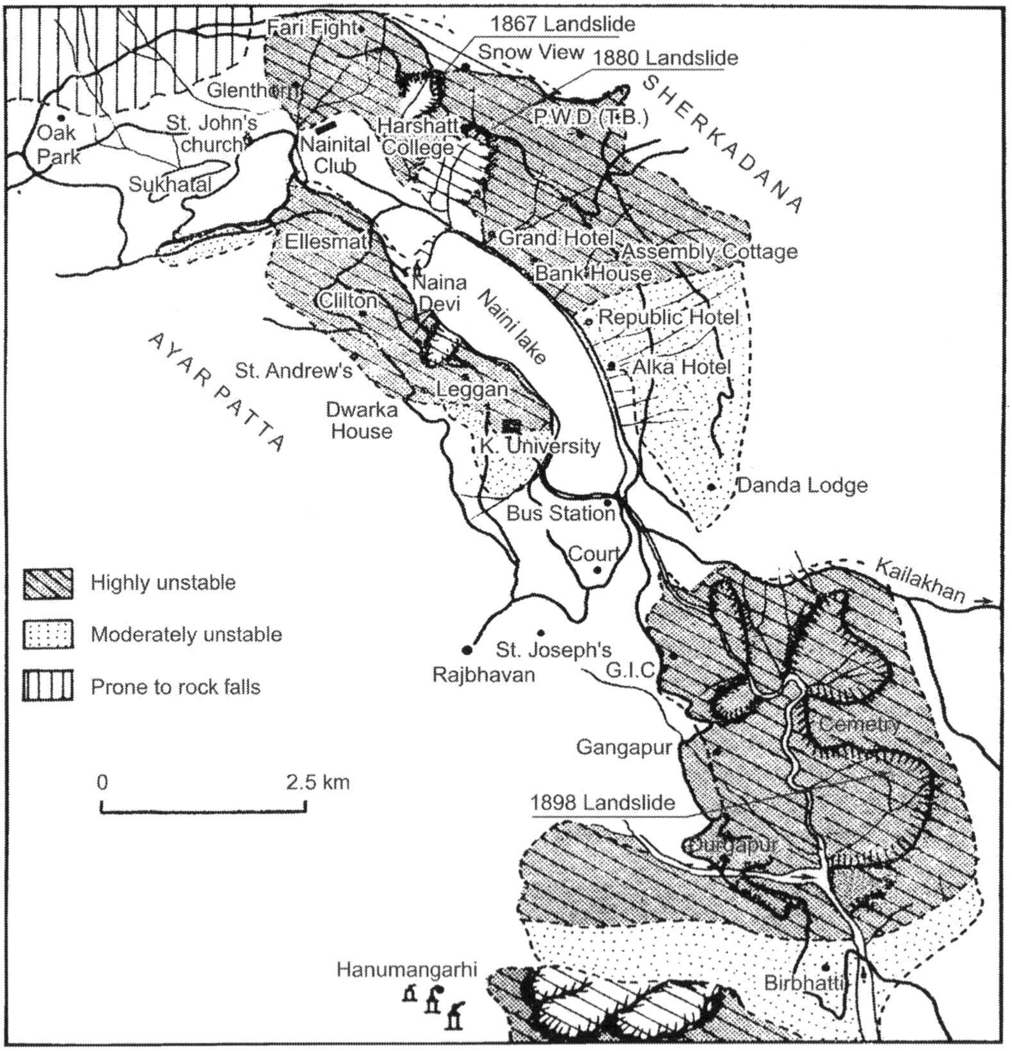

Fig. 11.20. Zones of instability in the township of Nainital and the Balia valley. (Based on B. M. Hukku, A. K. Srivastava & G. N. Jaitle, 1974; and A. K. Sharma, 1981).

The Kathmandu-Kakani belt—through which the road to Trisuli passes—receives annual rainfall 1100 to 2400 mm/yr. The steeply dipping gneisses, schists and quartzites forming steep slopes are in a highly weathered condition. Three major zones in terms of mass-movement are recognizable (Keinho et al., 1984). Intensive cultivation in the densely populated zone (250 to 300 persons/km^2) aggravated the naturally unstable conditions, as manifest in gullying, slumping and landsliding throughout the area. Significantly, slope failures were more common and extensive in non-irrigated terraces, whe absence of ameliorating activities by people created badland, leading to either abandonment or co version to pastures (Johnson et al., 1982). On the other hand, in the irrigated areas even though there were unlined ditches and stone-reinforced irrigation channels, there was extensive damage by gul formation. Human endeavours helped to restore the eroded and damaged terraces. The Kathmand Trisuli road cut through these terraces had greatly damaged the agricultural fields, and disrupted wate courses either by channelling runoff-water down the road bed, or by diverting from smaller channe to overburdened larger streams (Johnson et al., 1982). This exactly happened, and is happening in the whole of Uttarakhand, and perhaps elsewhere.

Hazard-Zoning in a Himalayan Town

As already stated, the township of Nainital in Uttarakhand is situated in the very vulnerable Ma Boundary Thrust zone— close to an active thrust and characterized by extreme rock deformation. has been repeatedly and severely ravaged by landslides of different kinds. The lake basin is still not free from the hazards. On the basis of creep movement, past occurrence of landslides, deformation of civil structures and geological conditions, three zones (Fig. 11.20) were identified in the early 1980s safe, moderately safe and very unsafe (Hukku et al., 1974; A. K. Sharma, 1981; and Valdiya,1988). would be most risky to make use of an unsafe area for settlement and engineering activities. Witnessing as the author does year after year, he finds no reason to revise this map. However, the risk factor h progressively increased.

Landslide Zones in the Western Ghat

In western India landslides are common along the Western Ghat on the steep (~60°) slopes overlooking the Konkan coast (Fig. 11.21). The Ghat scarps expose a succession of vesicular basalts interbedd with volcanic ash, red boles and agglomerates. The dense basalts forming the scarps and the ash beds make the gentler slopes. The mass-movements are generally confined to the plastic and semi-plas 20-m thick overburden material that has accumulated on the gently sloping (10°) vesicular lavas a volcaniclastic rocks. The movement takes place due to loss of shear strength at the interface of the volcaniclastic rocks or red boles with the lavas as and when rainwater infiltrates (S. C. Sharma, 197 Thigale, 1984). The rainfall is very heavy (>3000 mm/yr) in this belt. Steeply inclined shear faults and vertical joints provide easy access to the water, which triggers landslides. The landslide scarps are 60 to 180 m high and the amount of throw of the order of 200 to 400 m (Fig. 11.21). Since the impoundment of water behind the Koyna dam, the frequency of earthquake occurrence (*M* 2 to 5) has increased (especially in the Ratnagiri district), thus creating favourable conditions for landslides. In July, 1983 a 450-km long stretch of the Mumbai–Goa highway from Khed to Malwan (in Ratnag district) was seriously damaged in many places and as many as 62 villages ravaged, particularly in the Sangmaheshwar taluka, and the coastal villages in Malwan taluka in Sindhudurg district were laid waste (*The Times of India,* July14, 1983).

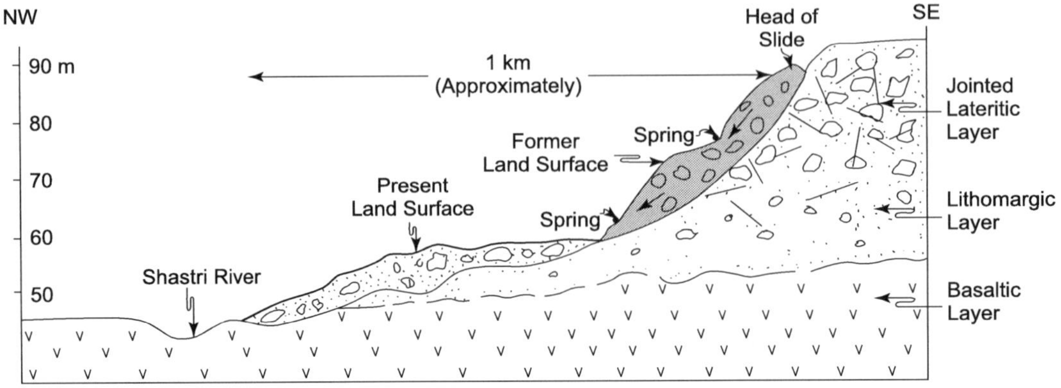

Fig. 11.21. Debris avalanches from a basaltic scarp in the Western Ghat belt (Sangmaheshwar taluka in Ratnagiri District) (After S. S. Thigale, 1984).

South Indian Landslide Zones

In South India landslides are common in the Nilgiri massif on the steep slopes of the horst that was uplifted to its present height (>2600 m) during Late Quaternary times. Demarcated by fault scarps, the massif in its upper largely deforested part has been ravaged by large landslides. The plateau is aracterized by a 20- to 30-m thick lateritic cap, and so yields easily to the impacts of rain, which is very intensive (Sheshagiri et al., 1982). On July 2, 1984 a massive slump, 6 km long and 200 m wide covered 50 ha of land in Meppadi, Wynad district, taking a toll of 15 lives. On November 9, 2001 ntinuous heavy rain brought down a 700-m high hillside, killing 36 persons of the Neyyatinkara lage 45 km north of Thiruvananthapuram in Kerala. Such destructive landslides have become far too quent and severe in recent years. Prepared on the basis of fieldwork and the satellite-imagery analysis mass-movement incidences the landslide-zoning map (Ramasamy and Muthukuar, 2008) shows that the dimension of landslides is increasing owing, primarily to human activities.

CONTROLLING LANDSLIDES

Strategy

As the principal cause of initiation or acceleration of mass-movements is water, this agent has to be prevented from entering into the affected area. It is not possible to prevent a landslide from occurring, but once initiated, it is certainly possible to minimize the hazard and damage by taking mely preventive measures. By constructing an efficient drainage network and making use of appropriate engineering practices the magnitude of risks from mass-movements can be minimized (Table 11.3). Many landslides are preceded by creep movements over a long period of time. This natural warning signal can be used to ensure that preventive and ameliorative measures are taken in time.

Table 11.3. Main Methods for Prevention or Control of Landslides (R. K. Bhandari, per comm., 1985)

Method	Different Uses of the Method
Excavation and Filling	(a) Unload upper parts of slope or load lower parts; achieve general flattening of slope; provid beams, etc.
	(b) Remove slipped materials; replace by the same material in a compacted state and with provision for drainage; alternatively, replace by better free-draining material.
	(c) Excavate the toe to induce sliding until flat, stable slopes are obtained.
Surface Drainage	(a) Divert surface water away from site; prevent entry of water into open joints, tension cracks etc. ensure drainage of cracks if possible.
	(b) Blanket slope with free draining material providing filters where necessary. This is particular useful for slopes subject to rapid drawdown.
Trench and Counterfort Drains	(a) Use gravel-filled trench drains aligned directly downslope and supplement by shallowe drains arranged in a chevron or herringbone pattern. While such drains are usually narrow, wider drains in which a bulldozer can operate are occasionally used.
	(b) Use counterfort drains, an earlier version of trench drains in which the invert is located beneath the slip surface, thus providing mechanical support in addition to drainage. These are also gravel-filled.
Cutoff or Interceptor Drains	Use cut off drain trenches, open or gravel-filled, running cross slope above the crest of the slop or uppermost scarp of the landslide. These are also called interceptor drains and may intercept only surface runoff when shallow or also groundwater flow when deep.
Horizontal Drains, Drainage Galleries, Drainage Well, etc.	(a) Use horizontal drains by drilling into slope and using porous or perforated liners. Usuall drains drilled at a small angle above horizontal lengths up to 100 m are not uncommon although much longer drains have been installed. In large slides horizontal drains may be used in combination with vertical drainage shafts or wells.
	(b) Drainage galleries maybe used in large slides if the cost can be justified. Additional drainag borings can be made through the sides, roof and floor of a drainage gallery.
	(c) Vertical drains connected to horizontal drains leading to gravity drainage flow are als sometimes used. In some cases vertical wells may discharge to a lower aquifer. Sometimes siphoning, automatically activated pumps or compressed air are used to drain water if gravit drainage is unsuccessful. Vertical drains may also serve as relief wells to reduce artesian pressures.
Electro-osmosis and Evapotranspiration	(a) In low permeability soils, water migrates from cathode to anode and may be removed with or without pumping. This method (electro-osmosis) has been used with success for stabilization of slopes during construction processes. Its use for permanent slope stabilization is uncommon, (b) Use of vegetation cover on slopes not only reduces erosion, but improves evapo-transpiration and consequently reduces pore pressures. At shallow depths roots of vegetation provide a strengthening effect also. While surface runoff and erosion are reduced infiltration is also reduced.
Restraining and Strengthening Structures Including Reinforced Earth	(a) Use of retaining walls or crib walls with base below probable slip surface. (b) Use of a wall of closely-spaced driven piles for shallow slides. (c) Use of anchored sheet pile or bored pile walls for deep seated slides. (d) Use of soil and rock anchors generally with prestressing and with or without retaining structures (very widely used for stabilization of translational rock slides) (e) Use of large diameter cylinder pile retaining walls. (f) Use of reinforced earth construction. This type of construction is increasingly being used with a great deal of success

Contd

Method	Different Uses of the Method
ime Column Method	The lime column method is used to stabilize soft clay down to a depth of up to 10 m. These lime columns are manufactured and installed in situ by mixing soft clay with unslaked lime. It is not necessary for columns to reach a firm base. The length should be such that the column goes below the probable slip surface so that load is transferred to soil both above and below the slip surface. The lime column method can also be used with advantage when high pore water pressures in silt or sand layers affect the stability of a slope because of drainage through the previously installed columns. The increase of shear strength due to lime columns also contributes to the stability of a slope. The risk of initiating a slide is small compared to that due to driving of timber or precast concrete piles.
ther Methods (not common use and some not fully evaluated)	(a) Grouting to reduce groundwater movement to area of interest.
	(b) Strengthening slip zone or shear zone by injecting cement grout usually aerated. (c) Deep soil stabilization by chemical means, e.g. introduction of quick lime, potassium chloride, etc. (d) Suppression of natural electro-osmosis by insertion of short-circuit electrodes. (e) Use of electro-osmotic anchors based on the principle that the pull-out resistance of anchors increases significantly when used as electrodes. (f) Ground freezing technique (considered to be an expensive method).
	(g) Heating or burning technique: excavation of wide counterfort drains, filling with alternate layers of clay spoil and coal, and igniting the coal thus baking the intermixed as well as adjacent clay and shales which are almost self-firing. Has also been used in non-carbonaceous London Clay in U.K. Loess soils have been stabilized by passing hot gases through tunnels and boreholes.
	(h) Use of blasting to disru pt slip surface and to improve drainage (a controversial perhaps even dangerous technique),
	(i) Bridging over an active or potential landslide area.
revention of Energy Release	Use of piles above (or outside) the limits of/area to reduce or discourage tendency of lateral stress release in sloping ground with high initial stresses. The concept of recoverable strain energy to explain progressive failure in overconsolidated clays and clay shales the successful use of this technique to arrest or prevent progressive failure, demonstrated
rosion Control	(a) Control of toe and surface erosion (where slope is under the action of sea, river, lake, reservoir, etc.) by use of concrete or crib walls, riprap, groynes, and spur dykes.
	(b) Control surface erosion due to runoff by use of surface drains, vegetation coyer, gabions, etc.
	(c) control of seepage erosion by use of inverted filters over the area of discharge or by interception with wells or sand drains at some distance away from the face.

The engineering technology involves provision of (1) surface and subsurface drainage, (2) removal unstable slope material and (3) the modification of slopes to restore stability and (4) the construction of retaining and associated walls. **A combination of measures is necessary to deliver the desired results** (Table 11.3).

Drainage Control and Dewatering Measures

he first step in a programme of slope stabilization is to get rid of water—the agent of provocation—by dewatering the affected area and directing the water away from the landslide area (Figs. 11.22

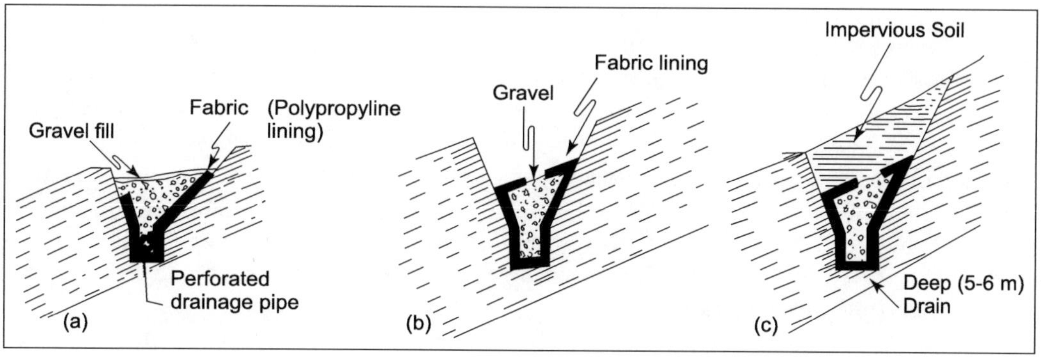

Fig. 11.22. The ditch dug to drain out water from the landslide-affected area is lined and then filled w gravel in order to save it from being clogged or chocked. If draining of only subsurface water intended, the gravel-fill is covered with impervious soil.(After Bhandari and Gupta, 1985).

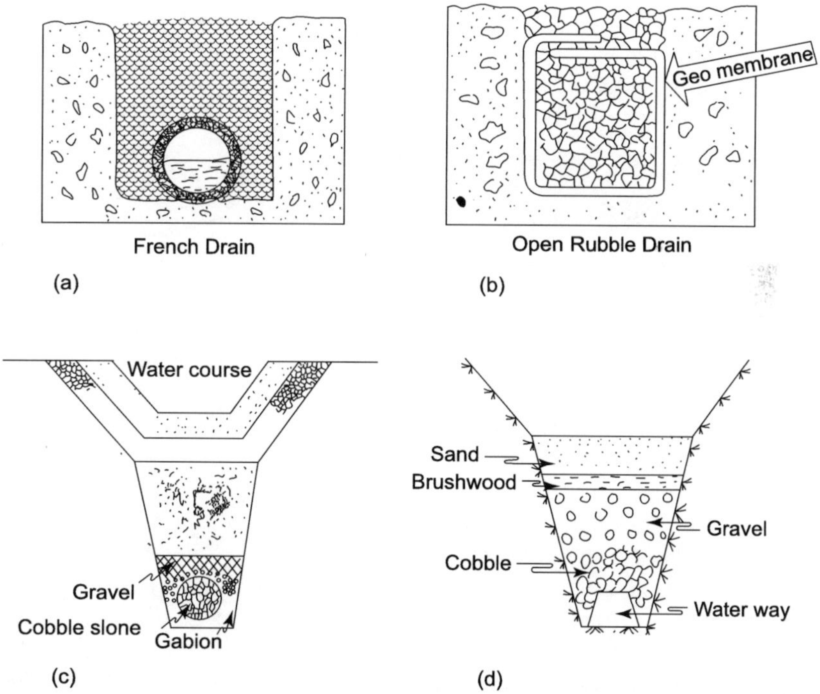

Fig. 11.23 Some typical surface and subsurface drains to ensure dewatering of the area affected by mass movement. (After R. K. Bhandari et al., 1985).

and 11.23). This is accomplished by constructing an elaborate drainage network. All streams a watercourses must be diverted around the crown of the slide or the potentially hazardous area through properly lined drains and ditches that have adequate gradient. The springs within the area are entrapped d likewise diverted through pipes into the drains. The drains within the affected area can be filled with avel and coarse sand (Fig. 11.23). A needle-shaped filter put on top checks the clogging by gravel m sediment-charged runoff. In special cases perforated pipes placed at the base serve as outlets. The ditches may be lined with polypropylene sheet to prevent seepage of water (Fig. 11.23).

Construction of a drainage network is accompanied by levelling the ground surface to eliminate all depressions where water can accumulate and create problem. **The cracks and openings are filled or plugged or grouted with cement or hydrated lime slurry or sodium silicate solution or bitumen emulsions. On critical spots the surface must be covered with impermeable layers of asphalt or ays to prevent infiltration of rainwater.**

Subsurface water is pumped out vigorously through tubewells or mechanized pumps put through drilled holes or adits. Galleries and tunnels can also be constructed if the situation becomes critical. But this is an expensive proposition. In the Tista valley in Sikkim, old telegraph poles (which were broken landslides) were perforated and used to drain out water from the debris mass (Harish Chandra, 1975). Perforated bamboo poles 2 to 3 cm or more in diameter and 8 to 10 m in length have proved useful in draining out subsurface water (Bhandari and Gupta, 1985). The dewatering operation effectively reduces the pressure buildup within the slide mass.

The muddy (clayey and/or silty) impermeable soils, such as loess threatened with instability (e.g. the Kashmir and Kathmandu valleys) are treated differently. In a rather expensive thermic treatment hot air is driven under pressure of 0.15 to 0.50 atmosphere through loose holes and tunnels in order to evaporate and drive out water from the interstitial space. This measure increases the shear strength of the soil.

Vegetating Damaged Slopes

An effective method of arresting erosion and a protecting damaged or exposed subsurface material is to cover it with a net of coir jute or synthetic yarn of about 2.5 to 1.5 cm opening (Figs 11.24b and 11.25), which will allow quick growth of vegetation. The area is fenced before the seed bed is prepared. The netting prevents the soil from breaking due to impacts of rain and wind, and allows hindered growth of vegetation. The soil is first seeded with quick-growing local grass. The root ps are planted in dug holes 15 to 20 cm apart—root to root, row to row, taking care that no turfs are planted in dug holes (Natarajan and Gupta, 1980). It is also admixed with an appropriate fertilizer leguminous plant or organic manure to promote quicker and richer vegetation growth. Chemical tilizers are applied—such as calcium ammonium nitrate if the soil is acidic—at the rate of 50 gm/100 (Bhandari and Gupta, 1984). The netting is firmly put in place by overlapping 5 to 8 cm fringes and pegged down with staples of 10 gauge steel wire 30 to 60 cm apart (Figs. 11.24b and 11.25). This netting eventually disintegrates to contribute to the store of manure. The seeded soil can also be covered with rous material such as chopped straw, shredded bark and wood pulp.

It would be desirable to plant self-generating stakes of quick-growing plants along the contours at some intervals (Fig. 17.24A) to strengthen the loose mass.

To ensure further protection of the slope and dislodgement of root slips in the seed beds, asphalt emulsion (mulch) can be sprayed (at the rate of 1 litre/m^2). The mulch eventually disintegrates into another fertilizers (Natarajan, 1978). The mulching is done just before rainfall. In extreme cases, plastic latex stabilizer is sprayed after seeding. A much simpler and more pragmatic method would be to mix the grass seed with a small proportion of rapidly growing stemmy plants such as sorghum, which acts as a nurse crop. Inoculation of seed with *Rhizobium* bacterial culture would ensure quick realization of the desired result. **The vertiver grass (*Vertiveria zazanoides*) known as "khaskhas" has proved to be a great stabilizer. The "kikiyu" shrubs is also a binder of loose material (Fig. 7.17).**

A number of grasses have been found useful for protecting slopes, such as 'nara' (*Arundo donax),* lemongrass, napier *(Pennisetum purpureum), kikiyu (Pennisetum clandestinum), 'gorda' (Chrysopogon fulvus), C. nepalensis,* etc. The choice depends on local conditions, including climate and soil composition. The legume *Pueraria hirsuta (P. thunbergiana)* with deep roots has been found very promising. It is planted in January and February.

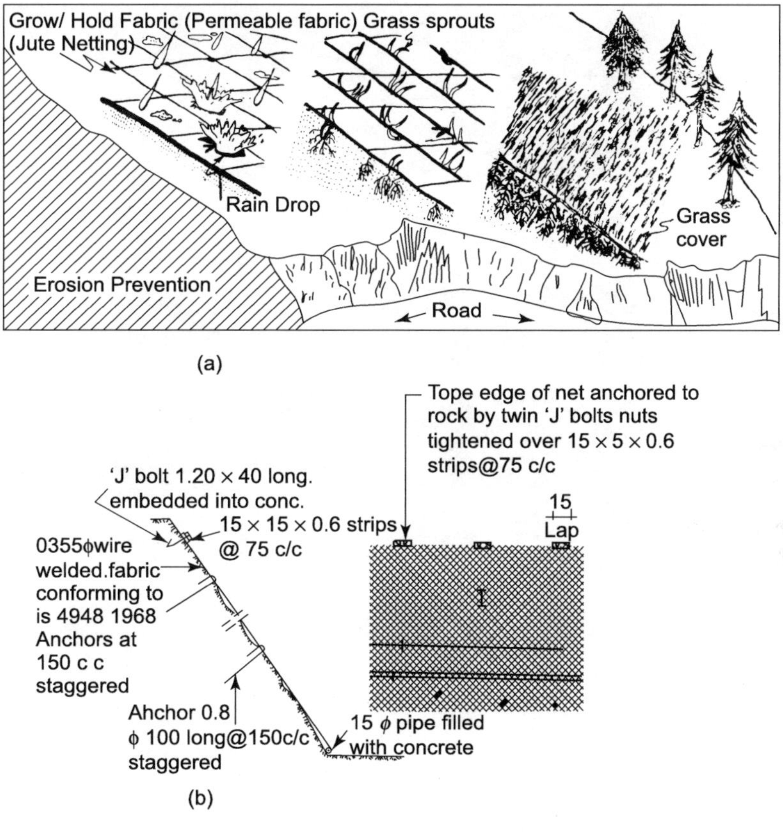

Fig. 11.24. (a) Putting vegetal cover on the landslide-ravaged slope of a road. Three stages of growth grass cover under the protection of jute-netting. (Based on T. K. Natarajan and S. C. Gupta 1980.) (b) Method of fixing coir net on the unstable slope. (From R. K. Bhandari et al., 1985.)

After partial stabilization has been achieved, the grass-covered area is planted with shrubs and tre with deeper roots. Not only do they dewater the soil but also increase its shear strength through their deeply penetrating roots. Among the shrubs recommended (Tejwani et al., 1975; R. K. Gupta, 197 are 'beshram' *(Ipomoea carnea),* 'shamalu' *(Vitex negundo)* and *Jatropha curcas.* Poplar, birch, adlar d oak trees are very useful. Trees such as 'khair' *(Acacia catechu),* 'gainthi' *(Boehemeria rugulosa),* angara' or 'mandar' *(Erythrina suberosa),* 'shisham' *(Dalbergia sissoo), Wendlandia exserta,* willow *(Salex tetrasperma), Bombax ceina, Woodfordia fruticosa, Lannea coromandalis, L. grandis, Moringa pterygosperma* (in phosphatic soil), *Bauhinia retusa* (in calcareous soil) have far-reaching effects on the bility of the slopes even though their growth is slow (R. K. Gupta, 1979).

The afforestation in the catchment of the Birahi, Garurga and Patalganga tributaries of the River aknanda in Uttarakhand resulted in the reduction of active landslide areas from 1409 ha in 1972 to 93 ha in 1995 (Kimothi and Juyal, 1996).

Fig. 11.25. Controlling movement of debris by jute netting. (Courtesy: Dr T. K. Natarajan, New Delhi).

Slope Modification

The stability of the slope can be increased by grading it by constructing benches, which not only reduce the slope but also serve as traps for falling, sliding and creeping material so that downslope damage is prevented (Fig. 11.26). But the reshaping of a slope involves appropriate disposal of the debris generated. The cost is also very high.

Another method of stabilizing a slope is to reduce the load at the head of the landslide and/or strengthen the base by enlarging the toe. **If only 4% of the slipping mass can be removed from the head and placed at the toe, the stability will increase by 10% (Zaruba and Mencl, 1969). This also implies removal of heavy structures (including buildings and big trees) and dumps from the top of the unstable area** (Fig. 11.3).

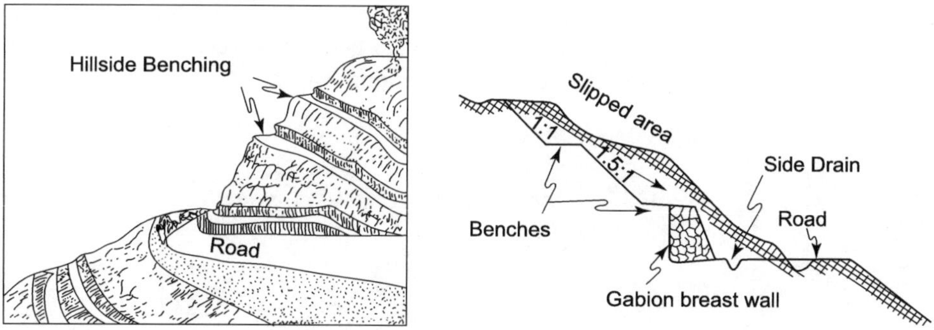

Fig. 11.26. Roadsides can be stabilized by modifying the slope such as developing benches, which redu the slope steepness and serve as traps for falling, slumping or creep debris.(Based on (a) G.C. Sharma, 1985; and (b) D.R. Coates, 1981, *Environmental Geology*, John Wiley, New York, 701 p., Fig. 11.28).

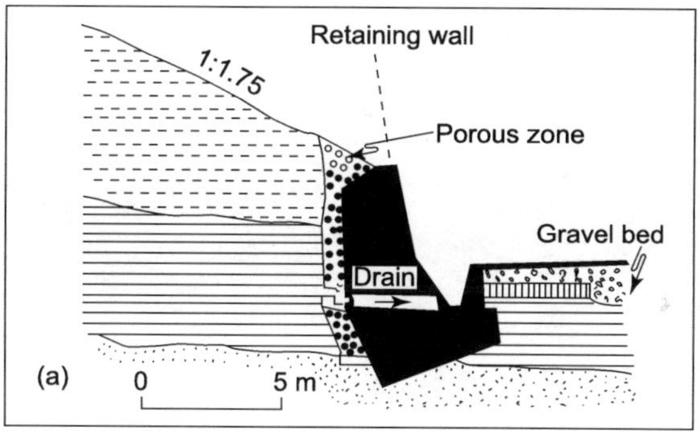

Fig. 11.27. Keyed retaining wall to provide support to the slope. Behind the wall the empty space is filled w gravel to provide passage of subsurface water.(Based on Q. Zaruba and V. Mencl, 1969).

Retaining Walls and Buttresses

To provide support at the base of the threatened slope or slid mass and to prevent toe erosion, a variety of retaining walls (Figs 11.27, 11.28 and 11.29) of concrete masonry or cribs can be constructed. A ***buttress wall*** is a heavy, low-height structure at the base that provides safeguards again toe erosion. If free space is available, the wall is built a little away from the toe and the intervening space filled with gravel and other permeable material to allow drainage of water. The free passage of wat prevents buildup of water pressure inside the slid mass behind the wall. The retaining and buttress walls must have weep holes and drop channels, if these are to be effective and long-lasting.

Another very effective, fast and economical method of preventing slope failure is the one adopted in deep cuts in the Tallahatti Channel of the Kalinadi project, northwestern Karnataka. *Pre-stressed anchor pads* (Fig. 11.28) are installed at the toe of the slope, each anchor weighing 147

tonnes. These are arranged in a grid pattern in three rows each and connected with one another by crossbeams (Upadhyay, 1980). The pre-cast concrete fensing panels (Ajit Singh, 1985) and soil reinforcing elements allow quick construction of the retaining wall.

Fig. 11.28. Pre-stressed anchor pads, each having a weight of 147 tonnes and mutually connected with cross-beams in the Talahatti Channel, Kalinadi project, Karnataka. (Courtesy: Dr U. S. Upadhyaya).

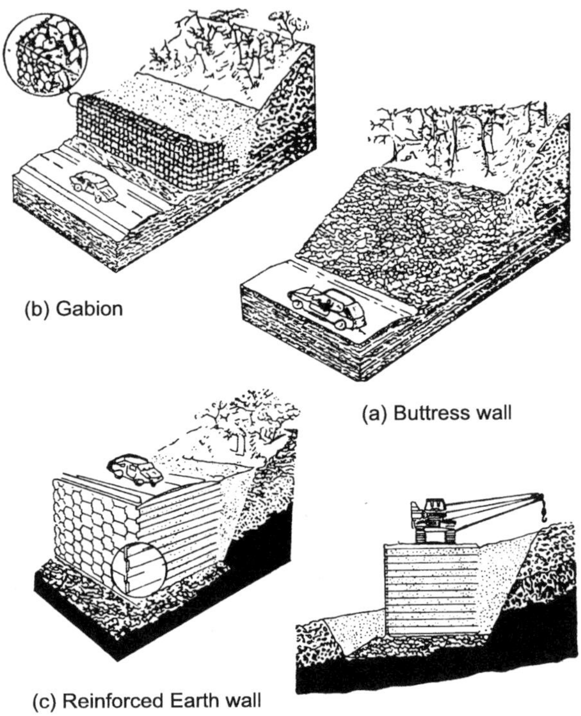

Fig. 11.29. Flexible retaining walls, including reinforced earth wall of modern concept.(After R.K. Bhandari et al., 1985.)

The ***cribb wall*** is constructed with wooden poles or pre-cast concrete units set up in cells which are ed with gravel or stone. The *gabions* are steel-mesh cages filled with gravels (Figs. 7.17, 7.18 and 11.29). **Baskets of gabions and cribb walls can support shallow sheets of landslide material. But ey are hardly effective in areas afflicted with creep movements or affected by debris flows.**

A row of deep piles connected by cross bars at a few levels and with the gabion baskets would prove much effective in arresting slides.

It is always desirable to put a layer of monofilter or polyfilter or polypropylene before emplacing the aining walls. These filters prevent loss of solid material of the structure. To stabilize embankments of reservoirs, canals and rivers, porous ***fibriforms*** are filled or injected with high-strength cement mortar *in* *situ.* Resistant to acids, alkalies and microbial activities, these fibriforms assume the shape conformab to the local microtopography and thus impart strength to the structure (Fig. 7.20).

The new technology of road construction has demonstrated the great merit of *reinforced ear retaining walls* (Fig. 11.29) in slopes prone to creeping and sinking (Bhandari et al., 1985). Light in weight and flexible in nature, these retaining walls can allow vertical settlement (as much as 2.5 cm 3 m) and horizontal displacement without impairing the structure.

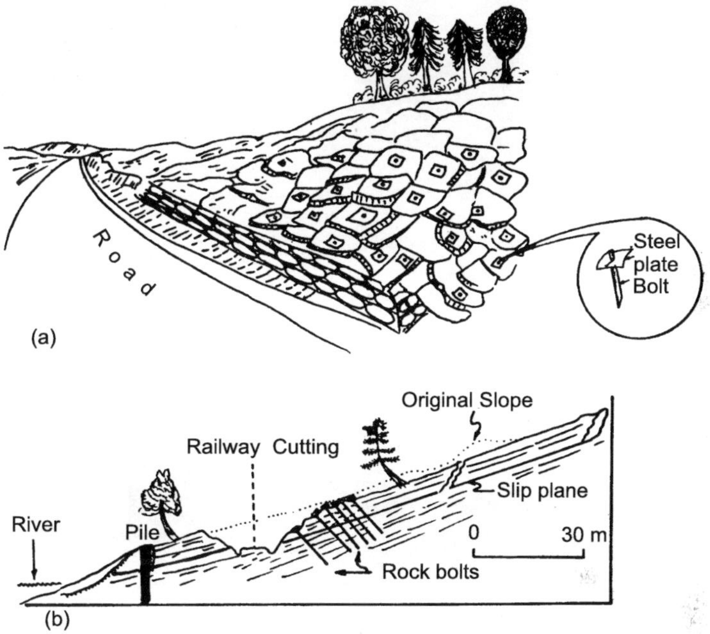

Fig. 11.30. Driving steel bolts through inclined strata can stabilize the slope, as on this road, (a) After D. Coates, 1981, *Environmental Geology*, John Wiley, New York, 701 p., Fig. 11-29. (b) After Zaru and Mencle, 1969.

Piles, Bolts and Shear Keys

Rows of steel or concrete piles are driven deep into the ground, in order to establish anchorage for non-plastic, unstable or slipping mass, particularly if the movement is taking place at shallow depths. Commonly, the piles are linked together by horizontal bars. This measure is supplemented sealing of joints, cleavages and cracks.

In mines and tunnels and on roadsides ***steel bolts*** are used to prevent blocks or sheets of rocks from falling or slipping (Figs 11.30 and 11.31). Holes are drilled through potentially unstable jointed or cleaved rock slabs and bolts fixed in them. Gneisses and schists in the Khuni Nala Slide (Fig. 11.1 between Ramban and Banihal (Kashmir) have been stabilized by driving steel bolts.

Stabilization of the slipping rocks can also be achieved by making prism-shaped holes compact filled with clays as done in the case of dams (Figs. 17.2 and 17.4).

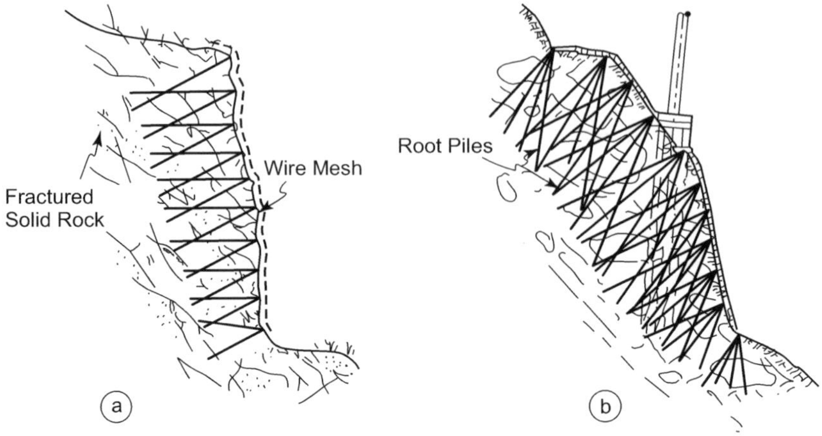

Fig. 11.31. Application of root piles minimizes hazards of rockfall and slope failure of certain types. The wire net prevents hazards on roads. (After R. K. Bhandari et al., 1985).

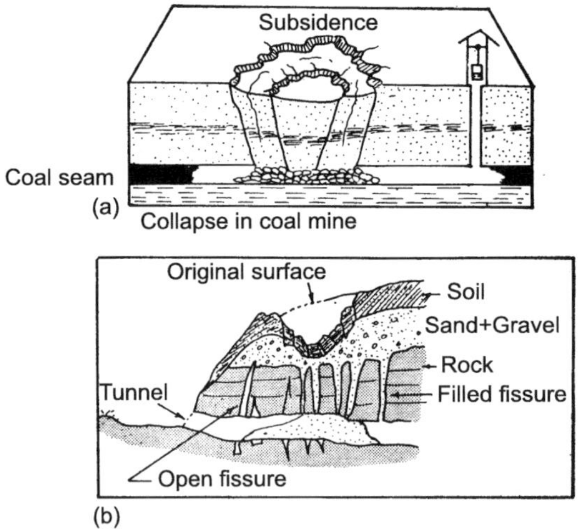

Fig. 11.32. Sagging of ground resulting from removal of coal from an underground mine. Tunnelling can also cause ground subsidence. (Based on P. D. Krynine and W.R. Judd, 1957.)

SUBSIDENCE AND SWELLING OF GROUND

Nature of Subsidence

pressions on an otherwise flat surface, an irregular pot-holed surface, wavy or undulating roads and wl-shaped water holes are signs of sinking of a part of the ground by the withdrawal of solid or id material from underground. Commonly, it is a slow process but the culmination may have tragic nsequences, such as witnessed in the Jharia coalfields in Jharkhand (Fig. 11.32) or in the historic city of Venice in Italy. The subsidence of the ground in Venice has invited encroachment of land by seawater, inundating a sizeable part of the beautiful city. This city is sinking at the rate of 2 mm/year (Y. Bock of Scripps Institute of Oceanography, San Diego) (*The Times of India,* March 23, 2012). In the rainy season the depressed parts of the Jharia coalfield become submerged by floodwater from the Barakar Riv which might someday swing its course towards the subsided area.

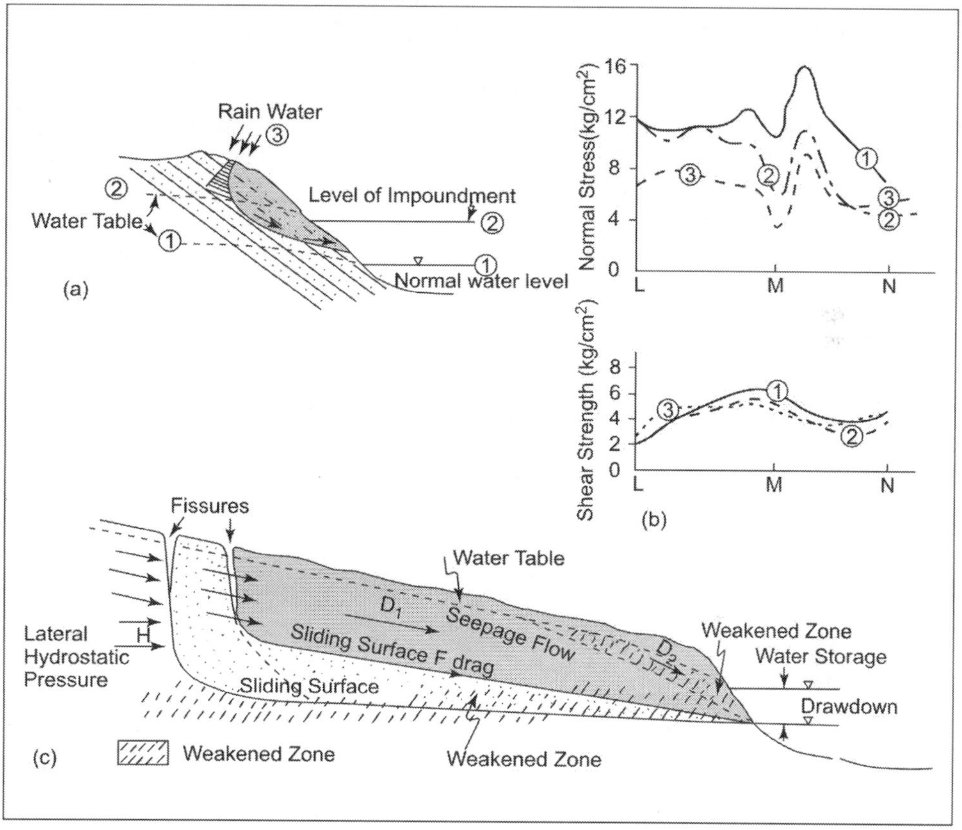

Fig. 11.33. Withdrawal of lateral support of water in a lake or reservoir following drawdown causes mass movements, and resultant subsidence if other conditions also exist. (Based on L.P. Petrov Jasjunas, 1982; and G.U. Dezhen and Wang Sijing, 1982.)

Causes and Mechanism of Subsidence

A number of causes of ground subsidence have been suggested, all related to the withdrawal of solid and or liquid material from deeper levels underground. As a consequence of the withdrawal water or liquid petroleum through pumping, the pore pressure within the liquid-bearing rocks diminish considerably. Simultaneously, the water table sinks. In the terrain of sediments and softer sedimentary rocks the soil-layer or aquifer becomes compressed vertically—particularly if a large number of we or pumps draw out heavily for a prolonged period of time (Fig. 5.10). Thousands of hectares of ground the oilfields of California have subsided for this reason—9 m in the central parts in the Inglewood field between 1937 and 1967; at the rate of 0.5 m/yr in the Nagata gasfield in Japan; and at the rate 25 cm/yr in the Po valley gasfield in Italy (Keller, 1982). The degree of subsidence depends on the uantity and rate of withdrawal of fluid and on the compressibility and depth of the layers from which e fluids are tapped (Bolt et al., 1975). Following reduction of pore pressure at a rate of 280 kg/cm^2 American oilfields—the grain-to-grain load became dominant, leading to compaction and eventual surface subsidence.

The removal of coal from the underground mines in Jharia caused a considerable surface subsidence varying from 6 to 25cm. The Baliapur-Jharia road sank 1.2 to 1.6 m, while the ground over e Kujama colliery went down by 2.4 m (Fig. 11.32). Withdrawal of water from lakes, reservoirs and tanks in unstable settings causes subsidence of marginal areas as a result of removal or reduction of lateral support to the slope (Fig. 11.33).

Road cutting and mining on slopes likewise bring about creep movement and attendant subsidence.

The drying up of soil also leads to subsidence. Small areas around brick kilns or belts adjacent to ws of large trees such as poplar, willow, eucalyptus and elm — which draw considerable water from e soil through evapotranspiration — may subside (Fig. 11.34). Draining away water from the marshes and bogs, which contain highly compressible organic matter, has led to subsidences. In eastern United ngdom, the subsidence was of the order of 5 to 10 m since the early nineteenth century following reclamation of the moors. This is also a problem in the marshy tracts of the Ganga Basin including ndarban and the southern suburbs of Kolkata.

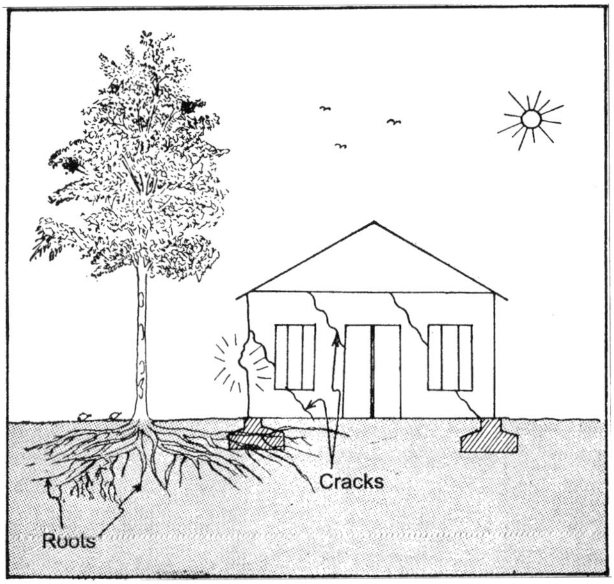

Fig. 11.34. Soil desiccation caused by the action of roots leads to shrinkage of clays and subsidence of ground, manifest in the development of cracks in buildings.

In the Himalayan region in the zones of intense crushing and pulverization of rocks and resultant formation of gouges in zones such as of the Main Boundary Thrust in the outer Himalayan belt, development of ***sagponds*** and related depressions is a common phenomenon. The subsidence is due to the compaction of the crushed material following infiltration of water or sometimes due neotectonic movements. A 70-m stretch of the road section near Dalhousie (Himachal Pradesh) subsid about 2 m between 1947 and 1953, the depression occurring on the highly fractured carbonaceous phyllites thrust over by schists, gneisses and granites (Krishnaswamy and Jain, 1977). The groundwat here had dissolved away the crushed material.

Preventive and Corrective Measures

Ground subsidence can be prevented or minimized by sand-stowing, i.e. filling the underground min with sand or unwanted debris in a phased manner (Chapter 6). Leaving pillars of ores to support t roof-beds controls the subsidence or collapse.

Water depletion caused by excessive withdrawal resulting in ground subsidence can be countered by resorting to artificial recharge of groundwater (Chapter 4) during the rainy seas using stormwater, floodwater or even drainwater. To compensate for the deficiency created by t withdrawal of water, oil or gas, such liquids as saline seawater, floodwater and drainwater ha been successfully introduced and subsidence checked.

Swelling of Ground Surface

Certain clays such as montmorillonite absorb water rapidly and expand consequently. The pressure this kind of clay within the soil causes swelling of the ground and resultant damage to structures, houses, bridges, dams, etc. Percolation of water from cesspools, sanitary tanks, swimming pools and ponds are known to have caused great distress to adjoining buildings.

The swelling of clays is particularly common in dry areas where there is a drastic change in the moisture content from summer to the rainy season. The Deccan and Malwa plateaus, consisting of basic lavas and volcanic ash is overlain by what is known as ‘regur’ soil. The ‘regur’ has swelling s• problems. The weathering of lavas gives rise to montmorillonite clays prone to this phenomenon. During the dry season the soils are impenetrably hard, locally glazed or shiny and characterized by shrinkage cracks. But in the rainy season this soil turns into a very sticky and clinging mass which swells into mini ridges and bulges. This explains why buildings have cracks in Maharashtra.

To lessen the problem of swelling of clays, either the soil is replaced by another suitable material or, it is irrigated to saturation point before construction starts. The ground can be loaded sufficient and heavily to counter the pressure of the swelling soil. Chemical treatment to convert expansive cla into non-swelling types involves injection of lime, flyash, phosphate and other organic componen Construction of networks of drains with wide aprons on the walls, and covering the area with impermeab material and keeping away pools and sanitary lines, can considerably minimize infiltration of water in expansive soils.

SUMMARY

Among the factors that govern the stability of hillslopes, the structural conditions of rocks and soils, t gradient of the slope and the amount of water that finds access underground play very crucial roles tiating and controlling landslides. Seismic shocks trigger and accelerate mass-movements on a larger scale but **the principal agent is always water. It not only acts as a lubricant, especially if mixed with clays or other slimy material, but also reduces their shear strength by building up pore pressure within soils or rocks.**

The mass-movements are manifest in the form of slow movements without breaking into discrete failure planes (creep), comparatively speedy sliding on discrete planes of failures at deeper levels (slumps d slides), or very rapid, almost spasmodic descent or flow of non-cohesive masses of rubble (debris alanches) or clay, silt and mud (mudflow). The incidence of landslides is therefore very frequent and their magnitude greater during rainy spells, and in earthquake prone belts.

The strategy for controlling the occurrence of landslides is therefore to stop the entry of water or remove it from the threatened or affected area by a system of surface or subsurface drainage, and by resorting to pumping. The second step is to place buttress wall or retaining wall at the toe of the slope or slide in order to check the movement of the dislocated or destabilized mass. Finally, **the damaged slope and its soil should be put under a protective cover of deep-root grasses, shrubs and trees.**

Measures for protecting the river banks from currents and the coasts from waves include construction deflecting spurs, crib- and gabion-wall. The withdrawal of fluids (water, petroleum, gas) from their underground habitat or of solid ores by underground mining causes the ground surface to sag or even llapse under certain circumstances. The subsided areas are exposed to the threat of flooding as ppened in the Jharia coalfield. Related to subsidence is the swelling of soils due to the presence of such clays as montmorillonite, which expands in volume as a result of absorption of water. This phenomenon adversely affects the stability of buildings and other structures and is very common in dry eas where there is a drastic seasonal change in soil water-content.

One remedial measure or two would not be effective in preventing mass-movements. It is always a package of measures that has to be taken, as done successfully in the Tista valley, Sikkim and the Nilgiri Hills, Tamil Nadu.

In order to check subsidence due to withdrawal of fluids from underground, attempts be made to charge groundwater by waters of floods, storms or drains. Underground mines can be filled with sand and debris or waste to prevent roof collapse and ground subsidence. Good drainage and keeping away leaking bodies or sources of water provide the best remedy against ground swelling.

12 Hazards in the Snowy Mountains

HAZARDS IN THE HIMALAYAN DOMAIN

In winter and early spring the strategic lines of transportation, trails of pilgrims and mountaineers and the defence installations in and across the Great Himalayan Range are often threatened with -borne masses of snow. Descending from the mountain tops wgfith savage speed and flowing e rivers of snow and ice, the avalanches engulf valleys for miles and destroy whatever comes in their ways.

Nestling in the lap of the Trishul (7122 m) at an elevation of 4778 m is the well-known glacial lley of Rupkund in Uttarakhand. Some 200-odd bodies of men and women pilgrims from Haryana Rajasthan, who were overtaken by a savage avalanche, presumably unseasonal, were found buried here in snow and ice. The holy shrines of Hemkund, Badrinath, Kedarnath, Gangotri and Yamunotri in tarkhand, Trilokinath in Himachal Pradesh and Amarnath in Kashmir continue to be threatened by alanches in winters. About 135 soldiers and civilian helpers encamped at Goma on the southwestern pe of the high ridge in the Siachin Glacier region were buried under more than 20 m-thick debris of ow in early April 2012 (*The Hindu*, April 8, 2012).

ln the Himalayan slopes the avalanches come down in waves—6 to 7 waves within short spans of to 24 days. In March 1979 about 237 people were killed in a day by avalanches that ravaged large eas in the Spiti and Lahul valleys in Himachal Pradesh. In the winter of 1982-1983, avalanches ok a toll of over 100 people in Jammu-Kashmir alone. An analysis done by the Snow and Avalanche udy Establishment of the Defence Research and Development Organization showed that on an erage 30 people die every year from these snowy mountain hazards in the Himalayan domain in e pre-1980s (Fig.12.1). Avalanches of catastrophic magnitude repeat themselves over a period of ars, the frequency of which is yet to be established for the Himalayan slopes. However, there is a rceptible increase in the number of fatality over time, particularly because our defence forces are nstrained to live and move around during snowy winters in the Himalayan region of snow and ice (Table 12.1).

Major avalanches in Kashmir come every year down the slopes of the Pir Panjal, pronouncedly in e months of January, February and March (Mohan Rao and Rangachary, 1983). The road to Leh and dakh becomes perilous during winter when nearly 70 % of the highway is affected (Fig. 12.1).

Table 12.1. Avalanche Disasters in India (Gupta et al, 2001)

State	Year	Damage
Himachal Pradesh	March 1978	30 people killed. Road and property damaged.
Lahaul and Spiti (Himachal Pradesh)	March 1979	237 people killed. Communication lines disrupted.
Jammu and Kashmir	December 1982	100 people killed. Communication lines disrupted.
Ladakh (Jammu & Kashmir)	December 1984	27 soldiers killed in an avalanche generated by artillery fire.
Himachal Pradesh	March 1991	An avalanche took place 4-5 times at Tinku, between January-March. The roads was blocked for 40 days in 1991.
Himachal Pradesh	September 1995	The avalanche caused descent of huge chunks of debris, which later melted to create a flood.

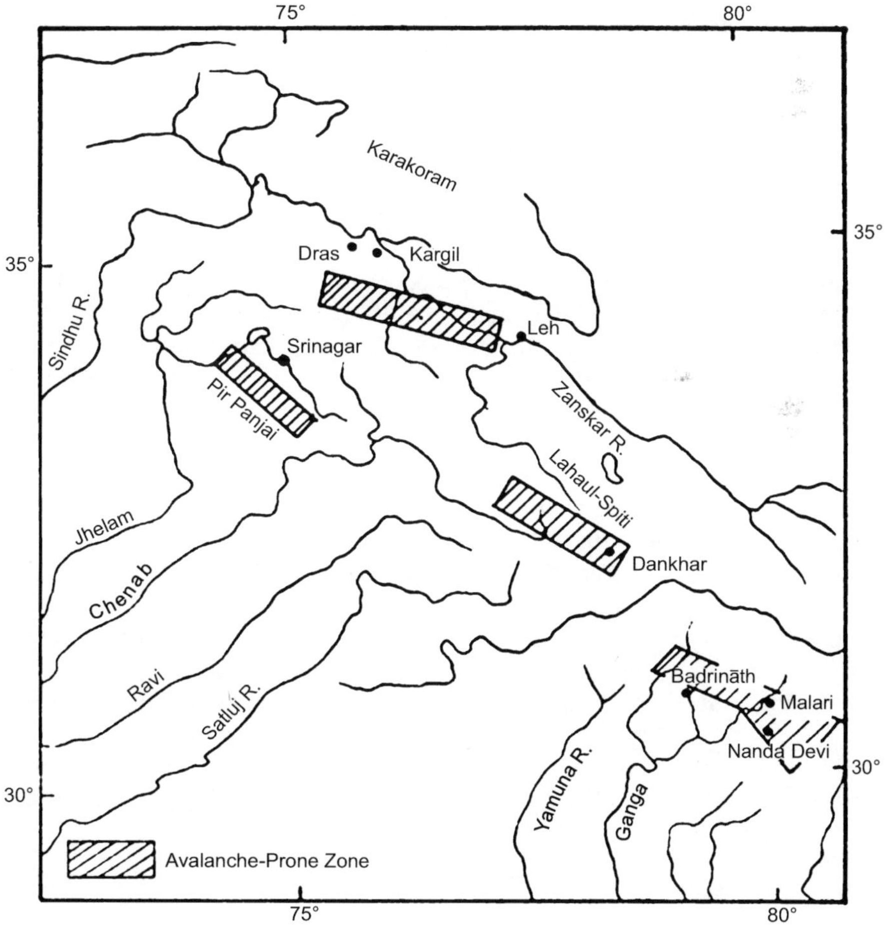

Fig. 12.1. Avalanche-prone areas in the Himalayan domain investigated by the Geological Survey of India and the Snow and Avalanche Studies Establishment of the Defence Research Organization.

TYPES OF AVALANCHES

Nature

asses of snow-and ice-crystals admixed with air descending from mountain-tops at very high speeds d flowing like rivers of snow and ice blocks in the valley below are known as avalanches. They occur winters and early springs, generally in the afternoons. The ice crystals are admixed with air and water pour so that an avalanche is a visco-elastic body having density between 20 kg/m^3 and 600 kg/m^3 that **moves down the precipitous slopes with a speed of 30 to 300 km/h.** While the flowing dry avalanches •ve with a speed of 20 to 40 metre per second (and the wet one slightly less at 10 to 30 m/s) the air-rne powder avalanches have a speed of 30 to 70 metre per second (Mohan Rao, per, comm., 1985). ne that struck the village Goskiar in Keylong (Himachal Pradesh) with destructive fury in March 1979 d a speed of 33 metre per second. **The thrust (impact pressure) of the heavy descending mass was also very high—5 to 50 tonnes per square metre. In the Kashmir Himalaya the pressure recorded was 20 to 30 t/m^2** (Mohan Rao and Rangachary, 1983). The air blast was stunningly high—0.5 t/m^2. nderstandably avalanches blow away everything, crush down trees and houses, and toss up vehicles as if they were toys.

On the mountain slope, several layers of snow are superimposed, each layer having its own distinctive operty (Fig. 12.2). The change in physical conditions such as the rise in temperature during the ys or in springs, the increase in wind velocity, the steepening of slopes resulting from progressive accumulation of snow, etc. cause appreciable thermodynamic changes in the snow, leading to mass-•vements of snow and ice.

In the Himalayan realm avalanches originate predominantly from elevations of 3500 to 4000 m. opes of 30 to 45° are most vulnerable for occurrence of avalanche although any slope from less than ° to more than 60° can experience this hazard. Wet avalanches are known to occur on slopes less an 15°. Convex slopes, naturally, are most vulnerable, particularly where these are bereft of trees and rubs or large rocks. Smooth grassy slopes are more prone to slides of snow masses. The north-facing slopes are affected more during mid-winters, while the south-facing slopes become dangerous in springs and during sunny afternoons.

An analysis of avalanche statistics over Kashmir and Himachal Pradesh (1965 to 1976) along with eteorological parameters showed that 72% of the avalanches (mainly loose snow) were caused by ry heavy snowfalls occurring over snow peaks; 15% may be attributed to temperature and rain (slab alanches) and the remaining 13% to wind. The effects of other parameters such as humidity appear to insignificant (Anon., 1984). **The average snowfall intensity propitious for triggering avalanches in this setting is more than or equal to 3 cm per hour and the minimum accumulation of 40 to 60 cm in 24 hours (Mohan Rao and Rangachary, 1983), which generally happens in February and March.**

Three types of avalanches are recognizable in the Himalayan slopes—loose-snow avalanches, slab alanches and airborne-powder avalanches. These three types can be put into two categories—storm-duced avalanches and delayed-action avalanches (K.C. Agarwal, 1979).

Fresh-Snow Avalanches

The ***loose-snow avalanches, also called fresh-snow avalanches*** consist of loose crystals of snow mixed with air. The loose aggregate is set in motion by snow-storms. It is the topographic drop in the

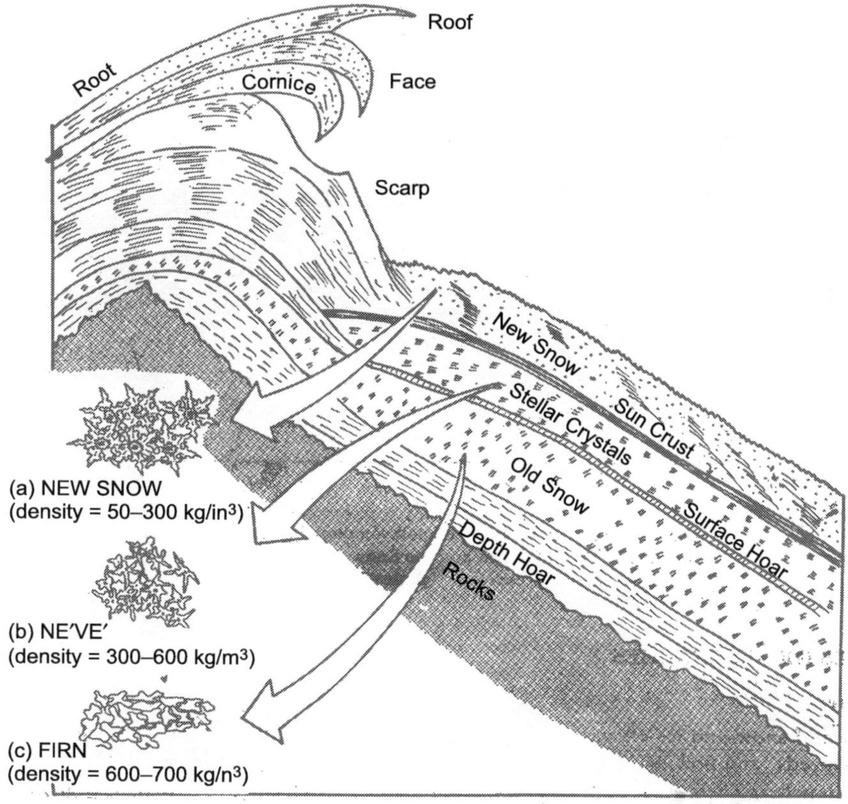

Fig. 12.2. Development of conditions propitious for triggering of avalanches. A storm has piled up snow masses into heaps with overhangs or cornices. These overhangs will crumble, move down and become air-borne, giving rise to fresh-snow avalanches. (After David Cuppe, 1982.) Inset (left bottom) shows change in the morphology of snow crystals due to load metamorphism. (From W. M. Marsh in the lower left and J. Dozier, 1981).

path that makes them air-borne. The billowy, powdery mass (>100 million m^3) of dry snow hurtles down the steep slopes (Figs. 12.3A, 12.5); gathering speed upto 300km/h, developing very strong pressure a creating shock waves in the atmosphere of the valleys below. They occur throughout the snow seaso either during or immediately after snow-storms. They are generally small in size and the impact pressu is also low.

Slab Avalanches

In some situations the snow may continue to pile up, with the consequence that the lower layers get hardened due to load metamorphism and converted into cohesive layers made of compact aggrega of rounded ice crystals (Fig. 12.4). This compact mass is attached to the slope with varying degree firmness. The weakening of the bond due to the presence of empty space or unfrozen ice, the shoc generate by earthquakes, explosion or movements of vehicles or skiers leads to failure of the ice laye

The dismembered layers of ice break down into slabs and blocks along fractures, as they slip and hurtle wn in the form of slab avalanches (Figs. 12.3b, 12.5b and 12.6). Belonging to the category of ***delayed avalanches*** these are frequently witnessed in the valleys of Ladakh, particularly on the convex and uth-facing sunny slopes (Raina et al., 1982). They are often dangerous and unpredictable, generating they do very high impact pressures.

A
B

Figure 12.3. (a) Fresh-snow avalanche in the Nubra valley, Ladakh, in Kashmir. (b) Slab avalanche developing on the slope, Kashmir. (Courtesy: Dr S.S. Sharma, Snow & Avalanche Studies Establishment, Chandigarh) (From Valdiya, 2004) (See colour Fig. 12.3)

Hoar Avalanches

n the northern slopes of the Great Himalayan Range—not exposed to sunshine and thus characterized extremely low temperature gradient across the snow layers—the formation of large pyramidal hollow

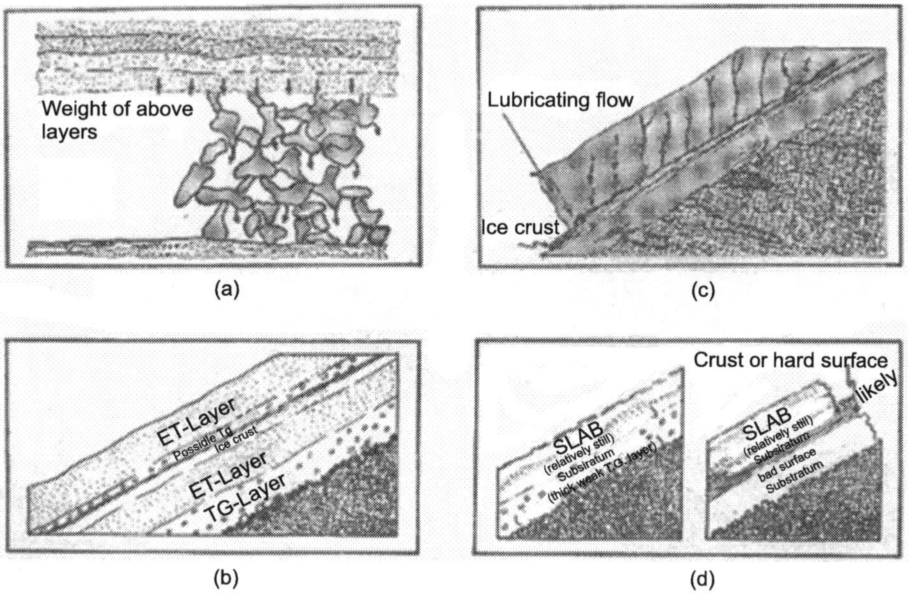

Fig. 12.4. (a) Snow layers undergoing metamorphism due to compression as the weight of piling snow grows. (b) Layering of snow pack on the slope due to temperature-gradient (TG) metamorphism in response to strong temperature gradient and equitemperature (ET) metamorphism resulting from knitting together of snow into a tight texture. (c) Lubrication of ice-layers resting on the hardened snow pack or on the ground surface is instrumental in destablization. (d) A thin weak layer sandwiched between stiff slabs and hard surface or crust is highly unstable. (After R.I. Perla and M. Martinelli, 1978, *Avalanche Handbook,* U.S. Department of Agriculture and Forest, Forest Service, Washington, 254 p. Figs. 58, 65, 70 and 124.)

crystals is very common. The brittle cohesionless depth hoar crystals crumble under the weight of upp layers, giving rise to air-borne ***depth-hoar avalanches*** triggered during snowfalls. The snow cloud causes whirling of snow into the air and is associated with air blasts which can achieve high velociti of destructive potential.

On the contrary, during cold clear nights the snow absorbs like a black body—all radiation availab in the atmosphere, but simultaneously emits maximum radiation, thus leading to cooling by mo than 10°C (Cuppe, 1982). As a result the immediately overlying water vapour in the air condenses in vertical ice crystals on the surface forming surface hoar. The crumbling of these crystals due to increasi weight culminates in the sliding of snow mass resulting in ***surface hoar air-borne avalanches.***

Factors that cause Avalanches

Three major parameters are involved in the triggering of avalanches—air temperature, wind veloci and slope gradient. The mountain slope is the most important factor—the steeper the slope, the great the potential—but only upto the limit of 60°. For beyond this limit the slope cannot allow accumulati of snow. The slopes ranging between 30° and 45° are very prone to avalanches in the Ladakh a

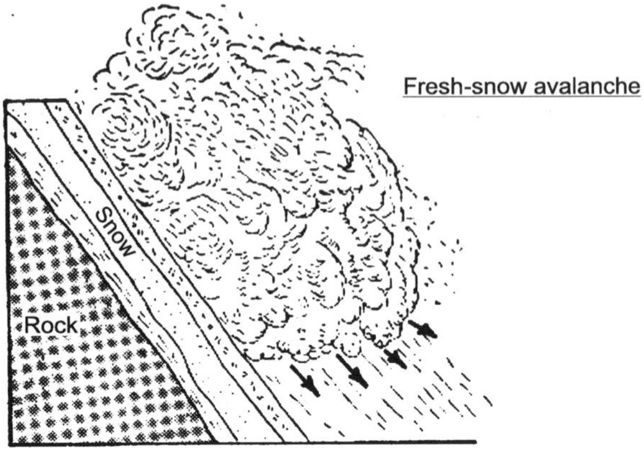

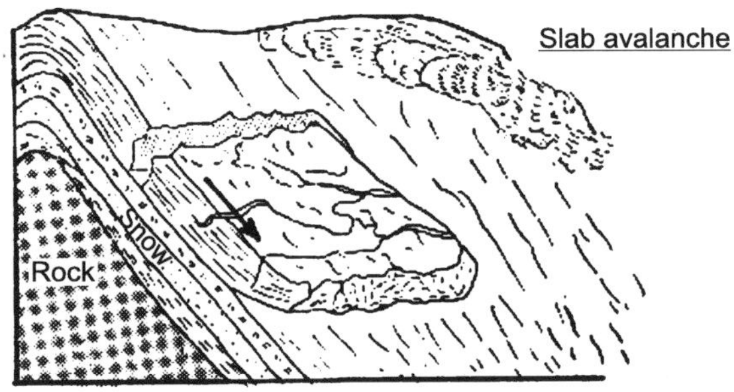

Fig. 12.5. (a) Storm-induced airborne fresh-snow (powder) avalanches moving with velocity varying from 30 to 300 km/h and having a thrust of 5 to 50 tonnes/m^2. (After D. Cuppe, 1982.) (b) Delayed slab avalanche resulting from weakening of bonds of ice crystals and development of fractures at the head. As the slide moves down the slope, it breaks down into blocks and fragments. Such avalanches occur after snowstorms during sunny afternoons and in springs. (After David Cuppe, 1982.)

machal regions. The vegetal cover has a restraining influence on the movement of snow masses. The rren mountains are thus very prone to this hazard, as witnessed in the Badarinath area in the upper aches of the Alaknanda River (Fig. 12.7).

The second factor influencing occurrence of avalanches is the temperature of air and its diurnal riation. In Kashmir, where the snow pack is 3 to 5 m thick, the diurnal variation is of the order of 6 to °C, the maximum day temperature in sunny days being 5 to 10°C. This temperature variation is very nducive to generation of avalanches (Raina et al., 1982). The southern slopes of the Great Himalayan nge are exposed to maximum solar radiation, promoting melting of snow and weakening of cover ring the day, leading to delayed slab avalanches. Rainwater also causes pronounced thermodynamic anges resulting in the formation of very rounded, smooth and convex grains of ice, commonly occurg in clusters. The smaller ones are the first to melt. The meltwater percolates through interstitial ace and accumulates above the frozen waterproof layers. This water is responsible for large-scale alanches.

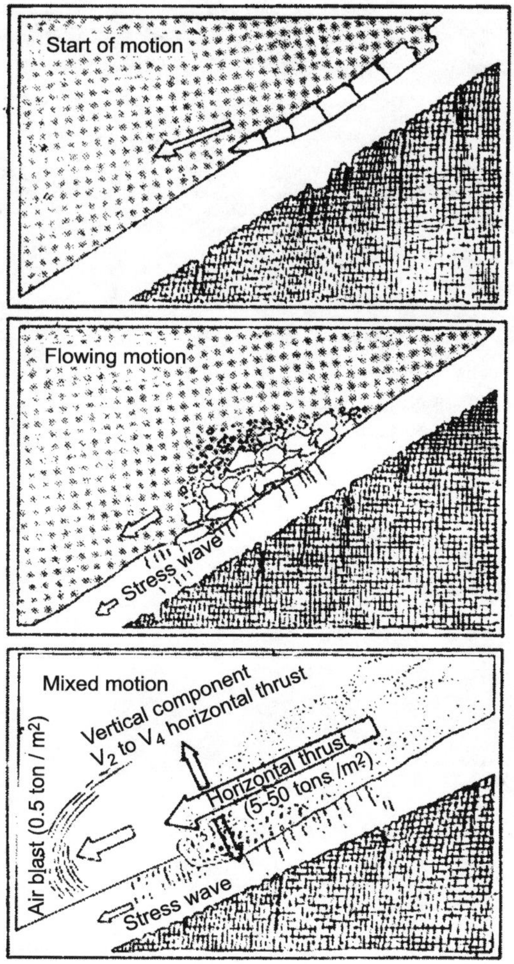

Fig. 12.6. Process by which the slab avalanche changes into a powder avalanche and resultant generation of a severe storm which rushes down the valley. (From R.I. Perla, and M. Martinelli, 1976, *Avalanc Handbook,* U.S. Department of Agriculture, Handbook 489, Washington Fig. 117).

Wind velocity is the third and very crucial factor in triggering fresh snow powder avalanches. T wind blows the snow mass to the leeward side; sometimes giving rise to overhangs and cornices (F 12.2), as very commonly seen in Ladakh, in Upper Sind (between Gagangir and Shitkari) on the oth side of the Zojila Pass and in the Jhelam basin, where the wind blowing with a velocity of 15 km/hr the valleys produces eddies and whirl winds. There were 37 avalanches within a stretch of 13 km in year in these regions once in the early 1980s (Raina et al., 1982).

The earthquake shocks trigger avalanches of great dimension over extensive areas as happen in Spiti and Lahul during the January 1975 earthquake. In 1970 an earthquake triggered catastroph avalanches in Peru that obliterated the township of Yungay, killing 18,000 people.

Fig. 12.7. Avalanche-struck Badrinath in Uttarakhand. It was a mild avalanche. (Courtesy: Prof. M.P.S. Bisht, Garhwal University, Srinagar) (See colour Fig. 12.7)

Glacial-Lake Outburst Hazards

lated to the hazards of the snowy mountains are the glacial lakes formed behind unstable moraine ms. The bursting of these lakes threaten to bring great hazard as the water levels of the lakes rise due rapid melting of snow and ice (Fig. 22.4). The outburst of such lakes discharge at the rate of as much 30,000 m^3 per sccond, thc rcsulting floods damaging up to the distance of 200 km downstream.

Using satellite imagery for surveillance has proved effective in Nepal and Bhutan. In 1994 the rsting of the Lugge Tso Glacial Lake in Bhutan caused a destructive flood, wiping out Namche Power ant and extensive damage (Mool, 2001). The glacial outburst (Fig. 22.4) in the Phochu river in 1994 as equally destructive (Joshi, 2001).

MITIGATION OF AVALANCHE HAZARDS

Strategy

One way of coping with the hazard of avalanches is to reduce their intensity and destructiveness by raisi barriers or some such engineering structures in their path, or diverting their course. **Precipitating their descent in chosen periods of time is yet another way of ensuring safety of traffic on the tracks.**

The Snow and Avalanche Study Establishment achieved considerable success in forcasting t occurrence of avalanches on strategic roads. The algorithm developed for predicting the time avalanches has met with 60 to 75% success along the Leh–Kargil highway and in the Siachin Glacı Valley (Bhutyani et al, 2007).

Pre-emptive Artificial Triggering

A dreadful idea grew in the head of an Austrian officer E. Aichinger when he saw, during the Fi World War in 1916, that the gunfire from the embattled Italian and Austrian armies in a valley below t Dolomite Mountain was causing avalanches. He bade his gunners to direct their shellfire at the snow covered slopes. More than 18,000 troops of both the sides perished under the avalanches that we triggered off by his command. A similar event occurred in December 1987 in the Siachin Glacier whe artillery gunfire during military skirmishes triggered an avalanche that killed 27 soldiers.

This experiment has been used for **triggering pre-emptive avalanches of small size by repeatedly hurling explosive projectiles on the snowy slopes and blowing off the overhanging cornices and thus releasing potential avalanches without allowing them to grow to an unmanageable size. This is done at chosen times and under safe conditions—thus forestalling the possibility of naturally occurring unpredictable avalanches. Artillery recoil-less launchers and air-guns have been used very commonly.**

In the Nubra Valley in the Siachin region, light-weight portable infantry weapons are now used prompt avalanches in order to prevent the possibility of naturally occurring unpredictable avalanch (Sharma, 2004). In the Jammu-Srinagar highway in order to pre-empt avalanches explosives are fired to 3 m above the snow surface through *Avadhav Vishphotak Vahan* developed by the Snow & Avalanc Studies Establishment (Sharma, 2002).

Avalanche-Preventing Structures

In the upper part of the slopes where avalanches commonly develop and get triggered, structur normal to the slope direction are constructed. The retaining structures (Fig. 12.8) in the upper part the slope prevent buildup of stresses in snow masses in the formation zone and thus prevent occurren of avalanche. They check deposition of snow on the leeward slopes and thus minimize the load build and consequent glide and creep of the snow mass. ***Terraces*** on the slope accomplish much the same result. ***Retaining walls*** of steel and concrete appreciably help avert rupturing of snow heels that give rı to slab avalanches. Below the tree-line the ***snow drift control*** structures include ***snow bridges*** or ***snow rakes*** of timber or aluminium or steel tubes or prestressed concrete and ***snow nets*** of wires (Fig. 12 or nylon ropes (Madan, 1977). They help in changing the wind-flow characteristics and in preventi concentrated accumulation of snow on vulnerable slopes.

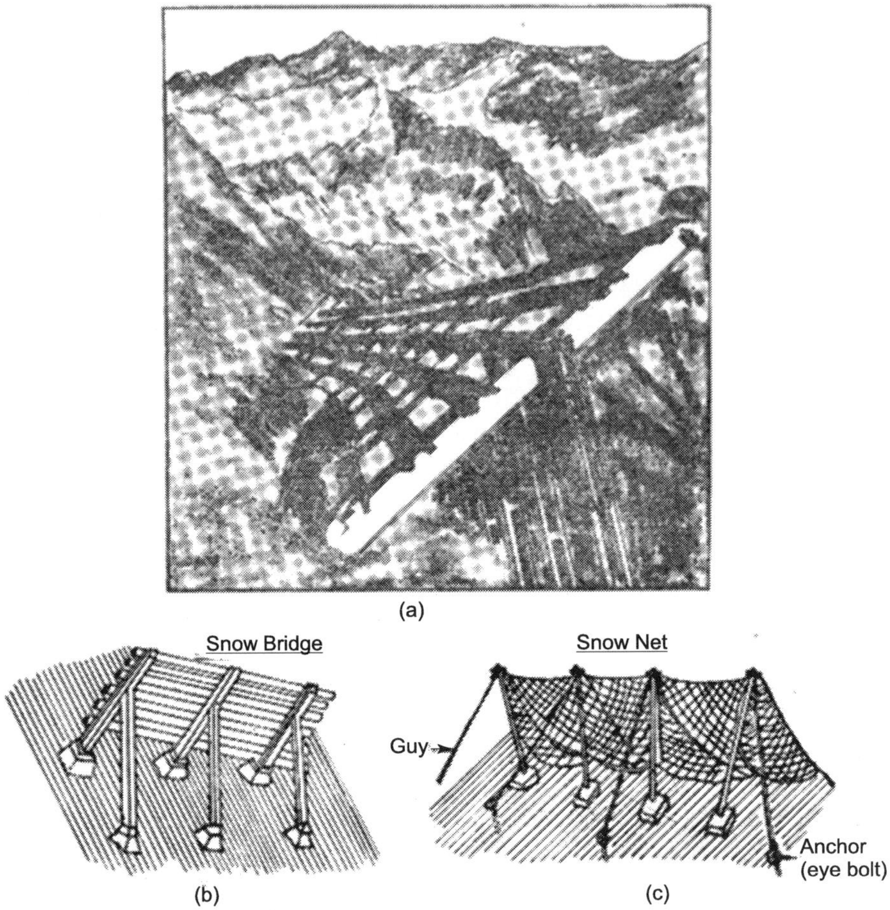

Fig. 12.8. One method of mitigating avalanche hazards is to build structures that prevent buildup of stresses in snow masses in the formation zone, and thus prevent avalanche occurrence.(a) Steel retaining structures in the upper part of the slope. (After B.A. Bolt, H.L. Horn, G.A. Macdonald and R.F. Scott/ 1975, Fig. 6.2 b.) (b) Snow bridge in the upper part of slope. (c) Snow net installed in the avalanche zone. (b and c After A.S. Madan, 1977).

Diversionery Structures

ong the Srinagar–Leh highway, effective measures to combat snow hazards have been taken. In dition to the dams and mounds constructed in the valleys, ***snow galleries*** protect the highway by guiding d passing on the avalanches over them. The ***guidewalls*** built upon the galleries limit the width of flow of the snow mass. ***The jet roof*** (Fig. 12.9) causes the wind to blow swiftly and prevent accumulation of snow and ice. ***Avalanche ramps*** (Madan 1977) allow passage over installations and buildings (Fig. 12.9). ***Avalanche wedges*** (Figs. 12.10 and 12.11) deflect the snow, break the speed of the descending masses of ow and ice and thus provide protection to buildings and installations. **Underground construction is**

the best measure against avalanches in chronically affected areas. The Banihal tunnel on the Jamm Srinagar highway was constructed to avoid endemic avalanches on the slopes of the Pir Panjal Range.

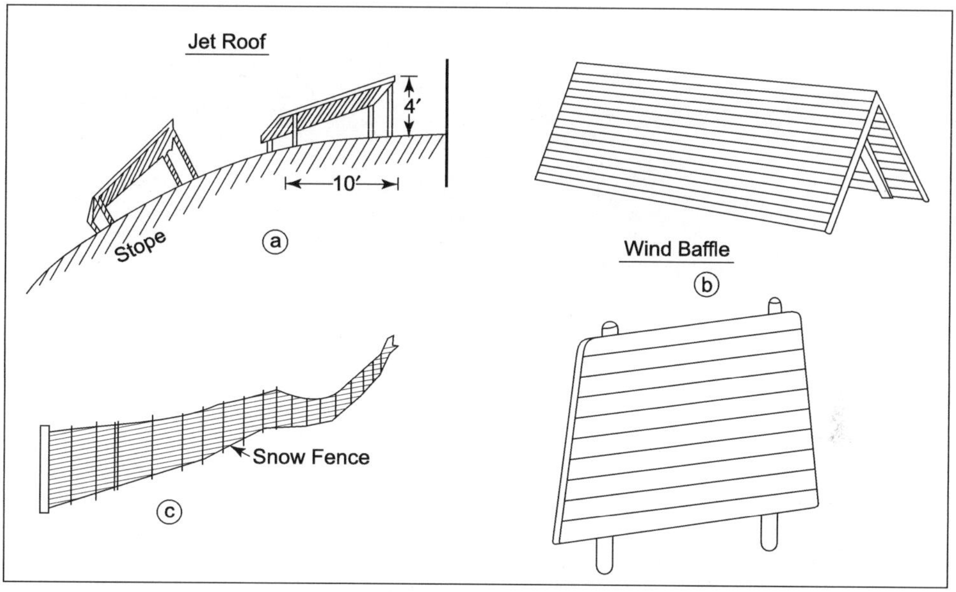

Fig. 12.9. Snow-drift control structures help in changing the wind flow characteristics and in preventi concentrated accumulation of snow on vulnerable avalanche slopes. (After A.S. Madan, 1977).

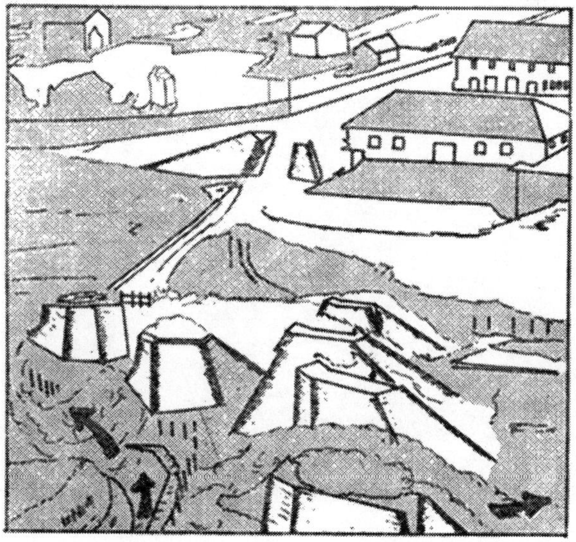

Fig. 12 10. Concrete wedges not only serve as a brake but deflect the flowing mass of snow and thus prote buildings and structures in a village.

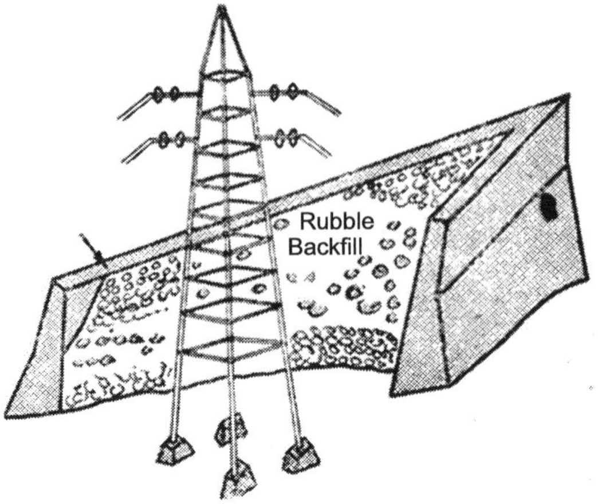

Fig. 12.11. Closer view of a concrete wedge protecting an electric installation.

Fig. 12.12. Avalanches have posed constant threat to our strategic roads in Kashmir particularly in Ladakh during winters. Along with blocks of ice come uprooted trees and shrubs, burying the road. (Courtesy: Dr M.R. Bhutiyani, SASE, Chandigarh) (See colour Fig. 12.12)

SUMMARY

Great masses of snow and ice called avalanches move with savage velocity, varying from 30 to 3 km/h in the snowy realm. They are triggered as a result of thermodynamic changes following variati in physical conditions such as the temperature of air. The velocity of wind, and the gradient of the slop on which the snow accumulates. The steeper the slope, the greater will be the mass-movements. T maximum amount of snow accumulates on slopes between 30° and 45°. Unforested bare slopes a more prone to movement of the fresh snow masses. The wind redistributes the snow, piling it up in heaps and thus creating imbalances—leading to mass-movement or fresh-snow avalanches. Earthqua shocks trigger avalanches on a very large scale. Diurnal variations of temperature of the order of to 10°C are conducive to the development of avalanches, particularly of the delayed slab-type. In t Himalaya, the south-facing slopes receiving a comparatively larger amount of solar radiation are mo prone to avalanches, particularly in the afternoons of early springs.

Four types of avalanches are recognizable. Fresh-snow avalanches are induced by storms during immediately after heavy snowfall, causing snow mass to slip down the slope-drops and become airborn Delayed avalanches occur after the snow masses have undergone load metamorphism and conseque compaction. The weakening of the bond of ice crystals and attendant development of fractures at t head may precipitate movement of sheets and their eventual fragmentation which moves down as a sl avalanche.

There are two ways of mitigating the hazards of avalanches. One is to trigger smaller avalanches artificially at chosen times under controlled conditions—before the snow masses grow big enou to cause natural unpredictable avalanches. The artificial triggering is done by hurling explosi projectiles through guns. The other way is to deflect or divert the moving mass by building win baffles, snow fences, snow nets, braking barriers, etc. so that the snow masses do not accumula to dangerous proportions. Retaining walls, fences and avalanche-wedges can ward off hazards to buildings and installations.

13 Floods and Flood Management

THE INDIAN SCENE

More than 40 m ha of land in India is prone to floods. On an average 10 million ha (3.3%) and 5 m ha •pped areas supporting about 113 million population is annually affected by floods. Until December 1980 floods had killed more than 18,000 people (*The Hindustan Times,* December 15, 1981). The do-Gangetic Plains supporting quite more than 40% of the country's population suffer maximum mage (Fig. 13.1); and here 60% (15 m ha) of cultivated land used to be inundated in spite of 12,265 m of embankments and 21,810 km of drainage channel till then. While about 6 m ha of land were fected by floods in 1953, little less than 12 m ha were ravaged in 1980. The population affected by e calamity of floods increased from 3.6 million in 1960 to 7 million in 1978. **One of the reasons for this phenomenal increase in damage is the fact that factories and housing colonies and settlements re and are increasingly located in land subject floods, that is in the floodways.** The constraints of nduse thus have a direct bearing on increased flood hazards. **Our excessive reliance on, and recourse civil engineering works to cope with the hazards—instead of restoring the degraded lands ecological health and containing erosion and checking deforestation—is another cause of the lamities of spates.**

Flood hazard is a global phenomenon. In the period 1874–1979 nearly 10,700 people died due to ods the world over (Coates, 1981). It is depressing to realize that compared to the number of people fected by floods 18 million in Brazil, 42 million in Bangladesh, 49 million in Indonesia, and 103 llion in China, more than 113 million Indians are exposed to the fury of floods (Thomas and Kale, 11). The Huangho of China and the Kosi of Bihar are particularly notorious for wildly recurrent ods and wayward behaviour.

It is more than obvious that India, like the rest of South Asia, is witnessing increasing trend in short spells of ***extreme rainfall events*** (Kale, 2011). The frequency of very heavy rainfall events increased er mid-1970s, particularly in central India (Goswami et al., 2006) as well as in eastern and northeastern dia.

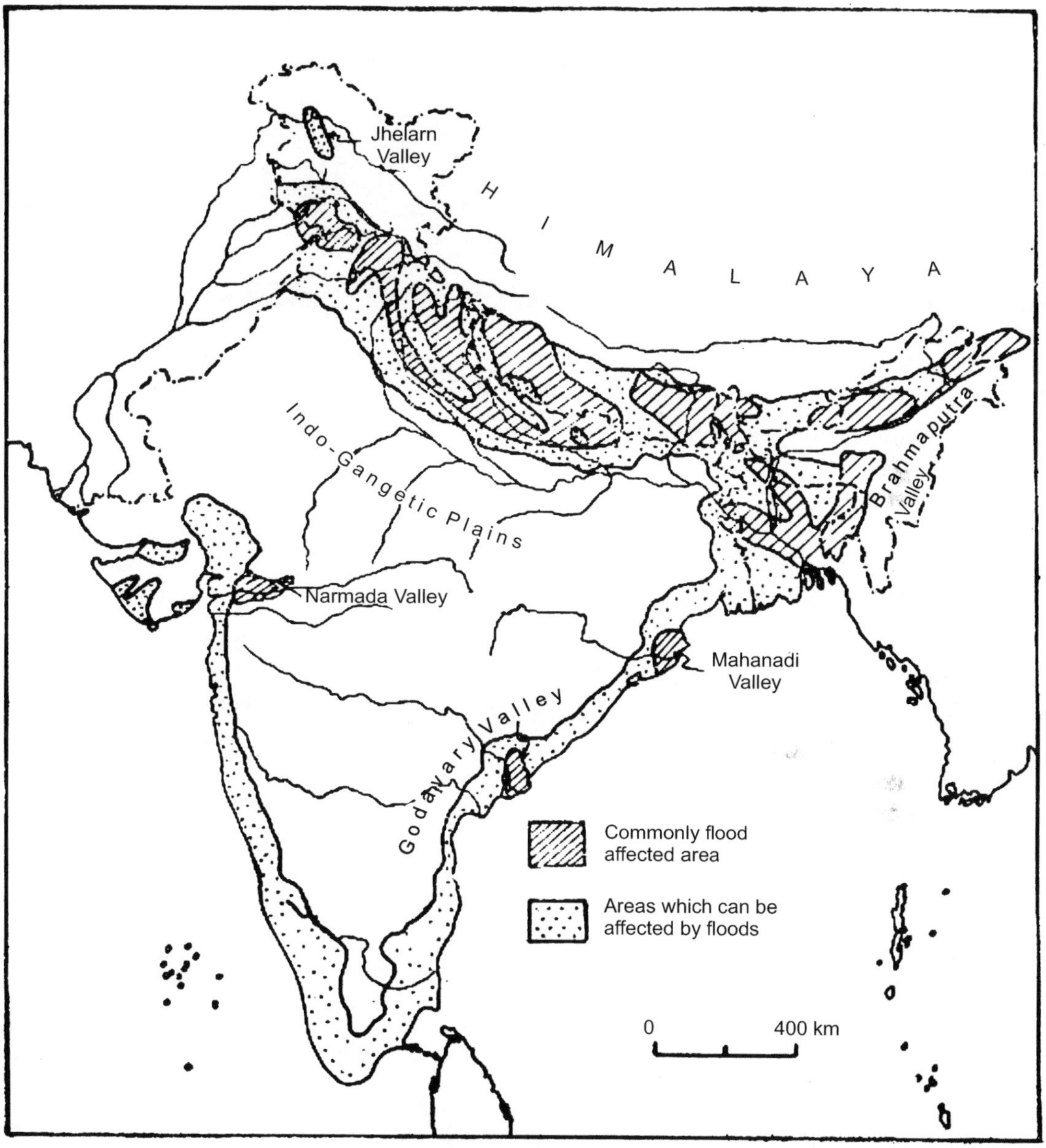

Fig. 13.1. About 40 m ha of land in India is prone to floods and 5 m ha of cropped areas are annua affected—60% being in the Indo-Gangetic Plains, which are densely populated. (After D. C. Das 1977).

Table 13.1. Devastating Floods in the Period 1980–1996 (Gupta et al, 2001)

Duration	Area affected	Synoptic systems
17–23 July 1980	East Uttar Pradesh	Low pressure area
7–27 August 1980	Central & southwest Uttar Pradesh	Low depression
–10 September 1980	Southeast Uttar Pradesh	Land depression
18–24 September 1980	South Odisha, Andhra Pradesh and central Uttar Pradesh	Deep depression
–15 July 1981	Gujarat	Low pressure area
16–12 July 1981	Rajasthan	Low pressure area
9–29 July 1981	Uttar Pradesh	Low pressure area
6–12 August 1981	East Uttar Pradesh	Cyclonic storm
3–9 September 1981	East Uttar Pradesh	Land depression
9–25 August 1982	East Madhya Pradesh	Pronounced low pressure
28–31 August 1982	North Odisha	Depression
0 August–3 September 1982	Uttar Pradesh	Land depression
0–23 June 1983	Gujarat	Land depression
11–17 August 1983	West Maharashtra	Through off the Maharashtra coast
18–31 August 1983	Northwest Uttar Pradesh	Low pressure area
5–19 September 1983	Marathwada, west Maharashtra	Low pressure area
8–14 September 1983	Southeast Uttar Pradesh	Land depression
21–27 June 1984	West Bengal	Pronounced low pressure
28 June– 11 July 1984	Uttar Pradesh	Low pressure area
3 August–5 September 1985	Bihar	Low pressure area
2–25 September 1985	East Uttar Pradesh	Land depression
–18 September 1985	Bihar	Pronounced low pressure
17–23 July 1986	Bihar	Land depression
7–20 August 1986	North Andhra Pradesh	Deep depression
23–29 July 1987	Bihar	Low pressure area
0 July–20 August 1987	North Bengal	Cyclonic circulation
3–16 September 1987	Bihar plateau	Pronounced low pressure
5–19 July 1988	North Gujarat	Cyclonic storm
5 August–7 September 1988	North Gujarat	Cyclonic storm
21–28 September 1988	Panjab	Low pressure area
19–26 July 1989	Coastal Andhra Pradesh	Depression
13–26 July 1989	Western Maharashtra	Depression
1–6 July 1990	Northeast Rajasthan	Low pressure area
16–29 August 1990	Vidarbha	Depression
23–31 August 1990	Gujarat	Depression
5–31 July 1991	Vidarbha	Deep depression
–13 September 1991	North Bengal	Cyclonic circulation

Contd.

Duration	Area affected	Synoptic systems
16–22 July 1992	Gujarat	Cyclonic circulation
10–16 September 1992	Jammu and Kashmir	Cyclonic circulation
10–16 September 1992	East Uttar Pradesh	Pronounced low pressure
1–14 July 1993	Gujarat	Cyclonic circulation
8–14 July 1993	Panjab	Cyclonic circulation
15–20 July 1993	Jharkhand	Pronounced low pressure
15-21 July 1993	North Bengal	Pronounced low pressure
9–15 September 1993	Uttar Pradesh	Pronounced low pressure
14–27 July 1994	Gujarat	Low pressure area
1–7 September 1994	Odish	Low pressure area
1–7 September 1994	Vidarbha	Pronounced low pressure
17–23 August 1995	Bihar	Low pressure area
5–15 September 1995	Haryana	Low pressure area
20–26 June 1996	Rajasthan	Low pressure area
25 July–7 August 1996	Bihar	Low pressure area

CAUSES OF FLOODS

Basic Causes

Whenever the magnitude of flow of water exceeds the carrying capacity of the channel within its bank the excess water overflows on to the flood plains and causes floods. Thus, two factors are critical responsible for the occurrence of floods :

(1) The flow of excess water resulting from increased runoff in the catchment are, and
(2) **Reduction in the carrying capacity of the channels as a consequence of accumulation of sediments and or the obstructions built on floodways.**

Excess Flows

There can be several causes of excess flow—heavy rainfall and cloudburst (Fig. 13.5), the melting snows on a large scale with attendant bursting of dams built of ice blocks, sudden and excess release impounded water behind dams, and bursting of man-made or landslide debris built dams.

A rainfall of 10 cm on just one square kilometre area means 100,000 m^3 of water. The inabilı of the ground to absorb a sizeable part of this amount of water would naturally cause excess ***overland flow or runoff.*** The larger the catchment area and greater the magnitude of rainfall, the greater would the volume of overland flow (Fig. 13.2). **Excessive rain within a short period is called a cloudbur** Cloudbursts are very common in the Himalaya, Odisha and central and western India (Fig. 13.3 There was clear correlation (Table 13.2) between extreme events (excessive rainfall) and occurren of flood disasters (Kale, 2011). The frequency of very heavy rainfall events increased after mid-197 particularly over central India, and in eastern and northeastern India (B.N. Goswami et al, 2008). 1979 within a span of one day, there was 37 to 50 cm rainfall in Saurashtra. Earlier in 1978 the Mor area in Saurashtra (Fig. 13.4) experienced an even worse tragedy of deluge and disaster (with 14 ople dead), as the Machu Dam II collapsed following heavy, incessant rains. Some other examples of ods (in 1967, 1969) are those of the Luni River in the dryland of Rajasthan, and in the valley of the •si River of Kumaun in Uttarakhand in 1978 and 2010.

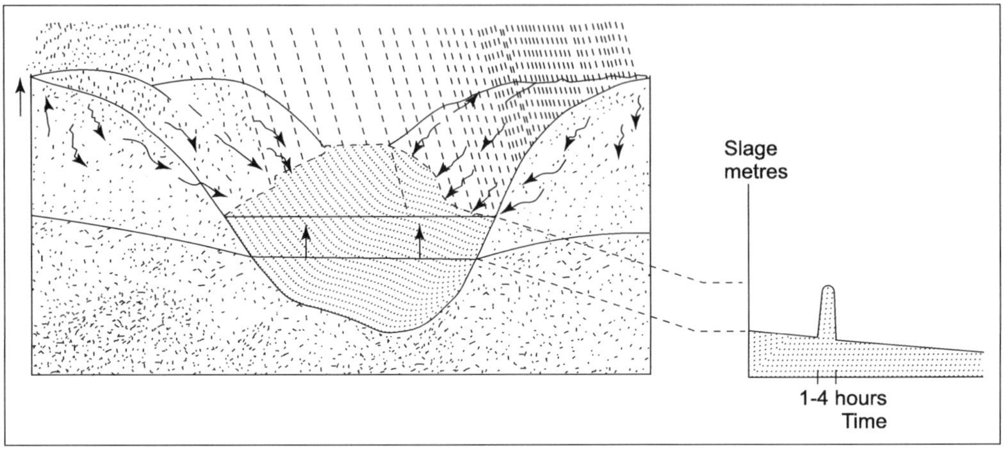

Fig. 13.2. The larger the catchment area and greater the magnitude of rainfall, the greater would be the overland (runoff) flow. The ground is not able to infiltrate excess rainfall, resulting in greater overland flow. (Reproduced with permission from W. M. Marsh and J. Dozier, 1981, *Landscape,* Addi-son-Wesley, Reading, 637 p., Fig. 13.13.)

Table 13.2. Flood Events having Recurrence Interval of >100 Years in the Indian Subcontinent Between 85 and 2009 (After Thomas and Kale, 2011)

Country	Date	No. of Deaths	No. of Persons Displaced	Area Affected (km^2)
Pakistan	August 12, 1997	165	8,36,300	2,76,900
India	October 29, 1999	9803	1,00,00,000	1,91,900
India-Bangladesh	September 18, 2000	1,468	2,40,00,000	1,55,000
India-Bangladesh	July 20, 2008	142	2,25,000	2,83,800
Pakistan	August 2, 2008	35	2,00,000	32,490
India-Nepal	August 18, 2008	400	1,00,00,000	1,63,700
India	September 22, 2008	2,400		4,64,300
ri Lanka	November 22, 2008	61	1,71,000	1,03,400

On the basis of a **climate modeling study, Lal et al., (2009) predicted that during the winter months there will be 5 to 25% less rainfall and in the summer season the monsoon rainfall will t only be 10 to 15% more than the normal, but also quite variable and spacially irregular in ntral India. In other words, there will be more spells of intense rainfall and frequent instances extreme rainfall. The prediction is coming true, particularly in the central sector of the malayan belt, where there is pronounced decline in winter precipitation (both rain and snow)**

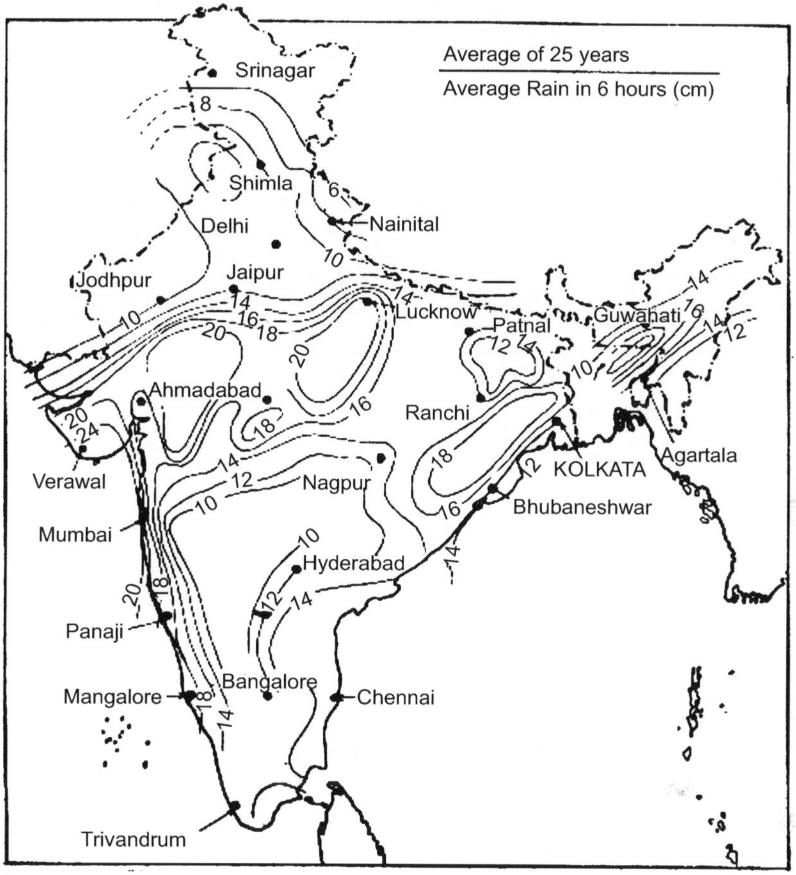

Based upon Survey of India map with the permission of the Surveyor General of India. The territorial waters of .India extend into the sea to a distance of twelve nautical miles measured from the appropriate base line. Responsibility for the correctness of internal details Shown on the map rests with the publisher Government of India Copyright 1986.

Fig. 13.3. In the early 1980s excessive rainfall within a short period of time (cloudbursts) in different parts of the country expressed in terms of an average of 25 years. (Based on S. D. S. Abbi, 1981).

and prolonged spells of drought in summer broken by cloud bursts in localized areas. Moreov there is perceptible shifting of the seasons. This implies that the hazard of flood will aggrava with passage of time.

The **bursting of dams** causes catastrophic floods such as of the Machu Dam in Saurashtra whi destroyed the whole township of Morvi on August 11, 1979. Earlier in 1961 the bursting of the Pansh and Khadakvasala dams in the Mula River destroyed a part of the city of Pune in Maharashtra. Followi a cloudburst in the Great Himalayan Range on August 6, 1978, the valley of Kanodia was blocked ov its 4-km stretch by a colossal volume of morainic debris that slipped from the mountain slope. A 3-k long and 3-m deep lake was formed. As the debris mass moved downstream, it blocked the passa of the Bhagirathi (Ganga) River forming another mighty dam. The bursting of this dam 14 hours lat generated calamitous floods that wiped away the hamlets of Gangnani and Dabrani on the pilgrim pa

Gangotri, reactivated multitudes of huge fossil (very old) landslide cones, and caused widespread voc right down to Uttarkashi.

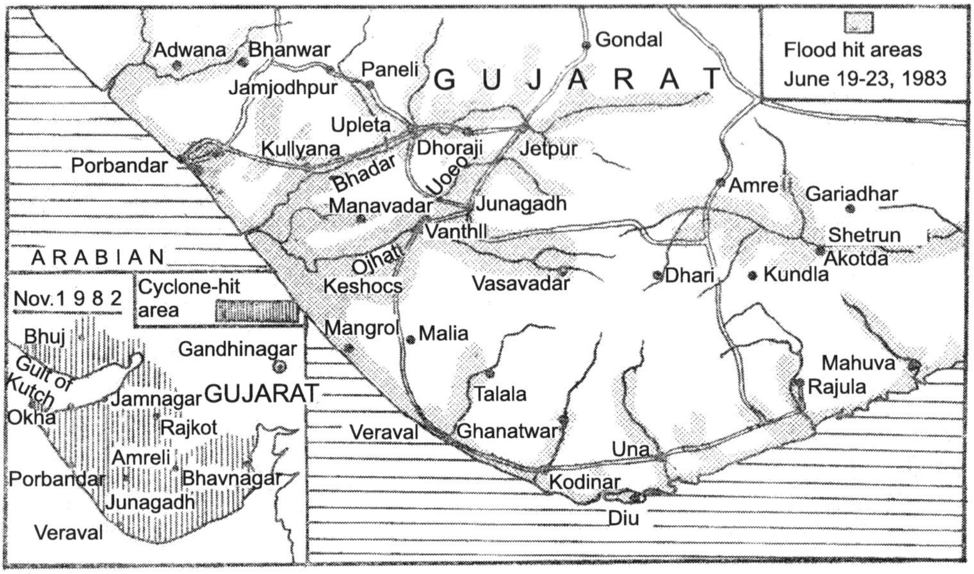

Fig. 13.4. Cyclones and tidal waves accompanied by torrential rains caused inundation of a large part of Saurashtra in November 1982 and June 1983.

A 130-m high dam formed of the debris generated by massive landslides in the Palang Zambo River Arunachal Pradesh gave way on June 10, 2000 and the powerful resulting flood washed away even big bridges, killed 30 people and rendered 30,000 people homeless (in Valdiya, 2004). In 1994 the rsting of the Lugge Tso Glacial Lake in Bhutan caused a destructive flood, wiping out Namche Power ant and extensive damage (Mool, 2001). The glacial outburst (Fig. 22.4) in the Pochhu river in 1994 as equally destructive (Joshi, 2001).

Rise in the sea level in the coastal belts following tidal waves or hurricanes bring about devastating ods. Cyclones also bring heavy downpours for prolonged period as happened in two days in November 82 and then in June 1983 in Saurashtra, resulting in overflowing of 27 dams—some by over 2 m.

Reduced Carrying (holding) Capacity of Rivers

e capacity of channels conveying water is diminished by (1) the accumulation of sediments derived m severe erosion in the catchment, (2) the formation of cones and fans of landslide debris (3) such pediments as bridges with piers and protruding abutments and (4) the occupation of the floodway for bitation, commercial activities, or industrial development and so on.

A floodway is designated as the area that can convey water of the 100-year flood (1% chance of currence) without increasing the height of water-surface in that floodway more than one foot at y point (L. Palmer quoted by Coates, 1976). To an engineer, however, **a floodway is a waterway** nstructed for getting relief from river floods (S. V. Chitale, per. comm., 1985). **The occupation of**

floodways has always led to tragic consequences whenever the water-level has risen above the sa level.

It is well-known that the carrying capacity of a river is governed by, among other factors, dischar of water (including sediment load) that it carries. Therefore, any excess of water and sediment wou easily upset the balance, resulting in spilling over the banks. The sediment in movement with water deposited in the channel either on the bed (to form riffles) or on bends (giving rise to pointbars Fig. 13.1 to form fluvial deposits. These deposits restrict the passage of water and thus reduce the carrying capac of the channel. It may be pointed out that the proportion of bedload—sediment carried by rivers at th basal level—of the Brahmaputra in Assam is 17%, and of the Ganga–Brahmaputra in Bangladesh 38 (D.C. Goswami, 1985). This results in spreading of the flood water on the adjacent plain. **The extensi flooding in eastern Uttar Pradesh and northern Bihar, especially by the Gandak, Narayani and K• rivers is primarily due to the reduced carrying capacity of the rivers choked with sediments deriv from accelerated erosion in the Himalayan region and inadequacy of slope in the flood plains.**

Runoff *Versus* Infiltration

The flood discharge of a stream depends on the amount of runoff or the ground (overland) flow rainwater. As already explained, the runoff is determined by the amount of infiltration of water in the soil, which in turn depends on the nature and extent of vegetation, texture of soil, and length a steepness of slope (Figs 5.4 and 13.5). Unless the permeable soil and/or vegetable litter carpeting t soil absorbs rainwater in substantial proportions, a large part of the precipitation will flow overla down the slope. For example, the litter-covered under-forest ground absorbed rainwater at the rate 38 mm/hr, compared to 19 mm/hr by agricultural fields (Ghosh et al., 1980). When the rate of rainf is higher than the infiltration capacity of the ground, there will be overland flow, thus contributing excessive streamflows—the floods. In the case of permeable soil there is greater percolation of wat leading to a rise of the water table and the sprouting of springs and seepages lower down the slop thus augmenting the surface flow through this subsurface path arriving at the stream not far behind t surface runoff (Fig. 5.3).

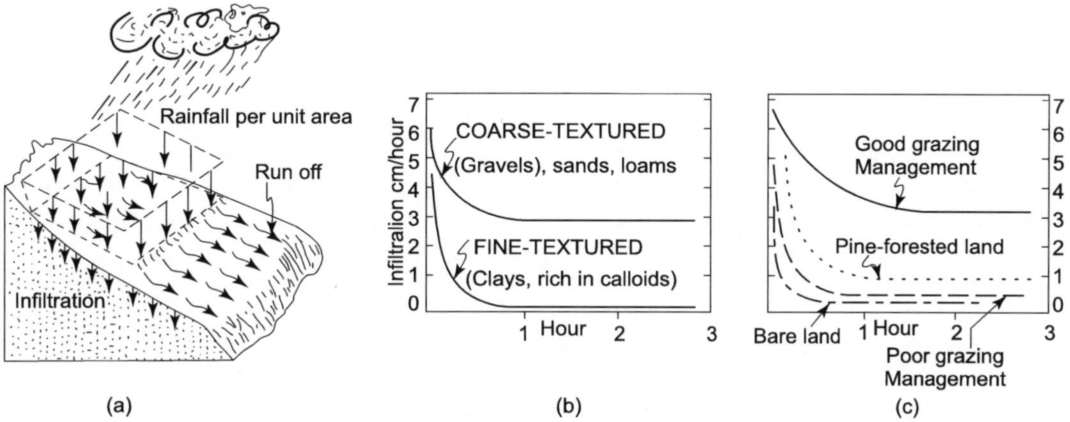

Fig. 13.5. The overland flow (runoff), that contributes to floods, depends on the magnitude of infiltration in the soil. The infiltration depends on the texture of soils, the landuse and the nature and extent vegetation that covers the soil.(Based on A. N. Strahler, 1971).

The nature and extent of vegetation have strong control on the proportion of the runoff water. e vegetation that generates plentiful litter permits greater infiltration, and consequently a lesser mount of runoff. In the forested watersheds of the Nilgiri massif, for example, the runoff was 50 to % less than from the treeless open catchment (Table 13.3). Likewise, in an experimental forest in hradun, the peakflow recorded in the tilled land was 64 mm/hr, in the eucalyptus farm 0.5 times less, meadows 2.5 times less, in abandoned fields 3.5 times less, (Fig 13.5) and in grazing land 10.5 times s (Singh, 1982). The burning, cutting and overgrazing increased the peak discharges by, respectively, %, 34% and 32%, in an experimental forest near Chandigarh. Clearfelling in a watershed forest used 28% increase in the total runoff, and 73% increase in the peak flow (Mathur et al., 1976). **It ll be quite obvious that deforestation (including burning, clearfelling, overgrazing) contributes nsiderably to the augmentation of floods. In view of the fact that about 76% of the land surface the country has lost its good forest cover, one can easily understand the phenomenal increase in e recurrence and severity of floods.**

Table 13.3. Runoff *Versus* Vegetal Cover in the Nilgiri Mountains (Data from S. Singh, 1981)

Forest Cover	Runoff
90%	35 mm/hr
57%	63 mm/hr
13%	113 mm/hr
Cultivated benches	37%
Degraded grassland	71%
Natural forests	59%

The rapid pace of urbanization entailing roofing, paving and asphalted-road construction, has nsiderably reduced infiltration and augmented surface runoff, causing local floods in practically big cities like Delhi, Kolkata, Bengaluru and Mumbai. In a city with 40% impervious surface d as much served by storm sewers, the occurrence of floods is three times higher than when banization had not overtaken it (Keller, 1982).

Sediment Filled in Channels

e amount of sediment carried by stream water in suspension or by traction is governed by the formula that **if the velocity doubles, the amount of sediment particles carried would increase 64 times. In flooded river, the velocity increases manifold, the sediment-carrying capacity thus increasing veral hundred times.** The amount of sediments carried by a river—which it erodes from catchment eas as well as from its own banks—is an index of the degree of erosion (Table 13.4).

In Assam, the deposition of large volumes of sediments in the Brahmaputra between Pasighat and uwahati and its tributaries such as Puthimari has resulted in considerable decrease in the carrying pacity of the river. In the 8 to 10 km wide channel of the Brahmaputra the formation of large bars and ands testify to this fact. Embankments built to the aggregate length of 3642 km with 599 km long ainage channels have not been able to control the recurrent floods in Assam (S. Goswami, 1998).

With great volumes of sediments flowing down the Himalayan rivers (Tables 13.4 and 7.5), the beds these rivers are rising rapidly. In southeastern Nepal, the beds of the rivers in the 'Bhabhar' belt was rising at the rate of 15 to 30 cm/yr more than thirty years ago. **The bed of the Kosi in Bihar flowi within considerably raised levees is now at a higher level than the floodplain (Fig. 13.7). Excess discharge causes spillover or overtopping of levees and widespread inundation.**

Table 13.4. Sediment Yield of Some Himalayan Rivers in the Early 1970s (Data from G.P. Gupta, 1979)

River	Point of Measurement	Sediment Yield (ha-m/100 km^2/yr)
Ganga System	Farakka	4.33
Tista	Anderson Bridge	98.40
Arun	Tribeni	11.91
Sunkosi	Tribeni	27.30
Tamar	Tribeni	60.76
Kamla	Chisapani	28.72
Saptakosi	Sonakhambi	16.32
Trisuli	Trisuli Bridge	28.00
Brahmaputra System	Pandu	7.81
Buhri Dihing	Khowang	17.73
Debang	Jiagaon	26.34
Dikrong	Dikrongghat	25.95
Lohit	Digarughat	34.20
Manas	Mathanguri	7.85
Noadihing	Nansari	14.86
Ranga	Rangaghat	8.29
Dihang	Rangaghat	7.95
Subansiri	Bhimpuraghat	10.91
Sindhu System		
Chendi	Konthan	25.20
Marusundar	Tillar	11.67
Juh	Panchtirthi	79.24

In northeastern Uttar Pradesh and adjoining Bihar, yet another factor is playing a powerful role. is the tectonic activity. While the Lesser Himalayan terrane of the Nepal Himalaya is raising at the ra of 3 to 4 mm/yr (Nakata, 1989) or 2 to 3 mm/yr in the spatial wavelength of 25 to 35 km (Jackson a Bilham, 1994), the northern part of Indo-Gangetic Plain in the front of it is subsiding at the rate of 0 to 0.3 mm/yr (Sinha and Friend, 1991). The uplift of the Nepal Himalaya is accelerating the rate denudation which is more than 6 mm/yr (Arita and Ganzawa, 1997), with attendant greater generati of sediments and their transportation to the plains during floods. The subsidence of the northern part the Kosi–Gandak floodplains is resulting in the decrease of the gradient of channel beds, slackening flow of water, and consequent deposition of sediments in the channels. The Bagmati River, for examp is being filled at the rate of 9.3 cm/yr (Sinha and Jain, 1998). Similarly, in Assam the Brahmaput channel aggraded 16 to 21 cm in just eight years, from 1971 to 1979. The deposition of a large volum of sediments has resulted in the formation of large and small sandbars and islands within the 8 to 10 k wide channel of the Brahmaputra. This explains why Assam is repeatedly ravaged by uncontrollab floods (Goswami, 1998).

In the period between 1738 and 1964, the Kosi shifted 167 km westward primarily in consequence of otectonism to which the Indo-Gangetic Plains are subject, and secondarily due to heavy sedimentation the channel and building up of alluvial fans, resulting in breaking of water through the banks during ods and thus causing destruction of 15,000 km^2 of fertile fields. The breaching of levees causes the Kosi change its course. In fact most rivers of the northeastern part of the Indo-Gangetic Plains including e Tista, Gandak, Narayani and Ganga have been changing their courses (Valdiya, 2010), causing nsiderable damage year after year. The abandoned valleys, being rich in moisture, underground water d mineral nutrients, are reclaimed for cultivation and even occupied for settlement. During floods the aters return to their abandoned courses, bringing about colossal damages, as happened in 2008 when e consistently westward-migrating Kosi changed its course and flowed through its abandoned channel ht through the middle of its alluvial megafan, bringing unimaginable devastation (Table 13.1).

The second major effect of the large amount of sediment transportation by floodwater is the silting of the floodplains. With greater and greater sedimentary material contributed by gully erosion rticularly in deforested tracts and wastelands, the debris derived from excavation for mining and road nstruction and from recurrent landslides, the finer sediment spread on the floodplains is no more rich organic nutrients. The result is diminished productivity of agricultural land.

FLOOD HAZARDS IN INDIA

Mountain Barrier and Rainfall

India the Indo-Gangetic Plains suffer greatest damage from floods. The reason is quite obvious—the fty barrier of the Himalaya mountain is instrumental in holding back the rain-clouds from moving rthwards, and causing their precipitation in torrents. The axis of the low-pressure trough lies close the foothills, resulting in heavy rainfalls in the Himalayan front (Fig. 13.6). Likewise, the barrier of e Sahyadri-Nilgiri-Annaimalai massif is responsible for heavy rainfall in the Western Ghat and the •nkan-Malabar coastal belt.

However, it is the mid-topospheric low-pressure system between 2 to 6 km developing over Saurashtra d Kachchh which is responsible for heavy rains over northern Maharashtra.

Brahmaputra Basin

the Brahmaputra valley vast tracts remain submerged under flood waters for about three months in a ar in spite of the width of the river channel being 6 to 10 km. Following the 1950 earthquake, which bstantially modified the channel gradient, about 2.9 m ha of land were submerged in 1954 (Nath, 82). In the upper reaches the north-bank tributaries change their courses frequently. The deposition of ormous volumes of debris at the confluences of the tributaries forces the Brahmaputra currents to flow uthwards, resulting in persistent and severe erosion along the south bank. During the period 1923 to 54, the 355-km long stretch of the bank between Kobo and Dhubri was considerably eroded.

Floods in the Dryland of Western India

recent years the desert areas in western India, which have an average annual rainfall of 15 cm, ve been ravaged by catastrophic floods even though sporadic and localized in extent. **With *their developed drainage systems,* and bereft of vegetal protection, these barren wastes are unable to cope with the excess water when cloudbursts or heavy rains suddenly pour down water in**

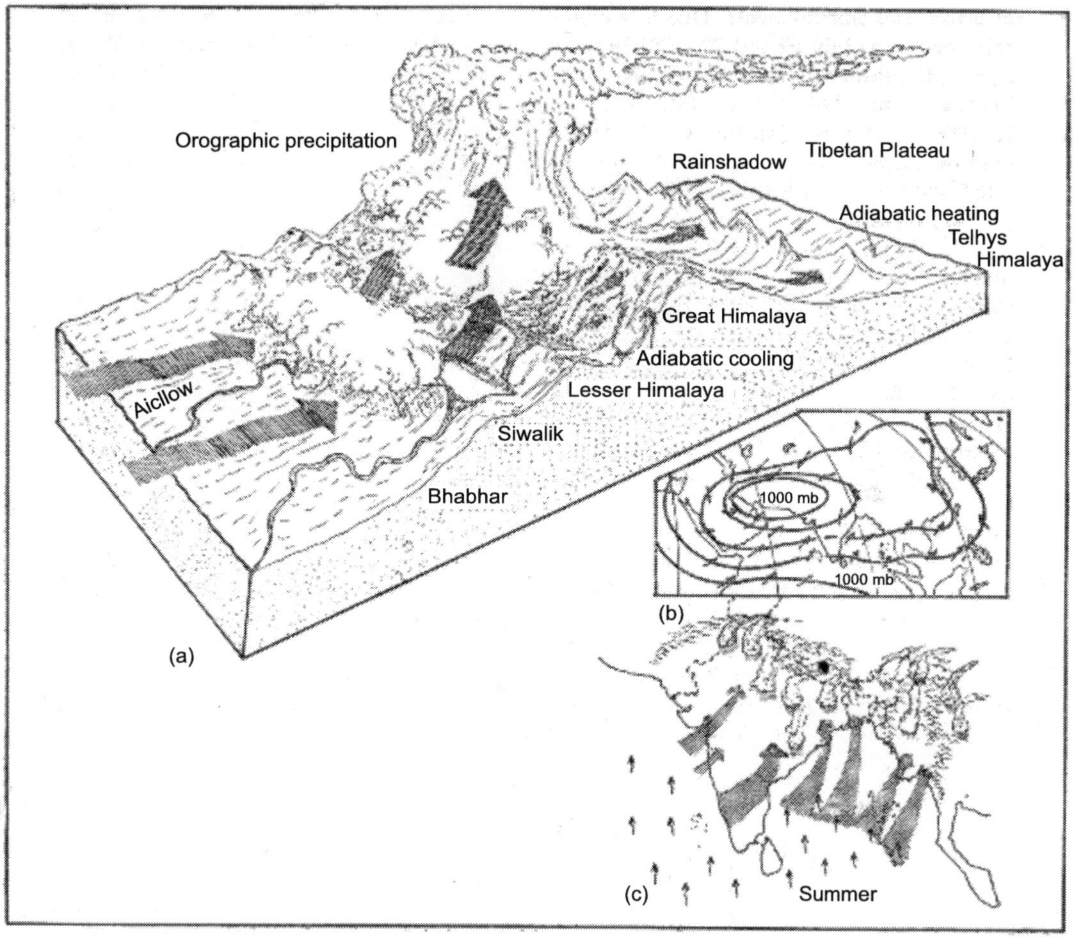

Fig. 13.6. The lofty Himalayan mountain barrier is responsible for holding back moisture-laden clouds and precipitating them so that the Himalayan rivers carry great volumes of water. The insets explain the meteorological cause of rainfall in the Indian subcontinent. (Modified after Figs. 7.10 and 8 W. M. Marsh and J. Dozier, 1981).

torrents. There is thus spilling over the banks, overtopping and breaching of levees and embankmen Due to a cluster of convective clouds, there was 82 cm of rain between July 10 and 20, 1981 in t districts in Rajasthan—Bikaner, Nagaur, Tonk, Jaipur, Bharatpur, Sikar, Bundi, Kota, Sawaimadhop and Ganganagar. In Jaipur district alone, 144 persons lost their lives, about 19,030 houses were destroy and 1,15,000 ha of good agricultural land damaged due to deep gully erosion and spreading over of s following disastrous flood. Several dams and embankments were destroyed and as many as 16,000 we filled with silt (Seth et al., 1982). The change in the courses of the Dhond and Morel rivers resulted wiping out of the entire village of Hingonia in Tonk district.

A well-marked low-pressure monsoon depression caused unprecedented heavy rainfall for fi days from 15th to 19th July, 1979 in the Luni basin in desertic western Rajasthan. Resting on t rd pan or bedrock the thin soil got so saturated very quickly that the runoff overwhelmed the arid aste, destroying 93,526 dwellings in 1026 villages, killing 10,700 heads of livestock, taking a toll 493 humans and rendering 55,867 ha of land unproductive. The flood waters carried sediments in avy concentration—40.2 gram/litre (compared to the normal 0.86 g/l) due to scouring erosion on the nks and in the catchment area (Sharma and Vangani, 1982). **Floods in the desert land are clearly a nsequence of excessive runoff and under-developed drainage system. Such events have become ite common in recent years.**

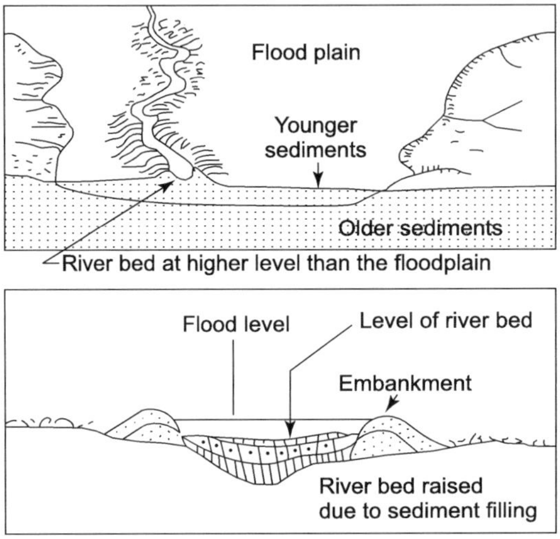

Fig. 13.7. In the Kosi–Gandak domain, where voluminous deposition of sediments is progressively increasing, the making of higher embankments has aggravated the situation. For, the beds of the river channel now lie at a higher level than the floodplain. Overtopping of and breaches in the levees bring about disastrous flooding. (From Valdiya, 2004, 2011).

MANAGEMENT OF FLOODS

Strategy

he management of floods implies not letting the excess runoff water flow suddenly and *intensely* the drainage network. The measures taken are intended to reduce the proportion of runoff increasing infiltration, spreading the excess water that manages to get into the channels and sposing it of through channels or canals. It is a common misconception that damming rivers can ate flood hazards. According to K. L. Rao (1975) even if all possible dams were to be constructed India and Nepal, only about 8 to 10% of the total flow would be stored, the remaining 80 to % would still be free to flow and make floods during excessive discharges. The situation has orsened since the early Nineteen Nineties. There is thus a strong case for conserving flood flows by eans other than surface storage behind dams in rivers.

The various measures of combating flood problems are discussed.

Table 13.5. Indian Rivers in which More than 10 Severe Floods Occurred in the Period 1969-81 (Based on data from D. N. Dhar et al., 1981.)

River	Point of Measurement	State	No. of Floods	Flood level above danger level (m)	
				Ordinarily	Highest
Betwa	Matatila Dam	Uttar Pradesh	16	2.0 to 2.9	2.9
Ghaghara	Chaukaghat	"	19	1.0 to 1.9	1.9
Ganga	Balia	"	10	1.0 to 1.9	2.9
Ganga	Farakka	Bengal	14	1.0 to 1.9	2.9
Kosi	Kuersella	Bihar	13	1.0 to 1.9	2.9
BurhiGandak	Samastipur	,,	13	1.0 to 1.9	2.9
BurhiGandak	Khagaria	"	27	1.0 to 1.9	2.9
Punpun	Supalpur	"	26	1.0 to 2.9	4.9
Son	Japala	"	10	1.0 to 1.9	1.9
Son	Koelwar	"	12	1.0 to 1.9	2.9
Brahmaputra	Noamatighat	Assam	11	1.0 to 1.9	1.9
Subansiri	Badatighat	Assam	25	1.0 to 2.9	3.9
Puthimari	N.T. Road	"	16	1.0 to 1.9	2.9
Rapti	Birdsghat	Uttar Pradesh	7	1.0 to 1.9	1.9
Manas	N.T. Road	Assam	10	1.0 to 1.9	1.9
Dhansiri	Numaligarh	Assam	21	1.0 to 1.9	2.9
Subarnrekha	Rajghat	Orissa	5	1.0 to 1.9	7.0
Narmada	Garudeshwar	Gujarat	14	1.0 to 3.9	17.9
Narmada	Bharuch	"	13	1.0 to 2.9	5.9
Tapi	Burhanpur	Maharashtra	7	1.0 to 4.9	7.6
Yamuna	Delhi	Delhi	4	1.0 to 1.9	2.9
Yamuna	Pratappur	Uttar Pradesh	4	1.0 to 1.9	5.2
Gomati	Jaunpur	"	8	2.0 to 2.9	3.9
Godavari	Bhadrachalam	Andhra Pradesh	3	1.0 to 1.9	1.9
Mahi	Vasol	Gujarat	3	1.9 to 2.9	3.9

Treatment of Watersheds

The most effective measure of flood management involves reduction of runoff by inducing a increasing infiltration into the ground in the catchment area. As already pointed out, the infiltration considerably augmented if the soil is permeable and porous, or the ground is covered with humus-ri soil, particularly if the ground is carpeted with vegetal litter which forms under forests of deciduo trees. A dense vegetal cover not only promotes considerable infiltration, but also causes a large amou of water to be disposed of by evapotranspiration. **The best measure of flood management is thus**

ing the catchment under afforestation with trees that generate larger amounts of litter. Pines and eucalyptus are certainly not in this category of trees (Chapter 7).

Terracing and strip-cropping also enhance infiltration and reduce the amount of flood runoff.

Reservoirs and Detention Basins

reduce flood discharges in rivers, reservoirs and detention basins (Figs. 7.15 and 13.8) are developed putting small check dams or major dams such as the Hirakud Dam. across the Mahanadi, the Panchet, aithon, Tilaiya, Konar and Aiyar dams in the Damodar basin, the Ukai Dam in the Tapi valley, and so . A portion of the flood water is retained behind the dams, the larger part being allowed to flow down e channel under controlled conditions. **If the objective of the dam is to contain floods, the reservoir •uld have to be kept only partially filled, if not empty, or at least the level lowered during times great flood potential. However, this is seldom done and therefore the so-called 'multipurpose' servoirs have failed to abate flood hazards.**

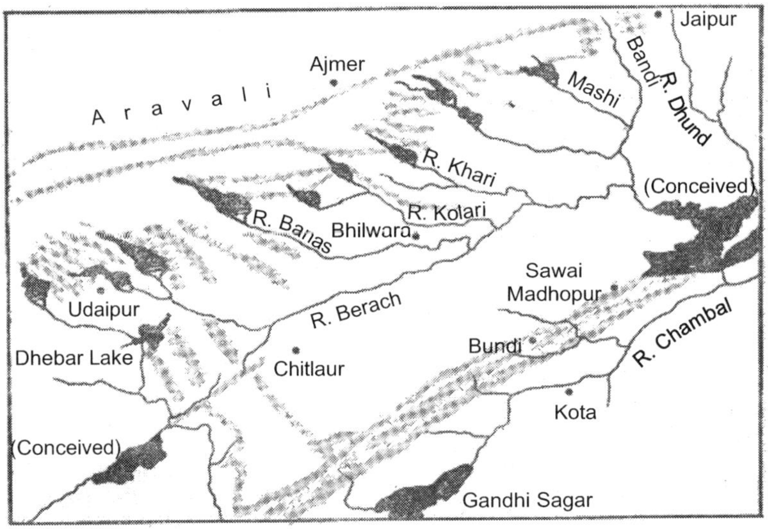

Fig. 13.8. A conceptual diagram showing smaller dams in tributary streams can check flood-flows in the sub-basins and thus minimize flood hazards in the main river, such as the Chambal in Rajasthan.

The construction of numerous smaller dams in tributary streams would check flood flows from e sub-basins and considerably minimize or even eliminate flood hazards in the basin of the main ver (Fig. 13.8). Ponds, tanks and other surface storage structures in the upper reaches of the streams d rivers are very useful in not only containing floods effectively but also in harvesting water for dry asons. A larger number of ponds and smaller reservoirs obviate the need for a larger one in one place e breaching of which may cause flash floods).

In the plains, natural depressions such as marshes and moors can very well serve as reservoirs to eck floods. Old quarries and mines can also serve as detention basins.

Water Spreading

Flood water can be distributed thinly over areas such as paddy fields (Figs. 5.26a and 13.9) and des dry lands, where a part of the water is lost due to evaporation and quite a proportion infiltrates into t soil. There are 5.8 m ha of paddy fields in the country. If half this area is inundated under 20-cm de flood water, nearly 5 million ha-m water can be prevented from flowing down the channels to becom floods.

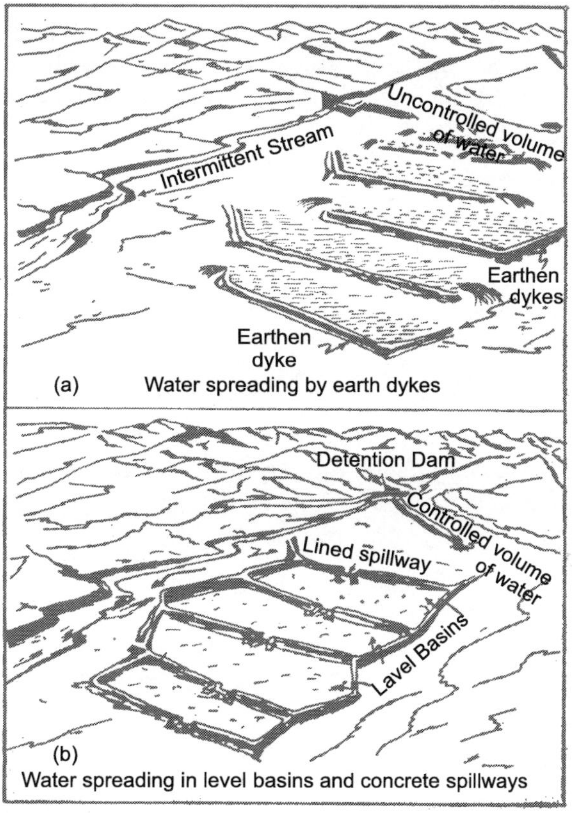

Fig. 13.9. Severity of floods can be minimized if the excess water is spread on fields, particularly pad fields and in wastelands. (After D. Hillal, 1982, *Negev. Land and Water and Life,* Praeger, New York, 163 p.).

Floodwaters can likewise be spread in marshes, lakes and depressions such as those dug along t railway lines and highways.

Groundwater Recharge

The simplest, cheapest and possibly the most effective method of reducing flow on the groun would be to artificially induce infiltration through a large number of dug wells(Fig. 5.27) sun along the beds of ephemeral streams (Rama, 1971) and monsoon kharif channels (Chaturvedi et a

81; 1983). Pumping-in can augment and quicken infiltration of excess flows. If during the dry season, ensive pumping is carried out along the rivers, streams, abandoned channels, canals and irrigation annels, the water table will sink considerably (which is tantamount to emptying the underground ervoir) so that a large space is made available for storing flood waters induced to infiltrate through charge wells (Fig. 5.27). The tubewells are located at intervals of 0.4 km, eight on either side of the annel, each pumping at the rate of 0.1 m^3/s (Chaturvedi et al., 1983). A network of dug wells can also rve the purpose quite well. The quantity of water that would be stored underground in this manner will pend on the thickness and holding capacity of the permeable aquifer. The lateral extent of the aquifers the river basins (such as in the Indo-Gangetic Plains) being virtually limitless, and the thickness not considerable, enormous volumes of excess water can be very gainfully stored underground and the zards of floods abated. This method of flood control is doubtless the most effective, most useful, and ast harmful to the environment.

Stream Channelization

network of canals is quite capable of reducing flood hazards, although the use of canals for flood ntrol is not a commonly adopted or accepted measure. **The canals serve as temporary storage. ater is held in the channels as the floodwaves move downstream. The severity is thus reduced.** ne can judge the extent of relief achieved from the fact that in a flood in 1937, the channels of the Ohio ver in U.S.A. stored floodwater nearly 8.3 times the total capacity of Lake Mead behind the Hoover am on the Colorado.

Channel improvement (by deepening, widening, straightening, lining and clearing out of vegetation d debris) increases the flood conveyance capacity of the rivers (Fig. 13.10). These measures are pplemented by bank stabilization through construction of ripraps, dykes or spurs (Figs. 7.16 and 7.19) d planting of deep-root trees on the embankments.

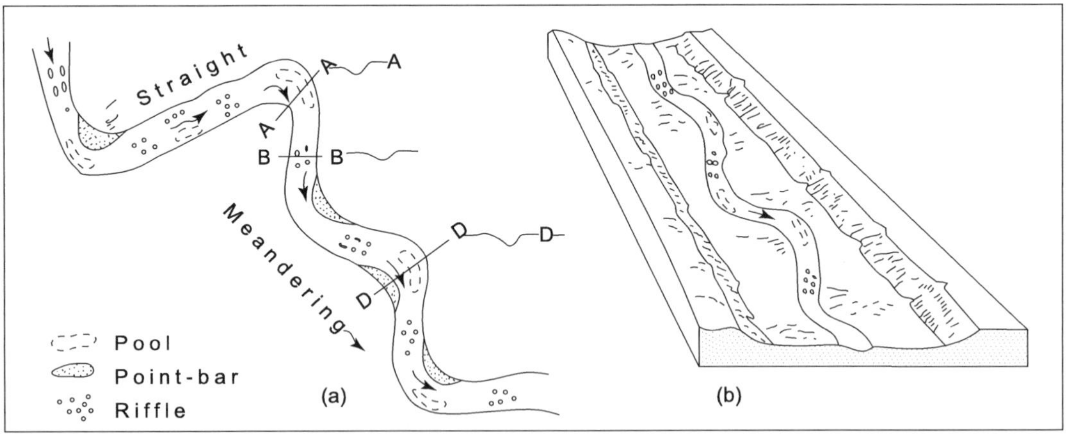

Fig. 13.10. Channel improvement by straightening, widening, deepening, clearing away debris and vegetation, and paving the floor which increase the carrying capacity of the river. Embankments prevent spillover of floodwaters. The river must be allowed to meander within its belt, which means that the embankment must be quite away from the banks. (Based on E. A. Keller, 1976, *Environmental Geology,* Bell and Howell, Columbus, 488 p., Fig. 3.24, 9.24.).

There is no denying that **channelization creates a number of environmental problems related disruption of riverine ecology and equilibrium between flowing water, sediment concentration an rate of erosion.** The erosion becomes brisker and the channel consequently enlarged and deepened. T deepening may set in a chain of degradational activities upstream. The swifter flow may adversely affe aquatic fauna and flora used to a quieter pace of life in the ecosystem characterized by pools, riffles a pointbars (Fig. 13.10). This would lead in some cases to a reduction in biological productivity. T replacement of the pools, for example, by long stretches of riffles causes the elimination of breedi areas, and the higher velocity in channels may be too much for the endurance of organisms. Thus in t Ruby River, Montana (U.S.A.) trout fish productivity fell from 133 lb/1000 ft of channel to 67 lb/10 ft when the bank was stabilized by riprap (Keller, 1982). There is yet another danger. Channelization the upstream part may increase flood hazard downstream, for the floodwater would now quickly rea downstream— much faster than the lower reaches can discharge their own water. Also, the flood pea of the tributaries may now coincide with that of the main river and thus aggravate the situation.

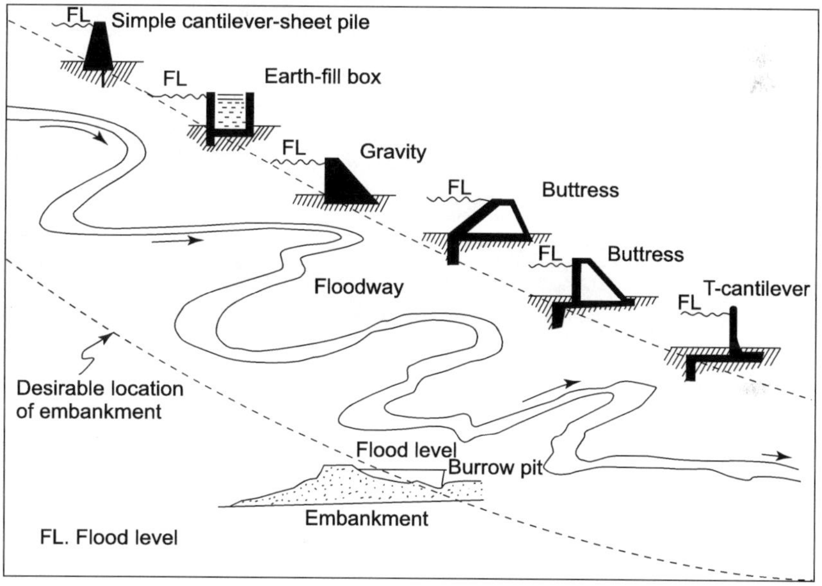

Fig. 13.11. Embankments to contain flood fury must be built quite beyond the flood fringe so that the river spate is free to swing in its meander belt. Cross sections of various models of embankment are shown along the left-bank side. (Valdiya, 2011).

Channelization of the floodwaters would prove more effective. If the excess discharge can distributed through a network of canals linked with rivulets and streams, the hazards of floo will considerably diminish. Not only would these canals and streams provide passage to excessi discharges, but also serve as temporary storage of floodwater for a short duration. If one were to consid the extent of the network of canals interlinked with streams and rivulets, one would realize that the to of aggregate storage would be considerable. Desilting of channels periodically, if not regularly, wou also provide some relief (Valdiya, 2011).

Flood Embankments

uilding embankments or 'bunds' parallel to the river banks (Fig. 13.10) to prevent spilling of excess ater appears to be widely adopted in flood management in India. This is evident from the fact that tween 1954 and 1979 nearly 12,265 km aggregate length of embankments were constructed—246 m on the banks of the Kosi, 249 km along the Bagmati, 208 km on the bank of the Mahananda and 317 m along the Burhi-Gandaki in northern Bihar alone. In the Brahmaputra valley in Assam 2 to 3 m high mbankments are constructed 1.6 km away from the river edge (Fig. 13.11).

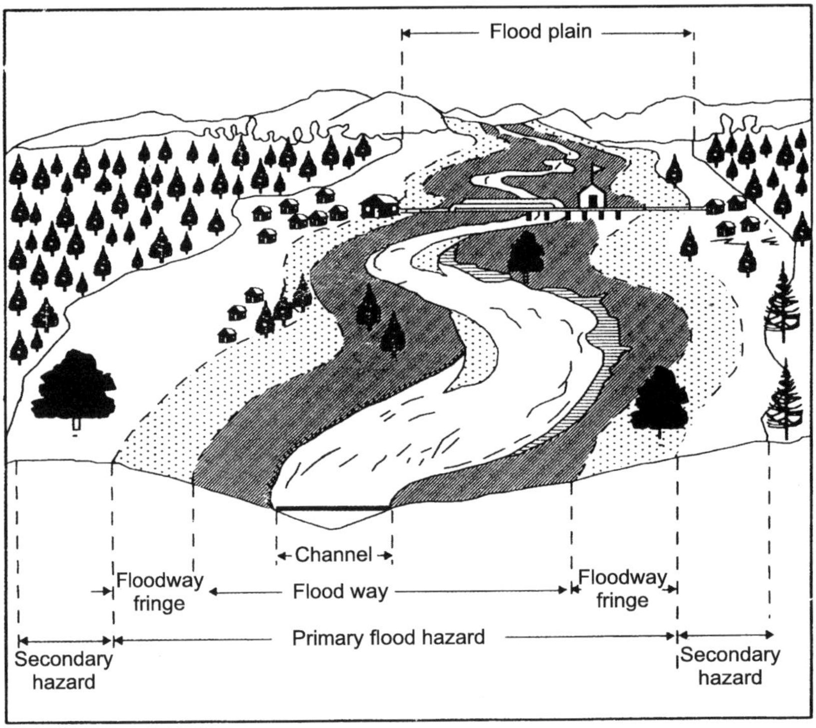

Fig. 13.12. Floodways, (which are covered by at least one foot above the height of normal floods at least once in 100 years), are frequently ravaged by floods and therefore should be kept free of all constructions to enable uninhibited flow of flood waters. In the flood fringe, which is inundated to a lesser extent by the 100-year floods, construction can be permitted if appropriate protective measures including flood proofing of buildings are taken. (From Valdiya, 2011).

However, it is time to realize that building embankments or levees on the banks of the channels not the solution to the problem of flood hazards. Embankments away from the channel on the *gher edge of the floodways* offer a better option (Fig. 13.12). The flood management programme ust take into consideration the fact that **it is natural for rivers to compulsively meander in their belts and therefore they will have to be provided with adequate corridors for meandering without or hindrance. Any attempt to diminish the width of this corridor (floodway) and curb their freedom to meander would prove counterproductive** (Valdiya, 2011).

Hazards-Zoning and Landuse Regulations

Flood management planning requires information on floodplains, particularly the identification floodways in relation to landuse (Fig. 13.13). This map (Coates, 1976) shows

(i) areas or belts that are recurrently inundated,
(ii) areas where construction cannot be permitted, and
(iii) areas where recharge of groundwater is taking place (with the objective that no polluted flo water is allowed to infiltrate.)

In order to cope with the hazards of floods, it is imperative that floodways of all major rivers a identified and precisely delineated in the 1:50,000 scale topographic maps (Fig 13.14). Remote sensi and study of aerial maps will be helpful. The second step is to make bypasses for excess floodwate through canals skirting the settlements on the floodway. We must have and implement strictly t floodway regulations.

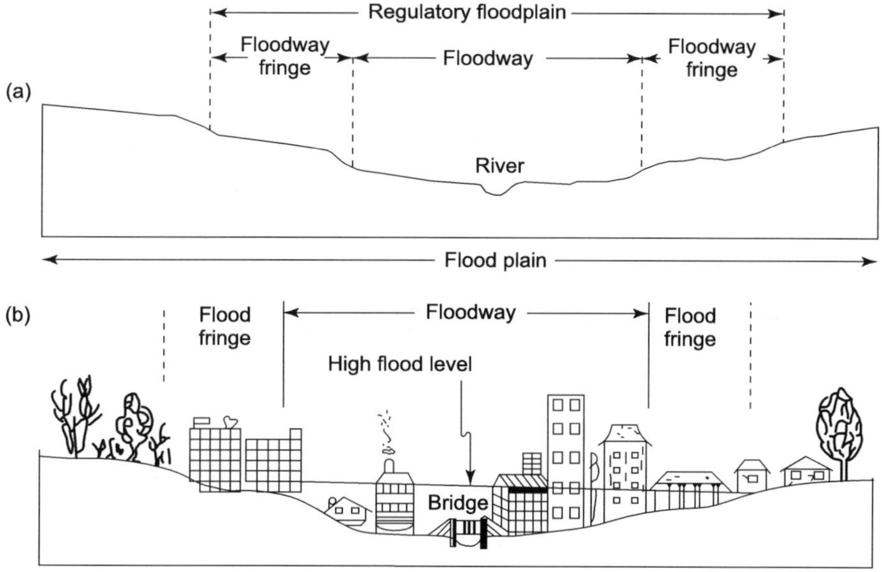

Fig. 13.13. (a) Diagrammatic cross section of a river basin showing the channel, floodway and flood fring The floodway is the vulnerable part of the floodplain. (b) Man-made constructions in the floodw considerably impede the flood discharge.

The ***floodway*** is inundated with at least one foot deep flood water at least once in a 100-year peri (with 1% chance of recurrence) (Fig. 13.12). It is this belt which is subject to frequent floods (Figs. 13. and 13.13), and therefore must be kept open to allow flood waters to flow without inhibition. Constructi must be banned in the floodway. **All kinds of construction such as residential buildings, commercı and industrial complexes, road embankments and bridges impede the free flow of floodwat moving in its own waterway. Whereas the railway bridges span the floodways and are therefo less vulnerable to damage and less problematic, the road bridges span just the river channels**

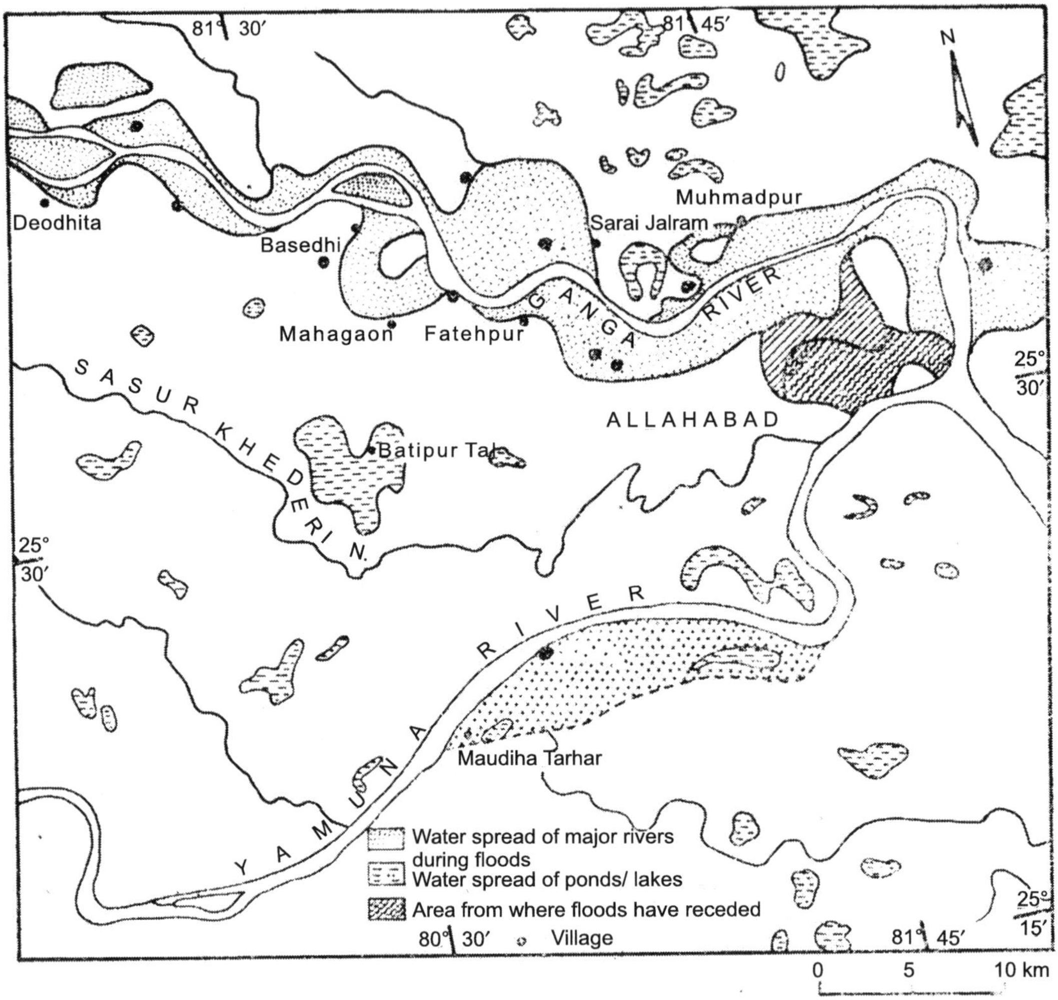

Fig. 13.14. Flood-hazard zoning map showing the floodways of the Ganga and Yamuna rivers just before the "Sangam" (R. S. Chaturvedi and R. Mohan, 1983).

e relatively narrower normal passage during drier seasons. Bridges with closely spaced piers or th projecting abutments restrict this passage. The road embankments, with which the bridges e invariably linked, are like dams across the floodways, the bridges serving as open sluice gates. he small and large houses that have been built in the floodways have compounded the problem not allowing free flow of rivers in spate.

Since there will be more frequent spells of intense and extreme rainfall, the people occupying e floodways have to brace for the disasters that the floods will bring. Whatever we do, the vers will reoccupy their natural waterways — the floodways — during spells of heavy discharges aldiya, 2011).

If need be, recourse to legislative measures to restrict engineering, industrial and economic activitı in these zones will have to be taken. These measures include not permitting construction of buildin or rigorous prescription for designs and flood proofing, aquisition by the government of the vulnerab land for developing it into a greenbelt or wildlife sanctuary, etc. Alternatively, the development rig of flood-prone land is bought by the government, although the original owner may retain it as a far but may not use it for commercial or residential or industrial purposes. Government assistance, ba loans, insurance against hazards, and civic facilities including telephones, bus service, and water a electricity supplies should be denied to those who occupy the declared flood-hazard zone despite t ban. In the identification of flood-prone areas, the natural water regime of the river and the histo of past occurrences of floods is closely studied (Figs. 13.13 and 13.14). The ***flood fringes,*** which a covered by the 100-year flood discharges, are protected by engineering structures in order to allo construction of flood-proof buildings.

Admittedly, it would be extremely difficult to enforce regulations in a time(beginning of 21st centur when "de-notification of land" has become a trade of those who are in positions of power, and t intervention of power-wielders would make disincentive measures unsuccessful.

Flood Forecasting and Warning

Forecasting floods and giving timely warning to the threatened people and relief agencies is an essent part of rational flood management. This calls for the development of an efficient network of warni systems. The pre-requisite is monitoring (i) the amount of rainwater, (ii) the volume of discharge a velocity of flow, (iii) flood height, depth of flooding, and (iv) the duration of inundation and the ar likely to be inundated.

Table 13.6. Warning Stage and Highest Recorded Flood Levels in the Early 1970s (After K.L. Rao, 1975)

River	Recording Station	Warning Stage (m)	Above Sea Level Highest Recorded Stage	
			Flood Stage (m)	Date
Ravi	Madhopur	349.4	349.8	5.10.55
Beas	Mandi	213.7	215.2	6.10.55
Satluj	Harike	210.3	211.2	6.10.55
Yamuna	Tajewala	382.2	385.9	1924
Yamuna	Delhi	204.8	206.4	15.10.56
Ganga	Garhmukteshwar	198.7	199.0	17.6.63
Ganga	Varanasi	71.3	74.7	1882
Ganga	Patna	48.6	49.6	6.8.71
Ganga	Narpur	21.0	22.5	26.8.55
Ramganga	Raini	210.9	211.8	6.8.69
Gandak	Bhainsala	96.0	96.7	22.8.61
Burhi Gandak	Samastipur	46.0	46.8	26.8.71

Contd

River	Recording Station	Warning Stage (m)	Above Sea Level Highest Recorded Stage	
			Flood Stage (m)	Date
Rapti	Birdsghat	73.8	76.6	1925
Bagmati	Hayaghat	45.7	46.9	16.8.70
Mahananda	English Bazar	24.2	24.8	1938
Kosi	Barahkshetra	130.5	131.8	5.10.68
Tista	Jalpaiguri	86.1	87.2	5.10.68
Torsa	Cooch Bihar	40.4	42.6	20.7.60
Barak	Silchar	20.7	21.9	17.6.69
Manas	Mathanguri	99.1	100.6	25.5.62
Burhi Dihing	Kharang	102.1	103.4	31.8.58
Subansiri	Bhoniparghat	96.2	96.9	9.8.57
Brahmaputra	Dibrugarh	104.2	105.5	17.7.69
Brahmaputra	Guwahati	49.7	51.1	23.8.62
Brahmaputra	Dhubri	28.6	29.8	1.8.72
Imphal	Imphal	781.8	783.0	18.6.63
Brahmani	Jaunpur	20.1	22.8	1960
Mahanadi	Naraj	21.0	27.9	1925
Subarnrekha	Rajghat	4.6	6.2	1943
Baitarni	Akhuapudi	19.2	21.9	16.8.60
Godavari	Dhonkswaram	16.7	17.7	16.8.53
Krishna	Vijayawada	18.1	24.3	7.10.03
Vamsadhara	Golta	24.9	35.7	12.7.60
Pennar	Dam Site	537.3	537.4	21.10.62
Kaveri	Mettur Dan	240.8	242.2	20.8.61
Bhavani	Khodiar	220.0	224.9	26.10.36
Tapi	Hope Bridge	29.0	31.1	6.8.68
Narmada	Bharuch	7.6	12.6	7.9.70
Mahi	Kadna	100.5	104.9	15.9.59

SUMMARY

bout 40 million ha of land in India is prone to floods and 10 million ha annually ravaged by inundation. e areal extent of flooding and the magnitude of damage have been phenomenally increasing in the last decades. Whenever the flow of excess water exceeds the carrying capacity of channels within the nks, flooding takes place. The causes of excess flow in the channels are heavy rainfall, particularly in orter spells of time, extensive melting of snow and ice, bursting of natural or man-made dams, and rm surges along coasts resulting from piling up of waves due to high wind.

The excess flow or discharge in a stream results from increased runoff due to reduced infiltration in the soil of the catchment areas. There is thus a greater overland flow when the infiltration is reduced d to decimation of the forest ecosystem with its multistories of grasses, shrubs and trees. Such forests a invariably associated with litter mass which have a great capacity for absorbing rainwater and reduci overland flow. The loss of porosity and permeability of soil such as resulting from stamping by grazi livestock in catchment areas, alkalinization-salinization in irrigated fields in flood plains and coveri of grounds by impervious asphalt or concrete-pavements and/or roofs in the process of urbanizati greatly reduce infiltration, and consequently augment overland flows that make floods. With only 24 of the surface of the country being covered with forests, more than 16.03 m ha of agricultural fiel being salinized-acidified-alkalinized, and large tracts being encroached upon for urbanization, it but natural that the frequency and severity of floods are increasing phenomenally. **An urban area with 40% impervious surface and storm sewers will have three times as many floods as befo urbanization.**

One of the results of floods is **increased erosion and greater transportation of sediments, leadi to a rise of river beds and consequent reduction in the carrying (conveying) capacity of the channels. Rivers like the Kosi, Tista, Jaldhaka, Gandak, Narayani, etc. frequently break throu their levees or man-made embankments and change their courses because of the rise of their be resulting from deposition of excess sediments.** During floods, the overflowing water returns to t abandoned channels and cause enormous damage. The sediments spread on the flood-plains also cau loss of productivity of the soils. For a large proportion of the sediments derived from gully erosion a landslides and excavation for mining and roads do not contain humus and mineral nutrients.

Of the 40 m ha of fertile lands affected by floods, more than 60% lie in the Indo-Gangetic Plain where the population is densely packed. In the Brahmaputra valley a vast tract is submerged for mo than three months in a year during the monsoon, and excessive accumulation of debris brought tributaries from the north has promoted rapid erosion on the southern bank. In recent years, even t semi-arid and arid terrains of Rajasthan and Gujarat have been ravaged by destructive floods, primarı because of inadequate carrying capacity of their undeveloped drainage systems and little infiltration rainwater into the dried up soils, owing to the absence of vegetation.

The flood management plan involves reduction of runoff by increasing infiltration throu appropriate afforestation in the catchment areas, reducing and controlling discharges by develo ing detention basins and reservoirs by constructing a series of checkdams and minor dams in su basins, inducing artificial recharge of the groundwater extensively through intensive pumpi and dugwells along dry streams, rivers, canals and channels, spreading and distributing excess wat thinly over such areas as paddy fields, depressions and desert tracts, diverting of possible exce water through channels or canals to areas and rivers deficient in water, and building embankments prevent spilling of flood waters.

It is imperative to have flood hazard-zoning maps, and plans of preparedness. **Encroachment of floodways should not only be discouraged but banned.** An effective system of flood-forecasting a warning through networks of stations is an essential part of flood management.

14 Coastal Hazards – Cyclones and Tsunamis

HAZARDS ON INDIAN COASTS

Vulnerable Belts

The eastern coastal belt embracing Bangladesh, Bengal, Odisha, Andhra Pradesh and Tamil Nadu (Figs. 14.1, 14.2) is very vulnerable to repeated hazards of surging sea waves and heavy downpours that company spirally moving fierce winds, mostly in the months of October, November and December (Tables 14.1, 14.2 and 14.3). Cyclones are endemic on the Andhra coast, one hitting every second year. Seven major cyclonic disasters hit Bangladesh within a short period of less than a decade—once in

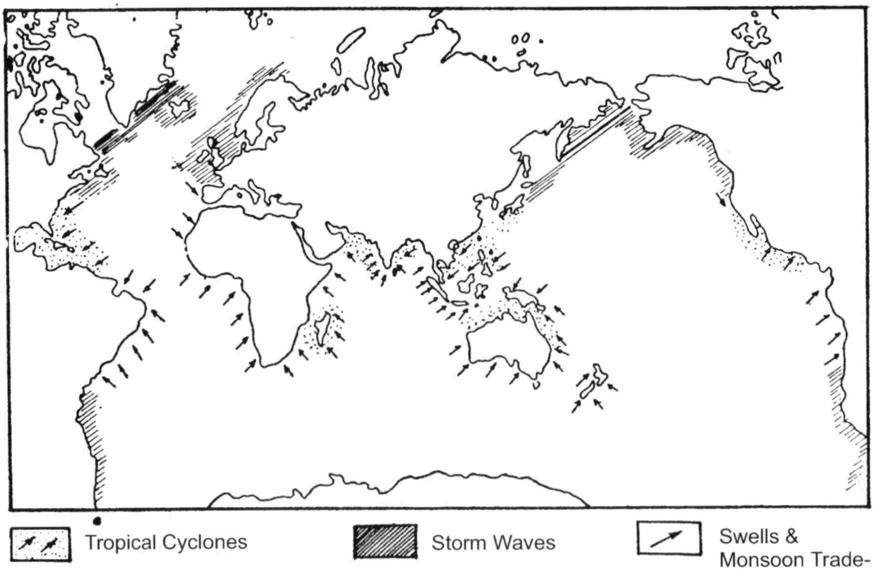

Fig. 14.1. Global coastal belts affected and locally ravaged by tropical cyclones and storm waves. (From Bloom, 1978.)

1960, twice each in 1961, 1963 and 1965. The cyclone that struck the coastal belt of India on November 13, 1970 took a toll of 200,000 human lives and over 800,000 heads of livestock, destroying more than 20,000 houses, 80% of the standing paddy crop, and 65% of the fishing capacity in 9000 localities. the tropical cyclone of November 19 to 20, 1977 more than 8500 people and 30,000 heads of cattle and 244,000 goats living on the Andhra coastal belt were killed. The October 17 to 29, 1999 cyclones that ravaged the coast of Odisha, had the wind speed of 260 to 300 km/hour, creating a surge waves as high as 8 m. Over 20,000 people were killed, 200,000 trees uprooted, 2 million hectares of land severly damaged, 520 km roads washed away, and salinity problem was created on the 3 to 4 km wide coastal area, mostly in the cultivated fields. In some areas the saline sea water went 15 km deep inland throu inlets, destroying 90% cropland and devastating 320,000 hectares of land (*in* Valdiya, 2004).

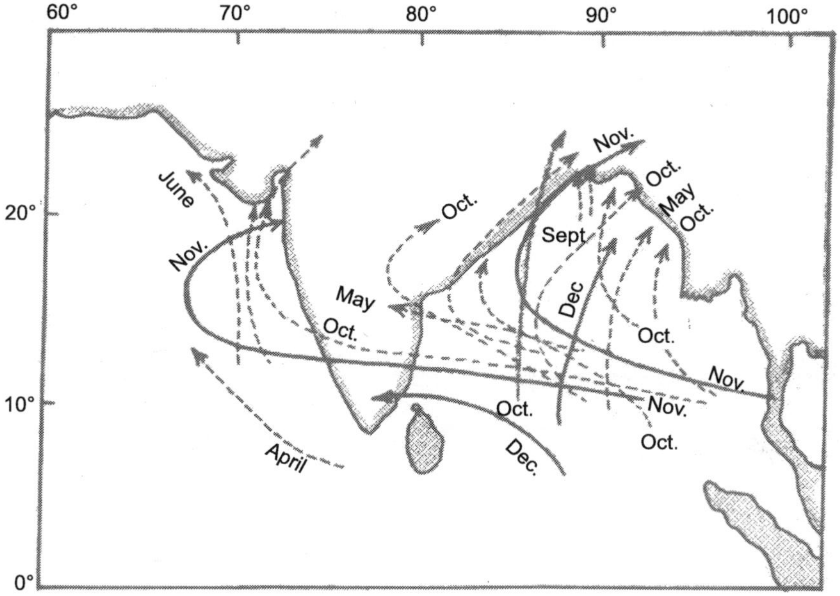

Fig. 14.2. Tracks of tropical cyclones (normalized) in the Bay of Bengal, Arabian Sea and Indian subcontinent (After P. Koteshwaram, 1974).

About a 25-km wide belt of the eastern coast of India is vulnerable to cyclonic hazards, the risks being particularly severe in the mouths of rivers. The cyclone-prone districts (Fig. 14.2) include Midnapur and 24-Parganas in Bengal, Cuttack, Ganjam and Puri in Odisha, East Godavari, Krishna, Nellore, Prakasam and Srikakulam in Andhra Pradesh and Chinglepet and Tanjavur in Tamil Nadu (Table 14.1).

Although there has been no change in the trends of cyclones originating over the Bay of Bengal, **there is nearly two-fold increase in the frequency of *intense* cycles during May and November** (Singh et al, 2001; Unnikrishnan, 2006; Niyas et al, 2009). These cyclones give rise to high storm surges along the coast.

Table 14.1. Frequency of Cyclone Landfalling in Different Cyclone-Prone Districts in the Period 1889-1970 ased on A.K. Saha et al., 1984)

East Coast		West Coast	
State/District	**Frequency**	**State/District**	**Frequency**
Bengal		***Kerala***	
Midnapur	11	Khozhikode	2
24-Pargana	21	Kannnur	1
		Ernakulam	1
Odisha		Malappuram	1
Balasore	11	***Karnataka***	
Cuttack	11	South Kanara	2
Ganjam	5		
Puri	6	***Maharashtra***	
Andhra Pradesh		Kolaba	4
East Godavari	8	Ratnagiri	3
Guntur	3	Thane	3
Krishna	12		
Nellore-Prakasam	11	***Gujarat***	
Srikakulam	13	Bhavnagar	1
Visakhapatnam	6	Bulsar	1
Tamil Nadu		Bharuch	1
S.Arcot-Puducherry	4	Jamnagar	1
Chingelpet	15	Junagadh	7
Ramanathapuram	2	Kachchh	3
Tanjavur-Pudukottai	11		
Tirunelveli	2		
Total	152	Total	33

Frequency of Cyclones

On an average, three to four severe cyclonic storms form in the Bay of Bengal, mostly from April to ne in the pre-monsoon period and September to December in the post-monsoon period (Table 14.2). It may be noted that **cyclonic storms are five times more frequent in the Bay of Bengal than in the Arabian Sea** (Fig. 14.2). This is quite evident from the hazards that the eastern coast faces year after year, compared to those borne by the west coast where northern Gujarat and Saurashtra are the most vulnerable areas.

On a global scale 80 to 100 severe cyclones form in the world's oceans each year.

Storm Surges

The passage of cyclones causes a sudden reduction of atmospheric pressure over the coastal belt—as much as 60 to 100 millibars — giving rise to storm surges (Table 14.3). Indeed, both cyclones and tsunamis bring in their wake high ***storm surges,*** **manifest in a sudden and marked rise of sea level. They may reach a height of 5 to 6 m above the normal sea level, the highest so for recorded of 12 m accompanied the 1876 cyclone that ravaged Bakerganj in Bangladesh. More than 250,000 peop were drowned to death. In the year 1998 the Kachchh coast in Gujarat fell victim to 18 to 20 high storm surges created by a powerful cyclone which had the wind speed of 235 km/hr along the Kandla coast.**

Table 14.2. Seasonal Frequency of Cyclones in India in the Period 1891 to 1982 Indicating the Time Occurrence of Hazards (After A.K. Saha et al., 1984)

Month	East Coast	West Coast
January	4	0
February	0	0
March	2	0
April	11	2
May	15	4
June	32	6
July	33	3
August	27	2
September	23	5
October	40	13
November	40	7
December	22	4
Total	249	46

More than 90% of the loss suffered by the coastal settlements is due to storm surges as they flo the terrain extensively, sweep away houses and cultural features, destroy farms and fields and cau widespread salinity in soil and water resources. In late November 1984 a cyclone on the Andhra coast destroyed a little less than a lakh houses and another lakh were partially damaged, and in large areas the waves threw a thick cover of sand (in some places as much as one metre), thus rendering barren more than 2000 ha of agricultural farms. As already stated, due to the supercyclone of October 1999, 320,000 ha of land was damaged, destroying of 90% of cropland.

CYCLONES AND THEIR GENESIS

Nature

Spirally moving tropical storms are known as ***hurricanes*** in western countries and ***typhoons*** in eastern Asia. In larger cyclonic storms having diameters of 60 to 160 km the air rotates counterclockwise with a speed of more than 200 km/hr at the periphery, while the centre (eye) remains calm (Fig. 14.3). **The highest wind speed recorded in the Indian subcontinent is 300 km/hr when the Odisha coa was hit by a supercyclone in October 17 to 29, 1999. They move forward with a speed of abo 25 km/hr and have a life-span of nine days on the average.** However, they spend their destructive fury completely within 24 hours after landfalling. But there are cases of cyclones tracking back to the sea, getting renourishment, and then returning to strike again, as happened during the 1978 cyclone in Bengal, and the 1999 supercyclone in Odisha.

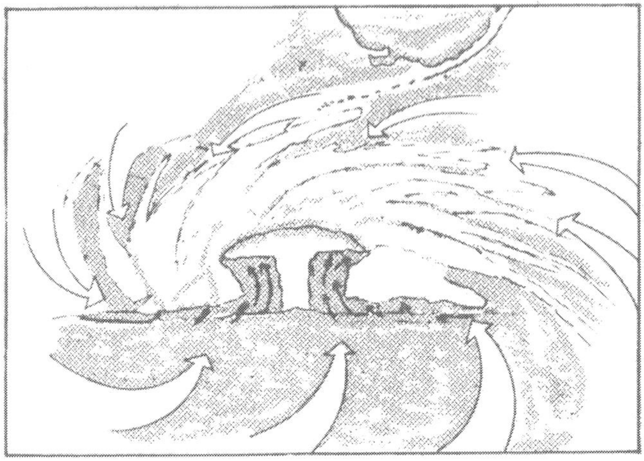

Fig. 14.3 A. Nature and dimension of a tropical cyclone. It is also known as a hurricane and typhoon. An entirely different kind of storm occurs on land. It is the tornado which is smaller in size but fiercer in velocity. (After W.M. Marsh and J. Dozier, 1981, *Landscape,* Addison-Wesley, Reading, 637 p., Fig. 8.17.)

Fig. 14.3 B. The storm that ravaged the coast of Saurashtra and parts of Gujarat on June 8,1998 (From *Current Science,* 78(3), October 2000). (Courtesy: Dr A.S. Rajawat, Space Applications Centre, Ahmadabad). (From Valdiya 2004). (See colour Fig. 14.3 B)

Tornadoes that strike the central plains of U.S.A. very frequently are not sea-borne coastal hazards, and are much smaller in dimension—0.5 to 1.5 km in diameter but they rotate very violently with speeds as high as 800 km/hr and may stretch as high as 25 km into the sky.

Heavy Rains and Floods

Exceptionally heavy rains accompany cyclonic storms—150 to 250 mm, or even 1000 mm locally — in the short duration of the passage of the storm. During the late November 1984 cyclone, as against t normal rainfall of 285 mm for the whole of November, the downpour was 600 mm in just five days and 1100 mm in the Sullurpet, Gudur and Naidupet areas in Andhra Pradesh. In the coastal belt adjacent to the Bay of Bengal, the rainfall during these events varies from 75 to 300 mm per day.

Fierce wind and heavy rain are compounded by floods, which may rise as high as 14 metres an inundate vast areas stretching deep inside the land through riverways. The October 1964 cyclone caused flooding as far inside as Acra in the Midnapur district via the Hooghly River. In the 1970 cyclo that took a toll of 200,000 people and 800,000 heads of livestock, a vast tract of land was submerged as deep as 6 m. The cyclonic storm that swept the Saurashtra coast on November 10, 1982 brought 18 to 20 m high storm surges invading the coastal belt, killing 200 people and rendering thousands homeless in the district of Amreli, Bhavnagar and Junagadh.

Genesis

In the tropical belt (5 to 15°N) sea water gets heated upto 27°C or more, so that the air above the warm water develops low pressure areas, as happens in the South Andaman Sea, South Bay of Bengal and Southeastern Arabian Sea. The low pressure area remains stationary for three to four days and draws energy from the sea surface. As the pressure in the centre falls as much as 20 millibar, the wind speed increases to 40 km/hr and cloud bands start spiralling around the centre, causing squalls. Since in the centre the pressure falls (even by 3 to 5 mb) the warmed up air rises as high as 12,000 m or more, inducing the wind from the surrounding areas to rush inwards creating spirally moving storms (Figure 14.3 A,B). These storms, turn at a speed of 100 to 120 km/hr, anticlockwise in the northern hemisphere but clockwise in the southern. The vortex of the air spirals inwards toward the centre where the pressure is minimum and contains only light wind or may even be quite calm—called the ***eye of the storm***.

The cyclone then moves landwards—to areas of lower pressure (Figs. 14.2 and 14.4). T outcome is extremely heavy rainfall exceeding 70 to 300 mm/day.

Geological Changes and Ecological Impacts

During the November 1977 cyclone in a 125-km long stretch of the Andhra coast between Lan and Janavripatnam the beach-dune complex was breached at several places (e.g. Machhalipatnam). Sediments were exposed in a number of places. The erosion was accompanied by formation of lobate sandbars and waterlogging in parts of the deltaic lowlands between Machhalipatnam and Machhugunta. The aquatic life was severely disturbed, and the groundwater considerably affected. The May 1979 cyclone, likewise, greatly modified the beach between Nellore and Vadarevu where it invaded 20 inland.

Table 14.3. Some Severe Cyclones in India (Data from A.K. Saha et al., 1984; Gupta et al., 2000)

Date	Landfall speed (km/hr)	Maximum Wind point	Minimum Pressure	Cyclone speed	Storm surge height	Loss of life
ct. 5-10, 1964	Contai (Bengal)					50,000 (drowning)
						30,000 (epidemics)
ov. 24–Dec. 1, 1876	Bakerganj (Bangladesh)	243	919	21		10,000 (drowning)
						100,000 (epidemics)
Sept.19–23,1885	False Point (Odisha)				3 to 12	
ct.15-16, 1847	Contai (Bengal)	226			3	75,000+6,000 Cattle heads
ov. 3, 1946	Narasapuram (A.P.)	130	965	11	3 to 5	750 + 30,000 cattle heads
ec. 20-24, 1964	Rameswaram (T.N.)	169	970	8	3 to 5	500
ec. 4-6, 1972	N. of Cuddalore (T.N.)	150	97	11		80 + 141 cattle heads
Sept.8-11, 1976	E. of Contai (Bengal)	160	972	17	2 to 3	10 + 40,000 cattle heads
ov. 14-20, 1977	N. of Chirala (A.P.)	259	941		5	8547+574200 cattle heads
May 5-13, 1979	Kilakkarai (T.N.)	222		3 to 5		700 + 300000 cattle heads
Sept. 1985	Odisha					84 + 2,600 cattle heads
ov. 1987	Andhra Pradesh					50 + 25,800 cattle heads
						8,400 houses
ne 1989	Odisha					61 + 27,000 cattle heads 14,000 houses
May 1990	Andhra Pradesh					928 + 14,000 houses
ov. 1991	Tamil Nadu					185 + 540 cattle heads
pril 1993	West Bengal					>100 lives
ct. 1999	Odisha					200,000 trees uprooted, 2 million ha land severely damaged, 90% cropland destroyed

Fig. 14.4. Odisha cyclone on October 29, 1999. (Courtesy: Professor B.N. Goswami, IITM, Pune) (See colour Fig. 14.4)

Cyclonic storms have appreciably reshaped the landforms of the Odisha-Andhra coast, particularly in the deltas of the Mahanadi, Godavari and Krishna rivers (Vaidyanathan, 1977, 1983). Even in normal times, the waves and tides accomplish considerable transportation of sediments by littoral drift at the rate of 0.53 to 1.0 m t/yr in the Visakhapatnam coast (Vaidyanathan, 1975), and 1.5 m t/yr along the Paradip coast in Odisha. As a result of accumulation of drifting sediments behind artificial impedimen such as a light houses, very many large sand dunes had formed e.g. the 40-m high Pudimadaka dune and the 90-m high Bhimli dune (Prudhvi Raju and Krishna Bhagvan, 1984).

Cyclonic storms give rise to spectacular landforms, and very suddenly. The November 1977 cyclone affected 320 km^2 area of the Krishna delta, while the 1969 event caused the disappearance of the entire hamlet north of Gangavaram behind the sand dune which was reshaped in just 12 hours.

As already mentioned, not only do the surface water bodies become polluted (with saline sea waters), and the agriculture soils rendered saline, the croplands of the coastal belts also may become covered with beach sands—enough to make the fields useless and requiring expensive remedial measures. the 1977 Andhra cyclone more than 1.46 million ha of cropland were severely damaged in this manner. The October 1999 supercyclone created salinity problem on a 3 to 4 km wide coastal area and destroyed 90% cropland (Table 14.3). In the wake of inundation followed epidemics causing more death from epidemics than from drowning.

TSUNAMIS AND DIMENSION OF HAZARDS

Origin

Seismic sea waves or tsunamis are generated by a sudden displacement of crust along under-ater faults on the sea-floor, particularly in the oceanic trenches that girdle the Pacific Ocean d the Bay of Bengal in the Indian Ocean. The oceanic trenches are elongate, narrow and deep depressions into which the oceanic crust plunges intermittently. Unlocking of the subduction zone generates earthquakes. Accompanied by large magnitude earthquakes, **the abrupt displacement of e seafloor along a fault or thrust creates a void (Fig. 14.5 upper). To fill this void, seawater from sides rushes in with terrific speed, colliding and then turning back in the form of sea waves ig. 14.5 lower) that may travel with a speed as high as 900 km/hr.** Their wavelength is very large—

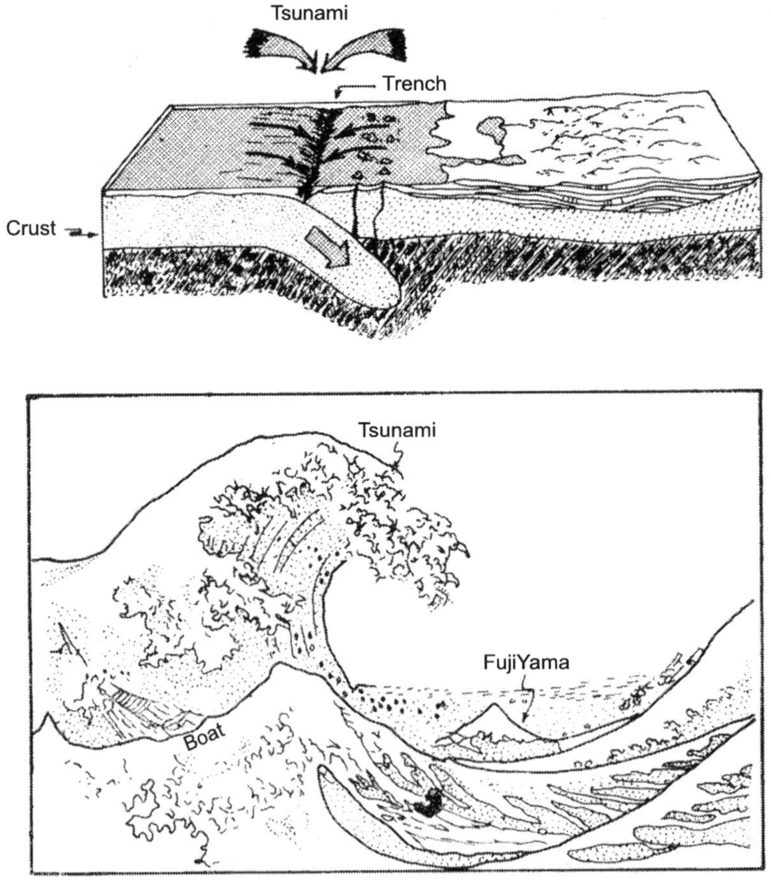

Fig. 14.5. Displacement of the crust due to subduction causes a large volume of water to rush towards the void thus formed, the collision of inrushing waves and their turning back as seismic sea waves.

100 to 200 km, but their amplitude is very small—merely 1 to 2 m in the high seas. These waves travel very far and cause havoc across the oceans thousands of kilometres wide. The earthquake that originated in the Oman Sea off the coast of Balochistan on November 27,1945 created 12 m high waves along the Makran coast and struck 1600 km at far off Mumbai where the waves rose 2 m high, and even affected coasts as far as Sri Lanka and Madagascar islands.

Tsunamis thus originate in the offshore belt of the Aleutian islands, Kamchatka Peninsula, Japanese islands, Chilean coasts, and Alaska. In the Indian Ocean the Andaman-Indonesia Island Arc is flanked the west by what is known as the Java Trench. The east-west trending coastal hill ranges of Balochistan is fronted by the Oman Trench (Table 14.4).

Table 14.4. Some Great Tsunamis in the World's Oceans

Date	Source Region	Visual Run-up height (m)	Reports from
August 13, 1868	Peru, Chile	>10	Africa, New Zealand, Hawaii
June 15, 1896	Honshu, Japan	24	Sanrika, Japan
March 3, 1933	Honshu, Japan	>20	Sanrika, Japan
April 11, 1946	Aleutian	10	Wainaku, Hawaii
May 23, 1960	Chile	>10	Waiakea, Hawaii
Nov. 27, 1945	Makran, Pakistan	12	Mumbai, Sri Lanka, Malagasy
Dec. 26, 2004	Tamil Nadu coast	>7	Nagapattinam, Sri Lanka
March 11, 2011	Honshu, Japan	>10	Tohoku, Tohako-Oki, Fukushima

Seismic Sea Waves

More than 80% of the total seismic energy produced on earth is concentrated in the circum-Pacific b where about 200 earthquakes of magnitude up to 6 occur every year. Many of these earthquakes occur deep under the sea in oceanic trenches. The stronger of them generate powerful seismic sea waves called ***tsunamis*** (Fig. 14.5). Tsunamis have ravaged the circum-Pacific coasts frequently and destructively.

The east coast in Tamil Nadu has been ravaged periodically by tsunamis. There is mention of major coastal hazard in the *Valmiki Ramayan* at the time when the "Ram Setu" was being built link Bharatvarsh with Lanka (Valdiya, 2012). In the historical Chola period around 1000 yr B.P tsunami crippled the ancient port of Kaveripattinam (Rajendran et al., 2011). There were 2.5 m hi waves at Chennai when an earthquake of magnitude M_w 7.9 hit the Andaman Islands in 1888. The 1941 earthquake (M_w 7.7) also generated destructive waves along the east coast, and the disasters brought by the 2004 tsunami caused by the Sumatra earthquake of M_w 9.3—due to rupturing in the 1200 km long strength of the Andaman–Java subduction—is too well-known to be recounted.

Submarine sediment slides causing abrupt displacement of seawater may also generate tsunamis as happened in 1933 in the Sugami Bay, Japan, and in 1964 in Yaldez. In 1958 a colossal avalanche in the Lituya Bay (Alaska) caused a major seawave. Both the submarine slides and snow avalanch were related to seismic shocks. In 1883 the Krakatau Caldera, an island in the Indonesian archipelago, collapsed following an explosive volcanic outburst. The cauldron collapse led to the generation of a catastrophic tsunami—40 m high seawaves rolled onto the Java and Sumatra coasts, killing 36,000 people. The surge waves were detected as far away as the English Channel in the United Kingdom.

Fig. 14.6. The December 26, 2004 tsunami hits the marine beach in Chennai. (Courtesy: Professor B. N. Goswami, I.I.T.M. Pune) (See colour Fig. 14.6)

Facing the east coast of Peninsular India is the Andaman island arc. Its western front is, as already ted, delineated by an oceanic trench that has been completely filled up with the sediments delivered the Bay of Bengal by the Ganga and Brahmaputra. This sediment-filled oceanic trench is tectonically extremely active.

The March 11, 2011 earthquake (M_w 9) originating from the Tohoku-Oki Trench off the east coast of Honshu, Japan, brought the calamitous tsunami that nearly destroyed the Fukushima nuclear power plant, among the many disasters that wrecked Japan.

Impacts

As already stated, waves having low amplitude (1 to 2 m) and long wavelengths (100 to 200 km) travel with a high speed of 800 to 900 km/hr in the high seas. But as they reach the shallow coasts, their speed diminishes progressively—180 km/hr in 40 m as deep water and 50 km/hr in the 20 m depth. The amplitude on the other hand increases and the waves may rise to a height more than 15 m (locally even upto 30 m). Obviously, the destruction due to these high waves can be colossal in magnitude. e high surge waves that destroyed the city of Lisbon on November 1, 1755, killing 60,000 people, rose upto 10 m high. The impact of this tsunami was felt across the Atlantic in the eastern seaboard of the U.S.A., the Azore islands, and the West Indies islands. The coastal areas of Sweden, Holland and Britain were severely hit. The May 22, 1960 seismic seawaves that originated in the Concepcion Trench hile) travelled through the 10,600 km wide Pacific Ocean at a speed of 700 km/hr, sweeping clean the buildings of many blocks in the city of Hilo in the Hawaii islands.

Seismic sea waves have disastrous effects also on aquatic and plant life in the near-shore environment. The onslaught of waves is preceded by a withdrawal of coastal waters, leaving the reefs dry, thus causing extensive fish mortality. The rushing-rampaging waves then erode away coastal terrains with their clif beaches and dunes.

COASTAL HAZARD MITIGATION MEASURES

Identification of Hazard-Prone Belts

The first step in a coastal hazard mitigation programme is the identification and demarcation on ma of the areas and belts prone to cyclones and attendant surge storms. In coastal India a 24-km wide belt in the following districts was identified as a hazard zone by the Cyclone Review Committee (Saha al., 1984):

Tamil Nadu	: Chingleput, Tanjavur
Andhra Pradesh	: East Godavari, Krishna, Nellore, Prakasam, Srikakulam
Odisha	: Balasore, Cuttack, Ganjam, Puri
Bengal	: Midnapur, 24-Pargana
Gujarat	: Junagadh, Amreli, Bhavnagar

The wind damage may extend over a much wider belt upto 100 km inland. The tidal bores may sweep deep inside the land through the mouths of rivers, as happened in the Roopnarayan River in Beng (1864) and the Mahanadi River in Odisha (1920), and in many rivers in 1999.

The Coastal Vulnerability Index Atlas prepared by Indian National Centre for Ocean Information, Hyderabad, 2012, shows that 638 km stretch of the Tamil Nadu coast embracing Kanchipuram and Chennai districts, and 37 km stretch of the Andhra Pradesh coast between East Godavari and Visakhapatnam districts are very high risk areas (*The Hindu,* May 18, 2012).

Developing Warning Systems

Based in ground observatories, the meteorologists in ships or on oceanographic research vessels take note of several cyclone indicators for prediction: (i) the eye and the surrounding walls of cloud together with characteristic cloud structures, (ii) the drop in pressure, and (iii) the maximum wind velocity. These parameters are monitored and the cyclone field mapped (with pressure and speed values). The directi of the cyclone is tracked in order to predict whether and where and when precisely it would strike the land. Also required is the advance knowledge of the speed of winds and magnitude of the storm surges they would generate.

Aircraft can be used to probe the cyclones at higher altitudes and measure various meteorological parameters, as done in U.S.A. and Japan. The aircraft-probing provides the most accurate and detailed information of the eyes and the intensities. There is a need for reconnaissance investigation at levels 12 km, 6 km, and 1 km above sea level. **Satellite imageries have provided very valuable information, and helped in determining the direction of the cyclones.** "By observing the rotation of major cloud patterns over 24 hours by polar orbiting satellites, it is possible to visualize the deflections of cyclo tracts" (Saha et al., 1984). **Through the geosynchronous INSAT-IB satellite, cyclones can continually monitored and their behavioural pattern determined.**

The most efficient and effective method of forecasting cyclone tracks is using the technique of radar to track them continually. The S-band radar has been found to be most useful in watching and tracking cyclones.

Tsunamis take 15 to 20 minutes to reach the land, while the seismic waves from earthquakes take only about one minute to cover that distance. This fact forms the basis for organizing warning systems, as is done in Russia, Japan and U.S.A. The seismic wave travel-time charts for different urces around the Pacific with lines representing the time of travel from the source area to the position on the chart is a useful measure in the warning system. It takes 10 hours for the tsunamis to travel from Chile to Japan. There is thus ample time for alert and preparedness.

Reduction of Risks in Vulnerable Areas

Advance warning of the impending disaster or hazard allows people to evacuate the threatened areas and reach safer places. Before the 1979 Andhra cyclone, the entire coast was evacuated so that the human casualty was minimum. **But warning alone is not enough; people must have adequate perception of zards and a willingness to adjust. Even though Bangladesh, Odisha and the Andhra coasts have been repeatedly devastated by cyclones for thousands of years, the people have made little efforts cope with the hazard—there are no integrated plans for adjustments and there never have been any attempts at modifying dwellings.** In Bangladesh people perch themselves on tree tops, climb up double-storey buildings or occupy embankments of roads and railway lines.

In Andhra Pradesh, fishermen have made significant adaptations in their huts to cope with the high winds. Their huts are made round in plan, and the thatched roofs are steeply pitched (Fig. 14.7). The pillars are thickly painted with asphalt or covered with polythene at the ground level to check attack from termites. The anti-cyclone features in a hut also include a central anchor post, metal corner straps fix frame members to corner posts, and beams. These huts are built on raised mounds (platforms) mud as a safety measure against flooding (Fig. 14.7a). **However, by and large people seem to be indifferent to the need of incorporating structural improvements. Moreover, the settlements remain dangerously close to the sea.**

Round or polygonal shapes of houses, which cause reduction of wind pressure, are most suitable. If rectangular, their lengths should be made parallel to the direction of prevailing winds, and their roof steep, hip-type (Fig. 14.7), the pitch of the roof being greater than 30°, which reduces wind load. The overhangs should be as small as possible or should have provision for the passage of wind such as vents louvers. Buildings which are built on raised floors need to be anchored at the corners to the ground through galvanized steel.

Settlements must be quite far from the coast, and located behind natural windbreaks such as sand dunes or clumps or rows of trees or hedges to reduce the impact of winds. To stabilize the w on the leeward sides, small gaps are left in the windbreaks. This calls for building up of a massive wall of forests of tall and deep-rooted trees all along the coast in the cyclone-prone belt. As in the case of windbreaks in the deserts, these walls of forests should contain multiple rows of appropriate trees, some fast-growing and others slow-maturing. A massive effort at afforesting the coastline would be doubly rewarding, for these forests would also provide fuel and fodder to the people.

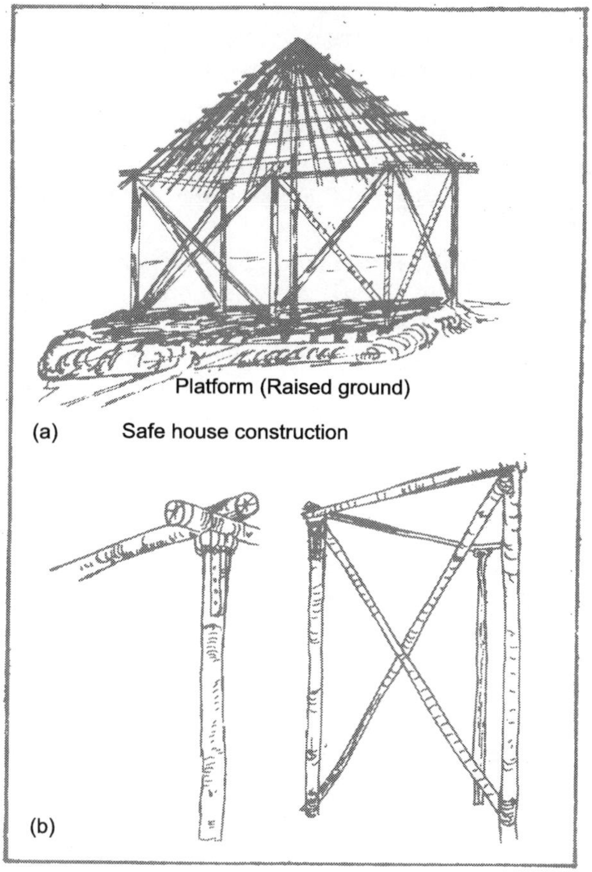

Fig. 14.7. Adaptation in structural plan of dwellings to strengthen them against the onslaught of fierce wind surge waves and flooding. (After A. Davies, 1984).

SUMMARY

Spirally moving storms (cyclones/hurricanes or typhoons) developing in the Bay of Bengal and Arabian Sea in the tropical belt have been striking the coastal belts of Bangladesh, Bengal, Odisha, Andhra Pradesh, Tamil Nadu and Saurashtra year after year, particularly in the months of May-June and October-November-December. Districts Midnapur, 24-Pargana, Balasore, Cuttack, Ganjam, Puri, East Godava Krishna, Nellore, Prakasam, Srikakulam, Chinglepet, Tanjavur, Bhavnagar, Amreli and Junagadh have been identified as cyclone-prone areas. These storms with wind velocity exceeding 100 to 120 km/ bring with them extremely heavy rainfall, often in the order of 150 to 250 mm within the duration of the passage of the cyclone over the belt. There is widespread damage from fierce winds, heavy downpou and sudden and deep inundation. To compound the tragedy, the cyclones induce sudden and marked rise of sea level—called storm surges—as high as 12 m—which not only flood the coastal belt but block t draining out of rainwater and river waters. On an average, 80 to 100 severe cyclones occur every year all over the world, and about 10 ravage the coasts of India.

Seismic sea waves or tsunamis are not very common on the coasts of India, but have caused enormous struction on the circum-Pacific coastal belts. The tsunamis result from sudden sinking of the ocean or following displacement along faults in the oceanic trenches. These waves of great wavelength and mall amplitude travel with a terrific speed of 800 to 900 km/hr and strike the coasts as 15 to 25 m high surge waves with speed upwards of 60 km/hr. The destruction along the coasts is thus catastrophic, both to human habitations and aquatic ecosystems.

Coastal hazards management involves developing an effective and efficient system of monitoring the movements of cyclones through satellites, radar, and aircraft and giving advance warning to enable the ople to evacuate threatened areas. Adopting modifications in the construction of a design that would engthen structures to withstand the onslaught of fierce winds, surge waves and floods, and developing a massive wall of multistoried forests along the coast are two very important measures that should receive greater attention from the governments and the public alike.

The most effective and doubly rewarding measure would be to build massive walls of forests all along the vulnerable coast.

15 Volcanic Hazards

VOLCANIC BELT GIRDLING THE INDIAN SUBCONTINENT

One natural hazard from which the **continental** Indian is free is the volcanic activity. But nearer home, the Andaman-Nicobar island arc, demarcated by an active subduction zone, includes a chain of sea-mounts (Alcock, Sewell, Narcondam) and recently alive volcanoe like the Barren Island (Fig. 15.1). A subduction zone is an elongate, narrow, long oceanic trench into which plunges the oceanic crust along with its sedimentary pile and islands. The Barren Island is a part of the Andaman–Indonesia Island Arc, a product of the tectonics of the subduction zone. The neighbouring islands of Indonesia — the th most populated nation with nearly 150 million people contain more active and potentially active lcanoes than any other region on earth. More than 135 volcanoes have been pouring out fire, ash and molten material during the Holocene period, and some of them have been very violent. The 1982-1983 eruption of the Galunggung in West Java caused widespread destruction and marked socio-economic nsequences on more than half a million people (Sudradjat and Tilling, 1984). Over 20,000 people re killed when the 4909-m high Navado Del Ruiz volcano in Colombia, South America, poured out h in the morning of November 13, 1985. Mixed up with snow and rain, the ash flow smothered the city of Armero, 48 km away. The intensity of the explosion was second only to the 1883 Krakatau explosion Indonesia, which released energy equivalent to that released by 30 hydrogen bombs.

BARREN ISLAND

The Barren Island represents possibly a wreck of a former larger volcano erupting andesitic and rnblende-dacitic lavas and cinders and forming a 3-km wide amphitheatre-like crater with two 300-m gh cones inside. Captain Blair had witnessed in 1775 glowing cinders and blocks being hurled up from e crater (Wadia, 1975). Rising 2300 m above the floor of the sea, the Barren Island volcano erupted 1832, in March to October, 1991, December 1994 to May 1995, and on 28 May 2005, the last one ejecting juvenile gases, ashes and incandescent fragments of basaltic composition from a sub-aerial en vent — a conical to oval-shaped depression (Bandopadhyay et al., 2006). Earlier in 1991 and 1995, there was eruption from several vents and deposition of huge quantity of pyroclasts and lava. The uption formed a lofty column 500 m in height (Fig. 15.4).

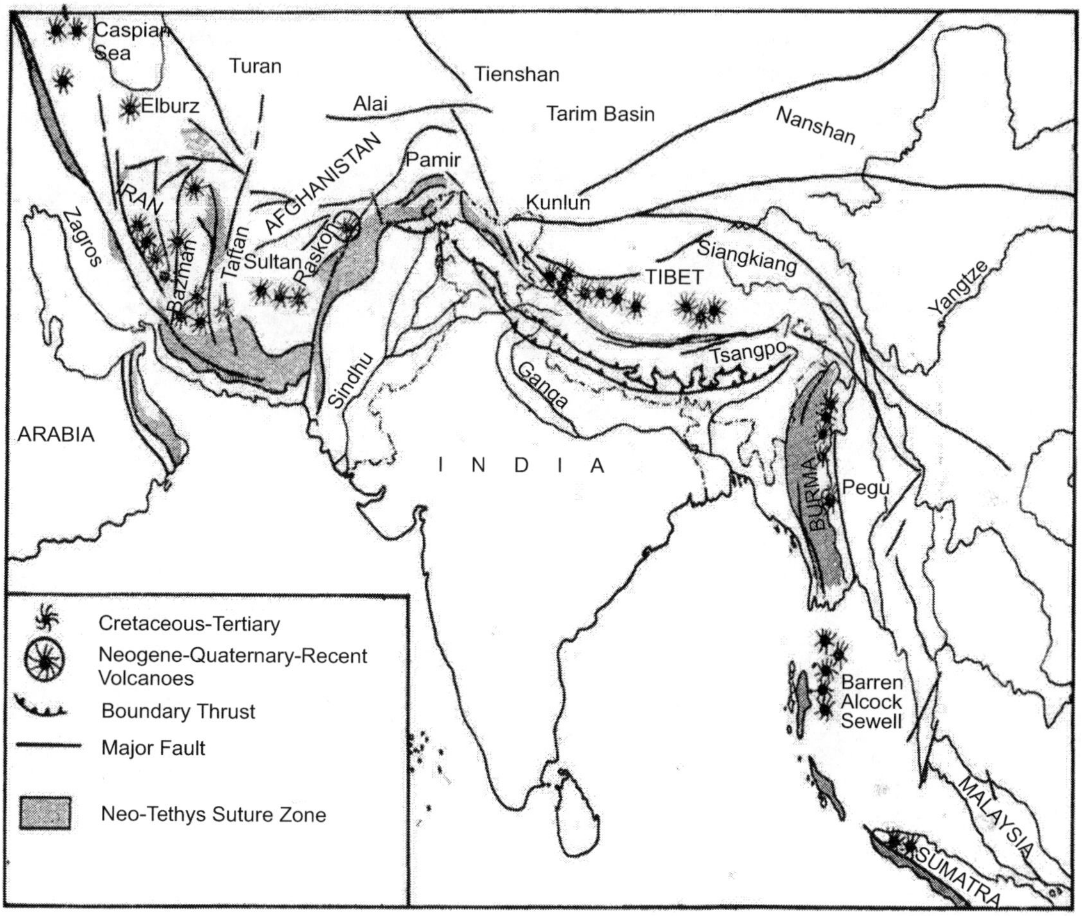

The territorial waters of India extend into the sea to a distance of twelve nautical miles measured from the appropriate base line. Responsibility for the correctness of internal facts shown on the map rests with the publisher. The topographical details within India are based upon Survey of India map, with the permission of the Surveyor General of India. The copyright 1986 of which vests with the Government of India.

Fig. 15.1. The belt of Neogene to Quaternary (2.0 to 20 million year old) volcanoes frames the Indian subcontinent. This belt represents the tectonic junction of Indian and Asian crustal plates, and an ancient zone of crustal sinking or subduction. (Based on J. Stockhn, 1980).

In the northward extension of the Andaman Island arc in the central Myanmar basin there is a line of volcanoes in the Pegu Yoma — the Thautalon, the Loimye and the Mount Popa — which used to pour andesitic lavas until the Quaternary times (Fig. 15.1). North of the line of the Tsangpo–Sindhu valleys—which delineates the northern tectonic boundary of the Indian plate—there is a belt of 20 2 million year old Neogene to Quaternary volcanoes that erupted lavas of andesite, dacite and latite as well as ignimbrite (representing glowing avalanches). To the west in Balochistan there is another belt of volcanoes—Raskoh, Chagai (Cretaceous–Tertiary), Koh-i-Sultan, Taftan and Bazman (Quaternar Recent). This andesitic volcanic belt lies about 100 km north of the plate boundary demarcated by t Oman Trench representing a subduction zone parallel to the Makran coast (Valdiya, 1984, 2010).

should be obvious that the Indian subcontinent is girdled on three sides by a narrow belt of volcanoes which were active until the Quaternary times. The belt in fact marks the boundary of the Indian plate against the Asian plate and forms part of the past subduction complex (Fig. 15.2).

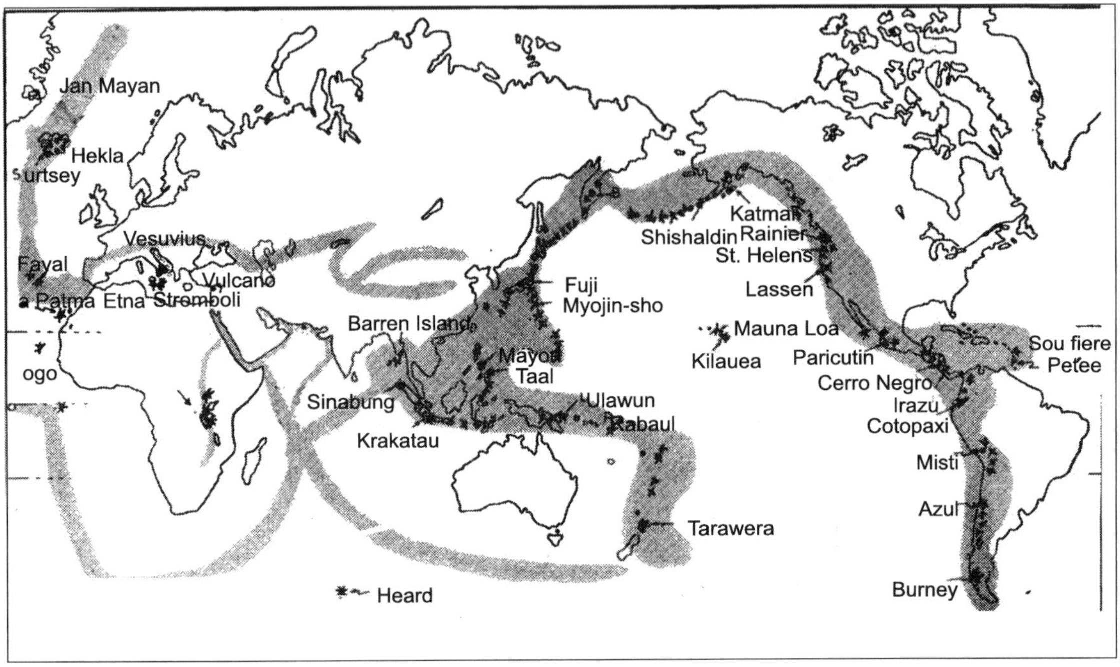

Fig. 15.2. Belt of volcanic hazards showing distribution of some of the major volcanoes. There are more than 500 active volcanoes in the world. The shaded area indicates the belt of high seismicity. (After R. I. Tilling, 1982).

Volcanic activity is restricted to narrow belts lining the junctions of crustal plates (Figs. 15.2 and .3), such as around the Pacific Ocean in which they form a ring of fire (75 to 80%), or mid-oceanic ges in the Indian, Atlantic and Pacific oceans. The deeply fractured or faulted zones in the continents ch as the East African rift zones, which represent the initial or aborted stage of separation of the ntinents under tension, are loci of volcanism. In the distant past (70 to 60 million years ago) the Narmada valley and the Konkan-Malabar belts — both dissected by multiplicity of deep faults — were similar zones of great volcanoes pouring out lavas and pyroclastics. These volcanoes gave rise to the upendous Deccan Volcanics that cover the whole of Maharashtra and large parts of Madhya Pradesh d Gujarat— the Deccan and Malwa plateaus.

ORIGIN AND TYPES OF VOLCANIC ACTIVITY

Volcanic belts mark the sites of impingement of convection plumes and convection currents beneath the crust. The impact of the plumes and currents (***hot spots***) produces tension in the crust resulting in its itting apart and the rise of hot molten material (Fig. 15.3 a). Reaching the surface quietly, the molten

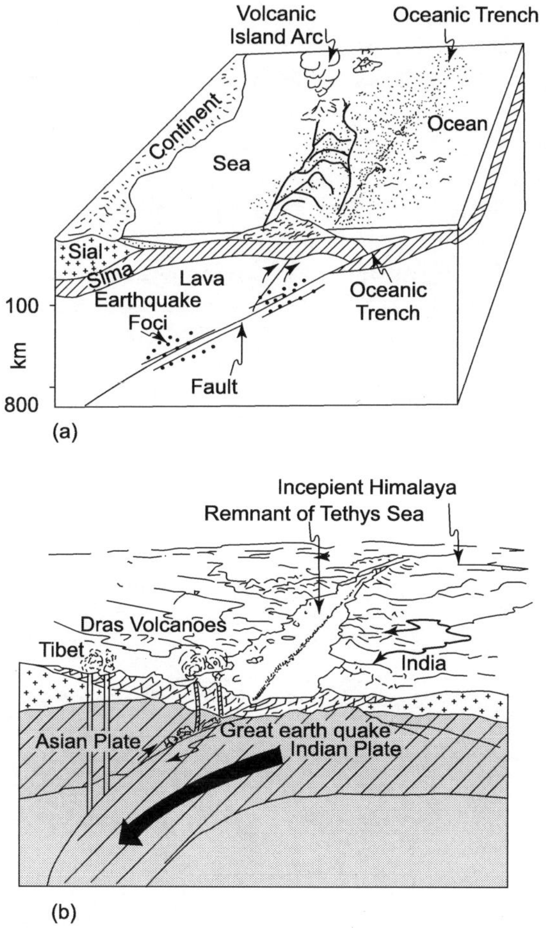

Fig. 15.3. (a) The origin of volcanic activity is related to sinking and subduction of the crustal plates into mantle, followed by melting of crustal material and then its rise to the surface through fissur and faults. The Barren Island volcano is located in the northern sector of the Andaman–Indonesia Island Arc. (b) Volcanic activities are also related to the splitting of crustal plates, such as the Java-Andaman trench. (Modified after W.M. Marsh and J. Dozier, 1981. *Landscape*, Addison-Wesle Reading-637 p., Fig. 23.12.)

material called ***lava*** forms—depending on the viscosity localized cones (e.g. Sumatran volcanoes)—broad dome-shaped shields (Hawaiian volcanoes like Mauna Loa) or spreads far such as in the case of the Deccan Volcanics of Maharashtra and Malwa.

The mobile lavas spread over gently sloping ground, flow down the valleys and low-lying areas b their speed is slow enough to allow threatened people to get out of their paths. However, if the basaltic lavas emerge beneath the water, the latter would flash into steam and make the eruption more explosi than if on land. This happened in 1963 off Iceland (R. S. Yeats, per. comm., 1985). The volcanic roc thus produced are hyaloclastite, which are very common on the ocean floor.

The second type of volcanic activity is explosive in nature, entailing the generation and emplacement of fragmented material (pyroclastics)—blocks of bubbling pumice, granules of cinder, sand-sized lapilli d very fine-grained ash — over wide areas around the orifice. The violence of an explosion may blow off the whole top and disperse the fragment and ash hundreds of kilometres around, as happened in 1883 when the Krakatau (one of the Indonesian islands) blew off its head. The pyroclastics form steep-sided cinder cones, such as Mount Vesuvius in Italy. The hot pyroclastics blow off the vent and •ve down the sides of the volcanoes as a fiery mass. This hot moving mass engulfs and traps air, heats it, and consequently becomes an increasingly hot, mobile and **swift-flowing avalanche called *nuees ardentes* (Fig. 15.5). These 'glowing avalanches' moving with a speed of the order of 50 to 100 km/ and having a temperature of several hundred degrees centigrade, are terribly destructive**. They become so intensely hot that fragmental materials fuse and get welded together into what is called the ignimbrite or welded tuff.

The ancient city of Pompeii in Italy with its population of 16,000 was buried 4 metres deep under e fine pyroclastics when Mount Vesuvius sent down waves of glowing avalanches in A.D. 79. Mount lee destroyed the city of Saint Pierre on the island Martinique in 1904.

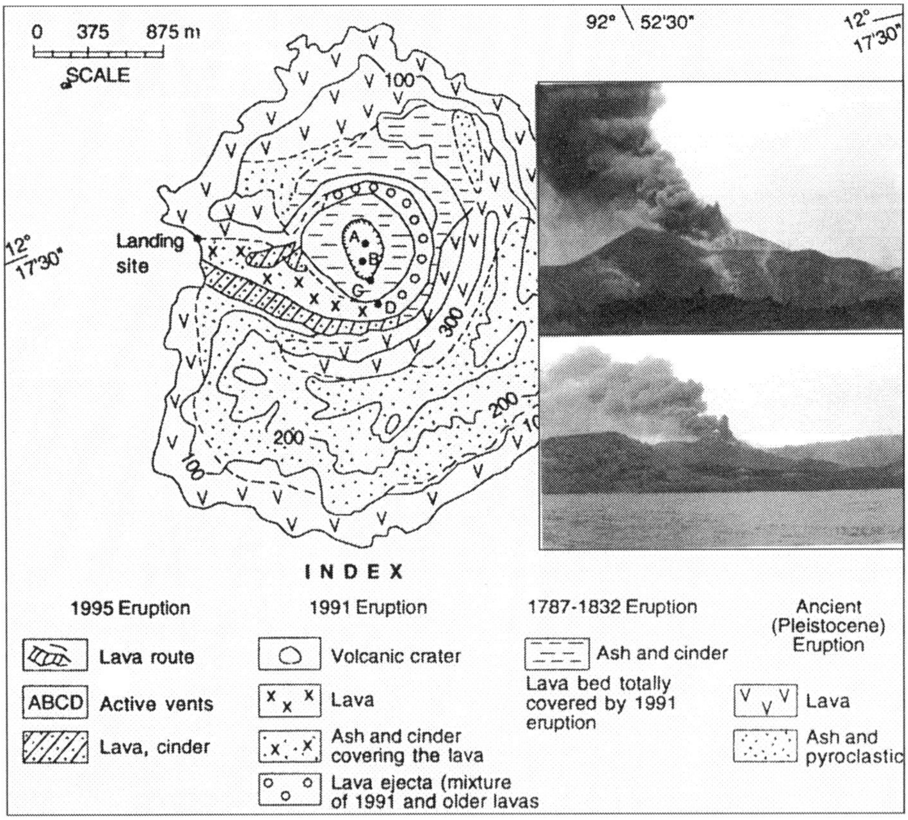

Fig. 15.4. Geological map of the Barren Island, showing the emplacement of lavas, ashes and pyroclastics. (From Haldar and Luhr, 2003) Inset shows the crater — an elongate depression with a rim characterized by saddles. (From Bandopadhyay et al., 2006). (See colour Fig. 15.4)

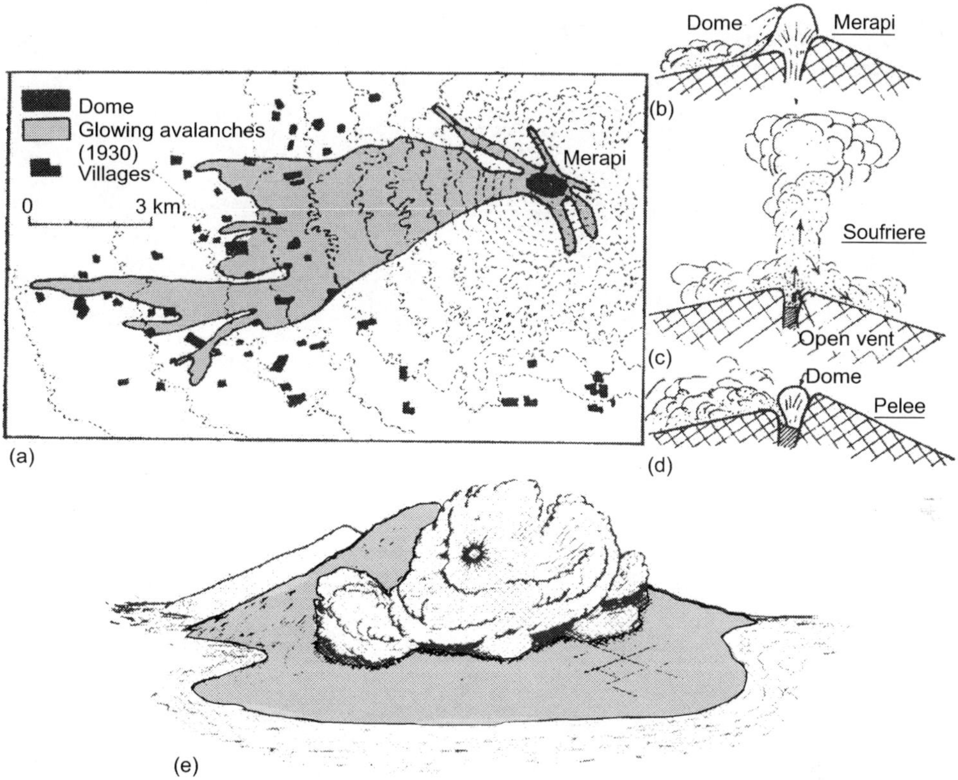

Fig. 15.5. (a) The spread of the glowing avalanches (*nuees ardentes*) from the crater of Merapi volcano Indonesia. The different types of hot material flow rapidly down the slope as a mass of burni gases and volcanic ashes. (After G. A. Macdonald in B. A. Bolt, H. L. Horn, G. A. Macdonald and R. F. Scott, 1975, *Geological Hazards*, Springer, New York, 328 p., Figs. 2.14 and 2.13.) (b) sketch of the glowing avalanches. (After W. M. Marsh and J. Dozier, 1981, *Landscape*, Addison Wesley, Reading, 637 p., Fig. 24.16.).

NATURE OF VOLCANIC HAZARDS

Burning and Incendiarism

The glowing avalanches (***nuees ardentes***) are by far the most hazardous of all natural phenomena. Everything that comes in their paths meets with complete destruction. Forests and settlements are la waste.

Mud-Flooding

Explosive volcanism is often accompanied by heavy rains. The rain water mixes with the abundant a and finer pyroclasts and gives rise to **volcanic mudflows** or ***lahars***. The lahars move down the slope

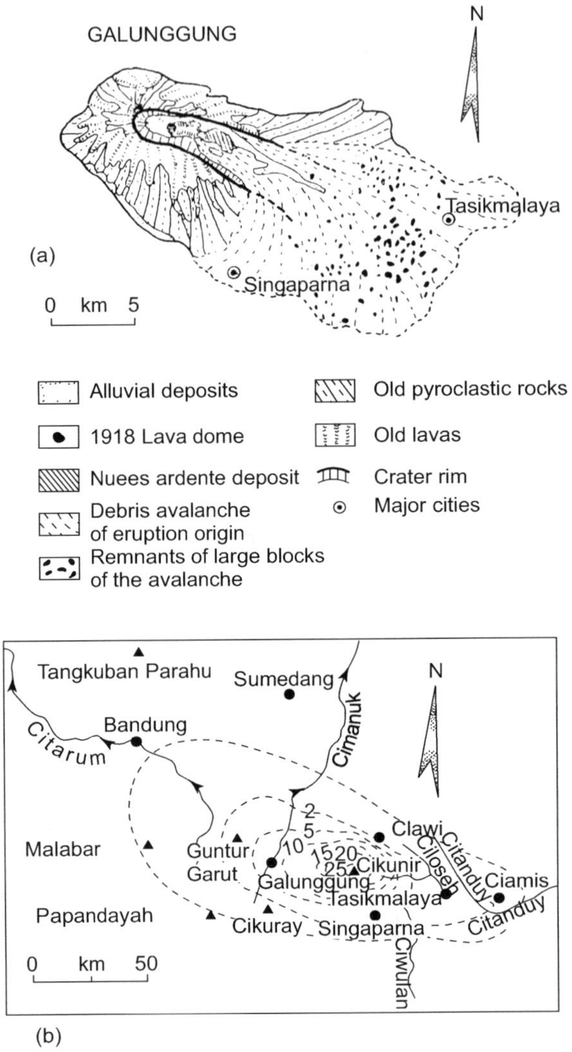

Fig. 15.6. The Galunggung volcano in western Java (Indonesia) was active from April 1982 to August 1983. More than 10,000 km^2 area was blanketed with a one cm or more thick mass of pyroclastics amounting to more than 300 million cubic metres. The isopach map shows accumulation of the volcanic ejectments as on September 14, 1982. The contours show thickness in centimetres. Circles represent main cities, and triangles the volcanoes. (After A. Sudradjat and R. I. Tilling, 1982, *Episodes,* 7:13-19).

uch faster than flood water, their speed being in the order of 90 to 100 km/hr and spreading to 100 to 0 km from the source. The damage to agriculture and settlements is not only extensive, but also very severe, as discernible time and again in the islands of Java, Sumatra and the Philippines. The explosive uption of Galunggung (April 1982 to August 1983) not only covered more than 10,000 km^2 area with one centimetre or more thick pyroclastics (300 million m^3), and spread ash over as distant areas as Bandung and Jakarta, but also caused great destruction of structures and agricultural fields, particular in the upper reaches of the Cibanjeran and Cikunir rivers, flattening all trees within one kilometre the crater (Fig. 15.6) (Sudradjat and Tilling, 1984). The threat persisted even after the waning of t eruption, as rain-induced cold lahars during the rainy season flooded the countryside with 120 milli cubic metres of the ejecta dumped by *nuees ardentes* in the catchment. The eruption thus touched the lives of more than half a million people—uprooted 35,000 from their houses, rendered unproducti over 94,000 ha of cropland and threatened 87,000 ha of agricultural fields with cold lahars.

Contamination

In the Galunggung event ponds for fishery were destroyed, pastures and poultry farms suffered great and fresh-water sources were heavily contaminated and distribution systems interrupted. About 3 million people and animals living within a radius of 150 km of the volcanoes were exposed to respirato diseases caused by prolonged and/or heavy exposure to ash and gases (Sudradjat and Tilling, 1984).

Thus, hazards are posed by three major processes—contamination of the air and the wat supply in the catchment by volcanic ashes, destruction of pastures rendering forages unpalatab to livestock, and destruction of agricultural fields and structures.

PREDICTION OF VOLCANIC ERUPTIONS

Signs of Awakening of Volcanoes

There are several precursors to resumption of volcanic activity—or awakening of a volcano—which may be taken as the basis for forecasting, warning and preparing for coping with hazards. The study of past history of volcanic activities in the region in conjunction with an analysis of various signs helps predicting the location, periodicity and nature of the impending event.

Microseismicity

The pressure of rising magma and compressed gases not only causes reopening of old fissures a fractures, but may give rise to new ones. Seismicity is the manifestation of fracturing and attendant movements. **Moderate to small earthquakes precede volcanic activity by hours, days, weeks even years**. In the 1977-1982 activity of the Usu Volcano in Japan, the precursory microseismic swarm occurred for 30 hours, their magnitude increasing progressively until the central part of the summ crater upheaved and the northeastern rim overthrust 180 metres northwards (Yokoyama, 1984). In som cases the seismicity considerably or almost completely diminishes when the eruption begins. And in other cases the ground shaking may continue during or after the event. **The seismicity pattern th varies from place to place and from time to time**.

Ground Deformation

Associated with seismicity is the structural deformation of the area as manifest in bulging u or doming of the ground surface. The tiltmeters placed in arrays close to the centre of the activity, efficiently indicate changes in the slope of the surface of the volcano, and the magnitude of rate deformation. **In the case of the Usu (Japan), the doming was evident as early as five months before e main event. Saint Helens (U.S.A.) began to dome up about six weeks before the May 18, 1980 uption** (R. S. Yeats, per. comm., 1985). The summit of the Kilauea Volcano in the Hawaii islands g. 15.2) was uplifted 0.9 m in the period 1919–1921, but then sank down 1.1 m after the magma amber was depleted of its content during the 1924 eruption (Bolt et al., 1975).

For monitoring activities of the Ganglunggung volcano, distance measurements on several lines was ne, using laser-ranging instruments and permanent retro-reflector targets. This setup monitors inflation deflation of the volcano—in the order of 20 mm for 4 to 5 km length. Some eruptions were predicted by measuring contraction of the lines. Ground deformation is 2 to 3 times faster than preceding tectonic earthquakes or accompanying tectonic deformation.

Rise in Heatflow

e rise in molten material and heated gases is manifest in higher heatflow in the volcanic areas. **riodic infrared photography obtained by aircrafts provide very dependable indications coming volcanic events as realized in the case of the May 1980 eruption of Saint Helens in regon (U.S.A.). Photographs taken before the dawn of 18th May indicated great increase in heat oduction which, in retrospect, was a clue to the great eruption a few hours later** (R. S. Yeats, per. comm., 1985). The temperature of spring water steadily increases, the fumaroles (which emit hot volatiles) become more active and become hotter, and the crater-lake water may boil as the temperature es by 11°C or more as seen in some Japanese craters. However, the Indonesian craters showed no perceptible rise in the temperature of water. (The temperature rise in itself cannot be taken as a reliable basis for predicting volcanic eruption, but only as a useful parameter for alerting observers).

Gaseous Emission

here was marked increase in SO_2 emission before and during many (but not all) eruptions of the alunggung volcano which then declined from 500 to 1200 tonnes/day in August to 300 t/d in October (Sudradjat and Tilling, 1984). The ***fumaroles*** emit many gases including CO_2, CO, SO_2, SO_3, H_2S, HCl, H—all very harmful to life. Gaseous hot springs called ***solfataras*** give off sulphurous gases (SO_2, SO_3) and the ***mofettas*** (the *fumaroles* emitting CO_2 and CO), which combining with atmospheric water give rise to acid rains. These can be easily monitored. As magmatic gases rise from the depth of the crust to e vent, their composition undergoes changes, as is evident from the fumeroles. This is reflected in the •dification of the composition of the groundwater held in aquifers. Thus the water of springs and wells would evince marked compositional changes. In some of the Japanese fumeroles, there was increase the proportion of HCl, SO_2, sulphur and chloride just before the eruption. But this kind of change is certainly not a rule. The SO_2 emission suggests that gases are coming directly from the magma without tensive interaction with groundwater (e.g. Saint Helens in 1980). The H_2S emissions occur when the groundwater interaction with SO_2 is enough to convert it into sulphuric acid.

Electromagnetic Changes

The gravimetric and electromagnetic properties of the ground show perceptible change prior to the volcanic eruption as demonstrated in the investigation carried out in the volcanic terrains of Kamchatka Russia particularly before the 1966 eruption.

Animal Behaviour

Animals become extraordinarily agitated as early as three to four days before the beginning the eruption. Dogs run around excitedly, dig holes and sniff, and cattle move away from the sı weeks before the onset of volcanism. The strange behaviours of animals can be taken as precurs indications of volcanic activity.

MITIGATION OF VOLCANIC HAZARD

Keystones of Mitigation Programme

The keystones in the programme of coping with volcanic hazards are:

(i) Assessment of thc activc volcanic areas on the basis of past history, present tectonic setting and prevalent geophysical conditions and natural phenomena, such as springs, fumeroles, etc.;
(ii) Monitoring precursor indicators to identify volcanic activity; and
(iii) Hazard preparedness programme including civil defence measures and practical steps to be taken against the hazard by different agencies and organizations.

Prediction and Warning

An analysis of the long history of volcanic and associated activities should allow experts to recognize the pattern and periodicity of expected volcanism. **The prediction cannot be based on one or tw precursory indicators, for one sign alone is never reliable**. Several different monitoring methods are correlated—seismic surveillance, ground-deformation, gas emission, aircraft-based remote geothermic surveys, analysis of animal behaviours, etc., in order to forecast the events, as was done successfully in the case of the 1980 eruption of Saint Helens.

If evacuation is not feasible, the following measures of combating the hazards can be taken:

Diversion of Volcanic Flows and Glowing Avalanches

The streams of lavas and lahars are best diverted, and certainly not stopped, by constructing walls of requisite height, width and strength, set diagonal to the slope on topographically strategic positions. Broad-based and weighty, these barriers are constructed of loose stone blocks. They are known to have successfully withstood the thrust of lava flows.

Digging channels can also divert the lava and lahar flows. To minimize damage from lahars, dyk are constructed along river banks. The dykes may be constructed of piled boulders wrapped around by wire mesh. Sandpockets are placed to trap loose finer-size material and check dams built to preve sliding of larger blocks.

The smaller glowing avalanches filling of valley courses can be diverted by building high barrı walls, although there will always be spillover and overtopping. Since *nuees ardentes* move by turbulent flow and can cross over surfaces of considerable relief, as seen at Saint Helens, this measure common may not be effective. Another strategy for containing glowing avalanches and lava floods and lahars to allow them to fill the reservoirs behind dams.

Spraying and Bombing Lava Flows

raying water on the advancing lava flows has helped their quicker cooling and congealing as done in awaii and Iceland. Bombing can also halt movement of the lavas flowing through tunnels and tubes. e bombing breaks and shatters the tunnel roofs, impedes the flow, and releases dissolved gases, ndering the lavas more viscous, and eventually halts their advance (D. R. Crandell and H. H. Baldron in Tank, 1953). Bombing also breaks natural levees of the surface channels and thus leads to spreading and eventual congealing of the lavas. There is one snag in this method. The gaseous and ash clouds may be too thick to allow accurate hit at the target.

Combating Gaseous Hazards

Acid gases are very harmful to both plants and animals. The acid rain destroys foliage, seedlings and nder plants. Particularly dangerous are CO_2, CO and the gases which cause bleaching of foliage. Being heavy, these gases accumulate at lower levels as pools in valleys, and thus form pockets of death. The various ways of protecting oneself from gases include wearing gas-masks and covering the face with cloth moistened with vinegar, urine or water.

SUMMARY

ncommon though extremely destructive natural phenomena, the volcanic activities do not pose hazard to any part of the **continental** India. The volcanic spectacle witnessed occurred in the oceanic part India occurred in 1803, 1991, 1994, 1995 and 2005 in the tiny Barren Island east of the Andaman chain. The melting the crust sinking into the mantle in the subduction zone (represented by oceanic trenches) and the impingement of rising convection plumes against the crust causes melting of rocks d formation of lavas and gases. When a fault, fissure or some kind of opening taps the molten mass, volcanic activities take place. These activities are restricted to the plate junctions such as the coastal belt dling the Pacific Ocean or where the crustal plate has split up, such as the East African rift valley. The Neogene-Quaternary volcanic belt through the central Myanmar basin, the Tsangpo-Sindhu valleys north of the Himalaya, and Balochistan, denote an ancient plate junction.

Volcanic activity is manifest in quiet emplacements of hot mobile lavas that flow down the valleys or spread on the terrain. It may be a violent explosion, ejecting bigger and smaller fragments of rocks d ashes and gases. The blown off fine material (ash) engulfs the air, heats it greatly, and the red hot ass moves down the slope with terrific speed as glowing avalanche of very destructive nature. The companying rain may convert the volcanic ash and finer ejectments into mudflows called lahars. **hile the glowing avalanches burn down forests and settlements, the lahars wipe away whatever mes in their way and cover fields and pastures with volcanic mud. The gaseous emissions create spiratory problems and pollution, and the ashfalls contaminate sources and supplies of water.**

There are several precursory indicators of the awakening of volcanoes—high degrees of croseismicity, bulging up or doming of ground surface, higher heat flow (manifest also in the rise of temperature of the water of springs and wells), change in the composition of groundwater due incoming gases such as HCl, SO_2, Cl, CO_2, CO, etc., variation in geomagnetic fields, and erratic behaviour of animals.

The study of the past history of volcanic activities in the region helps predicting location, periodicity and the nature of impending volcanism. Engineering measures to cope with the hazards inclu construction of heavy, robust and strong walls and channels to divert the flow of lavas, glowi avalanches and diagonally lahars. Spraying water and bombing lava tunnels and tubes slow down the speed and eventually halt the movement. Lavas, lahars and glowing avalanches can also be contained by impounding them in reservoirs behind dams in valleys.

16 Scope of Engineering Geology and Geotechnics

ROLE OF GEOLOGISTS IN ENGINEERING

The development of various utility structures and facilities useful to the society requires planning, designing and construction by engineers. These structures stand on the surface of and emplaced within the earth; and all kinds of earth materials are used in their making. **The application of the knowledge of geology to understand various natural conditions that affect safety, soundness, capability and economy of engineering works is called *engineering geology*.** The location, design and size of engineering structures are governed by lithological and structural conditions, particularly the geometry and the dimension of structural discontinuities and the nature, the thickness and the disposition of fillings. Their safety and longevity depends on geodynamic conditions such as seismicity, movements of groundwater and surface water, bearing- and shear-strength of the soil, etc. **An important consideration relevant is information on whether or not the area for engineering works is prone to natural hazards; and if it is, what would be the magnitude of risk and the frequency of occurrence of the anticipated hazard**. In hazard-prone zones, many geological parameters need independent investigation. For instance, for a seismic hazard the magnitude and frequency of damaging ground-waves depend on whether the underlying material is massive, layered, fractured or fragmented rocks, or consolidated or semi-consolidated sediment, or dry or wet soil.

The success of engineering works depends on their compatibility with their geological environment, and on an understanding of the geological processes that would affect them. This understanding leads to the formulation of safety measures. Requisite geological information is us indispensable to any cost-benefit analysis. The inadequacy of information or failure to interpret it correctly would not only cause delay and escalation of cost, but also lead to faulty performance or even failure. In the Chhibro-Khodri tunnel in the Tons Valley (western Uttarakhand), for example, the highly squeezed and shattered masses of rocks of the fault zone caused a delay of more than six years, mainly because of inadequate realization from surface mapping of the complicated geological conditions of the area (Chaplow, 1983) characterized by the imbricate or schuppen structure between the tectonically active Main Boundary Thrust and Nahan Thrust (Valdiya, 1980, 1984, 1998, 2010).

An additional and **very crucial dimension of the engineering geologist's assignment is the assessment of the impact on land, soil, water and other ecological systems of construction activities related to various engineering projects**. Interference with the ecological system by engineering works could bring about changes that could adversely affect the stability and integrity of the area. **There is thus an imperative need for prudent application of geological principles through multilayered checks and controls, and multiple options instead of the single-option approach of engineers**.

PARADIGM OF GEOTECHNICAL INVESTIGATIONS

Multi-phase Investigation

Geotechnical investigation is a multi-phase programme (Table 16.1). A preliminary study is carried out to provide a broad picture of the geological setting and the generalized range of problems, including the hazards likely to be encountered. In the case of projects for water-resource development, collection of hydrogeological-hydrological data is done at the preliminary stage.

Inseparably linked with the reconnaissance investigation is the environmental impact assessment, embracing an analysis of the consequences that would follow from various engineering activities. The appraisal of adverse effects on the ecological balance, on the society and on the psychology of the people to be affected would be needed in order to **seek alternative options on locations, layouts and designs of engineering structures**.

The feasibility investigation involves a detailed study of all geological and geophysical parameters, particularly the stability of the foundation and availability of construction material. In-depth studies are required not only to find a suitable site, the precise location of the proposed structure and select appropriate design, but also to work out the schedule of the project programme, keeping in mind the geological constraints.

The construction-stage investigation encompasses continual monitoring of changes of the behaviour of the ground and its response to the loads and other stresses generated as a result of engineering activities. A study of the features unearthed as a result of excavation and drilling, determination of bearing strength of the foundation rocks and soil, probing the interior through adits, shafts and intensive drilling, geomorphological study for identification of risky spots prone to natural or induced hazard and measurements of various geophysical parameters are carried out in order to formulate strategy to strengthen the weaker spots, or find out the need, if any, for readjustment and/or modifications of t proposed project. **This is the stage of the project programme at which the efforts of engineers are coordinated with the studies carried out by geologists in order to reduce delay, ensure stability of the structure and minimize damage to the ecosystem**.

According to Joshi (1981), a record of 140 major dams in India, showed a weighted mean period of 10 years of delay (the range being 4 to 46 years). Only six dams were completed within the acceptable period of six years—Koyna (1956 to 1961), Gandhi Sagar (1954 to 1961), Upper Bhavani (1959 to 1965), Matatila (1953 to 1958), Talkable (1958 to 1964) and Menahedi (1973 to 1978). To give a very few specific example, the study of the Jarlapur reservoir of the Machkud hydro-electric project indicat that the rate of sedimentation at 0.76 million m^3/yr was much faster than the assumed rate of 0.17 milli m^3/yr, so that the dead storage (7.1 million m^3) would be wiped out in just nine years. When the dam was 30 feet high, a decision was taken to raise its height by another 10 feet in order to prolong the life of the reservoir to 100 years (Swamy, 1981). Likewise, the alignment and layout of the Ukai Dam in the Tapi valley were significantly changed when a shear zone was detected in the foundation. The revelati that the bed of the Yamuna river is filled with 30-m thick sediment—instead of the estimated 10 m—Vyasi, necessitated not only a change of design of the dam but also caused a considerable delay in the implementation of the project (Framji and Krishna, 1981). The Supa Dam in Karnataka was redesigned considerably when a 24 to 50 m thick zone of weathered granite was discovered (Ramachandran and Ghosh, 1981). It was again in the construction-stage investigation that buried channels were discovered in the Bargi and Mahanadi rivers, necessitating construction of concrete diaphragms to stop underground seepage of water.

Table 16.1 Methodology for Site Investigation for Water-Resource Development Projects (After R. Chaplow, 1981)

Phase of Investigation	Purpose	Geotechnical Investigation		Approximate Duration	Anticipated Output and Result
		Location	Operations		
(1) Preliminary Preliminary Study	(a) Identification of major geological hazards	Regional Studies	(a) Review of existing data including maps, aerial photographs, reports.	5-8 months	(a) Identification of possible project for further study.
	(b) Identification of possible project(s)		(b) Site reconnaissance.		(b) Identification of general conditions and specially main potential geotechnical problems.
			(c) Commencement of geological mapping.		(c) Designing detailed site investigations and preparation of contract documents.
			(d) Limited subsurface investigations.		
		Award of Main Site Investigation Contract			
(2) Feasibility Study	(a) Confirmation of feasibility of project	(a) Regional studies, progressively concentrating on local areas of alternative projects, particularly areas of major geotechnical significance.	(a) Continuation of geological mapping.	9-12 months, but actual time depends upon geotechnical complexity of site and amount of resources available for investigation.	(a) Identification of preferred project for development.
	(b) Selection of general arrangement of component parts of single preferred project.		(b) Subsurface site investigation — geophysical surveys — boreholes — exploratory excavations— in-situ testing —laboratory testing—instrumentation		(b) Confirmation of technical feasibility of construction.

Contd

Phase of Investigation	Purpose	Geotechnical Investigation		Approximate Duration	Anticipated Output and Result
		Location	Operations		
			(c) Commencement of special studies (e.g. seismicity-monitoring)		(c) Providing geotechnical data sufficient to permit production of preliminary designs of component parts of project,
					(d) Identification of geotechnical aspects requiring detailed study and modifying investigation requirements accordingly.
(3) Design Study	(a) Producing predictive geotechnical model of site conditions of sufficient reliability.	Site-specific studies of locations of structures and areas likely to affect performance of the project or slope stabilizations.	(a) Completion of geological mapping	20-30 months minimum from commencement of preliminary studies.	(a) Providing geotechnical data and parameters for detailed project design and preparation of construction control documents,
	(b) Selecting geotechnical design parameters.		(b) Concluding sub-surface site investigation contract		(b) Identification of areas of geotechnical uncertainty and design. Further investigation for completion during initial stage of construction.
			(c) Continuation or conclusion of special studies (e.g. seismicity monitoring)		
Commencement of Main Construction Operations					
(4) Confirmatory Study	(a) Resolving remaining geotechnical uncertainties,	Site-specific studies in areas of geotechnical uncertainties.	As necessary to resolve uncertainties.	First 3-6 months of construction.	Confirmation of modification of geotechnical design assumptions.

Contd

Phase of Investigation	Purpose	Geotechnical Investigation		Approximate Duration	Anticipated Output and Result
		Location	Operations		
	(b) Confirmation of validity of predictive geotechnical model of site conditions.				
(5) Construction Investigation	(a) Monitoring and recording geotechnical conditions encountered during construction.	Site-specific studies in all excavations and in areas around site to monitor response of the ground to construction operations.	(a) Geological logging of all excavations.	Extending throughout the construction period but generally concentrated in early stages when excavations are done and ground treatment operations in progress.	(a) Confirmation or modification of geotechnical design assumptions.
	(b) Investigation of particular areas to clarify ground conditions.		(b) Other investigations as found necessary during construction.		(b) Production of records of as found ground conditions.
	(c) Recommending modifications to design or construction procedures in response to as-found ground conditions.				
Completion of Construction and Commencement of Operation of Project					
(6) Operation Monitoring	(a) Monitoring performance of the operating project.	Site-specific studies.	(a) Recording and interpreting the result of project instrumentation.	Throughout life of project on a regular basis but particularly during first five years of operation	(a) Definition of behaviour of ground in response to operation of project.
			(b) Visually monitoring behaviour of project area, specially slopes.		(b) Recommending any required remedial work.

Demarcation of Hazard-Prone Areas

entification and delimitation of potential hazard zones through interpretation of aerial photographs and ground-mapping are important aspects of geotechnical investigations. In the Himalayan and the Sahyadri regions, both have irregularly developed soils and are characterized by neotectonic activities causing weakening of the structural framework. Hazardous events like mass-movement and earthquakes place considerable constraints on the type and design of engineering structures. Terrain evaluation is thus an indispensable exercise in investigating the susceptibility of slopes to failure as witnessed in the valleys of the Tista in Sikkim-Darjiling and the Giri and Beas in Himachal Pradesh. At the Pong Dam site in the Beas valley, intake structures for the tunnel had to be introduced on the left-abutment hillslope. In the Giri-Bata project site the powerhouse had to be constructed underground in view of endemic slope failures and massive landslides. The cavernous nature of the Kalhel limestone in the valley of Siul (a tributary of the Ravi) led to abandonment of the project for fear of extensive seepage of impounded water. In the Kamthikheri (Pench valley) project in central India, the cavernous marble foundation had to be sealed by grouting to check seepage and failure (Ramachandran and Ghosh, 1981).

The Obra Dam on the Rihand River in Uttar Pradesh, the 42-m Kamthikhere Dam on the Pench River in the Nagpur district (Maharashtra) and the 73-m Bajaj Sagar Dam on the Mahi River in Gujarat required very special treatment (Sundaraiya and Wakpanjar, 1996). A concrete diaphragm of 0.6 m thick and 3 m apart across the entire thickness of the foundation rock put deep into the underlying beds, and grouting of the weakened roads, buildings of a grout curtain along the central line had to be done.

The need for in-depth geotechnical investigations for the alignment of roads and rail-lines is being increasingly felt. The unsoundness of road alignments, the recurrent landslides on roads, and the colossal damage to forests, hydrological system and agriculture in the mountains call for comprehensive pre-construction stage investigations.

ENVIRONMENTAL IMPACT ASSESSMENT

Multidisciplinary Team Work

An environmental impact assessment is an exercise designed to identify and predict effects of actions taken by engineers on the biological and geophysical environment and on man's health and well-being (Munn, 1977). This assessment is carried out at the same time as the geotechnical investigations. The assessments are, however, done at successive stages of the geotechnical investigations. **It is essential that the field investigation for impact evaluation be carried out by personnel independent of vest interests and free from bureaucratic pressures or constraints**. The study calls for involvement of competent workers specializing in such fields as engineering geology, hydrology, ecology and socı economics and for their collective team work.

Imperatives of Impact Evaluation

The imperatives of evaluation processes are the following:

1. Articulation on a scientific basis of the socio-economic significance of the project,
2. **Expression of commitment for sharing expected benefits between the communities adverse affected by the project (such as deprived of homes, farms, etc.) and the people for whom the project has been planned.**
3. Exposition of the full gamut of environmental changes that are likely to take place as a consequence of engineering activities for the project.
4. Indication of sociological, economic and cultural problems that the disruption due to engineering activities would generate to the traditional way of life of an established community.
5. **Suggestions about alternative possibilities on the basis of a comparison of effects of various projects**—that is, suggesting different ways of realizing the goal in a socially desirable, technically feasible and ecologically sound way.

6. **Appraisal of the financial burden of restoring the natural environment and resettling the displaced/affected people, including providing them with essential community services.**

Cost-Benefit Analysis

is assessment is indeed an exercise for fixing the cost of every possible impact on the environment, mputing the financial burden of remedial measures, determining prices of living and non-living resources that would be damaged or lost, attaching monetary values to such intangible fallouts as mental distress and trauma of the adversely affected people (e.g. the uprooted communities living upstream of the proposed dam or the people exposed to heavy pollution generated by a chemical plant). To recapitulate, the items under costs should include the following:

(i) Loss of all natural resources such as forests, pastures, wildlife, aquatic fauna, mineral deposits and soils.
(ii) Loss of yields of agricultural farms due to submergence resulting from waterlogging or alkalinity-salinity of over-irrigated fields.
(iii) Loss of fertility of aquatic fauna (from industrial pollution).
(iv) Exodus of wildlife and aquatic fauna.
(v) Instability of hillsides leading to mass-movements and attendant damages due to excavation for roads, canals, foundations and mining.
(vi) Cost of afforestation to compensate the loss of forests resulting from inundation, clear-felling, road-cutting, poaching, etc.

One of the most crucial requirements of this appraisal programme is seeking the viewpoints, values and environmental perceptions of the people who would be affected by the development projects. There should be a machinery for wide-ranging and periodic dialogue between the sponsors, the affected, the researchers, and the decision-makers. Such attempts at consultation with e affected people would minimize the psychological injury inflicted and the attendant unfortunate consequences. **The planners must see that the benefit goes to all—even to those who have been adversely affected.** ***Giving monetary compensation for the loss of land, home and other property is rtainly not a sharing of benefits of the project executed: it is merely buying peace, momentarily.***

DEVELOPMENT WITHOUT DESTRUCTION

The principal objective of all engineering works is to develop or augment facilities, provide energy and water, and build pathways of transportation and communication, so that the society harvests their nefits. If any activity reduces these benefits, or causes distress even to a sector of the community, it will be a counter-productive exercise. The methods of engineering work should therefore ensure development without causing destruction of the natural environment as well as the socio-cultural fabric of the society.

Leopold Aldo (1941) summarized the situation that prevailed in the U.S.A. The same situation prevails in India today; "The sober citizen who would never submit his watch or his motor to amateur tampering, freely submits his lakes to the draining, filling, dredging, pollution, abilization, mosquito control, algae control, swimmer's itch control and the planting of any fish able to swim. So also with rivers. We constrict them with levees and dams, and then flush them with edging, channelization and floods, and fill them with silts of bad farming. The willingness of the public to accept and pay for these contradictory tamperings with the natural order arises, I think, from at least three fallacies in thought. First, each of these tamperings is regarded as a separate project because it is carried out by a separate bureau or profession and expertly executed because its proponents are trained, **each in his narrow field. The public does not know that bureaus and professions may cancel one anoth** and that **expertness may cancel understanding**. Second, any construction mechanism is assumed to be superior to a natural one. Third, we perceive organic behaviour only in those organisms which we have built and do not realize that tampering with a part may effect the whole. We do not yet know that this is true of soils and water. **Thus, man too wise to tolerate hasty tinkering with our political constitution, accepts without qualm the most radical amendment to our biotic constitution."**

It is the responsibility of the engineering geologist to see that such tinkering with environmental order is prevented, or kept to the minimum.

17

Dams and Reservoirs: Geotechnical Considerations and Environmental Impact

PURPOSES OF DAMS

Multipurpose Dams

order to store water for irrigation, generation of electricity, and moderation of flood, commonly solid barriers of partly or wholly impervious material are constructed across rivers, preferably at places where the valleys are narrow. Most of the dams built serve more than one purpose. These are ***multi-purpose dams***. Of the 1554-odd major dams in India in 1979 only 17 had been identified as flood-control projects (Agrawal et al., 1982), for their principal objectives were irrigation and power-generation (Table 17.1).

best the dams can hold back flood water only upto a level. A ***diversion dam, barrage*** or *weir* serves to divert or channelize water from the river through canals. ***Check dams*** are constructed on tributaries nging sediments from degraded uplands, for controlling inflow of debris and sediments eroded from the catchment area.

Table 17.1. Major Dams of the Indian Subcontinent until the Beginning of 1980s

Dam	River (State)	Year of completion	Dam Height/ length (m)	Purpose I = Irrigation E = Electricity F = Flood	Gross Reservoir Capacity (million m^3)
Bhakra	Satluj (Himachal Pr.)	1963	226/518	IE	9621
Idukki	Periyar (Kerala)	1974	169/366	EF	1996
Koyna	Koyna (Maharashtra)	1963	104/853	EI	2796
Srisailam	Krishna (Andhra Pr.)	1982	143/512	E	8722
Nagarjun Sagar	Krishna (Andhra Pr.)	1965	125/4865	EIF	11550
Cheruthoni	Cheruthoni (Kerala)	1974	138/650	EF	1996
Sardar Sarovar	Narmada (Gujarat)		137/1210	IEF	9492
Pong	Beas (Haryana)	1978	133/1950	IE	8570
Hirakud	Mahanadi (Odisha)	1956	59/4800	IF	8105
Thein	Ravi (Himachal Pr.)		157/878	EI	2670
Mangla	Jhelam (Pakistan)	1967	138/3335	IEF	6358
Tarbela	Sindhu (Pakistan)	1976	143/2743	IEF	13690
Kalagarh	Ramganga (Uttarakhand)	1978	126/626	IF	2443

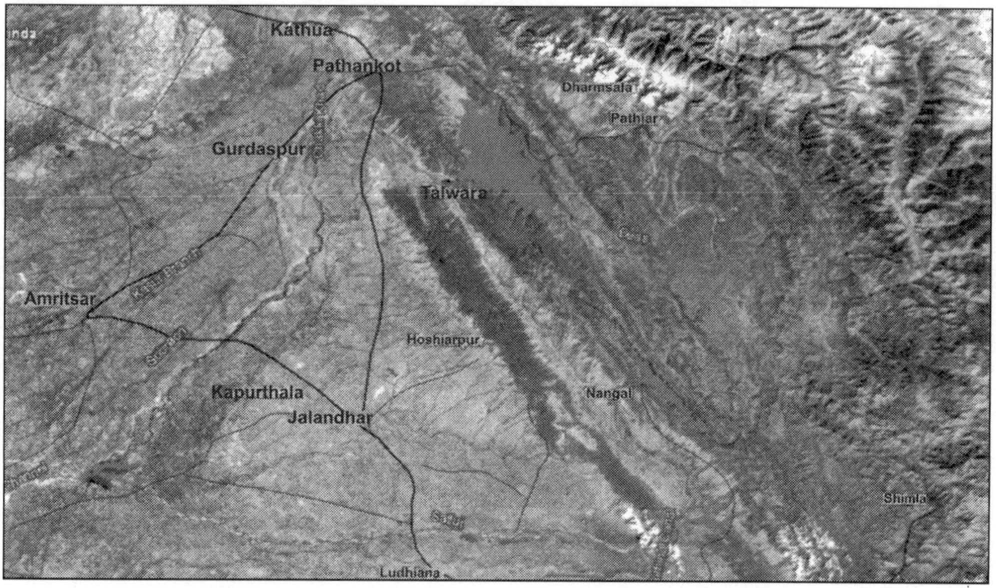

Fig. 17.1. Satellite photo of the reservoirs behind the Bhakra Dam on the Satluj River (right) and the Pong Dam on the Beas River (left) in Himachal Pradesh. (Photo *courtesy:* Dr J. Bhatt, Haldwani). (See colour Fig. 17.1)

Since a dam is an artificial creation designed to impound (Fig. 17.1) or divert water of a riv and change the hydrologic regime, its construction by large-scale excavation has a considerable impact on the environment. Keeping in view the geological and geomorphic conditions of the area, the choice of site of a dam is of crucial importance even as its type should be compatible with the geological condition and in harmony with natural environment.

Selection of Sites and Types

The primary requisites for selecting a site for a dam, apart from adequate stream discharge, are the presence of narrow course, preferably a gorge (Fig. 17.1) with suitable foundation conditions and the availability of construction material. The geological conditions of the foundation, the nature of construction material available and the seismicity of the area govern the choice of the type of dam.

If the valley is very narrow and the rocks exposed on the steeper valley walls and the river-bed are strong enough to take the transferred pressures, an ***arch dam*** can be built (Fig. 17.2c). Designed to withstand the pressure by arch action, the dam is composed of a single or multiple concrete wall, curved in plan with its convex face upstream. The pressure of the water in this situation is transferred to the abutment through each horizontal arch ring. The abutments must therefore be of very strong and immovable rocks. Only the weight of the structure is transferred to the base. Should the foundation rock be permeable, allowing seepage (which will initiate uplift action), an arch dam is the obvious choice. It may be emphasized that an arch dam can be constructed even at those sites where the foundation rocks are not very strong, provided that these are adequately treated for the control of leakage, and the abutments are unquestionably strong. An example of this type of dam is the 169-m high Idukki Dam across the narrow valley of the Periyar River in the Southern Sahyadri Range, Kerala.

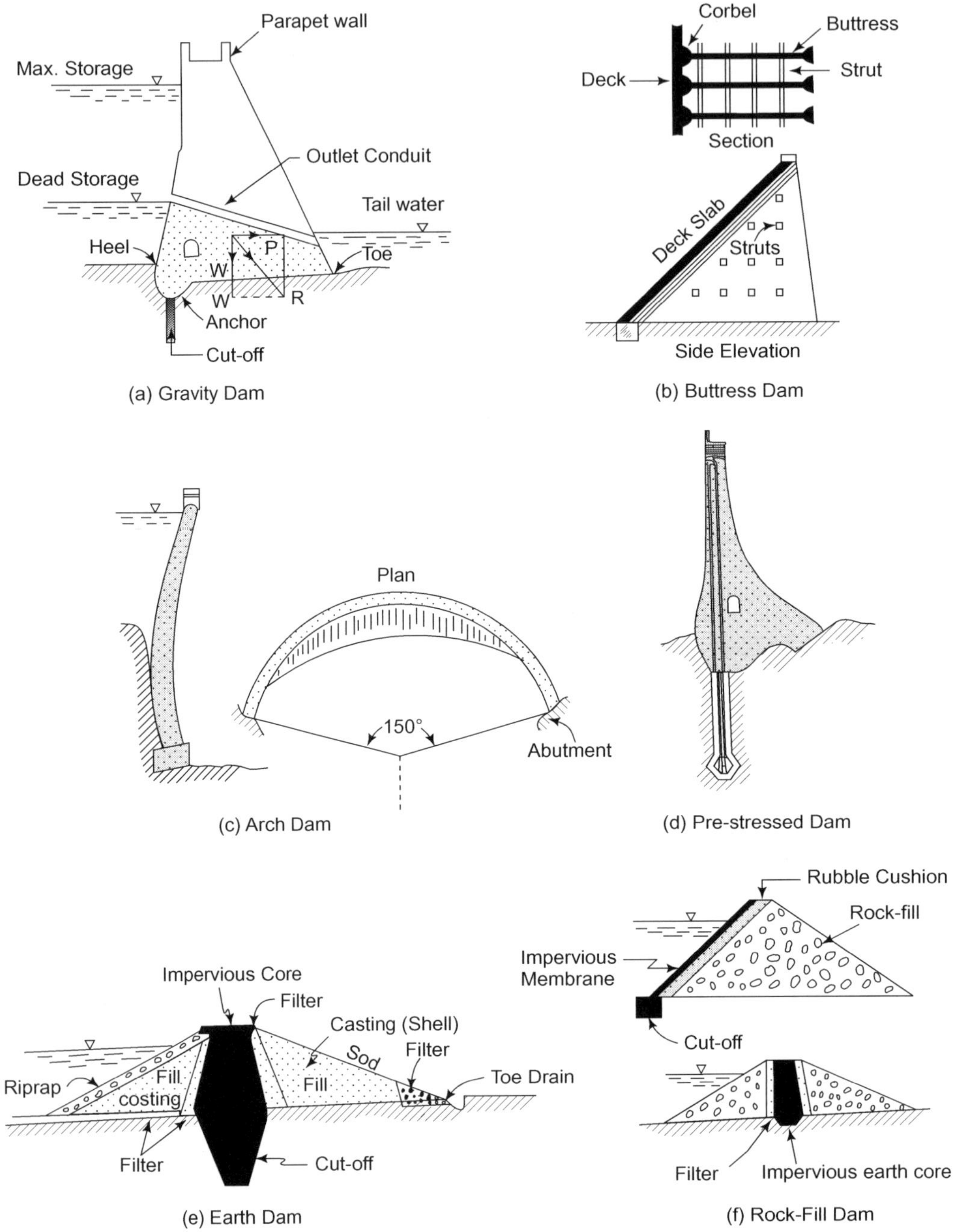

Fig. 17.2. Various types of dams. The cutoff (shown in black) is made up of impermeable mud and clay. (Based on D. P. Krynine and W. R. Judd, 1957, *Principles of Engineering Geology and Geotechnics*, McGraw-Hill, New York, 730 p., Figs. 15.1-15.2.

If the valley is wide and the foundation rocks have varying strengths, a ***buttress dam*** can be constructed. It consists of an upstream-sloping deck of reinforced concrete that bears the load of impounded water, and vertical buttresses aligned normal to the plane of this slab. The vertical buttresses transmit the load of the deck and the impounded water and accumulated sediment to the foundation (Fig. 17.2b). The pressure per unit area on the foundation rocks being very high, their shear strength should be high. The location of the buttress dam can be planned in such a way that it does not rest on weaker members.

Adapatable to all conditions and locations are the ***gravity dams*** which owe their stability to the weight of their sheer bulk. The height of the gravity dam depends on the shear strength and bearing capacity of the foundation rocks or material. If the valley is narrow and the foundation rocks strong, a masonry or concrete gravity dam is preferred (Fig. 17.2 a). The 226-m high Bhakra Dam across the Satluj River (Figs. 17.1, 17.6 and 17.7) in the Siwalik terrain, and the 104-m high Koyna Dam in the Deccan volcanic terrain (Figs. 17.12 b and 17.19) are straight gravity concrete dams, while the 125-m high Nagarjuna Sagar Dam and the 143-m Srisailam Dam (Fig. 17.10) on the Krishna River rest on the Archaean Peninsular Gneisses and Proterozoic quartzites. They are masonry gravity dams. Where the foundation cannot bear the load of the heavy structure of the dam or where there is alternation of weak and strong rocks in the foundation or where pressure of the impounded water and sediment, the shocks of earthquakes are high and if the valley is quite wide (>150 m) ***rockfill or earth dams*** (with hard clay cores) (Fig. 17.2 f) are constructed, as has been done in the Ramganga River near Kalagarh in the Siwalik terrain and the 261-m high Tehri Dam across the Bhagirathi River (Fig. 17.8) in the terrain of the Precambrian phyllites. In cases where the foundation is constituted of unconsolidated or poorly consolidated material of very weak shear strength and characterized by permeability such as valley-fi or alluvial accumulations, ***earth dams*** of height less than 30 m are suitable (Fig. 17.2e). The Farakka Dam on the Ganga in Bengal is an example. Resting on 35 m thick clay-silt and sand the Farakka, like all earth dams, has a wide base—five times its height—to ensure stability.

Some of the dams in wide valleys are composite in structures—partly masonry or concrete, and partly rockfill or earthen, as exemplified by the 59-m high and 26-km long Hirakud Dam on the Mahana River flowing through the Archaean granites, quartzites and slates. The 62-m high Bhavani Dam Tamil Nadu is founded on Archaean Gneisses, granulites and schists is a composite dam.

STABILITY AND FAILURE OF DAMS

Forces Acting on Dams

The stability of gravity dams, as already stated, is derived from their weight—the heavier the structure, the greater would be its stability. The vertical static forces (Fig. 17.3) acting downwards are (i) the weight of the dam structure and associated superimposed features (W) and (ii) the weights of water (W_w) and sediments (W_s) lying on the upstream side of the dam. Counter-acting the total weight is the buoyancy force or uplift action of the water, which is equal to the weight of the water displaced by the part of the dam that remains submerged. (It is equal to $1/2m \times H \times B$, where m = unit weight of water, H = depth of water in the reservoir, and B is base width of the dam. It acts at a distance $B/3$ from the upstream end). If the water manages to get under the dam in the foundation, the pore-pressure will act upwards to reduce the vertical downward force, and thus undermine the stability.

The impounded water exerts very strong horizontal pressure (P) on the dam (the magnitude being ½ mH^2 facting at $H/3$ distance from the base (Fig. 17.3). This horizontal pressure of the impounded water is reinforced by the forward thrust of the accumulated sediments (*Ps*). The upper portion of the dam is subjected to the impact of waves (*Pw*) generated when the wind blows over the reservoir water. Counter-acting the total forward thrust ($P + Ps + Pw$) is the pressure of the tail water (*Pt*) accumulated in front (toe) of the dam. Added to the horizontal destabilizing forces are the shocks generated by earthquakes which cause dynamic increase in the water pressure (Ohri and Ahuja, 1981).

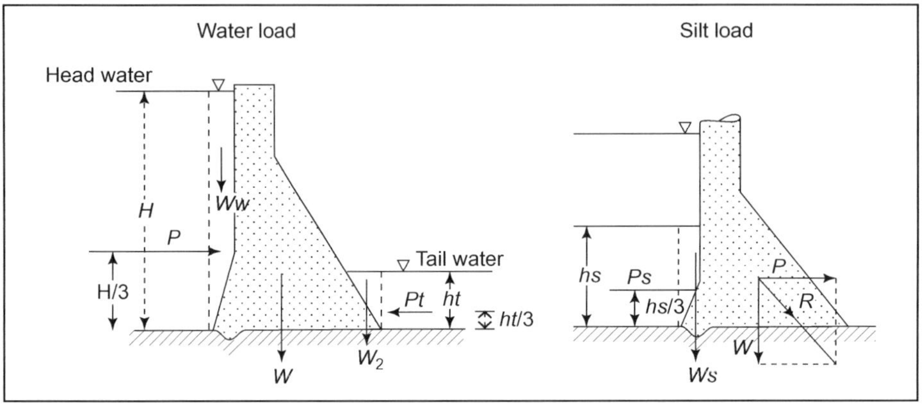

Fig. 17.3. Forces acting on a gravity dam. W = weight of dam; W_w = weight of water lying on the sloping face of the dam; W_s = Weight of sediment accumulated on the sloping face; H = depth of water impounded; hs = thickness of accumulated sediments; ht = column of tail water in front of the toe; P = lateral forward pressure of impounded water; Ps = forward thrust of accumulated sediments; R = resultant of total vertical and horizontal (lateral) forces. (After Krynine and Judd, 1957, Figs. 15.15,15.16).

To summarize, **the horizontal pressure, including thrust of waves and seismic shocks tend to stabilize the dam, while the vertical pressure resulting from weight imparts stability. So long as the resultant of the horizontal and vertical forces intersect the foundation at a distance *within the base of the dam*, it remains stable (Fig. 17.3). In other words, the smaller the ratio of the horizontal and vertical pressures the greater is the stability**. The value of this ratio should be less than the efficient of friction of the dam on its foundation rocks (0.7 to 0.6) or soils (0.5 to 0.4).

Causes and Mechanism of Dam Failure

ny force that reduces the magnitude of the vertical force acting downwards, such as the uplift tion of water which was able to infiltrate in the foundation material, or increasing the intensity the horizontal force such as wave actions, floods, and earthquakes, will weaken the stability of e dam and may cause failure. If the coefficient of friction between the dam and its foundation rocks decreases, such as happens when the latter becomes slippery due to the presence of montmorillonite ay and infiltrated water, the dam may slide along its base. Thus, the foundations of clays or fine silts and weathered rocks are likely to provide conditions conducive to dam failure. This is initially anifest in the formation of mud flows near the heel of the dam due to outward movement of softened, mobilized material. The strategy for preventing sliding is to increase the strength of the foundation material by taking a series of steps—roughening the surface and keying the structure to certain depths. The stabilization of dam is done by constructing a ***keywall* or *cutoff trench*** at the heel of the dam (Figs 17.2f and 17.4b). To prevent or minimize leakage, elastic (clay) **cut off walls, diaphragm or cores** are constructed beneath the dam structure. Such structures lengthen the path of seepage (Fig. 17.2 a, d, e, f; Fig. 17.4b). This has been done in the Pong and Kalagarh dams founded on the mudstone-sandstone alternation of the Siwalik. By giving upstream slope to the base of the dam (Fig. 17.5 a-d) and providing horizontal joints, the stability of the dam can be increased.

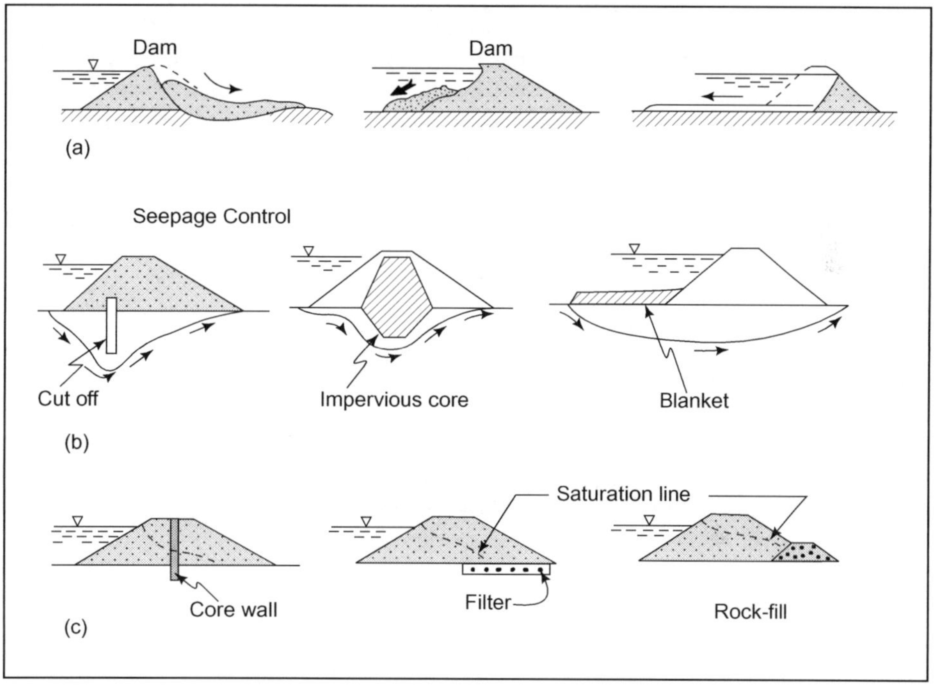

Fig. 17.4. (a) Mechanisms of failure of earth dams. (b) Methods of controlling seepage—cut-off walls, deep keying, impervious core and impervious aprons. (c) Construction of filter adjacent to heel a toe. (From D. P. Krynine and W. R. Judd, 1957. *Principles of Engineering Geology and Geotcchnic* McGraw-Hill, New York, 730 p., Figs. 16.22 and 16.26).

When the resultant force intersects the foundation outside the base of the dam, the dam topples or overturns. This can happen when the reservoir is roughed up by shock waves generated by avalanches, massive landslides or earthquakes. A disastrous flood emanated from the failure of the 168-m hi Vaiont Dam in Italy on September 2–3, 1963 due to a colossal landslide following incessant rain and an earthquake on the unstable slope of Mount Toe (Fig. 17.26).

The *uplift action* of water that manages to get under the dam through percolation is perhaps t most potent of all the destabilizing forces. The interstitial water under the base reduces the frictional resistance of the ground under the dam. When it gets linked to the reservoir water, the high hydrostatic pressure that develops is directly proportionate to the hydrostatic head of the reservoir water. During ods, the water level rises and the forces countering the downward directed vertical forces becomes very strong, leading to toppling of the dam. The 26-m high masonry arch dam on Tigra River (Madhya Pradesh) failed in 1919 due to sliding of the dam on its foundation resulting from uplift action of the water which had seeped under the dam through multiple joints in the sandstone-shale alternation dipping gently downstream. The breach was 396 m long, and the dam broke up into huge blocks.

Water gets under the base of the dam through pore spaces in sands and gravels (e.g. under Dehar Dam across the Beas River in the Upper Siwalik Boulder Beds), through caverns and cavities in carbonate rocks (e.g. Salal Dam on the Chenab River in the Jammu Limestone, and Kamthikheri Dam in Madhya Pradesh on the Sausar Limestone), via vesicles, pipes and lava-tunnels in the volcanics (e.g. many dams in the Deccan Plateau) and through open joints and fractures in igneous and metamorphic rocks. The Pandoh Dam on the Beas River located on the Chail phyllites was originally conceived as a concrete m. It had to be shifted from its original location and made a 76 m high rock-fill dam when it was found that the foundation rocks were highly jointed and fractured.

One-third of the failures of eathern dams have resulted from seepage (Fig 17.4). The failure of the great Teton Dam in Idaho (U.S.A.) on June 5, 1976 which damaged one million hectares of land d killed 1600 heads of cattle, is attributed to piping—an internal erosion of impervious fine-grained material of the core section near the right abutment (Flog, 1979*)*. Erosion of the upstream face of the earth dam at its toe is responsible for many a failure (Fig. 17.4). Rapid drawdown—lowering of the water level due to withdrawal—in reservoirs is tantamount to withdrawal of lateral support that imparts stability. This causes erosion and landslide on the hillslopes adjoining the reservoir and damage to the upstream face of the dam and embankment.

Overtopping of the dam—overflowing of water following rise in water level during_floods—is the major cause of the failure of a large number of dams in Rajasthan and Gujarat. The tragedy that overtook Morvi (Saurashtra) in 1978 was caused by overtopping and the resultant failure of the Machu Dam following excessively heavy and prolonged downpour. The failure of the Viswasmitri Dam due to overtopping caused extensive damage to the city of Vadodara. Nearly half the city of Pune was seriously damaged when the Panshet and Khadakwasla dams failed one after the other in 1963 following excessive rain which caused overtopping of the dams. Such examples are very common in the semi-arid region of western India, where unusually heavy rains have caused overtopping of dams.

GEOLOGICAL CONDITIONS OF FOUNDATION MATERIAL

Stability Factors

The stability of a dam depends on the shear strength of the foundation rocks and or soils. It is therefore imperative to obtain full information on the structure and composition of the foundation material through a variety of methodologies including seismic sounding, core drilling, water percolation tests for permeability-porosity, jacking tests for modulus of elasticity, bearing strength, etc. The foundation rocks and or soils should have enough shear strength to resist deformation on imposition of loads, and must be resistant to disintegration, decomposition and erosion. The strength of the rocks and or soils depends not only on composition and texture but also on their structure.

Since considerable drilling is done in the foundation, permeability tests of the foundation medium can be done. The permeability is expressed in terms of centimetre per second. It is commonly believed at for dams upto the height of 15 m, the foundation would be sufficiently watertight if its permeability does not exceed 30.4 cm/year.

Structure of Bedded Rocks

It is ideal to have foundation rocks of uniform physical strength. For, alternating hard and soft permeable and impervious rocks create problems of differential settling and leakage. The planes of bedding, foliation, cleavage and joint provide pathways to seeping water, as do porous and permeable layers in sedimentary succession.

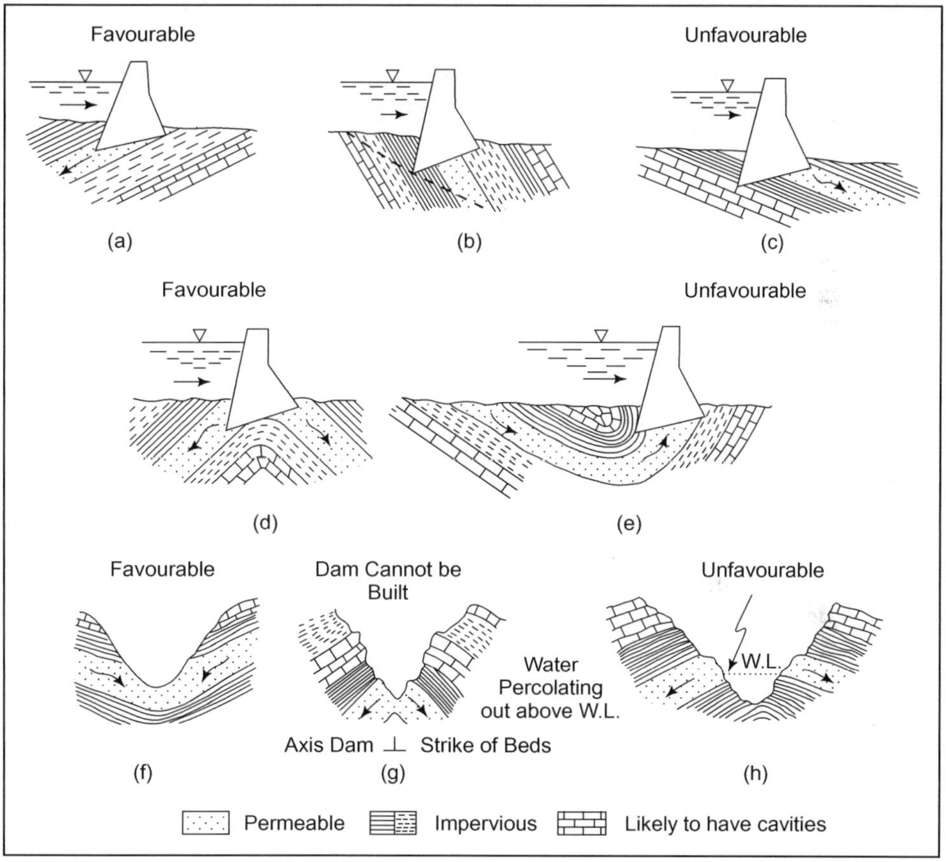

Fig. 17.5. Taking into consideration the fold structures and lithology (alternation of permeable and impervious horizons) the sites of dams are chosen. Since uplift action of the water that manages to get under the dam through seepage is the most potent of all destabilizing forces, the infiltration of water h to be stopped or made minimum by placing the dam in such a way that an impervious bed forms an apron at the heel.

It is always safer to align the dam, as far as possible, parallel to the strike of the beds, because an oblique orientation may bring in rocks of different physical characteristics into the foundation and cause problems of differential settlement or permit seepage through more permeable beds coming into the foundation. **The dam should be located in a manner that the impervious layers form aprons on**

the edges of the dam, particularly on its heel. This will prevent seepage (Figs. 17.4 and 17.5). Take the example of the Bhakra Dam. It is located on the alternating sandstone and mudstone succession of the Siwalik, and has aprons of mudstone both at its heel and toe —76 m wide at the toe (Figs. 17.6 and 17.7). Admittedly, this was not a desirable setting. Although the clay aprons did check leakage, they created more problems of differential foundation settlement. Very extensive treatment had to be resorted as pointed out elsewhere, involving deep excavation and backfilling with concrete.

The condition when the beds striking across the valley dip upstream is very favourable (Figs. 17.4 and 17.5). Downstream dipping steep beds provide comparatively better condition. But the gently downstream dipping beds provide unfavourable sites, for the beds induce the tendency to slide, particularly if clayey or shaly seams are present. The Bhakra Dam is located on the 50° to 70° downstream-pping beds (Figs. 17.1 and 17.6). The 133-m high, 1700-m long earth-fill Pong Dam is built on the gently upstream-dipping (10° to 13° NNE/N). Upper Siwalik is constituted of alternating sandstone and plastic mudstone seams, the latter having very low shear strength. The dam is positioned on the southwestern limb of the synclinally folded Middle and Upper Siwalik succession close to the active Satlitta Thrust dipping 20° to 30°N. Development of artesian conditions in the reservoir has been remedied through ief wells in the synclinal part. The 126-m high rockfill Kalagarh Dam on the Ramganga River in Uttarakhand is likewise located obliquely on the upstream dipping (50°) Middle Siwalik sandstone.

The synclinal crestal parts are less favoured than the anticlinal for the location of dams. In ch cases prevention of seepage and water-tightness can be achieved by placing an artificial apron or blanket of clays, cement, concrete, gunite or impervious mud on the upstream side of the dam, or by constructing cutoffs (Fig. 17.4b).

Shear Zones and Faults

ear zones are characterized by multiple fissures and splitting, crushing and granulation of rocks. **The ear-zone rocks thus have low bearing strength and may locally provide passage to infiltrating-eping water.** The Main Boundary Thrust zone in the outer Himalaya evinces these characteristics and problems although it has not come in the way of any of the dams in terms of foundations. **Faults ossing the valley do not pose much risk, unless they evince neotectonic activity and related smicity. It is risky to place the dam on the fault line or shear zone, especially if it crosses the ngth of the dam obliquely**.

The fault breccia may need special treatment against leakage by grouting. The crushed and sheared material should be excavated and replaced with clay or concrete. A grout curtain may also improve the condition. The 226-m high concrete gravity Bhakra Dam is located very close to a thrust (Fig. 17.6) and a crossfault. The faulting has caused repetition and elimination of the claystone band at several places. On either side of this middle claystone bed are shear zones and the one on the downstream side display higher degree of shearing (Ramachandran and Gangopadhyay, 1972). The shear zone material was cavated 45 m deep and filled with pre-cooled concrete. The foundation was strengthened by cement grouting down to a depth of 9 to 15 m. A 90-m deep grout curtain was provided near the heel, besides a 15 m thick concrete slab connecting the heel to the sandstone member (Fig. 17.7 inset).

In the Bhagirathi valley, the downstream-dipping Chandpur phyllites are characterized by narrow but numerous shear zones (Fig. 17.8), generally parallel to the foliation planes (Pant et al., 1982). The shear zones are 10 to 30 cm wide, but locally as much as 11 to 22 m (Navani et al., 1991). On this kind foundation is built the 261-m high rock-fill Tehri Dam. Considerable grouting and replacement of shear-zone material by concrete was done to strengthen the foundation.

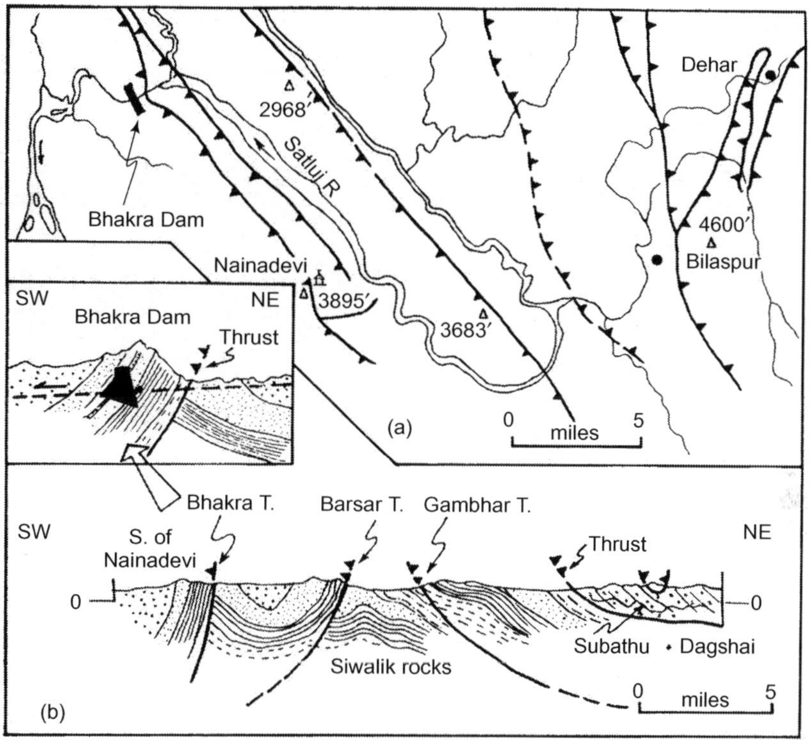

Fig. 17.6. Map and sections of a part of the Satluj basin where the Bhakra Dam is located. Note the location of the dam on the Nainadevi Ridge, south of the SW-dipping Bhakra Thrust which has placed the Lower Siwalik rocks over the Middle Siwalik succession. (After P. C. Hazra, V. S. Krishnaswam and J. Swaminath, 1959).

In southern India the lower Bhavani Dam (compare: Fig. 17.9) had to be shifted 12 m downstream when a band of treacherous talcose schist related to the shear zone was discovered. The masonry Kulamadu Dam, a part of the Idukki project is founded upon a 4-m wide shear zone in the graphite schist. The shear zone had to be treated with grouting to a depth of 9 m before building the dam. The 27- to 30-m wide shear zone in the Archaean rocks in the Mathupally valley had to be excavated to a depth of 40 m below the river bed, resulting in the doubling of the originally proposed 45 m height (Jhingran and Balasundram, 1966).

The foundation area of the Nagarjuna Sagar Dam (Fig. 17.10 b-c) across the Krishna River is characterized by not only a 5 m wide fault zone in the gneissic granite on the left abutment, which had to be excavated out and backfilled with concrete, but also a 1.5 m wide shear zone across the dam ax which had to be treated by constructing a 8.5 m deep cut-off trench near the heel and many radial and axial grout holes up to a depth of 73 m, in addition to blanketing the exposed quartzites and slates with impervious clays (Srinivasan et al., 1972). Under the 143-m high Srisailam masonry dam (Fig. 17.10a) horizontally disposed 0.2 to 0.9 m thick shear zones at a depth of 13 to 20 m below the bed of the Krishna River are characterized by gougey shale and very cavernous pockets. The right-bank quartzite

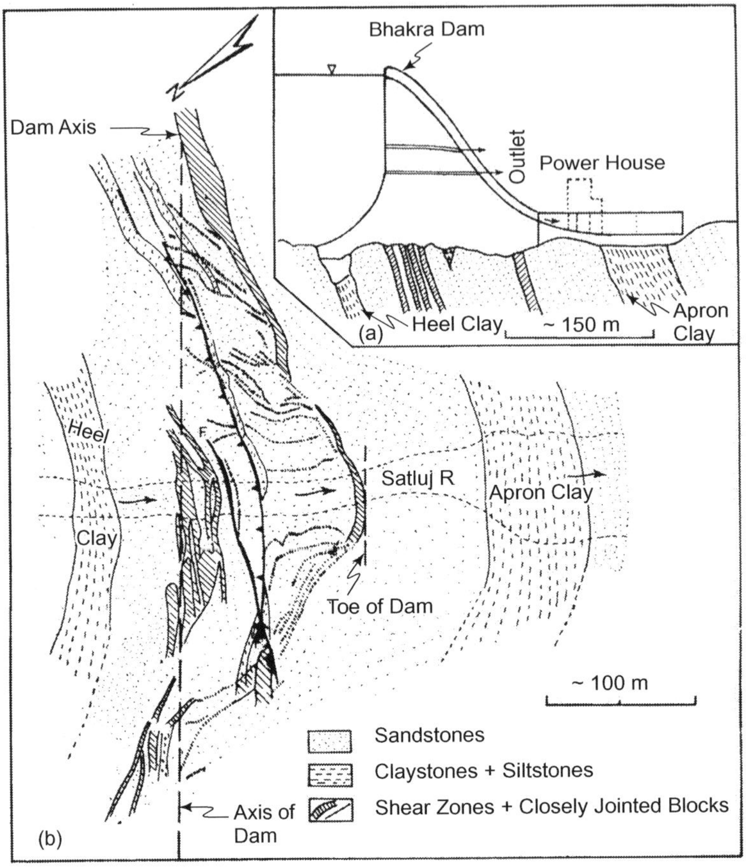

Fig. 17.7. Geological map of the Bhakra Dam site showing a thick shear zone and closely jointed blocks related to a fault that has attenuated and locally eliminated the clay horizon in the middle of the dam site. The section shows the dam heel keyed inside the impervious clay beds, and the toe connected with the concrete apron on the downstream clay bed. (After S. K. Shome, (Ed.), 1983, *Geotechnical Features of Major Dams in India,* Ind. Soc. Engineering Geol., Kolkata, 114 p.).

is extremely pitted and porous (Motwani, 1981). The construction of the 68-m high earth-and-masonry composite Ukai Dam on the Tapi River revealed 15 to 30 m wide three shear shear zones trending east-west in an overstepping manner. Grout curtains upto the depth of 15 m below the cut-off trenches and extensive treatment with chemical grouting (using sodium silicate-sodium aluminate mix) had to be done to strengthen the foundation material. The 104-m high masonry Koyna Dam (Figs. 17.12 and 17.19) is situated in the valley revealing 2 to 4 m wide N–S to NNE–SSW trending shear zones at a depth of 13 m below the bed of the river in the terrain of the Deccan Volcanics.

The faults and fault zones are likewise treated for potential instability. The excavation for the Peechi Dam in Kerala showed the left bank abutment to be fractured, the fractures being related to faults. Chemical and cement grouting were done to prevent possible leakage.

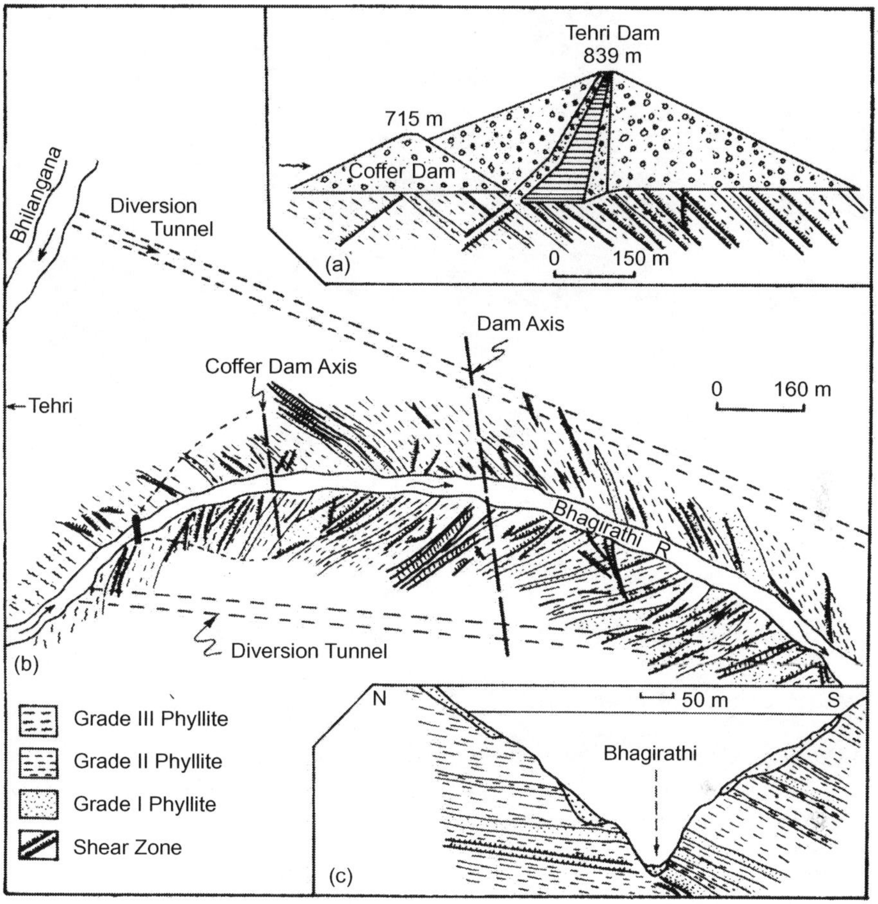

Fig. 17.8. Detailed geological map and sections of the Tehri Dam site in the Bhagirathi valley, showing shear zones and three grades of the Chandpur phyllites (grading done on the basis of deformation features and resultant bearing strength). The rock-fill dam has an earth core. (Simplified after Pant, V. Kumar and U. P. Gupta, 1983).

A large number of dams have been built or are being built on the Narmada River (Fig. 17.11). The river flows in a rift valley—an elongate depression formed between two nearly parallel or reverse faul The seismicity of the dam together with offsetting and displacement of recent landforms indicate that the rift boundary faults are quite active. In the Sardar Sarovar area, apart from a ENE—WNW trending dolerite dyke, there is a 7-m wide zone of steep reverse faults (Pancholi, 1997). On the site of the Navagam Dam on the Narmada River (Gujarat) a number of NE—SW trending steep faults and dolerite dykes cut the succession of the sedimentary rocks of the Bagh Formation capped with Deccan lavas. The brecciated rocks of the 15 m wide fault zone running along the river and dipping 50°N, have very low crushing strength of 48 psi, thus requiring excavation to a depth of 15 m (and backfilling wı concrete) and intensive grouting upto a depth of 30 m (Srinivasan et al., 1972). On the site of the 59 m

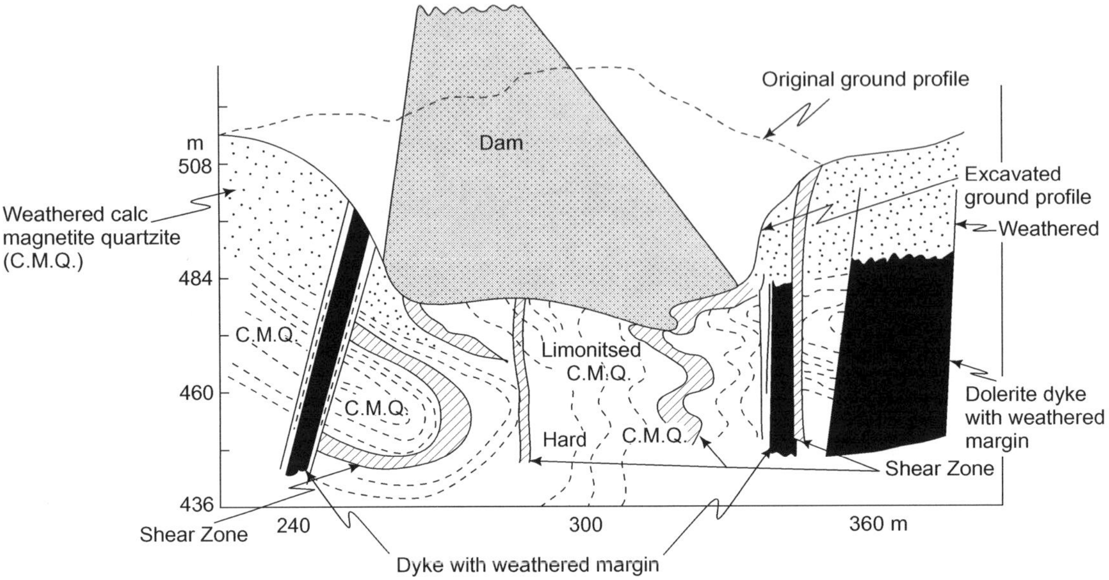

Fig. 17.9. Cross-section of the 102-m Supa Dam in the Kalinadi River flowing through the deeply weathered Dharwar terrain in Karnataka. (After V. S. Krishnaswamy, R. K. Sundaram, K. C. C. Raju, D. N. Seshagiri and A. N. Agrawal, 1982.)

gh Hirakud Dam in Odisha, in addition to a 6 m wide shear zone parallel to the foliation of schists, there is a 4.5 m wide fault zone on the right bank parallel to the valley. Characterized by a 0.6 to 1.5 m ck clay gouge and crushed shale, the fault zone had to be dug upto a depth of 3 m and backfilled with ncrete before the dam was founded (Hukku et al., 1972).

Composition and Texture of Foundation Material

n loose material, it is dangerous to build a dam higher than 7.5 m and unwise to risk a dam moderate height of 9 to 22.5 m (Fox, 1935). Poorly consolidated sediments like mud, silt and nd have very low bearing strength. They become still weaker when moistened. If exposure to air leads to shrinkage of the clays, the contact with water or moisture may cause swelling, particularly when montmorillonite is present. Either way, the stability of the dam is affected. The clay material is therefore removed to the level where moisture never dries out completely and is replaced by concrete. The exposed surface is then sprayed with emulsions of gunite or asphalt before pouring concrete. Poorly cemented porous sandstones, limestones and tuffs have smaller bearing strength. It is never safe to build heavy and high (> 20 m) structures on the foundation of such rocks as mentioned above.

Alternation of hard (competent) and soft (incompetent) rocks in any succession, such as sandstone-shale, dolomite-salt, gneiss-schist, do not provide safe conditions for the foundation of ms. The weaker rocks are vulnerable to deformation and differential movements. The six important ms in the Outer Himalaya—in the Tista, Kosi, Ramganga, Satluj (Bhakra), Beas (Pong) and Ravi (Thein) valleys—are all located on the Siwalik sedimentary succession characterized by alternation of ndstone and mudstone or gravel and having by and large low competency, the unconfined compressive

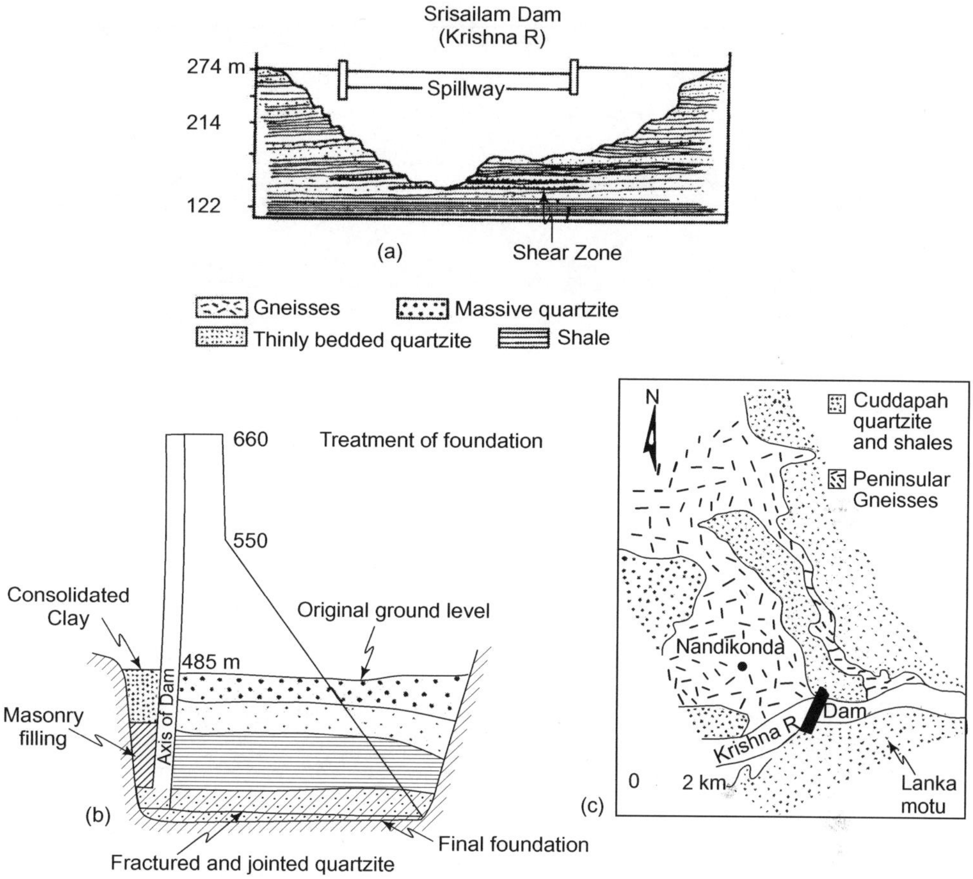

Fig. 17.10. (a) Cross-section of the Krishna valley at the Srisailam Dam site (Andhra Pradesh) showing almost horizontal disposition of shear zones in the 'Cuddapah' succession. (b) and (c) Cross section and map of the Nagarjunasagar Dam site, Krishna River (Andhra Pradesh) (a and c after S.K. Shome (Ed.), 1982; and b after R.V. Chalapathy Rao, 1972.)

strength being between 50 and 300 kg/cm^2 (Ramachandran and Ghosh, 1983). For this reason, the band beset with piping condition under the Pong Dam (Beas valley) had to be completely removed in the key trench upto a depth of 11 m. Additionally, 30 m deep grout curtains had to be put.

If free from structural infirmities like fissures, fractures and joints, such rocks as granites, dolerite basalts, amphibolites, granulites, gneisses, quartzites, well-indurated sandstones and massive dolomites have sufficiently high bearing strength, and make very good foundation rocks. But **schists, phyllites and slates, characterized as they are by foliation planes, are not strong; they are rather treacherous, even when intercalated with competent, massive rocks. Such rocks are removed from the foundati before the dam, or for that matter when any heavy structure is constructed**.

Under the Koyna Dam thin layers of massive basalt are underlain by tuff and breccia (10 to 13 m), tuffaceous ash and red bole (1.5 to 2 m) (Fig. 17.12b). The tuff has a high absorption character and

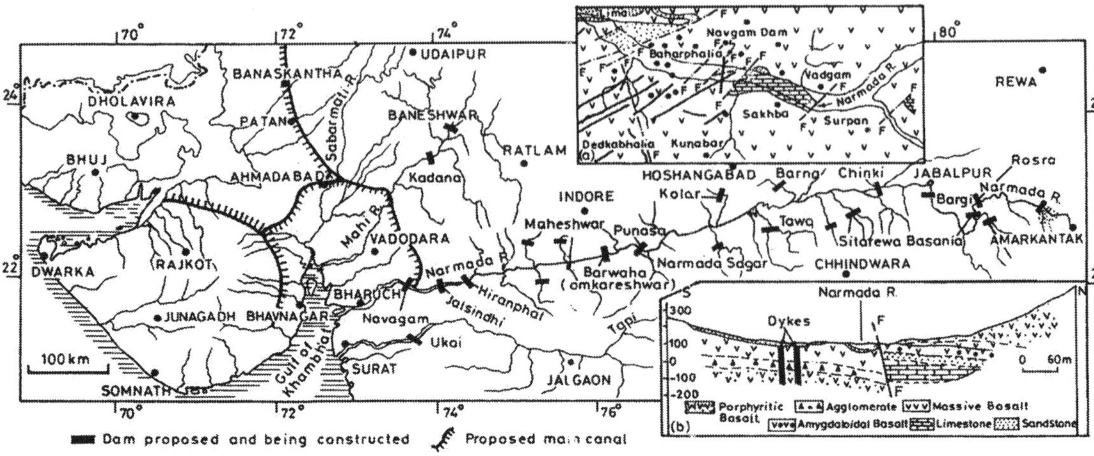

Fig. 17.11. Damming the Narmada and its many tributaries has resulted in the augmentation of storage of the river water. The water stored behind the Navagam dam has brought about greenery in northern Gujarat and Saurashtra. The upper inset shows the outline of the geology of the Navagam dam site, and the lower inset (cross-section) the structural make up of the Narmada valley. (From Valdiya, 2004; After Srinivasan et al., 1982).

hibits erratic variation in elasticity modulus and compressive strength. The lava flows are characterized by jointing (including columnar), pipes, tunnels, vesicles and cavities, and are thus particularly leaky and treacherous, This situation required deeper digging upto a depth of 18 m, with a provision of deep shafts on the upstream side.

Limestones are usually cavernous with numerous solution channels extending to considerable pths. These channels not only provide paths of escape for water, but often cause collapse and us destabilize the setting. Under the 42-m high Kamthikeri dam on the Pench River (60 km NNE of Nagpur) the 'Sausar' marble is extremely cavernous. The cavernous parts of the 30 m thick marble band **d to be dug out and backfilled with concrete, and intensive cement grouting carried out (Patil, 1982).**

The cavernous nature of limestones, dolomites and marbles created serious problem of safety of quite a few dams. The Obra Dam on the Rihand river in Uttar Pradesh and the 42-m Bajaj Sagar Dam on the Mahi river in Gujarat faced this problem (Sundaraiya and Wakpanjar, 1996). Construction of concrete diaphragm 0.6 m thick and 3 m apart across the entire thickness of the foundation rock, entrenched deep into the rock beds, together with grouting solved the problem.

In the terrain of the Vaishnodevi (Jammu) limestone in southwestern Kashmir, the bed of the Beas had to be dug upto 35 m below the level of the river bed to remove the mass of karstic dolomite and dolomitic limestone riddled with sink holes, tunnels and breccias before the Salal Dam could be constructed (Dhawan and Patni, 2011).

In Assam the Kopli Dam flooded the karstic terrain of limestones and shales upstream of the impoundment. Extensive treatment was done to prevent the reservoir water from sinking and leaking (Nanda and Mukherjee, 2009).

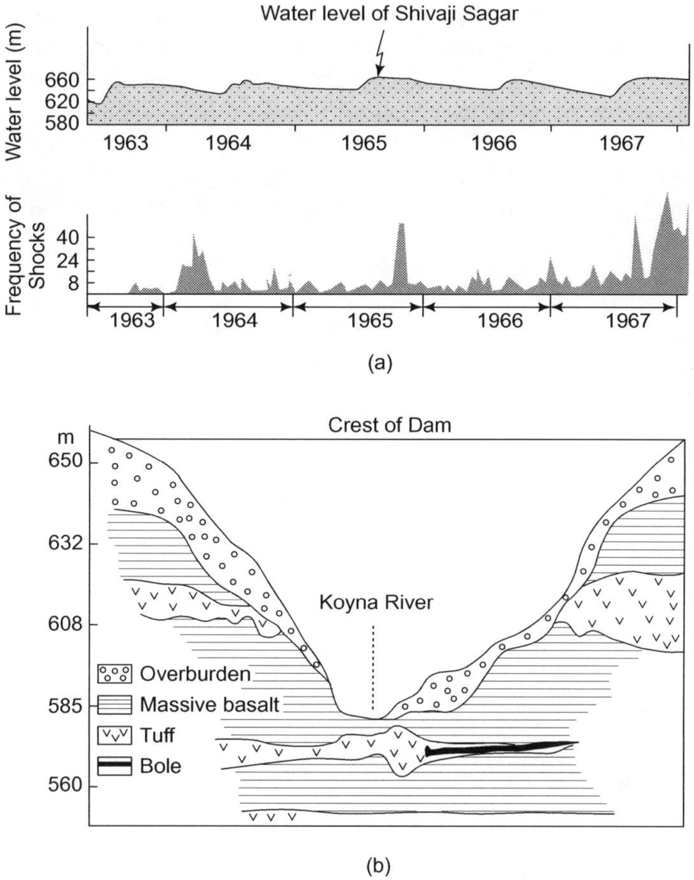

Fig 17.12. (a) The relationship of frequency of earthquakes and maximum level attained by the impounded water in the Shivaji Sagar. (After S. K. Guha, P. D. Gosavi, M. M. Verma, S. P. Agarwal, J. G. Padale, and S. C. Marwad, 1968.) (b) Simplified cross-section of the Koyna Dam site in t Deccan Volcanics terrain (Maharashtra). The foundation of the dam had to be dug deeper than the basal tuffacous horizon in order to get over the problem of weakness of the foundation material. (From J. P. Rothe, 1973).

Weathered Foundation Rocks

Weathered rocks pose serious problems of instability and require special treatment. The granitic rocks of central India (Fig. 17.13b), for example, are deeply weathered upto a depth of 50 m in some places (D. K. Ghosh, per. comm., 1983). In southern India the situation is not different (Figs. 17.9 and 17.13). The weathering of rocks gives rise to formation of clays and gritty soils where the weathering is complete. The clays are not only treacherously slippery, but also have a strong propensity for swelling and shrinking as already stated, thus generating strong stresses in the foundation.

The Linganmakhi Dam in the Sharavathi River in Karnataka is located on a deeply weathered granite-eiss with amphibolite and schists. Beneath this 54-m high rockfill dam there is a 20 m thick weathered clayey material in the river bed and on the right bank (Fig. 17.13a). This situation required constructing a 33 m deep pile foundation (Srinivasan et al., 1972). In the construction of the concrete gravity dam on the Rihand River in southeastern Uttar Pradesh, the 26 m thick weathered material mantling the ijawar' gneisses, schists and amphibolite on the right bank and of the 1 to 6 m soil in the overflow sections had to be replaced by concrete (Fig. 17.13b). The Tilaiya Dam of the Damodar valley -project on the Barakar River is located on the deeply weathered Gondwana sedimentary rocks, the thickness of e profile of weathering varying from 3.3 to 4.5 m.

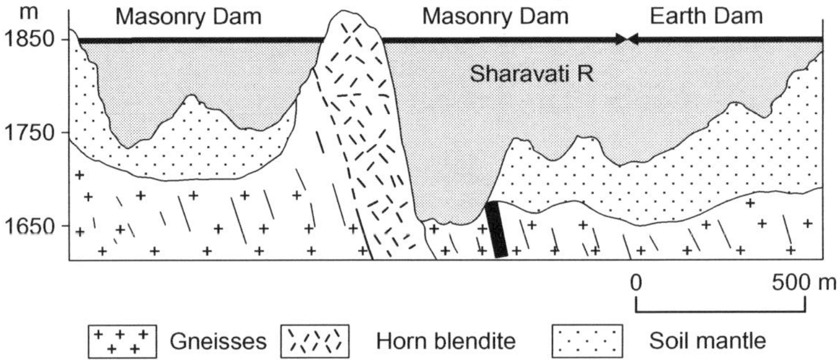

(a)

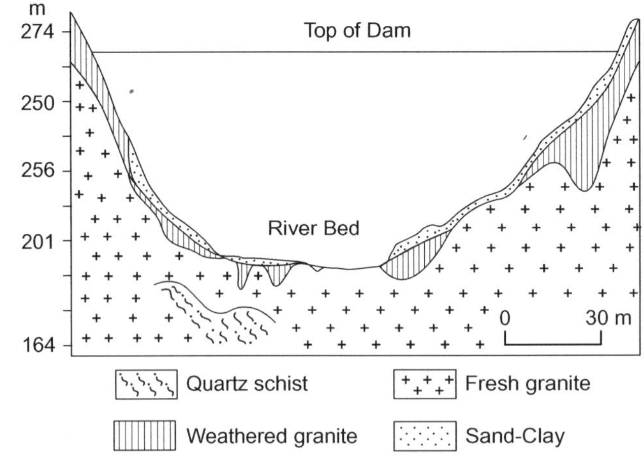

(b)

Fig. 17.13. (a) Cross-section of the Sharavathy valley at Linganmakhi Dam site (Karnataka) showing a thick mantle of soil formed by deep weathering of Archaean gneisses and amphibolite. (After Y. K. Murthy and B. S. Srivastava, 1972). (b) Cross-section of the Rihand Dam site in the 'Bijawar' terrain, southeastern Uttar Predesh.

Hidden Topography: Buried Channels

Careful exploration, including drilling and seismic sounding, have revealed a variety of concealed underground landforms in a number of places, both in the Peninsular and Himalayan regions. Buri channels or epigenetic valleys abandoned by the rivers in the past and now filled with fluvial depos have been recognized in the upper reaches of the Parbati valley in Himachal Pradesh, at Tapovan a Utyasu in the Alaknanda valley and at Maneri in the Bhagirathi Valley in Uttarakhand (G. Pant, 1972). The former channel-fills contain numerous large boulders closely packed in narrow gorges. **The buried channels allow free passage out to the impounded water if not checked by putting undergroun dams or concrete diaphragms** (Fig. 17.14).

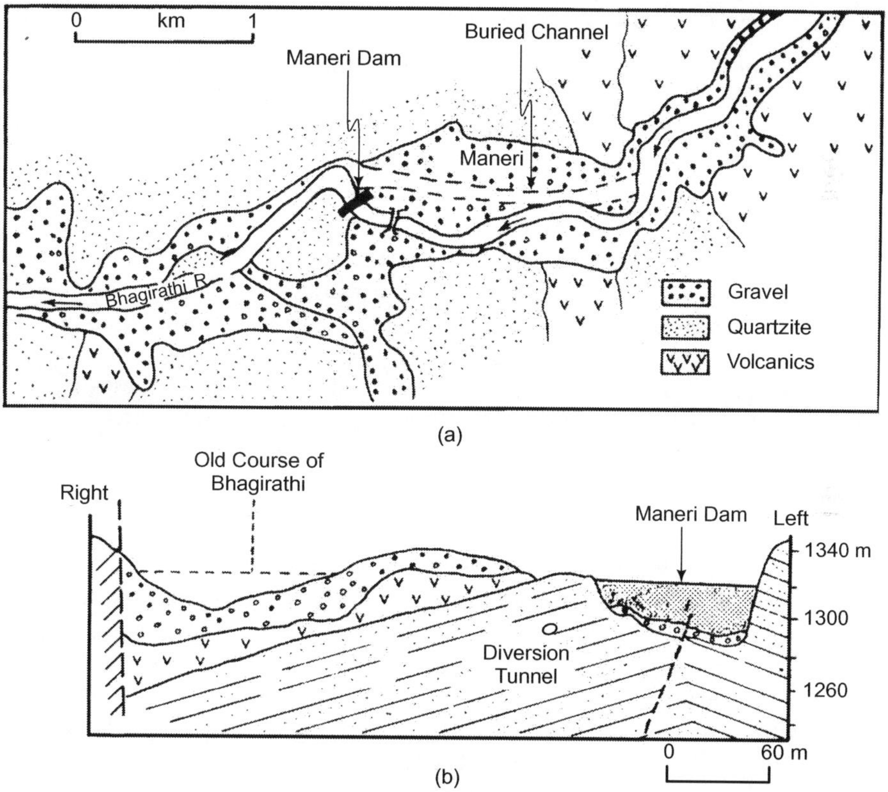

Fig. 17.14. Geological map and cross-section of the Maneri Dam site in the Bhagirathi valley (western Uttarakhand). The abandoned buried channel (old course) is filled up with fluvial debris. (Af G. Pant, 1975).

In the Srisailam Dam in the Cuddapah country of quartzites and slates in Andhra Pradesh (Fig. 17.10a), a 90 m wide and 13 m deep channel, filled 6 to 7 m with boulders close to the left abutment h to be subsurfacially dammed. In the Bargi and Mahanadi valleys in Odisha, concrete diaphragms were put to intercept the flow of groundwater through the buried channels. At the Obra dam in southeaste

Uttar Pradesh, beneath a thick overburden in the Rihand valley, a 49-m deep buried gorge coinciding with a fault zone was revealed near the contact of the Vindhyan and Bijawar formations. The leakage (10 to 12 cusec) through this gorge as well as through the cavernous 'Kajrahat' limestone of the Vindhyan Group was contained by a 23 m deep grout curtain and by 15-m deep concrete diaphragms put 3 m apart d grouting of the foundation and channel fill. Beneath the Moosakhand Dam in the Karmanasa River (southeast of Varanasi), a buried channel in the 'Kaimur' sandstone of the Vindhyan Group towards the left bank caused considerable seepage under the artesian head at the toe of the dam.

SEISMICITY CONDITIONS

Seismological Investigation

round motion due to earthquakes has a significant impact on the stability of dams. Earthquake ocks cause (i) slope failures and landslides, (ii) loss of the contact of the dam with its abutment, ı) sliding of the dam itself on the weakened slippery foundation, (iv) overtopping of the dam e to waves in the impounded water generated as a result of landslides, rockfalls, avalanches and so on. The faults may become active, with resultant displacement of the ground, and thus seriously impair or damage the structure. It is therefore imperative that an appraisal of seismic hazard potential of the area selected for the dam is made. **Recognition and delineation of active faults, and analysis of historical records of past occurrences are among the many important aspects of seismological investigations to be carried out**. Where seismological records are not available, monitoring devices are customarily installed to measure the magnitude of events, and examine the seismic behaviour of each feature. Additionally, determinations are made of ground strain, dynamic properties of foundation materials, modulus of elasticity and liquefaction potential of foundation soil or sediment. Accurate piezometric levels are determined to study the potential buildup of porewater pressure.

Seismicity Factor

All the above-mentioned factors are taken into consideration to compute the seismicity factor for dams. ell-built dams founded on firm ground are capable of withstanding moderate earthquake shaking, with a peak acceleration of say 0.2 g produced by earthquakes of $M \geq 7$. Earthen dams built on solid rocks suffered no damage even when the shaking was as high as 0.35 to 0.8 g from earthquakes *of* 8.25 (Meigh and Skipped, 1981). Dams founded on semi-consolidated, unconsolidated, cohesionless and water-saturated material are, however, badly affected by seismic shaking. The pore water pressure greatly contributes to the instability and eventual failure of the dams (Fig. 17.4).

Though the Koyna Dam was provided for earthquake acceleration of 0.05 g, the 1967-earthquake generated horizontal acceleration of 0.42 g, and vertical acceleration of 0.34 g (Mehta and Pradhan, 1972). It has built-in jointing in the structure of the dam that saved it from failure. The seismicity factor provided for dams is 0.66 g for horizontal acceleration and 0.33 g for vertical acceleration (Ramachandran and hosh, 1983). Most of the Himalayan dams are located close to faults and thrusts, some of them active.

The Tehri Dam in Uttarakhand has been designed to withstand peak ground acceleration of 0.446 g that would be generated by an earthquake of *M* 7.2. In the event of an earthquake of $M \geq 8$, the ectral acceleration in the field area would be as high as 1.0 g (Khattri, 1995). Apart from the ground celeration, the passage of ground wave may cause alternate opening and closing of fissures and joints in the foundation of the dam. This might allow ingress of water and thereby increase the pore pressure underneath the dam (Valdiya, 1997). Moreover, the earthquakes trigger landslides on a larger scale, filli the reservoir behind the dam with debris masses. This is already seen in the proximity of the dam.

GEOLOGY OF RESERVOIR AREAS

Benefits and Blacklashes

The formation of a reservoir behind a dam brings many benefits to the society—the stored water is us for irrigation and generation of electricity, for fisheries, recreation, navigation, industries, and to conta or moderate floods.

However, economic benefits apart, the environmental and socio-economic consequences of artific impoundment of water are very serious. The change from the riverine to lacustrine environment has far-reaching impacts on the ecology of streams. Certain geological events which are not beneficial may triggered. The submergence of forests, grasslands and agricultural lands brings about drastic changes in the biological productivity of various ecosystems. The drowning of settlements and agricultural fiel has traumatic effects on the uprooted communities, their lifestyles and culture.

The geological problems related to reservoirs are three-fold—the leakage of water and attenda problems, the sedimentation in the man-made lakes and consequent reduction of their lifespan and the seismicity induced by the impoundment of great volumes of water.

Leakage and Rise of Water Table

Water-tightness of the reservoir basins (Fig. 17.5) is of great importance in water-resource development programmes, particularly if the stored water is meant for irrigation. The presence of impervious rocks or soils provides insurance against percolation losses.

The position of the water table is the first thing to be studied. As already pointed out in Chapter S the water table roughly follows the surface topography, being deeper under higher places and closer to the surface in valleys. Its steep gradient is indicative of the lightness of texture of the ground material, while a gentler gradient is discernible in rocks or soils characterized by openings or high porosity. The impounded water of the reservoir infiltrates and causes a rise in the water table. When the level of wat in the reservoir is higher than the highest point (crest) of the water table, the ground water spills over the hump and there is thus unimpeded leakage (Fig. 17.15). As this causes loss of water, admittedly no dam is planned where the reservoir water may rise above the hump of the groundwater.

The volume of leakage depends obviously on the permeability of the rocks. The unconsolidated gravelly deposits, fluvial, alluvial and colluvial sediments, glacial drifts, soluble salts, cavernous lim

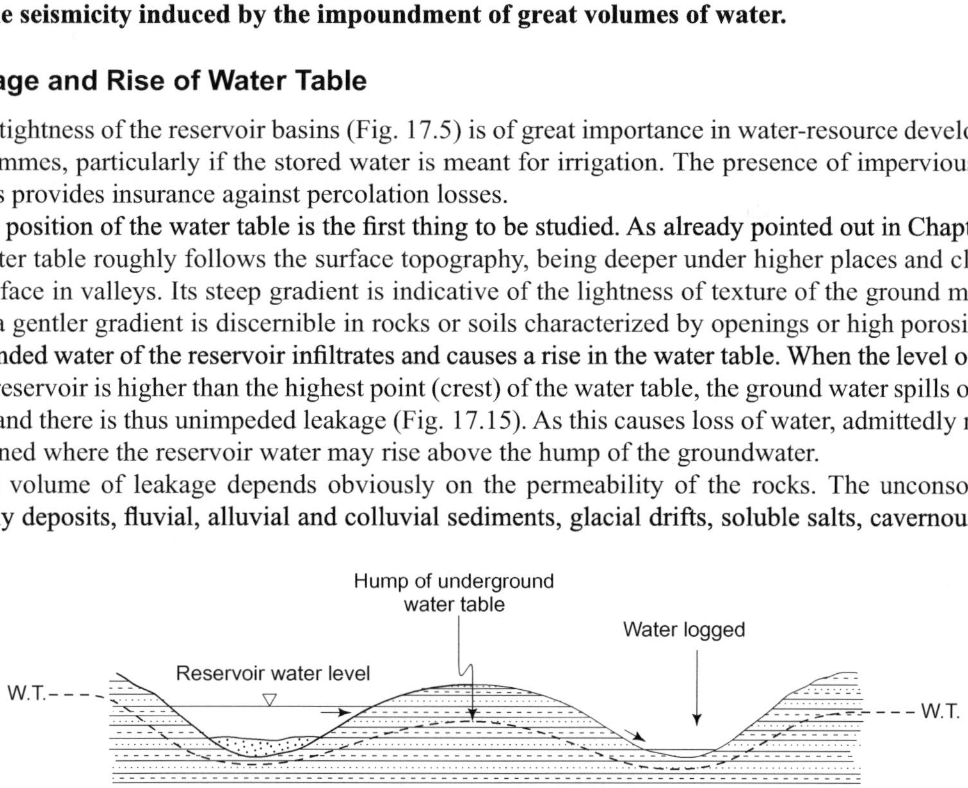

Fig. 17.15. When the level of impounded water rises above the hump of the underground water table there is seepage. This may cause waterlogging in the vicinity of the dam.

nes-dolomites, vesicular and scoriaceous lavas, and fissured-fractured crystalline rocks cause profuse leakage. Apart from the loss of stored water, there is waterlogging in the low-lying terrain outside the dams and embankments—leading to development of marshy conditions.

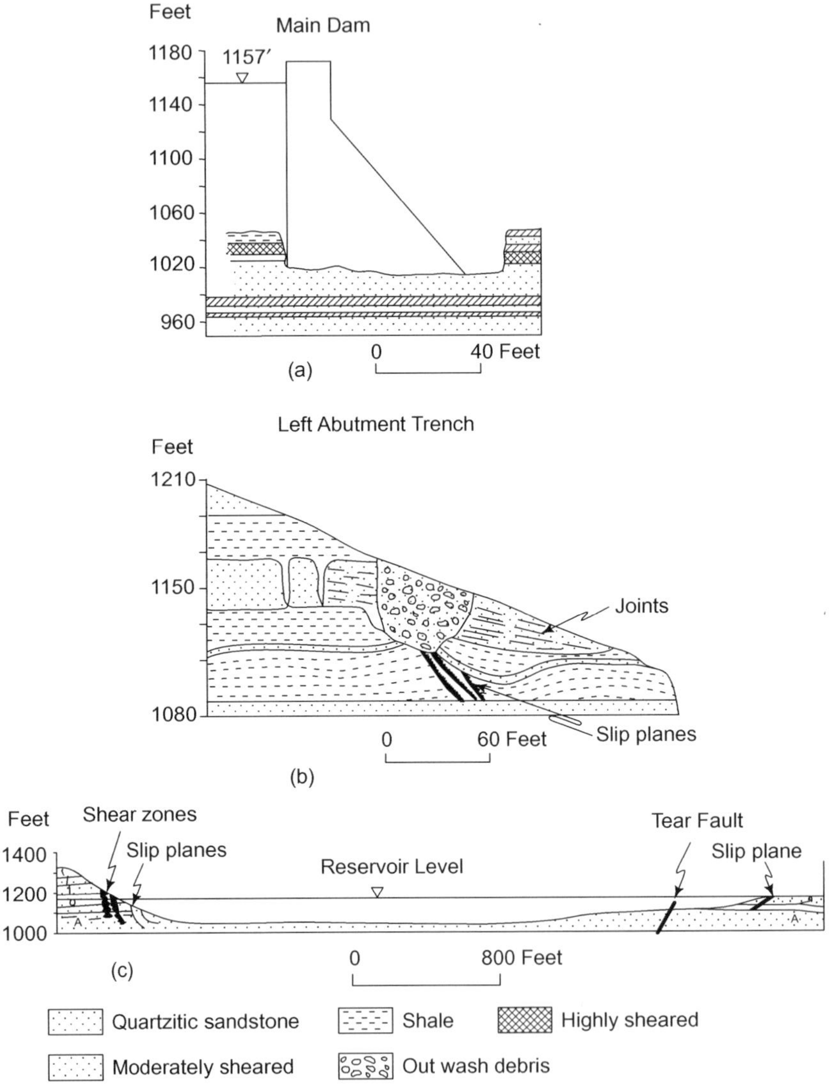

Fig. 17.16. Cross-sections of the site of the Rana Pratap Dam on the Chambal River, Rajasthan. (After S. N. Chaturvedi and Z. Ashraf, 1972.)

The 24-m high Dudhwa Dam in Madhya Pradesh resting on weathered granite and 2 to 6 m thick clay-sand beds was plagued with serious leakage in 1962 when the sand layer was opened by quarrying. The reservoir had to be emptied completely so that remedial measures could be taken, such as covering porous-permeable beds with a concrete apron throughout their submerged extent. Similarly, sand-boils and alarming seepage through the 6 m thick subsurface residual soil of weathered granite below the 15 m high Hirakud Dam were contained by putting an impervious blanket upstream and loading t embankment to check the seepage (Mehta and Pradhan, 1972). The Gandhi Sagar lake behind the 63 m high Kota Dam in the Chambal River created a serious problem of waterlogging as water leaked through transverse joints in the very gently downstream-dipping alternation of sandstone and shale of the Bhander Formation (Vindhyan Group). Grout curtains upto a depth of 20 m and concrete plugs on the left abutment were installed to check the seepage. Upstream near Chittaurgarh, the seepage from the Rana Pratap Sagar (Fig. 17.16) was checked by putting grout curtains at the heel through the clay layers of the Kaimur-Rewa formations and prestressing the foundation by placing extra load of concrete at the toe of the dam (Chaturvedi and Ashraf, 1972). The rock anchors were 3.5 m deep and spaced 1.5 m apart and placed downstream to prevent sliding.

In Meghalaya the 67 m high dam across the Kopili River and another 38 m high dam on a subsidiary stream Umrong in the Precambrian gneissic terrain unconformably overlain by 'Sylhat Limestone' created a problem of a high degree of seepage. The Sylhat Limestone is characterized by sink holes, cavities and solution channels—some of which are more than one kilometre in length and 20 m in diameter—and two prominent sets of vertical joints and several faults (Ramachandran and Gangopadhyay, 1972). Monitoring the movement of groundwater by use of radioactive tracers showed that seepage was taking place at a fast rate of one centimetre per second (cm/s) from the Umrong to the Langlai valley, the gross loss being of the order of 8.5 m^3/s from a one million square metre area.

Sedimentation and Loss of Storage Capacity

Sedimentation in reservoirs is a very serious problem, for it considerably reduces the life of the man-made lakes. Streams flowing into the reservoirs inevitably bring high loads of sedimen from the catchment areas, which of late have been exposed to accelerated erosion. The faster an deeper the erosion, the shorter will be the life of the reservoir.

The quickened pace of erosion in the catchment areas and resultant larger input of sedimenta debris at the rate 1.5 to 3 times more than the rate anticipated at the designing stage has result in progressive decrease of the water-holding capacities of reservoirs—at the rate of 0.30 to 0.90% annually (Lagwankar et al., 1994). The all-India rate of siltation varies between 0.15 m-ha/100 km^2/yr and 27.85 ha-m/100 km^2/yr. **An estimate done in the 1980s indicates that the economically usef lifespans of most of the man-made reservoirs have been reduced to half or a quarter of what w originally calculated by the planners of the dams** (Fig. 17.17, Table 17.2). The Pandoh reservoir in the Beas River lost 2.4 million m^3 of its total gross storage capacity of 41 million m^3 in the period 1977–1996.

The water impoundment or storage by damming of rivers causes, among other things, entrapment of sediments; so that the water that flows out of a reservoir is devoid of bed-load sediments and t suspended sedimentary particles are also not much in quantity. The water discharged from the reservoir therefore has stronger ability to erode deeper into the channel beds. In other words, the channels show accelerated erosion downstream of dams. This gives rise to the gully formation or deepening of the channels. **Due to damming in the upper reaches, the *sediment-starved rivers* in Kerala do not have enough sedimentary load to dump at their mouths, and are unable to contribute to the building of beaches.** It may be stated that the sediment deposited at the river mouths provide material for beaches.

In the absence of ***beach nourishment*** by rivers, the Kerala coast has become increasingly vulnerable to coastal erosion.

Table 17.2. Sediment Accumulation in Some Reservoirs (After Lagwankar et al., 1994).

Reservoir (River)	Year of Impoundment	Rate of Sedimentation (ha-m/100 km^2/yr)			Percentage of Annual Loss of Storage Capacity
		Assumed	Observed	Year	Gross
Pong (Beas)	1974	4.29	27.85	1986	0.41
Bhakra (Satluj)	1958	4.29	5.66	1987	0.32
Kalagarh (Ramganga)	1974	4.25	22.94	1986	0.33
Mayurakshi (Mayurakshi)	1955	3.75	6.83	1975	0.51
Matatila (Betwa)	1956	1.33	5.45	1990	0.71
Gandhi Sagar (Chambal)	1960	3.57	8.96	1976	0.29
Ukai (Tapi)	1974	1.49	7.16	1984	0.53
Koyna (Krishna)	1961	5.05	8.10	1986	0.025
Nizam Sagar (Marjina)	1930	2.38	4.89	1975	1.41
Tungabhadra (Tungabhadra)	1953	4.29	6.48	1985	0.49
Panchet (Damodar)	1956	6.67	5.88	1985	0.40
Maithon (Barakar)	1956	1.62	13.02	1971	0.50
Hirakud (Mahanadi)	1957	2.50	6.62	1984	0.61

A glaring example of the brisk decline of lakes is the premature silting-up of the reservoir behind the Ichari Dam on the Tons River (Fig. 17.17). Within a very short span of a little over two years (1975 1977) the reservoir was filled with sediments upto the level of the crest of the spillway 160 m high above the riverbed (Mohan and Tyagi, 1983). The total sediment transported into the lake was of the order of 29 million m in 1978–1979. The lake in the Jaldhaka River in northern Bengal behind a dam s likewise been filled up completely by sediments.

In the Govind Sagar behind the Bhakra Dam there was heavy sedimentation in the reach 24 to 26 m above the dam, forming a hump and resulting in maximum capacity loss between the elevation of 427 and 457 m above mean sea level. The average sediment deposition during the period 1959–1964 was 36.86 million m^3 and between 1959–1969 about 39.3 m m^3, 65% being in the dead-storage and 35% in the live storage estimated (Rao and Palta, 1973). As a result of this unchecked sedimentation the lifespan of the reservoir would be 103 years instead of the expected 250 years as estimated in the early 1970s (Singhota and Aggarwal, 1974). The rate of sedimentation behind the 12 m high Kalagarh Dam in the Ramganga was 1.90 ha m/km^2/yr of the catchment area so that within the stipulated lifespan of 100 years, about 5900 ha m of sediments would accumulate in the lake, reducing its life to 76 years (Mithal, 1982). According to another estimate, its lifespan has been reduced from 185 to 48 years.

How quickly a reservoir can be filled with sediments eroded from the hills of a tectonically resurgent gion is exemplified by the Kulekhani Reservoir in Nepal. The sediment loads of rivers and their tributary streams are deposited in the standing body of waters of the reservoirs, forming deltas where the streams meet the reservoirs. The deltas thus formed grew progressively downstream, eventually overwhelming the whole of the reservoirs due primarily to exceptionally large quantities of sediment delivered to the lakes. The Kulekhani Reservoir behind a 150-m high dam in southcentral Nepal is

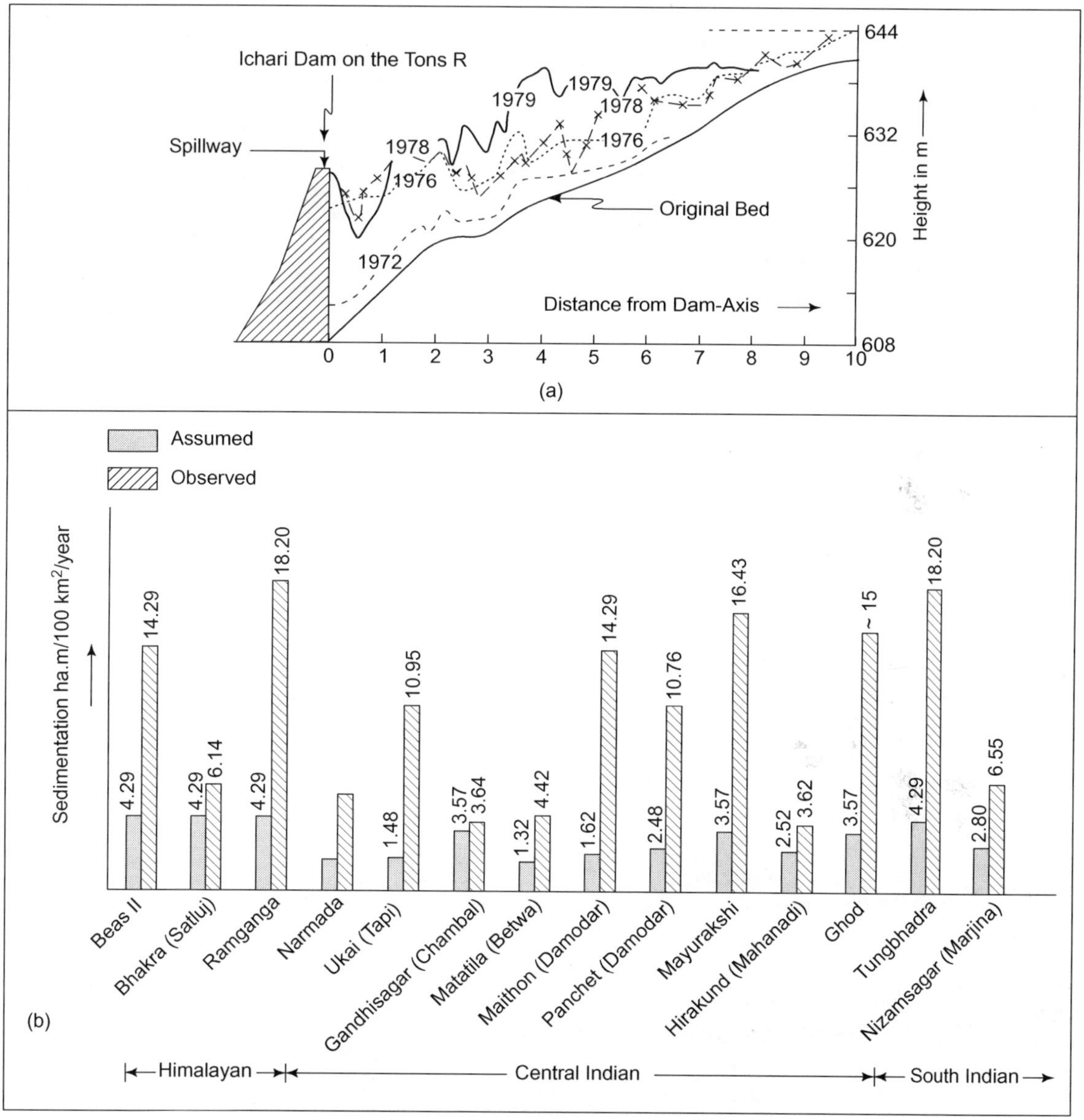

Fig. 17.17. (a) Excessive sedimentation behind the Ichari Dam in the Tons valley (western border Uttarakhand). The various lines show positions of bed levels in different years. In about two years, the reservoir was almost completely filled with sediments. (After J. Mohan and S. S. Tia 1982). (b) Assumed and observed rates of accumulation of sediments behind some Indian dams. (From R. K. Gupta, 1979).

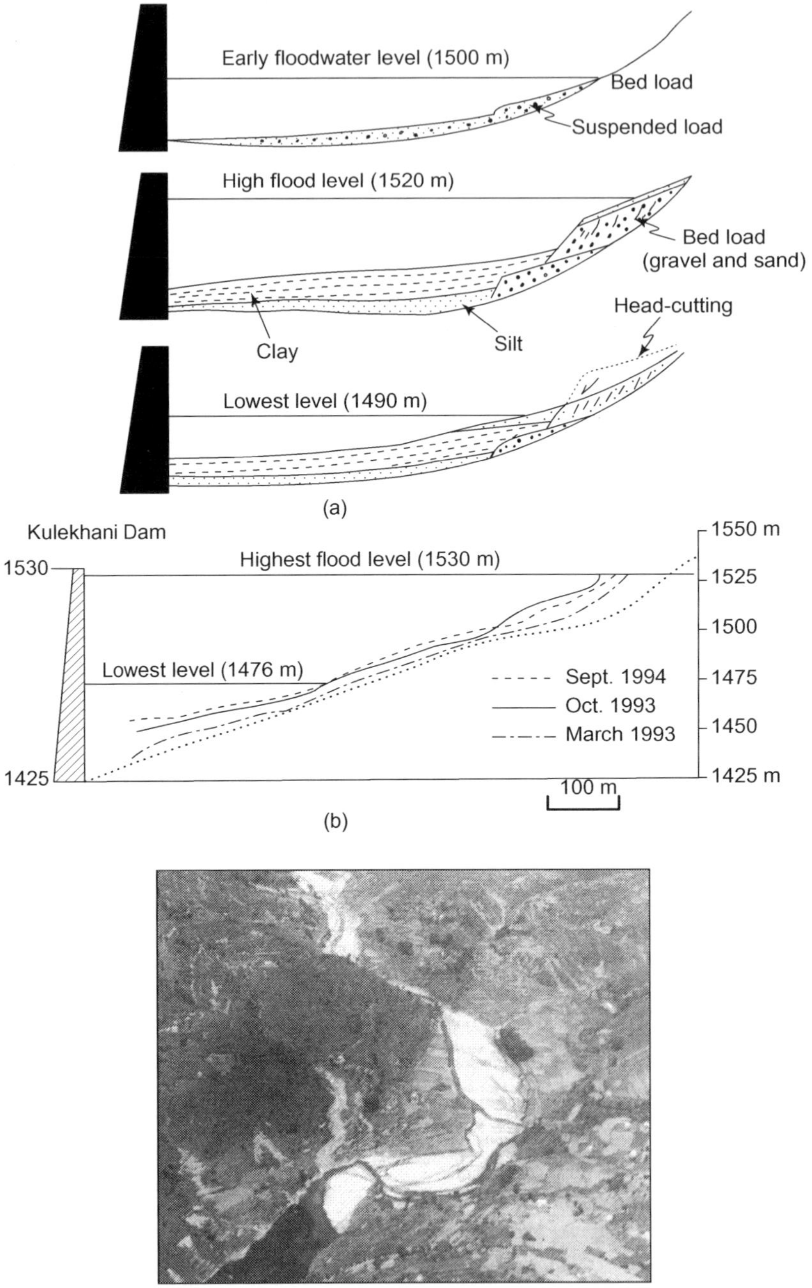

Fig. 17.18. (a) Southwest of Kathmandu in Nepal the Kulekhani reservoir behind a dam is being filled at a very fast rate. Notice the pattern of sediment accumulation. (b) Photograph showing the above phenomena. Note the growing delta of sediments within the reservoir in the valley. (After Galay et al, 1995 in Valdiya 2004). (See colour Fig. 17.18)

progressively losing its water-holding capacity due to rapid rate of sediment accumulation as the delta grows (Fig. 17.18). The original storage capacity of 85.3 million m^3 in 1982 dwindled to 78.3 million m by July 1993 due to sediment that was being eroded from the catchment at the rate of 1700 $m^3/km^2/yr$. During a single rainstorm on July 19, 1993, about 4.8 million m^3 of sediment was deposited, the rate of erosion in the catchment being 31,700 $m^3/km^2/yr$ (Galay et al., 1995).

In the Peninsular India the situation is not less serious. The storage capacity of the Neyyar reservoir in Kerala declined by 28.8 million m^3 in 38 years at the annual rate of 0.71% (Suresh et al., 2000). The Matatila reservoir in the Betwa River basin had lost 10.5% of its storage capacity by the end of 1979 at an annual rate of 0.80% (Singhal, 1982) (Table 17.2). The Ukai reservoir on the Tapi River lost 5.8% of its gross capacity at the rate of 0.83% per year between 1972 and 1979, the rate of sedimentation being as high as 11.19 ha-m/100 km^2/yr in catchment area against the assumed rate of 1.37 ha-m/100 km^2/yr. Consequently, the useful life of the Ukai reservoir might be reduced from the expected 273 years to 122 years (Sharma et al., 1983).

As already stated **the consequence of the quickened pace of erosion and attendant larger inp of sediments in the artificial lakes is that their useful life spans have been reduced to half or even quarter of what was originally estimated**. The fast sedimentation in lakes is a universal phenomenon. In U.S.A. every year about 1,235 billion m^3 of sediments were deposited in 1105 reservoirs causing on an average annually 3.5% loss of storage of smaller lakes, 0.16% of larger lakes, and 2.7% of the medium-sized reservoirs (Glymph, 1973). In central Europe in 19 reservoirs with a capacity of 1.49 to 23.15 billion m^3, the rate of storage depletion was 0.5% per year. Lake Nasser on the Nile River (Egypt) trapped 13 million m^3 of sediments annually. The Shehmen reservoir in Taiwan lost more than 45% of its storage capacity within five years of its commissioning in 1965. More dramatic was the case Anchaieaya River in Columbia which was filled to 25% of its capacity in just 21 months of its fillin and completely lost its storage capacity in a mere seven years (Coates, 1981).

World over the reservoir capacity has been decreasing at the rate of 0.8% per year, so that b 2050, 64% of the man-made reservoirs all over the world would have considerably diminish storage capacity due to accumulation of sediments (D.E. Walling, *U.N. World Water Development Report* No.3, 2010).

Another implication of the reservoirs trapping the sediments is that due to the lack of adequa amount of sediments delivered by rivers to beaches, the shoreline will be retreating.

Reservoir Filling and Induced Seismicity

A few months after the start in 1962 of the filling of the Shivaji Sagar behind the Koyna Dam, seism shocks accompanied by hollow muffled sounds were recorded, the frequency of which increased by t middle of 1963. The epicentres were all located in the neighbourhood of the dam or under the reservoir (Fig. 17.19) stretching to a distance of 40 km upstream (Guha et al., 1982; Narain and Gupta, 1968). Evidently, the filling of the lake water and its pressing down the floor with its load generating ne stresses underground, was responsible for this seismicity. It is noticed that the stresses build up to the peak slowly over several years after the initiation of the filling, and then there is a sudden failure of t ground, releasing these stresses in the form of earthquakes. This happened in the Shivaji Sagar area on December 10, 1967 when an earthquake of M 6.4 ravaged the area, killing 200 people, causing injury to 1500 and affecting as distant a place as Mumbai 250 km away. The major aftershocks on December 24, 1967 and October 29, 1968 were of magnitude 5.5 and 5.4, respectively. From 1962 to 1980 about 50,000 seismic events were recorded in this area. The induced seismicity continues.

A rise of more than 110 m above mean sea level (i.e. >50 m filling) of water level in Bhatsa (Khardi) lake 40 km north-east of Mumbai generated two strong shocks of M 4.0 (on 17.8.1983) and 4.8 (on 15.9.1983), in addition to including intense microseismicity (Patil et al., 1984), the epicentre being within a 5 × 7 km area.

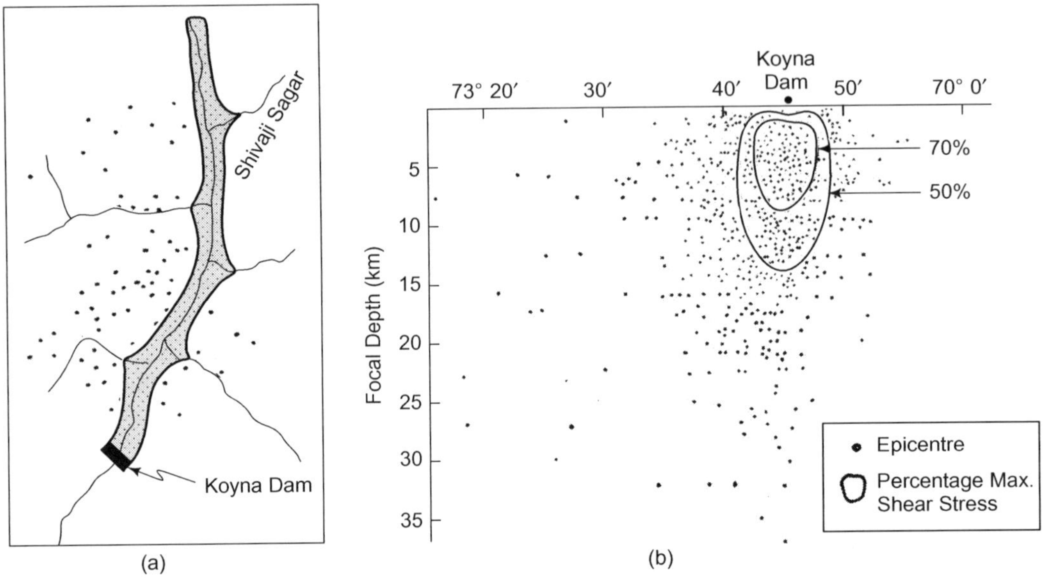

Fig. 17.19. Reservoir-induced seismicity in the Shivaji Sagar region, Koyna River, Maharashtra. Redrawn with permission from H. K. Gupta and B. K. Rastogi, 1975. *Dam and Earthquakes,* Elsevier, Amsterdam, 215 p., Fig. 43).

A similar incidence occurred in Lake Mead behind the Hoover Dam in the Colorado River in U.S.A. e first shock was felt in September 1936, about a year after the filling started. This was followed by 100 shocks in 1937 when the lake level of 330 m was reached, and the principal shock of *M* 5 occurred on May 4, 1939—nine months after the normal height of 355 had been attained (and when the volume was 35 billion m^3). In the next 10 years 6000 shocks were recorded in the area (Coates, 1981).

Analysis of seismic data by Gupta and Rastogi (1976) and Simpson (1976) from a number of reservoir sites all over the world established beyond doubt that seismicity is induced by and increases soon after e impounding of water (Fig. 17.21) even in essentially seismic areas such as Hsinfangkiang in China (1961, *M* 6.1), Kariba in Zimbabwe (1963, *M* 5.8) Kremasta in Greece (1966, *M* 6.3), Vaiont in Italy (1963). Monteynard in France *(M* 4.5), and Lake Mead in U.S.A. 1939, *M* 5) (Fig. 17.20, Table 17.3). **here is a positive correlation between the height of the water-column in the reservoirs and the smicity induced—6 out of 20 (30%) reservoirs with water height between 150 and 250 m have perienced RIS. And where the height of the water column is between 90 and 120, the RIS was t in only 6% of cases** (H. K. Gupta, 1984). It was thus noticed that as the water level rose, the degree of seismicity also increased and the strongest shock was registered when the level attained the highest point (Fig. 17.20), particularly higher than 100 m.

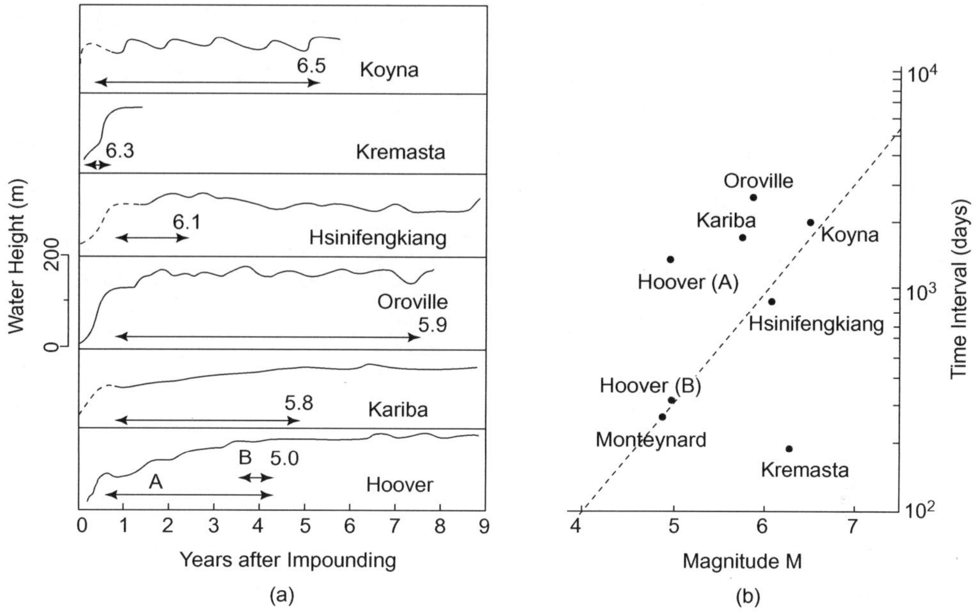

Fig. 17.20. (a) Relationship between the maximum height attained by the impounded water and the time of occurrence of induced seismicity. The arrows indicate time intervals. (b) Interval of time between the completion of reservoir filling and the last occurrence of an earthquake in relation magnitude. (After D. W. Simpson, 1976, *Engineering Geology*, 10 : 123-150).

Table 17.3. Reservoir-Induced Seismicity (RIS) in India (S.K. Guha, 1982)

Dam	Location (Coordinates)	Max. Depths of Reservoir Water (m)	Earthquake Magnitude (M)	Reservoir Induced Seismicity
Ghirni	18.4N:76.8E	15	2.0	III
Idukki	9.4N:77E	160	3.5	II
Khardi (Bhatsa)	19.6N:73.3E	52	4.8	II
Kinnersani	17.7N: 80.7E	62	5.3	II
Koyna	17.6N: 73.8E	100	7.0	I
Mangalam	10.6N: 76.5E	28	3.0	III
Mula	19.4N : 74.6E	44	1.0	III
Parambikulam	10.4N : 76.8E	66	3.0	III
Sharavathy	14.1N: 76.8E	38	2.0	III
Sholayar	10.3N:78E	59	2.0	M
Ukai	21.2N:73.7E	64	3.0	III

Seismicity: I = Intense, II = Moderate, III = Low (microseismicity).

A number of points about the RIS may be emphasized.

First, the occurrence of RIS is critically related to the height of the column (> 100 m) rather an to the volume of impounded water. In Shivaji Sagar whenever high levels (> 100 m) were ached following rapid filling, and this level was retained for a longer period, larger earthquakes ($M >$ 4) occurred at the rate of one every four months, in addition to many small-magnitude events.

Second, the reservoir filling does not change the maximum intensity of the naturally occurring rthquakes, it increases only the frequency of shocks.

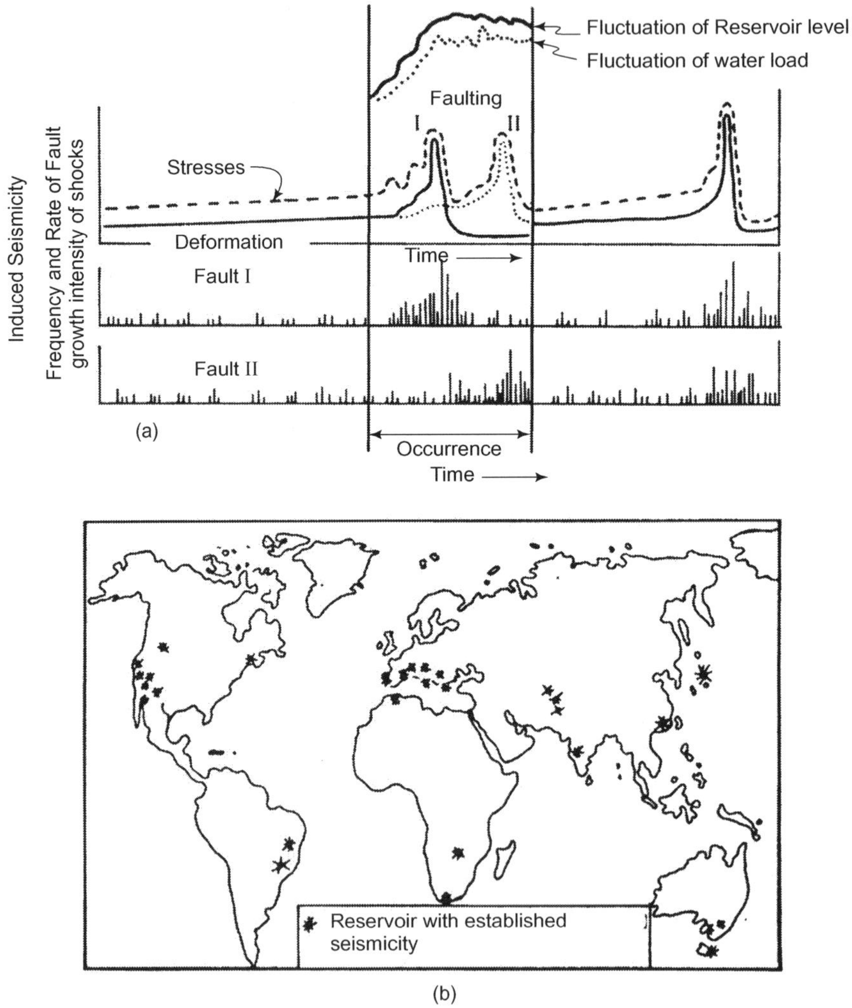

Fig. 17.21. (a) Relation of the frequency of induced-earthquakes, growth of fault, and effect of the reservoir filling. (After Nikolaev, 1974) (b) Places where reservoir-induced earthquakes of appreciable magnitude occurred.

Third, the RIS occurs only in the tectonically stressed unstable zones (Nikolaev, 1974). **It is a sult of reactivation of faults when the ground is overloaded with water** (Fig. 17.21). Impounded water finding its way by infiltration through fissures, shear zones, cavities and joints into the fault zon reduce the frictional resistance and normal force—the effective stress operating perpendicular to the fault plane—causing the movements (Fig. 17.22). The fluid pore pressure for a lake of 100 m depth only 10 bars, but combined with the effect of overloading (vertical stress) it would trigger off seismicity, if the stored energy were close to the failure point (H. K. Gupta, 1984).

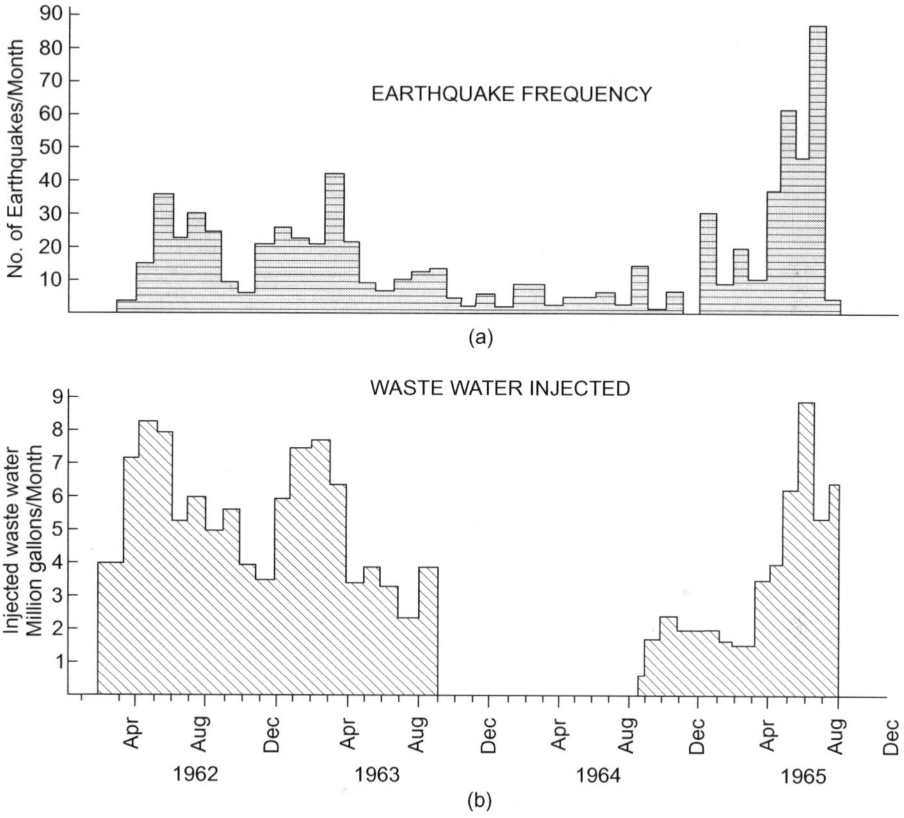

Fig. 17.22. Amount and rate of water injected underground through wells have clear relation with the frequency of seismicity induced in the Denver area, U.S.A. (Redrawn with permission from D. M. Evans in R. W. Tank, 1973, *Focus on Environmental Geology,* Oxford University Press, New York, 534 p., Fig. 3.)

Fourth, the presence of soft rocks, absence of fractures and the homogeneity of rocks do n favour accumulation of stresses and therefore preclude occurrence of the RIS. The brittleness of the rocks is a necessary condition. This explains why the reservoirs developed in the Siwalik-Murree belt, characterised by alteration of sandstone and mudstone, in the Himalayan region (such as at Tarbe Mangla, Pong, Bhakra, Kalagarh, etc. did not induce any seismicity.

To summarize, the RIS depends (Gupta and Rastogi, 1974; Huang Nain, 1982) on the (i) rate increase of water level, (ii) duration of high water level maintained, (iii) maximum height attained,

) deeper infiltration of the impounded water through passages provided by faults, fractures, joints, shear zones, and cavities, (v) existence of deeper tectonic condition such as stress buildup in pre-existing fault planes and weak zones, and (vi) brittleness of rocks such as quartzites, granites, basalts, etc. **Only all the last three conditions are present would there be an earthquake.**

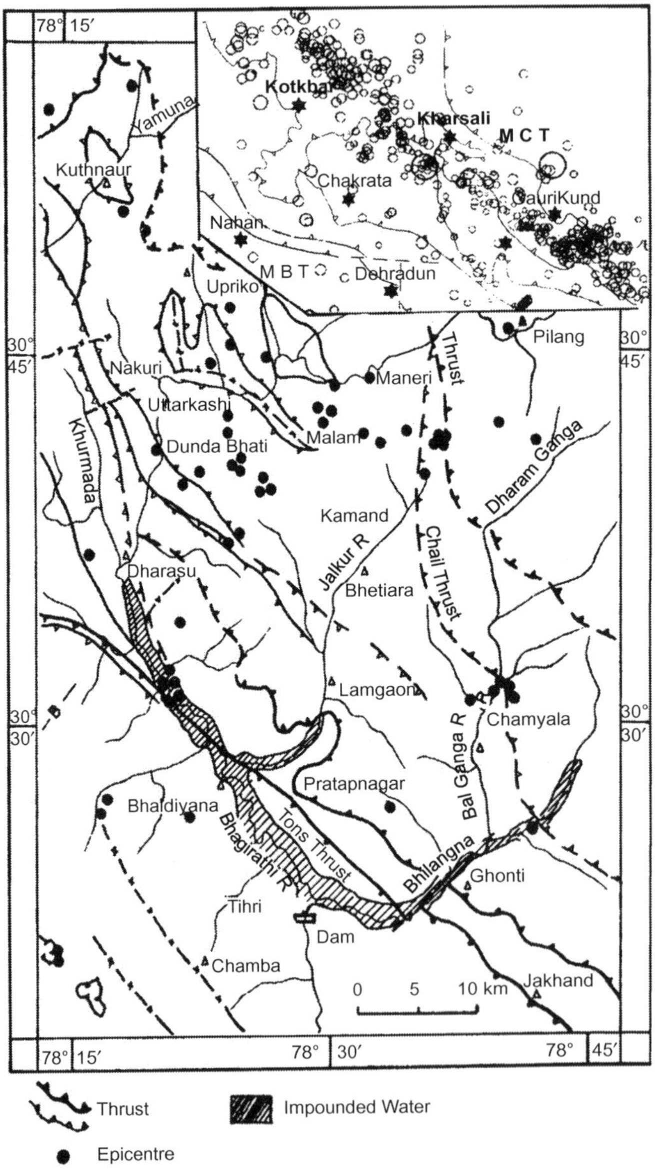

g. 17.23. Microseismicity (M < 3) restricted to the area 20 km NW of the Tehri Dam site in the Bhagirathi valley. (After R.C. Agrawal and P.N. Agrawal, 1982; geological map after Valdiya, 1980). The inset shows seismicity in western Uttarakhand (Garhwal). (From: Ajay Paul, 2011)

In a large part of the fault-riven Lesser Himalayan Nepal and adjoining Uttarakhand and Darjili regions high stresses prevail (Gupta et al., 1979; Gupta, 1984). The impoundment of large volumes of water to heights greater than 100 m might in all likelihood would increase the seismicity of the region. The many faults and thrusts (characterized by zones of shearing) would provide an easy path of infiltration of water. The transverse (tear) faults are particularly prominent, and some of them have be neotectonically active.

There was no induced seismicity in the case of the reservoirs behind the dams at Bhakra (Chaudhury and Srivastava, 1978), Pong (Srivastava and Dube, 1982) and Tarbela in Pakistan (Jacobs et al., 1975) because one of the prerequisites of RIS, namely the presence of brittle rocks, is absent in the Siwalik-Murree zone constituted of alternating plastic claystone-mudstone and soft sandstones (the larger part of the reservoirs lying over the terrain of these rocks). Significantly, in the vicinity of the dams on t Dihing and Subansiri rivers in Arunachal Pradesh, there was slightly higher microseismicity during the short period of survey (Srivastava and Dube, 1982). The geological conditions here are quite different

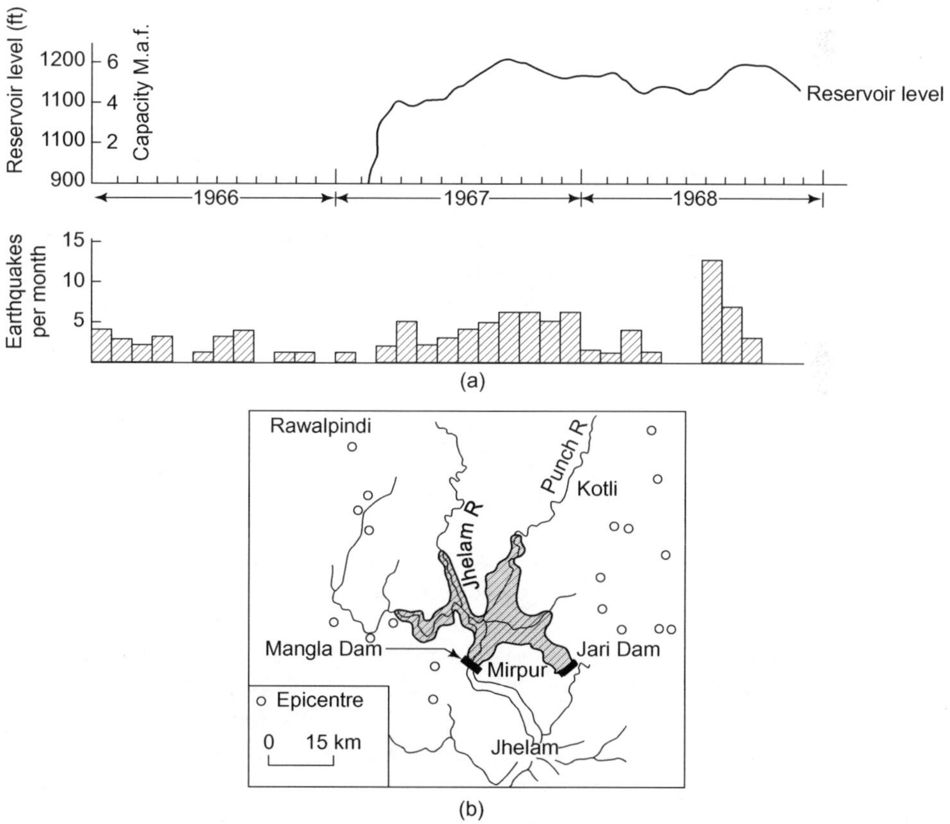

Fig. 17.24. Relationship between the maximum level of water in the reservoir behind the Mangla Dam on the Jhelam River (Pakistan) and the frequency of induced seismicity. (Redrawn with permission from H. K. Gupta and B. K. Rastogi, 1975. *Dam and Earthquakes,* Elsevier, Amsterdam, 215 p., Figs. 77-78.)

m those of the Siwalik zone in the northwestern Himalaya. The area of Mangla Dam in the Jhelam River in Pakistan (Fig. 17.24) also registered increase in seismicity of *M* 1 to 2 by a factor of *2* after the impoundment, but then this could be due to better recording capabilities deployed after impoundments. ter studies have indicated decrease in seismicity at Mangla (H. K. Gupta, per. comm., 1981).

ENVIRONMENTAL IMPACT OF WATER IMPOUNDMENT

ams and Environment

e socio-economic benefits of reservoir projects are many and too well-known. Farmers get assured supply of water for irrigation in a regulated manner and thus secure freedom from vagaries of unreliable, d often insufficient rain. They are thus able to increase production in farms. The electricity generated meets the varied and growing needs of cities, towns and industries. Floods are moderated by storing rain water. Navigation can be developed on the rivers following regulated courses, and so on.

The Narmada water has now reached the arid region of western Rajasthan (Fig. 17.11). Flowing through the 486-km long canal in Gujarat and 56-km reach in Jalor district in Rajasthan the Narmada water has brought greenery over 2.46 ha lakh hectares in over 233 villages in the Jalore and Barmer districts (*The Hindu*, February 29, 2012). What is heartening is that there is compulsion to use the water ough drip irrigation. The Indira Gandhi Canal bringing water from the Himalayan rivers has made the uthwestern Haryana and northwestern Rajasthan a blooming land of prolific crops.

A dam across a stream is not merely an engineering structure built to provide benefits to e people, but also an artificial creation that considerably modifies ecosystems and hydrologic gimes in areas both upstream and downstream.

It is therefore imperative that at the planning and implementation stages, the problems and fallouts that are likely to occur due to the construction of a dam are taken into account in order that timely and effective measures are taken to eliminate or minimize them, **and also to ensure that the benefits of e project are shared by the people adversely affected in the upstream areas with those living in stant towns and cities of the command area**. A wide range of changes are produced—the normal w pattern in equilibrium with the ecosystems is upset, the geological and geodynamic conditions are altered, the riverine and terrestrial ecosystems are replaced by lacustrine ecosystem, etc. **Dams are erefore not to be viewed within the restricted framework of technical feasibility and economic gains, but are also to be assessed in terms of the environmental changes they produce and the cio-cultural impacts they would have on the communities affected.**

The various dams constructed, and being planned, are undeniably sound in terms of technical sign and construction. However, so far no comprehensive evaluation of the effects of impoundment on varied aspects of the natural environment and the fabric of society has been seriously attempted. Also there are no provisions built into the framework of plans for the elimination or mitigation of negative environmental and socio-cultural effects. **If the project is to be really beneficial and aimed the welfare of the people, there must be an unambiguous enunciation of the ways and means balancing or compensating each negative effect in the upstream area with a positive one in e downstream region. There must also be a clear suggestion of alternatives entailing lesser mage to the environment and causing less distress to the people of the lake basin. Commonweal should be the major consideration, not capital cost and technical feasibility, even though these are portant factors.**

Erosion and Landslides in Reservoir Basins

The impoundment of water causes the groundwater table to rise, creating new and recurrently changing gradients of the water table, as water is filled and withdrawn from the reservoir. The rise of the wat table reduces the shear strength of rocks and soils, and its fluctuation produces drastic alteration in t pore-water pressure. This phenomenon promotes landslides along the banks particularly if the slopes are already in a state of instability (Figs. 11.34 and 17.25, 17.26). One of the greatest tragedies occurred on September 9, 1963 on the site of the 168-m high Vaiont Dam in the Italian Alps when more than 224 million/m^3 (4 m t) of rockmass slid down the mountain slope at a rate of 25 to 30 metres per second within a short duration of 30 to 60 seconds, filling the reservoir almost completely and generating 7-minute catastrophic flood that killed 2600 people within a 2 km downstream stretch (Fig. 17.26).

Evaporation Loss and Microclimate Changes

In a warmer country such as India, large amounts of reservoir water are lost through evaporation particularly during the drier seasons. The larger the surface areas, the greater the loss. The loss of water thus defeats the very purpose of storing it. In the Govind Sagar it is assumed that 12.3 million m equivalent to 1.2 m column of water—having an irrigation potential of 8,00,000 ha—is lost annually by evaporation (Rao and Palta, 1973). Lake Nasser in Egypt annually looses 15 billion m^3 of water, which is about 14% of the annual input, that is, more than half of the incremental gain from constructing the storage facility (Fahim, 1981). Likewise, in the American south-west the evaporation amounts to 3 m of water column of the reservoir (Coates, 1981). The loss from evaporation is immense in a hot country like India.

The water body also produces microclimate modification in the lake basin and adjoining area Studies carried out in the Pechora-Vychenda-Kame reservoir in the erstwhile U.S.S.R. indicated that the areas where the climate was affected was nearly four times the area of the lake basin (P. P. Micklin in Detwyler, 1971). The affected zone was cooled in early summer, but warmed towards its end. The summers became shorter and more humid, and the winters less severe. The net loss of heat now available to plants has affected their growth in farms and forests.

Changes in Quality of Water

The condition of water from flowing to stagnation undergoes a qualitative change. Among the ma things, the decay of organic matter washed down the forested slopes and derived from the submerged vegetation promotes vigorous growth of organisms feeding on the products of organic decay. The stagnant inlets and embayments of reservoirs thus become breeding grounds of germ-harbouring, disease-carrying organisms especially during drawdown periods.

Since the formation of the Nagarjuna Sagar in Andhra Pradesh the incidence of fluorosis (Fig. 21.2)—terrible crippling bone disease "genu valgum"—has increased considerably in the command region irrigated by the reservoir water. The disease has also appeared in parts of Tamil Nadu and Karnataka where irrigation dams have been built, such as at Parambikulam, Aliyar and Hospet. Obviously, t fluorine-rich water derived from weathering of fluorite-bearing pegmatites of the terrain is the real caus Likewise, in Raichur district (Karnataka) malaria has spread since the coming of the Tungabhadra Dam. The incidence of filaria and Japanese encephalitis has similarly increased in areas in South India whe reservoir water is used. The Egyptian experience in Lake Nasser and its command area of the increase in schistosomiasis is a strong warning. It is a debilitating disease caused by tiny worms and acquired

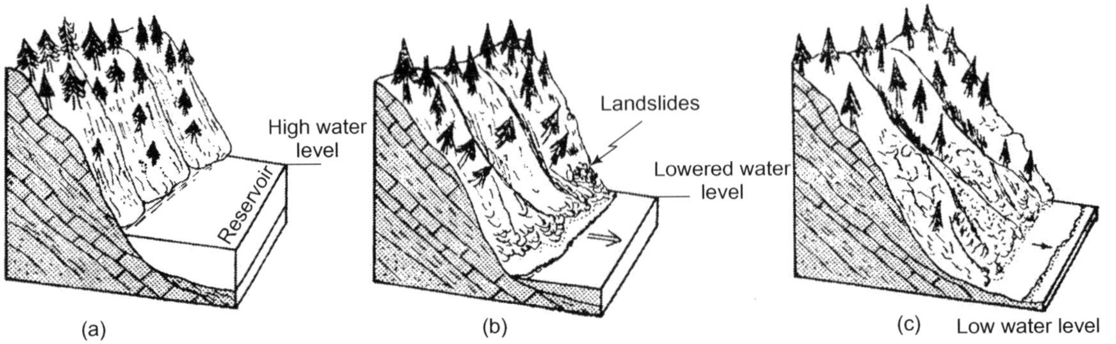

g. 17.25. The lowering of the water level of the reservoir is tantamount to withdrawal of lateral support to the hillside, leading eventually to landslides. (Based on W. M. Marsh and J. Dozier, 1981, *Landscape,* Addison-Wesley, Reading, 637 p., Fig. 26.20.)

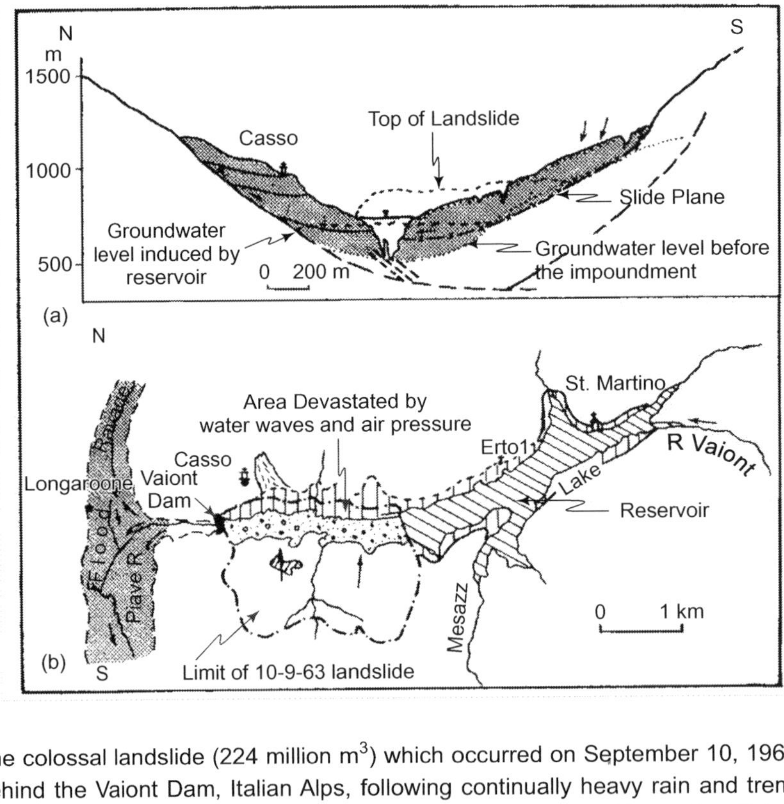

g. 17.26. The colossal landslide (224 million m^3) which occurred on September 10, 1963 in the reservoir behind the Vaiont Dam, Italian Alps, following continually heavy rain and tremors of reservoir-induced earthquakes. The flood that followed took a toll of 2600 people within a stretch of 2 km downstream. (Based on G. A. Kiersch in R. W. Tank, 1973).

through direct contact with water infested with larvae of the worm. Here the rate of incidence of t more severe variety increased from 3.2% in 1935 to 73% in 1977 (Fahim, 1981).

Destruction of Upstream Flora and Fauna

The submergence due to impoundment of water causes loss of forests (Fig. 17.27, Table 17.4), pastures, agricultural fields and also the flight of wildlife. According to an estimation (Bhumbla and Khare, 198 nearly 16,000 km^2 of forests were lost in the process of water-resource development or 479,100 ha by 1979 (Bowendar, 1982). Adding the agricultural fields, the extent of land submerged amounted to 2 m ha.

Besides, there is felling of trees on a large scale by the immigrant labour force for fuel and for building huts. The wanton destruction of forests in the Western Ghat and the Nilgiri as a consequence of the building of dams and reservoirs was an eye opener (Gadgil, 1979). With this phenomenon is related the exodus of fauna and the loss of biological diversity and the resulting invasion of pest weeds. In the environmental appraisal, therefore, genetic cost-accounting should be given a central place. This involves quantitative estimation of the genera/species to be affected and calculation of the time and energy taken by the plant population concerned to regenerate.

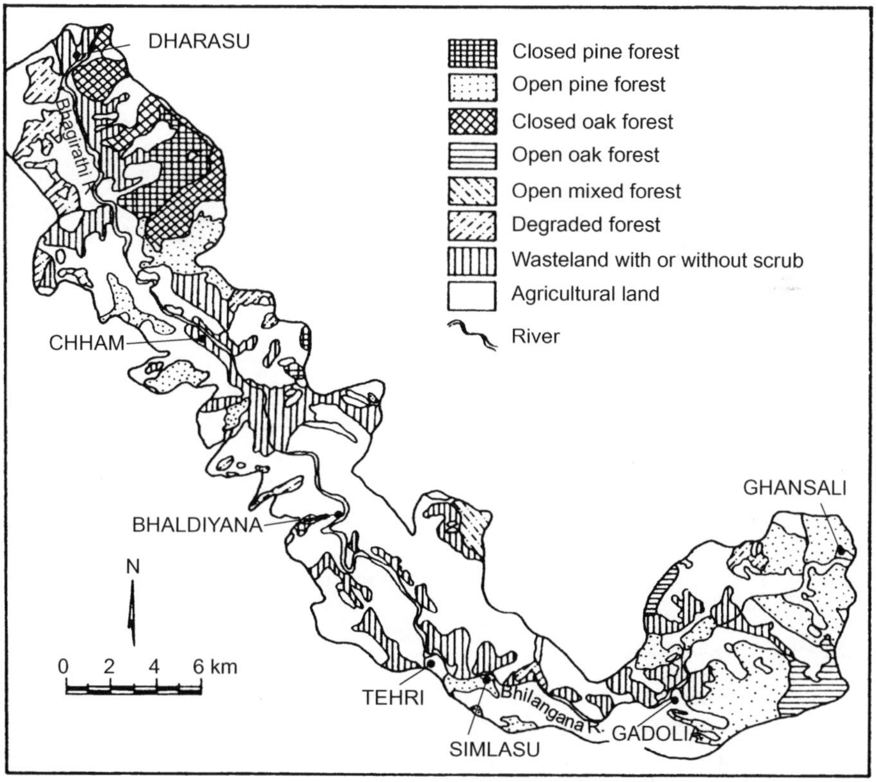

Fig. 17.27. Loss of forests and valuable cultivated agricultural lands in the valleys of the Bhagirathi and Bhilangana rivers, due to the construction of the 261-m high Tehri Dam in western Uttarakhand. (After Garg et al., 1995 in Valdiya, 2004).

Table 17.4 Forest Area Lost in a Time Span of nearly 4 Decades during 1951–1973 on Account of Water Resource Development and Road Construction (C. Sharma, 1977)

State	Forest Area Lost due to River Valley Projects (ha)	Forest Lost due to Road Construction (ha)
Andhra Pradesh	27,900	
Assam	19,700	6,600
Jharkhand	1,300	1,100
Gujarat	34,700	300
Himachal Pradesh	7,600	1,100
Jammu & Kashmir	100	200
Karnataka	28,700	1,700
Kerala	7,800	200
Madhya Pradesh	69,200	400
Maharashtra	10,300	32,800
Odisha	32,500	800
Rajasthan	14500	300
Tamil Nadu	45,100	100
Uttar Pradesh and Uttarakhand	93,200	3,400
Andaman-Nicobar		500
Goa-Daman-Diu		2,700
Arunadhal Pradesh	100	400
Bengal	1,300	1,900
All India	4,01,400 (11.8%)	54,700 (1.6%)

Aquatic communities are variably affected by the change from the riverine to lacustrine ecology. In the initial stage of impoundment due to superabundance of organic matter (which releases rich nutrients) there are vigorous microbial activities leading to explosion of some organisms, including some ecies of fish and weeds such as the water hyacinth. In Govind Sagar there was a rapid increase in the pulation of the "gid" fish (*Labeodero*), but disappearence of the spawning grounds caused dwindling or even elimination of other forms. The "gugli" (*Schyzothorex*) became rare, 'mahseer' (*Torputitord*) and 'kunni" (*Labeodero diechylus*) declined considerably (Rao and Palta, 1973).

Submergence of Mineral Deposits

The loss of mineral resources due to inundation is all too obvious. The Rihand Dam, for example, has submerged minor deposits of abrasives south of the border of the Mirzapur district (Uttar Pradesh). It may be emphasized that in the category of mineral deposits are included quarries of construction material as well.

Table 17.5. Extent of Displacement of People due to Building of Dams in the Last 50 Years (Data from: Indian Institute of Public Administration, 1988; and other sources)

Dam	No. of Persons Displaced (Approximate)
Kamthikhera	890
Ozarkhad	1,350
Bagh	1,370
Karanjwan	1,600
Palkhed	1,720
	2,260
Bennithora	3,000
Kenher	7,080
Kayadhu	8,860
Kabini	11,250
Hemavathy	11,600
Nagarjunasagar	13,230
Krishnarajasagar	15,000
Isapur	15,590
Dhimbe	18,000
Hirakud	18,000
Dham	19,730
Almatti	20,000
Mahi Bajaj Sagar	27,320
Warna	29,300
Koyna	30,000
Gohira	31,000
Hidkarl	31,300
Bhima	35,070
Bhakra	36,000
Tehri	86,000
Narayanapura	48,120
Gandhisagar	51,510
Ukai	52,000
Srisailam	52,050
Rihand	55,000
Balimola	60,000
Majal Gaon	65,300
Nizamsagar	67,440
Sriramsagar	75,090
Ukai	80,000

Con

Dam	No. of Persons Displaced (Approximate)
Tilaya	13,450
Maithon	28,030
Panchet	41,460
Tungabhadra	54,450
Kangsabati Kumari	1,25,000
Pong	16,000
Subarnrekha	1,20,000
Koel	98,000
Sardar Sarovar	4,00,000
Navagam	1,50,000
Maheswar	4,00,000
Bargi	70,000

ownstream Erosion

Devoid of suspended silt, the water released from the reservoir has a strong propensity for eroding the stream bed in order to satisfy its capacity of bed material. The stream bed is deepened, and lowering of the base level induces a new cycle of erosion and attendant geomorphic changes in some of the downstream tributaries. While the area immediately downstream suffers from erosion, further downstream there is oking of the channels. For example, at Yuma 560 km downstream of the Hoover Dam, the stream d was lowered by 3 m, while at Needles in California, silting of the river bed caused floods— which never occurred prior to the formation of the dam (Coates, 1981). Likewise, the Glen Canyon Dam on the Colorado was responsible for as much as 6 m deep erosion of the river bed. It was estimated that by 1985 the Nile River in Egypt downstream of the Aswan Dam would be deepened by 3 m (Fahim, 1981). The construction of the Akosombo Dam in Ghana accelerated erosion in the neighbouring Togo, where a town Kete on the coast lost to the sea hundreds of houses accommodating 10,000 people. **Similar studies for the Himalayan and other dams in India are yet to be carried out.**

It is not the deepening of the main stream that matters much, rather the initiation of a new cle of erosion in the entire drainage network that would manifest in the form of gullying of the tchment.

Quite another type of problem is seen in unlined canals fed by Govind Sagar. In the absence of silt in the canal water there was profuse growth of aquatic weeds since 1963, so much so that the channels were choked with algal growth (Rao and Palta, 1973). Costly and time-consuming measures for controlling the weeds had to be taken up.

allout of Canal Irrigation

This has been dealt with at length in Chapter Five. Over-irrigation of the command area causes extensive waterlogging in the moist regions and salinization-alkalinization in the dry regions (Table 17.6). More than necessary water causes rise of the groundwater table, which reaching close to the surface, retard plant growth, creates problems in foundations and converts depressions into marshes or pools. Capillary action in drier areas brings up salts and carbonates of calcium and sodium from the depth of the ground, ndering the soil sterile and unfit for cultivation for all time.

The round-the-year irrigation since the early Nineteen Seventies from Lake Nasser had waterlogged 80% of the productive areas, rendering 50% completely useless, and raising the water table from 15 m underground to 3 m. In Cairo the water table had come dangerously close to the surface, merely 81 cm below the ground level by the end of 1970s (Fahim, 1981). What has happened in the irrigated areas in Panjab and Haryana has been already described.

Uprooting of Human Communities

The inundation of a large part of the catchment area results in the destruction of the best part of the croplands in the mountains and hills—usually the fertile terraces—and moist floodways in the plai which are comparatively densely populated tracts. The submergence of valuable fields and settlemen not only causes uprooting of large communities but also puts a halt to food production of the area, thus countering the very objective of the water resource development project.

Table 17.6. Problem of Waterlogging and Salinization Associated with Water Impoundment by Mid 1980s (After Indian Institute of Public Administration, 1988).

Reservoir	Areas Affected (in ha)	
	Waterlogged	Salinized
Jaisamand	–	606
Bhakra	–	1,71,000
Paithan	215	660
Vani Vilas Sagar	276	–
Siddheshwar	558	4990
Malaprabha	1,146	139
Hidkal	3,521	289
Yeldari	9,595	316
Tawa	14,000	–
Rana Pratap Sagar, Stage I	16,402	–
Sriramasagar	20,000	–
Tungabhadra	41,000	81,000
Nagarjunasagar	57,000	23,000
Gandhisagar	70,000	64,838
Total	2,33,713	3,46,838

The Pong Dam in Himachal Pradesh rendered 16,000 families (each family presumably with s to eight persons) homeless. The Tehri Dam displaced 1,20,000 people from their hearths and homes (only 32,000 when the census was first done at the project planning stage), and the Navagam Da uprooted 1,50,000 people living in 326 villages. Even smaller dams of 44 to 55 m height in the Koel River (Jharkhand) caused displacement of 4000-odd adivasi (tribal) families living in 101 villages. **To compound the tragedy, no rehabilitation plans for them have been prepared so far. In South Ind on the contrary, the development of 295 reservoirs covering 12,00,000 ha of land did not crea any problem of uprooting and relocation, because they were imaginatively built in uninhabit terrain by engineers who had an appreciation of the sufferings of the people.**

Resettlement of Displaced People

an-made lakes resulting in uprooting of people cause disruption of the established pattern of social organization and engender acute psychological impact—distress, grief and trauma. As pointed out by Miller and Armstong (1982) the relocation of the uprooted communities not only involves investment of capital, but also deep understanding of the outlooks and attitudes, the culture and customs, the sensitivities and aspirations, and the real needs of the relocatees on whom a new life-style in a different physiographic setting and perhaps under different climate conditions is being imposed. With all physical hardships that the uprooting and resettlement cause, there is the breakdown of kinship ties and neighbourhood bonds in the new resettlements. All this generates a feeling of impotent rage and inevitable hostility. **As Thayer Scudder poignantly states, "it is indeed hard to imagine more dramatic way to illustrate impotence than to forcibly eject people from their preferred bitat against their will." This feeling of utter helplessness, mixed with rage, produces a crisis of identity—both cultural and individual.**

The resettlement schemes should be taken as an opportunity to raise the living standard of the displaced people, improve their social organization, and bring them into the national mainstream. Three important points must be taken into consideration:

(1) In the period intervening between the census carried out at the planning stage and the completion of the dam—which may be as long as a decade or more—the population of people threatened with displacement would have grown considerably. Many who were children would have grown into adults, having their own families. Therefore, the number of the people to be resettled is generally much larger than originally estimated. In the relocation schemes the additional numbers must be provided for.

(2) The diverse needs of different categories of people of varied age groups, avocations and professions must be taken into consideration.

(3) In the computation of compensation not only should the prices of land and property be included but also the cost of physically moving the displaced to the resettlement colony.

In view of the wide range of impacts of execution of water-resource development projects, it is necessary to embark on a comprehensive appraisal of environmental changes likely to take place as suggested in Table 17.5.

ALTERNATIVES TO BIG DAMS

reater Benefits from Smaller Dams

Realizing the damage on natural and social environments that ensues from the construction of large dams, and recognizing the great need for power generation and irrigation, it would be prudent to go in for larger number of smaller dams—10 to 30 m in height and with 50,000 to 1,00,000 m^3 reservoir capacity. Each could generate 5,000 KW of electricity. Apart from lower cost of construction and maintenance, shorter duration of completion time, and lower loss of electricity in long-distance transmission, they would not create too many environmental problems.

A comparison of the benefits of a big dam with those of a series of smaller dams located in the **tributaries in the same region** (Odum, 1959) **showed that the latter not only cost less but bring**

greater benefits without damaging or impairing the environment, croplands and settlemen (Table 17.8).

Even the detailed **investigations sponsored by the Planning Commission** (Table 17.9) **show that small dams not only cost less and start yielding benefits quite early, but also bring great benefits at the expense of far less damage to the environment, croplands and settlements** (Singh, 1993). There is no denying that the cost of the electricity generated from smaller dams would be high, but if one were to think of the value of the lost natural resources, the cost of restoration of damaged land and the huge expenditure—and enormous distress—on the relocation or rehabilitation of the uprooted people, the cost of electricity generated by small hydel project is more than compensated (Valdiya, 1992, 1997, 2004). **Two examples are sufficient to show this. The 28-m high and 95-m long dam acro** Jhelam River near Buniyar in the Baramula district of Kashmir has the capacity of generating 480 MW **of electricity. The 67-m high Nathpa–Jhakri Dam in Himachal Pradesh caused the displacement of on** 52 families (Mande and Vohra, 1988).

Table 17.7. Environmental Impacts of Development Projects (From: R.E. Munn, 1975; *Environmental Impac Assessment: Principles and Procedures*, John Wiley, New York)

A. Proposed Project
1. Objectives of the Project
2. Location and Land Requirement
3. Existing and/or Proposed Facilities (Cultural, Material and Technical)
4. Construction Procedure and Schedule
5. Procedures of Operation and Maintenances, and Safety Measures Proposed
6. Planning Constraints
7. Possible Alternatives to the Proposed Project
B. Existing State of the Natural Environment
1. Land Features (Topography) and Landuses
2. Geological Setting and Foundation Conditions
3. Geological Hazards
4. Meteorological Condition and Hydrologic Regime
(a) Climate and Rainfall
(b) Surface Hydrology
(c) Underground Hydrology
(d) Water Quality and its Monitoring
5. Biotic Resources
(a) Vegetation
(b) Wildlife
(c) Fisheries and Aquatic Communities
(d) Genetic Resources
(e) Endangered Species
6. Non-renewable Resources

Con

7. Historical and Archaeological Heritages (Cultural Resources)
8. Criteria for Measuring the Impacts
C. Areas of Human Concern (Socio-economic Conditions)
1. Population and Occupational Status and Demographic Impacts
2. Employment Pattern: Existing and Expected
3. Economic Condition and Impacts
4. Social Values and Amenities and Health Condition
5. Problem of Resettlement
6. Influx of Labour and Consequent Socio-economic Problems
D. Environmental Impact of the Proposed Action
1. On Land Features and Uses
2. On Geological and Soil Conditions
3. On Ecosystems and Species (Terrestrial, Aquatic and Threatened)
4. On Air and Water
5. On Renewable Natural Resources
6. Waste Production and Disposal
E. Measures to Avoid or Mitigate Adverse Environmental Effects
1. Preventive Measures
2. Remedial Measures
(a) Measures
(b) Effectiveness
F. Alternatives to the Proposed Project
1. Feasibility of Alternative Project
2. Economic Comparison Through Cost-Benefit Analysis
G. Recommendations
1. Significant Environmental Impacts
2. Mitigation Measures
3. Alternative Projects.

Table 17.8. Comparison of a Big Dam with a Series of Smaller Dams (Data from Paterson in Odum, 1959).

	Mainstream Reservoir	Multiple Head-Water Reservoirs
Number of reservoirs	1	34
Drainage area (mi^2)	195	190
Flood storage (acre-feet)	52,000	59,100
Surface water for reaction (acres)	1,950	2,100
Flood pool (acres)	3,650	5,100
Bottom inundated (acres)	1,850	1,600
Bottom protected (acres)	3,371	8,080
Total Cost	$ 60,00,000	$ 19,83,000

Table 17.9. Expenditure on and Benefits from Major and Minor Irrigation Projects: Planning Commissi Statistics (S.Singh, 1993).

National Plan	Quality/Expenditure (million rupees)		Cumulative Potential (million hectares)	
	Major/Medium	Minor	Major/Medium	Minor
Pre-Plan benefits	–	–	9.7	12.9
First Plan	3,800	760	12.2	14.06
Second Plan	3,800	1,420	14.3	14.79
Third Plan	5,810	3,280	16.6	17.01
Annual Plan 1966–1969	4,340	3,260	18.1	19
Fourth Plan 1969–1974	12,370	5,130	20.7	23.5
Fifth Plan 1974–1978	24,420	6,310	24.82	27.3
Annual Plan 1978–1979	9,770	2,370	25.86	28.60
Annual Plan 1979–1980	10,790	2,600	26.6	30
Total	75,100	12,930	168.88	187.16

According to Satyajit Singh (1993) a big dam displaces on an average 44,182 people. Taking in consideration the size of modern dams and the growth of population in the past sixty years, t figure of displaced people per dam can be taken at 10,000. There were 3,300 big- and moderate-si dams in the country uptil about 1991. This means that 33 million people in India were uproot (Table 17.5) due to dam-building projects (Roy, 1999).

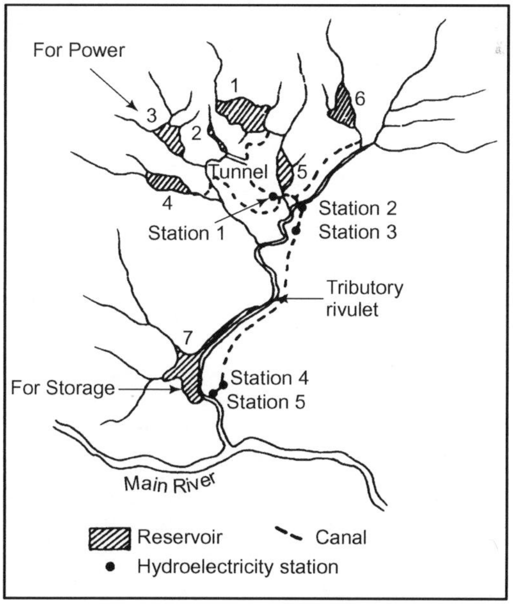

Fig. 17.29. The Liang Tian River Cascade Scheme profitably taps a small river in the mountainous Chi for storage of water for irrigation, flood moderation and electricity generation without creati environmental and social distress. (After Naibo, 1986 in Valdiya, 2004).

Long back, Panjab had shown the way for tapping hydroelectricity from even a canal flowing through gently sloping plains. About 300 KW of electricity generated was from a mere drop of 2.5 m in the western Yamuna Canal at Kakori in Sonipat district in a pilot plant. Together, thousands of such microhydel power plants in the myriads of streams, canals and rivers all over India are capable of meeting all the energy needs of the country.

A far more environment-friendly and socially less distressing way of harnessing the hydropower is what Zheng Naibo (1986) describes as **Cascade Development Scheme**, utilizing small rivers for multiple purpose. The Chinese have adopted this model, which costs less on engineering, provides eater benefit, and causes little environmental distress. A small hydropower station with a storage ervoir located in the upper reaches (Fig. 17.29) increases the firm capacity of the downstream station and reduces the amount of discarded water. All stations thus jointly maintain water balance between the tions. The scheme proves beneficial to the local community, and also results in the involvement of the local people in the construction.

crohydel Potential in India

Already in some states small hydropower projects (with station capacity of 25 MW) have been taken up. The aggregate installed capacity of small power units was 2953 MW by the end of December 2010. **A rvey showed that small hydroelectric power units can be built in 5718 sites in India which will ve a total capacity of 15,384 MW (Sukhatme, 2011).**

SUMMARY

The design of a dam built across a river depends not only on the purpose for which the project is dertaken (irrigation, power generation, flood control, etc.) and on the availability of constructional material, but also on the width of the valley, stream dynamics, and the geological conditions of the foundation. On a narrow gorge with weaker rocks on the bed but stronger on the valley slopes (abutments) an arch dam is preferred (e.g. Idukki in Kerala). In a wider valley without good foundation material, not uniformly good to bear the load, a multi-arch or buttress dam is suitable. Adaptable to all situations are gravity dams, which derive their stability from their own weight. Masonry and concrete gravity dams are useful when the valley is narrow and the ground strong enough to withstand the load of the structure. rthern or rockfill gravity dams are built where the foundation-grounds are not strong—either having low bearing strength (e.g. mudstone, cavernous limestones, unconsolidated sediments, gravels, etc.) or e beset with structural deficiencies (such as jointing, foliation, fracturing, faulting, brecciation, etc.). ch dams do not have a height of more than one-fifth of the width at their base.

A dam is subject to destabilizing lateral forces, comprising mainly lateral thrust or forward push of the impounded water, waves generated by earthquake shocks and avalanches or landslides, and horizontal ground motion due to earthquakes. To counteract this, the vertical force created by the load of the structure and weight of the impounded water and sediments accumulated on the face of the dam has to be increased. The problem arises when leakage and interstitial percolation permits reservoir water to enter under the base of the dam and exert upward pressure and thus create a condition of partial buoyancy.

The seeped-in water may also soften the foundation material and render it slippery. The consequent reduction of frictional resistance contributes to sliding of the dam. To prevent failure of the dam by toppling and sliding, it is imperative to eliminate all causes of leakage and infiltration of water und the dams by putting aprons of impervious clay, concrete or asphalt on the spot, structural feature or rock horizon exposed to water, or by sealing them by grouting. Constructing cut-off walls and trenches, grout curtains or concrete diaphragms effectively checks leakage. In the location of the dams, care is taken to see that impervious layers form aprons on the heel and toe and that the dip of these beds does not favour seepage.

There are three main problems related to the formation of reservoirs. The leakage of water due to rise of water level above the hump of the underground water table and through permeable rocks and or sediments such as cavernous limestones, vesicular or scoriacious basalts, unconsolidated gravels, etc. may defeat the very purpose of having the reservoir, and also cause waterlogging in the adjoining area. Accumulation of sediments derived from erosion in the catchment area results in decrease in the storage capacity and reduction to half, or even a quarter, of the expected lifespan of the reservoir. The fast rate of siltation in the man-made lakes in India, particularly in the Himalayan region is a matter of serio concern.

The occurrence of earthquakes induced by impoundment of water depends on (i) the maximum height ($\geq$100 m) of the water attained, the infiltration of water through cavities, fissures, faults, joints, (ii) t pore water pressure, and (iii) the existence of tectonically stressed ground—brittle rocks under strong stress resulting from neotectonism. Infiltration of water into this stressed zone stimulates movements a attendant seismicity. Unless all these conditions are met, there will be no reservoir induced seismicity.

Weighted against the potential benefits of reservoirs such as power generation, irrigation, flo control, recreation facilities and navigation, etc. are the many adverse consequences of the artific impoundment of water behind dams, viz. destruction of forests and pastures and the flight of wildlı due to submergence, loss of agricultural fields and uprooting of human communities, undesirab ecological changes, acceleration of erosion and landsliding in both the lake basin and downstream drainage network, increased occurrence of earthquakes, etc.

Both hydro-electricity and water from dammed reservoirs are only as cheap as they are because the environmental and social objections are seldom brought into account (Miller and Armostrong, 1982). "And when as much weight is placed on the protection of the environment and delicate ecologic system as on the meeting of economic needs, alternative less spectacular solutions to water problems, not entirely radical modifications of nature, may prove to be the wisest course" (P. P. Micklin in Detwyler, 1971).

18 Foundations of Buildings and Bridges

STRUCTURAL LOAD AND FOUNDATION STRENGTH

An Example from the Past

The Taj Mahal, the exquisite monument of immortal love, standing on the bank of the Yamuna River at Agra for more than three and a half centuries, is not only an architectural marvel but also has a superbly signed foundation. Taking into account the floods that recurrently hit the floodway of the Yamuna, d the unconsolidated nature of the alluvial sediments on which the edifice stands, **the designer of the Taj ensured proper distribution of its load through a well foundation (Fig. 18.1). Withstanding the ravages of time and natural hazards, it stands as a tribute to the geotechnical genius of the Indian architects of the past. The glory and stability of many such ancient monuments, including the Qutub Minar and Iron Pillar at Delhi, are mute testimony to the appropriate foundation conditions that ancient builders ensured for construction.**

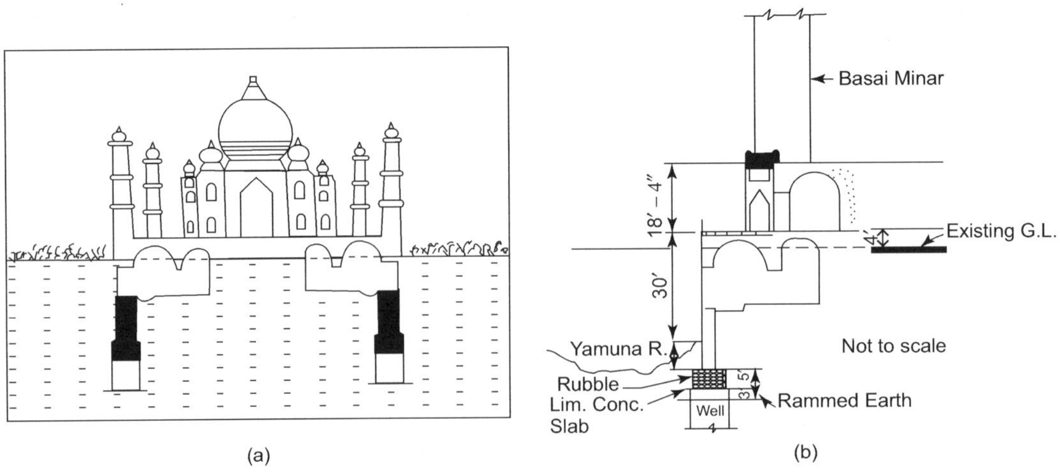

Fig. 18.1. The Taj Mahal, located on the alluvial sediments of the Yamuna River, stands on a well (pier) foundation. (From the emblem of the 4th International Engineering Geology Conference, 1982.) The *inset below* shows how the foundation beneath the Basai Minar (minarette) of the Taj transmits and distributes the load of the superstructure. (Redrawn with permission from R. F. Legget, 1962, *Geology and Redrawn Engineering,* McGraw-Hill, New York, 475 p.)

Forces Acting on Foundations

Whether it is a bridge or a building, the load of the structure is borne by the foundation. The load consists of its own weight (dead load) and the weights of things resting on it (live load). Then the are lateral forces generated by blowing wind or storms, rushing flood waters and ground accelerati due to earthquakes. The lateral forces push the structures sideways. The resultant of the lateral a vertical forces is transmitted to the foundation where it generates two types of stresses—compressi and shear. The former causes reduction in volume of the ground material under the foundation and consequent settlement leading to deformation and even plastic flowage. The innate strength — that the compressive and shear strengths—are commonly strong enough to counter-balance these forces deformation and change. **The stability of buildings and bridges depends principally on the strength of the material under the foundation to resist compression (reduction in volume) and shearing (deformation or plastic flowage), and secondarily on the manner in which the loads are transmitt to the ground.**

Intrinsic Strength of Foundation Material

The engineering properties of the material under the foundation—rocks, sediments or soils— **on which depends the stability of the structures, include shear strength, compressibility and bearing capacity, permeability, shrink-and-swell potential, erodibility and corrosion potential.**

The ***shear strength*** of a material depends on the tendency of the constituent particles to stick together (cohesion) and on the innate resistance the particles offer to all sorts of movements. This resistance called ***frictional resistance,*** which is generated as a result of interaction between individual particle The shape, size and density of packing of particles (texture) have a significant bearing on the friction resistance. **The entry of water_diminishes the frictional resistance, and consequently reduces the shear strength of the material under the foundation.** The sandstones with finer-sized constitue grains have higher shear strength than the coarser-grained sedimentary rocks. Higher porosity implı greater content of interstitial water and therefore lesser shear strength. Similarly, **the presence of discontinuities such as fissures, fractures, joints, and cleavages significantly diminishes the she strength of the rocks.** If the load exceeds the shear strength, there is either sliding along the plane of discontinuity or if the material is consolidated, there is plastic deformation or flowage. **It may be pointed out that the bearing capacity of any ground is always determined from the strength of the weakest member of the succession there.** The ***compressive strength*** represents the tendency or ability to resist decrease in volume due to consolidation and is a function of the elastic nature of the particles. Finer material like clays and silts have lesser compressive strength than the coarser material such as gravels and coarser sands. **The presence of organic matter and clays in the sandstone thus increases its compressibility—propensity for settling.** The compressive strength (Tables 18.1 a 18.2) is measured in terms of the stress required to break the specimen (unconfined at its end), a expressed in terms of pound per square inch (Krynine and Judd, 1957).

The *swell-and-shrink potential* exhibited by certain clays such as montmorillonite is related to the propensity of the clays to imbibe or chemically attract water. The water thus drawn insi the framework of the clay crystals forms layers of molecules alternating with layers constituted of S Al, and O. The development of layers of water molecules causes the clay crystals to expand, and th deprivation due to drying leads to shrinkage. It is stated that montmorillonite clays can expand fifte times their original volume, although soils seldom show more than 25% to 50% expansion (Kell 1981). **Even as little as 3% expansion is enough to destabilize the foundation.**

Table 18.1. Crushing Strength of Some Central and South Indian Rocks (After L. N. Vaidyanath and D. K. Ghosh, 1982).

Rock Kock	Crushing Strength in Dry Condition (kg/cm^2)
Archaean	
Granites	825 to 2262
Aplites	841 to 1342
Rhyolites	1153 to 1184
Dolerites	848 to 1222
Basalts (Sheared)	817 to 891
Amphibolties	848
Granulites	597 to 1571
Gneisses	636 to 1082
Marbles	608 to 761
Phyllites	1161 to 1244
Schists	77 tp 480
Quartzites	428 to 1870 to 3273
Cuddapah-Vindhyan Supergroup	
Limestones	1352 to 1570
Slates	480 to 742
Quartizities	836 to 1568
Porcellanite	1137 to 1149
Gondwana Supergroup	
Sandstones (soft)	35 to 54
Sandstones (moderately hard)	140 to 853
Sandstones (purple)	145 to 409
Sandstones (very hard)	471 to 1239
Shales	442 to 661
Deccan Volcanics	
Basalts (dense)	644 to 3007
Basalts (amygdaloidal)	118 to 766
Basalts (vesicular)	144 to 421
Volcanic Breccias	220 to 388
Tuffaceous Basalts	56 to 82
Tuffs	222 to 294
Laterites	76 to 384
Bauxites	296 to 611

The ***permeability of*** a material is its ability to allow passage of water without damaging its own ernal structure. It is measured in terms of the volume of water passing through the unit cross-section of the unit length of the specimen under study. The permeability depends on the interconnection of the pore-space and openings present within the material. The coarser the material and rarer the packing grains, the higher is the permeability. The gravel without finer material has very high permeability hundreds of $m^3/day/m^2$ in contrast to lower permeability of finer-grained cohesive material like clay—fraction of one $m^3/day/m^2$.

Table 18.2. Categorization of Rocks on the Basis of Compressive Strength (Krynine and Judd, 1957)

Category	Range of Compressive Strength (psi)	Rocks
I	>40,000	Basalt, Dolerite, Quartzite
II	40,000 to 25,000	Fine-grained Granite, Diorite, Basalt, Compact-cemented sandstone, Quartzite
III	25,000 to 10,000	Average sandstone, Coarse-grained granite, Gneisses
IV	10,000 to 5,000	Porous sandstone, Shales
V	<5,000	Tuffs, Chalk, Very porous sandstone, Siltstone

Closely related to the permeability is the **erodibility of the soil**—the ease with which water movi through the pore-spaces can remove away the constituent material. The cohesion and consolidation material are an insurance against easy erosion.

Some materials are prone to weathering or ***corrosion*** on the surface promoted by water and high temperature.

In brief, the stability of the foundation depends on three factors—the intrinsic shear and compressive strength, the degree of structural deformation (discontinuities) and the degree of weathering of the ground material.

GROUND CONDITIONS AND FOUNDATION DESIGNS

Ground Conditions

The ground may be made up of solid hard rock(s) very close to or exposed on the surface so that t foundation can be laid directly on it, as is done in such metropolitan cities as Bengaluru, Mumbai, Ne Delhi, Hyderabad, etc. In such cases evaluation of structural homogeneity or defects and bearing streng of the rocks is undertaken before emplacing the foundation of heavy structures. Commonplace or lighter structures can be placed without special treatment of the ground. In the case of weaker materials resti on stronger ones, it is the weaker mantle that will experience shear deformation and failure will resu On the contrary, if a stronger layer overlies a weaker material, the stronger member will spread the lo and thus reduce pressure on the weaker material. However, this aspect is dependent on the thickness the rigid layer and the load acting on it. Nevertheless, the shear failure by plastic distortion occurs only in the weaker material as the upper member bends downward if the load exceeds bending moment.

Differential settlement is the most serious problem, as it leads to distortion of structures. requires adequate remedial treatment of the ground. The safe limit of angular distortion due to settlement is 1 : 500 (Bell, 1982). The commonly accepted limit is 25 mm. The granular material shows settleme not more than 25 mm (rarely more than 50 mm), but in clayey material it is quite high, and m continue to take place for years. The clays deform easily under imposed shear stresses, as already explained, leading to differential settlement. If the ground is made up of a thick mantle of soil or loose consolidated sediment such as in the Indo-Gangetic Plains, the Mahanadi, the Krishna–Godavari, the averi and the Sabarmati Plains, the foundation should be such that it transfers and distributes the loads the structures through the overburden to the bedrocks (if they lie deep underneath) or to the hardened diments deeper down.

The presence of faults and shear zones do not necessarily mean an incurably unstable foundation. •wever, **weak material such as soft incompetent gouge and shattered material of the fault and shear zones have to be replaced by concrete or other competent materials before laying the foundation.** It is therefore necessary to undertake intensive exploration, including test-drilling and boring order to determine the underground conditions, including groundwater regime, bearing strength of the erburden material, etc. On the basis of the knowledge thus acquired, and taking into consideration the onomics of the project, the type or design of the foundation is selected (Figs. 18.2 and 18.3).

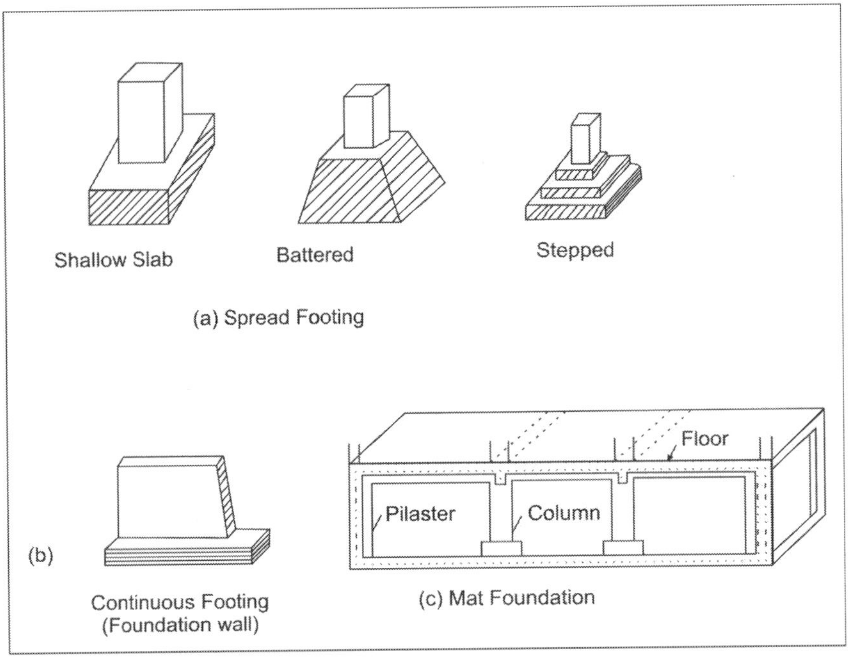

Fig. 18.2. The spread foundation and mat foundation ensure distribution of load over a larger area, and thus reduce the pressure per unit area to within the limit of the bearing capacity and shear strength of the foundation material. (Modified after D. P. Krynine and W. R. Judd, 1957, Figs. 13.2, 13.5).

Special Types of Foundations

The footing of the ***spread foundation*** takes up the weight of the structure and spreads it over a larger ea so that the load per unit area (tonnes/m^2) on the material under foundation is less (Figs. 18.2 a d b and 18.3). **Where the soil or overburden has low bearing strength, the foundation is made as a continuous reinforced concrete slab covering the entire area of the structure. This is a mat foundation. In case the soil is very weak and prone to plastic deformation in the presence of**

water, a box-type foundation is preferred. The overburden is excavated down to the depth of t bedrock—hardened horizon—and the space dug out is utilized for building as basements (Fig. 18 c). Another method of building foundations on very weak ground is to excavate continually benea the foundation as the height of the structure rises so that the structure progressively sinks to become a ***caisson foundation.*** The caisson foundation can also serve as an underground basement or even a storey (Legget, 1962).

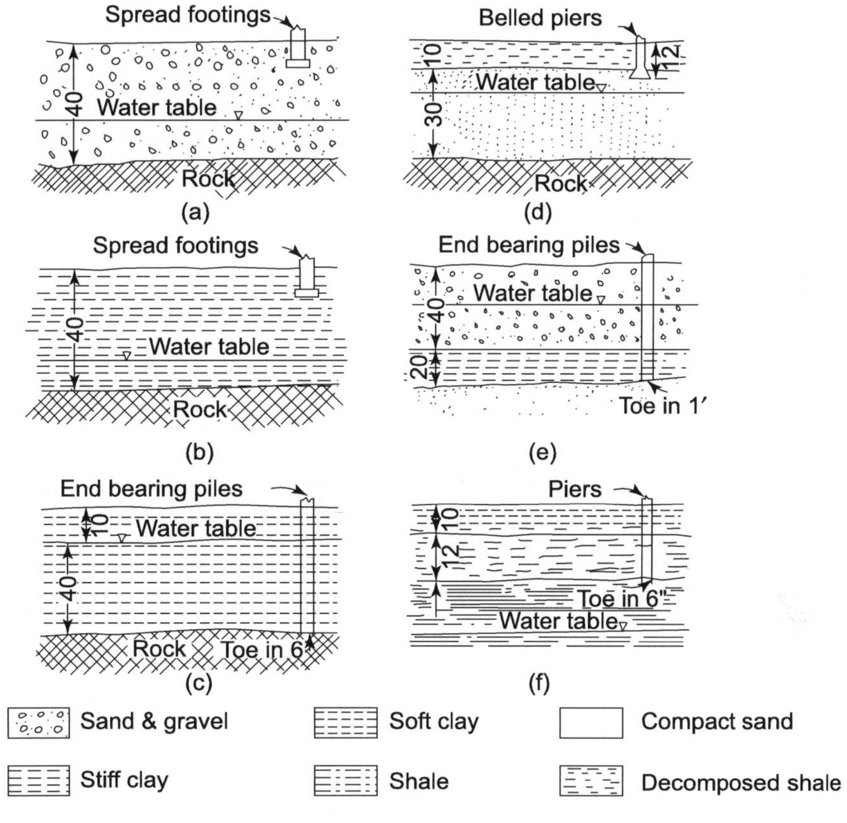

Fig. 18.3. The grain-size and density of packing (texture) of the sediment determines the design of the foundation. On firmer sand and gravel, and stiff clay, a spread foundation (a and b) is quite suitab Soft plastically yielding clays even if interlayered with stiff material call for a pile foundation (c and e), or piers (d and f). (Redrawn with permission from D. P. Krynine and W. R. Judd, 1957. *Principles of Engineering Geology, and Geotechnics,* McGraw-Hill, New York, 730 p., Fig. 13.12).

If the depth of the overburden is great, and the material consists of gravels and/or loose aggregate of sediment, *a pier foundation* or *well foundation* is made (Fig. 18.3 d, f). This is do by drilling holes (25 cm to 2 m) and filling the holes or wells with the concrete reinforced with bars steel. The pier transmits the load vertically downwards to the hardened horizon below and laterally the material on the periphery. **The foot of the piers can be advantageously enlarged to make it bell-**

shaped in order to spread the load (Fig. 18.2). The Taj Mahal provides the example of this kind of foundation.

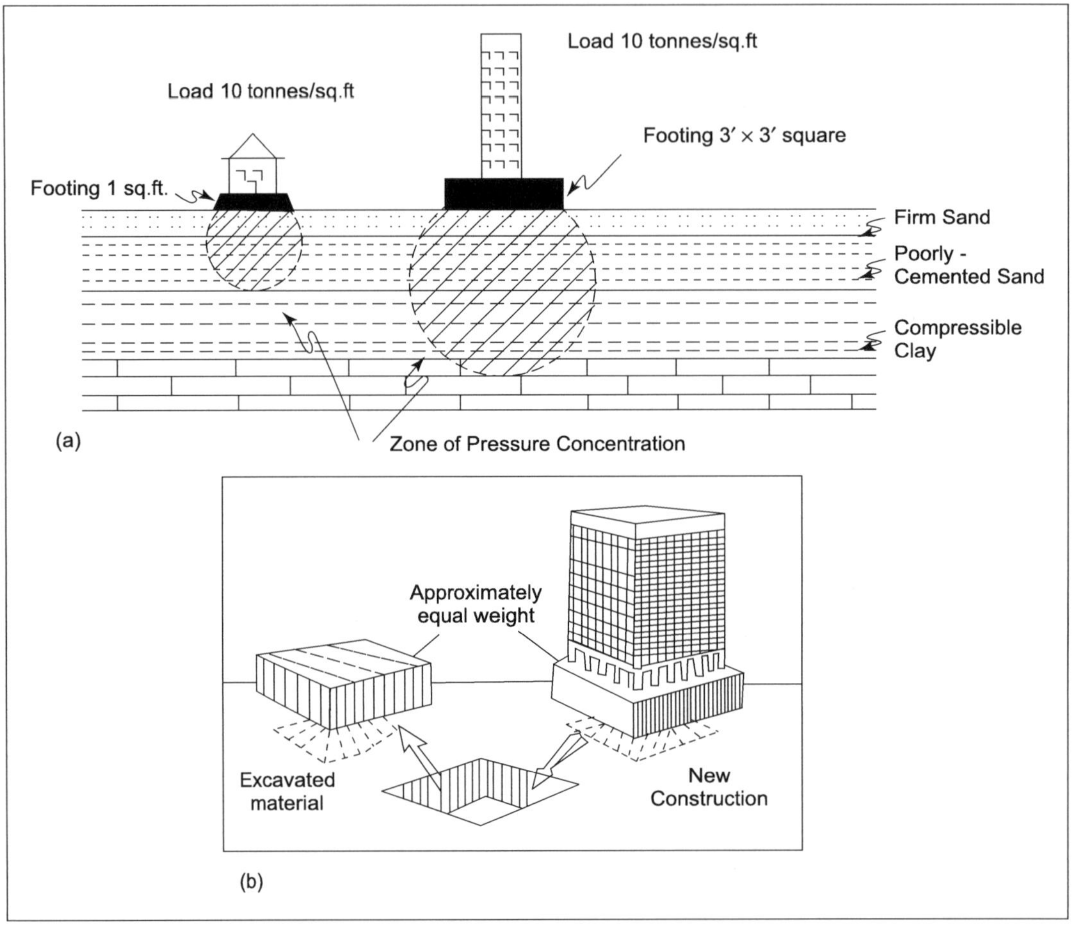

Fig. 18.4. (a) Distribution of the load of the structure on the foundation material of different compressive strengths would react differently to the load in the zone of pressure concentration. (Based on D. P. Krynine and W. R. Judd, 1957, Fig. 22.8). (b) Softer clays that cannot support heavy structures are excavated. The amount (weight) of excavation is equal to the weight of the building to be constructed. The building raised on the basement thus floats on the soil. (From a brochure published by the U.S. Geological Survey).

If the depth of the overburden is not very great and the soil is free from gravelly material, piles timber poles, steel or reinforced concrete (Fig. 18.3 c, e) are driven by hammering mechanically in der to make it reach the reasonably firm material—***point-bearing pile foundation***—or upto a depth here the adhesive soil can counter the load through frictional forces acting on the piles —***friction-pile foundation*** (Krynine and Judd, 1957). The transfer of the stress which steadily increases as the structure is built (Fig. 18.4a) depends on the nature of the overburden or soil. This calls for prior investigati of the soil properties—compressibility and safe bearing capacity against settlement and overstressi under the foundation (Legget, 1962). Particularly relevant is the investigation to find ways and mea of controlling settlement.

REMEDIAL MEASURES AGAINST SETTLEMENT

If an investigation indicates high settlement potentiality, it would be necessary to preload the grou adequately or render it compact by artificial means before placing the foundation. The ***vibrocompaction* entailing lessening intergranular space reduces settlement by more than 50 % (Bell, 1980) and increases shear strength substantially.** The poker vibrator or vibro-float can be penetrated into t soil to effect compaction of material lying even below the water tables. Another method of achievi dynamic compaction is repeated application of intense impacts, as the dropping of large weights (10 20 tonnes) from crowbar cranes from heights of 15 to 40 metres.

***Grouting* strengthens the rocks and soils considerably. Fluids such as cement slurry that set quickly are injected under great pressure into fissures, pores and cavities in the ground.** The grout holes are oriented in such a manner as to intercept as many discontinuities and holes as possible and upto a considerable depth. In water-saturated material, grouting with bitumen or resin is preferred. B the common grout is cement slurry.

However, if these efforts prove unavailing, the foundation should be made flexible characterized joints connecting the various blocks.

FOUNDATIONS ON SLOPES

In hilly and mountainous terrains, special precautions are needed to ensure stability against slope failure. The factors to be taken into consideration are: (1) the structure of the bedrocks, (2) the angle of slope a (3) the pattern of groundwater flow. Not only the removal of loose burden or soil is necessary, **the surface water and underground flow must be diverted at the very outset without fail. Grouting is resort to in order to seal the openings and fissures.** Likewise, the foliated, jointed and bedded rocks may ha to be bolted (Figs. 11.31 and 19.8b). The foundations are then keyed to the firm ground (Figs. 11.5 a 19.7). **Another very important measure is to cover the foundation with damp-proof coating and construct a drain on the side of the hillslope** (Fig. 18.5). The drain may have to be filled with grav and sand (Fig. 11.23) to enable the water to percolate down without being impeded by garbage or deb which commonly accumulate in open drains. This kind of foundation is recommended for such towns Udagamandalam, Mahabaleshwar, Dharamshala, Shimla, Mussoorie, Nainital and Darjiling.

GROUNDWATER AND FOUNDATION PROBLEMS

Swelling Clays and Pore-Water Pressure

Water-sensitive material like clays, loess and salts are particularly prone to plastic deformation and even liquefaction, leading to foundation failure in a situation where the ground water table approaches close to the surface—as in the waterlogged terrains in Haryana, Panjab and Delhi (Chapter Five). Clays such as montmorillonite that swell in the presence of water are a

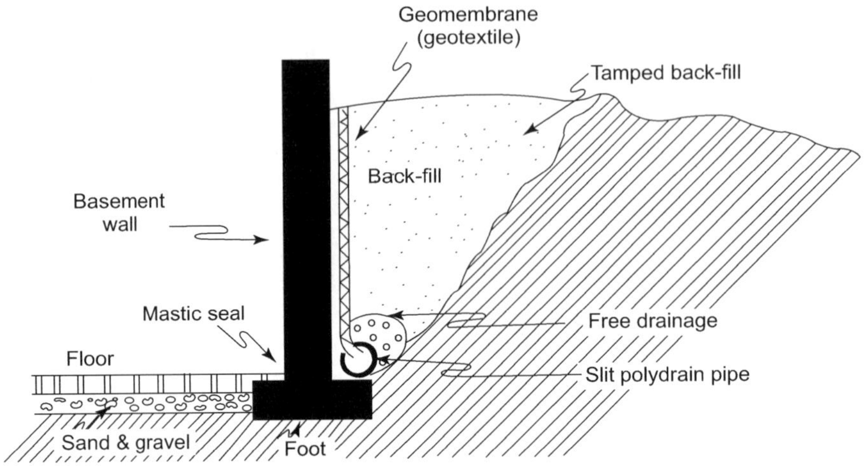

Fig. 18.5. On a slope in mountainous terrain special care has to be taken to prevent entry of groundwater into the foundation.

very treacherous component of the material below foundation. During drier seasons they shrink nsiderably. This explains why a very large number of houses and structures in the Deccan and Malwa ateaus have problems of differential settling as manifest in the cracks on walls and buckling of floors. similar situation is witnessed around lime kilns and brick kilns where heat dries up the clays, thus causing shrinkage.

It is a common knowledge that the water present in a soil or rock not only corrodes the foundation aterial but also reduces its shear strength. The infiltration of water builds up pore pressure, threatening the stability of the foundation. In areas underlain by soft clays that are prone to plastic deformation d swelling-shrinking, it is desirable to excavate the material equal in weight to the structure to be nstructed (Fig. 18.4) — the box-like basement "floats" on the soil. Since the weight of the clays under and around the structure is approximately the same as the structure, it does not settle or get deformed.

Internal Erosion

e percolation of the underground may also wash away finer or soluble material from the fabric of e soil rock. This commonly happens when the water table rises or lowers due to heavy pumping drawdown in reservoirs. The net result is internal erosion and differential settling. It is therefore essential to take into consideration the anticipated settlement problem and the foundation designed cordingly. There should be built-in provision for accommodating unavoidable settlement by jacking the foundation such as a three-point instead of four-point foundation and installing adjustable joint connections.

The leaning Tower of Pisa in Italy, the construction of which was started in 1174 and completed by 50, rests on a bed approximately 4 m thick of clayey sand underlain by about 6.4 m of sand and clay g. 18.6). Since its construction more than 650 years ago, the tower has been considerably displaced its total height of 54.6 m, the movement being of the order of 1.4 mm/yr. The clay layer at the depth 8.5 m below the foundation is responsible for this displacement.

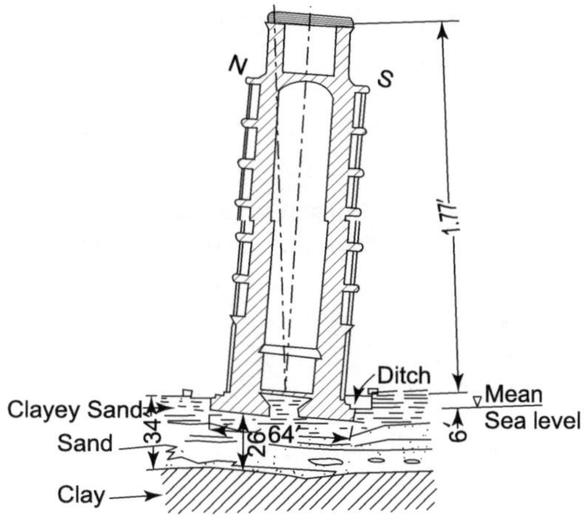

Fig. 18.6. Plastic deformation (yielding) of the soil layer under the Tower of Pisa (Italy) is responsible for differential settlement and considerable lateral displacement of the foundation and tilting of the tower. (Reproduced with permission from D. P. Krynine and W. R. Judd, 1957, Fig. 17.31.)

FOUNDATIONS OF BRIDGES

Special Problems

Forming an integral part of road or rail track, a bridge is located on considerations of, or requirements imposed by, the grade and alignment of the route. The problem before the geologist is therefore not to suggest a suitable site or alternatives for bridges, but to bring out the conditions of rocks or/and soils selected for placement of the terminal supports (abutments) and intermediate supports (pier The geologist suggests the type of bridge on the basis of existing geological conditions and prevailing dynamics of the water of the river to be spanned. There are a number of problems peculiar to bridges

(1) As a sizeable part of the ground is concealed under water, the knowledge of the geological conditi of the foundation can be gained only by drilling and extrapolation based on an in-depth study the banks.
(2) The valley is filled with sediments of all kinds — gravels, sands, clays and muck, depending the stage of maturity of the river. These sediments have variable bearing capacity.
(3) The load (weight) of the bridge and its supports are acting on a restricted area. Commonly, th load is transmitted through the structural design to the abutments so that the latter have to be very strong and stable.
(4) The bridge structure stands on waterway—indeed as an obstruction—and withstands the action the flowing water, which is swift and scouring during floods. The magnitude of erosion caused the water flowing through the reduced cross-sectional area under the bridge depends on veloci freak discharge, high-water period, and eddy generation. One of the considerations is the likeliho of the submergence of the bridge during floods, and the resultant threat of destabilization. T 370-m long bridge on the Narmada River near Jabalpur, for example, used to be submerged ye after year.

5) In wide valleys filled with sediments, there is always a possibility of the existence of buried channels, pot holes, cavities and depressions. In-depth investigations, therefore, have to be carried out to identify and locate them. The bridge on the Bilaspur-Jagdalpur highway (Madhya Pradesh), for example, passes through a karstic terrain in which flows the Sheonath River near Bilaspur, and the Indravati River near Jagdalpur. The bridge sites reveal all these infirmities pronouncedly.

Thus, a bridge must be safeguarded against erosion by rivers, damage by landslides and ckfalls, and foundation failures due to the presence of weakness planes or pockets of infirmities. This calls for thorough prior appraisal of the overall foundation conditions, including presence or herwise of pockets of sulphides and salts, cavities, shattered rocks, presence of harder rocks over aker layers, rock heterogeneity, zones of weathering, etc. The safety of bridges depends on the entification of these features and effective remedial measures.

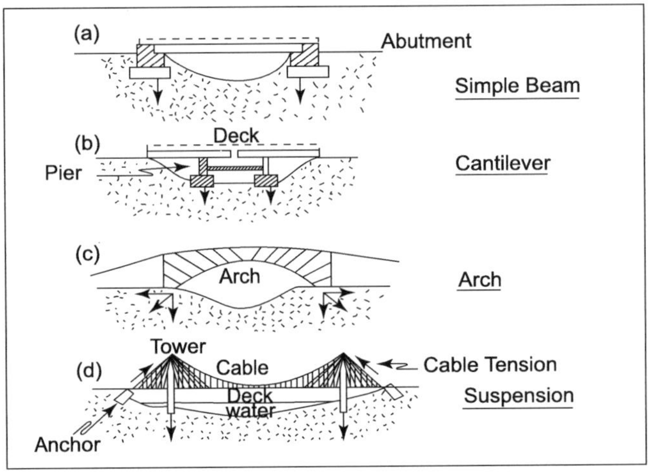

Fig. 18.7. Four types of bridges, their designs depending on the condition of the foundation material. (Redrawn with permission from D. Liverson, 1980. *Geology and Urban Environment,* Oxford University Press, New York, 120 p., Fig. 6.14).

Load Factor and Bridge Types

•t only the magnitude of forces transmitted to the foundation but also the manner in which the forces can be transmitted, determine the design of the bridge foundation. If the forces are to be transmitted rtically to the firm stable abutment, a ***simple beam bridge*** is the choice. In a ***cantilever bridge*** the rces are transmitted to piers standing on the sound foundation (Fig. 18.7). If the abutment is located a poorer foundation, the piers have to take the load. The wider the river, the deeper should be the netration of the piers. The examples are the 96-m long Salal bridge on the Chenab River and the huni Nala bridge on the Jammu-Srinagar highway.

In addition to the vertical load, there is a horizontal thrust exerting pressure on the abutments in the ***arch bridge*** and ***rigid-frame bridge.*** *In* such bridges the stability is ensured by anchoring the superstructure through cables to the rocks or concrete blocks on the abutment. This involves inclined tension as in a ***suspension bridge*** (Fig. 18.7).

Stability of Bridges

The stability of a bridge depends on the bearing capacity of the material of the foundation ground. Effective measures have to be taken against the possibility of settlement of the foundation as a consequence of load of the bridge structure plus the weights of rolling stocks. Unequal settlement pos a serious threat to the stability. **Spread foundation would ensure distribution of loads and reduction per-unit-area pressure and thus diminish the possibility of settlement.** Compacted coarse sands and confined compacted fine sand form good ground for the foundation for piers. On the other hand, soft much finer sediments such as silt and clays create problems, including failure of bridges. The piers the Dholpur bridge on the Chambal River are founded on the horizontal limestone interbedded wı sandstone and purple siltstone and claystone, and with a capping of subrecent conglomerate. The s• conglomerate was not able to stand the imposed load of 10 kg/cm^2, leading eventually to the failure of the bridge (Mathur, 1982).

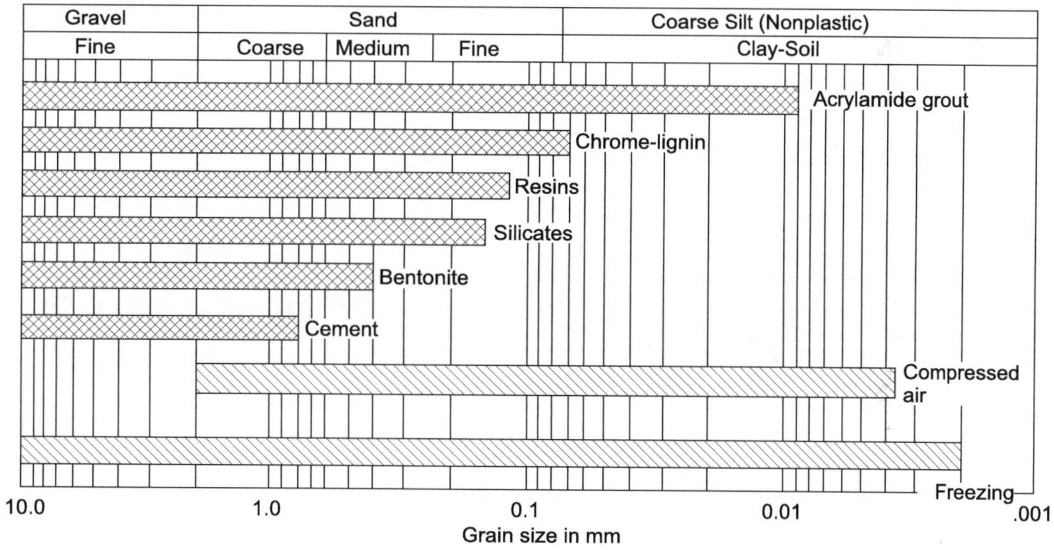

Fig. 18.8. Weaker foundation material can be treated chemically by means of grouting. The nature of t grout depends on grain-size of the foundation soil and or sediment. (Redrawn with permission from F. G. Bell, 1980, *Engineering Geology and Geotechnics,* Newnes-Butterworths, London, 497 p.,Fig. 5.14).

Taking the foundation deep to the bedrocks (limestone), which are capable of taking a load 6 kg/cm^2 would not have created this problem. **If a soft layer is encountered in shallower depths, a pile foundation is the obvious choice—more so if the bedrocks are overlain by such yielding material as sediments. In cases like this a differential settling of 10 to 15 cm is unavoidable and spread footing has to be designed.** The materials of the foundation of low bearing strength a consolidated through artificial means—chemical and cement grouting, mechanical compaction, e (Fig. 18.8).

Other measures of ensuring stability of bridges involve planning against the pressure of running ater, particularly during floods, the impact of drifting wood or ice blocks, storms, shocks of passing hicles including trains and seismic shocks. During floods the scouring action is very accentuated. The placement of riprap around the piers and abutments is a palliative measure. The abutments have to be feguarded against such hazards as mass-movements, rockfalls, and avalanches.

SUMMARY

e foundations of buildings and bridges transmit the load of the structure to the ground, which must have e capacity or strength to withstand the pressure without impairment. The principal qualities imparting rinsic strength to the material under the foundation are compressive strength (bearing capacity) and ear strength. These strengths are derived from the cohesion of the constituents—their ability to stick together and resist all kinds of movements or dislocation and deformation. These qualities depend on the textufcre and structure of the foundation. The presence of structural discontinuities, defects and res or cavities that allow entry of water and occurrence of swelling-shrinking and plastically yielding aterial (such as clays) appreciably reduce the intrinsic strength of the material of the ground under the foundation. The foundation design ensures the distribution of load over a larger area so that there is minimum pressure per unit area. Where the material under the foundation has lesser strength, a pile foundation is preferred.

There is little choice on the location of the bridge site, as it is governed by alignment and grade of the route. The design of the bridge depends on the bearing strength and shear strength of the material of the ground under foundation. If the abutment rocks are strong and stable, the load is transmitted through e structure of the arch and suspension bridges, and where the load has to be transmitted vertically wnwards, piers are driven deep if the bedrock is overlain by soft and yielding material.

In mountainous terrain special care is needed to counter the pressure exerted on the foundation by material that slides down unstable slopes, and to divert the surface and underground water.

In all cases, the ingress of water through permeable horizons and or fissures and openings has to be checked, for it is the water which reduces the shear and compressive strengths of the material of e foundation ground, and also may cause swelling of certain clays or even bring about liquefaction, erode away cement and finer constituents, thus weakening the texture and integrity of the material. hemical treatment (grouting) appreciably improves the strength of the ground under the foundation.

19 Geotechnical Considerations in Road and Canal Construction and Environmental Implications

ROUTE ALIGNMENT

Roads connecting urban centres must pass through places of habitation or special interest, wherever they are located if they are to serve the needs of the people. Likewise, canals are dug to divert water from rivers and reservoirs to the places where it is needed. There is thus little or no option for selecting the general alignments of routes of transportation (motor roads and railway lines) and of water conveyance (canals). All that the geologists have to do is to identify the geological and environmental problems and the hazards that are likely to be encountered along the chosen pathways. **In the Himalayan region, for example, the network of roads has been built and planned without taking into consideration the geological conditions and the adverse impact on the environment. One of the disastrous consequences is the reactivation of the many old landslide fans and cones these roads cross, bringing about immeasurable miseries.**

Road construction should be preceded by geotechnical investigation to examine the bearing capacity and shear strength of the roadbed substrate, the stability of the slope of the cuts, and the conditions of the groundwater that may have effects on the stability of the road or canal. Topographic and geological maps supplemented by satellite images or aerial photographs are very useful in suggesting alignments and identifying potential areas of instability, natural hazards and adverse hydrological conditions. In as much as the construction of roads and canals causes drastic changes in the natural set-up and therefore the ecological balance, the environmental implications of these development projects are of a very serious nature. This is also true in the plains where the roads (particularly "kachcha") are equally troublesome, because of their erosional impact. Cartroads, for example, initiate gully erosion in alluvial tracts and arid-semiarid terrain.

HILLSIDE CUTS: STABILITY PROBLEMS

Stability Factors

The main problem relating to alignment is safety from hazards of mass-movements on the hillslopes and ground subsidence in the plains. An excavation for a road or a canal is called a cut. **It is the stability of the slope of the cut that is of paramount importance.** This depends on (Figs 19.1 and 19.2) (i) gradient, height and steepness of the wall of the cut, (ii) alignment of the road with respect to the structure (dip of strata, foliation, cleavages, joints or faults and their orientations, (iii) lithological composition with particular reference to density and cohesion of the rocks being cut, (iv) flow pattern the underground water, and (v) nature of vegetal cover and vegetation types.

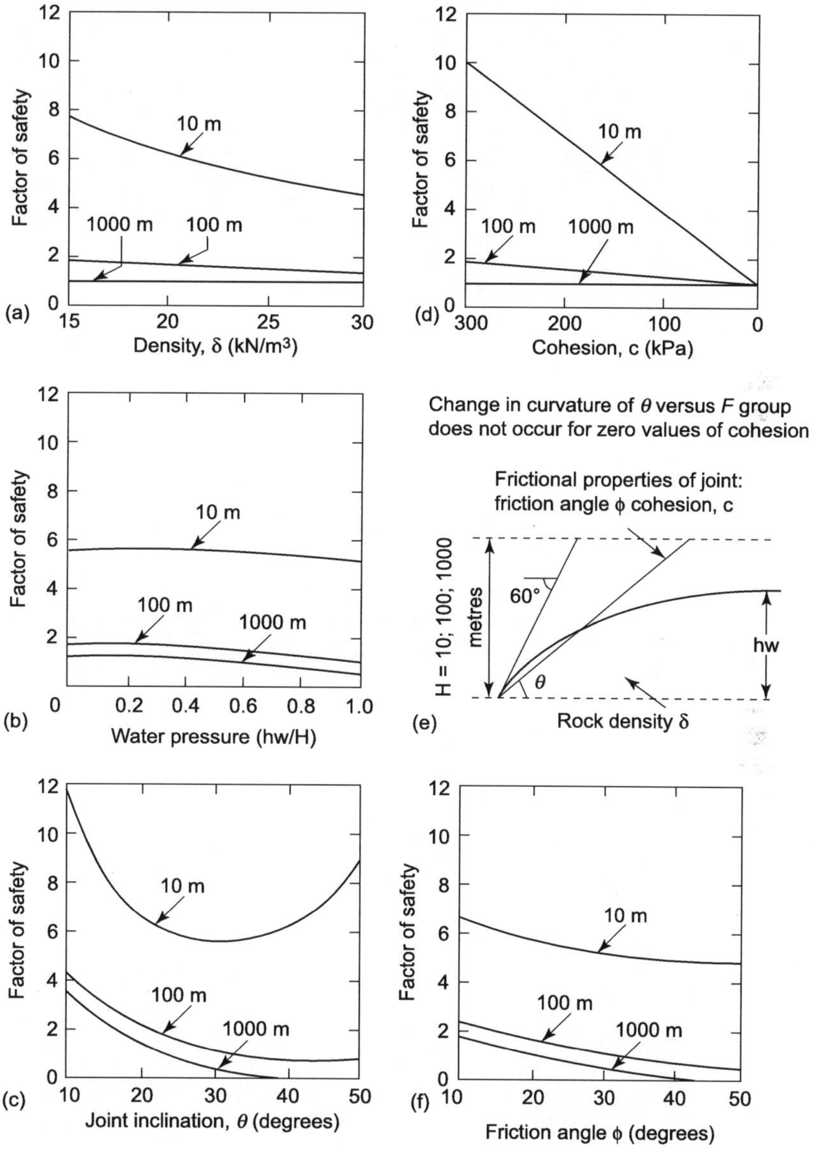

Fig. 19.1. In order to ensure stability and safety of the walls of excavations for roads or canals, a number of sensitive parameters have to be taken into consideration. The safety factor depends on density, cohesion, groundwater conditions, and inclination of joints, cleavages and bedding planes. (Redrawn with permission from F. G. Bell, 1980. *Engineering Geology and Geotechnics,* Newnes-Butterworths, London, 497 p., Fig. 3.5).

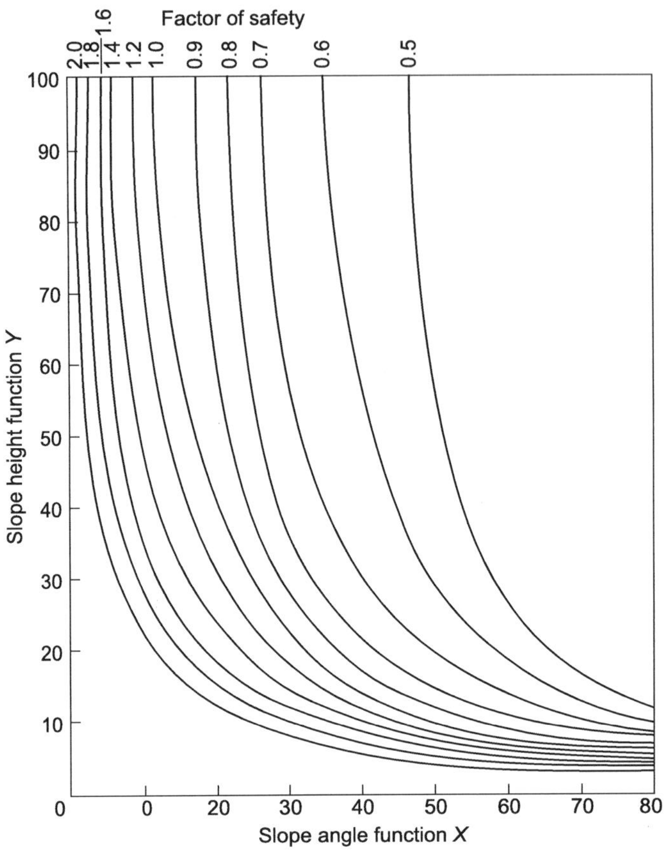

Fig. 19.2. The safety factor in excavation is a function of slope, height and slope angle. (Redrawn with permission from F. G. Bell, 1980. *Engineering Geology and Geotechnics,* Newnes-Butterworths, London, 497 p., Fig. 3.9 a.)

Steepness and Height of Cut-Walls

Excavations expose rocks and soils to the processes of uninhibited weathering. Landslides that follow road-cutting expose much larger areas to disintegration and decomposition. The result is the loss of intrinsic strength of the cut material. The exposure also allows percolation of water deep inside the rocks, thus promoting instability. Pressure release further aggravates instability. The height and slope of the sloping wall of the cut should be kept approximately proportional to the shear strength of the rocks. The greater the shear strength, the steeper and longer the slope of the wall can be (Fig. 19.2), as seen in granites, gabbro, dolerite, hornfelse, granulites, massive dolomites, quartzites etc. provided that they are free from weathering and jointing. On the other hand, **the slope of the wall should be between 0.5 : 1 and 0.25 : 1 in non-cohesive or unindurated sandstones and shales. In earthy or silty terrains, as in the Karewa (Kashmir) and Kathmandu valleys, or along the terraces in river valleys in the Himalayan region, the common slope values should be 1.5:1, 1.75 : 1 and 2:1. Fissility of rocks duces their shear strength, for the planes of parting serve as pathways of water infiltration, and**

hence the slopes of cut walls are made gentler in slates, phyllites, schists and gneisses. In case of sandy clays or clayey sands the slope of the cuts should be about 1.5 : 1. (Table 19.1, Fig. 19.3).

Table 19.1 Average Slope Values for Excavations for Roads (After A. H. Zubairi and P. D. Agarwal, 1985)

Rock Type	Horizontal:Vertical Ratio
Granites, basalts, other lavas	0.25 : 1 to 0.50 : 1
Gneisses, schists and marbles	0.25 : 1 to 0.50 : 1
Massive sandstones and limestones	0.25 : 1 to 0.50 : 1
Slates	0.50 : 1 to 0.75 : 1
Interbedded sandstones, shales, limestones	0.50 : 1 to 0.75 : 1
Massive claystones and siltstones	0.75 : 1 to 1:1

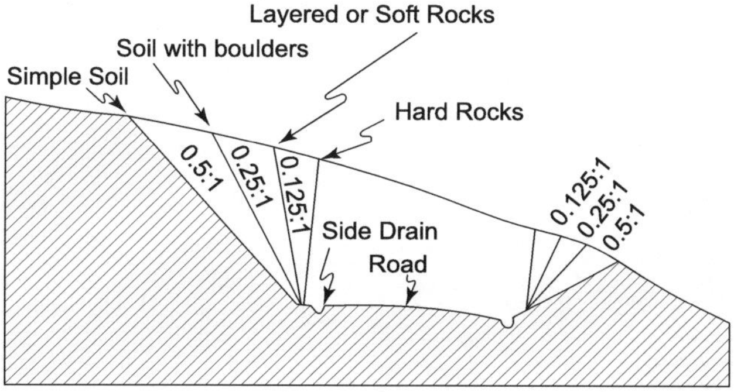

Fig. 19.3. In the excavation of road cuts, the angle of the slope is a crucial factor. It varies from soil to soil and rock to rock. (After S. C. Pal and R. C Das, 1952).

The slope angle greatly influences the strength of the force that causes slope failure—the steep the slope, the greater the chances of failure. Erosion, landsliding and excavation steepen the slope and thus create conditions conducive to its failure. Furthermore, removal of material by excavation considerably reduces lateral support, and thus promotes downslope movement, particularly if the bedding planes or cleavage/joint planes dip downslope (Fig. 11.7). **The dumping of debris is tantamount to adding load on an already destabilized slope. All activities of road-cutting thus promote failure of the slope and attendant mass-movements.**

If the slope is very high, it will be advisable to break it by developing horizontal beams or benches above the cut (Fig. 11.27). This would reduce the slope, and lessen the load of rocks on the slope. The benches also serve to drain away water and trap the slide debris or rockfalls. Cuts parallel to the strike are less stable than those which are normal to it. **The situation would be unfavourable if the beds or foliations and joints were to dip in the direction of the slope in the cuts paralleling the strike. And if water finds its way through the planes of discontinuities (cleavages, joints, fracture shear zones) there would be sliding of rocks** (Fig. 19.5), as happens so frequently in the Himalayan roads (Fig. 19.4). The situation can be ameliorated by making the slope non-symmetrical—that is gentler in the uphill part.

Fig. 19.4. Construction of a road in an unstable slope in Kumaun Himalaya, triggered massive and recurrent landslides, extensively killing vegetation. (Courtesy: Dr V.C. Tewari, Wadia Institute of Himalayan Geology, Dehradun). (See colour Fig. 19.4)

LOW-COST ROADS IN MOUNTAINS

Protective work at the early stage of road construction, particularly with respect to slope designs, and resort to new road technology involving application of geotextiles or geomembranes as impermeable ers and fabrics as filter sheets on retaining structures, rock-bolting and dowelling, reinforced earth structures, adequate drainage on the highway slopes, gabion mattresses and gabion walls etc. would prove a less costly highway venture. Considering the extremely high cost of recurrent maintenance and the economic implications of frequent and often protracted road closures during the rainy season, liberal investments in protective work would prove very beneficial in the long run.

The primary consideration in route alignment is the choice of mature and stabler slopes, and avoidance as far as possible of fans or cones of scree (talus) and landslide debris and of belts of hillside instability. Where it is necessary to cross such unstable zones, the height of the road cut should be minimum (<35 m); and maximum use be made of robust rocks at depth for erecting walls. **It is better to have a retained cut than a retained fill** (Fookes et al., 1985). The general gradient of the road should be 1 : 15 (4° 46′), the maximum permissible being 1 : 10 (5° 43′), while the radius of curvature of bends should not exceed 20 m (Fookes et al., 1985). It is not advisable to have the retaining structures higher than 15 m, and their aggregate length more than 20% of the total length of the road (Fookes et al., 1985).

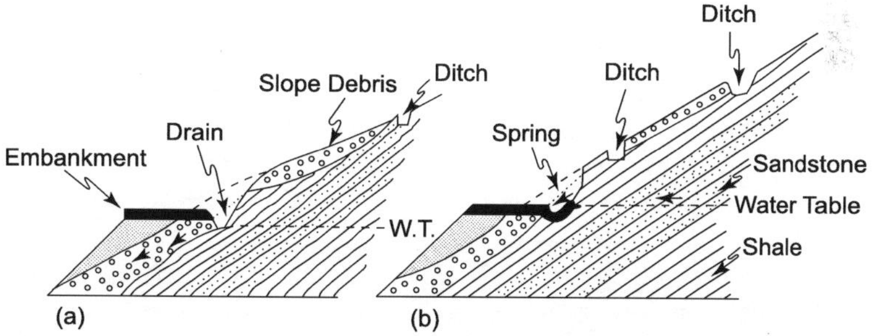

Fig. 19.5. Excavations for roads invariably cause disruption of groundwater. (a) The groundwater flowi (seeping) out weakens the base of the road, which is largely a fill. (b) It is drained out by constructi ditches higher up the slope, and a deeper drain with lining on the hillside of the road cut. (Redrawn with permission from Q. Zaruba and V. Mencl, 1976. *Engineering Geology,* Elsevier, Amsterdam, 504 p., Fig. 11.4).

Where the road has to traverse through the scree/landslide debris, it is advisable to cut the road either in the upper part where the accumulation of clasts is thin, or in the lower concave (toe) part constituted of bigger clasts which are comparatively stabler and less prone to rockfalls. It is not advisable to have hairpin bends on unstable hillslopes prone to mass-movements. Alluvial terraces must be adequately protected against toe erosion and flood hazards.

Gabion walls have proved four times cheaper than masonry or concrete walls (Fookes et al., 1985). Gabion walls up to the height of 20 m on talus slopes along the Dharan-Dhankuta highway in eastern Nepal have been found very effective in checking mass-movements.

Construction of protective ditches on the *natural slopes,* combined with brush-fencing (wattling) and establishment of fast growing vegetal cover on the road slopes will considerably diminish the problem of slope failure. Side drains should be led into *culverts that should have an original-bed gradient so that there is no deposition of debris in the channel. Wider and shallower ditches are preferable to steep-walled and narrow ones.* Ditches lined with heavy-duty polythene are not only less costly, but also more flexible than the masonry or concrete drains. Problematic gullies can be tackled constructing multiple check dams across them, and more successfully by providing well-bedded stone-ching keyed together (with a long axis perpendicular to the flow direction) or by a gabion mattress emplaced deeply in the gully bed and the gabion aprons on the banks.

ENVIRONMENTAL IMPACTS

•ad construction has a considerable impact on the environment—not only old landslides are reactivated, but also are triggered new ones in structurally vulnerable zones. The water table may be lowered, resulting in diminished spring discharges and diversion of stream courses and attendant acceleration of erosion. The blasting operation may cause widening of joints and fissures and resultant instability of the hillslope.

Groundwater Disruption

The cutting of a hillslope results in the interception of groundwater and consequent development of springs and seepages at the base of the wall of the cut. This beheading of the acquifer system leads to the lowering of the water table and the resultant drying up of springs and wells on the slope (Figs. 19.5 and 19.6) and groundwater starvation down-slope. Admittedly, this problem is quite minor, and can be medied easily. The lowering of the water table induces infiltration of water from the streams or rivers in the area, which in consequence would themselves have diminished discharges.

Another consequence of road cutting is the pollution of groundwater due to infiltration of polluted ater through fissures and openings (Fig. 19.6). The pollutants are derived from fumes and droppings from rolling stocks, particularly petroleum-run vehicles. The problem is serious where the cuts sink deeper than the water-yielding horizons or aquifers.

Production of Debris

The making of a cut involves the production of a sizeable volume of debris. In the Himalayan terrain where the slopes are steep, the hillside walls of the cuts are 3 to 5 m high, and the width about 4 m, the amount of debris generated varies from 40,000 to 80,000 m^3 per kilometre (on the assumption that one cubic metre (m^3) of rock gives rise to 1.4 to 2 m^3 of debris). In the early 1980s the 44,000 km long (aggregate) road network both metalled and non-metalled in the entire Himalaya must have generated approximately 2640 million m^3 of debris (Valdiya, 1985). Data related to the period 2004 to 2008 obtained from *Road Transport Year Book 2009,* Ministry of Road Transportation & Highways shows the aggregate length of the roads in the Himalayan belt Kashmir to Mizoram (in 2008) is more than 21,000 km. The more than 21,000-odd km length of the roads in the Indian states of the Himalayan region (in 2008) must have generated over 12,600 million cubic metres of debris ! In addition to this colossal volume of debris, there is debris formed by recurrent landslides during the rainy season. As already pointed out (Chapter Eleven), in the roads constructed and maintained by the Border Roads Organization, the annual rate of debris generation from seasonal mass-movements was 724 m^3/km/yr in the Jammu-Srinagar highway (Kashmir), 411 m^3/km/yr in the Tanakpur-Tawaghat road (Kumaun), and 591 m^3/km/yr in the Kameng highway (Arunachal Pradesh) (Patnaik, 1982). If the average is taken at 550 m^3/km/yr, every year about 24 million m^3 of debris was produced by mass-movements on the ads in the entire Himalayan region in the mid 1980s (Valdiya, 1985). Today the amount would be five times more.

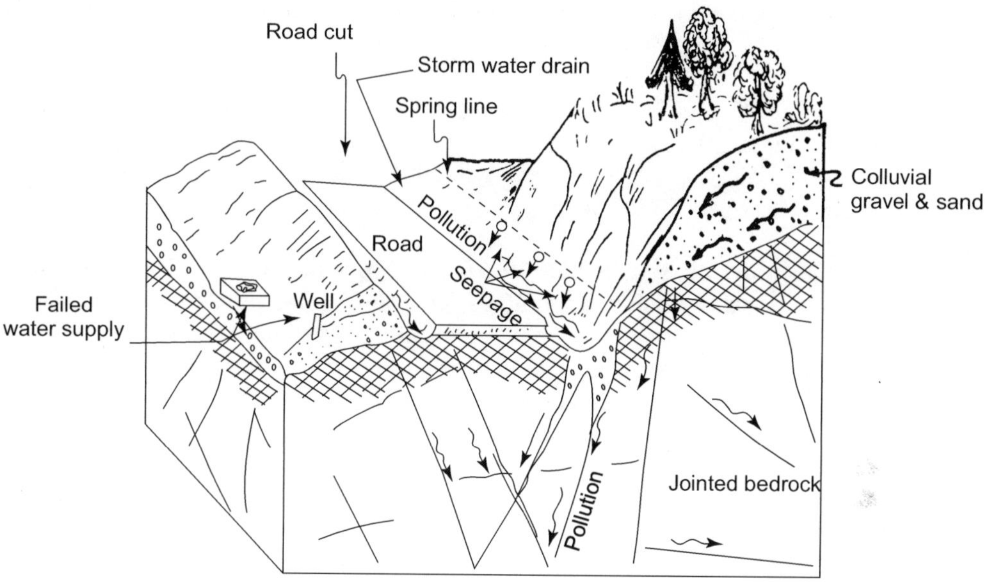

(a)

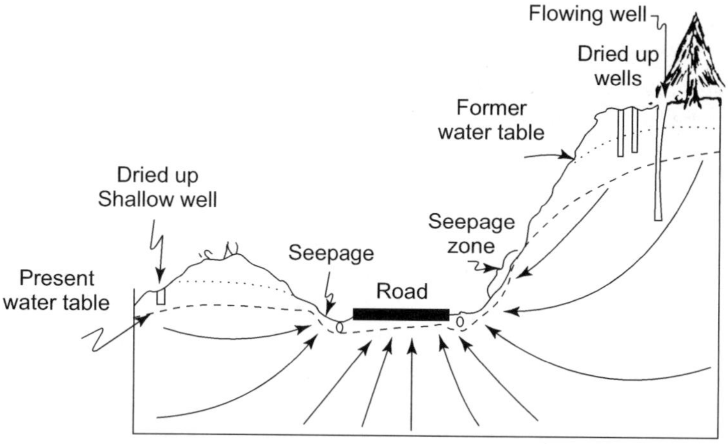

(b)

Fig. 19.6. Excavation on a hillslope means intercepting the water table of the groundwater, leading seepage (spring line) in the cut and drying up of springs and shallow wells on the hillslope. There is also the danger of pollutants finding their way into the groundwater. (Redrawn with permissi from R. R. Parizek, 1971, in D. R. Coates, 1981. *Environmental Geology,* John Wiley, New York 701 p., Figs 13.9 and 13.12).

Even if considered on a local scale, the annual debris generation on roads is colossal. On the Dehradun-Tihri road (Western Uttarakhand), for example, there were 10 medium-sized slides, each generating about 500 m^3 of debris, and one chronic landslide for every 10 km stretch of the road. Thus in the mid-1970s every kilometre stretch of the road annually produced 5000 m^3 of landslide debris, despite all efforts to control it (Bansal and Mathur, 1977). In the then mining belt in the Mussoorie Hills on a stretch of 6 km road, the per kilometre outfall was 1100 m^3 in the rainy season of 1978 (Haigh, 79). Out of the 470 outfalls identified in the early 1980s, 148 slides had their origin on the slopes with a tree cover of 40% or less, and 118 on slopes with 60% or more tree cover. Between Ghansali and Tilwara, north of the Ganga valley by Srinagar, about 102 landslides were counted within a stretch of 11 km (Negi, 1982).

The debris is invariably allowed to slide down the slopes into the stream beds. This causes enormous damage to the downslope vegetation including forests (Fig. 19.4) **and pastures and agricultural farms and to the river or stream regime.** The agricultural terraces occur generally at lower altitudes, so that the avalanches of debris cause widespread destruction of valuable arable land. The enormity of the loss suffered by individual farmers can be appreciated when one realizes the fact that in Nepal and Uttarakhand the per capita landholding was 0.1 to 0.2 ha (Rieger, 1977; Moddie, 1981) and that 68% of the individual landholdings were smaller than 3 ha (K. Singh, 1977). The population density is much higher now and the per capita land holding far less than what it was in the late 1970s. Even if a patch of this petty holding (a fraction of one ha) is damaged or lost, the very survival of the landowner is in danger. The implication is obvious—unstoppable outmigration and even exodus from very large parts of the Himalayan region. **It is imperative to stop this malpractice and to force road-builders to dispose off the debris in an environmentally sound manner.**

The destruction of valuable forest has been equally serious. According to an estimate, excavation kills one tree in every 500 trees standing on the hillslope (Haigh, 1982, 1984). A perusal of Table 17.4 would demonstrate that in the period 1951 to 1973, nearly 54,700 ha or 1.6% of valuable forests were lost on account of road construction in different parts of the country (L. C. Sharma, 1977). **The extent of damage to vegetal life in the Himalayan region since 1973 is quite staggering.**

An example will highlight the environmental implication of indiscriminate road construction. The construction of the Dhankuta–Dolaghat road in the Sunkosi valley in eastern Nepal triggered numerous slides, adding to the already very high bedload of the tributaries of the Sunkosi (Jha, 1982). The fans debris brought by the tributaries filled the valley of the main river. This encroachment caused undercutting of the opposite bank, leading to eventual collapse of that slope as well. Thus, there was extensive slope failures on both the banks—the failures progressively expanding up the deforested slopes. If this goes on unchecked, in course of time the hillsides would be deprived of the water in their springs and shallow wells as a result of the lowering of the water table. Even the water of the Sunkosi er seems to have sunk in the debris that choked the channel. Examples of drastic channel modification due to road construction are numerous throughout the Lesser Himalayan region in Uttarakhand and Himachal Pradesh. Most of the streams have become wider, and locally grown into ravines.

This will be the scenario for the entire Lesser Himalayan region from Jammu to Arunachal Pradesh and beyond, if engineers continue to construct roads in the manner they do.

Admittedly, not all the landslides are recurrent. Some slides occur only once, the conditions of instability can be corrected, and the incidence of slope failure diminished in number and extent with the passage of time—the slide zone heals up to stability. Thus on the old roads only a few landslides remain unhealed, and may occur as chronic cases.

Debris Disposal

There are a number of methods to dispose off debris. Debris can be used to build robust parapet walls all along the stretch of the road on the outer margin. Apart from utilizing the debris, the construction of the parapet wall will provide safeguards against accidents. The debris can be transported for use elsewhere or stacked appropriately for use as construction material for making building blocks, as ballast, as fill to develop playing grounds, parks, orchards and building site including bus stations. The debris can be stacked behind specially built retaining walls in the concave bends of road alignment, taking due precaution against infiltration of water.

The flat wide sites formed by the debris can serve as parking places, tree-farms for fuelwood a fodder, flower gardens, orchards and playgrounds—if these are closer to settlements. In this mann many villages and towns along the roads would have suitable space for playgrounds, recreation parks or farms for fuelwood and fodder. The roads would then become a multi-purpose feature.

Unscientific Methods of Excavation

The liberal use of explosives for excavation is greatly weakening the hillsides, particularly in the unstable zones, and ancient landslide deposits through which roads have been cut. Vibrations caused by blasting of rocks and passage of heavy vehicles are capable of reactivating old (fossil unstable landslides) as witnessed in the Bhagirathi and Alaknanda valleys in a number of places. The vibrations generated by blasting have opened up fissures in fractured rocks and thus facilitated entry of the destabilizing agent the water—causing loss of shear strength of rocks. **It may be remembered that when the explosives in a drill hole are detonated, the pressure of gases generated may exceed 1,00,000 atmospheres so that the zone around the hole is completely shattered and the tensile cracking spreads far and wide.** The limit of the ground velocity (shock) upto which the rocks remain undamaged (intact) is 50 mm per second. This calls for right spacing of short holes, keeping in view the strength, density and structure of the rocks. As a rule, the spacing must be between 0.75 and 1.25 times the thickness of the overburden.

Road development involves stream-crossing and culvert construction. Ephemeral streams bringing debris loads on the roads can be ignored, but **the problem of draining out a concentrated flo of rainwater on the roadbeds is quite serious and cannot be overlooked. The roadbed water concentrates into powerful tenors and carves out gullies, particularly on bare debris fills an dumps on the slopes. These gullies later expand and undermine the entire road segment, thus creating unseen hazards to traffic. This is happening on roads all over India. The effective remedy is construction of appreciably wide drainage ditches on the inner side of roads in the hills which can be linked with the culverts and streams and must *always be kept free* of litter and debris.** Most of the hill roads have drainage ditches choked with debris, for there is no—locally absolute absence of—attempt to clear them regularly.

In view of all these environmental problems of construction of roads and canals, it is necessary to carry out environment-impact appraisal before embarking upon a road or canal building project (Table 19.2).

Table 19.2. Evaluation of Environmental Impacts of Road-Building (From: R.E. Munn, 1977; *Environmental Impact Assessment: Principles and Procedures,* John Wiley, New York)

A. General Setting of Region	
	1. Topography
	2. Geology
	3. Climatic condition
	4. Existing network of roads and paths
	5. Route alignment of proposed road
B. Objectives of the Road Project	
	1. Size of the area and population it will serve
	2. Volume of traffic expected in the first, second and third years
C. Technical Features	
	1. Length of road
	2. Width and type of road
	3. Slopes of hills it will cut through, and heights of road cut
D. State of the Environment of the Belt through which the Road will Pass	
	1. Hillside-stability and problem of erosion
	2. Seismicity of the region
	3. Identification of hazard-prone zones and hazard-zoning map
	4. Faunal communities
	5. Flora—extent and types
	6. Parts of road prone to flood or landslide hazard
E. Impacts of Road Construction on the Environment	
	1. On stability of hillsides
	2. On hydrological conditions—surface and underground waters
	3. On forest ecosystem
	4. On farms and settlements
F. Measures Proposed to Mitigate I Minimize Adverse Impacts	
	1. Balancing cut and fill
	2. Disposing of debris
	3. Reclaiming pits and securing embankments
	4. Preventing mass-movements
	5. Containing blasting
	6. Landscaping route belt
G. Socio-economic Concerns	
	1. Employment potential
	2. Displacement and/or exodus of people

MEASURES FOR STABILIZING WALLS OF ROAD CUTS

Various measures for stabilizing hillslopes have been discussed at length in Chapter Eleven. These will be recapitulated here.

Dewatering and Drainage

It is advisable to avoid zones afflicted with instability, and prone to creep or related mass-movemen and find an alternative route across the valley. Old landslide debris cones should be avoided as far as possible. If the location of the route in a geologically unfavourable area becomes unavoidable, the vulnerable spots should be carefully identified on the map, hairpin bends kept to a minimum, and t heights of the cut kept less than the critical limit (R. S. Dixit, per. comm., 1985). Also, half-cutting and half-filling will minimize the problem generated by the debris produced. Likewise, **big boulders and blocks providing natural toe support should not be removed. If a breast wall or toe wall is needed to provide this kind of support, the construction of the wall should be taken up simultaneous with the excavation of the road cut.** It would also be desirable to seal or grout the cracks and fissures on t slopes well before the first rain—at the construction stage itself.

The most effective and economical method of minimizing and controlling slope failure of the walls of road cuts is to develop a system of drains in order to divert runoff water from the slopes. The herringbone type of drainage has been found to be very efficient. A ***catchwater drainage ditch*** (Figs. 11.23 and 19.5) at the top of the excavated slope would prevent surface water from entering the cut and causing mass-movement and erosion. It is to be linked with the ***interceptor drain*** at the foot of the slope of the cut (Fig. 19.5). In the threatened sites deep trenches extending beneath the potential failure zone are lined with filter fabrics and filled with coarse granular material to form a counterpart dra (Bell, 1980). While allowing free passage of water, the granular fill acts like a supporting buttress.

An example from the Deccan plateau may be cited. In the sensitive Borghat section of the Mumbai–Pune highway, three important measures were suggested for the prevention of damage from water: (i) construction of ***catchwater drains*** of sufficient capacity and longitudinal slope to allow free flow surface water coming from the adjoining land, (ii) draining out surface water from the carriage way to prevent damage to the bitumen-pavement and percolation to the substrate, and (ii) adequate provision of subsurface drainage to relieve pore water pressure and internal erosion (Rane and Sukumar, 1985).

Since water-saturated soils are highly prone to slumping when excavated, it is desirable to lower the water table. This can be done by pumping out underground water through a series of open wells or tubewells. Underground water must be prevented from entering the zone of instability. This can be done by making a ***cut-off*** *(a* trench backfilled with asphalt or concrete) or ***grout curtain,*** the two forming impermeable underground barriers. A bentonite slurry-filled trench also serves as an impervious barri particularly if cement with a retarding agent is incorporated into the slurry.

Simultaneously, all cracks and holes on the surface are sealed with cement, bitumen or clay in order to prevent infiltration of surface waters including stormwater.

Reinforcement of Slopes

The various engineering measures for providing support to a slope include construction of ***buttress*** and *retaining walls* at the foot of the slope (Figs. 11.28 and 19.7). **The construction of these walls should be done *immediately after the excavation is complete.* Since a retaining wall has to withstand**

tremendous and unequal pressures and load, the width must be large enough to make it stable. rthermore, it is keyed deep upto the firmer horizon to which it transmits the load. There should be adequate provision for draining out water lodged inside the soil or rocks through weep holes and pipes in order to relieve buildup of pore pressure. ***Crib walls,*** preferably constructed from reinforced ncrete or steel units and set up in cells filled with gravel, provide good support to the wall or slope. ***Gabion walls*** of strong wire mesh enclosing stone pieces are quite good, particularly to prevent erosion (Figs 7.20, 7.19, 11.27). **Piles have been found effective in checking slides in the case of rocks but quite useless where the soil is a loose aggregate.**

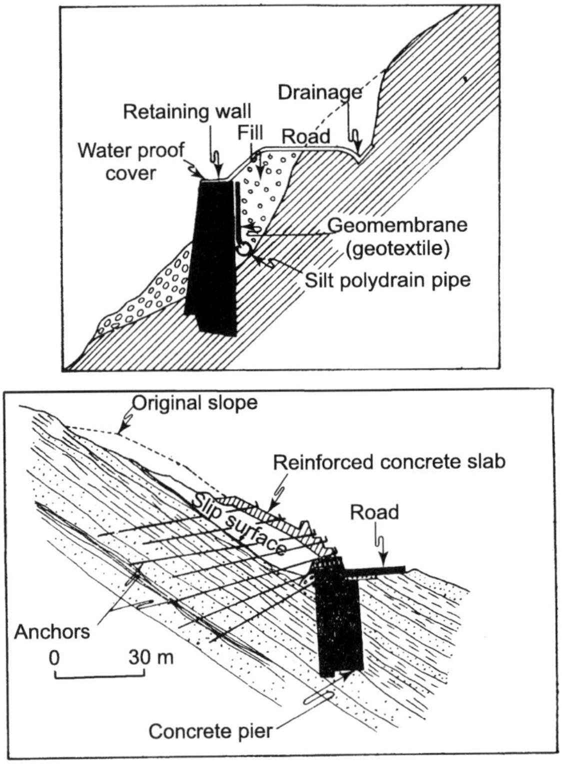

Fig. 19.7. Retaining walls keyed deep upto the firm beds provide support at the toe. The gravel mass filled behind the retaining wall allows accumulation of underground water without causing buildup of water pressure. This water is drained out through weep holes. The slope above the cut is reinforced by installing anchors and/or by covering with concrete slabs. (After Q. Zaruba and V. Mencl, 1976. *Engineering Geology,* Elsevier, Amsterdam, Fig. 11.5).

Wide fissures and fractures and cavities can be filled with concrete or masonry, not only to seal the passage of water but also to provide a sort of anchor. *Steel dowells* (Bell, 1980), about 2 m ng are fixed into holes driven across the bedded rocks beneath the slip surface, and grouted in place. Application of coatings of gunite (3.5 parts sand : 1 part cement by weight) not only seals the surface ainst infiltration of water, but also prevents dislodgement and consequent slipping of the soil. If injected under high pressure with gunite, the weaker rocks and soils become compacted and stabilized. The gunite coatings can be reinforced with wire mesh or rock bolts. An important precautionary measure is to make adequate provision for draining out underground water through the gunite coating, otherwise the pressure buildup will become too strong for the stability of the slope. They provide stability to the slope of the wall exposing rocks (Fig. 19.8 b). Another very effective device is the rock anchor, more than 3 m in length and installed often in groups (Figs. 19.7, 19.8 and 11.32).

***Rock bolts* (Figs. 11.31 and 19.8) prevent slipping and fall of jointed and foliated rocks by providing additional strength on critical planes of weakness within the rock masses. The bolts should be inclined (not normal) to the potential plane(s) of failure and penetrate as deep as much as 8 m. The bolts can be linked together by light steel sections or steel mesh to support the rock face (Fig. 19.8 a). The *nets of steel wire* suspended from the rock face provide safety against hazards from rockfall.**

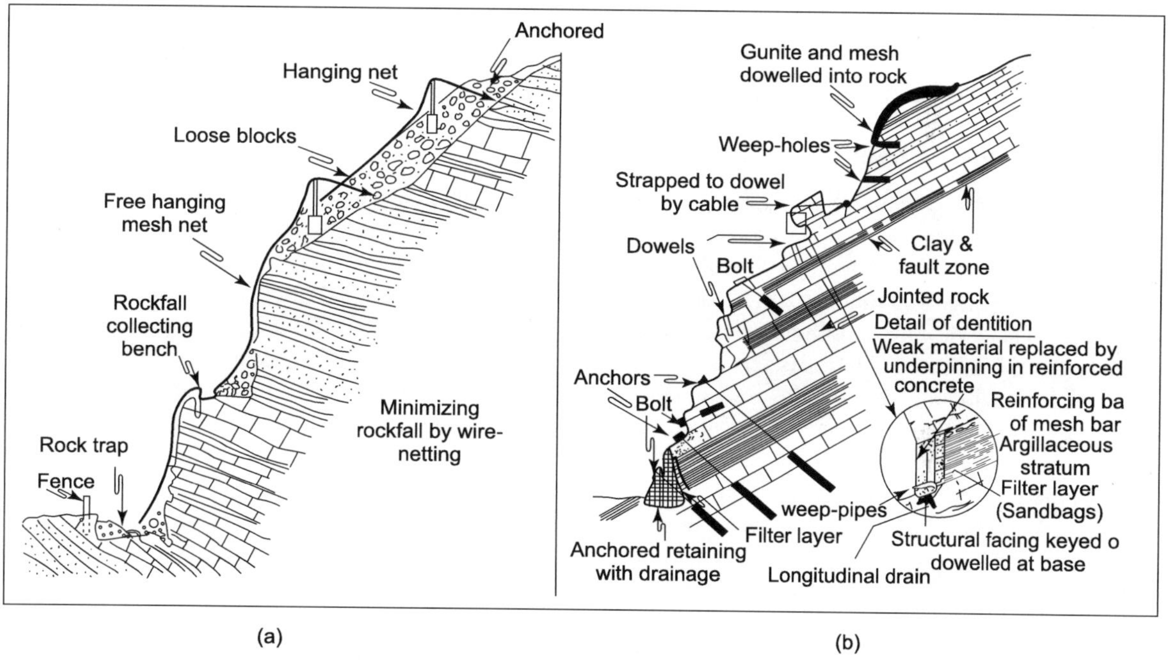

Fig. 19.8. (a) Suspending wire netting from the hillslope minimizes risks of rockfalls. Benches serve as traps besides modifying the slope to reduce the pressure on the road cut. (b) Rockfalls from jointed and fissured or foliated rocks can be minimized by installing steel dowells, rock bolts and anchors fixing dentition, and by covering parts of the surface with gunite and wire-mesh dowelled into t rocks. (Redrawn with permission from P. G. Fookes and M. Sweeny in F. G. Bell, 1980, *Engineering Geology and Geotechnics,* Newnes-Butterworths, London, 197 p., Fig. 3.23 a).

Afforestation

As already stated in Chapters Four, Seven and Eleven, **planting of trees that have high evapotranspiration propensity and deep roots substantially improves the stability of slopes (Fig 11.25), particularly if constituted of loose, unconsolidated material such as in a slumped mass or**

alluvial landform. Shrubs such as vertiver and kikiyu and grasses that thrust their roots deep and wide and hold together the slope material are as effective as the reinforcing steel rods, and change the loose mass into a resilient mat. However, care has to be exercised to avoid a situation where the trees put their roots into the fracture or joint planes and widen them.

Slope Modification

The costliest and possibly the least practical method of reducing the instability of hill-slopes is to divide them into segments by cutting benches about 3 to 5 m wide and less than 1 to 2 m high. In addition to providing drainage to runoff water and serving as traps to falling/sliding debris, these benches increase the stability of the steeper slope (Fig. 11.27). As already pointed out, the construction of benches reduces the slope and also lessens the weight of rocks on the slope. This measure diminishes the strength of the force causing slope failure.

CONSTRUCTION OF ROADS AND CANALS IN FLAT COUNTRY

Embankments: Stability and Failure

flatter terrains or plains the development of roads and canals involves the construction of earthen or ckfill embankments. The material used for the embankment is adequately compacted before covering with a multi-layered pavement constituted of differently processed road metals. The embankments are ilt on firm ground so that there is no problem of subsidence and related deformation (Figs. 19.9 a and 19.10 a). The three causes of failure of the embankments are (i) the unconsolidated nature of the material of the ground below foundation (such as clays, peats, loess, gravels) which do not have a bearing strength fficient to withstand the load of rolling stocks on the roads, (ii) limited thickness of the soft material resting upon hard compact sediments or rocks, such as along shores of ponds, lakes and seas, and (iii) alternations of hard and soft material—particularly soft layers sandwiched between hard ones—in the foundation. These situations are conducive to differential settlement, and attendant deformation of the embankment manifest in bulging, wrinkling and spreading (Fig. 19.9 a). The deformation is accentuated by water which gets into the foundation. **Road embankments without vegetal or concrete or tarmac cover are prone to gullying by water concentrated on the roadbed. Piping or tunnel erosion is the result of water seeping in through openings and intergranular pores. The pore water, as already pointed out, washes and eventually produces a network of tunnels. The collapse of these tunnels causes havoc on the roads and in the adjoining areas.**

Remedial measures involve intercepting and draining out subsurface water through drainage ditches. Under high and heavy embankments the softer material from the foundation must be removed completely and replaced by coarse sand and gravel. The compressive and shear strength of the ground material is increased by compaction effected by mechanical or chemical methods, including admixing the soil with cement, bitumen or chemicals two-three years before the pavement is laid.

Drainage Problems

Preliminary surveys of groundwater conditions in soils is carried out before constructing roads. The groundwater, particularly that which moves up through capillary action, creates problems in the foundation and in the subgrade immediately below the pavement. The subgrade may become wet or even waterlogged as has happened in many parts of Haryana, Panjab and Uttar Pradesh. The degree of saturation is normally higher under concrete pavements, compared to that under asphalt or clay covers, for obvious reasons. The water under pavement may lift it up, causing bulges on the roads.

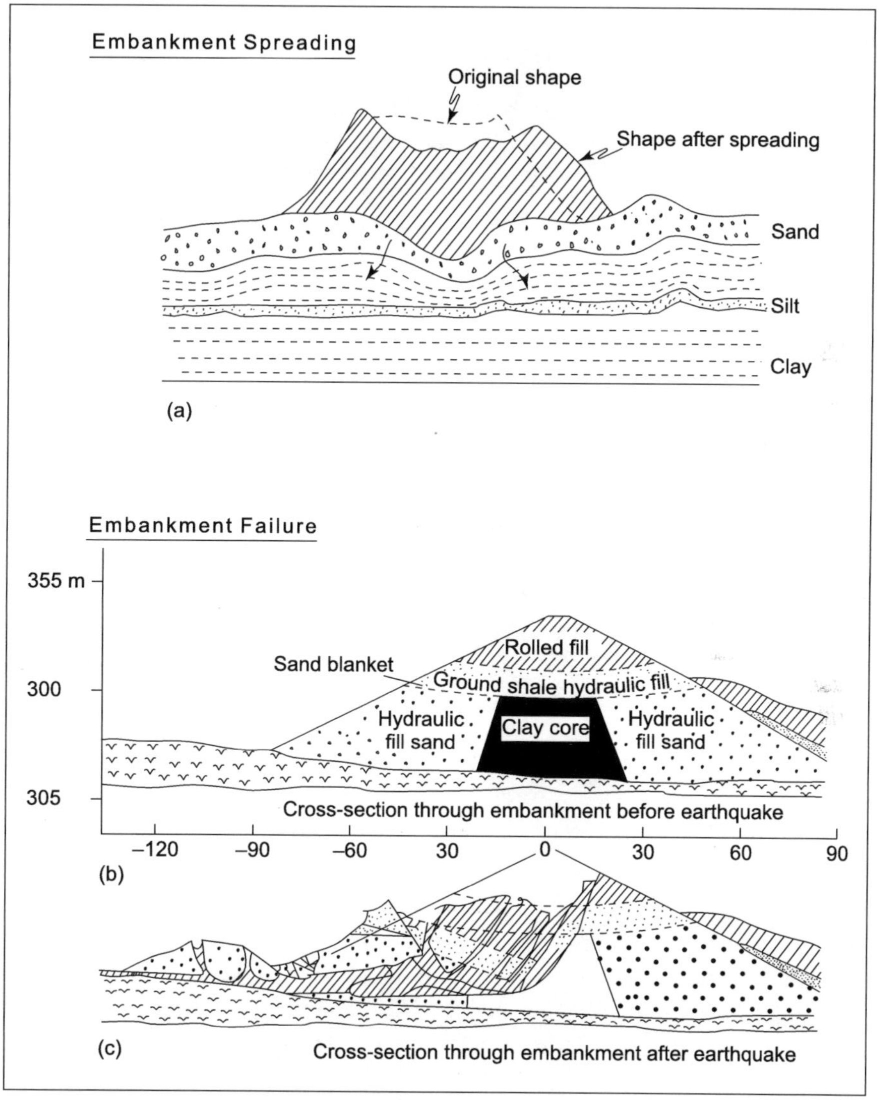

Fig. 19.9. (a) Alternation of competent and incompetent (the latter sensitive to water and pressure) layers beneath the embankment may lead to spreading and failure.(Based on P. D. Krynine and W. Judd, 1957. *Principles of Engineering Geology and Geotechnics,* McGraw-Hill, New York, 730 p., Fig. 16.10.) (b) Earthquake shocks are responsible for failure of the earth core in rock-embankments. Failure results from liquefaction of hydraulic sand-fill at the base of the embankme (After F. G. Bell, 1980, Fig. 11.17).

The remedial measures involve construction of longitudinal ditches and gutters and linking culverts through the embankments. In waterlogged terrain and in areas where there are springs, special endeavours to drain out water through deep ditches and lowering of the water table through dug wells and pumping need to be done.

SPECIAL PROBLEMS RELATED TO CANALS

The construction of canals involves not only excavating channels but also in the plains, building up embankments on both sides. Almost the same technology is applied in this case as in road embankments. The geotechnical and environmental problems are also very much the same.

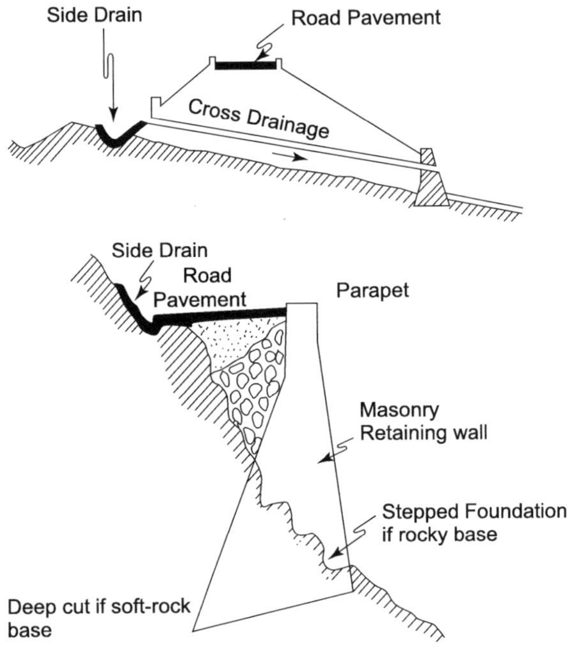

Fig. 19.10. (a) The safety of the earthen embankments for roads from erosion by storm- and flood-waters is ensured by providing drainage across the embankments, side drain on one side and buttress wall on the other. (b) The retaining wall in the road section passing through a steep slope is erected on a stepped foundation. The filling is done, with rubble or gravel in the lower and outer part and with sand and mud in the upper part. (After V. S. Rane and A. W. Kulkami, 1984.)

Environmental Changes

e construction of canals entails diversion of stream water through cuts on the hillsides or on flat plains, and/or building embankments or all of these works combined. The various problems arising from these activities have already been dealt with. There are two special problems that cause environmental anges—seepage of water and erosion in channels due to flow of water down the predetermined (rather prescribed) grade of the canal. In the mountainous regions the canals have to be safeguarded against landslides. Take the example of the 12.5 km long Lastar canal between Mayali and Badani in the Tihri district (western Uttarakhand), which suffered from 200 slides, each 50 m wide (Negi, 1982).

The diversion of streamwater through the canal causes alteration of the entire ecosystem, resulting in inevitable physical and biological changes in the affected stream, particularly in the faunal and floral communities living downstream of the barrage or dam, for their lives have no to be readjusted to the regime of diminished water. The curves of the pools and the riffles in betwe (Fig. 13.11) have their own ecosystems. Used to these ecosystems, the living communities find sw flowing water in straight-course canals very uncomfortable and even a threat to their survival.

The unlined canals of the Bhakra-Nangal canal system once faced a difficult problem of proliferati of weed. The water of a natural stream often and commonly carries suspended sediment, which is deleterious to weed growth. Freed of the suspended sediment due to impoundment of water behind the dam, the canal water favours a rich growth of weeds as already explained in Chapter Seventeen. Uncontrolled weed growth seriously impairs the carrying capacity of the canals.

Channel-bed Erosion

Water flowing straight through a predetermined gradient and without natural impediments is uniform swift. The result is appreciably greater scouring, compared to that in a natural channel. On the other hand, where the gradient is less, there will be accumulation of sediments leading to reduction in the carrying capacity of the channel. **It is therefore advisable not to have high gradient and steeper walls of cuts.**

An example of the problem resulting from channel-bed erosion may be cited (M. Haigh per, com., 1985). The Greenes Canal in Arizona (U.S.A.) constructed to divert the Santa Cruz River towards a storage reservoir was originally (in 1911) 1.5 m deep and 4 to 6 m wide. Floodwaters deepened it into a 3 to 4 deep trench. In 1985 it was 6 to 10 m deep and locally a 100 m wide ravine, fringed on both sides by a kilometre-wide belt of tributary gullies. The man-induced changes, including intensive pumping of groundwater leading to lowering of the water table to a depth of 100 m and consequent subsidence (by as much as 4 m) of the central part of the basin, played havoc with the drainage of the areas—the streams were diverted to the fissures and lost their waters, and the Greenes Canal became a dry ravi through which flows only the flood water. Therefore, while constructing canals through arid, semi-ar and alluvial tracts, care has to be exercised to ensure the canals do not cause appreciable channel-bed erosion.

Seepage

More than 25% of the water passing through unlined canals is lost by seepage well before reaching its destination. It is lost through cracks and pores in the bed as well as embankments. In the Bhagirathi valley in Tihri district (Uttarakhand), the minor canals were loosing 25 to 35% of their water (Negi, 1982). In the irrigated plains of Uttar Pradesh, Haryana and Panjab the seepage was locally as high as 40% in the early 1980s.

The impact of the seepage is manifest in the waterlogging and salinization-alkalinization of valuable agricultural lands, as explained at length in Chapter Five. Thus canal irrigation has created more problems than it has solved.

SUMMARY

Construction of roads and canals involves excavation on hill slopes and construction of embankments flat country. On hillside cuts the main problem relates to the stability of the hillslope, which depends on the structural condition and lithology. Excavation entails removal of natural support, leading to slope failure. It also produces great masses of debris that is allowed to slide downslope, causing struction of forests and agricultural fields and irreparable damage to both the surface and sub-surface water systems. Road construction triggers and promotes landslides on an extensive scale and is responsible for interception of the water table of groundwater that causes drying up of springs and shallow wells.

In order to ensure stability, the principal concern should be to make the slope of the cut less than the angle of repose and drain out water from the slope above the road and from the road t. Safe disposal of debris—by making appropriate use of it—is the most important exercise in road technology. Roads and canals should be in harmony with the environment and appear as "an integral part of the total landscape, rather than an offensive intruder."

In the plains or flatter terrains the development of roads and canals involves building embankments. Embankments should be constructed on ground that has requisite bearing strength and is free from layers or pockets of sediments sensitive to water and pressures load. The settlement of the embankments and consequent failure can be prevented by strengthening both the ground and the embankments through compaction and or chemical grouting.

The main problems relating to canals are seepage and channel-bed erosion, both of which can be nimized by lining the canals with asphalt, granite, impervious clays, pavingstone or flagstone.

20 Geotechnics of Tunnels

IMPERATIVE NEED

Tunnels serve to carry water for the generation of electricity, as the conduits for water supply for varied needs, and as the short passages on transportation routes. **The alignment of a tunnel as well as the methodology and programme of its excavation depend on accurate information of the geological setting of the tract.** The knowledge of competence, hardness and workability of the rock and or sediment enables engineers to select the type of explosives to be used for excavation and plan the system of supports to be installed on crucial spots. Comprehensive geological investigation is therefore a prerequisite before embarking on tunnel operation. The effective way of illustrating vital geological information is to make a series of longitudinal cross-sections along the proposed alignment, and to represent the conditions obtaining.

STRESS CONDITIONS IN TUNNELS

Rocks at depths, particularly deeper than 600 m, are under the great load of the overlying mass of rocks and or sediments (Fig. 20.1). The stresses so generated tend to give rise to strain, which is manifest in the form of displacement of material. The displacement takes place only if space is available for movement. Tunnelling provides that space. Small displacement is apparent from heaving up or bump up of the ground, and bending and or 'bowing' of steel support structures within tunnels. These happen when locked up stresses are released. Swift movements are expressed in explosive spalling off of rock material, occurring as it does when the residual stresses exceed the strength of rocks and or sediments.

Even without stored stresses, excavation causes movements of surrounding rocks and or soil material into the openings. However, there exists a system of self-balancing shearing stresses which makes the material draw inwards. In this phenomenon, known as ***arching***, there is neither generation nor elimination of pressure—rather there is simple redistribution of stresses (Krynine and Judd, 1957). It involves radial stresses acting normal to the wall of the tunnel, and circumferential stresses acting parallel to the circumference (Fig. 20.1). The radial stress at the wall is zero, but progressively increases inward. **The circumferential stress at any spot is twice that which acted there before the excavation started; and it decreases inwards (Krynine and Judd, 1957). The tunnel wall will not fall until the circumferential stress equals the compressive strength of the rock** (Schultz and Cleaves, 1955). So long as the shear strength of the material above the tunnel opening is more than the corresponding shearing stress resulting from the weight of the overburden, there will be no failure. A reinforced lining is therefore provided to divide and distribute the weight of the overburden (Fig. 20.1).

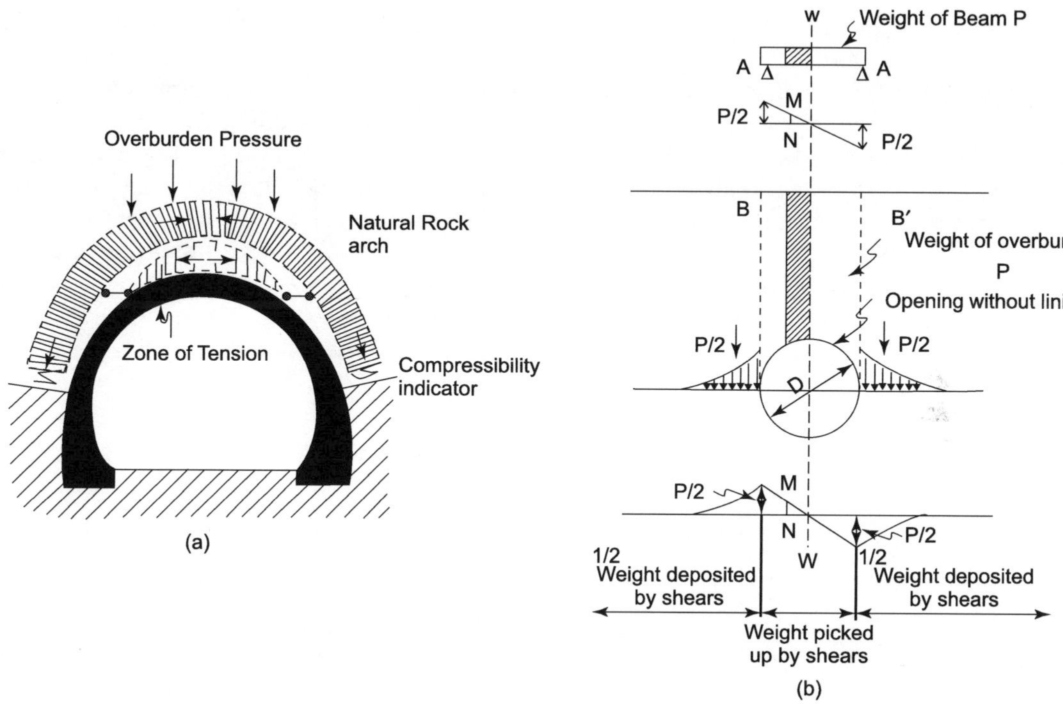

Fig. 20.1. (a) A zone of tension develops above the tunnel excavation and below the induced rock arch that is subject to overburden pressure. (After W. Zaruba and Mencl, 1976). (b) Shear diagrams showing relationship between shear stress and arching. (After D. P. Krynine and W. K. Judd, 1957, Fig. 9.7, with due permission).

In structurally deformed—faulted, fissured, jointed or foliated and or folded—rocks the arch acti around the tunnel is not always sufficiently strong to prevent failures, so that supports have to provided to ensure integrity and stability. The arch action is generally well developed to ensure stability in igneous rocks with sparingly developed joints, thickly bedded horizontal or very gently dipping ha sedimentary rocks and in volcanic rocks striking parallel to the tunnel axis, or very steeply dipping and vertical beds striking normal to the tunnel axis (Fig. 20.2).

INFLUENCE OF GEOLOGICAL CONDITIONS

Stratified Rocks

The stratification of rocks, whether sedimentary or metamorphic, causes distribution of pressu generated by the load of overburden. However, bedding planes and foliation or cleavage planes have certain inherent weaknesses. Jointing in rocks augments this weakness—particularly where these are perpendicular to the stratification.

If the principal planes of weakness (e.g. bedding, foliation, cleavage and joints) are horizontal and wide-spaced, a ***bridging effect*** is discernible (Fig. 20.2). If the spacing of transverse joints is more an the width of the tunnel, the joint blocks act like a solid beam subject only to a bending movement. long as this bending movement does not exceed the flexural strength of the rocks, the roof will remain stable even without a support (Schulz and Cleaves, 1955). On the contrary, **if the joints are closely spaced, and particularly if the beds are thinner, there will be possible overbreak**—the, breakage rarely continuing beyond a vertical distance equal to half the width of the tunnel above the top of the circular pay line. Where the beds are laminated thinly and dip 5° to 10° or less, roof spalling ay occur—particularly when jointing is developed (Schulz and Cleaves, 1955). Fortunately, the joint blocks are interlocked or held together with cementing material and therefore there are lesser hazards. The blocks of rocks must overcome frictional resistance, which gets reduced if slippery layers are present along them, and if the lubricating action of water is noticeable.

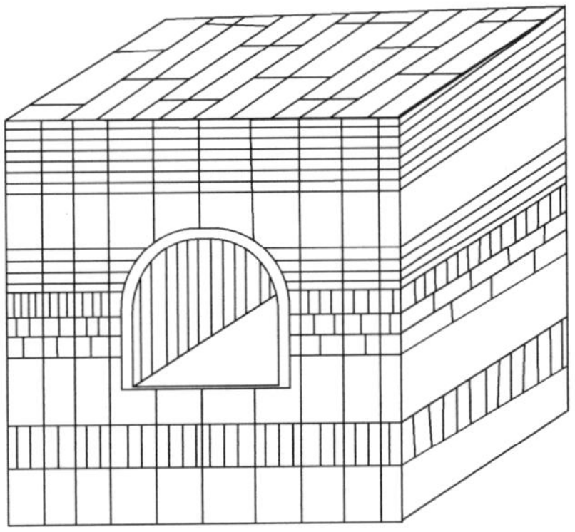

Fig. 20.2. A tunnel driven through horizontally or gently dipping beds of sedimentary/metamorphic tocks would not have any problem if the spacing of joints or fissures is wider than the width of the tunnel, but would experience fallout and overbreak if the joints or fissures are closely spaced. (After Krynine and Judd, 1957).

If a tunnel is *driven normal to the strike* of the steeply dipping or vertical beds having non-uniform conditions (Fig. 20.3), problems of seepage and overbreak would arise (Fig. 20.3b), ***particularly when hard and soft rocks alternate.*** In the Ramganga tunnel at Kalagarh, which cuts ough the alternation of highly jointed sandstones and maroon mudstones of the Siwalik, incidences of roof collapse plagued the excavation programme for quite some time owing to weak rocks and water flow. Usually, the very steeply dipping and vertical beds with spans equal to the widths of the cross-section, act as girders so that the sides of the tunnels remain stable.

If a tunnel is excavated *along the strike* of the vertical or inclined beds (Fig. 20.4) the conditions of stability are less satisfactory even if the mass of rocks above the roof is held up by frictional resistance along the bedding planes, which are uneven. If a tunnel were driven entirely through one rock, the weight of the rock would be brought to bear on the lining. Seepage would take place if the rock is open-textured (Fig. 20.4 b). Admittedly, the weight of the rock will depend on its physical characteristics and geometry of planes of discontinuities with respect to the alignment of the tunnel. The failures are generally due to shearing of rocks. **But if the tunnel is partly through harder and partly**

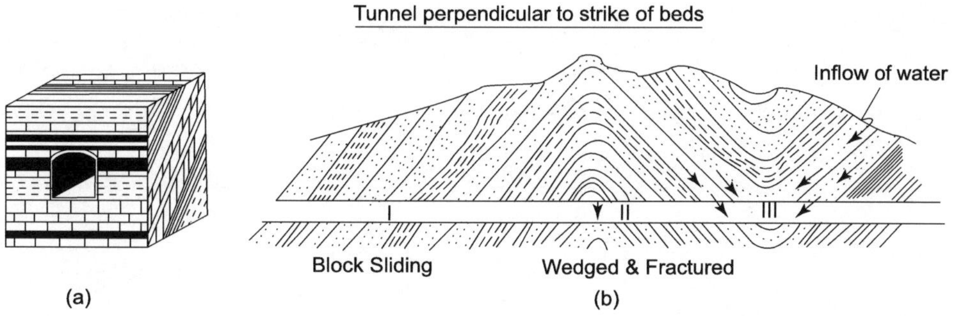

Fig. 20.3. Tunnel driven normal to the strike of steeply dipping or vertical strata may encounter problems of seepage and fallouts, particularly if softer and harder rocks alternate rapidly.*Inset* after Krynine and Judd, 1957.

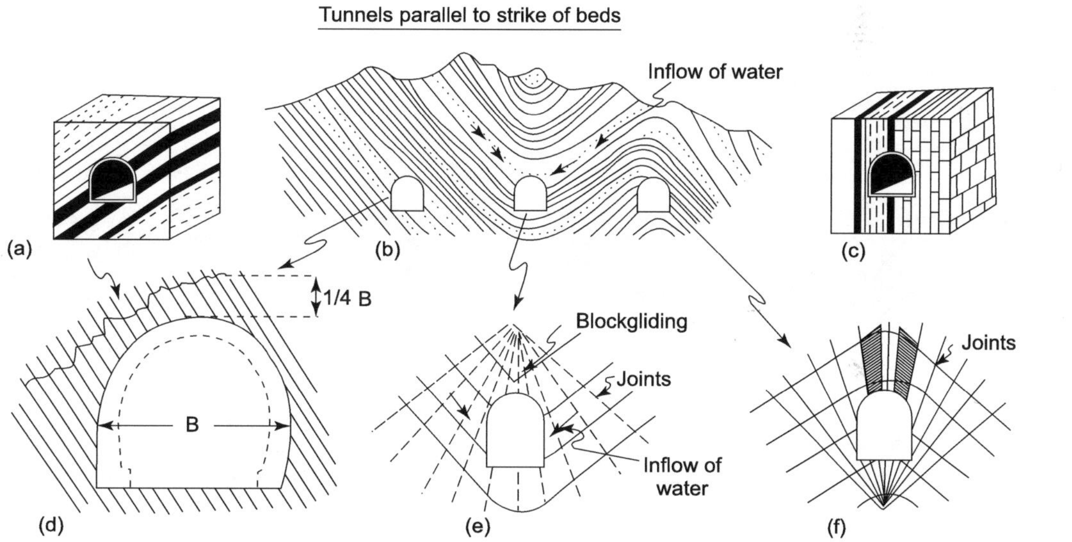

Fig. 20.4. (a) (b) (c) Tunnels driven parallel to the strike of the layered rocks encounter serious problems of instability and seepage, particularly if the beds are steeply dipping or vertical or if there is alternation of softer and harder rocks. (d) and (e) In synclinal cores the pressure is high and joints converge upwards so that the blocks bound by joint planes are liable to drop down just as interstitial water would seep or flow into the tunnel. (f) In the anticlinal cores the pressure is les the seepage is minimal and the chance of sliding down of rocks as joints converge downwards are lesser. Block diagrams in insets after Krynine and Judd, 1957.

through softer rocks, there will be a danger of sliding along planes of contact, and of seepage of water if the tunnel is driven below the water table. The 339-m long headrace tunnel of the Jaldhaka Project in the Darjiling Hills made parallel to the strike of a succession of schists and sheared gneiss to carry water at the rate of 7 cubic metre per second, experienced frequent failures as the roof collapsed and water rushed into the tunnel at the rate of 227 litre per metre. The huge chimney-like roof collapse was associated with a 6 m wide shear zone in quartzite. The upshot of these mishaps was the suspension of work for 11 months (Sen Sharma and Chowdhury, 1972; Hukku et al., 1972; Srikant and Roy, 1975). In such a situation "eccentric loads are imposed on the supports which must be properly backpacked and designed to carry much load" (Schultz and Cleaves, 1955).

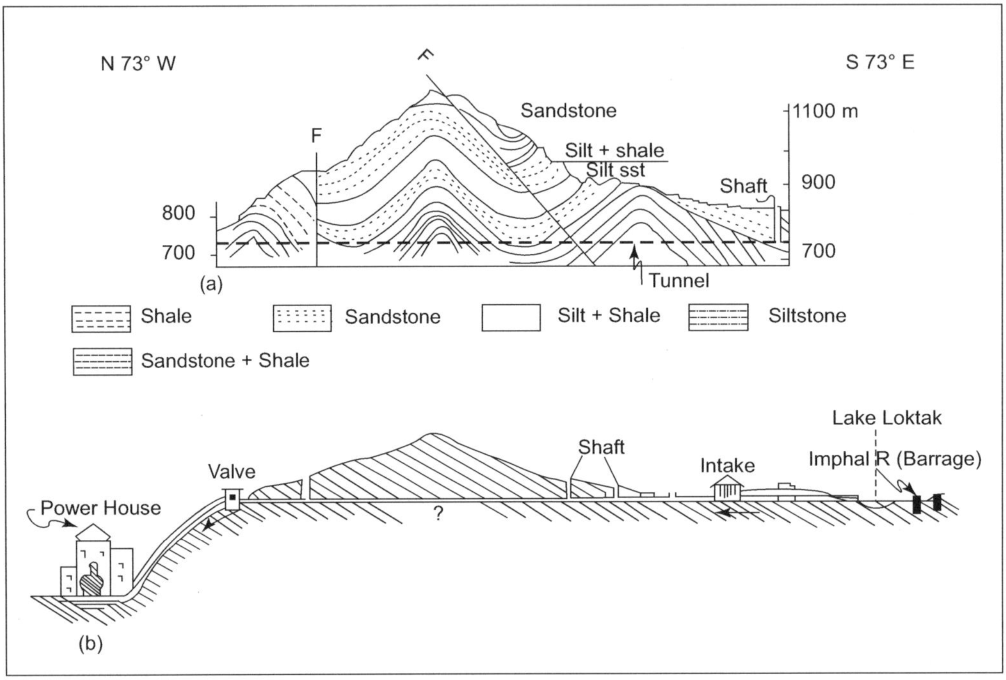

Fig. 20.5. The tunnel of the Loktak hydro-electric project in Manipur is driven across a Tertiary succession of sandstones and shales. The highly jointed synclinal core was beset with the problems of gushing out of sands, onrush of water, seepage of methane gas, and fallout of rocks. The ground surface witnessed landslides of tragic proportion which wiped away everything. (After Chaudhury and Chattopadhyay, 1983).

Influence of Joints

e cores of folds are commonly fractured and jointed (Fig. 20.4 e, f). In synclinal cores the compressive essure is high, the planes of joint and fracture converge upwards (Fig. 20.4). **Therefore, in the synclinal cores the wedge-shaped blocks—very similar to inverted keystone—are liable to drop down.** Simultaneously, the groundwater would readily flow into the tunnels. The 6.6 km long and 3.8 wide headrace tunnel of the Loktak project in Manipur (Fig. 20.5) traverses a N—S trending syncline made up of Tertiary sandstones and shales. The rocks dip 25—50° E or W. The sandstones are highly friable, very splintery shales and the rock conditions change practically every metre. **Consequently ere were numerous seepages of inflammable methane gas that was held in pockets, particularly in the vicinity of the fault planes** (Mehrotra, 1982).

Joints parallel to the axis of the tunnel are treacherous, particularly if they are inclined more than 45°. This situation is conducive to fallouts and roof collapse. The flat-lying joints may also cause dislodgeme from the roof, especially if the dip of the beds in the tunnel, aligned normal to the strike, is less than 15°. **While closely spaced joints are responsible for instability, the wide-spaced joints—wider apart than the width of the tunnel—pose no problems. Where joints are randomly oriented, the joint blocks are unable to move and hence there is no problem of rockfall or of roof crumbling.**

Faulting and Zones of Crushing

The faults are always associated with crushed or sheared rocks. The presence of such zones create unequal distribution of stress within a tunnel (Fig. 20.6), requiring special treatment and safeguards against hazards. **The rocks may be so severely battered and shattered in these fault zones—which can be quite thick—that they are prone to sliding, falling and even flooding by seeping wat The whole crushed material may locally gush out into the tunnel if located below the water table.** Further, the bearing capacity of rocks is so considerably reduced that even a slight addition of load of the overburden may cause destabilization.

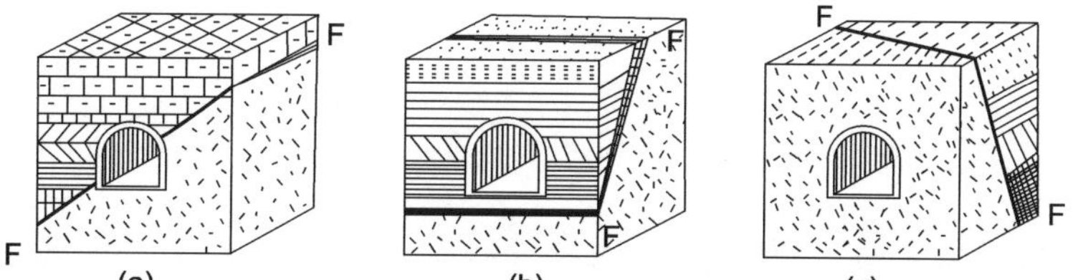

Fig. 20.6. Location of a tunnel with respect to faults. (Redrawn with permission from D. P. Krynine and R. Judd, 1957. *Principles of Engineering Geology and Geotechnics,* McGraw-Hill, New York, 730 p., Fig. 9.12).

Fault gouges containing swelling clays pose a serious problem of break-in and bulging of walls so that the tunnel would require support during its construction. The Beas-Satluj link tunn between Pandoh and Baggi in Himachal Pradesh (Fig. 20.7 a) charged with high hydrostatic pressure (100 to 300 psi) encountered many blowouts and collapse in the crushed zones, within the Precambrian quartzites and slates, dolomites, and particularly in the two shear zones within the Chail granite and associated talc schist (K.N. Srivastava et al., 1972; A. K. Srivastava, 1972). In one of the shear zone the entire assembly was pushed back 60 metres. The phenomenon led to twisting and buckling of ribs in parts of the tunnel where the thickness of the overburden exceeded 1400 m.

In the Chhibro-Khodri tunnel (Fig. 20.7 b) driven across the Main Boundary Thrust and Nahan Thru in the Tons valley in Uttarakhand (Krishnaswamy et al., 1970; Mehta and Sanganeria, 1972; Mehta al., 1972), the problem of blowouts and gushing out of water was encountered in the extremely sheared and crushed Tertiary rocks sandwiched between two thrusts. To overcome this problem, a kink was given in the tunnel alignment in order to reduce the width of the intra-thrust zone from 700 to 750 m along the original straight tunnel-line to just 250 m. The diameter of the tunnel in the intra-thrust zone w increased to 9 m from the original 7 m (in Shome (Ed.), 1982). Likewise, in the Giri-Bata tunnel, whi cuts across the 640 m wide zone of crushed gougey rocks between the Main Boundary Thrust and t Nahan Thrust (Mehta and Dua, 1975; Ghosh, 1975), the plastic flow of gougey material and the tilti and distortion of timber supports, concrete sleepers and steel ribs were experienced in parts of the tunnel where the stresses reached 35 kg/cm^2 (Krishnaswamy et ah, 1970). In order to reduce the thickness of the oblem zone, the tunnel had to be curved (Madhavan, 1982; Tyagi and Sharma, 1982).

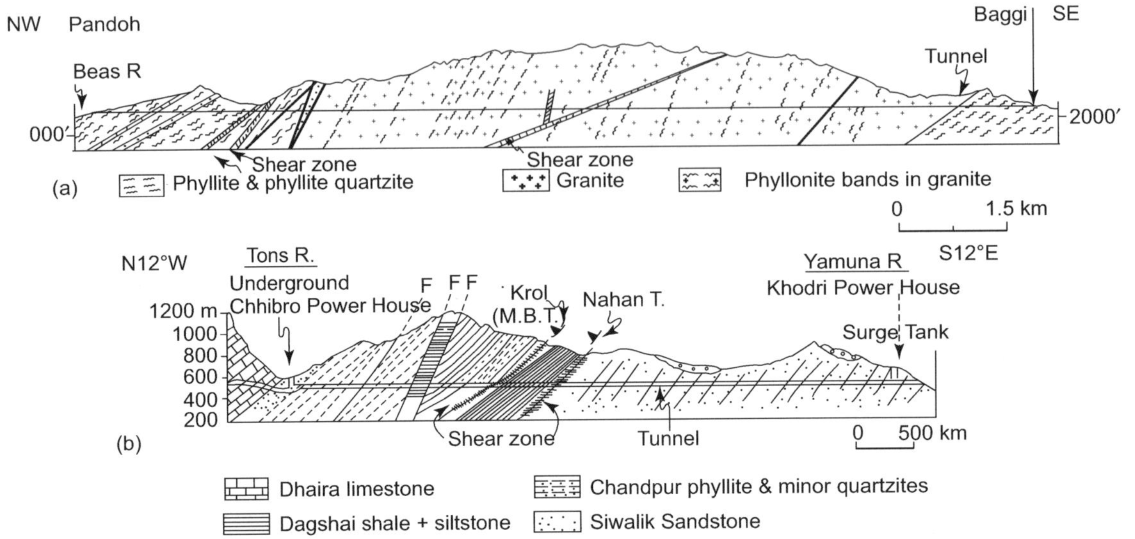

Fig. 20.7. (a) Cross-section of the mountain range through which the Pandoh-Baggi tunnel of the Satluj-Beas link project (Himachal Pradesh) is driven. The shear zone characterized by mylonitic granite and sericite-quartz schist created serious problems during tunneling. (From S. K. Shome (Ed.), 1982.) (b) The Chhibro-Khodri tunnel in the Tons valley (western Uttarakhand Himalaya) reveals the structure of the intra-thrust zone—between the active Nahan and Main Boundary thrusts. Problems of seepage gushing out of sands and distortion of support structures were encountered during tunnelling. (After A. K. Dube, J, L. Jethwa, B. Singh, Bhawani Singh, and R. S. Mithal, 1982.)

In the tunnel through the Pir Panjal Range in the Uri area (Kashmir) the zone of the Panjal Thrust is characterized by severe deformation of sedimentary and metamorphic rocks. The tunnelling faced the problem of squeezing out of rocks, requiring multiple drift methods and spiling in order to provide fficient stand-up time by building rock support and reinforcement (Stille et al., 1998).

The 27-km long Nathpa-Jakhri Tunnel across high-grade metamorphic rocks in Kinnaur (Himachal adesh) ensuring a drop of 450 m onto the Satluj River, cuts through 1,500-m wide shear zone ar Ratanpur and a fault near Dajkhas. The problem of distortion in the tunnel supporting system as encountered in the shear zone. Multi-drift tunnelling methods, coupled with drainage and rock nforcement had to be resorted (Gupta, 1998). The faults brought out 40° to 60°C hot water at the rate of 100 litres/second.

In the Ramganga tunnel roof collapse and displacement and distortion of steel support were countered in the 13 m thick zone of sheared clay (Mandwal, 1975).

Collapse of roofs, squeezing or bursting out of broken rock masses (resulting in deformation of pport system), vertible flood of underground water together with emission of inflammable methane gas, and related incidences were encountered in the making of the tunnels for the Loktak Project Manipur (Fig. 20.5), the Ranganadi-Dikrong River Project in western Arunachal Pradesh (Nand a Mukerjee, 2009), the Nathpa-Jhakri and the Parbati project in Himachal Pradesh and the Dulhasti Proje in the Kishtwar district in Kashmir (Kadkade, 2006; Dhawan and Patni, 2009). Perhaps most difficu was the tunneling for dropping the Dhauli River water into the Kali River in northeastern Pithoraga district in Uttarakhand. The 5.4 km long tunnel through the thrust-bound overturned tight synclinal fo made up of mylonitized porphyritic granites, phyllonites and schists, encountered all the problem that shear zones and severely deformed and fractured rocks create besides the inflow of hot water and ve high heat upto 42°C (Dhawan and Patni, 2011). The north-facing steep slope of the Chhiplakot mas covered thickly and overwhelmingly with huge blocks and enormous mass of chaotic debris loosened by recurrent slumps, slides and debris avalanche through ages had to be treated intensively to stabilize the unstable slope (Dhawan and Patni, 2011) at great cost.

It may be emphasized that if the crushed rocks of the shear zone are dry, the arch action is able to transfer the load to the side walls so that the roof support does not experience much addition of weight.

Neotectonic Activities on Faults

The movements on faults disrupt the tunnel and its lining and may even cause offsetting of the tunnel itself. The sudden release of stresses may disrupt everything, including the support. In the belts intersected by numerous faults, the possibility of movement taking place has to be taken into consideration. It will be safer to shift the alignment of the tunnels from areas that are tectonically active. The three major tunne in the Himalayan region mentioned above cut through the neotectonically active Main Boundary Thru and the Nahan Thrust. Geodetic measurements have proven this fact: the country south of the MB in the Tons valley is moving southwest compared to the northern block, the southward component of movement being 0.64 cm/yr and western component 0.27 cm/yr (Ansari et al., 1976). Along the Nah Thrust the movement is at the rate of 0.92 cm/yr (Sinvhal et al., 1976). The Shali Thrust encountered near Joginder Nagar in the Uhl hydel project is likewise active—the horizontal movement was of t order of 0.5 to 4 cm and vertical 1.5 cm in the observation period of two years (Arur and Raja, 1982 Provisions have been made in these tunnels to accommodate movements that may take place.

Tunnelling Through Massive Rocks

Massive rocks being commonly affected by jointing exhibit conditions similar to those of stratifi rocks. Usually, the blocks between joints are intimately interlocked, so that strong frictional resistan prevents their falling. However, weathering may plague these massive rocks, and the formation of clays may make them not only permeable but more compressible as well. Clays commonly swell and shrink, and the water infiltrating through fissures and joints aggravates the instability.

Tunnels Through Unconsolidated Sediments

Unconsolidated finer grained sediments such as sands, silts and clays have lower compressive streng and higher sensitivity to water. Dry clays like many other materials (excepting cohesionless sand show what is known as ***ground arching and bridge action***. It takes some time before the roof of the clay breaks down as excavation proceeds—the delay is called ***standup time*** and may range from several hours to several days, depending on the content of water and intrinsic strength of the sediments. The standup time is the duration in which an opening is created and failure of the overlying mass takes place. It varies from material to material. Roof support has to be provided before the expiry of the standup time.

Stiff clays, marls and loess have slightly higher tensile strength, but have a network of closely spaced hair-cracks, which is responsible for disintegration of clay columns on application of load on dehydration. **In the presence of water, the montmorillonite clay may swell as much as 10 times its original volume, and thus cause development of strong pressure on the tunnel lining and support system.** Lessening or removal of the load may lead to increase in volume of consolidated clays lying at depths and thus generate pressures. These were the problems which were encountered in the construction of the subway (metro) in the alluvial sediments under the city of Kolkata. The saturated nature of the sediments in the proximity of the Hooghly River and the presence of entombed organic matter representing ancient amps considerably aggravated the problem. Discontinuous, isolated pockets of sand within clays drain out small quantities of water into the tunnel, and large sand bodies (aquifers) bring large volumes of underground water in the tunnels.

CHANGES IN WATER REGIME

Problems

The opening of a tunnel alters the water regime perceptibly. The tunnel is indeed an underground drain to remove the interstitial underground water (Fig. 20.8). The extent and rate of removal of water through the tunnel depend on the permeability of material—there is onrush of water when the tunnel ikes a permeable zone. The tunnel connecting two reservoirs of the Kundah project in Tamil Nadu as flooded with a sudden rush of water and sands. Nearly 30 m below the fault there was a depression 5 m deep and 37 m wide) filled with water-saturated sediments (Hukku et al., 1972). The onrush of ater is mostly encountered along shear zones, joints and other planes of structural discontinuities, i.e. nes of secondary permeability. The Banihal tunnel across the Pir Panjal (Kashmir), constituted of ongly folded and faulted limestones, slates, conglomerate and lava, is frequently flooded with water gushing out in torrents. The Salal tunnel through the Jammu Limestone in the Chenab valley, and the ira-Siul project area in the Ravi valley in western Himachal Pradesh encounter cavernous limestones d dolomites. Obviously, the problem of flooding is serious in tunnels, particularly in cores of folds, as seen in the Loktak tunnel (Fig. 20.5) and the Sundernagar-Slapper tunnel (Fig. 20.7 a).

The environmental problem arising from excessive draining out of undergound water through tunnels is that the groundwater table is lowered in the area. Consequently, wells and springs dry, and there is dwindling or even disappearance of water in surface streams flowing over permeable zones. This was noticed in the Pandoh-Slapper tunnel section of the Beas-Satluj link project, where 50% of the springs and streams were reported to have dried up following the making of the tunnel (*Annual Report,* 1984-85, Department of Environment, New Delhi, pp. 79-82.)

Remedial Measures

Perhaps the most effective remedial measure to overcome the problem of seepage would be to lower the ater table below the level of the tunnel through pumping and deep dug wells (Fig. 20.8 b), provided that it does not cause ground settlement. This phenomenon would create a serious problem if the tunnel passes underneath habitations. Likewise, horizontal drainage in the form of bored wells drilled from the deep shafts provide relief from a seepage problem. An ejector system is applied to dewater soils a sediments. The ejector produces a vacuum around the tunnel and thus draws out water from the fine grained sediments. Admittedly, these suggestions are not practical in the mountainous regions, such as the Himalaya.

The second measure is to block the passage of water by sealing and grouting voids, fissures, fracture joints, open cleavages and cavities on the ground surface. The watershed can be managed to dra out rainwater and minimize infiltration into the ground. In the disaster-struck Loktak project area Manipur not only a number of 20 m deep shafts were sunk to drain out water but also many trenches a contour-drains on the hillslope were dug and check dams on all the streams constructed.

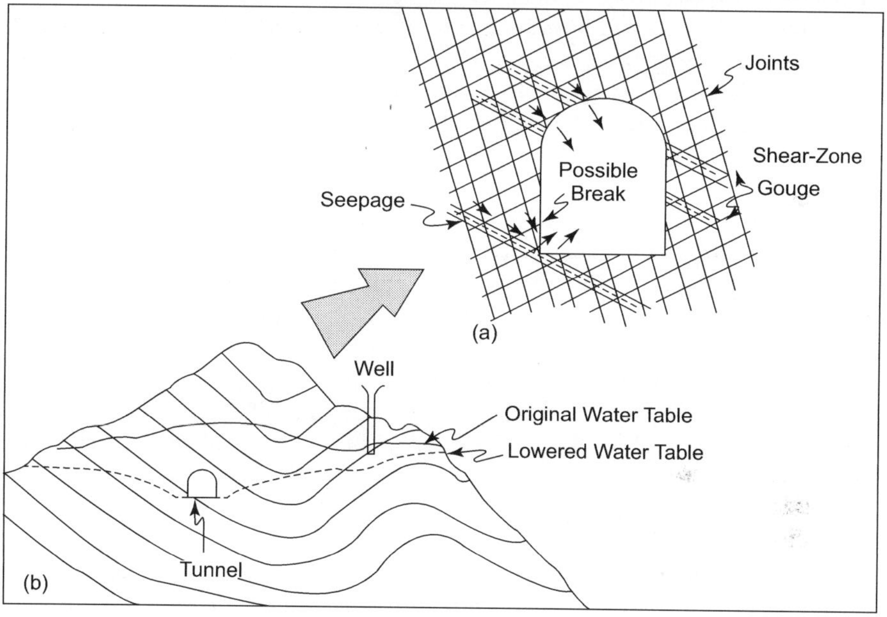

Fig. 20.8. (a) The joints, even in massive rocks, serve as pathways of percolating water and seepage tunnels. The gougey shear zones are particularly vulnerable to break-ins. (b) The tunnels ser as underground drains, leading to lowering of the water table and drying up of springs and wells on the ground surface.

SUMMARY

The principal concern of an engineering geologist involved in planning a tunnel along a proposed alignment is the identification and location of structural discontinuities or defects, and lithologic properties that may cause caving-in, overbreak and seepage of water. A self-balancing system of shear stresses generates arching action that keeps the rocks and or sediments from falling so long as the shear strength of the material above the opening is higher than the corresponding shear stress. A lining is provided to divide the weight of the overburden between the lining and the system of shears around the tunnel.

The bedding, cleavage and foliation planes have an inherent weakness. The joints and fissures perpendicular to these planes aggravate instability. Where the principal planes of weakness are rizontal, the presence of the bridging effect will ensure stability, although transverse joints, if closely spaced, may also induce overbreak. If the tunnel is driven normal to the strike of a succession of steeply dipping or vertical beds, non-uniform conditions related to load and incidence of seepage may be encountered, although the magnitude of the problem may not be serious. On the other hand if the tunnel is driven along the strike of beds or foliations, very unsatisfactory conditions such as overbreak and seepage are likely to be developed, particularly where soft rocks alternate rapidly with hard rocks, and where the succession is broken by faults or fractures. The rocks of the cores of folds e highly jointed and cleaved. In synclinal cores, blocks bounded by fractures and or joints are likely to drop down; and the seepage under artesian condition may be very serious. In the anticlinal cores the conditions are comparatively better.

The faults associated with shearing and crushing of rocks are particularly vulnerable and prone onrush of water and finer sediment. The pressure of gushing water and sediments may be strong enough to deform or break tunnel supports. Active faults are very hazardous in all respects, and must be avoided.

Unconsolidated fine sediments like sands, silts and clays have lower shear strength, lower permeability, higher compressibility and greater sensitivity to water. The hydration of clays causes swelling and resultant generation of pressure in tunnels and on their linings. The lowering of pressure may also cause expansion of the unconsolidated clays. Even the stiff clays due to presence of close-spaced microscopic cracks disintegrate when put under the load. Therefore, provision of support has to be provided at the very early stages of tunnelling through all kinds of clayey masses.

Opening up a tunnel is virtually like opening an underground drain. The result is a substantial eration of the groundwater regime. Not only are directions and rates of water movement changed, the water table also is lowered leading to drying up of wells and springs, and dwindling of water in surface eams flowing over permeable horizons.

In order to minimize the problem of seepage in tunnels a simple method is to sink wells and intensively mp out water so that the water table sinks below the level of the tunnel. Grouting of fissures, joints and ults, filling up depressions in cavities and constructing a network of drainage on the ground surface to drain out storm and surface waters are ecologically sound methods of overcoming the problem.

21 Problems of Pollution and Wastes

Humans of flesh and bone will not be much impressed by the fact that a few of their contemporaries can explore the moon, programme their dreams and use robots as slaves, if the planet earth has become unfit for everyday life. They will not long continue to be interested in space acrobatics if they have to watch with their feet deep in garbage and their eyes half blinded by smog.

— Rene Dubes

WHAT IS POLLUTION

human population has grown phenomenally, the magnitude and ramification of pollution have also creased—to monstrous proportions in many areas. As a consequence, the environment is adversely fected considerably, leading to impairment of health and ruination of aesthetic state of landscape several parts of India. In small quantities (measurable in milligrams per litre (mg/l)), the common ses such as carbon dioxide and nitrogen and the nutrients such as phosphates and nitrates are useful aterials. But when their proportion rises beyond certain limits, the ecosystems get imbalanced, and us inhospitable. Heavy metals like lead, zinc, mercury and cadmium measurable in terms of parts per llion (ppm) or billion (ppb) are harmful, particularly when their amount exceeds threshold values. ne ppm is equal to one cubic centimetre of ink dissolved in one cubic metre of water).

Different living beings have different levels of sensitivity or tolerance to the presence of these aterials in the ecosystems. Whether the pollution effect will be harmful or not is therefore a subjective atter.

Pollution is manifest in the expanding dumps of garbage and wastes, in the intractable mification of effluents in water, and the thickening canopy of smoke, dust and gases in the atmosphere. The air over urban and industrial centres is no longer breathable, the rivers and eams flowing past cities and towns have been fouled beyond measure by sewage and toxicants, e agricultural soils have been poisoned, and the landscape is marred by heaps of ugly garbage.

While garbage is an individual problem directly related to his level of cultural development d aesthetic sense, sewage is a measure of municipal efficiency, and the pollution of rivers and atmosphere is indicative of the degree of environmental perception of governments.

Needless to state, all the three are equally responsible for creating a situation in which, as Tom hrer put, "you brush your teeth with the best toothpaste, then rinse your mouth with industrial waste water".

SOURCES OF CONTAMINATION OF WATER

Among the many sources of water pollution the principal ones are:

(i) **Discharges of domestic and municipal sewage,**

(ii) **Effluents from industries,**

(iii) **Use of agricultural chemicals such as fertilizers, pesticides and insecticides**

(iv) Accidental or/and deliberate release of oil in the ocean water,

(v) Then there is natural pollution of groundwater due to salts and minerals derived from rocks a soils and land.

In this chapter only a few cases are mentioned—not to portray the status of all kinds of pollution, b just to demonstrate what is or are responsible for the deterioration of the quality of water and air.

(1) In 1981 the urban communities in India every day generated 24,752 million litres of waste wat while the rural sector produced 49,777 million litres—that is 74,529 million litres per day in t whole country of 68.3 million (Sundaresan, 1982). In 2012 when our population is one billion a 21 millions, one can estimate the amount of the production of sewage. Much of the content of t sewage is organic matter.

(2) A village dweller in 1980 on an average generated per day 60 gram of oxygen-demanding was so that 2,880 tonnes of biochemical oxygen demand (BOD) load was discharged every day the rural masses (Deodhar and Nath, 1981). The extent of pathogenic pollution by pathogen microbes can be gauged from the fact that even one gram of faeces contains millions of micr organisms, many of them disease carriers such as faecal coliforms. In urban centres the situati was and is still worse (Tabels 21.1 and 21.2) as the pollution is concentrated and the natural abilı of self-purification of the ecosystems (water) has become ineffectual.

(3) Industries, which are burgeoning at a terrifically fast rate, daily produced 55,000 million m^3 waste water—68.5 m m^3 of the effluents discharged in rivers and streams in the early 198 (Deodhar and Nath, 1981). Today it is several times more. Thus **all our industrial cities including Mumbai, Delhi, Kanpur, Kolkata, etc. have been contaminating without check the country life lines** (Tables 21.1 and 21.2).

(4) Indian pesticides industry produced in 1999, approximately 90,000 tonnes of pesticides belongi to about 200 different types. The major consumer was and continues to be the agricultural sect The result is obvious. **Even the groundwater is now polluted with the undesirable content of these harmful chemicals banned elsewhere in the world.** In Ahmadabad, for example, the me

Table 21.1. High Fluorine Content in Drinking Water in Affected Areas (From Jha and Jha, 1982)

State	Fluorine Content (ppm)
Assam	12.5
Bihar	1.2
Uttar Pradesh	13.5
Delhi	4.0
Panjab	23.0
Andhra Pradesh	11.8
Tamil Nadu	6.4

levels of DDT and HCH were as high as 4.74 mg/l and 256.9 mg/l., respectively (Report of Indian Council of Medical Research, 2000). In 1984 Bhopal in Madhya Pradesh suffered from one of the world's worst industrial disasters, when an accidental leak of methyl-isocyanate (MIC) (used for making an insecticides) at the Union Carbide plant killed 2,500 people and crippled for life thousands of people, who continue to struggle to get justice and to survive.

GROUNDWATER CONTAMINATION

Natural Pollution

some areas in Andhra Pradesh, Tamil Nadu, Karnataka, Gujarat, Rajasthan, Panjab and Uttar Pradesh g. 21.1 and Table 21.1) a disease has been affecting several hundred million people leading to •ttling and pitting of teeth, decay of bones causing deformation of the body frame, or excessive bone

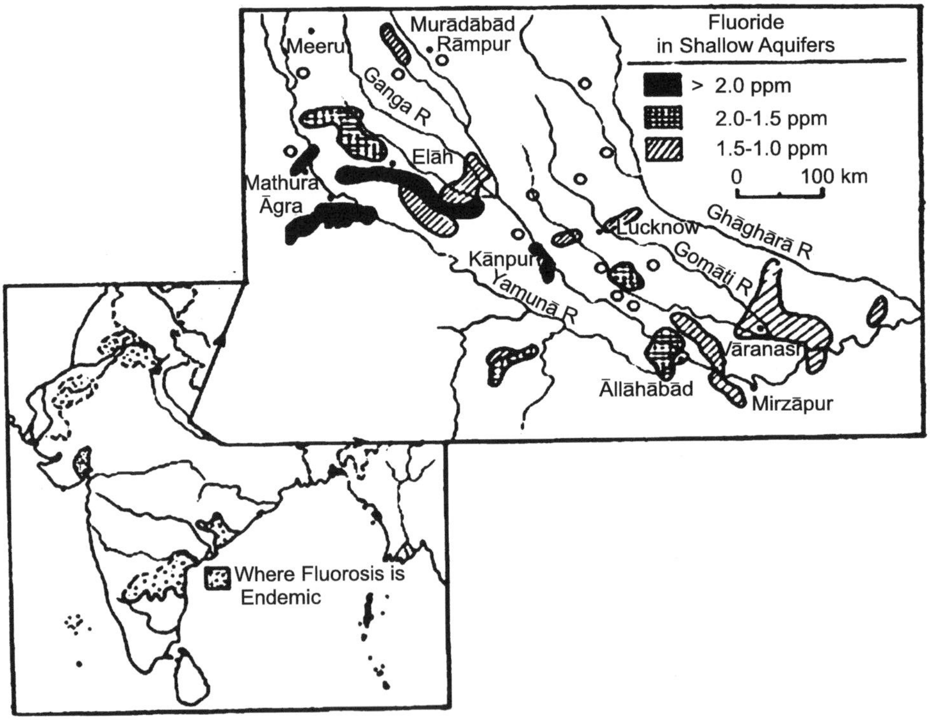

Based upon Survey of India map with the permission of the Surveyor General of India. The territorial waters of India extend into the sea to a distance of twelve nautical miles measured from the appropriate base line. Responsibility for the correctness of internal details shown on the map rests with the publisher Government of India Copyright 1986

Fig. 21.1. *Left*: Several hundred millions people in different parts of the country in the early 1980s were afflicted with the crippling bone disease-fluorosis, resulting from excessive (>1.5 ppm or 1.0 mg/1) fluoride content in water. There has been no attempt worthwhile to reduce the pollution. (After Anil Agrawal *et. al.*1982). *Right*: The groundwater in many areas in Uttar Pradesh has a higher content of fluoride. (Modified after M. C. Gupta and M. M. Gaumat, 1983).

formation in the periosteum and calcification of ligament tendon and aorta leading to immobilizati (Fig. 21.2), in addition to abdominal colic, constipation, anemia and insomnia. It is attributed to a hi content (>1.5 ppm or 1.0 mg/l) of fluorine in the groundwater. In Panjab and western Uttar Pradesh t dental syndrome of mottling is more prevalant, while in south India osteofluorosis affecting bones a skeletal structure is encountered (C.R. Krishna Murti, per. comm., 1985). Tens of villages of Prakasa Nalgonda and Guntur districts in Andhra Pradesh have been crippled (with "bent back") for life fluorosis affecting mostly children. The building of the Nagarjunasagar Dam across the Krishna rais the water table in the command area, bringing the fluorine-rich water—with its higher content molybdenum, Zn, Cu, Mg, Cd—into shallow wells and the subsoil. The molybdenum found its way in sorghum, cropped by the poor farmers for personal consumption. Molybdenum displaces copper in t body and seriously aggravates the disease. Over 75% of the victims belong to the age group 10 to years, the deformation setting in at the age of 6 to 7 years.

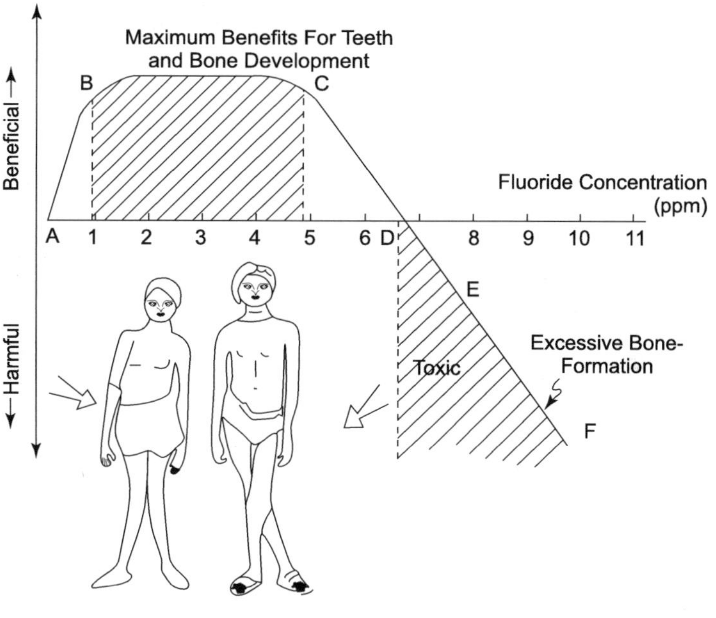

Fig. 21.2. The fluoride dose response curve. (After E. A. Keller, 1976, *Environmental Geology*, Bell and Howell, Columbus, 488 p., Fig. 11.4.) While small quantities may be beneficial to bone grow larger amounts cause excessive bone growth, deformation of the body frame and other debilitating effects such as manifest in these boys from Andhra Pradesh.

The fluorine content is higher (>8 ppm) in the groundwaters of districts Krishna, Karimganj, Anantp (Andhra Pradesh) and Betul (Madhya Pradesh) (Table 21.2), the chief source of fluorine being t minerals fluorite (CaF_2) and fluorapatite ($(Ca_5PO_4)F_3$) derived from the rocks pegmatites, granites a dolomites. In the Ganganagar, northwestern Jodhpur, eastern Nagaur, western Sirohi (Rajasthan) a Naliya (Kachchh) districts, isolated pockets had as high a content of fluorine as 8.5 to 19.0 ppm (Bisw and Saha, 1983). In the Amreli district in Gujarat nearly two thousand people were afflicted until abo 1981, particularly in Liliya and Lathi talukas, where the Deccan basalts contain higher fluorine (4.5 to m). In the early 1980s the content of fluoride (Figs 21.1, 21.3 and 21.4) in village Bairi on the rural nge of Kanpur (U.P.) was 2.70 mg/litre, while in the wells at Vinayakpur it was 0.55 to 1.80 mg/litre. l wells of Bairi and about 50% of Vinayakpur wells had water with fluoride content higher than the erance limit, leading to dental fluorosis and caries (Kayastha and Kumra, 1980). The well waters Bansi, Dumariyaganj (Basti), Milak (Rampur), Nanpura (Bahraich) and Biswan (Sitapur) in Uttar adesh (Fig. 21.3 and 21.4) were found to contain high levels of Mn, more than 200 mg/l (some as gh as 1000 mg/l)—leading to manganism, a disease resembling Parkinson's disease (Sondhi, 1983). In e Ghazipur, Varanasi and Mirzapur districts of Uttar Pradesh, groundwater derived through tubewells, ntained the arsenic concentration above the 10 µg/l limit in 60% cases, and above 50 µg/l in 20% ses (Shah, 2010).

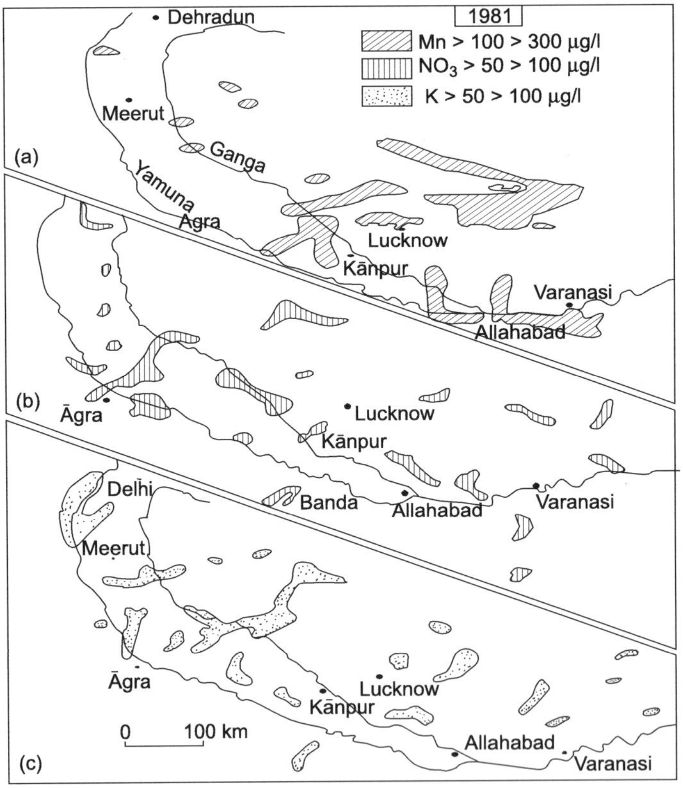

Fig. 21.3. In 1981 the groundwater (in dug wells) of Uttar Pradesh showed high levels (although still within permissible limits) of Mn, K, NO_3—the last due to excessive application of nitrogen fertilizers in agricultural fields. (Modified after (a) T. S. Sandhu, 1983; (b) Adarsh Kumar, 1983; and (c) D. K. Goel, 1983.)

Induced Pollution

many parts of Uttar Pradesh the groundwater in the early 1980s contained more than 200 mg/l nitrate (the safe limit being 45 to 50 mg/l) (Figs. 21.3 and 21.4) such as in Badshahpur in Jaunpur

(300 mg/l), Mirzapur, Saharanpur (300 mg/l), Bangarmau in Unnao (390 mg/l), Birdha in Hamirp (434 mg/l), Bansi in Jhansi (558 mg/l), Hastinapur in Meerut (694 mg/l) (Adarsh Kumar, 1983). **This is attributed to excessive application of nitrogen fertilizers, as evident from the fact that only tho aquifers are affected which underlay the fields treated with fertilizers.** The result is a strange infa disease causing blue colouring of the skin, and the incidence of goitre. The incidence of goitre is relat directly to iodine deficiency, but its appearance in the areas having nitrate-rich water bespeaks of t need for caution in over-use of nitrate fertilizers. Until about the early 1980s approximately 30 to 70% the people in 10 districts of the Tarai belt in Uttar Pradesh including Gonda, Gorakhpur, Deoria were the grip of this malady. The districts of Sidhi, Sarguja, Shahdol and Raigarh constitute the goitre belt eastern Madhya Pradesh in the early 1980s.

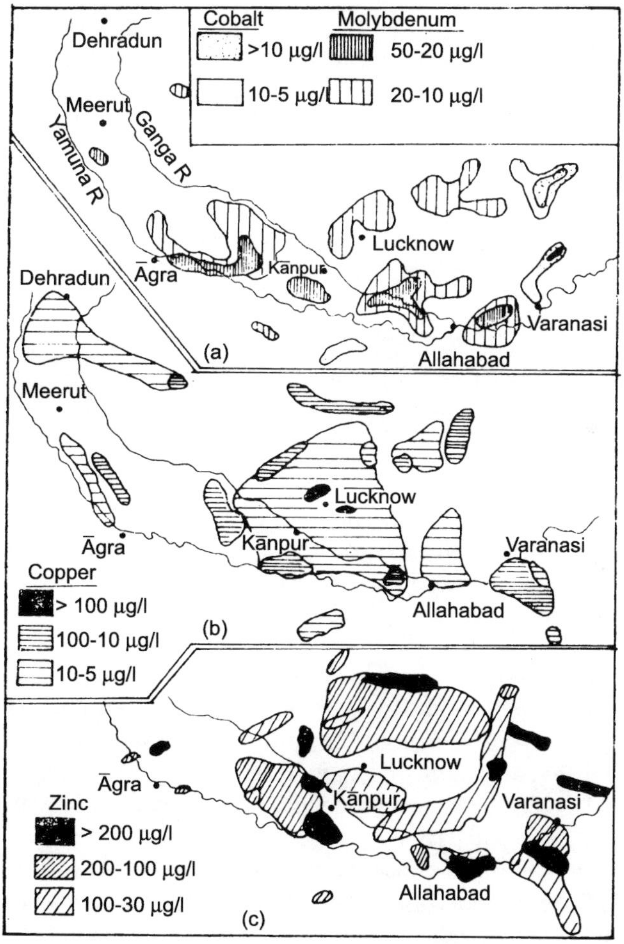

Fig. 21.4. The 1981 occurrence and distribution of pollutant trace elements such as cobalt, molybdenum, copper and zinc in the groundwater in dug-wells of Uttar Pradesh. (Modified after B. K. Hand 1983.)

The injection of waste water from industries of all kinds into wells as a measure of disposal is aught with hazardous consequences. Particulate matter and pathogen bacteria are not known to •ve far (>30 m), but the chemical pollutants including trace elements move quite freely far and deep and can thus cause serious underground water pollution (Fig. 21.5).

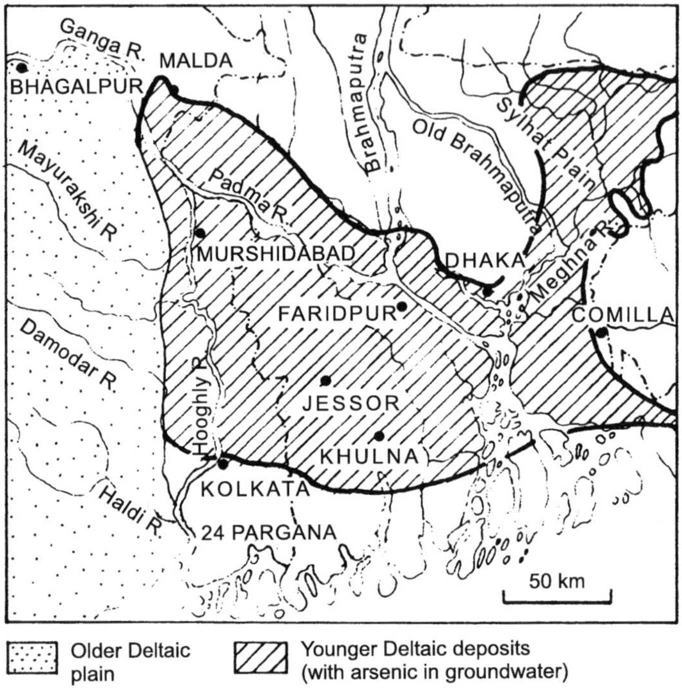

Fig. 21.5. Arsenic poisoning in well waters coming from the 50 to 80 metre deep aquifer in the floodplains of the Bhagirathi and Padma rivers in Bengal and Bangladesh. (Based on Mallick and Rajagopalan, 1998) (From Valdiya, 2004).

Sometimes municipal or domestic sewage finds its way to aquifers as a result of infiltration g. 21.6). In the early 1980s in the industrial city of Kanpur, for example, the shallow aquifers (50 to 5 m deep) showed cyclic rise of content of nitrate (during the rainy season) from 1 to 12 ppm to as uch as 45 ppm or locally even 70 ppm, of sodium to 15 to 175 ppm or even 1000 ppm, of chloride to much as 180 ppm, and of bicarbonate to 300 to 900 ppm, although still within the tolerance limits in 81 (Baveja, 1982) (Figs. 21.3 and 21.4).

Industrial effluents containing high concentrations of cyanide (1.82 to 7.55 mg/l), chromium (2.12 3.10 mg/l), arsenic (0.04 to 0.07 mg/l), lead (0.03 to 0.06 mg/l) have infiltrated into the aquifer in the eas where wools are dyed in the Ludhiana district in Panjab, causing pollution of underground water yanide 0.22 to 3.25 mg/l, chromium, 0.02 to 1.32 mg/l, arsenic 0.01 to 0.06 mg/l, and lead 0.01 to 0.06 g/l) above the permissible limit (Singh and Bhayan, 1982). The discharge from the Vaniyambadi (Tamil adu) tannery wastes containing inorganic salts into the Palar river (which has no surface flow) caused nsiderable pollution in subsurface water over a long stretch of the river (Table 21.3) as indicated by the rise in well-water of the content of total dissolved salts from 640 to 740 mg/l and of chlorides fro 88 to 180 mg/l (Sastry, 1984) in the early 1980s.

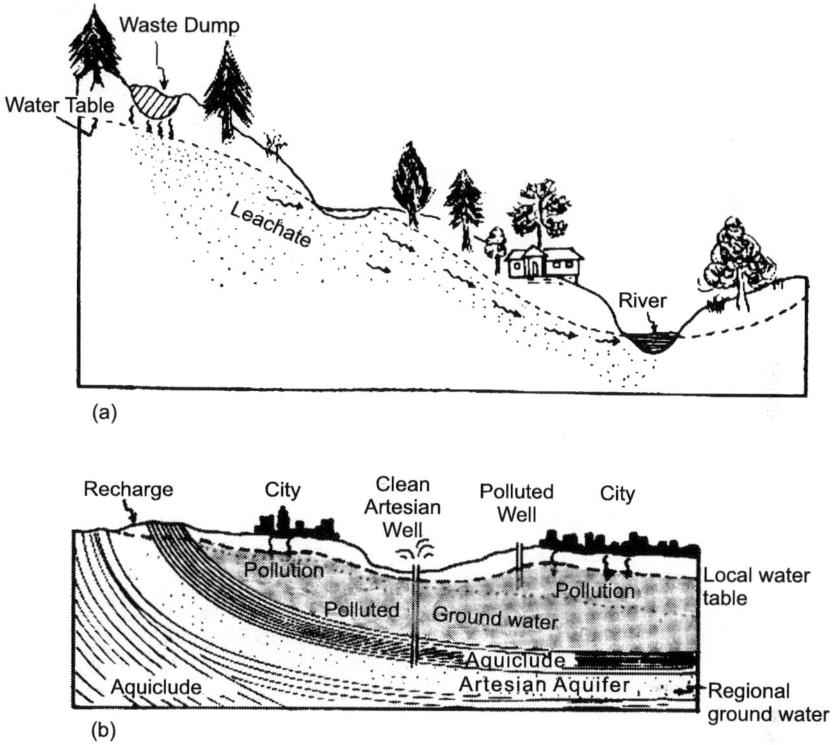

Fig. 21.6. (a) An unsealed or improperly covered waste dump located at a higher altitude up the slope is capable of causing pollution through leachate derived from waste. (After W. J. Schneider in R. Tank (Ed.), 1976, *Focus in Environmental Geology,* Oxford Univ. Press, New York, 538 p., F 1, 482. p.) (b) The waters of shallow, local aquifers become polluted by careless or indifferent disposal of urban seewage.

More than 600,000 people of about 400 villages in Bangladesh and West Bengal (Figure 21. are grievously affected—terminal illness and body disfigurement—due to the arsenic poisonin 20 to 70 times more than the prescribed safe limit. The poisoned water comes from 50 to 80 deep aquifers, pumped out through tubewells. Over-withdrawal of water has exposed the 50 to 80 de aquifer to oxidization, and the oxidized material leaches out through carriers of iron hydroxides deriv from ferruginous coatings around sand grains (Mallick and Rajagopal, 1996; Khan et al., 2003).

In the Kopanur and Chinnnamoolathara in Chittur taluk of the Palakkad district in Kerala, the wat of the dugwells showed fluoride concentration ranging from 0.2 ppm to 5.75 ppm, the deeper aquife showing concentration as high as 3.12 ppm (Shaji et al., 2007).

Report – 2011)

River	No. of Monitoring locations	Year	Observed Range of Water Quality Parametres						
			pH	Conductivity (µmhos/cm)	DO (mg/l)	BOD (mg/l)	COD (mg/l)	Total Coliform (MPN/100 ml)	Faecal Coliform (MPN/100 ml)
Ganga	34	2002	6.4–9.0	19–2720	2.7–11.5	0.5–16.8	1–30	300–25 × 10^5	20–11 × 10^5
	39	2005	6.1–9	23–1696	3.2–12.8	0.1–15.2	1.0–37.6	13–45 × 10^5	13–11 × 10^5
	57	2009	6.5–8.9	68–4460	4.3–11.2	0.2–16	2–34.2	2–65 × 10^4	0–4 × 10^3
Yamuna	23	2002	6.7–9.8	56–1959	0.1–22.7	1.0–36	1–112	27–26.3 × 10^6	11–17.2 × 10^5
	23	2005	6.8–9.1	90–2290	0.5–17.3	0.8–59	1–180	14–307 × 10^5	11–52 × 10^5
	27	2009	7.0–8.8	80–3040	0.0–17.9	0.2–103	2–115	4–23 × 105	9–21 × 10^8
Sabarmati	3	2002	2.9–8.6	269–13530	0.6–7.9	0.8–475	4–1794	210–28 × 10^5	28–28 × 105
	9	2006	6.79–8.67	256–3970	0.2–14.7	0.8–293	9–825	9–10 × 10^3	4–11 × 105
	9	2009	7.3–8.7	253–3200	0–8.9	1.2–46	36–44	240–46000	80–15000
Mahi	7	2002	7.1–9.2	175–5720	0.2–8.5	0.1–3.0	9–163	3–2400	3–75
	9	2005	7.5–9	182–7080	4.1–11.1	0.2–5.9	3–18	3–14 × 10^3	2–1 × 10^3
	9	2009	7.1–10	160–766	3.5–8.6	0.1–4.0	6–13	3–170	0–90
Tapi	10	2002	7.4–9.0	76–700	4.8–8.8	0.6–10.0	8–40	40–2100	2–210
	13	2005	7.2–9.4	186–1084	4–8.4	1–25.1	–	2–46 × 10^4	2–15 × 10^4
	14	2009	6.6–8.9	173–45400	3.7–8.2	0.6–12	6–68	14–39000	0–14000
Narmada	14	2002	6.9–9.3	102–1341	5.8–9.8	0.1–3.8	6–47	9–2400	2–64
	15	2005	3.3–9	190–1746	4.8–10.9	0.6–4.5	12–18.3	3–2400	2–210
	21	2009	6.5–8.9	178–1930	4.2–11.5	0.2–3.0	6–60	2–1600	0.90
Godavari	11	2002	7.0–9.0	118–1400	3.1–10.9	0.5–7.8	3–96	8–5260	2–3640
	18	2005	6.7–9.1	121–1300	0.8–8.7	0.5–20	4.0–80	2–33 × 10^3	1–10 × 10^3
	35	2009	6–9.2	115–3169	3.2–12.3	0.0–26	0.4–64	5–16000	0–340
Krishna	17	2002	6.8–9.5	28–11050	2.9–10.9	0.2–10.0	3–88	17–33300	3–1 × 10^3
	21	2005	6.5–9.9	69–43300	1.4–8.8	0.4–40	4–44	17.84 × 10^3	1–34 × 10^3
	22	2009	6.7–9.0	75–19960	0–12.6	0.3–9.6	4.0–45	8–170000	0–1400
Kaveri	20	2002	2.0–9.2	31–53100	0.1–12.6	0.1–26.6	30	39–16 × 10^3	2–28 × 10^3
	20	2005	6.2–9.5	28–48700	0.3–9.8	1–12	16–96	2–9500	1–3 × 10^3
	20	2009	6.5–8.9	65–81800	1.5–10.3	0.1–17	12–48	7–9200	2–5400

Contd.

River	No. of Monitoring locations	Year	Observed Range of Water Quality Parametres						
			pH	Conductivity (μmhos/cm)	DO (mg/l)	BOD (mg/l)	COD (mg/l)	Total Coliform (MPN/100 ml)	Faecal Coliform (MPN/100 ml)
Mahanadi	16	2002	7.3–8.9	114–15940	1.3–10.4	1.0–7.6	7–39	15–30000	50–17000
	21	2005	6.1–8.7	75–36279	4.5–10	0.2–16	4–150	$3–92 \times 10^3$	$78–54 \times 10^3$
	22	2009	17.39	6.7–8.8	0.2–11	0.2–7.1	3.2–28	5–1600000	110–160000
Baitarni	5	2002	7.3–8.3	54–78400	6.8–9.3	0.2–6.8	–	900–22000	700–11000
	5	2005	7–8.6	68–42257	5.2–8.8	0.4–4.3	12.9–20.4	$790–24 \times 10^3$	$3330–1 \times 10^3$
	5	2009	6.7–8.4	69–28400	6.1–9.0	0.6–3.4	3.9–22.2	630–5400	230–2800
Beas	19	2002	7.1–8.7	53–517	5.2–11.5	0.3–5.0	1–13	2–2400	2–1600
	19	2005	7–8.8	54–395	4.8–13	0.2–10	1.8–22	$2–11 \times 10^3$	2–1100
	23	2009	7.1–8.5	46–338	6.4–11.8	0.1–4.3	1.5–7.6	7–2400	2–1600
Ghaggar	15	2002	7.0–9.5	320–1012	2.6–9.6	1–180	4–560	43–14000	9–2500
	19	2005	7–9	21–2682	2.2–8.9	1–626	96–1600	$43–15 \times 10^4$	$14–5 \times 10^4$
	19	2009	6–8.5	350–2660	0.8–7.5	2.5–593	0.0–15.8	$500–1 \times 10^6$	$900–11 \times 10^4$
Kali East	2	2002	7.2–8.7	24–1930	6.7–11.9	1.9–67.0	66–421	$2100–48 \times 10^6$	$10 \times 104–36 \times 10^4$
	2	2005	7.4–8.4	23–1730	1.7–10.6	2–136	48–492	$7500–18 \times 10^6$	$2300–122 \times 10^4$
	2	2009	7.3–9.2	72–2124	0.0–13.3	3.6–353	140–140	$2800–35 \times 10^7$	$400–4 \times 10^7$
Subarnarekha	6	2002	6.5–8.0	113–355	5.2–8.5	0.2–12.0	4–96	150–1800	70–541
	6	2005	6.8–8.3	130–405	5.5–8.6	1.0–4.7	4–38	110–1400	78–700
	12	2009	6.4–8.4	164–717	4.0–8.5	0.4–6.3	4–104	280–2400	70–1300
Brahmaputra	6	2002	6.5–9.0	104–684	1.1–10.5	0.1–3.9	6–11	360–240000	300–24000
	10	2005	5.9–7.6	20–408	2–10.5	0.3–6.2	–	$300–24 \times 10^4$	$150–24 \times 10^4$
	10	2009	6.1–8.1	69–303	4.4–10.5	0.3–5.4	4.3–13.5	1–24000	0–1100
Pennar	4	2002	7.5–8.7	364–978	6.0–9.3	1.0–2.9	14–16	–	–
	5	2005	7.8–8.8	447–2340	3.8–8	0.8–2.8	12–13.2	16–2790	2–35
	5	2009	6.6–8.6	307–2450	3.3–10	0.7–3.9	5–24	9–3500	2–140
Satluj	20	2002	6.8–8.8	131–819	3.8–11.4	0.1–45.0	1–80	8–35000	2–3500
	21	2005	7.1–8.3	150–818	2.8–14.2	0.1–40	2.8–60	$1–35 \times 10^4$	$1–11 \times 10^4$
	22	2009	6.3–8.5	124–932	0.6–11.4	0.1–55	1.4–128	4–250000	0–110000

POLLUTION OF RIVER WATERS

Sources

l major rivers—which meet 85% of the drinking water needs of the country— are without an exception lluted, and some are dangerously poisoned due to toxins released by industries (Table 21.2).

It is disconcerting to realize that 4400 km aggregate length of the 14 major rivers of India t of nearly 6000 km stretch is forbiddingly polluted—with the dissolved oxygen content being much below the required level. This is the consequence of discharge of untreated municipal and dustrial effluents that goes on unchecked—and without shame. Not only the oxygen (largely lized in the decomposition of the organic pollutants) is far lower than the desirable limit (Table 21.3), t there is also addition of toxic elements like lead, zinc, chromium, copper, cobalt, etc. These toxic ments then go into the framework of the clays in the river beds, leading to the wider damage to the getation growing on the flood plains. The industrial pollution is more dangerous, because the effluents ntain quite many toxic elements (Table 21.7).

The rivers are thus heavily loaded with organic wastes, predominantly animal manure, human creta, decaying vegetable matter, and micro-organisms including pathogens. The perusal of Tables .2 to 21.6 will show how greatly the Ganga (Tables 21.4 and 21.5) and its tributaries such as Yamuna able 21.6) are polluted. For example, in 2003 the 22 km stretch of the Yamuna (Table 21.6) received wage at the rate of 1800 million litres per day. How fast are our cities polluting rivers can be judged m the affidavit recently submitted to the Supreme Court by the Central Pollution Control Board. Until 10 nearly 500-km stretch of the Yamuna was polluted, but by the early 2012 (in just two years) the lhi drains polluted additional 100 km stretch of the river—from Wazirabad in the north to Juhita in e south (Nivedita Khandekar, *Hindustan Times*, May 30, 2012). Thus 600-km stretch of the Yamuna is ghly polluted. Between Kanpur and Unnao the toxic elements within the alluvial sediments increased the Ganga by the factor of 14.7 chromium, 12.2 tin, 3.2 zinc, 2.6 copper, 1.6 nickle, and 3.6 organic rbon (Ansari et al., 1999).

Table 21.3. Broad Range of Nutrient Contents (mg/l) of Domestic Waste Water in Indian Cities (B.B. Sundaresan, 1981)

ype	Total Nitrogen (N)	Total Phosphate (P_2O_5)	Total Potash (K_2O)
Raw Sewage	60 to 70	20 to 25	40 to 45
Primary treated sewage	60 to 70	20 to 25	40 to 45
Secondary treated sewage	15 to 20	15 to 20	35 to 40

The Ganga at Varanasi—where 60,000 people take a dip every day—is polluted by inflow of sewage the rate of 200 million l/day in 2004), industrial wastes, plastic bags, animal caracasses and unclaimed man dead bodies, resulting in biochemical oxygen demand is as high as 20 mg/l, and faecal coliform unts around 1.5 million/100 ml (Mishra, 2005). The Ganga stinks as it leaves the city.

The daily sewage generation of Mumbai in 1982 was of the order of 1365 million tonnes, and the ahim creek alone (Fig. 21.7) daily carried 325 million litres of domestic effluents and 22 million res of industrial effluents (*Scavenger,* 13 (1): 22-25, 1983). The water of the Ulhas River draining the dustrial area of Mumbai was and continues to be so heavy with toxic elements like Pb, Zn, Cr, Cu, Cd, . that these elements enter the framework of even the riverbed clay minerals such as montmorillonite, te, kaolinite (Sahu and Mukerjee, 1983). The waters of the Kallu stream and the Thane creek likewise contained high amounts of toxic Hg, Cd, Pb and Cu, which later found their way into vegetables a fodder grasses grown in the floodplain. Table 21.8 shows that condition has worsened very much.

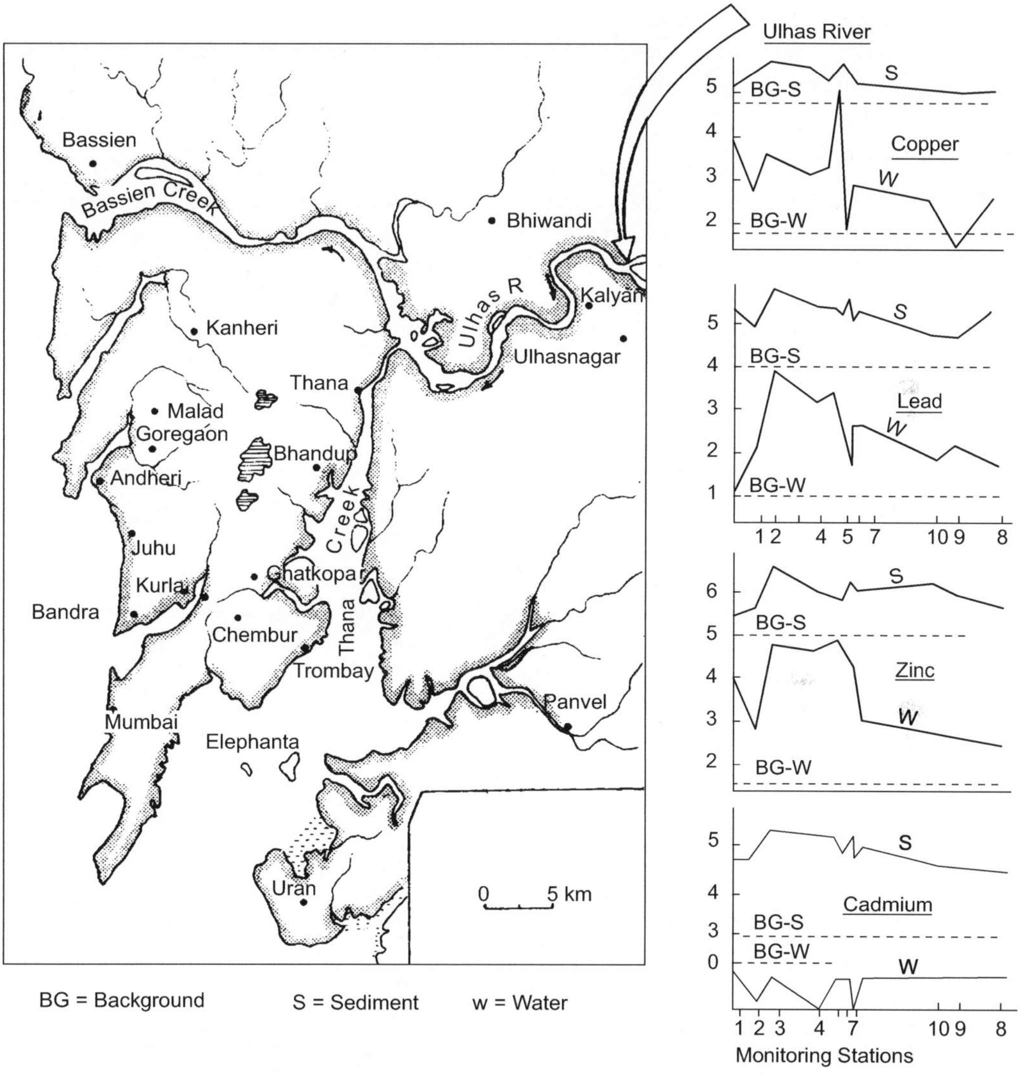

Fig. 21.7. State of water quality in the Mumbai region in the early 1980s. Not only the water, but also t clays of the sediments of the Ulhas River flowing through the industrial suburbs of Mumbai a polluted with heavy metals. (Graphs after K. C. Sahu and S. Mukerjee, 1983).

All the ten major rivers of Kerala are heavily polluted—the various industries dumping their tox effluents at the rate of 500,000 m^3/day in the early 1980s. The Chaliyar is the most polluted river, water being characterized by high contents of TiO_2, and compounds of Zn, Pb, Cr, Cu, Mg, Mo pl alkalies discharged by industries manufacturing and using dyes and pigments. One rayon factory avoor in the Kozhikode district alone added 54.8 million litres of effluents daily (*The Times of India,* y 20, 1983).

The Zuari river in Goa is quite polluted, both by discharges from the agro-chemical industry and by discriminate dumping of iron ores on the land and in the river.

It is thus apparent that municipal sewage and industrial effluents are the two major causes of pollution of river waters (Tables 21.3 and 21.7).

Parameters of Pollution

e normally dissolved oxygen (DO) in water degrades the wastes, and thus under normal conditions ngs about purification (Fig. 21.8). **However, if the quantity of waste becomes high, the dissolved ygen (DO) is not sufficient and not able to effect biodegradation and purification. The oxygen water is consumed by the micro-organisms feeding on organic pollutants (Fig. 21.8). This is called the *biochemical oxygen demand* (BOD) measurable in milligrams per litre (mg/l) of ygen consumed over a period of five days at a temperature of 20°C. The higher the number of cro-organisms, the greater the consumption of dissolved oxygen, and the greater the extent of eutrophication.**

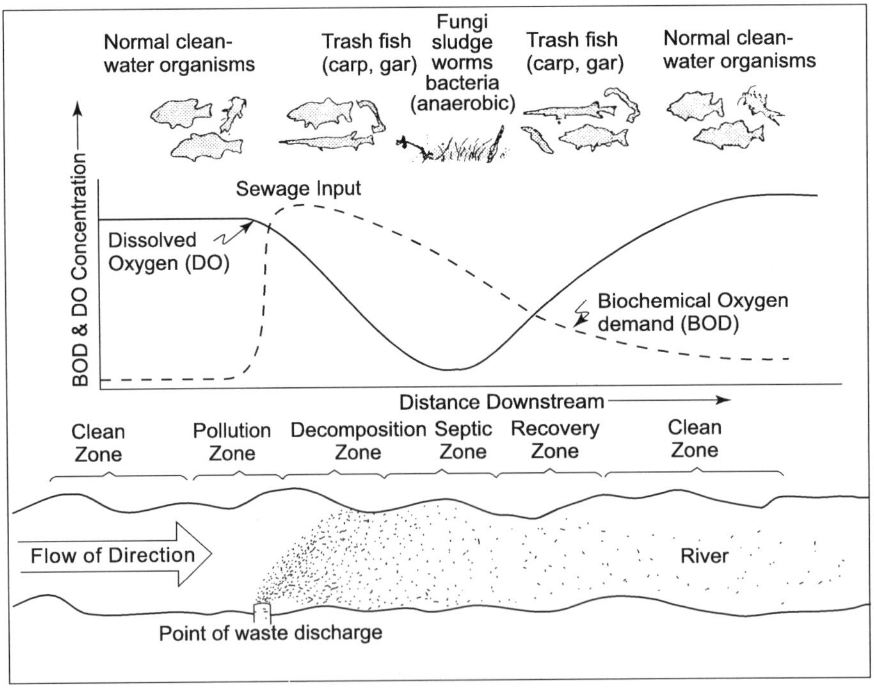

Fig. 21.8. A sewer drain of a city causing pollution of river water. The organic pollutants make a great demand on the dissolved oxygen (DO) for their biodegradation. If the oxygen content is high enough, the water is able to recover quickly—purify itself. This is indicated by the composition of the organisms living in the river. (After G. T. Miller and P. Armstrong, 1982, *Living in the Environment*, Wadsworth, Belmont, 491 p., Fig. 17.2).

Domestic and municipal sewage contain high quantities of phosphate, nitrogenous compound potassium and carbon (Table 21.3). They serve as fertilizers, and stimulate excessive growth of alg and other aquatic weeds which further overburden the river water with pollutants. The result is a furth depletion of oxygen from an already oxygen-deficient ecosystem. The death and decay of these wee make the waters murky and still more polluted and stinking. Many of the detergents, pesticides a related organic compounds including DDT are not degradable by bacteria, so that the waters becom hazardously toxic in view of the fact that India is annually producing and using more than 14,000 tonn of DDT and 41,000 t of BHC.

The threshold value of pathogens is 200 cells per millilitre of water. The rivers contain ma times higher amounts of bacteria, viruses and coliforms, particularly where the sewage of t cities is discharged into the rivers without treatment. This explains why 60 to 80 % of the diseas afflicting Indians are related to infection in the alimentary canal, such as diarrhoea, dysenter typhoid, jaundice, cholera, etc.

The effluents from industrial units and power plants carry various salts, metal compounds and fine divided metals that increase the acidity or alkalinity and toxicity of water (Table 21.7).

Table 21.4. State of Pollution of the Ganga River (Source: Central Pollution Control Board) A-14011/1/200 MON, dated 14/1/2010

Pollution Point	D.O. (mg/l)			pH			Conductivity (μmhos/cm)			BOD (mg/l)			Faecal Coliform (MPN/100 mL)			Total Coliform (MPN/100 mL)		
	Min.	Max.	Mean	Min.	Max.	Mean	Min.	Max.	Mean	Min.	Max.	Mean	Min.	Max.	Mean	Min.	Max.	Mean
Gangotri	7.2	10.1	8.6							0.0	2.0	1.0				2	340	
Haridwar	6.0	9.2	7.6							2.2	5.4	3.8						
Kanpur	1.6	9.9	5.6	7.4	8.8	8.0	67	860	479	3.8	21.0	8.3	1500	46000	16818	15000	240000	94250
Allahabad (Sangam)	6.5	9.7	8.0	8.1	8.4	8.3	270	758	469	3.3	5.4	4.0	1700	11000	5167	3500	17000	88
Varanasi	6.5	7.9	7.3	8.4	8.9	8.6	264	397	353	9.2	12.4	10.5	18000	1100007	45000	94000	220000	144833
Dakshi-neshwar (Kolkata)	4.1	8.2	6.1	7.4	8.2	7.8	201	378	316	1.4	6.0	4.2	35000	850000	283333	55000	1400000	6054

Tragedy of the Ganga

The Ganga is no more the life-giving, life-supporting river it was. It has degenerated into a b sewer—the dirtiest of all rivers (Tables 21.1, 21.4 and 21.5).

Having been dammed at Tehri in western Uttarakhand, **the Ganga descends onto the plains, on to be robbed of its water by huge diversions through the Upper Ganga Canal at Haridwar (whi reduces its discharge to mere 15 billion m^3/yr) and then by the Lower Ganga Canal near Aligar** That leaves so little water in the Ganga that the dry-season discharge at Kanpur is merely 90 to 386 m second, at Allahabad 279 to 997 m^3/second, and at Varanasi 278 to 1160 m^3/second (Das, 2011, citi CPCB Report 2009). Despite being joined by a number of tributary rivers, the Ganga is progressive polluted due to heavy discharges at the rate of 3000 million litres per day from towns and cities, whi boast of having facility of sewage treatment, varying from 13.5% in small cities to 27.8 to 50.4% big cities—329 million kilolitres (Das, 2011). Obviously nearly 50% of waste waters are discharg treated into this lifeline of central Indo-Gangetic Plain. Over 1.3 billion litres of sewage, 260 million res of industrial waste, runoff from 6 million tonnes of fertilizers and 9000 tonnes of pesticides used in riculture, and very large quantities of solid waste are daily released into the Ganga (Das, 2011, citing Markandeya and M.N. Murty, 2000).

Table 21.5. Quality of Water of the Ganga at Different Localities (Based on graphics from Central Pollution Control Board, 2011)

Locality	Parameters		
	DO (mg/l)	BOD (mg/l)	Faecal Coliform (MPN/100 ml)
Gangotri	9.0	2.0	>10,000
Rishikesh (upstream)	8.6	2.2	>10
Haridwar (downstream)	6.1	5.6	>12
Garhmukteshwar	8.5	2.6	~1100
annauj (downstream)	7.1	4.1	~1200
anpur (downstream at Jajmau)	5.6	8.3	>5000
Allahabad (downstream at Sangam)	8.0	4.0	~1200
aranasi (downstream at Malviya Bridge)	7.3	10.5	>12,000
Patna (downstream at Ganga Bridge)	7.4	2.6	~11000
okama (downstream)	8.5	2.6	~10,000
Bhagalpur	8.8	2.4	~8000
Behrampur	7.4	1.7	~30,000
olkata (Howrah-Shibpur)	6.0	1.9	~1,00,000
olkata (Diamond Harbour)	6.1	1.6	10,000

Table 21.6. Pollution of the Yamuna (as on Feb. 6, 2012) Caused by the Capital of India—Delhi (Source: lhi Pollution Central Committee Report # DPCC/W/Y/11-12/091–099

Locations	pH	COD	BOD (mg/l)	DO (mg/l)	Total Coliform (MPN/100 ml)	Faecal Coliform (MPN/100 ml)
Water quality criteria ('C' Class)	6.0–9.0	-	3 (max)	4 (min)	5000	-
At Palla	8.4	16	2.2	7.6	1100	730
At Surghat (downstream of Wazirabad)	9.0	36	5.4	5.7	4400	3500
Khajori Paltoon Pool (downstream Najafgarh drain)	7.6	380	110	-	240000	110000
Kudesia Ghat	7.6	140	30	-	46000	29000
ITO Bridge	7.5	260	38	-	110000	46000
At Nizamuddin Bridge	7.7	172	36	-	39000	29000
Agra Canal (Okhla)	7.6	180	37	-	35000	23000
After meeting Shahdara Drain ownstream Okhla Barrage)	7.6	176	40	-	240000	150000
At Agra Canal (Jaitpur)	7.7	160	36	-	160000	75000

Table 21.7. Broad Range of Water Pollutants from Some Major Indian Industries (Based on N. S. Deodh and K. J. Nath, 1981; B. B. Sundaresan, 1982)

Industry	Waste Water Generated (million m^3/yr)	Pollutants			Others
		BOD (mg/l)	COD (mg/l)	Suspended Solids (mg/l)	
Paper and pulp	450	160 to 165	610 to 725	375 to 410	Chlorides
Cotton textile	1530	200 to 350	400 to 800	350 to 550	Chlorides
Refinery	15	100 to 300	320	200 to 400	Phenol, Sulphides, chromium, Phosphate, Oil, H_2S, RSH, NH_3
Nitrogen Fertilizers	52				Nitrogen, Phosphate, Arsenic, Fluoride
Coke oven	9	200 to 1200	1000 to 3000	50 to 500	Phenol, Cyanide, Tar, Oil, NH_3, CN, CNS
Sugar	225	300 to 2000	600 to 4000	220 to 800	Nitrogen
Distillery	22	15000 to 40000	3000 to 8000	4500 to 12000	Potash, Nitrogen, Cl, SO_4
Pesticides		0 to 7000	3000 to 10000		Org.Nit. Sulphates
Tannery		1700	3500	1900	Cr, Sulphides, Tannin
Dairy		816	1340	690	Oil, Grease
Varicose rayon			210	200	Zinc, Polysulphide

The pollution is evident from the point it emerges into the plains at Rishikesh. As already stated, t municipal sewage and industrial effluents of 48 cities and 66 large towns into the Ganga are discharge In spite of its inherent ability for self-purification (10 to 25 times faster than other rivers) through qui coagulation of colloidal nutrients, it has in many segments indeed become "a biological nightmare", A. K. Sharma (1981) put it.

Amongst the major culprits at Kanpur are the tanneries and textile mills discharging 37.15 milli gallons per day of waste water, containing among other things, faecal coliform count 11,00,007 p 100 millilitre, and BOD 12.4 mg/litre (Das, 2011). The holy city of Varanasi dumps so much pollutan into the holy Ganga that the BOD level on the average is 10.5 mg/l, the faecal coliform count 4500 p 100 millilitre, and the total coliform count 1,44,833 per 100 ml (Das, 2011). Kanpur city discharg sewage and effluents from tanneries, textile mills, jute mills, and many pharmaceutical and chemic industries. The toxic metals added to the Ganga is Cd 0.05, Cr 6.45, Cu 0.88, Fe 8.80, Mn 0.55, Ni 0.2 Pb 0.19 and Zn 1.82 mg/L (Das, 2011). At Allahabad in the late 1970s the fertilizer factory at Phulp dumped 5500 m^3 of effluents everyday, enough to make the Ganga water fatal to fish and other aqua organisms for the 16-km stretch of the river. The bacterial count at the Kumbh Mela site in Allahaba during Mela days from 10 to 15 January in late 1980s rose to 9 million/litre against the normal 50,0 to 20,000/litre. The BOD value is 12.4 mg/l and the FCC 11,00,007 per 100 millilitre (Das, 2011). T celebrated temple city of Varanasi every year in the early 1980s dumped as many as 3000 half-bur human bodies, 6000 caracasses, 140 to 200 tonnes of flesh, 200 to 300 tonnes ash (by burning 110 tonnes of firewood), in addition to 925 million tonnes of domestic sewage and industrial effluents eve y (Agrawal et al., 1982). As already stated at present so much sewage inflow gets into the Ganga at ranasi that the BOD is as high as 20 mg/litre and the faecal coliform count is ~1.5 million/100 ml at e downstream end (Mishra, 2005). Downstream of Varanasi the Ganga turns black and septic, full of am and bubbling methane gas. The toxic metals added through wastes is Cd 0.16, Cr 8.12, Cu 0.16, 3.22, Mn 0.47, Ni 0.14, Pb 0.15 and Zn 1.58 mg/l (Das, 2011).

In the early 1980s the six major drains of Patna everyday used to discharge 6.6 million litres of w sewage, sullage and sludge, resulting in alarming bacteriological pollution (Ghose and Sharma, 84), the MPN count being 25 million per litre, the faecal coliform to streptococci ratio higher than e, coliform count 6,00,000/litre and streptococci 48,000/litre and the total bacteria 135 million/litre ilgrami and Datta-Munshi, 1984).

Downstream from Patna upto Bhagalpur the Ganga is polluted at several points with industrial fluents: the refinery at Barauni, the tannery of Mokamah and the distillaries, at the rate of respectively m l/day, 10.8 m l/day and 1.85 m l/day in the early 1980s. The pollution threatened the survival of *Hilsa illisia*, other fish and the Ganga dolphin (*Platanista gangetica*) among others. The bacterial count as 6,00,000 coliforms per millilitre and 48,000 streptococci per litre, the total bacterial density being .2/million per litre compared to the threshold limit of 200,000 cells per millilitre downstream of hagalpur some thirty years ago.

Taking into consideration these facts of pollution, the Ganga water can no longer be described as e-giving and holy.

Pollution from Mining Activities

ning and ancillary activities also contribute considerably to river pollution, as already discussed in hapter Six. It may be recapitulated that in the Bailadila iron mines site at Kirandih in Bastar district hattisgarh) more than 12 million tonnes of ore fines had accumulated in the period 1977 to 1980. The ns have been washing the wastes downstream, turning the river red (Anon, 1982).

The Tiri stream, which drains the lead mines at Zawar in Rajasthan, used to be polluted for 15 km wnstream. As a result of lead and zinc pollution, not only was the aquatic fauna adversely affected, t even the timber plants and other vegetation had registered perceptible decline. *Wringtonia tinctonia* and *W. tomentosa,* for example, were eliminated. A similar situation was discernible around Jhamarkotra osphorite mines near Udaipur and in the Khetri copper mines in Rajasthan.

The Damodar flowing through the coal-mining belt in Jharkhand is one of the very heavily polluted ers. Not only have the industries (such as the fertilizer plants of Sindri), power plants (Bokaro), and ores of coal washeries been discharging their effluents but also the many coal mines are contributing llutants in a dangerously reckless manner. Exposure of fresh rocks due to coal mining initiates athering, with inevitable oxidation and hydration of such minerals as pyrite, marcasite, ankerite (and derite and limonite) which produce sulphuric acid and soluble sulphates and oxides causing pollution water. These substances have adversely affected flora and aquatic life in the Damodar and its many butaries.

In the Singrauli Coalfields in south-eastern Uttar Pradesh, the water—which accumulated in pressions caused by opencast mining—and the streams draining the area showed higher concentra-n of Ca, Mg, Na, K, Fe and Al, derived obviously from coal seams and associated rocks (Philips et , 1985).

CONTAMINATION OF LAKE WATERS

General Pattern

The standing bodies of waters of ponds, tanks, reservoirs and lakes are not able to revive or puri themselves when their dissolved oxygen is utilized in oxygen-demanding biochemical processes whi take place due to the influx of organic wastes. The prolific growth due to presence of phosphate a nitrate nutrients and overcrowding of algae and weeds and their eventual death and decay creates t same situation. The result is the rapid setting in of anerobic conditions. Wholly bereft of oxygen a surcharged with foul-smelling H_2S, a lake suffers *eutrophication*—an environment of oxygen deficien and excess of hydrogen sulphide. This results in the destruction of its productivity (Fig. 21.9).

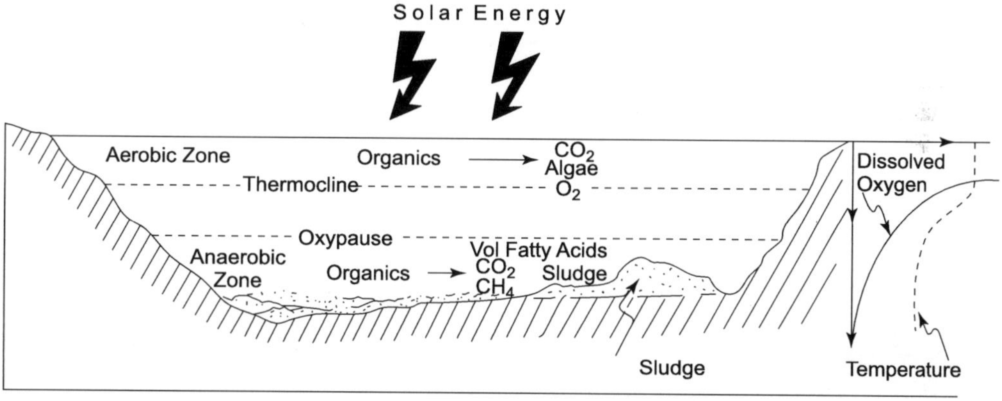

Fig. 21.9. In a standing body of water, such as a lake, an oxygen-deficient anaerobic condition develops the stagnant lower level, while the upper part may continue to have dissolved oxygen and warmth due to solar radiation.

Once a Well-Nourished Naini Lake

The Naini Lake (Fig. 21.10)—formed as a result of blockage of a stream in the geological past d to block movement along a major fault—was until about the beginning of the present century in advanced stage of eutrophication (Pant and Sharma, 1978; Pant et al., 1980; Sharma and Pant, 1982 This was manifest, as pointed out by Pant and his co-workers, in: (i) fairly reduced transparency (75 195 cm); (ii) rich crop (blooms) during warmer seasons of phytoplanktons, particularly microcysts; (i high non-algal suspended load in winter; (iv) prolific growth of zooplanktons such as *Bodo, Eutricha, Bursara, Epistylis,* and *Vorticella,* but decline in population of 'mahseer fish' (*Tor tor*) and large-sca mortality of small fish in winter when dissolved oxygen slumped down to 0.7 ppm (0.8 to 4 ppm) a CO_2 and NH_3-N values rose to, respectively, 22 ppm and 520 μg/l. The time of mortality of fish winter coincides with the periodic abundance of diatoms and myxophycea which causes depletion the main fish food (according to Das and Pande, 1978) (v) high content of phosphate (PO_n-P) above t critical level of 0.01 μg/l between 6.5 μg/l in winter and 70.00 μg/l in spring; (vi) eight-fold increa in ammoniacal nitrogen (NH_3-N) from 19 μg/l in 1954 to 156/μg/l in 1975: (vii) significant increase e concentration of nitrite-nitrogen (NO_2-N) and nitrate-nitrogen (NO_3-N), to 18.2/μg/l and 338/μg/l pectively in 1975; (viii) increase in hardness of water due to alkalinity from 4.00 ppm in 1954 to 400 m in 1975; and (ix) high concentration of free CO_2 as indicated by high pH 7.0 to 9.6.

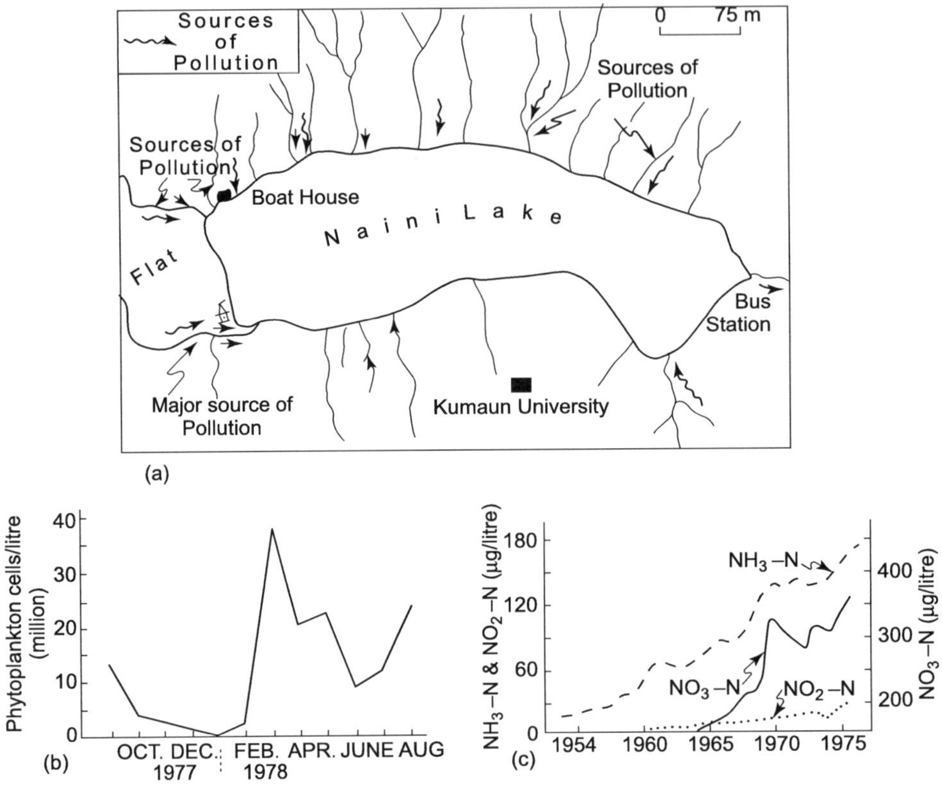

Fig. 21.10. The Naini Lake in Kumaun (Uttarakhand) was a well-nourished lake in terms of phosphatic and nitrogenous compounds derived from organic sewage from the many open drains, lakeside establishments and unchecked open defecation around the lake. The contents of these pollutants steadily rose since 1954, as was evident from the prolific growth of phytoplanktonic weeds. Now by periodic oxygenation (aeration), combined with removal of muck occasionally from the lake-side bottom, the lake has been restored to health. (From M. C. Pant, A. P. Sharma and P. C. Sharma, 1980 and P. K. Gupta and M. C. Pant, 1983).

The cause of eutrophication of the Naini lake was quite obvious (Fig. 21.10). A sizeable fraction the sewage of the burgeoning town used to, and continue to, find its way through the main drains o the lake, open defecation by the labour force and vagrants along the lake sides, discharges from ke-side catering establishments, a scattering of trash and foodstuff by uncultured people including urists. The accelerated rate of erosion in the northeastern hillside has also risen due to indiscriminate use-building, entailing generation of enormous volumes of debris which are allowed to move down e gutters. This brazen act added to the gradual deterioration of the lake, as evident from the fact that in the period 1960–1977 about 0.239 million m^3 of sediments was deposited in the lake at the rate of 0.0 mm^3/yr (Arun K. Sharma, 1981). The situation has changed markedly since the campaign was launch in the early first decade of this century—periodic aeration, and occasional removal of muck from t bottom, along sides and control in different discharges.

The Once Dying Dal Lake

The 12 km^2 shrunken Dal lake represents the remnant of a once (Pleistocene early Holocene) vast la that occupied the entire Kashmir valley. The shallow lake (maximum depth 6 m) has been encroach upon by a luxuriant growth of macrophytic vegetation (*Schizothorax*). Then it was progressive polluted by inputs of domestic garbage and sewage carrying (in the late 1970s) 16 tonnes of phosphor and 365 tonnes of nitrogen per year; by dumping of night soil and trash from then 5000 'shikara (houseboats) that floated on the lake; and by the mud brought by the streams from the adjoining area a very high rate of 80,000 tonnes/yr (V. Kaul and Raina, 1982). The dissolved oxygen content vari from 7 μg/l in spring to 5 to 4 μg/l in summer, the NO_3-N level varied from 7 μg/l in summer to μg/l in winter, NO_2-N content changed from 120 μg/l in summer to 440 μg/l in winter, and NH_3-N ro from 5 μg/l in summer to 15 μg/l in winter (Kaul and Pandit, 1979). Understandably, the fish populati declined following the damage and destruction of their spawning and breeding areas and prolific grow of weeds which generated 40,000 to 50,000 tonnes of organic matter annually. Even Wuler, Triga Tilwansar, Shieskar and Khanpur lakes in rural settings were in various states of eutrophication. In t early 1980s the orthophosphate concentration was high in Trigam (average 136 (μg/l), Wular (928 μg/ Tilwansar (60 μg/l), Khanpur (40 μg/l), Shieskar (928 μg/l in November), the transparency was low, t concentration of NH_3-N and NO_2-N chlorides, BOD, COD and H_2S quite high (Zutshi, 1984).

POLLUTION OF MARINE WATERS

Sources of Pollution

The coastal belts of oceans (Fig. 21.11)—which yield 90% of the world's marine food—are under fa increasing threat of pollution due to sewage discharges, spilling of petroleum oil from offshore oil-we that blow out, tanker wrecks and bilge, offshore drilling and various dock operations (Table 21.8). In t late 1970s the metropolis of Mumbai each day dumped 0.40 km^3 of sewage and 0.95 km^3 of industr effluents at the rate of 23 million litres into the sea (Sengupta, 1981). The total pollutants annually add to the sea amounted to more than 50 km^3. Today the situation is far worse as Table 21.8 brings out.

Pollution from Petroleum Oil

The second major cause of sea water pollution is petroleum oil. Nine-tenth of this quantity derived from offshore exploration and development of seabed mineral resources. Pipeline breache tanker accidents and offshore blowouts contribute more than 20%. Disposal from engines and machin (bilge) account for quite more than 6.6 %. The major tanker routes (such as from West Asia to differe countries) are heavily polluted. The giant oil slick releasing crude oil in the order of 50,000 to 30,0 barrels have been as long as 160 km and as wide as 15 km.

Table 21.8. Sources of Pollution of Coastal Water in India – 2009 (Source: Rahul Goswami in *Infochange News and Features*, September 2009)

City	Effluents Discharged (million litres per day)
Kolkata	------ (municipal) 7.0 ml/day (industrial)
Paradip	6.6 ml/day (municipal) 8.9 ml/day (industrial)
Visakhapatnam	68 ml/day (municipal) 55 ml/day (industrial)
Chennai	------ (municipal) 128 ml/day (industrial & power plants)
Puducherry	45 ml/day (municipal) 7 ml/day (industrial)
Cuddalore	13 ml/day (municipal) 7.5 ml/day (industrial)
Tuticorin	17.5 ml/day (municipal) 10.4 ml/day (industrial) 91.2 ml/day (aquacultural)
Kochi	36 ml/day (municipal) ------ ml/day (industrial)
Mangalore	28 ml/day (municipal) ------ (industrial)
Mumbai (Mahim)	830 ml/day (municipal) 93 ml/day (industrial)
Mumbai (Versova)	440 ml/day (municipal)
Mumbai (Bassein)	730 ml/day (municipal & industrial)
Hazira	----- ml/day (municipal) 60 ml/day (industrial)
Surat	300 ml/day (municipal) ------ (industrial)

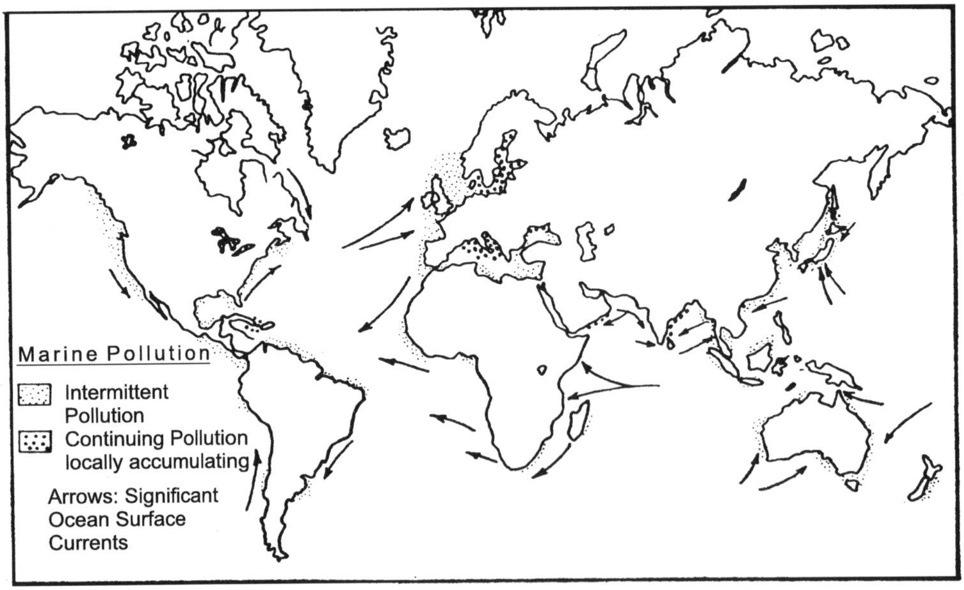

Fig. 21.11. The frequented routes of oil-tankers in the seas are heavily polluted with petroleum matter. The western coast of India is comparatively more polluted although still within safe limits. (Redrawn with permission from M. W. Holgate, M. Kassas and G. F. White, 1982, *The World Environment—1972–1982,* Tycooly, Dublin.)

The Dock Basin of Mumbai yielded 90,000 litres of recoverable waste oil in one year alone (*The Times of India*, October 2, 1984). A similar situation prevailed in the docks of Chennai and Vishakhapatna In the Indian coasts (Fig. 21.11) up to the depth of 20 m the petroleum hydrocarbon concentration w 0.2 to 0.05 ppm in the southern sector and Sri Lanka and 0.005 to 0.001 ppm in the eastern coasts in t late 1970s (Qasim and Sengupta, 1980). Pollution was comparably high in the western coast—even t planktons then contained higher concentrations of hydrocarbons (19.5 to 833 μg/g) and the sedimen were also affected (64.8 to 8.5 μg/g). The zooplanktons on the coastal belt were characterized by high values of heavy metals like Cu, Fe, Mn, Zn, Co, Ni (Qasim and Sengupta, 1980). It may be mention that as yet, the level of marine pollution along the Indian coasts is within safe limits even though t situation today is much worse than what it was in the year 1980.

The effects of oil pollution on the ecosystem are profound. It was realized that the lead concentrati of marine waters increased by a factor of 4 to 5 since the introduction of lead-based petroleum additiv The primary productivity of diatomaceous communities declined, while the growth of blue gre algae and flagellate increased (Miller and Armstrong, 1982). Highly toxic aromatic hydrocarbons a instrumental in causing the death of a number of aquatic organisms, especially the bottom-dwelli shellfish. Some of the hydrocarbons like the cancer-causing benzopyrene have got into the food cha through the fatty tissue of marine organisms. Oil slicks commonly attract sea-birds and cause ma mortality by poisoning.

TREATMENT OF POLLUTED WATERS

Domestic and Industrial Organic Waste Water

If there are no disease-carrying pathogens and toxic elements, the nutrient-rich sewages can be utiliz straight away for irrigation. **Removal of pathogens and toxic elements would make the waste wat more useful. The biological method of treatment involves holding the sewage in a tank, pond ditch for several days so that coarser particulate matters settle down and finer particles coagula through use of iron and aluminium salts or lime and caustic soda. The liquid is then filter through a mesh screen and allowed to trickle down through filters of sand or diatomaceous eart The filtered liquid fraction is then aerated by passing oxygen which also removes H_2O and CO This stimulates bacterial activity leading to biodegradation of fine or colloidal organic matt left in the liquid. This treated liquid can be utilized for all purposes, except drinking**. Resort chlorination in appropriate dosage kills pathogen bacteria. The muck collected from the settling tan can be used as manure.

A complementary process involves symbiotic activity of algae and bacteria in the presence sunlight, free oxygen and nutrients present in the waste water. The prolific algal growth consumes C● and excess pollutants (nutrients such as phosphates and nitrates) and releases abundant oxygen. T result is the cleaning of the waste water.

Toxic Wastes

Industrial effluents rich in toxic elements, including pesticides and detergents, are subjected physicochemical treatment—adsorption on activated carbon, neutralization, thermal reduction, and i exchange. Intense nuclear radiation, which generates highly reactive free radicals in water eliminat trace organic and inorganic pollutants including pesticides and pathogens (K. Krishnamurthy, 1980).

Water hyacinth *(Eichhornia crassifer)*—which doubles its biomass every eight days and covers to 40% of the water courses in India today—is capable of cleaning up metals by absorbing Cd, Ni, Cr, Zn, Cu, Fe present within the waste water (Bhattacharjee, 1983). **The weed can be used for nerating biogas and for making cement boards and paper.**

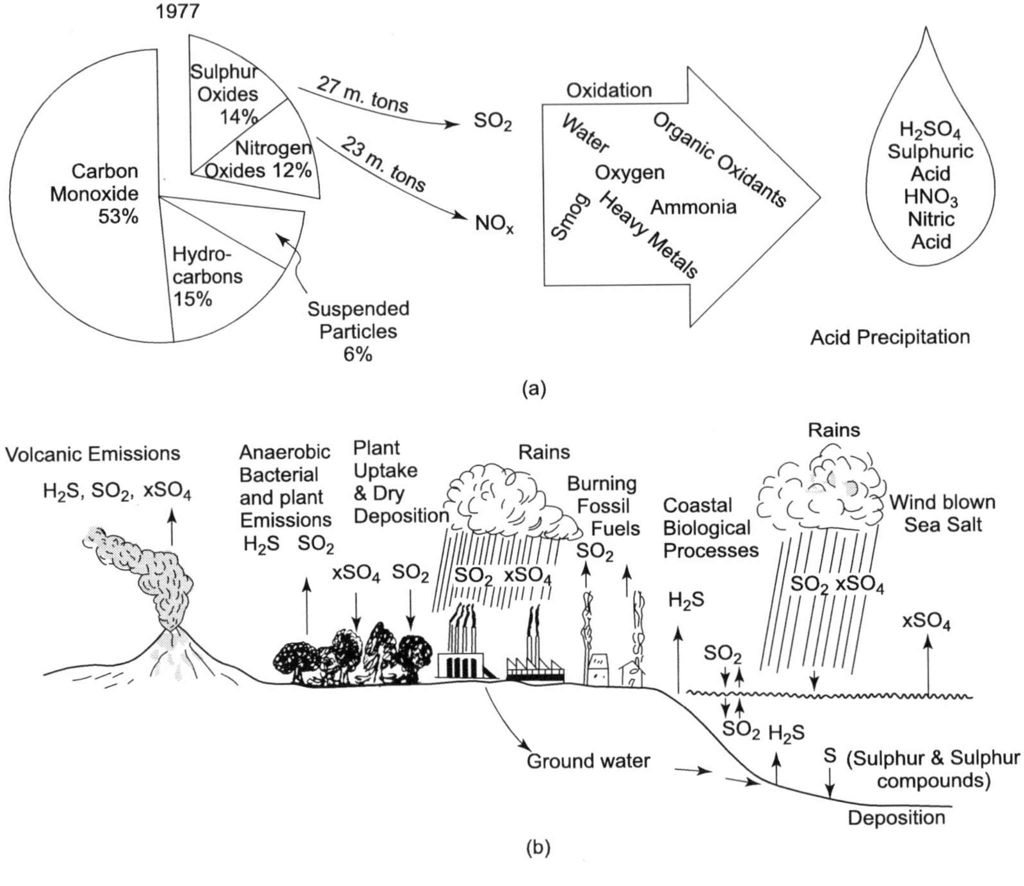

Fig. 21.12 (a) Amount of pollutants discharged into the atmosphere in 1977 and the formation of acid rains. (From E. A. Keller, 1982, *Environmental Geology,* Bell & Howell, London, 488 p., Fig. 13.35.) (b) Various sources of air pollution. (After D. R. Coates, 1981, *Environmental Geology,* John Wiley, New York, 701 p., Fig. 17.16.)

NATURE AND EFFECTS OF AIR POLLUTION

Nature

ties, towns, big villages and industries are incessantly breathing out fumes and gases, sometimes urious to life. The lower levels of the atmosphere have thus become heavily polluted (Fig. 21.12 b). is is in addition to what is being added by volcanoes, duststorms, forest fires and decaying organic matter in different parts of the earth. Much of the material put into the atmosphere comes down to t earth by rainfall at the rate of 130 to 525 tonnes/km^2/yr (Miller and Armstrong, 1982), and the fracti that remains in the atmosphere causes pollution and restricts transmission of solar radiation.

The sky over the Indo-Gangetic Plains and the Himalaya has become very hazy, more so duri the winters and summers. In winter, cold heavy air slides down the southern face of the Himalaya in the lower level of the Indo-Gangetic Plains and thus traps close to the ground the pollutant aeroso including soot (Rupakheti and Nakajima, 2011).

Table 21.9. Some Atmospheric Pollutants (From A. Agrawal et al., 1982)

Pollution	Principal Human Sources	Effects
Carbon dioxide	Fuel combustion for heating, transport, energy production.	No direct effect on people. May lead to increase in earth's temperature,
Carbon monoxide	Incomplete fuel combustion (as in motor vehicles).	Deprives tissues of oxygen. People with cardio-respirator diseases more sensitive.
Sulphur dioxide	Burning of sulphur-containing fuels like coal and oil.	Combined with smoke, increases risk and effects of respiratory diseases. Causes suffocation,irritation of throat and eyes. Combines with atmospheric water vapour to produce acid rain. Reduces crop yields. Leads to acidification of lakes and soils. Corrodes buildings.
Suspended particulate matter	Smoke from domestic, industrial and vehicular sources.	Possible toxic effects depend on specific corn position. Aggravates effects of sulphur dioxide. Reduces sunlight and visibility. Increases corrosion.
Oxides of nitrogen	Fuel combustion in motor vehicles and furnaces Forest fires.	Possible increase in acute respiratory infections and bronchitis and morbidity in children. Produces brown haze in city air. Causes corrosion.
Volatile hydrocarbons	Partial combustion of carbonaceous fuels, disposal of solid wastes.	React with other pollutants to produce eye irritants (acrolein aldehydes). Ethylene is harmful to plants. Aeros particles reduce visibility. May produce unpleasant odours
Oxidants and ozone	Emissions from motor vehicles, photochemical reactions of nitrogen oxides and reactive hydrocarbons.	Cause eye irritation and impaired pulmonaryfunction in diseased persons. Corrode mate-rials and reduce visibility Ozone is one of the most damaging pollutants of plants.

The most common types of pollutants (Table 21.9) include NO_2, NO_3, CO_2, CO, CH_4, NH_3, H_2 Cl_2, SO_2, and other volatile hydrocarbons and particles of carbon and tiny drops of water (0.01 to 1 microns) called aerosols. The maximum permissible limit of certain gases is as follows:

NH_3	100 ppm
CO_2	5000 ppm
Chlorine	1 ppm
Fluorine	0.1 ppm
HCl	5 ppm
H_2S	20 ppm
NO_2	5 ppm
SO_2	5 ppm
Ozone	0.1 ppm

The following reference limits have been set for the Indian cities (Sundaresan, 1982). Rise of these bstances above the respective limits would imply hazardous consequences.

Suspended particulate matter	150 μg/m^3 for 24 hours
Sulphur dioxide (SO_2)	80 μg/m^3 annual mean
Sulphation rate (SR)	0.5 μg SO_3/IOO cm^2/day
Dust fall (DF)	10 million tonnes/km^2/month

Effects

mospheric pollution affects climate on the surface of the earth, particularly the microclimate the affected areas. Breathing problems and respiratory ailments are the more serious nsequences. Carbon monoxide, emitted by automobiles, thermal power plants and refineries, duce oxygen supply in the blood vessels and creates cardiovascular problems. Some of the llutants such as sulphur compounds and polynucleic aromatic hydrocarbons may be mutagenic using genetic and chromosomatic abnormalities and even damaging internal organs and stimulating ncer.

Plants are much more sensitive to pollution than animals as manifest in diminished agricultural elds, early maturing of crops or delayed fruition of vegetables, or even annihilation of some species if exposed to prolonged and severe pollution (Khoshoo and Ahmad, 1981). These vegetal garies are related to changes in stromatal physiology, interference in the reproductive processes and arded seed germination. The increase in CO_2 concentration, for example, causes early flowering of matoes and cucumbers, but delays that of wheat, maize, sorghum, cotton, sunflower, etc. **This has a great impact on the agricultural economics of the world.**

Level of Air Pollution

e extent of pollution of a locality depends on the scale of its industrial-commercial activities, d associated vehicular traffic. With industrial development taking place at a fast rate, even small wns have caught up with big urban centres in the matter of producing pollutants. **Interestingly, winters the atmosphere of these cities is more polluted late at night or very early in the orning than during the days.** During windy summers and rainy seasons the winds decrease the llution level.

It was found that in 1979 the Mumbai metropolis daily emitted 1000 tonnes of pollutants, there ıng greater pollution over Lalbagh-Chembur-Trombay area (Fig. 21.7) with its fertilizer plants, ermal power plants, refineries and pharmaceutical and petrochemical units. The air over Mumbai ntained (in 1979) 38.4% CO, 34.4% SO_2, 9.8% NO_2-NO_3, 9.5% trace elements and 8% hydrocarbons *Scavenger,* 1983, 13(1):22—25). From Table 21.10 one can see that the condition today has not much proved.

In the early 1980s Delhi had extremely high quantities of pollutants—1400 tonnes of SO_2, hydro-rbons, CO_2-CO and NO_2-NO_3 emitted everyday by power plants, industries (200 tonnes/day), automo-es (466 tonnes/day) and domestic 'chulhas' (Usha Rai in *Times of India,* December 12, 1984). Cars, ses, trucks, motycycles, etc. were the biggest culprits, (contributing 40% of pollutants) annually discharging 60,000 t of CO_2-CO, 24000 tonnes of hydrocarbons, and 2,000 tonnes of NO_2-NO_3. Tab 21.10 shows the quality of air above Delhi at present among other cities of the country.

The contribution to the pollutant load made by domestic fuel is by no means insignificant—perha it is no less than that by fugitive emissions. Besides cattle dung and firewood burnt in 'chulhas' (whi do not permit proper burning) the coal used in homes produce a smoke that contains all comm pollutants along with significant amounts of polycyclic hydrocarbons (C. R. Krishna Murti, p comm., 1985).

In 1984 Bhopal (Table 21.10) in Madhya Pradesh was devastated by one of the world's worst industr disasters, when an accidental leak of methyl-isocyanate (MIC) (used for making an insecticide) at t Union Carbide plant killed 2,500 people and crippled for life thousands of people, who continue struggle to get justice (Report of Indian Council of Medical Research, 2000).

Air Pollution: Sources

Industries: **Fertilizer factories are as great generators of undesirable gases as the pesticide plants.** There are more than 100 large and medium fertilizer factories in the country emitting fluorine ga SO_2, SO_3, NO_2, NH_3, hydrocarbons and particulate matter into the atmosphere. Petroleum refineri and petrochemical industries emit SO_2, CO_2-CO, NO_x hydrocarbons, and trace elements like Ni a Ya. In late 1970s the Mathura Refinery, for instance, produced 25 to 30 tonnes of SO_2, 100 to 1 tonnes of CO_2-CO, 60 to 100 tonnes of hydrocarbons and 100 tonnes of nitric oxide besides harmf trace elements (Khoshoo and Ahmad, 1981). Many of the pollutants are carcinogenic and mutagen Polynucleic aromatic hydrocarbons are generated in the production of coke and gas in the refineries the manufacture of synthetics, alcohols and in wood-pyrolysis.

Halogenated hydrocarbons (fluorocarbon, chlorofluorocarbon trichlorofluoro-methane an dichlorofluoro-methane) are generated in the manufacture of detergents, refrigerants, propellan cosmetics, perfumes, dry cleaning fluids, solvents, etc. Vinyl chloride and its derivatives are produc in the plastic industry. Textile mills generate naphtha vapours, sulphuric acid gas, nitrogen oxide chlorine, formaldehyde and chlorine dioxide, etc., besides cotton dust which causes weakening of t respiratory function and bronchitis. A textile mill on an average spews out 1,25,000 litres of napt vapour daily.

Coal-fired thermal plants and steam engines contribute CO_2, CO, SO_2, soot and flyash. Ac rain (Fig. 21.12) is common in areas where the atmosphere has higher concentration of SO_2, such thermal power plants and refineries. A discharge of 25 tonnes of SO_2 daily will increase acidity from 1 to 200 mg/m^3 of air, such as happened in the Chembur-Trombay area in Mumbai thirty years ago. the industrial zone of Vishakapatnam the acid contained 2.0 to 5.9 ratio of SO_4+NO_3/NH_3+Ca with p exceeding it (Naidu and Kavitha, 2012).

Intimately related to industrial activities is the **usage of petroleum oil-driven automobiles, whi emit with CO_2-CO, NO_2, NO_3, hydrocarbons, complex lead compounds and aldehydes, in additi to black smoke rich in particulate matter. There are tens of million automobiles on the roads. T extent of pollution from them can be easily gauged, particularly during winters.**

Domestic Fuels

The burning of all kinds of fuels—wood, animal dung, agricultural wastes—generates soot, CO_2-C SO_2, hydrocarbons and small quantities of H_2S, NH_3, HCl, etc. (Table 21.11). **Thus almost eve**

Table 21.10. Air Quality over Some Cities of India in 2006, 2008 and 2010 (Source: Central Pollution Control Board, New Delhi, Report 2011)

ity	2006			2008			2010		
	SO_2	NO_2	RSPM	SO_2	NO_2	RSPM	SO_2	NO_2	RSPM
Agra	622	133	5	20	165	9	11	150	
Ahmedabad	**10**	**22**	**96**	**12**	**20**	**80**	**16**	**21**	**96**
Allahabad	-	-	-	8	35	128	5	24	218
Amritsar	12	31	-	15	36	-	14	36	218
Asansol	5	54	132	9	74	135	8	66	140
Bengaluru	**18**	**36**	**48**	**15**	**40**	**90**	**15**	**31**	**94**
Bhopal	5	13	59	7	15	93	7	15	59
Chennai	**7**	**10**	**57**	**6**	**9**	**48**	**9**	**15**	**116**
Coimbatore	11	34	42	5	28	55	6	28	75
Dhanbad	19	52	109	19	44	131	15	38	112
Delhi	9	43	136	5	45	198	5	55	259
Faridabad	10	22	155	13	25	139	18	30	164
Hyderabad	**5**	**27**	**81**	**6**	**27**	**87**	**5**	**25**	**81**
Jaipur	4	32	118	6	34	112	NA	NA	NA
Jabalpur	BDL	22	82	BDL	25	136	NA	NA	NA
Jamshedpur	38	51	128	37	51	172	36	48	154
Indore	6	13	109	9	17	174	14	18	120
anpur	7	21	180	7	23	209	7	34	208
ochi	BDL	BDL	53	5	19	40	4	11	36
Kolkata	**7**	**53**	**100**	**9**	**58**	**148**	**11**	**62**	**98**
ucknow	10	29	188	8	35	186	8	34	204
udhiana	16	39	282	10	39	251	9	32	229
adurai	10	27	36	10	23	41	11	25	47
eerut	-	-	-	10	42	115	8	48	170
Mumbai	**9**	**29**	**86**	**9**	**42**	**132**	**4**	**21**	**94**
Nagpur	7	24	70	8	32	98	7	29	86
Nashik	34	28	69	30	25	80	22	34	79
Patna	10	41	113	7	39	120	7	36	165
Pune	**25**	**42**	**133**	**22**	**38**	**99**	**26**	**36**	**65**
Rajkot	11	14	61	10	13	89	14	18	97
Surat	22	29	121	16	23	81	18	25	77
Vadodara	14	23	114	11	21	57	17	29	94
Varanasi	**16**	**19**	**100**	**16**	**19**	**106**	**18**	**20**	**NA**
Vijayawada	5	32	81	5	26	91	6	13	98
Visakhapatnam	11	32	99	10	31	87	7	16	69

$_2$: Sulphur Dioxide, NO_2 : Nitrous Oxide, RSPM : 60 microgram per cubic metre particulate matter, BDL : Below detectable level, NA : Not Available.

village, town or city in India is invariably surrounded by thick palls of polluted air during windle or breezeless days and nights.

Despite the cooking gas supplied to the masses, two thirds of households (49%—62.5%) in rural a 20.1% in urban—people continue to use firewood, crop residue cow-dung cakes or coal for cooki (*Census of India's Houselisting and Housing: Data Highlights*-2011) (*The Hindu,* March 14, 2012).

Table 21.11. Pollutants Generated by Burning of Domestic Fuels (Anil Agrawal et. al., 1982)

Pollutants	Fire Wood (kg/t)	Cattle Dung (kg/t)	Agricultural Wastes (kg/t)
CO_2-CO	1.63	0.69	1.59
SO_2	19.60	8.18	18.75
NO_2-NO_3	3.90	1.63	3.75
Hydrocarbons	23.50	9.81	22.05
H_2S	1.20	0.45	1.12
NH_3	1.20	0.45	1.12
HCl	1.20	0.45	1.12
Particulate matter	31.40	13.09	30.00
Total	**83.63**	**34.75**	**79.50**

Mining and Ancillary Industries

The obnoxious gases produced in mining operations are over and above the great amount of du generated by indiscriminate blasting and unscientific excavation. The sky over calcination plants carbonate-rock mines and carbonization plants in coal mines is considerably vitiated. In the early 198 over the Jharia coalfields the air was heavy with as much as 24 to 72 (even upto 187) μg/m^3 of SO_2, to 65 (shooting upto 147) μig/m^3 of NOx, 181 to 924 (rising to 1964) μg/m^3 of particulate matter, a 10 to 37 μg/m^3 of dust (Bagchi, 1982). This explains the high (10.8%) incidence of pneumoconio and antrucosis.

In the manganese mines in Chhindwara (Madhya Pradesh) the workers were affected with mangane poisoning resulting in mania, pathogenic laughter or even insanity. The problem was checked by str enforcement of wet-drilling techniques (Bagchi, 1981). In the mica mines of Jharkhand, the hi incidence of silicosis leading to turberculosis and other lung diseases is attributed to increased sili dust resulting from mica mining from pegmatite. In the Kolar gold mines cases of pneumoconio increased from 5 in 1978 to 86 in 1983. The ailment affects the liver, kidney, intestinal canals a nervous system. Asbestos dust is very harmful to the lungs (because it is responsible for the crippli illness asbestosis), and causes bronchial cancer. The dust particles work like a time bomb, and may ta as long as 25 to 30 years to cause an illness. The tremendous increase in global production of asbest from 0.67 million tonnes in 1940 to 5 m t in 1985, called for effective measures to combat the hazar One way of doing this is to reduce asbestos particles in the work places—mines and factories. Ind being poor in the production of this mineral (only 4% self-sufficiency in 1980) the problem has yet n assumed serious proportions.

DISPOSAL OF SOLID WASTES

Magnitude of Wastes Generated

the late 1970s a city dweller in India on an average generated 0.27 (±SD 0.11 to 0.08) kg of solid aste per day so that the 109 million people then concentrated in urban centres (1981) produced proximately 10.75 million tonnes of solid wastes—garbage, trash and other materials (Sundaresan, 82). Mumbai alone then produced 2000 tonnes of garbage every day (*Scavenger* 3(1): 22-23, 1983) d Delhi about 8000 tonnes/day. Table 21.12 shows how such solid wastes the cities are producing day. In the rural areas the solid waste produced is predominantly agricultural, and wholly disposable organic manure for the fields.

Fig. 21.13. In the late 1970s the global annual production of solid waste. (Redrawn with permission from H. M. Dix, 1981, *Environmental Pollution,* John Wiley, New York, 286 p., Fig. 9.1).

Table 21.12. Solid Waste Generated by Some Cities of India in 2005 (Based on NEERI Study; reported by Central Pollution Control Board)

City	Tonnes per day	Kg/person/day
Greater Kolkata	12,060	0.66
Greater Mumbai	11,645	0.51
Delhi	11,558	0.65
Chennai	6,404	0.71
Greater Hyderabad	5,154	0.65
Greater Bengaluru	3,501	0.45
Pune	2,724	0.53
Ahmadabad	2,636	0.42
Kanpur	1,839	0.49
Surat	1,815	0.47
Kochi	1,431	0.76

Contd.

City	Tonnes per day	Kg/person/day
Jaipur	1,426	0.45
Coimbatore	1,311	0.65
Greater Visakhapatnam	1,250	**0.67**
Ludhiana	**1,167**	0.61
Agra	**1,069**	**0.58**
Patna	989	0.42
Bhopal	919	0.46
Indore	908	0.43
Allahabad	**853**	**0.59**
Meerut	841	0.53
Nagpur	838	0.29
Jodhpur	825	**0.7**
Lucknow	**778**	0.25
Thane	**775**	0.45
Srinagar	**747**	**0.55**
Asansol	739	0.5
Varanasi	**739**	**0.45**
Vijayawada	**720**	0.5
Ghaziabad	**718**	0.54
Amritsar	**711**	**0.51**
Aurangabad	**702**	**0.57**
Faridabad	698	0.48
Dhanbad	654	0.45
Vadodara	634	0.31
Hubli-Dharwad	630	0.58
Mysore	**578**	0.52
Madurai	**568**	**0.34**
Tiruchirappalli	562	1.23
Jammu	559	0.66
Jalandhar	554	0.56
Solapur	551	0.46
Jamshedpur	539	0.35
Chandigarh	**509**	**0.46**
Bareilly	496	0.48
Gwalior	**477**	0.4
Aligarh	**472**	0.51
Puducherry	**470**	**0.67**
Tiruppur	462	0.61
Moradabad	**457**	0.52
Kozhikode	449	**0.37**
Gorakhpur	446	0.52

Contd.

City	Tonnes per day	Kg/person/day
Ajmer	**428**	**0.63**
Mangalore	424	0.57
Jabalpur	398	0.26
Bikaner	378	0.52
Bhubaneswar	373	0.41
Nashik	**345**	**0.22**
Ranchi	340	0.29
Rajkot	332	0.24
Raipur	331	0.34
Tiruvananthapuram	322	0.26
Saharanpur	321	0.51
Belgaum	315	0.45
Nellore	315	0.45
Bilaspur	311	0.67
Kolhapur	305	0.44
Jamnagar	302	0.39
Bokaro	275	0.4
Cuttack	**274**	**0.34**
Bhavnagar	266	0.37
Udaipur	264	0.49
Dehradun	**259**	**0.35**
Guwahati	258	0.23
Thrissur	241	0.2
Rajahmundry	238	0.42
Gaya	236	0.43
Patiala	223	0.5
Kurnool	222	0.47
Kakinada	221	0.42
Sagar	205	0.48
Bhagalpur	193	0.4
Thoothukkudi (Tuticorin)	191	0.57
Mathura	**178**	**0.4**
Ahmadnagar	170	0.36
Noida	168	0.4
Gurgaon	164	0.52
Tirupati	**164**	**0.39**
Nagercoil	164	0.57
Rohtak	162	0.4
Junagadh	156	0.45
Bhilwara	155	0.4
Cuddapah	150	0.41
Kharagpur	150	0.4

Contd.

City	Tonnes per day	Kg/person/day
Thanjavur	149	0.5
Darbhanga	148	0.4
Shillong	143	0.39
Puri	**143**	**0.65**
Karnal	130	0.43
Ratlam	130	0.4
Haridwar	**127**	**0.42**
Anand	120	0.4
Agartala	120	0.46
Palakkad	117	0.43
Faizabad	116	0.4
Bhusawal	103	0.4
Bharuch	101	0.41
Chitradurga	98	0.57
Haldwani-cum-Kathgodam	**98**	**0.45**
Aizawl	90	0.29
Chittoor	86	0.41
Ambala	80	0.42
Modinagar	80	0.41
Porbandar	79	0.29
Hoshiarpur	79	0.38
Jagadhri	77	0.55
Dibrugarh	76	0.4
Imphal	75	0.22
Chhindwara	69	0.33
Gandhinagar	68	0.25
Shimla	**62**	**0.31**
Navi Mumbai	**61**	**0.54**
Satara	61	0.41
Darjeeling	**51**	**0.34**
Dimapur	47	0.35
Kohima	21	0.19
Gangtok	20	0.5
Itanagar	**19**	**0.39**
	Total: 136,059	**Average: 0.5**

The 2011 Census data shows that on the average 18% of the households in India have two or mo married couples living together. This implies that ten or more persons comprise an average househo Every city-dwelling individual generates solid waste on the average 0.2 to 0.5 kg/day—that is 2 to 5 per household per day. The 78.86 million urban households produce 157.72 million kilograms or 3.9 million tonnes of solid waste per day! The 167.83 million rural households must be generating mo than 335.66 million kg per day (or 335.66 million tonnes) per day.

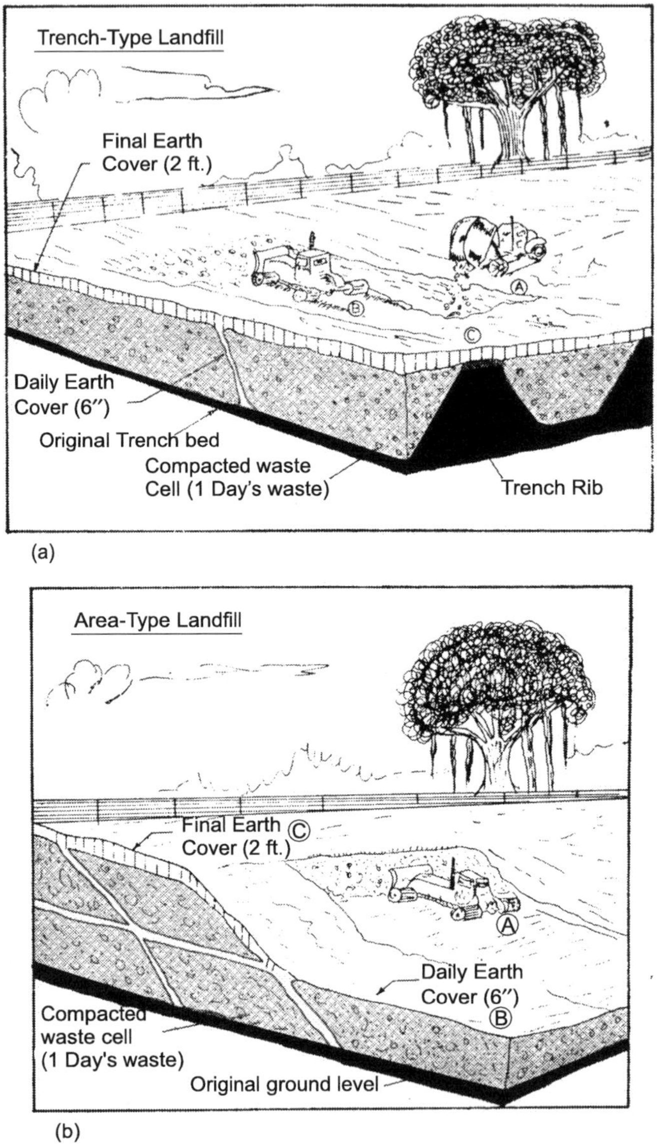

Fig. 21.14. Scientific technique of solid waste disposal in (a) trench-type and (b) area-type landfills. (Based on D. R. Coates, 1981, Figs. 20.20 and 20.21).

Over and above this, the defecation out in the open used to generate 51,000 to 715,000 tonnes/y in the late 1970s (Deodhar and Nath, 1981). The *Census of India's Houselisting and Housing: Data Highlights*-2011 (*The Hindu*, March 14,2012) shows that only 46.9% of the total 246.6 million useholds—30.7% in rural and 81.4% in urban households—in India have toilet facilities and **49.8%**

household people defecate out in the open. One can imagine the amount of excreta lying in t open—knowing that 625 million people (51% of the population of India) defecate in the open, also, mentioned in the World Health Organization report (*The Deccan Herald,* March 28, 2012 It could be of the order of 1 million to 10.5 million tonnes per day!

The third category of waste generated includes mining wastes and spoils. The generation of therm power involves production of huge quantities of ash (30 to 40% of the coal used). In 1980 all o thermal plants produced 16.6 million tonnes of ash (Sundaresan, 1982), some of which contained tox elements such as Zn (6%), Ba (12.2%), Cu (1.2%), As (0.02%), Va (0.08%), Mn (0.2%), etc. Likewis the fertilizer plants generated huge quantities of wastes (containing phospho-gypsum)—3.5 milli tonnes in 1980 (Fig. 21.17). Eighty per cent of this waste was released into the ecosystems.

Table 21.12 brings out the fact that no city, no town is behind in producing solid waste.

Scourge of Plastic Waste

A menace that stares darkly at the eyes of the Indians is the plastic waste comprising carry ba for groceries and vegetables, drinking water bottles and pouches, gutka and liquor pouches, shamp or detergent sachets, packets for biscuits salty snacks such as potato chips, peanuts, spices, and wh not (Table 21.13). Heaps, piles and mounds of their waste cover not only the metropolitan cities li Delhi, Kolkata, Mumbai, Bengaluru, Hyderabad and Chennai but also every small city, every town, a now every village. **Not amenable to biodegradation, and difficult to get rid of unless recycled, th ubiquitous dreadful waste is smothering growth of vegetation, preventing rainwater recharge, and killing hungry cattle which roam around the streets.**

If the production of plastic is not drastically cut, the land of India will be in the grip of wh Jug Suraiya humorously calls typically Indian thing *Indicus plasticus* (*The Times of India,* Mar 30, 2012).

It is only the governments which can control the rapid advance of the menace by directing t manufacturers to curb or stop the production and use.

Geological Condition for Sanitary Landfills

If recycling or incineration of the waste is not possible, it can best be deposited scientifically *sanitary landfills* in carefully chosen sites. The waste is spread systematically (Fig. 21.14) in thi layers, and then covered each day with 15 cm thick layers of earth, and subsequently compacted. T landfill should have an impermeable cover and inner liner of clays, or plastic or asphalt to preve infiltration of rain or surface water. Unless this is done—the landfill is made completely waterproof sealed—the infiltrating water would dissolve the soluble matter of the waste and produce leachate th would percolate down and out to pollute the groundwaters (Fig. 21.16). **The landfill has to be design in such a manner that downward percolating water or leachate moves only in one direction, an comes out from one point where it can be collected and properly treated. In addition to the factor rainfall, evaporation and runoff potential, the geological parameters are taken into considerati in selecting and designing the landfill in order to check the pollution of ground- and surfac waters from landfills.** The following factors are considered:

Topography and Slope: Uneven topography or slope steeper than 8 to 10% are not favourab because the weight of the waste can precipitate slope failure and release pollutants.

Table 21.13. In the late 1990s Percentage of Plastic in the Municipal Wastes Generated by Cities with pulation Larger than One Million (Source: Cental Pollution Control Board, 2000)

City	Percentage of Plastic
Indore	5.77
Bhopal	12.38
Dhanbad	5.56
Jabalpur	8.30
Jamshedpur	5.27
Patna	4.14
Ranchi	3.45
Bhubaneswar	5.70
Ahmadabad	5.29
Nashik	12.58
Raipur	7.07
Asansol	2.78
Bengaluru	9.72
Agartala	4.43
Agra	8.72
Allahabad	10.33
Daman	8.92
Faridabad	13.73
Lucknow	7.45
Meerut	54.48
Nagpur	7.45
Vadodara	7.58
Gandhinagar	6.40
Visakhapatnam	9.24
Dehradun	8.58
Ludhiana	8.27
Guwahati	10.04
Kohima	6.80

Drainage: Unless the area is well-drained and kept free from flooding, the landfill will cause ntamination of groundwater as a result of dissolution of such substances as phosphates, chlorides, Fe, u, Pb, etc. Moreover, the pore-water pressure buildup may trigger failure of the slope.

Soil Texture: Permeable soil or ground material or horizon or cover allowing water to enter the ndfill or letting the leachate and biogas to come out of it would create problems of pollution, nuisance even hazard from such gases as methane.

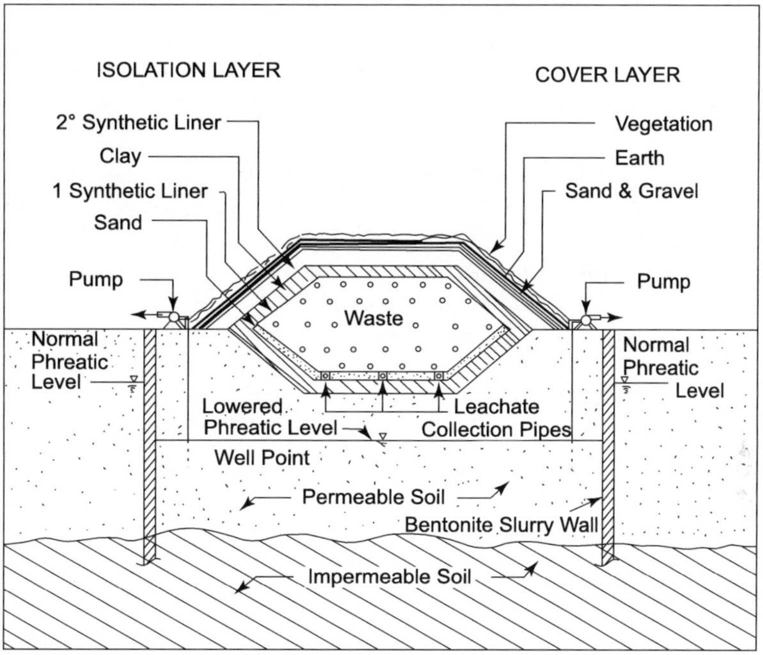

Fig. 21.15. The safe and sound method of disposal of chemically hazardous wastes—a conceptual design. (P. Pearce, 1983, Fig. 3).

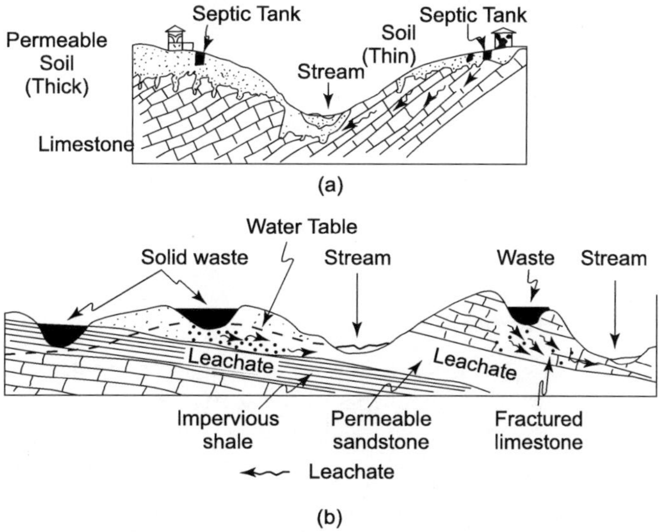

Fig 21 16. (a) Domestic sewage and septic tanks developed on permeable soil or rocks cause groundwater pollution as is happening in all towns in the Himalaya. (b) Leachate from solid waste dumped uncovered and unlined landfill causes pollution in areas of permeable rock and soil horizon.

Groundwater: Perhaps the most important factor is **the depth of the water table which must be ll below the base of the landfill. This will keep the waste free from being saturated with water and the formation of leachate. Failure to ensure this will cause groundwater pollution** (Fig. 21.16), in the case of the Nizamuddin (New Delhi) landfill, the base of which almost touches the water table ewari, 1983). It is therefore imperative to monitor the fluctuation of the water table in the area of ndfill, or where one is proposed. This calls for thorough understanding—in the rocky terrain—of the d-rock conditions such as bedding planes, joints, cleavages, fissures and faults which influence the nfiguration of the water table and the flow of underground water. In soils and unconsolidated sediment e porosity and permeability conditions are thoroughly investigated beforehand.

Conversion of Wastes into Resources

arly one third of the solid waste generated by the urban population is amenable to conversion into mpost, a valuable manure. The drier waste (trash) can be incinerated to generate biogas. **The process pyrolysis of waste involves thermal decomposition of the waste in an atmosphere of oxygen ficiency, leading to formation of oil, organic solids, and inorganic material.** Secondary materials ch as paper, glass and scrap-metals can be gainfully recycled. Thus, the solid waste can be converted o a variety of useful resources.

Recycling of Industrial Wastes

Yokohama (Japan) a plant since 1976 in addition to producing biogas, used to convert the municipal astes into toilet paper—46,000 rolls from the trash generated by 100,000 people. As a matter of fact e recycling of wastes increased in Japan from 16% in 1974 to 48% in 1978. Today it is very much gh. **India with its colossal amount of garbage can very well attempt recycling, at least a part of it. The wastes can be converted into paste and pressed into briquettes or pellets usable as fuel of gh calorific value.**

The slag produced in the making of steel—600 kg slag per tonne of steel produced—can be anulated, powdered and used in the making of portland cement. This will dispose of the waste dumps ry profitably.

Indian coal contains a high proportion (35%) of ash. A 10-tonne boiler using 600 t/day coal would oduce 250 t of ash (Chakrabarti, 1982). **The flyash is a very useful cementing material and is in great demand.**

Disposal of Radioactive Wastes

e fission of 1000 gram of U^{235} produces 999 gram of radioactive waste, which is really a mixture solid, liquid and gas. Some of the components of the wastes are dangerous, and some radioactive es have a half-life of hundreds or even thousands of years. In India the low and intermediate-level dioactive wastes in the form of slurry and stored in concrete-lined containers are embedded in shallow nd repositories (Fig. 21.17) with concrete walls. The high-level wastes are vitrified in borosilicate ass or synthetic rock, and then stored in concrete-encased steel tanks, 2 m deep. These containers are en stored for 20 years or more in deep underground stores, to be finally disposed of in salt beds or derground abandoned salt mines (Raja Ramanna et al., 2001).

The total radioactive waste generated in India is upwards of 2,73,000 m^3 (Ramana et al., 2001). e low-level and intermediate-level radioactive wastes are converted into slurry and put in concrete-lined containers, and then embedded in shallow land repositories having concrete walls. The high-lev radioactive waste, on the other hand is vitrified in borosilicate glass or synthetic rock and then stored concrete encased steel tanks at the depth of 3 metres. After remaining there for 20 years, the material then disposed of in salt beds or in abandoned mines (Ramana et al, 2001).

In concrete containers radioactive wastes would remain soluble for about 1000 years, by which tim they would automatically decay out completely (K. G. Vohra in *The Times of India,* January 8, 1983 It must be admitted that these methods do not provide foolproof prevention of leakage of gases a liquids. This explains why nuclear power plants were banned in the state of California since 1976, a the country (U.S.A.) has not gone for nuclear power plants after 1990. Sweden will close down all nuclear power plants in 30 years. Germany is doing the same. Japan also seems to be thinking in t same line.

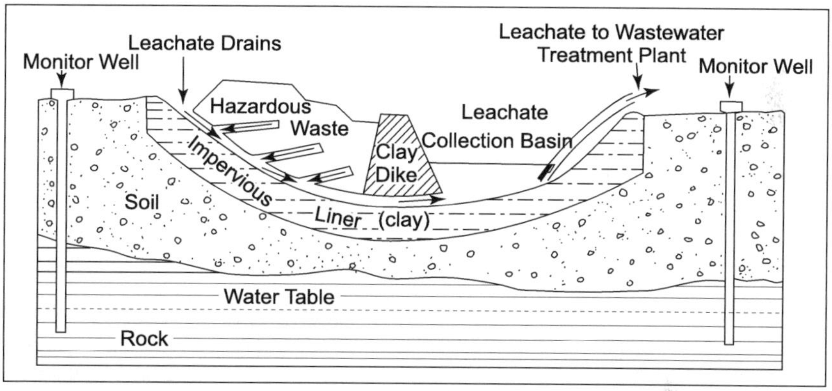

Fig. 21.17 A method of disposal of hazardous chemicals and radioactive wastes. The leachate may be collected at a point for proper treatment. (From P. Pearce, 1983, *Industry and Environment,* Sp. Issue, 4: 57-62, 1983.)

NOISE AS A FACTOR IN POLLUTION

Impacts of Noise

Noise has permeated the environment of our cities and towns to a distressful degree. Music through amplifiers or even religious prayer offered through loudspeakers has not only become nuisance but also a curse, causing health problem. The adverse effect of noise is beginning to tell the health of the people in a substantial way. It is a potential threat to the quality of life. The followi consequences may be mentioned:

(i) Prevention of sleep leading ultimately to insomnia and attendant fatigue.
(ii) Decrease in speech reception, communication distraction and diminished concetration, th adversely affecting job performance or efficiency.
(iii) Chronic physiologic disturbance including impaired hearing and sexual impotency.
(iv) In extreme cases there are irreparable cardiovascular, respiratory and neurologic damages, a quickening of the process of aging so that the life span is lessened. **According to Dr. Samu Rosen, on hearing a loud noise (Fig. 21.18), the blood vessels contract and impede the blo**

flow, and if the noise is continuously prolonged, the heart beat will increase and cause higher blood pressure and attendant heart troubles, suspension of breathing and stomach spasms. Recent research show that every 10 dB rise of noise increases 12% the risk of heart failure (*The Times of India,* June 29, 2012).

Table 21.15. Noise Pollution in Some Localities of India's Capital—Delhi during the years 2008 and 2009 (Source: Central Pollution Control Board)

Locality	June 2008		June 2009	
	Noise (Day Time)	Noise (Night time)	Noise (Day Time)	Noise (Night Time)
Ashok Vihar	61.7	52.6	60.1	50.1
Daryaganj	64	57.4	60.6	55.8
Defence Colony	60.1	53.8	59.8	53.7
Dwarka	60.9	52.3	58.1	54.5
Greater Kailash	63.1	62.2	59.2	49.7
Janakpuri	62.7	55.1	59.4	53.1
Karol Bagh	62.9	61.3	64	57.4
Lajpat Nagar	63.6	61.8	61.2	51.9
Moti Bagh	59.3	52.4	58.6	47.7
Mukherji Nagar	61.7	55.5	62.8	56.2
Paharganj	63.6	60.4	62.6	57.6
Patel Nagar	57.6	48.7	60.2	50.5
R.K. Puram	60.2	55.7	58.7	49.1
Rajpura Road	62.1	52.5	57.6	48.7
Sarita Vihar	69	59.1	63.6	54.3
Tilak Nagar	62.1	54.9	59.2	48.4
Vasant Kunj	62.9	58.6	59.5	55.8
Minimum	57.4	48.7	57.6	47.5
Maximum	69	62.2	64	60.8
Average	**61.47**	**54.96**	**60.64**	**60.54**

For Residential Area: Day time standard 55 dB; Night time standard 45 dB

Measurement of Noise and the Tolerance Limit

und intensity is measured in terms of pressure of sound waves on the ear drums, and the scale is garithmic. The unit is the decibel (dB). **Each decibel rise represents a tenfold increase in sound tensity. Since high-pitch tones are very disturbing, they are given more weight and expressed as BA (Figs. 21.18 and 21.19). Hearing loss begins to occur, with prolonged (eight hours or more) posure to noise levels of 80 to 90 dBA (Table 21.16). It becomes painful around 120 dBA.** Those ho are continually exposed to a noise level exceeding 80 dBA, run the risk of impaired hearing or even afness.

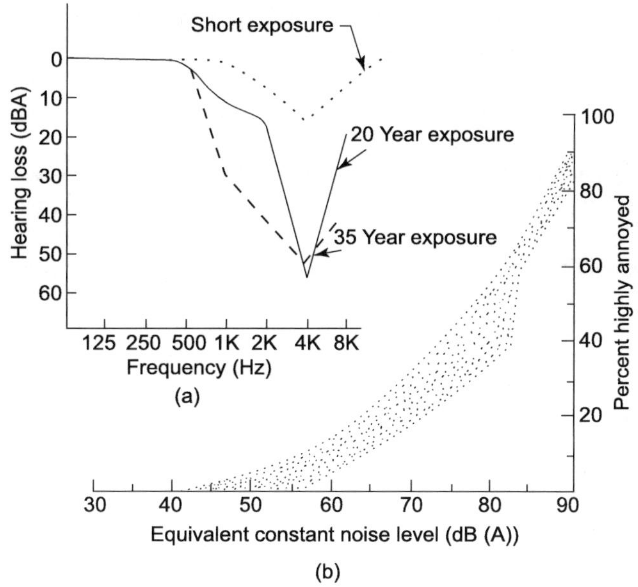

Fig. 21.18. (a) Hearing loss due to exposure to noise, particularly above 80 dBA. (Redrawn with permission from H. M. Dix, 1981, *Environmental Pollution,* John Wiley, New York, 286 p., Fig. 4.3.) Noise causes great annoyance, particularly when above 70 to 80 dBA. (A summary of 12 soc surveys, E. El-Hinnawi and A. K. Biswas, 1981).

Table 21.16. Average Noise Levels (dB[A]) in Five Metropolitan Cities in 2006 (Source: Compendium Environment Statistics India, 2007)

City	Day/Night	Industrial Area	Commercial Area	Residential Area	Silence Area
Kolkata	Day	78	82	79	79
	Night	67	75	65	65
Mumbai	Day	76	75	70	66
	Night	65	66	62	52
Chennai	Day	71	78	66	63
	Night	66	71	48	49
Bengaluru	Day	78	76	67	67
	Night	53	57	50	50
Delhi	Day				
	Night				

Our Noisier Cities

Mumbai, Delhi and Kolkata are amongst the noisiest cities in the world. In the early 1980s non-festival days in Delhi the sound levels reached more than 70 dBA (compared to a WHO standa ermissible) of 45 dBA). On the festival days, including the marriage season it rose to 90 dBA or more the 1980s. The Table 21.15 indicates how noisy has the capital of India become today. The normal ise level of Mumbai was 90 dBA, and industrial areas as high a level as 120 dBA. In the early 1980s e highest noise 84 dBA was recorded at V.T. Station, Dadar, around 6 p.m. (Fig. 21.19b), and the oper and Nanavati hospital areas near Mumbai airport recorded 70 to 74 dBA in normal times to 86 A when the aircrafts flew over (Dixit et al., 1982). At the airport the level rose to 98 dBA when the anes took off (Menon, 1983). In Table 21.16 is given the noise levels of some Indian cities. Everybody els that our cities have become unbearably noisy. **Indian homes too have also become very noisy, e level rising in most to above 60 dBA** (Kayastha and Kumra, 1980).

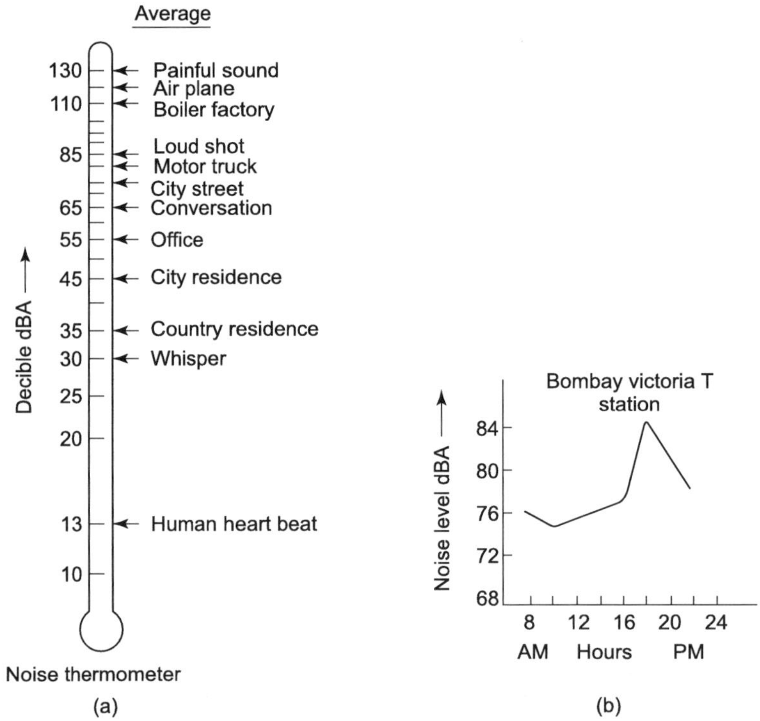

Fig. 21.19. The scale of noise generated by different human activities. (From M. S. Kasbekar, 1981). (b) The noise level used to be highest around 6 p.m. at Victoria Terminus Station in Mumbai in the early 1980s. (After G. R. Dixit, T. N. Mahadevan and R. K. Kapoor, 1982).

Noise Abatement

roduction of damping devices between the source and the receiver by putting sound absorbers, or lating the sound generators are two ways of minimizing the decibel level.

1) **New technology for less noisy pile-drivers, air-compressors, motor bikes, air conditioners, fans, etc. would help abate the distress.**
2) **Noise control standards for vehicles and aircrafts and factories need to be enforced.**

(3) **Heavy levy or tax on the use of sound amplifiers would perhaps be the most effective way dissuading people from generating unnecessary noise.**
(4) **Rubberized wheels in trains (as in Paris and Montreal) and soft horns (France).**
(5) **Complete ban on loudspeakers except for public addresses in designated places wou possibly minimize the noise level.**
(6) **Planting trees like casuarina, tamarind, banyan and neem within and around urban centr would considerably abate the noise pollution.**

Table 21.16. Maximum Permissible Noise Levels (H. M. Dix, 1981)

Situation	dBA
Road traffic near residential areas	70
Noise on sites of construction	70
Factory (where workers spend eight hours a day)	90
Threshold of pain (duration 20 s)	120
Maximum for impulse noise	135
Maximum for instantaneous noise	150
Ear-drum rupture	180
Lung damage	194

SUMMARY

Man is the biggest polluter. His activities have led to the generation of great amounts of pollutants of varied kinds and the introduction of poison in the food chain leading to impairment of health an deterioration in the aesthetic quality of the landscape. Both sewage from homes and municipalitı and effluents from industries are polluting water bodies such as rivers, lakes and the sea in the ear 1980s. The urban and rural communities daily generated more than one thousand million litres of was (polluted) water which finds way into these water bodies. Today we are generating far, far more. T extent of pathogenic pollution can be gauged from the fact that even one gram of faeces contains millio of micro-organisms, bacteria and viruses. Industries produce waste water characterized by organic a inorganic toxic matter—all discharged into rivers. All the 14 major rivers of the country are consequent polluted. **The Ganga is no more the life-giving river that it was.** It has degenerated into northe India's gutter, carrying all kinds of filth, garbage, sullage and effluents from 48 larger and 66 modera towns with their burgeoning industries and multiplying populations. In the late 1970s the holy city Varanasi added every year as many as 3000 half-burnt human bodies, 200 to 300 tonnes of ash, 140 to 2 tonnes of flesh, 6000 carcasses, besides more than 340,000 million litres of municipal sewage. The capi of India then contributed 109,500 million litres of sewage annually. The bacterial count of the Gan water at Patna showed a load of 600,000 cells of coliform per litre and 48,000 cells of streptococci p litre—6,48,000 cells/l against the permissible 200,000 cells/l in the early 1980s. With pollution remaini uncontrolled, we can guess how much more pollutants we are putting into the Ganga.

Not only rivers, but also the groundwaters have become polluted as a result of infiltrati of sewage and industrial effluents. This is evident from the quality of well water in cities an

industrial centres. The lakes have become too rich in organic matter including natural nutrients, thus owing various degrees of eutrophication. Some are too well-nourished in terms of pollutants to be eful and others are on the verge of death.

The coastal areas which supply 80% of the seafood in India are becoming increasingly polluted sewage discharged by cities and effluents from coastal industries. Mumbai, in the early 1980s, for ample, dumped 0.40 km^3 of sewage and 0.095 km^3 of industrial effluents every day so that the total llutant annually added amounts to 50 km^3. In fact, domestic municipal sewage from the coastal Indian pulation amounted to 35 km^3, solid wastes to 90 million tonnes, industrial effluents to 3.5 km^3, inland fluents to 0.14 km^3, pesticides to 12,500 tonnes and detergents to 31,250 tonnes. The western coast is mparatively more polluted than the eastern, although still within the safe limit.

Cities, towns and big villages are incessantly breathing out fumes and gases and other pollutants. **any of the pollutants are cancer-causing, some aggravate respiratory troubles and others are responsible for pathologic abnormalities.** This is in addition to natural pollutants such as dust emitted volcanoes and raised from deforested-degraded land by storms and tornadoes. The result is that the croclimate of cities has considerably deteriorated—smogs immobilize traffic during winters in most the cities in northern India. The sky over New Delhi is heavy with pollutants— in early 1980s 1400 nnes of SO_2, hydrocarbons and CO_2-CO, NO_2, NO_3 emitted everyday by power plants, industries, tomobiles and 'chulhas'. The burning of fuel-wood, dung and agricultural wastes generates soot, CO_2, O, NO_2-NO_3, SO_2, hydrocarbons, NH_3, H_2S and HCl. Thus almost every village or town wears a thick ll of polluted air, particularly on breezeless nights. Mining and mine-related industries are greatly gravating the problem of pollution as witnessed in the sky over coal-carbonization and carbonate-cination plants.

The urban population of India generates millions of tonnes of garbage, trash and other solid wastes nually. Rural folk release huge amount of faeces in their fields. A sizeable part of this solid waste can converted into manure, a part can be incinerated to produce biogas, and some parts could be recycled r use, such as fuel-briquettes and pellets from organic wastes, raw material for cement from slags of el, flyash from power plants as cementing material, etc.

The average level of noise in the capital of India is quite above 65 dB compared to the admissible andard of 50 dB. Other cities and towns are no less noisy. Noise has become not only a nuisance, t also is telling on the health of the people, particularly those who have high blood pressure and rdiovascular problems. Apart from causing hearing loss, the continual exposure to loud noises ove 80 to 90 dBA may cause blood vessels to contract, augment heart beat and blood-pressure, and produce breathing troubles. Technological innovation should be resorted to reduce noise from e noise-making machines and implements. A heavy levy or tax on the use of sound amplifiers eds to be imposed. Planting of trees within and around cities would considerably abate noise pollution.

22 Global Warming and Climate Change

INEVITABLE WARNING

We all know and strongly feel that the air above us is heavily polluted by gases and particulate matters, cluding dust, soot and aerosels. The particulate matter dispersed in the air reflects back or scatters away solar radiation and thus diminishes the heat input to the earth's surface. The 1815 eruption of an Indonesian volcano generated so dense a cloud of ash that complete darkness prevailed for three days over the Island of Madura, 800 km away, and the summer temperature of 1816 in England dropped by 2° to 3°C below the normal, resulting in the failure of crops. For the U.S.A., it was the year of no summer. The dust clouds raised from degraded-deforested tracts produce similar conditions.

Oil and coal-fired plants and engines have been adding particulate matter including soot and flyash in the air since the beginning of industrialization. It is estimated that human activities are responsible r the injection of dust, soot, oil spray, sulphates, salts and fluorides—in the 1970s at the rate of 300 million tonnes per year (5 to 45% of total atmospheric particulates). This was over and above the natural ections through volcanism and dust storms (773 to 2200 million tonnes per year). Approximately 50 70% of anthropogenic emission are sulphate particles (Miller and Armstrong, 1982). **The particles having a residence time of 1 to 5 years affect the global climate. But soot-like black materials absorb the solar heat and radiate it to the earth so that the earth's surface is warmed up. Both the processes work together.**

There is another aspect also.

CO_2 and Greenhouse Effect

The presence of CO_2 generated by burning of fuels and by various industries has a great impact on the global climate. The amount of CO_2 has been rising steadily at an exponential rate of 4% per decade from 0±10 ppm in pre-industrial times to 340 ppm in the early 1980s (Fig. 22.1) and to 390 ppm at present *The Times of India,* January 10, 2012). **This gas (CO_2), while lets the sunlight enter the atmosphere, inhibits or prevents the heat rays (infra-red) from returning back to the sky and thus causes a warming up of the atmosphere** (Fig. 22.2). **This is called the greenhouse effect.** The warming up of e earth became perceptible and is being felt by all since the 1990s (Table 22.1 and Fig. 22.2) as a result of inevitable build up of CO_2.

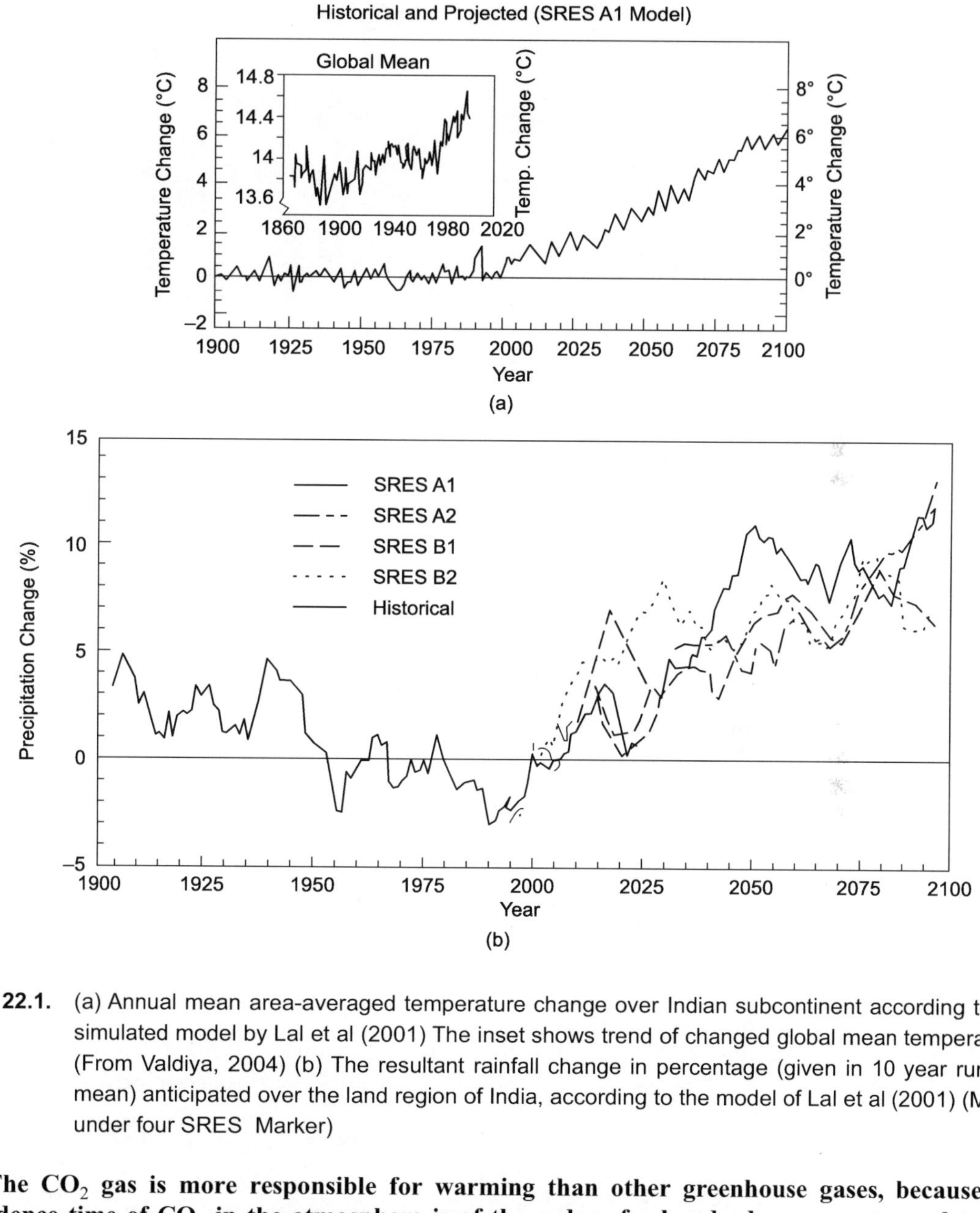

Fig. 22.1. (a) Annual mean area-averaged temperature change over Indian subcontinent according to the simulated model by Lal et al (2001) The inset shows trend of changed global mean temperature. (From Valdiya, 2004) (b) The resultant rainfall change in percentage (given in 10 year running mean) anticipated over the land region of India, according to the model of Lal et al (2001) (Model under four SRES Marker)

The CO_2 gas is more responsible for warming than other greenhouse gases, because the residence time of CO_2 in the atmosphere is of the order of a hundred years, compared to few weeks of sulphate aerosols (Srinivasan, 2008). Unfortunately the CO_2 content is rising at the rate considerably more than 4% per decade. This is having disastrous effect on global climate. Moreover, another greenhouse gas chlorofluorocarbon (CCl_2F_2) continues to be produced in large quantities India alone produced 1800 million tonnes in 2001 for coolants in refrigerators and air conditioners (N. Singh, 2002).

Table 22.1. Increase in CO_2 Content in Atmosphere and Rise of Average Temperature on the Earth's Surface (Based on data from B. Hileman, 1982, and others)

Year	Global Concentration of CO_2 (ppm)	Average Temperature Rise (°C)
1960	285	
1980	293	05 to 0.8
1967	320	
1970	330	
1980	335	0.1 to 0.2
2000	380 (375 to 384)	
2050		2
2100		3 to 4.5

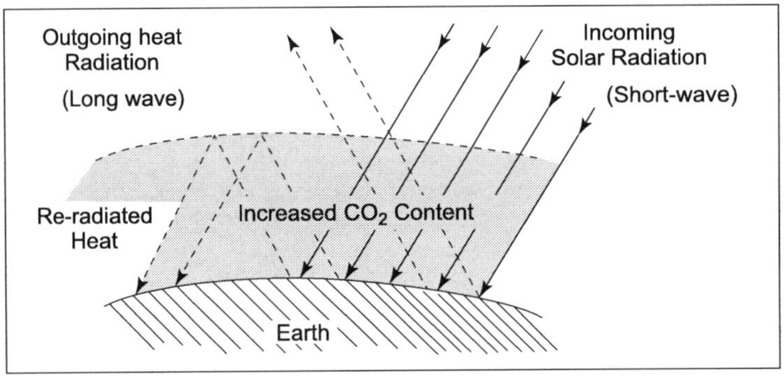

Fig. 22.2. The greenhouse effect of CO_2 in the atmosphere. The CO_2 gas lets in solar radiation but inhibits or prevents the escape of reflected heat, resulting in warming up of the air.

The other pollutants are the aerosols of dust, ash, soot and sea salt of the size 0.001 mm, which cover 10 million km^2 extent of the sky over northern Indian Ocean, the Bay of Bengal and the Indian landmass during winter months (Ramanathan et al, 2001).

The greenhouse gas emission (Table 22.2) from energy, industry, agriculture and waste sectors nstitute respectively 58%, 22%, 17% and 3% of the net 102 equivalent emission per day (S. L. Jain, 2012).

Large-scale deforestation greatly contributed to the rise of CO_2 content in the air, for the plants consume a considerable amount of this gas to generate oxygen through the process of photosynthesis or carbon assimilation. With less than a fourth of land left with forest cover in India, it is inevitable that the atmosphere is beginning to show a steep rise in CO_2 content. Since the oceanic water is also warming up, its capacity to dissolve and hold CO_2 in captivity is also diminishing.

Table 22.2. Greenhouse Gas (GHG) Emissions in India by Various Sectors in the Period 1994 and 2007 given in Million Tonnes of CO_2 Equivalent (MTC_{eq}) (Cited from Valdiya, 2004)

Sector	Emissions in 1994		Emission in 2007		CAGR (%)
	MTCEq	In(%)	MTCeq	IN(%)	
Electricity	355.03	28.4	719.30	37.8	5.6
Transport	80.28	6.4	142.04	7.5	4.5
Residential	78.89	6.3	137.84	7.2	4.4
Other energies	78.93	6.3	100.87	5.3	1.9
Cement	60.87	4.9	129.92	6.8	6.0
Iron and steel	90.53	7.2	117.32	6.2	2.0
Other industries	125.41	10.0	165.31	8.7	2.2
Agriculture	344.48	27.6	334.41	17.6	-0.2
Waste	23.23	1.9	57.73	3.0	7.3
Total without LULUCF	1251.95		1904.73		3.3
LULUCF	14.29		-177.03		
Total with LULUCF	1228.54		1727.71		2.9

Percentage emissions from each sector are with respect to total GHG emissions without LULUCF. 1994 and 2007 respectively.

Nitrogen Oxides and Ozone Layer

Nitrogen oxides that petroleum-fired engines (including mobikes, cars, buses, trucks, tractor aeroplanes), pesticides and fertilizer plants, refineries, etc., generate, are slowly nibbling t ozone layer lying at the height of 16 to 60 km. The ozone layer prevents or inhibits the entry of ultraviolet rays that cause skin diseases. The CO_2 is always accompanied by NH_3, CH_3 and chlorofluorocarbon (CCl_2F_2) and methane. **The amount of CCl_2F_2 increased to 0.3 ppb and CCl_2F to 0.2 ppb and methane increased at the rate of 1 to 3% per year until the end of 1970s. The trend has continued with greater intensity. These gases have damaging effect on ozone in the upper layer of the atmosphere. Small and big holes have been formed. The net result is the warming up of the atmosphere (Hileman, 1982). The possible impact of concentration of greenhouse gases together with acid mist and serious damage to the ozone layer is manifest in unprecedented decline in the forests of West Germany, Switzerland, Sweden, Austria, Italy and the Netherlands** (Blank, 1985 The acid fogs are known to leach out ions, and cause deterioration of plants (as evident in Poland, Czech Republic and Slovakia).

The holes in the ozone layer due to excess of CCL_2F_2 in the sky above Antarctica and Arctic are growing in size thus letting in more and more ultra-violet rays. The hole that was 1.1 million km^2 1979 grew to 29.9 million km^2 in 2010 over Antarctica (S.L. Jain, 2011).

Rate of Temperature Rise

The rate of global warming is 0.2°C per decade, the net impact being the rise of mean surface temperature by more than 0.7°C (Srinivasan, 2008). Latest report shows that from the record temperatures over a period of 37 years from 1973 to 2009, it is quite apparent that western India getting warmer at the rate of 0.13°C per decade (that is 0.013°C per year). This warming is attributed two factors—0.07°C/decade rise resulting from global climate warming and 0.06°C/decade increase e to landuse and land cover changes (Nayak and Mandal, 2012). But then other parts of the country must be warming up at different rates. At this rate, **the annual mean-area averaged surface warming over the Indian subcontinent would range between 1.18 and 1.00°C in the twenties, 2.87 and 63°C in the fifties, and 5.09 and 5.55°C in the eighties of this century, according to a modelling study (Lal et al, 2001). The temperature change would be more pronounced during winters and springs compared to that in summers and autumns.** The northern and central parts of India would perience greater warming (upto 3°C) and the southern India slightly less (2°C) by the end of this ntury (Lal et al, 2001).

The mean annual temperature for the whole country during 2010 was +0.93C above the average of e years 1961–1990, the months March, April and May being the warmest with 1.8C above normal (Annual Report, 2010 of India Meteorological Department).

IMPACT OF GLOBAL WARMING

Impact on Rainfall

The impact of surface warming will be pronounced to quite perceptible in rainfall pattern over the Indian subcontinent—5 to 25% less rainfall in winters and 10 to 15% more during the summers (Lal et al, 2001). There will be less rainfall and snowfall in the Himalaya during winters, and more frequent spells of intense rainfall, with more instances of extreme rainfall during summer rainy season (Lal et al, 2001). Rainfall is likely to become irregular in central India. The time of onset of monsoon will also became variable.

Melting of Ice Caps and Glacial Retreat

The melting of snow and ice has increased in the Himalaya, the melting producing melt water at the rate of 500 km^3/year. More than 67% of the Himalayan glaciers have retreated in the past two decades. However, in the last 100 years the Himalayan glaciers have behaved differently in different areas and at different intervals of times (Raina, 2011). In the Kashmir Himalaya the napani Glacier retreated by 500 m and the Siachin Glacier advanced by 700 m between 1862 and 09, then retreated by 400 m in 1929–1958 period, and hardly retreated thereafter (Raina, 2011). A research team from Grenoble University, France came out with the estimate of the Karakoram glaciers ıning thickness at the rate of 0.11 to 0.22 m/yr in the period 1999 to 2008 (*The Times of India,* April , 2012). In the Uttarakhand Himalaya the Gangotri Glacier (Fig. 22.3) retreated at the rate of 14.6 yr in the period 1935–1962, 46.4 m/yr in 1962–1990, 375 m/yr during 1990–1994, and 25 m/yr in e period 1994–1998 (Tangri, 2000). The Bhagirathi Kharak Glacier retreated at the rate of 7.42 m/yr ring 1962–2008. The Dokrani Glacier is retreating at the rate of 16.6 m/yr (Dobhal et al, 2004). The e of recession of the Satopanth and Bhagirath Kharak glaciers—between 1962 and 2006—was 22.88 year and 7.42 m/year respectively (Naninwal et al., 2008). The Pindari Glacier has retreated at the e of 8 to 10 m/yr since the recording was done in 1906. Satellite Resources-2 LISS 4 data show that e Milam Glacier, which receded 1328 m from 1954 to 2006, is now dammed at its front by a lake, the mming caused by accumulation of ice-coned moraines (Balu Govindha Raj et al., 2012). The snout has retreated further back. The Zemu Glacier in the Sikkim Himalaya retreated 44 m in 56 years i.e., 05 to 1961 (Vohra, 2000).

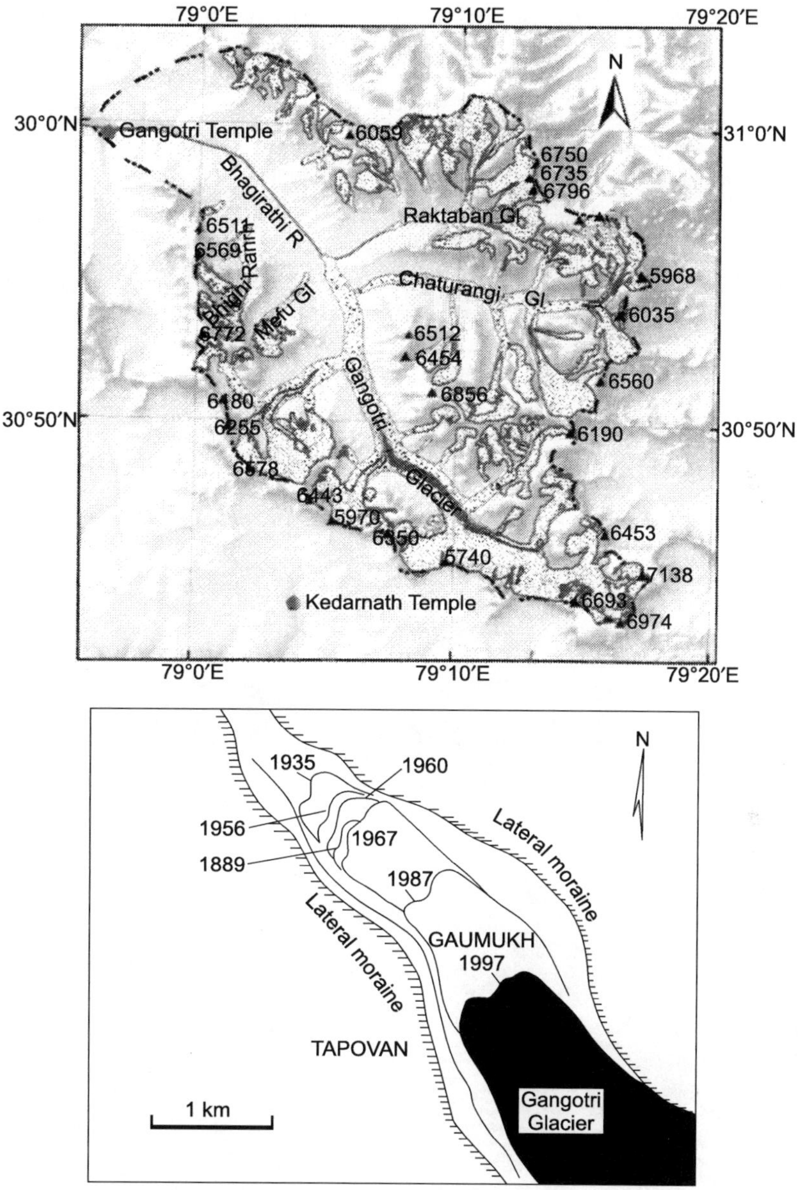

Fig. 22.3. The Gangotri Glacier, the source of the Bhagirathi River (the Ganga), has been retreating since the recording was first done in 1937. ((a) After Mukherjee and Sangewar, 2001; (b) From Valdiy 2004). (See colour Fig. 22.3)

It may be added that the shrinking of the ice cap and glaciers can also be attributed partially to the negative mass-balance due to considerable variation in snow precipitation (Gurung et al, 2011).

Shrinking of the ice cap has been very spectacular in Antarctica and Arctic. In Arctic the icepack melted considerably along the edges with some shelves disintegrating disastrously at the rate of 2 m/yr. The thickness of the icepack in less than 30 years decreased from the average 3.1 m to 1.8 m d its extent diminished by 6% in the period 1978–1996. The rate of reduction was 35370 km^2/yr by 2002. The situation in Antarctica is much worse.

(a)

(b)

Fig. 22.4. Fast melting of ice of the glaciers has created a serious hazard—the bursting of glacial lakes and resultant floods. (a) Rapshtreng glacial lake in Bhutan. To the left is the Lugge Tso in the Pho Chu basin. The dam-burst occurred in October 1994.(From Mool, 2001)(b) Tso Rolpa glacial lake in the Tamakosi basin Nepal. It is 12 km long and 55 m deep at part.(From Mool, 2001). (See colour Fig. 22.4)

Change in Atmospheric Circulation

There will be an impact on the circulation of ocean currents also. Since the polar region would be warmed two to three times more than lower-latitudinal zones, the temperature difference between the two belts will decrease. **This will cause 20% reduction in the wind velocity and consequent circulation of ocean currents. The net result will be appreciable modification in the global climate pattern.**

Impact on Vegetation

The cumulative effect of the rise of temperature and change in the rainfall will be considerable on the vegetation—weed species are likely to slowly displace the tree species, and the productivity crop plants, such as wheat and *Sorghum* would be affected (Rao et al, 1995). **The warming up of the earth's surface will also cause a shift in the agricultural zones and attendant disruption of food supply, and deterioration of freshwater supplies in semi-arid belts.** Climatic regions in temperate zones will be shifted towards the poles. The phenomenological events such as bud bursts, flowering a senescence as well as length of the growing seasons are likely to be affected greatly (Bawa et al., 201 This is already quite visible in the Himalaya all along its length and breadth.

Sea Level Rise

Estimates of the mean sea level rise at selected stations along the Indian coast shows the rate of sea level rise of 0.78 mm/yr at Mumbai, 1.14 mm/yr at Kochi, 0.65 mm/yr at Chennai, and 0.75 mm/yr Vishakhapatnam (Unnikrishnan et al, 2006). The sea level rise during the 20th century was close to 2 mm/yr, which is higher than what it was in the last half a century – 1.7 mm/yr (Shetye, 2011).

As a consequence of rise of the sea level, more than 5763 km^2 area along the Indian coast, including 1700 km^2 of prime agricultural land, would be submerged under water, causing the displacement of *eventually* 7.1 million people, particularly in Odisha, Bengal and Bangladesh. Already the island Ghora in the Sunderbans in Bengal has lost 41% of its land, resulting in displacement of 7000 peop (Nandini Mitra, 2007).

URBANIZATION AND CLIMATE OF CITIES

The impact of man's various activities is greatest in cities. The four factors that make the cities warmer upto 10°C more than the countryside (Fig. 22.5) are:

(i) Energy generation and release of heat due to burning of wastes
(ii) Concrete building, asphalt pavements and roads that trap heat
(iii) Obstruction in the circulation of breeze and wind, and
(iv) Presence of pollutants in concentrated form.

To this may be added the water vapour generated by cooling plants. The particulates serves as nuclei of condensation of water vapour producing cloudiness and fog. **The result of all this is that cities have become "heat islands" characterized by temperature 5° to 10°C higher than the surrounding countryside.** Within the region of capital of India, Sitaram Bazar, Bhikaji Cama Place, Connaug Place, Noida and Badarpur areas are heat islands where the temperatures are 8° to 10°C higher th the green areas within and around the metropolis—the Sitaram Bazar registering as high as 10.7 more temperate comparatively (Jayashree Nandi, *The Times of India,* May 30, 2012). **The moderate**

intensity urban heat island over city centre of Trivandum reaches the maximum value of 2.4°C over and above that of the countryside (Ansor et al, 2012).

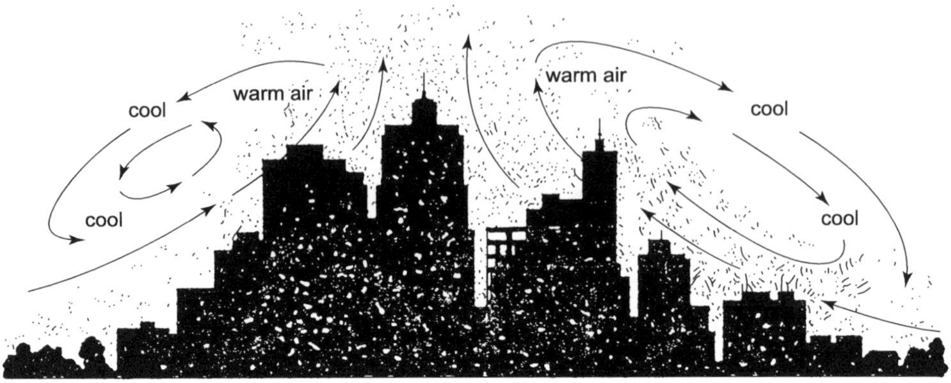

Fig. 22.5. The temperature over cities and industrial centres is as high as 10°C above the surrounding countryside. (Redrawn with permission from G. T. Miller and P. Armstrong, 1982, *Living in the Environment,* Wadsworth, Belamont, 491 p., Fig. 11.7).

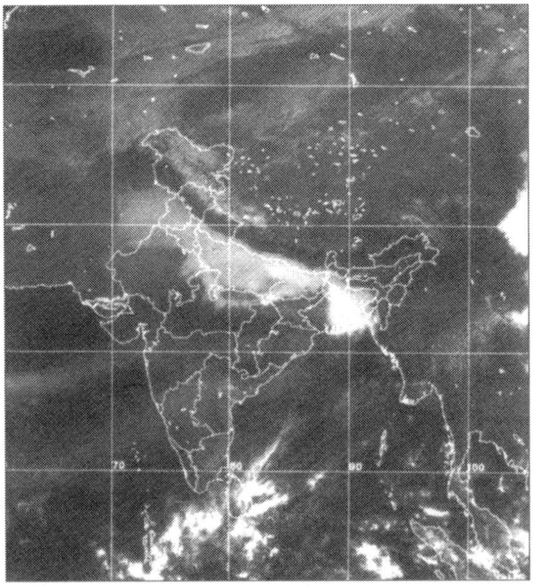

Fig 22.6. Dense fog in mid-January 2010 over the Indo-Gangetic Plains. (Ref: Kalpana-1, VIS Imagery) (See colour Fig. 22.6)

Soot, flyash and dust combine with sulphur oxides, carbon oxides and fog to give rise to industrial smog in cold and wet areas. The NO_2 may react with accompanying hydrocarbons in the presence of sunlight to form secondary pollutants called photochemical smog, a brownish haze

in dry warm areas. The smogs greatly reduce visibility and drastically curb aircraft and vehicular traffic as has been happening without let up year after year over larger part of the Indo–Gangetic Plai (Fig. 22.6), putting the rail and air traffic to utter chaos. In the over-irrigated swathes of agricultur lands the problem of fog has became extremely serious due to excess moisture in the air.

In certain topographical situations in winter under certain climatic conditions, dense cool air gets trapped under a layer of warm air (Fig. 22.7) in which pollutants accumulate to dangerous levels. this temperature inversion persists for a couple of days, the condition may become lethal. Fortunately these inversions do not last longer than a few hours. Although similar situations have not been so far witnessed anywhere in India, indiscriminate industrialization involving generation of pollutants **can create conditions conducive to the development of such pockets of death.** Safeguards against this kind of dangerous pollution will have to be taken in industrial and mining centres being developed in the mountainous region.

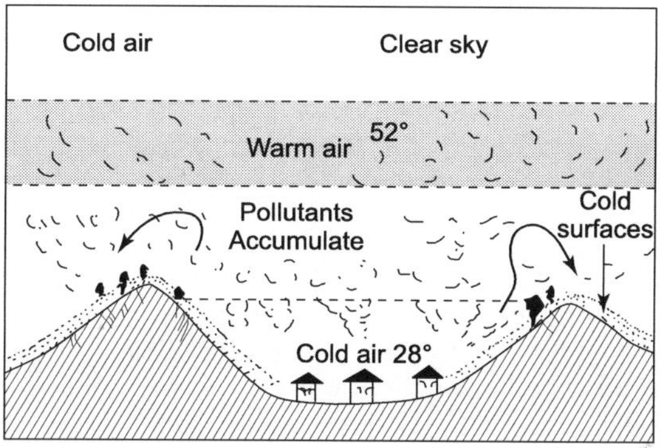

Fig. 22.7. In cities or industrial hubs located in areas of topographic peculiarities, temperature inversion may occur during winter, leading to concentration of pollution under a layer of warm air. (After Miller and Armstrong, 1982, Fig. 18.4).

SUMMARY

The global warming, and resultant climate changes are the consequences of the increase in the content of greenhouse gases in the atmosphere like carbon dioxide together with ammonia, methane, halogenat compounds of carbon, and particulate matter such as soot, soil sprays, sulphates, salts and dust, generated increasingly by energy sector (58%), industry sector (22%), agriculture sector (17%), and waste sect (3%). Therefore, there will be no check on the rise of atmospheric and surfacial temperatures, unless their production of greenhouse gases is curtailed or checked at their sources. The tens of millions of petroleum-fired vehicles on the roads produce excessively large amounts of dangerous gases a particulate matter.

In a situation where there is growing clamour for faster industrialization in all parts of our land, overwhelming is the demand for having personal motor vehicles, insatiable is the need for all kinds of consumer goods, and when the governments have little option but to yield to the wishes of the corporate world, the utterly helpless common man and the equally miserable marginalized experts can do ve little to improve the condition.

23 Renewable Sources of Energy

THE ENERGY CRISIS

The rate of energy consumption is determined by the economic and technological development of the country (Fig. 23.1). This is best illustrated when comparing the energy consumption of a highly developed country (USA) with that of a developing country like India. In 1981 India had 15% of the world population and consumed nearly 1.5% of the total energy generated, in comparison to the U.S.A. which with only 6.25% of the world population consumed nearly 33% of total global energy produced. In the late 1970s about 85% of the energy generated was consumed by industrialized countries and the remainder 15% by the non-industrial world (El-Hinnawi, 1981). Even as recent as 2003, when we compare the consumption of total primary energy of 9969 kg oil-equivalent by the USA, 5160 kg oil-equivalent by France, 1700 kg oil-equivalent by U.K. and 832 kg oil-equivalent by China, India's consumption was only 325 kg oil-equivalent (Narayanan, 2004). It would be much higher now. **But India is nowhere near the point of meeting its energy needs.**

The per capita energy consumption in the early 1980s was 300 kWh in developing countries in comparison to 6000 kWh in developed countries (Tolba, 1982). The demand for energy was then doubling every 14 years (Miller and Armstrong, 1982). Within India itself, about two-thirds of the total demand for energy (1983-1984) was for commercial use alone. The source of this energy was and continues to be coal, oil or electricity. The source for non-commercial purposes (Fig. 23.2, Table 23.1) re and continue to be firewood, charcoal, animal and vegetable wastes. The percentage of their uses varies from area to area. In the early 1980s, of the 194 million tonnes of coal replacement energy for non-commercial purposes, approximately 64% was generated by fuelwood and charcoal, 23% by cattle dung and 13% by vegetable wastes (Fig. 23.2).

Despite the cooking gas supplied to the masses, 49% of households—62.5% in rural and 20.1% in ban centres—continue to use firewood, crop residue, cowdung or coal for cooking (*The Census of India Houselisting and Housing: Data Highlights-2011, The Hindu,* March 14, 2012). The fuelwood consumption in the year 2004-2005 was 312 kg per capita in the rural area and 110 kg per capita in the urban (K.D. Singh, 2008). This implies that **86 million tonnes of woodmass was annually removed from forests primarily for cooking in the period 2004-2005.**

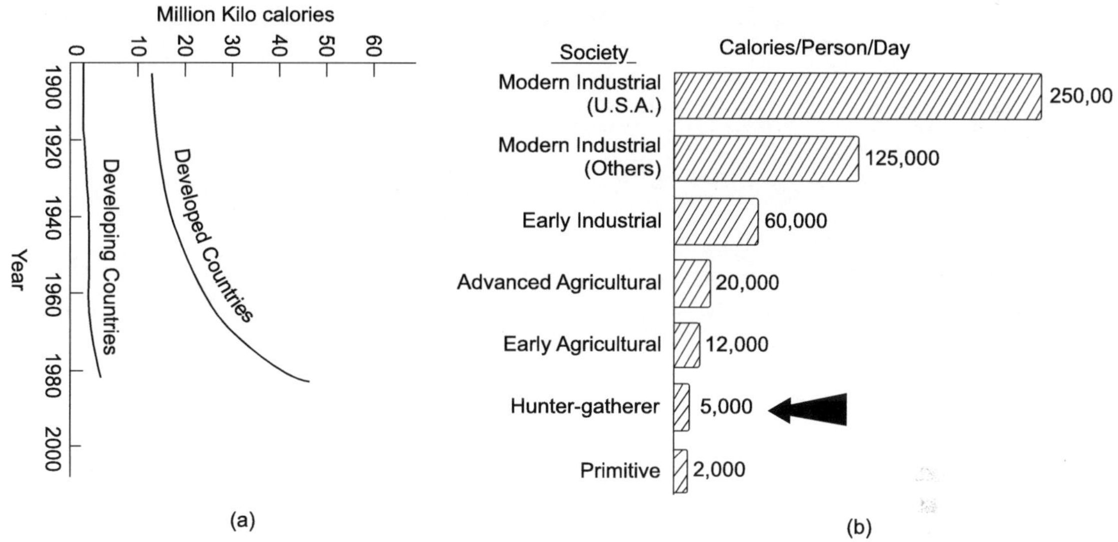

Fig. 23.1. (a) Per capita consumption of energy in developed and developing countries in the late 1970s. The demand for energy was then doubling every 14 years, (b) The amount and rate of consumption of energy is determined by the economic and technological progress of the country. (Based on G. T. Miller and P. Armstrong, 1982) *Living in the Environment,* Wadsworth, Belmont, 491 p., Fig. 14.2 and 14.1.)

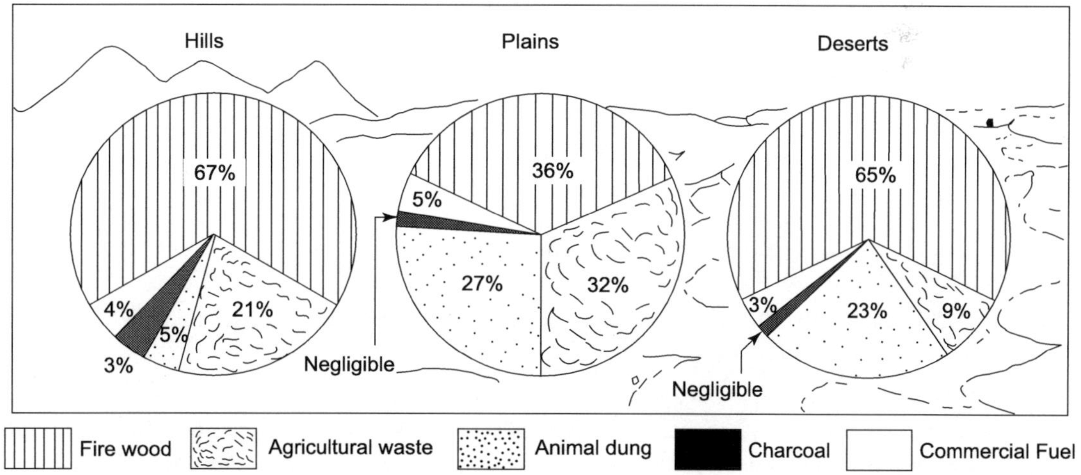

Fig. 23.2. The source of energy for domestic use varies from place to place in India. Pattern of energy consumption in 1975-1976 in northern India. (After Anil Agrawal, Ravi Chopra and Kalpana Sharma, 1982, *The State of India's Environment—1982,* Centre for Science and Environment, New Delhi, 189 p. Fig. 4 on 153 p.)

Table 23.1. Rural Domestic Fuel Consumption in India in the year 1978–1979 (Morgan and Moss, 1981; El-Hinnawi and Hashimi, 1983)

Type of Fuel	Percentage (%)
Fuel wood and charcoal	51.6
Dung cake	10.7
Soft coke	3.1
Agricultural waste	17.9
Kerosene oil	11.9
Gas	1.4
Electricity	3.3

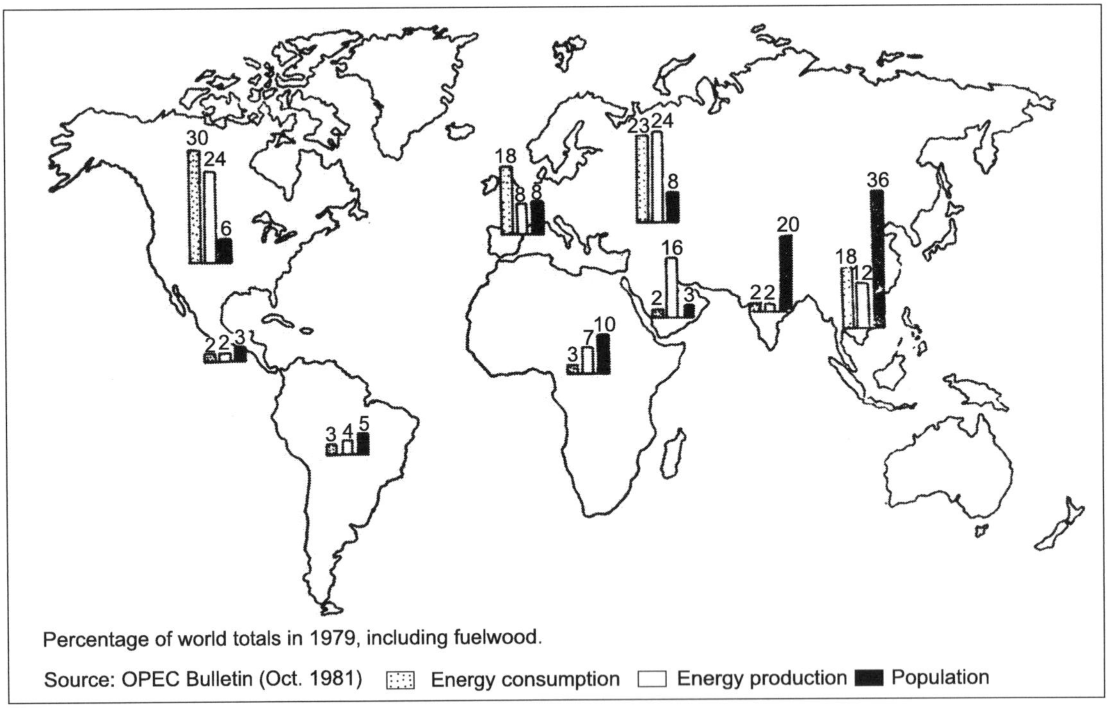

Fig. 23.3. Histograms indicating the distribution of population and production and consumption of energy in various subcontinental regions in the early 1980s. (Redrawn with permission from E. El-Hinnawi and M.H. Hashimi (Eds.), 1981. *Renewable Resources of Energy and the Environment,* Tycooly, Dublin, 210 p.).

There is greater crisis on the commercial-industrial energy front.

ENERGY OPTIONS

Non-renewable reserves of fossil fuel (coal, petroleum) for energy production are finite and th demands are dangerously outstripping generation. There is no denying that the burning of fossil fuels causes serious air and water pollution. Moreover, considerable environmental degradation (including scarification of land, mass-movements on slopes and deforestation and damage to hydrological system everywhere) is caused by the mining of these materials. Needless to state, our forests cannot be tapped further for meeting the wide demand for fuelwood. In view of these factors, it has become imperative to seek beyond the fossil fuels the renewable, preferably non-polluting sources of energy. Tables 23.2 and 23.3 show the potentials of various sources that can be tapped for energy.

It is evident from the Table 23.2 that against the total installed capacity of 5,74,984 MW, the annual electricity production could be 1229 TWh (Sukhatme, 2011). In 2009-2010 the supply was just 142.7 TWh. **Unable to meet the requirement of energy from the conventional sources of energy, such as coal-based thermal power plants, hydrothermal power plants, atomic energy plants, petroleum and natural gas, India has no option but to go in a very big way for tapping renewable sources of energy.**

Table 23.2. Estimating Total Potential of Renewable Energy Sources for Generating Electric Power in India (Sukhatme, 2011).

Source		Total installed capacity (MW)	Plant load factor	Annual electricity production (TWh)
Solar energy	Direct Thermal	125,000	0.20	219
	Photovoltaic	125,000	0.20	219
Indirect	Hydroelectric (large)	148,700	0.37	482
	Hydroelectric (small)	15,384	0.37	50
	Wind (on land)	65,000	0.16	91
	Wind (off shore)	65,000	0.16	91
	Biomass	23,000	0.30	60
	Wave energy	?	-	?
	Marine currents	?	-	?
	Ocean thermal energy conservation	?	-	?
	Tidal energy	7,900	-	17
Total		574,984		1229

In recent years deposits of gas hydrates containing methane have been discovered in the deep sea environment. For the future use, a technology needs to be developed to extract methane from the gas hydrates.

The generation of energy from the sun's rays, tides and oceanic currents, wind and hot springs, even if costly and geographically restricted in distribution, have no adverse effect on the environment. The colossal volume of biomass can be used to produce ten times the present total world energy consumption for all commercial purposes.

Table 23.3. Energy Potentials : Global Scale (From : Hinnawi and Biswas, 1981)

Global Energy Source	Annual Potential
Sun's radiation	1×10^{18} kWh
Near-surface wind energy	1200×10^{12} W
Ocean thermal gradient	50×10^{12} W
Wave flux	2.7×10^{12} W
Tides	6.0×10^{10} W
Hydro-electricity	2.2×10^{6} MW
Geothermal source	100 GW (e)
Biomass	171,000 million tonnes/yr

ENERGY FROM SOLAR RADIATION

Constraints and Basic Approach

On an average the earth's surface receives solar energy (Fig. 23.4) at the rate of 200 W/m^2/hour. In deserts such as the Thar in Rajasthan, the energy input exceeds 275 W/m^2/hr. However, a part of the

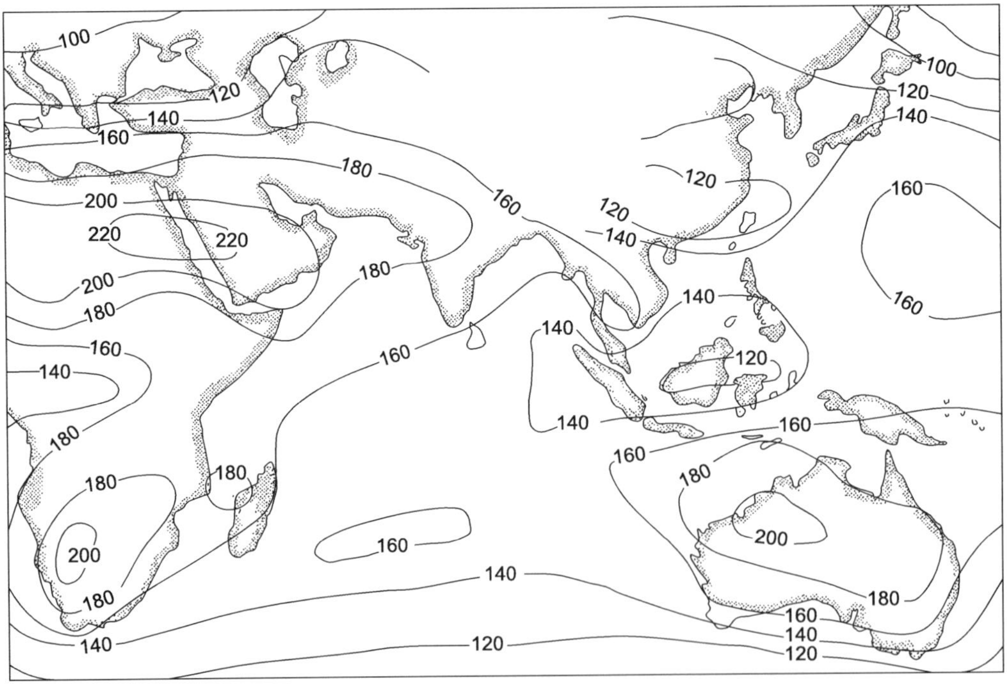

Global solar radiation (kcal/cm2/year)

Fig. 23.4. Global solar radiation distribution in Asia and Africa in terms of total solar radiation kilo-calories per unit (cm^2) area per year. (After W. M. Marsh and J. Dozier, 1981, *Landscape,* Addison-Wesley, Reading, 637 p., Fig, 6.9.)

energy received is lost through reflection, so that the net energy gained by the earth's surface is abo 80 to 90 $W/m^2/hr$. It is instructive to know that if only one per cent of this energy falling on the Sahara desert alone were to be converted into electricity, the solar radiation would be capable of supplying quite a part if not the entire energy needs of the world (Miller and Armstrong, 1982). And this without polluting the ecosystems. Technologically this has become feasible.

Solar energy can be harnessed directly as well as indirectly. The radiation can be converted directly into electricity during the day by photovoltaic cells and stored in batteries. A photovoltaic cell of silicon having a diameter of 8 cm normally generates electricity at the rate of 0.5 W/hr, and an array of several hundred photoelectric cells (Fig. 23.5) would be able to meet the needs of a family.

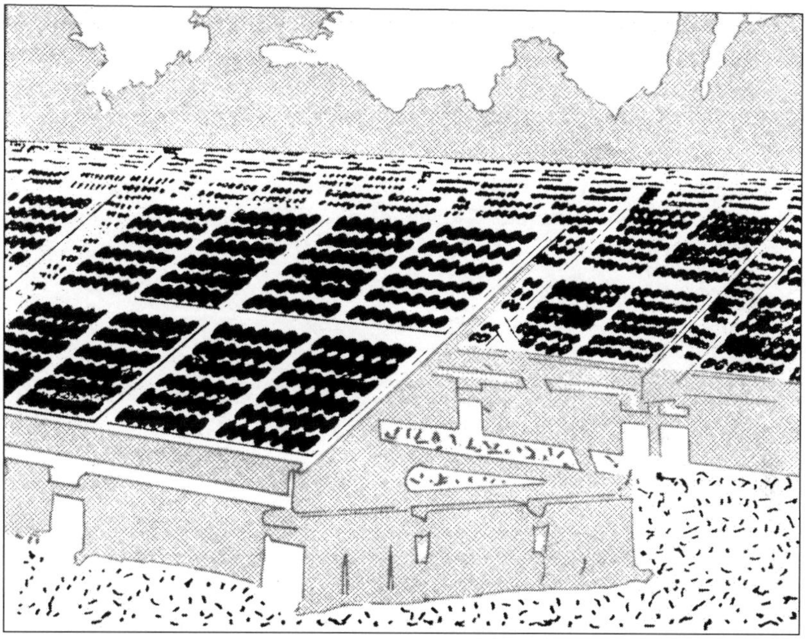

Fig. 23.5. The direct way of converting solar energy into electricity is to install an array of photovoltaic cells. This one consists of 13600 photovoltaic cells generating 15 kW of electricity. (After Leslie Wae, 1981, *Span,* October, p. 18.)

The indirect method of tapping solar energy involves concentrating the sun's rays by use of mirrors or blackened collectors on water to produce steam that runs an electricity generator. The heated water can be used for a variety of purposes, including warming homes during winters and in the colder regions. In one technique, collectors coated with black heat-absorbing substance are placed on roofs or open points and water circulated through them. In another technique, an array of mirrors that can be automatically manipulated to turn in the direction of the sun, concentrate the solar rays on overhead tanks (boilers) where steam is generated (Fig. 23.6). The steam is then fed into the turbine of the generator as is the case in thermal power plants.

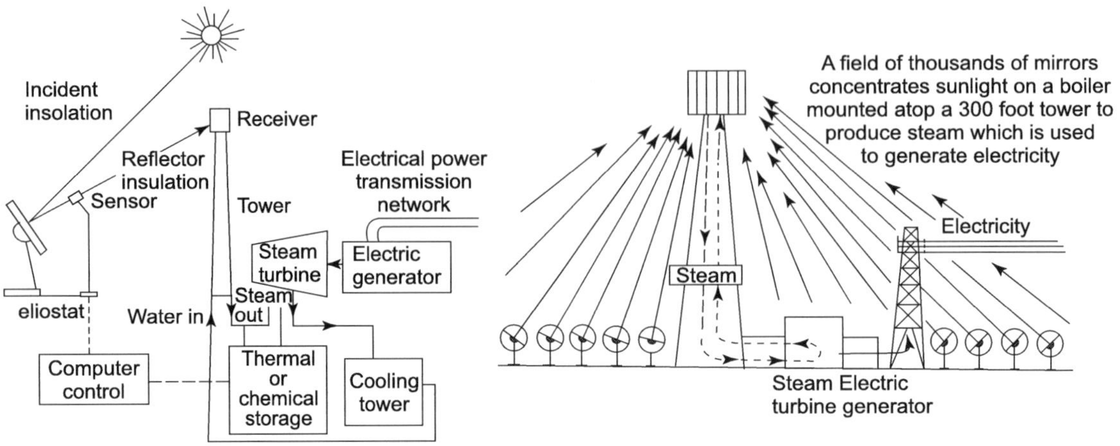

Fig. 23.6. The indirect way of converting solar energy into electricity is to concentrate solar radiation by a series of mirrors on an overhead tank (boiler) where steam is generated. The steam is then used to turn the turbine of the generator. The timing devices keep the mirrors in alignment with the sun. (Redrawn with permission from D. R. Coates, 1981, *Environmental Geology,* John Wiley, New York, 701 p., Fig. 7.13).

Solar Energy Use in India

Solar energy has been successfully used in various parts of the country to run tube-wells and rice mills, for lighting homes, streets and hospitals in towns and villages, and for heating water in hotels and homes, etc.

In India approximately 100 megawatt of electricity has been produced for localized use through photovoltaic conversion modules. A large-scale photovoltaic system requires nearly 4 ha of land for generating of one megawatt electricity. A number of grid-connected photovoltaic power plants can be installed in suitable locations in Rajasthan, Gujarat, Western Madhya Pradesh and Maharashtra. Operating 7 to 8 hours a day for atleast 250 days in a year, the annual energy production could be raised by 219 million megawatt (Sukhatme, 2011).

However, if one were to consider the amount of solar radiation received by the landmass of India for long durations in a year, it seems that the potential of solar energy generating electricity is quite higher than what Sukhatme estimates. The solar-thermal plant at Mathania in Rajasthan, for example, **ll generate 140 MWp in a solar field 22 hectares in area, implying that an average figure of 1.25 ha/** MW is enough for a solar PV farm (Narayanan, 2012). Thus, Sukhatme's estimate of 4 ha/MW is under-evaluation of the potential of solar energy. According to Narayanan (2012), India has the solar potential of well over 800 GWp which can generate 1400 TWh/year at a realistic plant load factor of 0.2.

Germany, the land which receives much less solar energy than India, and for much shorter duration in a year, has produced 22 GW solar power per hour—an amount equal to 20 nuclear power stations— from the government-mandated solar plants, through midday hours on Fridays and Saturdays and fed into the national grid on Saturdays. The solar plants thus meet nearly 50% of Germany's midday electricity requirement (*The Hindu,* May 28, 2012). Germany has not only established a world record, but also shown the world the way to meet the challenge of energy crisis.

India plans to build large grid-connected solar power plants with a commulative installed capacity of 20,000 MW_p by 2020. One among the 20 such megawatt-size plants in India, a grid-connected photovoltaic power plant at Yalesandra, district Kolar (Karnataka) produced 3.34 million kWh of electrical energy in the very first year (2010) of its life (Mitavachan et al., 2012). The Kolar and t Belgaum plants together are expected to produce energy at the rate of 4 million kilowatts per hour per year, provided that there is no inverter failure (Mitavachan et al., 2012).

Distribution of Solar Radiation in India

A comprehensive survey in the late 1970s by Mani (1981) showed that **every year over 3000 to 3200 hours of bright sunshine are available in Rajasthan, Gujarat, western Madhya Pradesh and northern Maharashtra, and 2800 to 2600 hours in the rest of the country excepting Kerala, Kashmir and Assam.** In the Gujarat-Rajasthan region more than 2000 kWh/m^2 are received as global solar radiation while Bengal, Assam and eastern Bihar receive less than 1700 kWh/m^2 (Mani, 1981). During winter (December to February) the duration of sunshine is more than 280 to 300 hours per month in the central parts of the country including Rajasthan, Gujarat, Maharashtra, Andhra Pradesh, and northern Karnataka. The lowest values of 200 hours per month are recorded in Kashmir, along the Himalayan foothills and northeastern Assam. In summer (May to June) the duration of sunshine exceeds 300 to 320 hours per month over the region extending from Panjab to northern Bihar and northern Maharashtra, while the southern Peninsula, Kashmir and Assam have less than 220 hours of sunshine per month. During this period in Rajasthan and adjoining Gujarat, the global solar radiation is of the order of 220 kWh/m^2/month, while over Assam and Kerala between 180 to 190 kWh/m^2/month (Mani, 1981).

With the onset of the monsoon there is a significant decrease in the duration of sunshine practical over the whole country, except Kashmir. The west coast of the Peninsular India, southern Madhya Pradesh and northern and eastern Maharashtra have about 100 hours of sunshine per month, while nearly half of the country comprising the Peninsula and northeastern India receives a little more or less than 150 hours monthly sunshine—Rajasthan 180 to 200 hours, Assam and Saurashtra 140 hours, Tamil Nadu 170 hours. In autumn the duration of sunshine increases to 280 to 300 hours per month in Rajasthan and Assam and less than 200 hours per month in eastern Peninsular India during the monsoon months.

Remote sensing through satellite 'Samir' shows the distribution of solar heat received by the surface of the Indian subcontinent in the months of July and August 1984 (Fig. 23.7).

HARNESSING THE WIND

Availability

In the coastal and high Himalayan regions wind speed on an average is 8 to 10 km/hr. **The energy of the wind is utilized to turn turbines of electric generators. If the wind speed is doubled, its electricity generating capacity is also increased.** The successful operation of hundreds of windmills in different parts of the country for the generation of electricity and for running pumps for irrigation has demonstrated that wind power can be tapped gainfully without causing any harm to the environment. Significantly, the average efficiency of windmills is 35% compared to 5 to 15% in sol units (Coates, 1981).

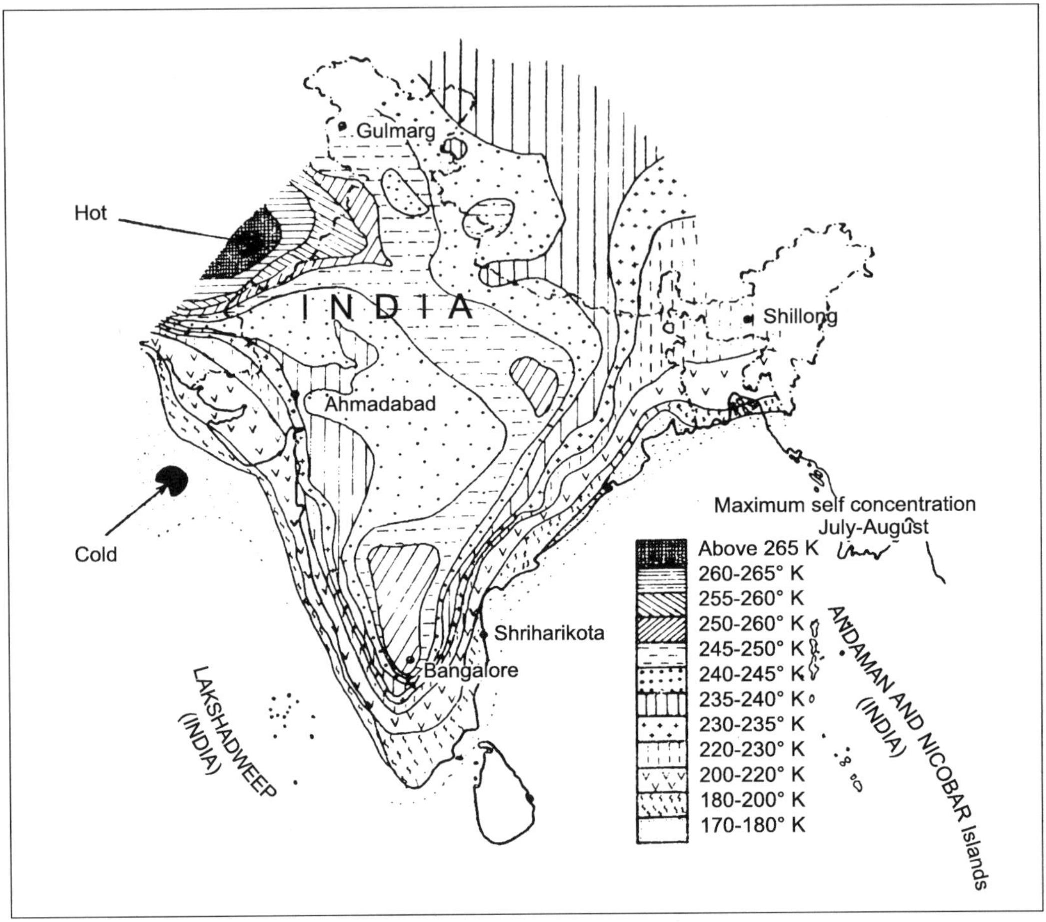

ed upon Survey of India map with the permission of the Surveyor General of India. The territorial waters of India extend into the sea to a distance of twelve nautical miles measured from the appropriate base line. Responsibility for the correctness of internal details shown on the map rests with the publisher Government of India Copyright 1986

Fig. 23.7. Areas receiving equal solar radiation, as remote sensed by satellite 'Samir' in the months July–August, 1984. (Redrawn from *Indian Space Programme,* ISRO, Bangalore, 24 p., Fig. on p. 12).

Methodology

The motion of the wind (Fig. 23.8) is conveyed to the dynamo to generate electricity. The electricity so generated can be utilized immediately, stored in batteries, or fed into the regional power grid. **A windmill system of 20 towers can supply the electricity needs of a small town. If the windspeed is 9 km/hr, a windmill with a 12-blade rotor having 5 m diameter can pump 2500 litres of water per hour, and if the velocity of the wind is 36 km/hr, the water pumping capacity rises to 23,000 litres per hour upto the height of 20 m** (*The Times of India,* September 16, 1983).

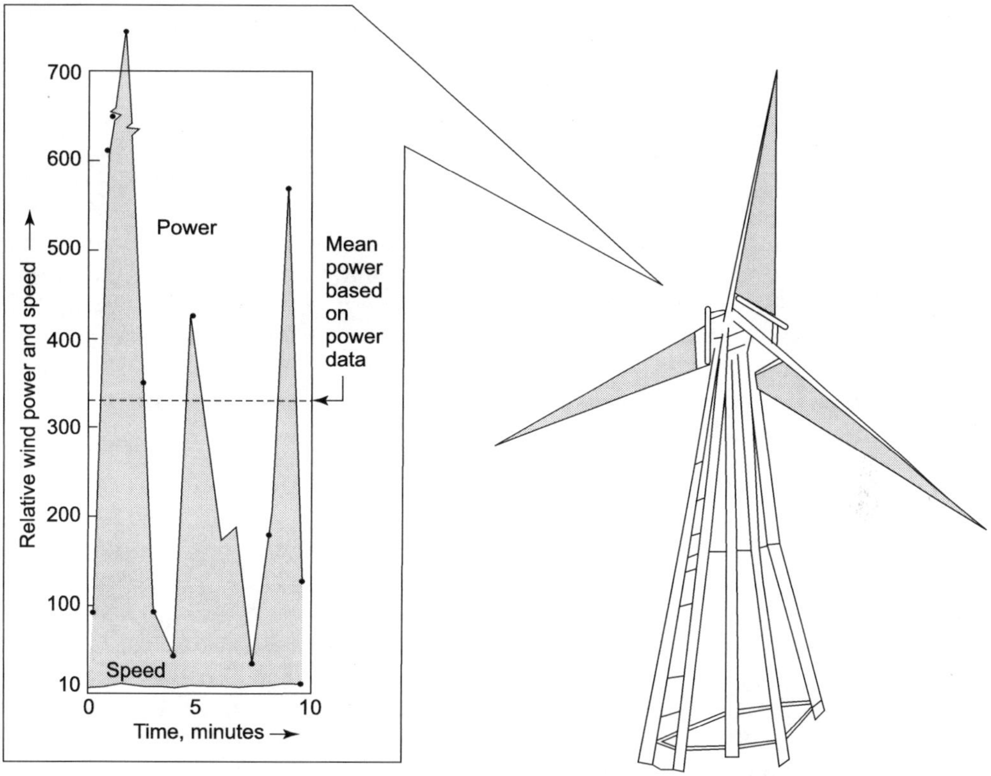

Fig. 23.8. (a) The wind-speed governs the amount of energy generated by a windmill. (Redrawn from W. M. Marsh and J. Dozier, 1981, *Landscape,* Addison-Wesley, Reading, 637 p., Fig. 30.6) (b) A 12-blade rotor with 5 m diameter at a wind-speed of 9 km/hr is capable of pumping water at the rate of 2500 litre/hour.

A better alternative would be to pump water into a reservoir or a huge overhead tank. The release of water of this reservoir/tank during windless periods would generate hydro-electricity as well as serve the purpose of irrigation. In this manner the variable supply of wind-generated electricity can be made more regular. Since the wind velocity is stronger over offshore belts, floati offshore windmills, with each unit linked to a turbine of 2 mW capacity can be installed in these belts. Hydro-electricity so produced can serve the purpose of irrigation. In this manner the variable supply of wind-generated electricity can be made more regular.

Potential in India

Daily mean windspeed measurements from 400 stations in India in the late 1970s showed (Mani, 1983) that Gujarat, parts of Rajasthan, western Madhya Pradesh, coastal belts north of 15°N latitude, southern parts of Tamil Nadu, the Andaman-Nicobar islands and Lakshadweep islands have high wind speed (Fig. 23.9). During the monsoon months the windspeed is uniformly high

over practically the whole of India, except the eastern coast from 10 to 15°N latitude, northern and central India (Fig. 23.10). About 20 potential sites for small to moderate machines operating windspeeds of 8 to 80 km/hr, with a rated output of less than 100 km/kW/hr were identified in the country by the end of 1970s.

In Gujarat the windspeed in Bhuj, Dohad, Dwarka, Rajkot, Verawal, Jamnagar and Bhavnagar is high. The Jamnagar-Bhavnagar belt experiences strongest wind speeds (10.5 to 18.3 km/hr) during the monsoon period (June to September) while the Dohad–Veraval coast has strong winds (3 to 5 km/hr) during winter (October to February) (Jagdeesh and Varshneya, 1982). The net estimated (in the late 1970s) wind power in Gujarat, according to Jagdeesh and Varshneya, was of the order of 26,700 million kWh. This was greater than the total electricity consumed in the whole state during 1978-1979 (11,000 million kWh). In the mountainous regions, the local wind blowing through gaps may attain a considerable speed. **Across the Great Himalaya in the Tethyan domain in Uttarakhand the wind ●ws fiercely after 11 A.M. until unset, practically all over the belt. This wind energy can be ofitably tapped and fed into a regional grid.**

Use of Wind Energy in India

There has been dramatic increase since 1992 in the generation of electricity by harnessing wind power in Kerala, Tamil Nadu, Andhra Pradesh, Karnataka, Madhya Pradesh, Rajasthan and Gujarat. A Spanish company has commissioned wind energy plants in Karnataka generating an aggregate 85.5 MW of energy—at Tuppadahalli in Chitradurge district, Arasingundi (13.2 MW), and Analur (16.3 MW) (*The Hindu*, June 7, 2012). This energy is enough to meet the needs of 3500 homes. The electricity produced in 2009-2010 was 18,187 GWh (Sukhatme, 2011). The estimated potential for wind resource on land is of the order of 65,000 MW. The potential in the offshore may be as much as that on land. The annual electricity generation would be around 91 TWh in each case (Sukhatme, 2010).

Recent assessment by Lawrence Berkeley National Laboratory in U.S.A (as reported in *Journal of Renewable Energy,* The Hindu, March 29, 2012) shows that the potential of generating ergy from the wind is 20 to 30 times higher than the official estimate of 102 gigawatt—if India were to install wind power plants over 1629 km^2 land, mostly along the western and eastern coasts of India.

The present author's observation in the last 40 years across and north of the snowy Great Himalayan Range makes him believe that the wind energy can be gainfully used on a large scale along the narrow lleys of south-flowing rivers that emerge north of the Great Himalayan Ranges where the wind speed is very strong after 11 A.M. until sunset. Tapping the power of wind in Himachal Pradesh, Uttarakhand and Sikkim will make northern India less dependent on oil and gas as sources of energy.

ENERGY FROM TIDES AND WAVES

Kinetic energy from the vertical motion of waves and the horizontal to–and–fro rolling movements of tides can be converted into electric energy in estuaries and bays, such as in the Gulf of Kachchh, in the estuaries and deltas of the Kaveri, Godavari and Ganga (Sundarban). The up-and-down motion of waves is made unidirectional by using a variety of mechanical devices (Fig. 23.11). Similarly, the movement waves and tides are rendered unidirectional by special valves installed with paddle-wheels fixed on a dam-like structure put across the bay or estuary (Fig. 23.12).

Tides have greater potential for generating electricity, particularly where the difference in low tide and high tide is large as witnessed in the White Sea and Okhotik Sea (Russia), La Ranee estuary (France), Severn estuary (U.K.), Bay of Fundy (Canada), and the Bays of Khambhat and Kachchh (India). The regular long-period motion of the ebbs and flows turns the generators through a devi which has lift translators with vanes to guide the tidal flow. In La Rance (France) 240,000 kWh electricity is generated by tapping the tidal power.

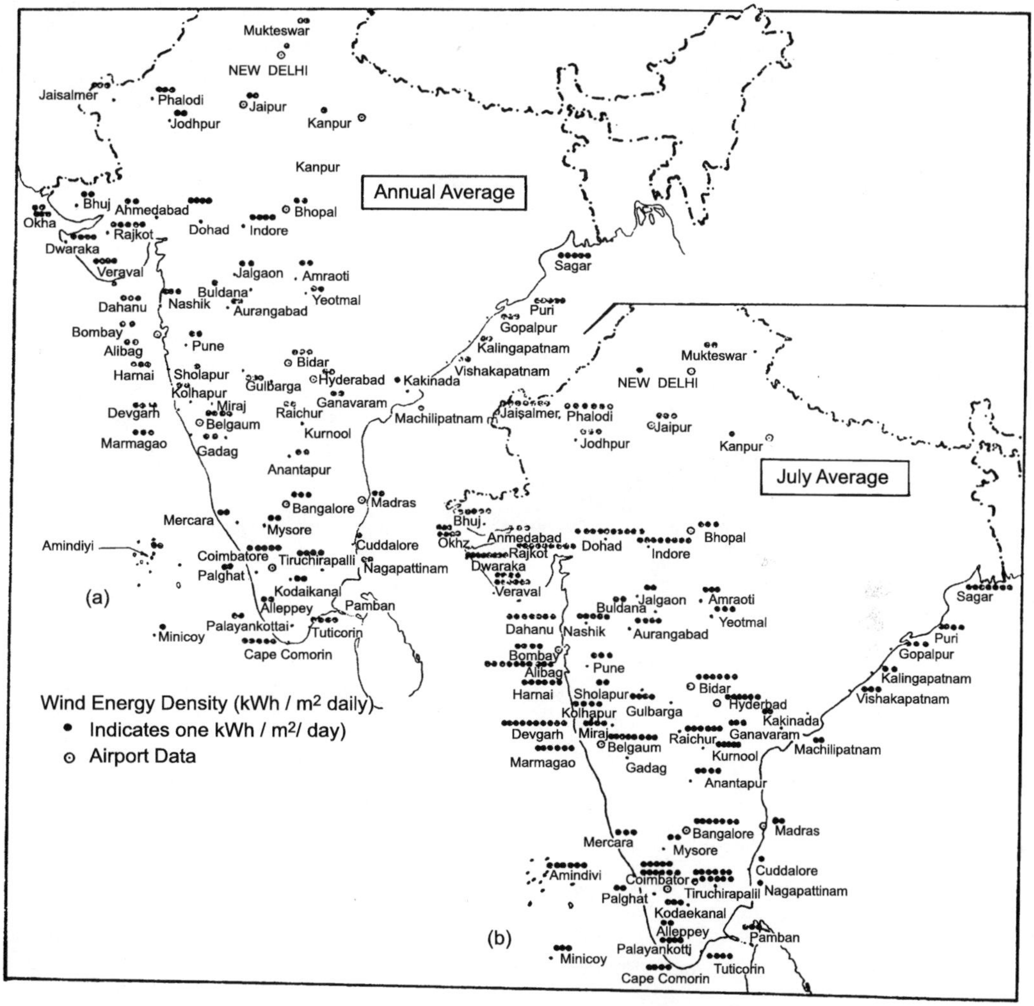

Based upon Survey of India map with the permission of the Surveyor General of India. The territorial waters of India extend into the sea distance of twelve nautical miles measured from the appropriate base line. Responsibly for the correctness of internal details shown on the maps rests with the publisher Government of India Copyright 1986.

Fig. 23.9. Wind energy density in different parts of the country in the early 1980s: (a) annual average, (b) average during the monsoon season. (A. Mani and D. A. Moolay, 1983, *Wind Energy Data for India,* Allied Publishers, New Delhi).

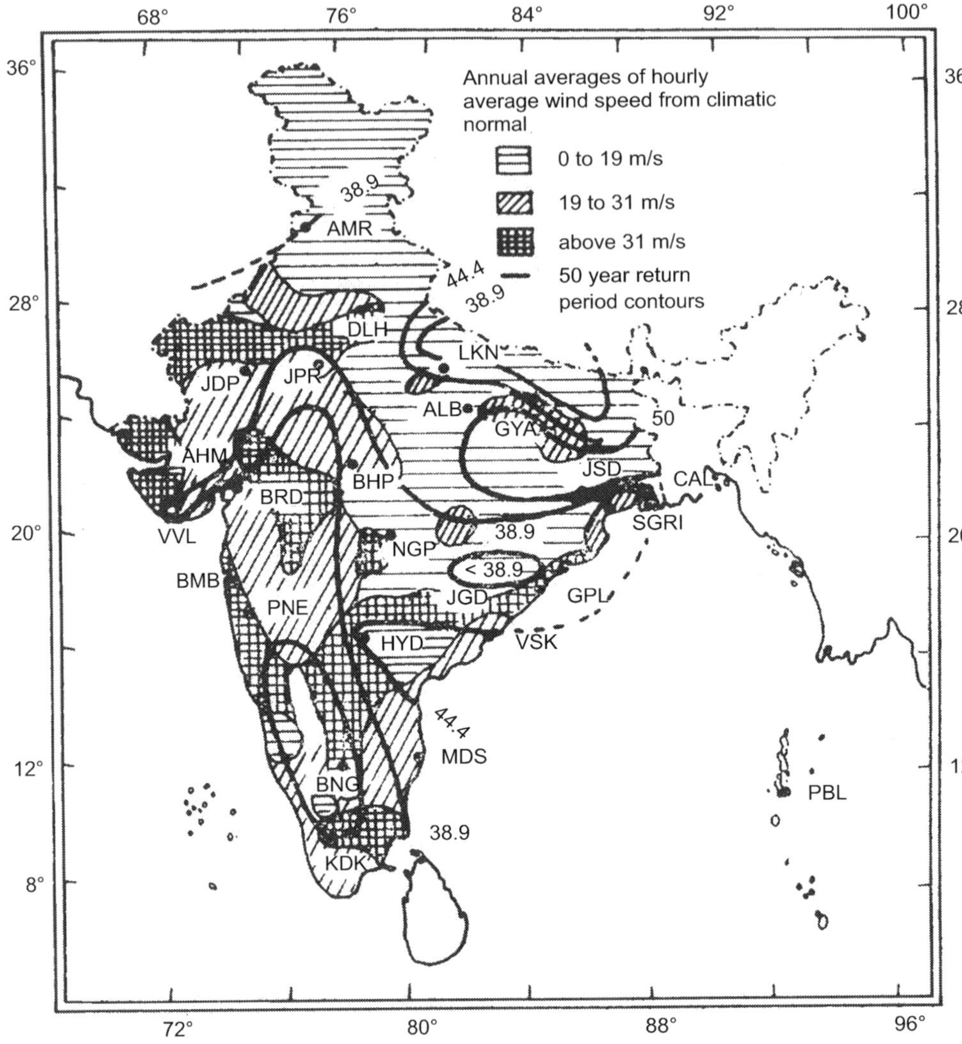

ed upon Survey of India map with the permission of the Surveyor General of India. The territorial waters of India extend into the sea to a distance of twelve nautical miles measured from the appropriate base line. Responsibility for the correctness of internal details shown on the map rests with the publisher © Government of India Copyright 1986

Fig. 23.10. Mean windspeeds (in metres/second) represented by contours, and 50 year-return extremes. (Redrawn from K. Narasimha and V. Shrinivasan, 1984, Fig. 5).

The waves keep the ocean surface waters perpetually in motion. The Salter Design makes use of huge vanes (20 to 40 in number) bobbing up and down due to the rolling wave motion (Fig. 23.12). This rolling motion is converted to turn a turbine through no-return valves. The ocean waves may be concentrated in converging channels, and the breaking waves maintain a head of water whose concentrated momentum can drive a turbine (Qasim, 1980).

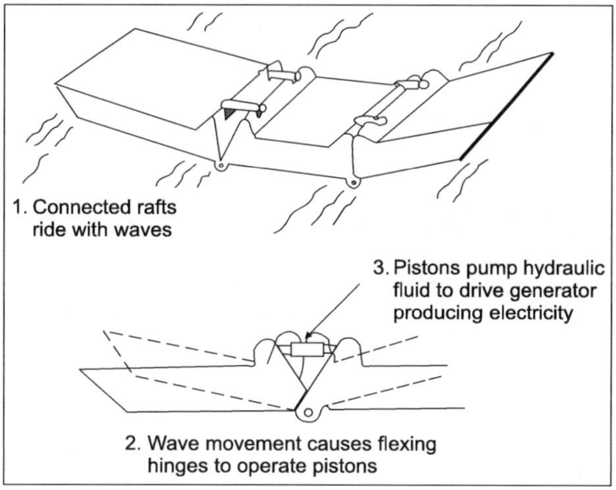

Fig. 23.11. The up-and-down motion of ocean waves is converted into a unidirectional motion with this mechanical device. The device turns the wheel of electricity-generating turbines. (Redrawn with permission from D. R. Coates, 1981, *Environmental Geology,* John Wiley, New York, 701 p., F 7.17).

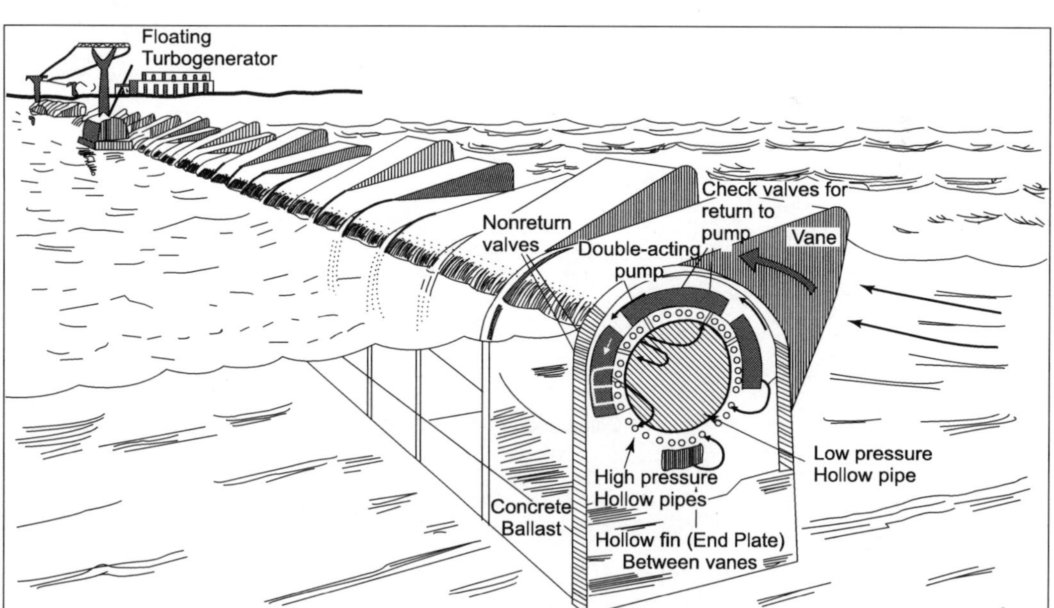

Fig. 23.12. The to-and-fro rolling motion of oceanic currents and tides is made unidirectional by valves installed on a dam-like structure put across the bay or estuary. (Redrawn from J. D. Isaac and R. Schmidt, 1980, *Science,* 207 (4428), Fig. 3).

The Gulf of Khambhat (where the average tidal range is 7 m) and the Gulf of Kachchh are two potential regions where the energy of the tides can be harnessed. It is estimated that the machines with the capacity of 7900 MW, generating electricity about 17 TWh/year can be installed (Sukhatme, 2011).

The oscillating water column system installed at the depth of 10 m off the coast of Trivandrum has been generating 18 KW of electric power (Sukhatme, 2011).

THERMAL AND SALINITY GRADIENTS IN SEA WATER

In the warm waters of the Arabian Sea and the Bay of Bengal, the surface-water temperature may rise to 28°C, while the deeper layers remain very cold. The temperature difference thus established can be utilized to operate ocean thermal energy conversion (OTEC) plants (Fig. 23.13). Liquid ammonia or some such liquid of low-vaporization point becomes vaporized when it is made to circulate from deeper

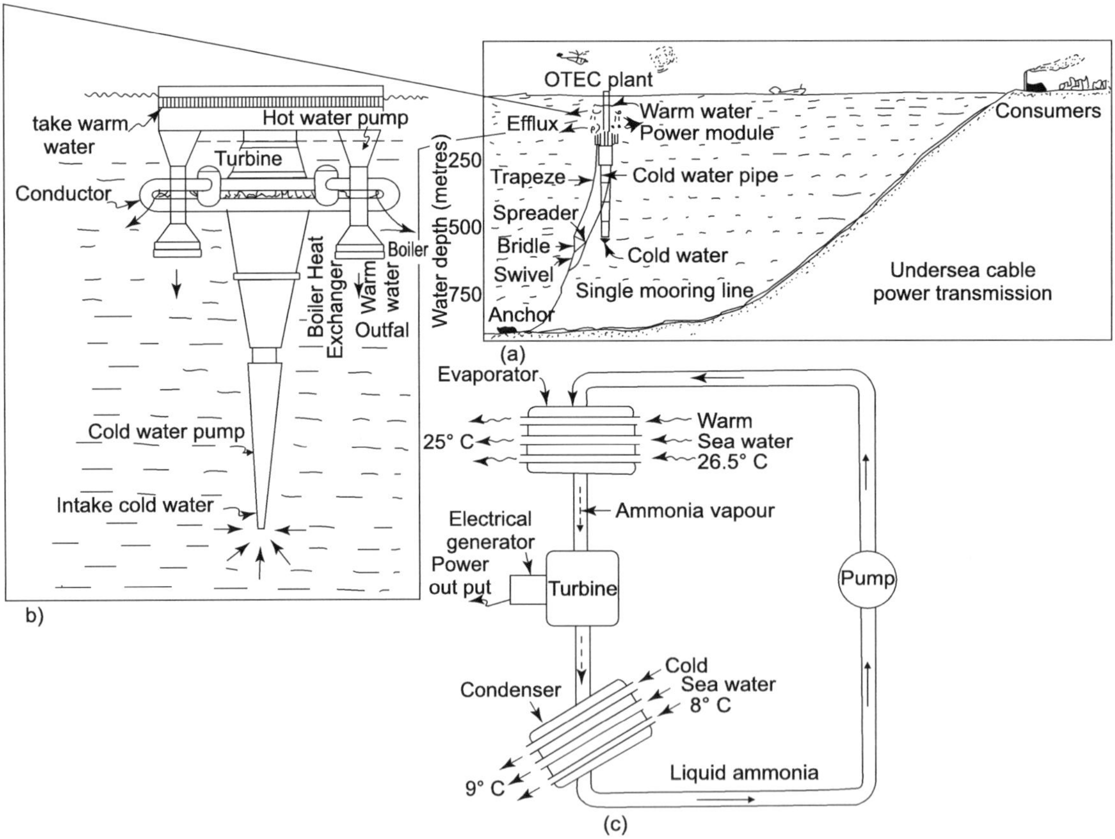

Fig. 23.13. (a) The ocean thermal energy conversion (OTEC) plant operates as a result of the difference of temperature of the water at the surface and at deeper levels (thermal gradient). The liquid ammonia gets vaporized. The vapour then turns the turbine. (After S. Z. Qasim, 1980, *Science Today*). **(b) Close-up view of the heat exchanger. (After D. R. Coates, 1981,** *Environmental Geology,* John Wiley, New York, 701 p., Fig. 7.19.) (c) The principle of the DTEC plant.(S. Z. Qasim, 1980).

cold levels to higher warm levels. The vapour then turns the turbines of the generator of electricity. Even though costly, the technology seems to be quite promising. Its efficiency has been increased by maki the heat exchanger larger in size, by proper choice of the working fluid, and by taking the cold-wat pipe to greater (>1000 m) depths (Qasim, 1980). The capacity of each OTEC plant can be 25 to 50 mW and can be installed in such places as the islands of Andaman-Nicobar, Lakshadweep and along coastal Tamil Nadu (such as Kulasekharapattinam in Tirunelveli district.

Similarly, the difference in salt concentrations in sea water at different depths—salinity gradient—can be used for generating power. The salinity gradient produces a difference in osmotic pressure—as much as 24 atmospheres where freshwater mingles with saline water. This osmotic pressure gradient provides the kinetic energy for electricity generation.

TAPPING GEOTHERMAL ENERGY

Availability

The thermal energy stored in the earth's crust upto a depth of 3 km is manifest in the form of rising geothermal gradients, particularly in some sedimentary basins such as the Sabarmati in Gujarat (where the heat flow is 1.7 to 1.9 HFU) and the upper Brahmaputra in Assam. The surface and groundwater th is able to reach great depths in such geothermally live basins get heated and even converted into steam. The steam-charged hot waters reappear on the earth's surface as hot springs and geysers (Fig. 23.15). Italy showed the way in 1904 for tapping the hydrothermal fields at Laderello and generating electrici of the order of 370 million watts (Fig. 23.14).

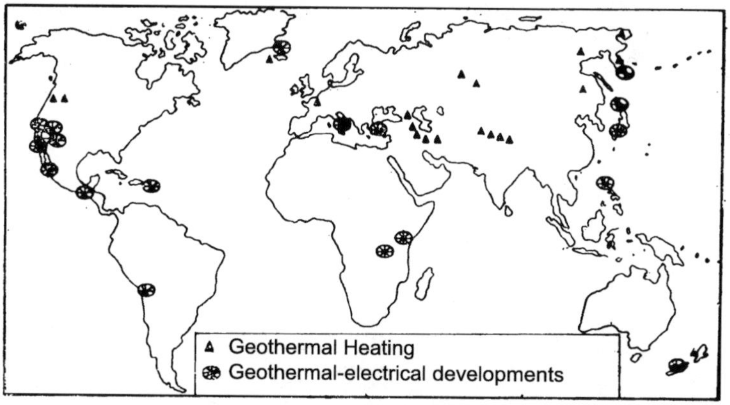

Fig. 23.14. Italy showed the way for tapping hydrothermal energy from hot springs. Now Hungary, Japan New Zealand, U.S.A. and quite many other countries are gainfully utilizing the energy of hot springs in a big way (Redrawn from G. Facca, 1980. In a publicity brochure circulated at the Intern. Geol. Congress at Paris).

Methodology

There are three ways in which the geothermal energy can be utilized. First, the hot waters and superheated dry steam can be directly conveyed to power plants to operate the generator turbines as done in coal-based thermal plants (Fig. 23.15). Second, the surface water is injected into

fractured rocks in the deep interior of the geothermal area through deep wells in order to heat and vaporize it; it is then brought back to run the turbines (Fig. 23.15). Third, the heat exchanger containing liquid having a low boiling point is immersed in the underground well. The liquid vaporizes and returns to the surface and runs the generator.

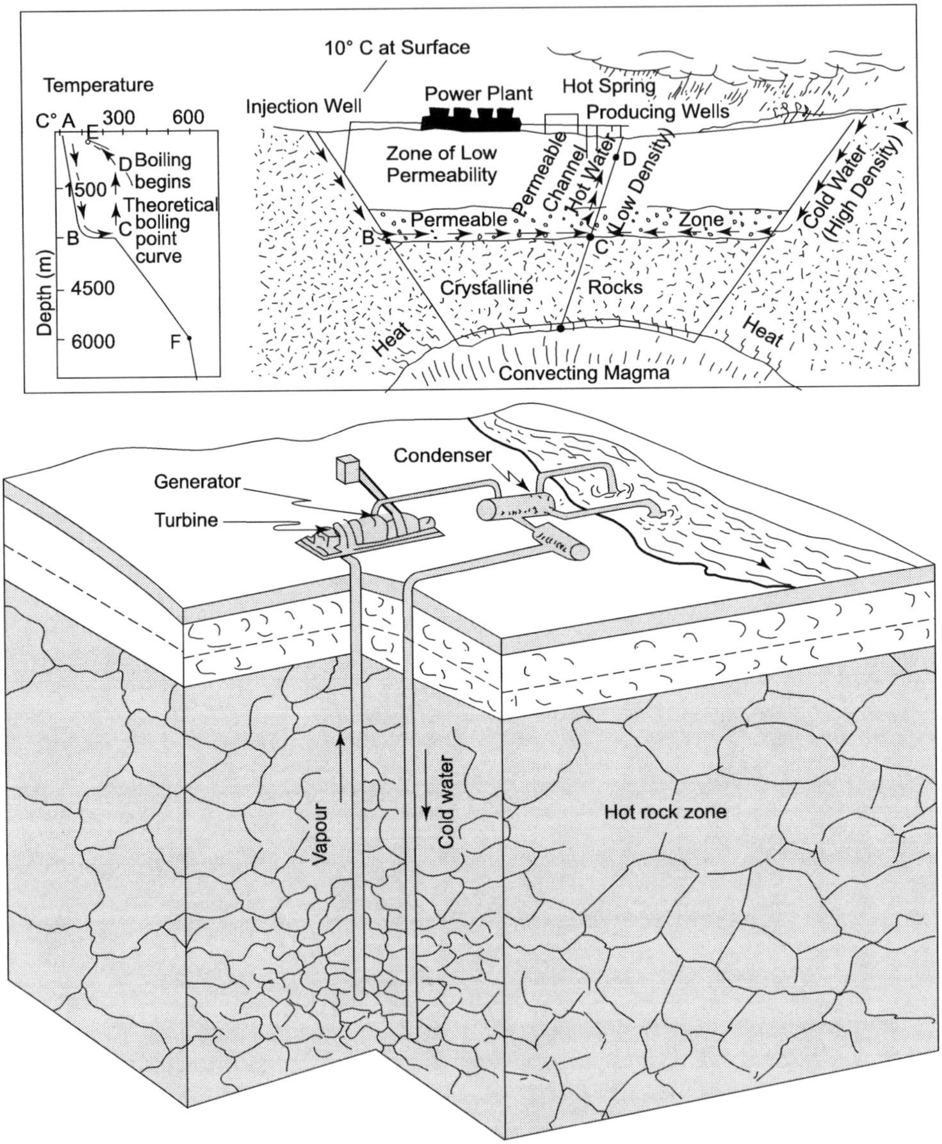

Fig. 23.15. (a) The superheated dry steam and hot water are utilized to operate the turbines of electricity generators. The surface water reaching great depth becomes hot and converted into steam. (After E. A. Keller, 1976. *Environmental Geology,* Bell and Howell, London, 488 p., Fig. 13.35). (b) In areas of high heat-flow, water injected through pumps is heated, vaporized and brought back to turn the turbines. (G. Facca, 1980).

Geothermal Resources in India

A comprehensive survey by Krishnaswamy and Shanker (1982) and Shanker (1988) reveals that **of the 340 thermal areas (Fig. 23.16 A and B)—113 of which storing an aggregate amount of 40 × 10 calories of heat—have the potential of generating 10,600 million watts of electricity (Table 23.3 This is equivalent to the energy generated from 5,730 million tonnes of coal or from 38,230 million barrels of oil. Only 46 of these hot springs have a temperature high enough (>150°C) to generate electricity by conventional methods, the cumulative power potential being 1838 million watts. There are 59 springs of intermediate temperature range (150° to 90°C), and having the potential of generating electricity through the use of the binary cycle.**

Table 23.3. Thermal Springs of the Indian Geothermal Belts of Energy Potential (After V.S. Krishnaswamy and Shanker, 1982; Ravi Shanker, 1988).

Geothermal Area	Number of Geothermal Areas	
	Total number of Reported Springs	Identified Hydrothermal System
Jammu and Kashmir	37	14
Himachal Pradesh	34	20
Haryana	3	1
Uttarakhand	37	14
Sikkim	7	2
Arunachal Pradesh	11	3
Assam	4	1
Meghalaya	1	
Andaman Nicobar Islands	1	
Odisha	11	3
West Bengal	6	1
Jharkhand	60	14
Madhya Pradesh	13	5
Rajasthan	8	5
Gujarat	13	8
Maharashtra	40	20
Andhra Pradesh	43	1
Tamil Nadu	2	
Karnataka	6	1
Kerala	3	

There are about 340 hot springs in ten geothermal fields (Figure 23.16 A and B) where the heat flo values vary between 30 and 468 million watts per square metre and the thermal gradient of 70 to 100°C/ km (Shanker, 1988). The important geothermal fields where temperatures exceed 150°C are in Pu and Chhumthang (Ladakh), Manikaran (Himachal Pradesh), Tattapani (Madhya Pradesh), and Jalgaon (Maharashtra), producing significant amount of energy.

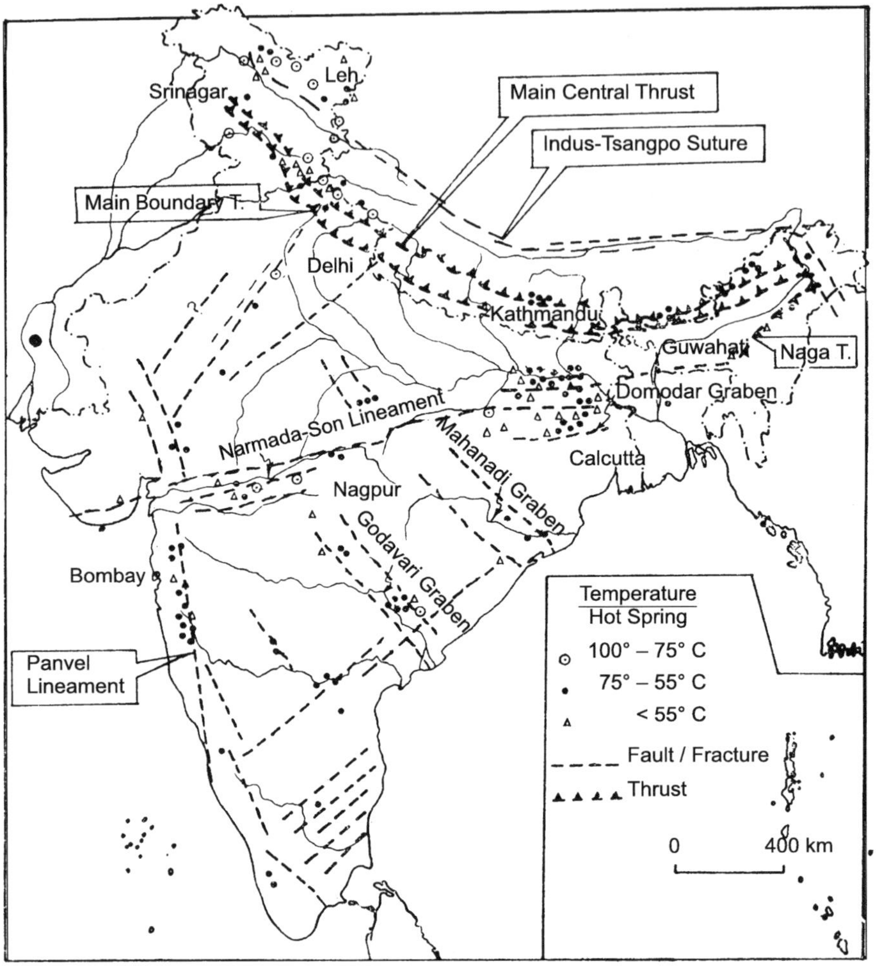

ed upon Survey of India map with the permission of the Surveyor General of India. The territorial waters of India extend into the sea to a distance of twelve nautical miles measured from the appropriate base line. Responsibility for the correctness of internal details shown on the map rests with the publisher. © Government of India Copyright 1986.

Fig. 23.16. (a) Of the 340 thermal areas in India 113 have been identified as having potential for generating energy. They have hydrothermal energy potential of 10,600 million watts equivalent to 5,730 million tonnes of coal or 38,230 million barrels of oil. (Based on S. Deb, 1977; V. S. Krishnaswamy and R. Shanker, 1982; and K. S. Valdiya, 1984).

The hot springs discovered on April 19, 2012 in the fractured granites at the Siachin Base Camp is producing sulphurous warm water at the rate of 12000 litres/hour (Asit Jolly, *India Today*, May 21, 2012). Seven kilometres away from the Sasoma thermal spring, the Siachin spring promises to bring comfort and hope to the army based 3540 m above the mean sea level.

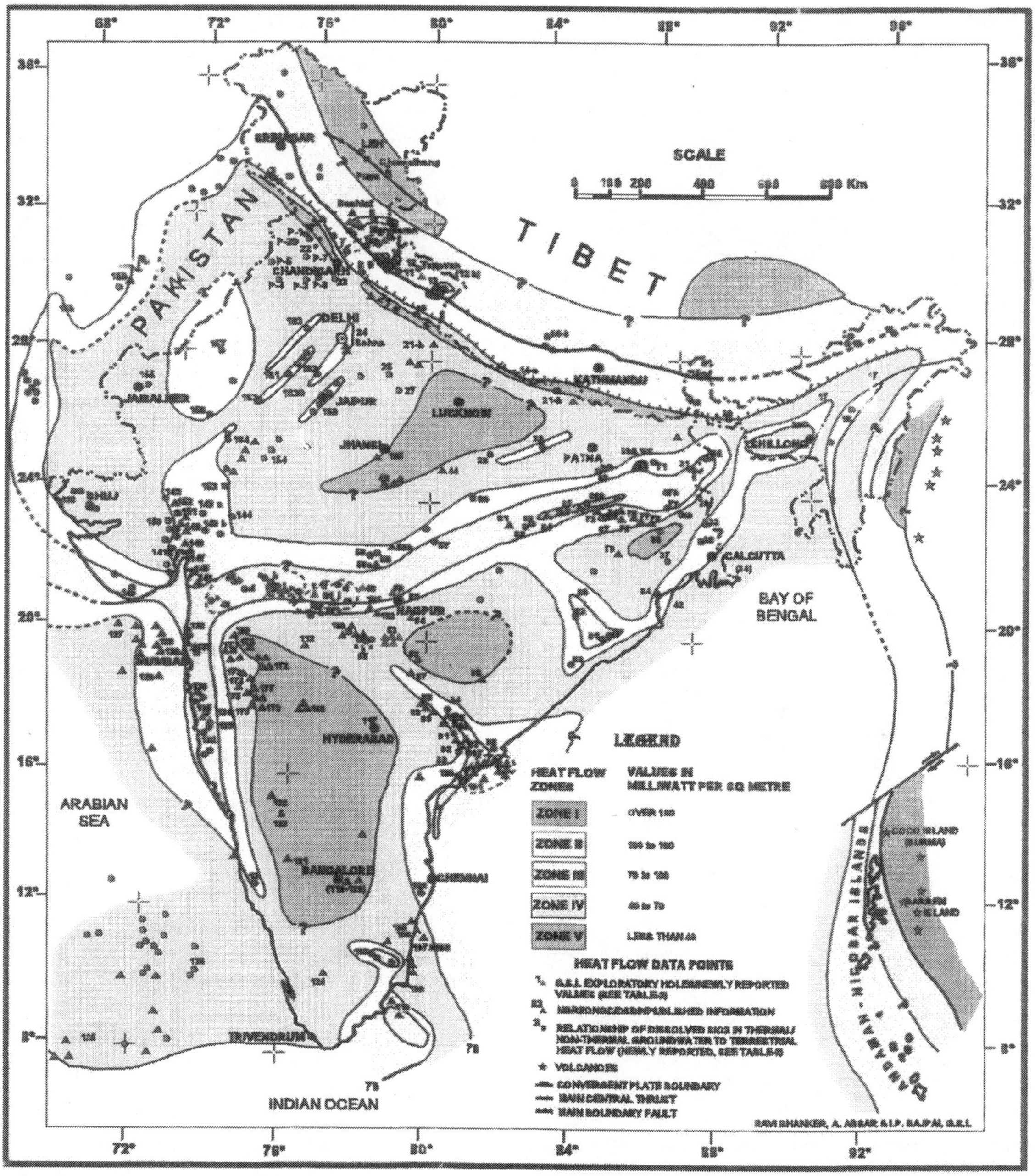

Fig. 23.16. (b) Heat-flow map of India showing belts of high heat flow and locations of geothermal sprin (From: Shanker, 1988). [See colour Fig. 23.16(b)]

The ten geothermal belts (Fig. 23.16, Table 23.3) are:

1. The Sindhu valley in Ladakh along the Indus-Tsangpo Suture Zone with its hot springs in Puga and Chamthang valleys, recording temperatures as high as 140°C. The Puga fields have a power potential of 400 kW per bore. A 7-kW power plant has been in operation since 1977.
2. The zone of the Main Central Thrust at the base of the Great Himalaya lithotectonic province, extending from Manikaran in Himachal Pradesh through Yamunotri, Gaurikund (Manadakini R.), Badarinath (Alaknand.R.), Tapovan (Western Dhauli R.), Sobla (Gori R.) north of Kathmandu, south-central Bhutan, to western Arunachal Pradesh. The Manikaran hot springs in the Parbati valley (Kullu), for example, have temperature as high as 93° to 96°C, with a power potential of about 450 kW per bore, if the flow is taken at a very conservative value of 400 m^3/h (Shridhar Sharma in *The Times of India,* August 12, 1984).
3. The Naga-Lushai belt in the zone of obduction in the Indo-Myanmar Ranges.
4. The Andaman-Nicobar belt in the immediate proximity of the active subduction zone.
5. The Mahanadi valley graben in Odisha.
6. The Damodar valley half-graben in Bengal-Jharkhand and adjoining northeastern Madhya Pradesh. The thermal springs such as Tatapani and Bakreshwar of this geothermal belt register temperatures varying from 70° to 67°C. The adjacent Chhota Nagpur (Ranchi) shield is characterized by 20 groups of hot springs in the Hazaribagh district, the important one being Surajkund with a temperature of 87°C (Deb, 1977).
7. The rift valleys of the Narmada and Tapi rivers which show a tendency towards neotectonic activities and seismicity.
8. The sediment-filled Sabarmati graben with its high thermal gradient. The 33 hot springs of Khambhat area extending over a 300 km tract, have a power generating capacity between 1850 and 2300 kW per bore (S. Sharma, *The Times of India,* August 12, 1984).
9. The Konkan coast of Maharashtra demarcated on the shelf riven by NNW-SSE trending boundary faults and a few N–S step faults on the coastal belt with a large number of thermal fields (S. Deb, 1977).
10. The Karnataka craton torn by deep faults and shear zones.

New Technology

Low-enthalpy resources (<150°C) need to be taken by resort to innovative drilling, and special power-neration technologies such as installing heat exchangers and binary fluids (Chandrasekharam and Bundschuh, 2008). Even though costly in the beginning, there are no hidden costs in the form of environmental mitigation and restoration.

ENERGY FROM BIOMASS

Principle and Potential

Biogas—from which energy can be generated—is formed as a result of fermentation of dung of cattle, human excreta, crop wastes, garbage, and aquatic weeds. Fermentation takes place at temperatures higher than 28°C. Hence **the biogas plants are efficient only under warmer climate condition. The fermentation of the biomass (Fig. 23.17), carried out by aerobic bacteria, produces hydrocarbon**

gases, predominant methane (70 to 60%) with CO_2 (30 to 40%), H_2S and N_2 in small quantities. The residue left behind in the tank is rich in nutrients and can be used as valuable manure.

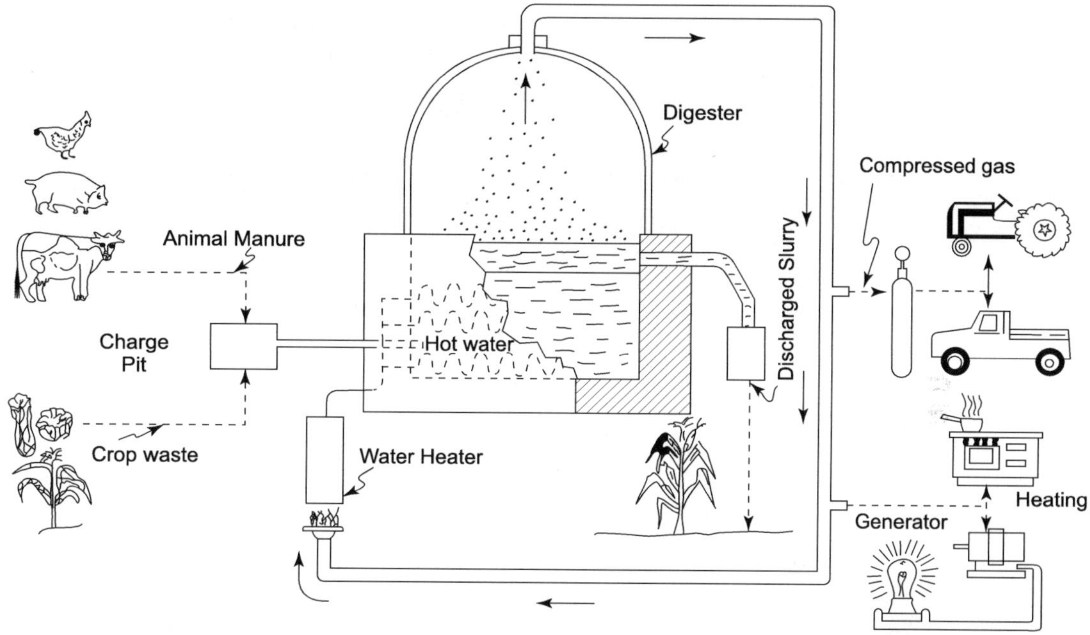

Fig. 23.17. Biogas is formed as a result of fermentation of cattle dung, aquatic weeds, bird droppings agricultural wastes and domestic garbage. The gas can be used for cooking, heating and lighting the home and for generating electricity. The residue is a valuable manure. (Redrawn with permission from D. R. Coates, 1981, *Environmental Geology,* John Wiley, New York, 701 p., Fig. 7.20).

The dung of an average cow can produce 280 litres or 0.3 m^3 of gas per day, so that **five heads cattle in a household can supply enough gas (1.5 m^3/day) to meet the varied needs of a family for cooking food, heating water, and by generating electricity for lighting rooms, running tubewells, etc.** (Fig. 23.18). The efficiency of the device can be increased by making use of such other ingredien as aquatic weeds, night soil, and crop wastes, and keeping the temperature between 28° and 48°C within the plant.

The *FAO Livestock Report*-2007 shows the population of livestock in India was 173 million heads in 2006. Now it must be quite higher than 175 million. India has enough cattle to produce dung to run tens of millions of biogas plants. **As far back as 1970s there was enough cow dung (575 to 630 million tonnes per year to run 18 million family-size plants of capacity 1.68 m^3/day and more than a million community-size (140 m^3/d) plants** (V. S. Kurup in *The Times of India,* August 6, 1981).

Added to this volume of dung, about 2880 tonnes of night soil existed in villages released daily by people in the open fields in the early 1980s (Deodhar and Nath, 1981). The 2011 Census showed th 49.8% household people defecate out in the open. Taking the whole of India, on an average 10 members make a household (*The Census of India Houselisting and Housing: Data Highlights-2011, The Hindu,* arch 14, 2012). One can imagine the amount of human excreta spread out in the fields and elsewhere. In 1981 about 683.3 million Indian people annually generated 18.6 to 26.1 million tonnes of excreta (Deodhar and Nath, 1981). By that yardstick, **today 1210 million people must be annually producing 32.9 million tonnes to 46.2 million tonnes of faeces, nearly half of which left out in the open. The excreta mixed with dung and or organic wastes can be used to produce biogas.**

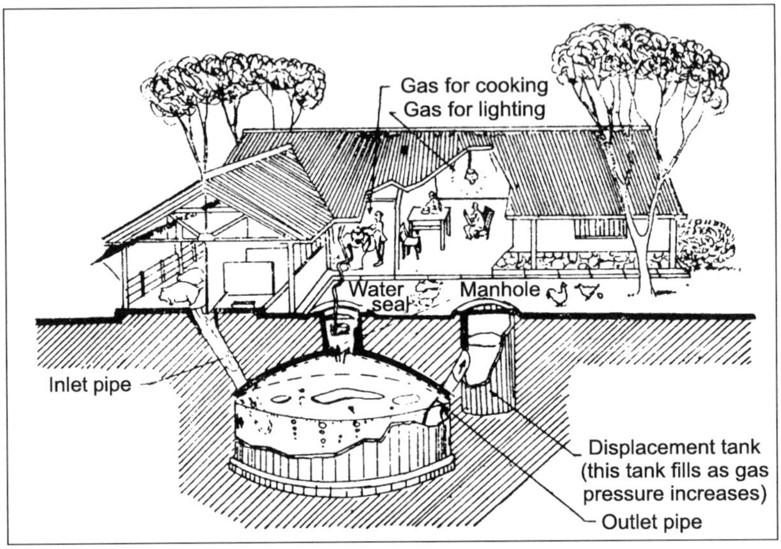

Fig. 22.18. A family-size biogas plant producing at the rate of 15 m^3/day can meet the varied needs of the household as in this Chinese home. (Anon, 1980, *Guidebook on Biogas Development,* U.N.O., New York, 127 p., Fig. 2).

The weeds that clog the channels of rivers, streams and canals constitute a very useful mix with animal dung. The agricultural wastes in the late 1970s amounted to about 230 to 240 million tones. **At present it is more than 300 million tonnes, and can produce as much biogas as 30 million tonnes of coal. Considering the various sources and the great volume of biomass that is available in the country, it can be deduced that the energy potential from biogas is 15 to 20 times greater than that of energy sources like coal and petroleum.**

Biogas Technology

The Khadi and Village Industries Commission pioneered the movement for constructing biogas plants in the country (Fig. 23.19). The original design has a capacity-range of 1.5 to 85 m^3/day of gas output, and the biomass used is cattle dung mixed with an equal amount of water and stirred into a slurry. The retention period is 50 to 55 days. In the second generation design (in addition to dung) night soil, water-hyacinth and rice straw in the ratio of 1:1: 2 are used and the temperature is kept at about 30°±2°C by connecting the tank with an external water jacket heated by solar radiation. **Every kilogram of this biomass-mix produces, on an average, 300 litres of biogas and the retention period is only 15 to 18 days** (S. K. Subramaniam, 1978). **A way of overcoming the problem of low temperature as in the**

Himalayan region or during winters in the plains is to cover the plants with plastic sheets and add molasses, algae, urea or urine to the slurry. By connecting toilets to the dung digesters, the Sulabh Shauchalaya organization was able to illuminate the streets of Patna and elsewhere.

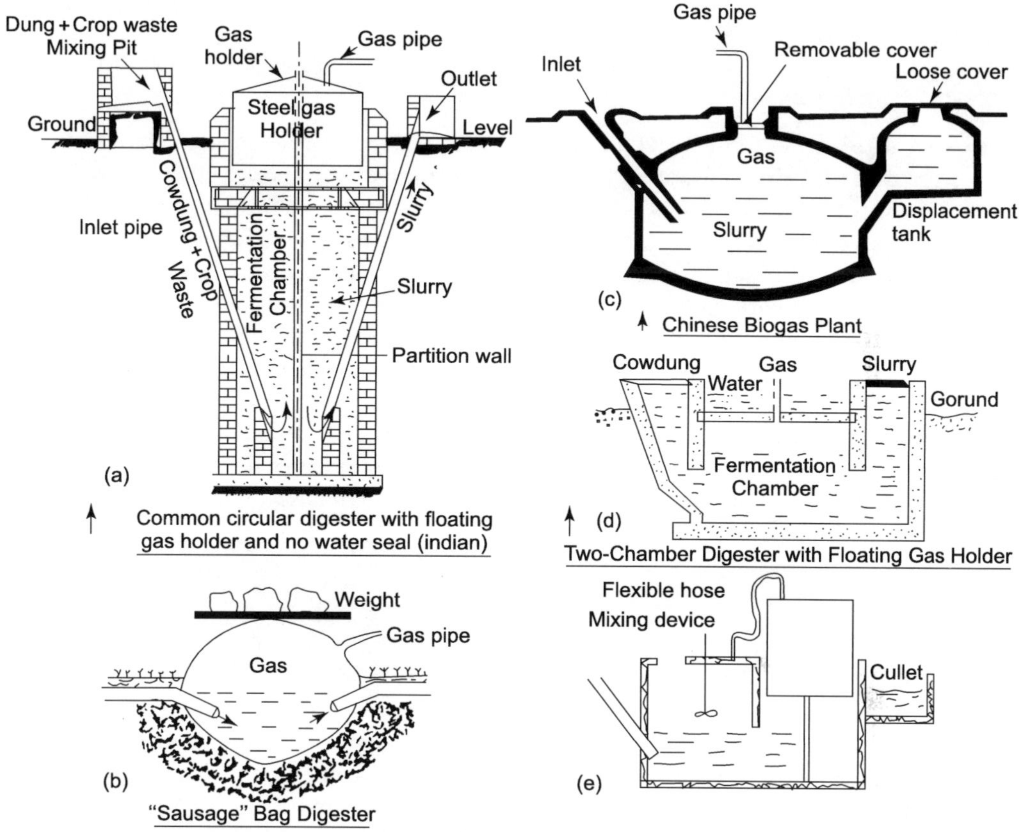

Fig. 23.19. Various types of biogas plants. (Redrawn from Anon, 1980, *Guidebook on Biogas Development,* Energy Resource Development series, No. 21, U.N.O., New York, 127 p., Figs. 5, 6,8,12.) (b) – (Chinese biogas plants. (After E. El-Hinnawi and A. K. Biswas, 1981).

As already mentioned, **weeds such as water hyacinth, water lettuce, salvinia, hydrilla, duck weeds, and algae have been found to be very useful supplements to the cattle dung. One hectare of water hyacinth—which has a growth rate of 8 to 16 t/day—is capable of producing 70,000 m of biogas containing 69% methane** (Khoshoo, 1982). A captive water hyacinth field, irrigated town sewage, can thus greatly augment the biogas production in areas where cattle dung is not easily available or is in short supply.

Municipal sewage in a concentrated state within closed chambers (digestors) generate sludge gas containing 53 to 68% of methane with O_2, N_2 and H_2S in about four weeks, the biogas potential being 9500 m^3/day per million of the population served by the sewage treatment plant (Rai, 1984).

India's total potential of energy generated by biomass is about 23,000 MW—18,000 MW from the combustion and 5000 MW from bagasse-based co-generation units in sugar mills. The annual electricity generated would be of the order of 60 TWh (Sukhatme, 2011).

ALTERNATIVE FUEL

Ethnol and Biodiesel

Fermentation of biomass under certain optimum conditions also generates alcohol, usable as a substitute for or as a mix with petrol in combustion machines. This has been demonstrated with resounding success by the Brazilians who are running a sizeable proportion of their motor vehicles th biofuels. Six vegetation types containing cellulose matter have been identified as energy plants—garcane, sugarbeet, sweet sorghum, maize, tapioca, and the mahua flowers. **The oil-bearing fruits of jatropha and *Pongamia pinnata* have been found to be a very fruitful source of biodiesel by resort a technology based on catalytic trans-esterification. Large-scale jatropha cultivation on about 11.2 million hectares of land could produce enough biodiesel "to substitute for fossil diesel to the extent of 20%"** (Bhardwaj et al., 2007). The other oil-bearing substances that can be used are palm oil d sunflower oil. The use of ethnol as a mix with petrol will greatly reduce consumption of petroleum and dependence on the import of crude oil. If India is to check the ever-rising curve of price hike and lessen the misery of the common man the import of petroleum will have to be reduced, and a substitute fuel or fuel-mix introduced. There is no other option.

Gas Hydrates

Large deposits of gas hydrates (methane entrapped within frozen molecules of water) have been entified on the continental margins of India—particularly offshore the Krishna-Godavari delta, as ll as in the eastern flank of the Andaman Islands. One cubic metre of gas hydrate formed in deep sea can produce 164 m^3 of methane (T. Kumar, 2009). The estimated amount of gas deposit is 2000 billion bic metres—sufficient to provide twice the energy resource compared to the conventional form of ergy (Ramana et al, 2007). The difficulty is that we have as yet not been able to develop a technology to tap this large reserve lying under the seawater.

NUCLEAR ENERGY OPTION

In order to meet nearly 10% of the country's total installed capacity the Atomic Energy Commission aims at installing nuclear power plants to generate 63,000 MWe by A.D. 2032 in order to meet a part of the country's total installed capacity (R.K. Sinha, Chairman, AEC, in *Hindustan Times,* May 1, 2012). The country is now in a position to produce all the fuel and heavy water requirements needed for longer term programme. According to Ramanna (1984), the nuclear radiation will not exceed 0.15% of the total radioactivities from various sources, including cosmic rays, to which the population is exposed. The 20 nuclear power plants in India at present together are generating 4780 MWe—merely 3% of the total installed capacity of power generation (Monto, 2010). It is hoped that by 2020 A.D., 35GWe nuclear energy would be available in the country.

One kilogram of natural uranium U^{238} generates as much energy as 35,000 kg of coal (Bose, 1981). The uranium mineral mined contains only 0.7% of the fissionable U^{235} isotope, which participates in the heat-generating nuclear fission. The principal problem is therefore to enrich the uranium in U^{235} content from 0.7% to 3% (Bose, 1981). The commercially exploitable uranium reserves in the early 1980s were about 49,000 tonnes of U_30_8 from various reserves located mostly in the Singhbhum district (Jharkhand) and very subordinately in Meghalaya, Madhya Pradesh and Karnataka. Some 7800 tonnes of U_30_8 can be recovered from the copper tailings in Roam-Rakha, Surdi and Mosabani. Copper Mines; and another 1750 tonnes of U_30_8 can be extracted from phosphoric acid from phosphate-based fertilizer plants (Ramanna, 1984). A 10,000 mWe nuclear power generation programme will require 37,000 tonnes of U_30_8 over the 25-year life-time of the reactors.

Table 23.5. Nuclear Power Programme of a Few Selected Countries in the early 1980s (After Raja Ramann 1984)

Country	Number of Operating and Under-construction Units	Total Capacity of Operating and Under-construction Units (mWe)	Share of Nuclear Capacity in Total Power Capacity	Future Programme	
				Projected Nuclear Capacity in mWe (year)	Projected share of Nuclear Power Capacity in Total Power Capacity
India	10	2,270	>3%	10,000 (2000)	10%
Japan	38	29,045	>12%	62,000 (2000)	27%
Erstwhile U.S.S.R	84	58,672	>7%	58,672 (1900)	NK
U.S.A.	130	119,053	>13%	119,053 (1990)	NK
U.K.	42	12,556	>10%		

In recent years moderate sized to very large deposits of uranium minerals have been discovered, particularly in the Cuddapah Basin in Andhra Pradesh, in Southern Meghalaya and in parts of Karnataka, making India less dependent on foreign suppliers. Our deposits of plutonium-bearing minerals are very large, perhaps largest in the world. The supply of nuclear fuel is thus assured.

The spent-fuel material contains unused U^{235} and plutonium Pu^{239}, the latter being a deadly pollutant (even 1/1000 g may cause death and 1/10,00,000 g lung cancer). Our scientists have got over the problem of disposing off nuclear wastes by making use of it as a fuel in the fast-breeder reactors, such as at Kalpakkam. India has also developed a mixed fuel obviating our dependence on enriched imported material.

SUMMARY

The consumption of energy is increasing at a galloping rate, thus threatening the conventional energy resource base. The world in the decade 1970–1979 consumed as much oil as it did in the previous 100 years. More than 75% of the total energy demand in India is for commercial purposes. And, of the energy generated for non-commercial non-industrial purposes, 64% is contributed by fuelwood and charcoal, 13% by dungs and 23% by vegetal wastes. The annual firewood consumption is far outstripping t annual productivity of our forests, resulting in their alarming degradation and destruction. Realizing that e fast-depleting reserves of fossil fuels are extremely finite and knowing that their extraction causes considerable damage to various ecosystems including water bodies, there is no option left but to seek alternative renewable sources of energy to meet the demands of our growing population.

Every square-metre area of the earth's surface receives on an average, 200 W of solar heat per hour. If only 1% of this energy could be converted into electricity, the world's entire energy needs would be met. Over 3000 to 3200 hours of bright sunshine (at the annual rate of 2000 kWh/m^2/yr) are received annually by the vast stretches of Rajasthan, Gujarat, western Madhya Pradesh and northern Maharashtra, and 2600 to 2800 hours (at the rate of 1700 kWh/m^2/yr) by the rest of the country, except Kerala, Kashmir and Assam. This energy can directly be converted into electricity by use of photovoltaic cells or utilized through collectors and mirrors to heat water and generate steam to run the turbines of generators.

The kinetic energy of wind can be used to run windmills to operate tubewells and to run dynamos to nerate electricity, which can be stored in batteries. The efficiency of a windmill is 35% as compared to 5 to 15% of the solar units. It is estimated that 20 million megawatts of wind-generated energy can be produced by utilizing the most promising sites all over the world. The variable supply of wind-nerated energy can be supplemented by using hydro-electric generators in reservoirs filled through pumps operated by windmills themselves during windy spells. Coastal Saurashtra in Gujarat, parts of Rajasthan and western Madhya Pradesh, the Great Himalayan–Tethyan domain of the Himalaya the central belt north of 15°N latitude, parts of Karnataka, Kerala and southern Tamil Nadu, and the islands of Andaman-Nicobar and Lakshadweep have a high potential for tapping wind energy, particularly during the monsoon season.

Recent surveys demonstrated that the potential of wind power for generating electricity is 20 to 30 mes more than what official figures show.

The vertical and horizontal motions of ocean waves, tides and currents, after being rendered unidirectional through mechanical devices, can be used to turn the turbines of generators installed in such places as the Gulf of Kachchh, the estuaries in Sundarban (Bengal) and the Kaveri delta rather than all along the coastal belts.

The thermal gradient obtaining between warm surface waters and cold deep-level water in the oceans can vaporize liquids like ammonia. The vapour is then fed into turbines to generate electricity. Similarly, salinity gradient generates osmotic pressure which can be utilized for power generation.

The thermal energy stored in the crust up to a depth of 3 km is manifest in the form of hot springs and in geothermal gradients in some areas, particularly in thick sedimentary basins like the upper Brahmaputra valley in Assam and the Sabarmati valley in Gujarat. The hot water and steam of thermal springs can be utilized directly, or the steam can be fed into turbines to generate electricity. Of the 113 areas of geothermal potential in India, 46 having temperature >150°C have a total capacity of generating together 1838 mW of power and 59 springs with a temperature range of 150 to 90°C can generate electricity by making use of the binary cycle. In addition to heating houses, and glass houses where vegetables can be grown, the energy can be used for many purposes. At Puga a 7-mW plant has been generating electricity since 1977, giving warmth and comfort to our defence and security forces, and bringing greenary in their vegetable gardens.

Fermentation of biomass comprising cattle dung, human excreta, agricultural wastes, domestic-municipal garbage, aquatic weeds (like water hyacinth and algae) has the potential of generating 15 to 20 times the power generated by coal and petroleum. Five heads of cattle in a household generate enough dung to run a family-size plant producing gas at the rate of 1.68 m^3/day. India has the potential to run tens of millions of family-size and more than a million community-size (140 m^3/day gas production) biogas plants. **At places where cattle dung is in short supply or not available, captive water hyacinth farms irrigated by municipal/domestic sewage can supply a more vigorous source of biogas.** Toilets connected to biogas plants would augment the production of gas as also the mixture of crop wastes such as rice straw and petroleum products.

Oil-bearing plants and fruits, such as jatropha, are being increasingly used to generate biofuels. **Large-scale jatropha cultivation of over 11 million hectare land alone would prove to be substitute for 20% of the diesel India uses today.**

One kilogram of natural uranium can generate as much energy as 35,000 kg of coal. Having sufficie to sizeable reserves of natural uranium and plutonium, and with its proven success in utilizing the mixed-oxides fuel, India can venture into an ambitious nuclear power programme provided that adequate safeguards are taken against accidental leakages and pollution.

References

Abbas, N. and Subramanian, V., 1984. Erosion and sediment transport in the Ganga River Basin. *Indian Jour. Hydrol., 69:* 173–82.

bbi, S.D.S., 1981. Flash floods and their warning. *Proc. Intern. Conf. Flood Disasters,* Ind. Nat. Sci. Acad., New Delhi, 1: 238–50.

brol, I.P., 1990. Problem soils in India, *In*: ***Problem Soils of Asia and the Pacific***, FAO, Bangkok, 153–165.

haryya, A., Sengupta, B. and Ghosh, R.N., 2000. Mid-1997 tidal surge of Bay of Bengal at Digha on the east coast of India: Its damage evaluation and mitigation, *Jour. Nepal Geol. Soc.*, 21, 75–86.

dyalkar, P. G., 1975. *Paleogeography, Framework of Sedimentation and Groundwater Potential of Cropland Alluvial Valleys of India.* Govt. Printing and Stationery Depot, Mumbai, 261 p.

dyalkar, P. G., 1984. Hard-rock hydrology: Indian scenario. *Proc. 71st Indian Science Cong.,* Ind. Sci. Cong. Assoc, Kolkata, 5(2): 1–52.

grawal, Anil, Narain, Sunita and Sen, Srabani (Eds) 1999. *The Citizens' Fifth Report* (2 parts), Centre for Science & Environment, New Delhi, 440 p. and 256 p.

grawal, A., Chopra, R. and Shukla, K., 1982. *The State of India's Environment—1982,* Centre for Science and Environment, New Delhi, 189 p.

grawal, K. C., 1979. Snow and Avalanche Study Establishment and its Activities. *Nanda News Bulletin,* Mountaineering Club, Nainital, 1(7): 2–8.

grawal, M. C., Agrawal, R. P. and Chaudhury, R. S., 1974. Infiltration characteristics of soils as related to soil physical properties. *Jour. Indian Soc. Soil Sci.,* 22: 285–89.

grawal, P. N. and Kumar, A., 1982. Microearthquake reporting for engineering application. In: B. B. S. Singhal (Ed.), *Engineering Geosciences,* Sarita Prakashan, Meerut, pp. 181–86.

grawal, R. C. and Agrawal, P. N., 1982. Monitoring of contemporary seismic activity for project sites. *Proc. Fourth Intern. Cong. Eng. Geol.* Oxford & IBH, New Delhi, pp. 19–32.

ggarwal, Yash P., Sykes, L. R., Armbruster, J. and Sbar, M. L., 1973. Premonitory changes in seismic velocities and prediction of earthquakes. *Nature,* 241, 101–104.

hmad, M., Ratan, S. and Dhar, B. B., 1996. The impact of processing of lead-zinc ores on the surrounding environment. *In*: *The Impact of Mining on the Environment*, A.A. Balkema, Rotterdam, 229–236.

hluwalia, P. P., 1983a. Salinity ingress prevention programme in coastal area of Saurashtra. *Ind. Jour.*

Engin. Geol., 12(2-3): 23–30.

Ahluwalia, P. P., 1983b. Monitoring and management of groundwater in saline coastal area of Saurasht *Proc. Sem. Assessment, Development and Management of Groundwater Resources,* Central Groundwater Board, New Delhi, 2: 253–62.

Ahmad, M. S., 1983. Groundwater exploration in Bundelkhand Granites in southern Banda distri U.P. *Proc. Sem. Assessment, Development and Management of Groundwater Resources,* Central Groundwater Board, New Delhi, 1: 141–50.

Allen, R., 1980. *How to Save the World,* Kogan Page, London, 150 p.

Anon., 1969. *Glossary of Terms Relating to River-Valley Projects,* Part III (Dams and Dam Section Indian Standards Inst., New Delhi, 29 p.

Anon., 1975. What should an environmental impact assessment do? In: Munn, R. E. (Ed.), *Environmental Impact Assessment—Principles and Procedures,* SCOPE Pub. No. 5, John Wiley, New Yor 190 p.

Anon. (Ed.), 1975. *Engineering Geology Case Histories,* Misc. Pub. No. 29, Geol. Surv. Ind., Calcut 256 p.

Anon., 1977. *Desertification and its Control,* Indian Council of Agricultural Research, New Del 358 p.

Anon. (Ed.), 1977. *Desertification, its Causes and Consequences,* Pergamon Press, Oxford, pp. 17–2

Anon., 1979. *Guidelines for Design of Large Earth and Rockfill Dams,* Indian Standards Institution, New Delhi, p. 26.

Anon., 1983. *Indian Mineral Industries at A Glance—1981,* Indian Bureau of Mines, Nagpur, 165 p.

Anon., 1983. *Bharatiya Antarisksha Karyakram,* Indian Space Research Organization, Bangalo 24 p.

Anon., 1983. Mapping of forest cover in India from satellite imagery 1972-75 and 1980-82: Summa Report, Nat. Remote Sensing Agency, Hyderabad, p. 9.

Anon., 1984. *Proceedings of the National Seminar on Quakes, Floods and Cyclones* (10-11 April, 1982 Gujarat Engineering Research Institute, Baroda, 171 p.

Anon., 1984. *State of the Art Report of Studies on Himalayan Glaciers,* Department of Science Technology, New Delhi, pp. 19–46.

Anon, 2009. *Indian Minerals Yearbook*, Indian Bureau of Mines, Nagpur, 80 chapters.

Anon, 2010. *Report of the Task Force for Himalayan and Hill Areas of India*, Planning Commission, Government of India, New Delhi, 112 p.

Anon, 2010. *Degrated and Wastelands of India: Status and Spatial Distribution*, National Council Agricultural Research and National Academy of Agricultural Sciences, New Delhi, 158 p.

Anonymous, 2000b. *Reclamation/Restoration Techniques and Strategy for Mined Areas*, Indian Bureau of Mines, Nagpur, 122 p.

Ansar, Shareekul, Dhanya, C. R., Thomas, G., Chandran, A., John, L., Prashanthi, S., Vishnu, R. a Zachariah, E.J., 2012. A study of urban/rural cooling rates in Thiruvananthapuram, Kerala, *Ind. Jour. Geophys. Union*, 16, 29–36.

Ansari, A. A., Singh, I. B. and Tobschall, H.J., 1999. Status of anthropogenically induced metal polluti in the Kanpur–Unnao industrial region of the Ganga Plains, *Environ. Geol.*, 38, 25–33.

Ansari, A. R., Chugh, R. S., Sinvhal, H., Khattri, K. N. and Gaur, V. K., 1976. Geodetic determination of earth strains and creep on the Krol Thrust in the Dak-Pathar area, Dehradun district. *Him. Geol,* 6: 323–37.

ita, K. and Ganzawa, Y., 1997a. Rate of erosion in Nepal, Lesser Himalaya, *Jour. Geography*, 106, 156–167.

ita, K. and Ganzawa, Y., 1997b. Thrust tectonics and uplift process of the Nepal Himalaya, revealed from fission-track ages (in Japanese with English abstract), *Jourl. Geol. Tokyo Geographical Soc. (Geology)*, 106, 156–167.

ur, M. G. and Rajal, B. S., 1981. Study of crustal movements across Shali Thrust in the area of Shanan (Joginder Nagar) in Himachal Pradesh. *Tectonophysics,* 71: 151.

ya, A. S., 2001. Case study in India: Diagnosis, repair and strengthening of damaged buildings. *In*: V.K. Sharma (Ed) *Disaster Management*, Indian Institute of Public Administration, New Delhi, 157–176.

hish, M., 1979. Agricultural economy of Kumaun hills: Threat of ecological disaster. *Economic & Political Weekly,* June 23.

Avasthi, S.C., 1966. A note on the reh problem in Uttar Pradesh. *Indian Mineral,* 20: 49–52.

asthi, S.C, 1977. A note on the possibility of artificial growth of salt petre. *Indian Mineral,* 31: 20–28.

ba, M., 1979. Beach erosion in Kerala—problems and some solutions, *Professional Paper No. 6,* Centre for Earth Science Studies, Marine Science Division, Trivandrum, 15 p.

bu, R., Tejwani, K. G., Agarwal, M. C. and Bhushan, L. S., 1978. Distribution of erosion index and iso-erodant map of India. *Ind. Jour. Soil Conservation,* 6(1): 1–12.

Bachkheti, N. D., 1981. Inaugural address, *Proc. National Sem. Forests and Environment,* Karnataka Forest Department, Bangalore, pp. 1–4.

gchi, S., 1981. Dust pollution with special reference to manganese and mica ores. In: A. K. Sharma and Archana Sharma (Eds.), *Impact of the Development of Science and Technology on Environment,* Ind. Sci. Cong. Assoc, Calcutta, pp. 132–34.

lasundaram, M. S. and Gangopadhyaya, S., 1972. Cavernous limestone and problem of reservoir leakage, Kopili Hydel Project, Assam. In: Krishnaswamy, V.S. and Mehta, P. N. (Eds.), *Proc. Symp. Geol. & Eng. Problems of River Valley Projects,* Ind. Soc. Eng. Geol., Calcutta, pp. 379–90.

li, Y. P. and Kanwar, R. L., 1977. Reclamability classification of ravines for agriculture. *Soil Conservation Digest,* 5(2): 40–47.

ndopadhyay, P.C., Mitra, S.K., Pal, Tapan and Raghav, S., 2006. The 2005 eruption on Barren Island, Andaman Sea, *Current Science*, 90, 620–622.

ndopadhyaya, J., Shiva, V., Ashis Ganju, M. N., Menon, A. G. K., Nadir, K. L., 1983. *The Doon Valley Ecosystem, Report on Natural Resources Utilization in Doon Valley,* Centre for the Study of Development Societies, Department of Environment, New Delhi, 94 p.

ndyopadhyaya, M., 1982. Geotechnical study of hillslope in Eastern Himalayas for environmental development and town planning. *Proc. Fourth Intern.Congr. Eng. Geol,* Oxford and IBH, New Delhi, 1(1): 335–42.

Banerjee, S. P., 1981. Environmental problem due to mining in Jharia coalfield. In: Desh Bandhu (Ed *Environmental Management,* Indian Environmental Society, New Delhi, pp. 81–95.

Bansal, R. C. and Mathur, H. N., 1976. Landslides—the nightmare of roads. *Soil Conservation Digest,* 4(1): 36–37.

Bansode, R. B., Mehta, P. N. and Prakash, I., 1982. Assessment and control of factors contributing siltation of Ukai Reservoir. *Proc. Fourth Intern. Congr. Eng. Geol,* Oxford & IBH, New Delhi, 7(4 19–24.

Barnett, A., Pyle, L. and Subramanian, S. K., 1978. *Biogas Technology in the Third World: A Multidisciplinary Review,* Intern. Development Research Centre, Ottawa, 132 p.

Barua, D.K., 1997. The characteristics and mobility of some major rivers in Bangladesh, *Water Nepal* 5, 109–127.

Battawar, H. B. and Rao, V. P., 1969. Effectiveness of crop cover for reducing runoff and soil losses. *Ind. Jour. Soil and Water Conservation,* 17(3-4): 39–50.

Bawa, Kamalji, S. and Markham, A., 1995. Climate change and tropical forests, *Trends in Ecology and Evolution*, 10, 348–349.

Bawa, Kamalji S., Primack, P.B. and Oommen, Meera, 2011, *Conservation Biology*, Universities Press, Hyderabad, 589 p.

Baweja, B. K., 1982. Effects of urbanisation and industrial waste on groundwater regime in Kanp metropolitan area, Uttar Pradesh. *In*: B. B. S. Singhal (Ed.), *Engineering Geosciences,* Sarita Prakashan, Meerut, pp. 143–55.

Bell, F. G., 1980. *Engineering Geology and Geotechnics,* Newnes Butterworths, London, 492 p.

Betz, F. Jr. (Ed.), 1975a. *Environmental Geology,* Halsted, Stroudsburg (Pennsylvania), 390 p.

Betz, F. Jr., 1975b. Landuse planning. In: F. Betz Jr. (Ed.), *Environmental Geology,* Halsted, Stroudsburg (Pennsylvania), pp. 35–39.

Bhambhri, R., Bolch, T. and Chaujar, R. K., 2012. Frontal recession of Gangotri Glacier, Garhw Himalaya, from 1965 to 2006, measured through high-resolution remote sensing data, *Current Science*, 102, 489–494.

Bhan, Suraj and Singh, Shamsher, 1994. Soil and water conservation scenario in India. In: L.S. Bhusha I.P. Abrol, and M.S.R. Mohan Rao (Eds) *Soil And Water Conservation* (Vol. 2), Ind. Asssoc. Soil Water Conservation, Dehradun, 883–901.

Bhandari, R. K., 2002. *Seismic Microzonation* (unpublished report), Centre for Disaster Mitigation a Management, Anna University, Chennai, 198 p.

Bhandari, R. K. and Gupta, C, 1985. Problems of landslides in the Himalaya and future directions. I J. S. Singh (Ed.), *Environmental Regeneration of Himalaya: Strategy and Concept,* Cent. Him. Env. Assoc. Nainital, pp. 39–57.

Bhandari, R. K., Mehrotra, G. S., Nainwal, H. C. and Raiwani, K. K., 1985. Hill roads and Himalay landslides. *Proc. Sem. Construction of Roads in Hill Areas,* Ind. Road Congress, New Delhi, p 123–41.

Bhardwaj, Anshu, Tongia, R. and Arunachalam, V. S., 2007. Scoping technology options for India's security: Part II coal to liquids and biodiesel, *Current Science*, 92, 1234–1250.

Bhargava, G. P., 1989. *Salt-affected Soils of India: A Source Book*, Oxford IBH, New Delhi, 261 p.

hattacharjee, A., 1983. You cannot fight that weed, can you? *Ecoforum,* 8(1): 12–19.

hattacharyya, T., Ram Babu, R., Sarkar, P., Mandal, C., Dhyani, B.L. and Nagar, A.P., 2007. Soil loss and crop productivity model in humid subtropical India, *Current Science*, 93, 1397–1403.

howmick, P. K. and Misra, Ravi, 2010. Indian oil and gass potential, *In*: *Glimpses of Geoscience Research in India*, Indian National Science Academy, New Delhi, 218–228.

Bhumbla, D. R., 1982. Canal irrigation and its impact on crop production and environment. *First Ind. Env. Congr.* (New Delhi) (Mimeo), 12 p.

Bhumbla, D. R. and Khare, A., 1984. *Estimate of Wastelands in India,* Society for Promotion of Wastelands Development, New Delhi, 16 p.

hutiyani, M. R., Kale, V. S. and Pawar, N. J., 2002. Variations in glaciohydrological characteristics of some southwestern Himalayan river basins in this century. *In*: H.C. Nainwal and G. Prasad (Eds) *Geodynamics And Environment Management of Himalaya*, H.N.B. Garhwal University, Srinagar, 130–138.

lgrami, K. S. and Datta-Munshi, J. S., 1984. Ecology of River Ganga, *Annual Report MAB Workshop,* Department of Environment, New Delhi (Mimeo.).

sht, K. S. and Sah, M. P., 1999. The devastating landslide of August 1998 in Ukhimath area, Rudraprayag district, Garhwal Himalaya, *Current Science*, 76, 481–484.

swas, M. R. and Biswas A. K. (Eds.), 1980. *Desertification Case Studies,* Pergamon, Oxford.

swas, A. B. and Saha, A. K., 1982. Groundwater resources of India, with special reference to their salinity and pollution hazards. In: C. K. Varshney (Ed.), *Water Pollution and Management Reviews,* South Asian Publishers, New Delhi, pp. 113–27.

swas, A. B. and Saha, A. K.,1985. Environmental hazards of the recession of piezometric surface of groundwater under Calcutta. *Proc. Ind. Nat. Sci. Acad.,* 51 A: 610–21.

swas, A. K., 1982. Environmental and subsurface water development. In: Anon. (Ed.), *Water for Human Consumption,* Tycooly, Dublin, pp. 375–95.

swas, A. K., 1985. The web of interdependence. *Science Age,* 3(7): 4–8.

ank, L. W., 1985. A new type of forest decline in Germany. *Nature,* 314: 311–14.

asco, F., 1975. Les mangroves des l'Inde, *Institut Francais, Pondicherry. Trav. Sec, Sci. Tech.,* 15: 175.

oom, A. L., 1978. *Geomorphology,* Prentice-Hall, New York.

lt, B. A., Horn, H. L., Macdonald, G. A. and Scott, R. F., 1975. *Geological Hazards,* Springer-Verlag, New York, 328 p.

rdoloi, B. N., 1976. The problem of rehabilitation of the Jhumiya families of the hill areas of Assam. In: Negi, S. S., *Environmental Problems of the Himalaya,* Bishan Singh Mahendrapal Singh, Dehradun, 1982, 188 p.

se, D. K., 1981. Accounting of nuclear power. *Economic & Political Weekly,* August 8, pp. 1313–18.

wonder, B., 1982. Deforestation in India. *Intern. Jour. Env. Studies,* 18: 223–36.

adshaw, A. D. and Chadwick, M. J., 1980. *The Restoration of Derelict Land,* Blackwell, Oxford, 317 p.

own, L. R., 2001. *Eco-economy: Building an Economy for the Earth*, Hyderabad, Orient Longman, 334 pp.

Bruun, P. and Nayak, B. U., 1980. *Manual on Protection and Control of Coastal Erosion in India,* National Institute of Oceanography, Panaji (Goa), 174 p.

Burdon, D. J., 1982. Hydrological conditions in the Middle-East. *Quart. Jour. Eng. GeoL,* 15: 71–82

Caine, N. and Mool, P. K., 1981. Channel geometry and flow estimates for two small mountain stream in the Middle Hills, Nepal. *Mountain Research and Development,* 1(3-4): 231–43.

Caine, N. and Mool, P. K., 1982. Landslides in the Kolpu Khola drainage: Middle Mountains, Nep *Mountain Research and Development,* 2(2): 157–73.

Caratini, G., Thanikaimani, G. and Tissot, C, 1980. Mangroves of India: Palynological study and rece history of the vegetation. *Proc. IV Intern. Palyn. Conf,* Lucknow, 3: 49–59.

Carson, B., 1985. Erosion and sedimentation processes in the Nepalese Himalaya, *ICIMOD Occasional Paper No. I,* ICIMOD, Kathmandu, 39 p.

Chaddha, D. K., 1999. Groundwater *In*: *Vision 2050*, Indian Water Resources Society, New Delhi, 8–14.

Chalapati Rao, R. V., 1972. Geology and foundation treatment of the Nagarjunasagar dam, Andh Pradesh. In: Krishnaswamy, V. S. and Mehta, P. N. (Eds)., *Symp. Geol. and Eng. Problems of River Valley Projects,* Ind. Soc. Eng. Geol., Calcutta, pp. 19–32.

Chandra, A., 2004. Emerging trend in oil and gas exploration India — a bright future ahead, *Indian Jour. Geophys*., 25, 3–11.

Chandra, D., Singh, R. M. and Singh, M. P., *Text Book of Coal (Indian Context)*, Tata Book Agenc Varanasi, 2000, 402 p.

Chandra, P. C., Ramakrishna, A., Singh, S.C and Reddy, P.H.P., 2002. Geophysical approach delineate natural fracture zones in hard rocks, *In*. Thangarajan et-al (Eds) *Sustainable Development and Management of Groundwater Resources in Semi-arid Regions With Special Reference to Hard Rocks,* Oxford & IBH, New Delhi, pp 75–84.

Chandra, H., 1975. Landslides in Teesta valley, *Proc. Sem. Landslides and Toe Erosion Problem with Special Reference to Himalayan Region,* Ind. Soc. Eng. Geol., Calcutta, pp. 147–64.

Chansarkar, R. A., 1975. Geologic and geomorphic factors in landslide investigations. *Sem. Landslides and Toe Erosion Problems with Special Reference to Himalayan Region,* Ind. Soc. Eng. Geol., Calcutta, pp. 66–72.

Chandrasekharam, D. and Bundschuh, J., 2008. *Low-Enthalpy Geothermal Resources for Power Generation*, CRC Press (Taylor & Francis Group), London.

Chaplow, R., 1981. Investigations of hydroelectric projects. *Proc. Sem. Accelerated Hydroelectric Development in India,* Central Board of Irrigation and Power, New Delhi, 1: 9-16.

Chaturvedi, A. N., 1982, *Poplar Farming in U.P., India,* Ambassador Press, New Delhi, 42 p.

Chaturvedi, A. N, 1985, *Firewood Farming on Degraded Lands.* U.P. Forest Bulletin No. 50, Fore Department, Lucknow, 52 p.

Chaturvedi, M. C., 1981. Flood management—new concepts, technology and planning approach. *Proc. Intern. Conf. Flood Disasters,* Indian National Science Academy, New Delhi, pp. 1–15.

Chaturvedi, M. C, Revelle, R. and Srivastava, V. U., 1975. Ganga water machine-induced groundwat recharge. In: *Water for Human Needs, Proc. 2nd World Congr. on Water Resources,* 4: 195–203.

haturvedi, R. S., Bhattacharya, D. C, Kamal, P., Krishnamurthy, J. and Sunder Raman, N.,1983. Integrating remote sensing techniques in groundwater exploration—a typical case study from Bundelkhand region in U.P. *Proc. Nat. Sym. Remote Sensing in Development and Management of Water Resources,* N.R.S.A., Hyderabad, pp. 267–76.

haturvedi, R. S. and Mohan, R., 1983. Delineating flood-affected areas of south U.P. using satellite remote sensing techniques. *Proc. Nat. Symp. Remote Sensing in Development and Management of Water Resource,* Indian Institute of Photointerpretation and Remote Sensing, Dehradun, pp. 79–87.

haturvedi, R. S., Sharma, R. S. and Tangri, A. K., 1983. Evaluation of hydrological regimes in Saryu Project area of Gonda and Bahraich districts, U.P., using remote sensing techniques. *Proc. Nat. Symp. Remote Sensing in Development and Management of Water Resources,* Ind. Inst, of Photo-interpretation and Remote Sensing, Dehradun, pp. 113–17.

haturvedi, S. N., 1975. A geological appraisal of slope instability and proposed remedial measures at Khuni Nala bridge site along National Highway-IA, Doda district, Jammu and Kashmir. *Proc. Sem. Landslides and Toe Erosion Problems with Special Reference to Himalayan Region,* Indian Soc. Eng. Geol., Calcutta, pp. 175–86.

haturvedi, S. N. and Ashraf, Z., 1972. Geological problems encountered during the construction of the Rana Pratap Sagar main and saddle dams, Chambal Valley Hydel Scheme, Rajasthan. In: Krishnaswamy, V. S. and Mehta, P. N. (Eds.), *Symp. Geol. & Eng. Problems of River Valley Projects,* Ind. Soc. Eng. Geol., Calcutta, pp. 129–46.

haubey, V. D. and Tanwar, B. S., 1982. Groundwater development in alluvial plains of north India. In: B. B. S. Singhal (Ed.), *Engineering Geosciences,* Sarita Prakashan, Meerut, pp. 96–110.

haudhry, Pradeep, Bohra, N. K. and Choudhary, K.R., 2011. Conserving biodiversity of community forests and rangelands of a hot arid region of India, *Land Use Policy*, 28, 506–513.

haudhury, A. K. and Chattopadhyay, B., 1982. Geological setup and prediction of tunnelling problems along the Loktak Hydel project, Manipur. *Proc. 4th Intern. Engin. Geol. Congs.,* 4(2): 93–100.

haudhuri, N. and Chakrabarti, S. P., 1984. Pollution potential of the Yamuna basin. In: C. R. Krishna Murti (Ed.), *Facets of Environmental Problems,* Indian Nat. Sci. Acad., New Delhi, pp. 1–14.

hen, W. P. and Molnar, P., 1977. Seismic moments of major earthquakes and the average rate of slip in Central Asia. *Jour. Geophy. Res.,* 82: 2945–69.

ates, D. R. (Ed.), 1973. *Environmental Geomorphology and Landscape Conservation* (Benchmark Papers in Geology), Dawden, Hutchinson & Ross, Stroudsburg (Pennsylvania), 360 p.

Coates, D. R., 1976. *Geomorphology and Engineering,* Dawden, Hutchinson and Ross, Stroudsburg (Pennsylvania), 360 p.

Coates, D. R., 1981. *Environmental Geology,* John Wiley, New York, 701 p.

burn, A. W., Hughes, R. E., Illi, D., Nash, D. F. T. and Spence, R. T. S., 1984. The construction and vulnerability to earthquakes of some building types in the northern areas of Pakistan. In: K. J. Miller (Ed.), *Intern. Karakoram Expedition,* Vol. I, Cambridge Univ. Press, London, pp. 229–52.

ggin-Brown, J. and Dey, A. K., 1955. *India's Mineral Wealth,* Oxford Univ. Press, New Delhi, 689 p.

leman, A., 1982. A chain reaction of land misuse in Britain. *Intern. Jour. Env. Studies,* 19: 91–95.

leman, J. M., 1969. Brahmaputra River: channel processes and sedimentation. *Sedimentary Geology,* 3:129–239.

Cupp, David, 1982. Avalanches: Battling the jauggernaut. *Nat. Geogr. Magazine,* 162(3): 2822305.

Dabral, B. G., Nath, P. and Swarup, R., 1963. Some preliminary investigation on rainfall interception leaf litter. *Ind. Forester,* 89(2): 112-16.

Das, D. C, 1977. Soil conservation practices and erosion control in India—a case study. *Soil Bulletin of F.A.O., 33:* 11–50.

Das, S., 2006. *Groundwater Overexploitation and Importance of Water Management in India — Vision 2025*, The Indian Geological Congress, Roorkee, 24 p.

Das, D. C, Bali, Y. P. and Kaul, R. N., 1981. Soil conservation in multipurpose river valley catchmen problems, programme, approach and effectiveness. *Ind. Jour. Soil & Conservation,* 9(i): 6–20.

Das, D. C. and Singh, Shamsher, 1981. Small storage works for erosion control and catchme improvement: mini case studies. In: Morgan, R. P. C. (Ed.), *Soil Conservation Problems and Prospects,* John Wiley, New York, pp. 325–50.

Das, D. K., Raju, J. S. and Kidwai, A. L., 1983. Chloride budget in a sub-basin of Betwa river basin central India—a tentative approach for evaluation of groundwater recharge. *Proc. Sem. Assessment, Development and Management of Groundwater Resources,* Central Groundwater Board, Delhi, 11–21.

Das, S. M. and Pandey, Jyotsna, 1978. Lake pollution as evidenced by physical, biological and chemic parameters in the Nainital Lake, U.P. *Proc. Nat. Sem. Resources Development and Environment in the Himalayan Region,* Department of Science and Technology, New Delhi, pp. 502–512.

Dasgupta, K., 1982. *Groundwater Problems and Investigations in Western Rajasthan.* Misc. Pub. N 49, Geol. Surv. India, Calcutta, pp. 179–86.

Dasgupta, N. N., 1982. Problems of fuel for domestic sector. *Jour. Mines, Metals and Fuels,* 3 176–79.

Das Hazra, P. C. Krishnaswamy, V. S. and Swaminath, J., 1957. Regional geology and seismicity of t Bhakhra Dam project. *Proc. First Symp. Earthquake Engineering,* Roorkee University, Roorkee, pp. 171–92.

Dasmann, R. F., 1959. *Environmental Conservation,* John Wiley, New York, 375 p.

Datt, G. K., 1983. Mapping the environment. *Intern. Jour. Environ. Studies,* 19: 25-33. Davis, I., 198 The vulnerability and reduction of damage-risk in small houses subject to natural hazards. *In:* J. Miller (Ed.), *International Karakoram Expedition,* Vol. I, Cambridge Univ. Press, London, p 290–307.

Davis, S. N., Reiton, P. H. and Pestrong, P., 1976. *Geology and Our Environment,* McGraw-Hill, Ne York.

Deb, S., 1977. Possibilities of harnessing geothermal energy in high heat-flow zones of Peninsular a Extra-peninsular India. *Bull. Oil & Nat. Gas Comm.,* 14(1-2): 1–13.

Deming, H. G., 1975. *Water, the Foundation of Opportunity,* Oxford Univ. Press, New York, 942 p.

Dendy, E. E., 1968. Sedimentation in nation's reservoirs. *Jour. Soil and Water Conservation,* 23(4 135–37.

Deng, Q., Jiang, Pu, Jones, L. M. and Molnar, P., 1981. A preliminary analysis of reported changes groundwater and anomalous animal behaviour before the 4th Feb., 1975 Haicheng earthquake. I *Earthquake Prediction, An International Review,* Amer. Geophy. Union, Washington, pp. 543–65 odhar, N. S. and Nath, K. J., 1981. Water pollution and its control in India. In: A. K. Sharma and Archana Sharma (Eds), *Impact of Development of Science and Technology on Environment,* Ind. Sci. Cong. Assoc, Calcutta, pp. 152–61.

sai, B. I. and Gupta, S. K., 1971. Artificial Recharge Experiment for Underground Storage of Water, based on Siphon Principle, Report Phy. Res. Lab., Ahmadabad, 83 p.

sh Bandhu, 1981. Environmental situation in India. In: Desh Bandhu (Ed.), *Environmental Management,* Indian Environmental Society, New Delhi, pp. 29–43.

sikachar, S.V., 1984. Oil and gass deposits in the deltaic sequence of sediments in Cambay Basin and Upper Assam Valley, *Jour. Geol. Soc. India*, 25, 200–220.

twyler, T. R. (Ed.), 1971. *Man's Impact on Environment,* McGraw-Hill, New York, 731 p.

y, A. K., 1978. Utilization of wastes of mineral industry. *The New Sketch,* January, pp. 1–2.

y, Dilip, 1977. Mountain-biosphere reserves—the great Himalayan biosphere. In: H. C. Rieger (Ed.), *Dialogue—Himalayan Mountain Ecosystem,* Max Mueller Bhawan, New Delhi, pp. 102–108.

hanaraju, R., Roy, M., Madhuparna and Vasude, S.G., 1993. Uranium mineralization in the southwestern part of the Cuddapah basin: A petromineralogical and geochemical study, *Jour. Geol. Soc. India*, 42, 135–149.

har, O. N., Ghose, G. C. and Kulkarni, A. K., 1981. A catalogue of major and devastating floods in India and their space-time distribution. *Proc. Intern. Conf. Flood Hazards,* Ind. Nat. Sci. Acad., New Delhi, 1 : 147–61.

har, S. K., 1985. Snow clearance operations in Zoji La (Ladakh). *Proc. Sem. Construction of Roads in Hill Areas,* Ind. Road Congress, New Delhi, VI: 1–10.

hawan, G. and Patni, B., 2011. Stability of spillway cutslope at Dhauliganga Hydroelectric Project, *In: Proceedings Seminar Slope Stability: Challenges in Infrastructure Project*, Central Board of Irrigation and Power, New Delhi.

hawan, G. and Patni, B., 2009. Construction-stage geotechnical problem of Dhauliganga Dam and their remedial measures, *Proceedings INDOROCK-2009*, New Delhi.

Dhruva Narayana, V. V., 1993. *Soil and Water Conservation Research in India*, Indian Council of Agricultural Research, New Delhi, 448 p.

bble, R. D., Nairn, I. A. and Neall, V. E., 1984. Volcanic hazards of North Islands, New Zealand—an overview. *Proc. Colloq. Earthquakes Geol. Hazards Predictions,* 27th Intern. Geol. Congr., Moscow, pp. 144–59.

x, H. M., 1981. *Environmental Pollution,* John Wiley, New York, p. 286.

xit, G. R., Mahadevan, T. N. and Kapoor, R. K., 1982. A noise pollution survey of Bombay. *Scavanger,* 12 (2): 20–25.

•bhal, D. P., Gergen, J. T. and Thayyen, R. J., 2004. Recession and morphogeometrical changes of Dokrani Glacier (1962–1995) Garhwal Himalaya, *Current Science*, 86, 692–701.

•ornkamp, J. C., Brunsden, D., Jones, O.K.C., Cooke, R. V. and Bush, P.R., 1979. Rapid geomorphological assessments for engineering. *Quat. Jour. Eng. Geol.,* 12(3): 189–204.

•uglas, I., 1981. Erosion risk assessment and land classification for soil conservation. In: R. P. C. Morgan (Ed.), *Soil Conservation Problems and Prospects,* John Wiley, New York, pp. 63–73.

Douglas, J., Sholto, Robert, A., Hart, J., Ranganathan, G. S., 1982. Designing a forest farm. *Forest Farming,* Natraj Publishers, Dehradun, 149 p.

Down, C. G. and Stocks, J., 1977. *Environmental Impacts of Mining,* Applied Science Pub., Londo 371 p.

Driscoll, E., 1981. The power of small dams. *Span,* August, pp. 15–17.

Dubos, Rene, 1972. *A God Within,* Charles Scribner & Sons, New York.

Dunne, T. and Leopold, L. B., 1978. *Water in Environmental Planning,* Freeman, SanFrancisco, 818 p.

Eckholm, E. P., 1976. *Losing Ground: Environmental Stress and World Food Prospects,* W. W. Norto New York, 223 p.

Eckholm, E., 1982. Coming down to earth. *Ecoforum,* 1(1): 1.

El-Hinnawi, E., 1981. *Environmental Impact of Production and Use of Energy,* Tycooly, Dublin, 322 p

El-Hinnawi, E. and Biswas, A. K. (Eds.), 1981. *Renewable Sources of Energy and the Environment,* Tycooly, Dublin, 219 p.

Evans, D. M., 1966. Man-made earthquakes in Denver. *Geotimes,* 10(9): 11–18.

Facca, G., 1980. *Future Geothermal Energy* (in a brochure distributed at the Intern. Geol. Congress he at Paris).

Fahim, H. M., 1981. *Dams, People and Development: The Aswan High Dam Case,* Pergamon, Ne York, 187 p.

Flawn, P. T., 1970. *Environmental Geology,* Harper and Row, New York, 313 p.

Flogg, Ch. G., 1979. Geological causes of dam breaks and failures. *Bull. Intern. Assoc. Eng. Geol,* 20:231–32.

Fox, C. S., 1935. *A Comprehensive Treatise on Engineering Geology,* Govt. of India Press, Calcut pp. 224–43.

Framji, K. K. and Krishna, H., 1981. Feasibility investigations for hydroelectric projects—perspectiv and suggestions. *Proc. Sem. Accelerated Hydroelectric Development in India,* Central Board Irrigation and Power, New Delhi, 1: 75–82.

Gadgil, Madhav, 1980. Wildlife resources of India. *Golden Jubilee Commemoration Volume,* Indian National Science Academy, New Delhi, pp. 1–28.

Gadgil, Madhav and Meher-Homji, V. M., 1982. Biological diversity. In: *State of Environment—Some Aspects,* National Committee on Environmental Planning, Department of Environment, New Del Chapter 5, pp. 1–33.

Galay, V.J., Okaji, T. and Nishino, K., 1995. Erosion from the Kulekhani watershed, Nepal during t July 1993 rainstorm, *In*: Schreier, H., Shah, P.B. and Brown, S. (Eds) *Challenges in Mountain Resource Management in Nepal*, ICIMOD, Kathmandu, 13–24.

Gandhi, Indira P., 1981. Inaugural address. In: A. K. Sharma and Archana Sharma (Eds.), *Impact of Development of Science and Technology on Environment,* Indian Sci. Cong. Assoc, Calcutta, pp. 1–4.

Gangopadhyay, B., 1985. A case history of Kaliyasaur landslide at km 147 on Rishikesh–Joshima road. *Proc. Sem. Construction of Roads in Hill Areas,* Ind. Road Congress, New Delhi, V: 21–32

Ganguli, A. K., 1983. Rationale of a project for development of an experimental urban domestic fu forest in Greater Bombay, *Scavenger,* 13(1): 2–8.

ansser, A., 1964. The Alps and the Himalayas. *Rep. 22nd Intern. Geol. Congr.,* 11: 387-99.

aur, V. K., 1984. Earthquake risk to Tehri Dam (unpublished Report), National Geophy. Res. Inst., Hyderabad.

hassemi, F., Jakeman, A.J. and Nix, H.A., 1995. *Salinization of Land and Water Resources*, Centre for Resources and Evironmental Studies, The Australian National University, Canberra, 335 p.

hildyal, B. P., 1981. Soil of the Garhwal and Kumaun Himalaya. In: J. S. Lal and A. D. Moddie (Eds.), *The Himalaya,* Oxford Univ. Press, New Delhi, pp. 120–37.

hose, A. K., 1982a. Environmental implications of rock blasting in mines and quarries. *Jour. Mines, Metals & Fuels,* Special Issue on Advances in Explosives—Usages and Blasting Techniques (January), pp. 33–39.

Ghose, A. K., 1982b. Environmental control and the mining industry. *Jour. Mines, Metals and Fuels,* 30: 167.

hose, N. C. and Sharma, C. B., 1983 Pollution load of drainwater entering the river Ganga at Patna, Bihar. From: Ph.D. Thesis of Sharma, Patna Univ., Patna.

hosh, D. K., 1978. Role of geomorphology in project planning of control area. *Proc. Symp. Morphology and Evaluation of Landforms,* pp. 153–67.

hosh, D. K., 1980a. Weathering of granites and their influence on dam foundation in central India. In: Sharma, H. S. (Ed.), *Perspectives in Geomorphology,* Concept Pub. Co., New Delhi, 3: 165–78.

hosh, D. K., 1980b. Geological influence of landslides and other mass-movements in the planning of Baira-Baledh-Siul link project, H. P., India. *Proc. Intern. Conf. Landslides,* New Delhi, pp. 19–23.

hosh, R. and Bathija, T.S., 2001. Need for speedy water resources development in India, *Ind. Jour. Power & River Valley Development*, Golden Jubilee Number, 73–77.

hosh, R. C, Subba Rao, B. K. and Ramola, B. C, 1980. Interception studies in 'Sal' *(Shorea robusta)* appice forest. *Ind. Forester,* 106: 513–14.

ymph, L. M., 1973. Summary: Sedimentation of reservoirs. In: Ackermann, W. E., White, G. F., and Worthington, E. W. (Eds.), *Man-made Lakes: Their Problems and Environmental Effects,* Amer. Geophy. Union, Washington, pp. 342–48.

•el, D. K., 1983. Pollution of groundwater by potassium in U.P. *Proc. Sem. Assessment, Development and Management of Groundwater Resources,* Central Groundwater Board, New Delhi, 2: 269–79.

•odbread, S. L. and Kuehl, S. A., 1998. Floodplain processes in Bengal Basin and storage of Ganges–Brahmaputra river sediments: An accretion study using 137Cs and 210 Pg geochronology, *Sedimentary Geology*, 121, 239–258.

•swami, B. N., Venugopal, V., Sengupta, P., Madhusudan, M.S. and Xavier, P.K., 2006. Increasing trend of extreme events over India in a warming environment, *Science*, 314(5804), 1442–1445.

•swami, D. C., 1985. Brahmaputra river in Assam: Physiography, basin denudation and channel aggradation, *Water Resource Research*, 27, 959–978.

•swami, D. C., 1998. Fluvial regime and flood hydrology of the Brahmaputra River, Assam, *In*: V.S. Kale (Ed) *Flood Studies in India*, Geological Society of India, Bangalore, 53–76.

•swami, S., 1998. Assam's dire need, *In*: *Souvenir National Workshop on Geodynamics of Northeastern India*, Dibrugarh University, Dibrugarh, pp. 36–41.

Govindha Raj, K., Kumar, K.V., Mishra, R. and Mukhtar, M.A., 2012. Recession of Milam Glaci Kumaun Himalaya, *Curr. Sci.*, 102, 1351–1352.

Grant, Rachel, A. and Halliday, Tim, 2010. Predicting unpredictable; evidence of pre-seismic anticipato behaviour in the common toad, *Jour. Zoology*, DOI: 10.1111/j.1469–7988.2010.00700.X.

Guha, S. K., 1982. Review of the present status of research on reservoir induced seismicity and relat phenomena. *Proc. Fourth Intern, Congr. Engineering Geology,* Oxford & IBH, New Delhi, 7(4 119–32.

Guha, S. K. and Bhattacharya, U., 1984. Studies on prediction of seismic risk in north-eastern Ind (20°-30°N, 90°-100°E) (unpublished report), Geol. Dept., Poona Univ., Pune.

Gupta, M. C., Sharma, V. K., Gupta, L. C. and Tamini, B. K., 2001. *Manual on Natural Disaster Management in India*, Department of Agriculture & Cooperation, Government of India, New Del 204 p.

Gupta, Anil K., Tyagi, P. and Sehgal, V. K., 2011. Drought disaster challenges and mitigation in Ind Strategic appraisal, *Current Science*, 100, 1795–1806.

Gupta, R. P., Kumar, S. and Samadder, R. K., 2011. Palaeochannels as reservoirs for watersh management, *Proc. 4th Intern. Conf. Research in Watersheds*, Fairbanks, A.K., 156–161.

Gupta, G. P., 1975. Sediment production—status report on data collection and utilization. *Soil Conservation Digest,* 3(2): 10–21.

Gupta, H. K., 1984. Seismicity in the vicinity of dams on Himalayan rivers and the problem of reservo induced earthquakes. *Jour. Geol. Soc. Ind.,* 25(2): 85–93.

Gupta, H. K. and Rastogi, B. K., 1975. *Dams and Earthquakes,* Elsevier, Amsterdam, 229 p.

Gupta, H. K., Rastogi, B. K. and Narain, H., 1973. Earthquakes in the Koyna region and comm features of reservoir-associated seismicity. *In*: Ackerman, W. E. et al. (Eds.), *Man-made Lakes Their Problems and Management of Groundwater Resources,* Central Groundwater Board, Ne Delhi, 1: 283–289.

Gupta, H. K., Rastogi, B. K. and Ramakrishna Rao, C. V., 1979. Reservoir-induced seismicity. *Geophy. Res. Bull,* 17: 391–404.

Gupta, M. M., 1983. Distribution of fluoride in shallow aquifers, U.P. *Proc. Sem. Assess. Develop, and Management of Groundwater Resources,* Central Groundwater Board, New Delhi, 1: 283–289.

Gupta, P. K. and Pant, M. C., 1983. Microbenthos of Lake Nainital (U.P., India) with particular referen to pollution. *Water, Air and Soil Pollution,* 19: 397–405.

Gupta, R. K., 1977. Watershed—the right approach to ecosystem management in the Himalayas. In: C. Rieger (Ed.), *Dialogue: Himalayan Mountain Ecosystem,* Max Mueller Bhawan, New Delhi, p 142–52.

Gupta, R. K., 1979. *Plants for Environmental Conservation,* Bishan Singh Mahendrapal Sing Dehradun, 247 p.

Gupta, R. K., 1983. *The Living Himalayas,* Today and Tomorrow's, New Delhi, 379 p. Gupta, R. K. a Prakash, I., 1975. *Environmental Analysis of Thar Desert,* English Book Depot, Dehradun, 484 p.

Gupta, S. K., 1983. Use of paddy fields in groundwater recharge basins. *Proc. Sem. Assessment, Development and Management of Groundwater Resources,* Central Groundwater Board, Ne Delhi, 2: 282–96.

upta, S. K. and Dave, V. K. S., 1982. Geological investigation of the Bhagirathi blockade of August, 1978. In: B. B. S. Singhal (Ed.), *Engineering Geosciences,* Sarita Prakashan, Meerut, pp. 50–64.

urung, D. R., Bajracharya, S. R., Immerzeel, W. and Kulkarni, A.K., 2011. Understanding dynamics of snow and glaciers in the Hindukush–Himalayan region, *In*: *Earth Observation and Climate Change*, ICIMOD, Kathmandu, 17–21.

aigh, Martin J., 1981. Floods and erosion in north India. *Himalaya: Man and Nature,* 5(1-2): 21–25.

aigh, M. J., 1982a. Soil erosion in India. *Discussion Papers in Geography,* Oxford Polytechnic, Oxford, 17: 1–124.

aigh, M. J., 1982b. Road development and rural stresses in the Indian Himalaya. *Nordia,* 16: 135–40.

aigh, M. J., 1984a. Deforestation and disasters in northern India. *Land Use Policy,* July, pp. 187–98.

aigh, M. J., 1914b. Ravine erosion and reclamation in India. *Geoforum,* 15(4): 543–61.

aigh, M. J., 1984c. Landslide prediction and highway maintenance in the Lesser Himalaya, India. *Z. Geomorph N.F.,* 51: 17–37.

aldar, D. and Luhr, J.F., 2003. The Barren Island volcanism during 1991 and 1994–95: Eruptive style and lava petrology, *Mem Geol.Soc. India*, 52, 313–338.

anda, B. K., 1983a. Occurrence and distribution of trace elements in natural waters of U.P. *Proc. Sem. Assessment, Development and Management of Groundwater Resources,* Central Groundwater Board, New Delhi, 1: 293–304.

anda, B. K., 1983b. The effect of fertilizer use on groundwater quality in the phreatic zone with special reference to U.P. *Proc. Sem. Assessment, Development and Management of Groundwater Resources,* Central Groundwater Board, New Delhi, 2: 415–25.

ardin, Garret, 1968. The tragedy of the common, *Science*, 162, 1243–1248.

ardin, G., 1985. *Filters Against Folly: How to Survive Despite Economists, Ecologists and the Merely **Eloquent***, Viking Press, New York.

ansen, J., Johnson, D., Zacis, A., Lebe Deff, S., Lee, P., Rind, D. and Russel. G., 1981. Climate impact of increasing CO_2. *Science,* 213 (4511): 957–66.

ardin, G., 1968. The tragedy of commons. *Science,* 162: 1243–48.

are Ram, P. and Rama Rao, R. V., 1983. Saltwater-freshwater interface at Vishakhapatnam coastal aquifer. *Proc. Sem. Assessment, Development and Management of Groundwater Resources,* Central Groundwater Board, New Delhi, 2: 263–73.

arikrishna, P. and Sudhakar Reddy, C., 2012. Assessment of increasing threat of forest fires in Rajasthan using multitemporal remote sensing data 2005–2010), *Curr. Sci.*, 102, 1288–1297.

ays, W. H., 1981. *Facing Geologic and Hydrologic Hazards: Earth Science Considerations.* Geol. Surv. Professional Paper 1240 B, U.S. Govt. Printing Office, Washington, 108 p.

ays, W. H., 1982. *Preparation for and Responding to a Damaging Earthquake in the Eastern United States: A Workshop,* U.S. Geol. Surv., Washington, 179 p.

im, A. and Gansser, A., 1939. Central Himalaya, Geological Observations of the Swiss Expedition, 1936. *Mem. Soc. Helv. Sci. Nat.,* 73(1): 1–245.

Hillel, Daniel, 1982. *Negev—Land, Water and Life in Desert Environment,* Praeger Pub., New York, 162 p.

Holdgate, M. W., Kassas, M. and White, G. E., 1982. *The World Environment*—1972-1982, John Wile New York, 637 p.

Howard, A. D. and Remson, I., 1978, *Geology in Environmental Planning,* McGraw-Hill, New Yor 428 p.

Hukku, B. M., 1972. On the control of leakage and piping through buried bedrock channels at t Moosakhand Dam site, Varanasi district, U.P. In: Krishnaswamy, V. S. and Mehta, P. N. (Eds), *Proc. Symp.Geol. & Eng. Problems of River Valley Projects,* Ind. Soc. Eng. Geol., Calcutta, pp. 435–3

Hukku, B. M. and Narula, P. L., 1975. A geotechnical assessment of the stability of slopes at Khuni Na and landslide at Nashri on national highway, N.W. Himalaya. *Sent. Landslides and Toe Erosion Problems with Special Reference to Himalayan Region,* Ind. Soc. Eng. Geol., Calcutta, pp. 165–7

Hukku, B. M., Raju, K. C. C. and Sen Sarma, S. B., 1972. Problems of engineering geology in Ind *Rec. Geol. Surv. India,* 104(2): 31–44.

Hukku, B. M., Srivastava, A. K. and Jaitle, G. N., 1974. Evolution of lakes around Nainital and proble of hillside instability, *Himalayan Geol.,* 4: 516–31.

Hukku, B. M., Srivastava, A. K. and Jaitle, G. N., 1976. Measurement of slope movements in Naini area. *Ind. Jour. Eng. Geol,* 1: 457–66.

Ikeya, M. et al., 1997. Pulsed electric field before Kobe and Izu earthquakes from seismically induc anomalous animal behaviour, *Episodes*, 20, 253–260.

Issac, J. D. and Schmidt, W. R., 1980. Ocean energy: Forms and prospects. *Science,* 207 (442 265–73.

Ives, J. D., 1982, Mapping of mountain hazards. *Impact,* 32(1): 79–88.

Ives, J. D. and Messerli, B., 1981. Mountain hazards mapping in Nepal: Introduction to an appli mountain research project. *Mountain Research and Development,* 1(3-4): 223–30.

Ives, J. D. and Messerli, B., 1984. Stability and instability of mountain ecosystems: Lessons learnt a recommendations for the future. *Mountain Research and Development,* 4(1): 63–71.

Jackson, M. G., 1985. A strategy for improving the productivity of livestock in the hills of Uttar Prades In: Singh, J. S. (Ed.), *Environmental Regeneration in Himalaya,* Centr. Him. Env. Assoc, Nainital, pp. 130–54.

Jacobs, K. H., Armbruster, J., Seeber, L., Pennington, W. and Farahatulla, S., 1979. Tarbela reservo Pakistan—a region of compressional tectonics with reduced seismicity upon initial reservoir fillin *Bull. Seism. Soc. Amer.,* 69: 1175–92.

Jacobson, T. and Adams, R. M., 1971. Salt and Silt in ancient Mesopotamia. In: Detwyler, T. R. (Ed *Man's Impact on Environment,* McGraw-Hill, New York, pp. 383–94.

Jagdeesh, A. and Varshneya, N. C., 1982. Utilization of wind energy in Gujarat State—A case study. *Ind. Jour. Power and River Valley Development,* 32(1-2): 7-10. Jerry, Hans, 1980. *The Soil Resource: Origin and Behaviour,* Springer-Verlag, New York, 377 p.

Jain, Sharad, K., 2011. Population rise and growing water scarcity in India — Revised estimates a required initiatives, *Current Science*, 101, 271–276.

Jain, Sharad, K., 2012. Sustainable water management in India considering likely climate and oth changes, *Current Science*, 102, 177–188.

Jain, S. L., 2011. Ozone hole over poles: Current status, *Mausam*, 62, 577–584.

hwa, J. L., Dube, A. K., Singh, B. and Mithal, R. S., 1982. Evaluation of methods for tunnel support design in squeezing-rock conditions. *Proc. Fourth Intern. Congr. in Eng. Geol.,* Ind. Soc. Eng. Geol., Oxford & IBH, New Delhi, 5(2): 121–34.

a, L. B. and Jha, M., 1982. Fluoride pollution in India. *Intern. Jour. Env. Studies,* 19: 225–30.

a, V. K., 1982. Highway alignment survey in mountainous areas and its impact on environment—a case example from Okhaldhunga area, Nepal. *Proc. Fourth Intern. Congr. Engineering Geol.,*Oxford & IBH, New Delhi, 1:403–412.

ngran, A. G. and Balasundaram, M. S., 1969. Broad consideration in selection of dam sites: some case histories. *Bull. Nat. Inst. Sci. India,* 40: 10–24.

hnson, K., Olson, E. A. and Manandhar, S., 1982. Environmental knowledge and response to natural hazards in mountainous Nepal. *Mountain Research and Development,* 2(2): 175–88.

nes, L. and Molnar, P., 1976. Frequency of foreshocks. *Nature,* 262: 677–79.

nes, L. M., 1995. *Pulling Down Roots in Earthquake Country*, Southern California Earthquake Centre, Los Angeles, 28 p.

shi, D. D., 2001. Geoenvironmental problems of the glacial belt of Bhutan High Himalaya and remedial measures: A case study of Raphstrong Lake in Lunana area, *Jour. Geol. Soc. Ind.*, 57, 65–76.

shi, V. K., 1981. Management of major hydroelectric projects. *Proc. Sem. Accelerated Hydroelectric Development in India,* Central Board of Irrigation and Power, New Delhi, 1: 60–66.

dson, S., 1968. Erosion of the land or what is happening to our continent? *Amer. Scientist,* 56: 356.

yal, G. P., Dadhwal, K. S., Katiyar, V. S., Joshi, P. and Arya, R. K., 1998. Mined area rehabilitation in Himalaya: Sahasradhara experience, *In*: P.B. Saxena (Ed) *Himalayan Environment and Sustainable Development*, Surya Intern. Publication, Dehradun, 254–263.

adkadi, D. G., 2006. Problems encountered during construction of head-race tunnel of Dulhasti Hydroelectric Power Project, *Proceedings Water and Energy International*, Central Board of Power & Irrigation, New Delhi.

adur, Sandesh and Bawa, Kamal, 2005. *Sahyadris: India's Western Ghats — a Vanishing Heritage*, ATREE, Bangalore, 240 p.

aila, K. L., Gaur, V. K. and Narain, H., 1972. Quantitative seismicity map of India. *Bull. Seism. Soc. Am.,* 62: 1119–32.

aila, K. L. and Sarkar, D., 1978. Atlas of isoseismal maps of major earthquakes in India. *Geophysical Research Bulletin,* 16(4): 233–67.

akkar, Y. P. and Bhatnagar, N. C, 1983. Effect of electroplating wastes on groundwater in Ludhiana, Punjab. *Proc. Sem. Assessment, Development and Management of Groundwater Resources,* Central Groundwater Board, New Delhi, 2: 435–46.

ale, Vishwas S., 1998. Monsoon floods in India: A hydrogeomorphic perspective, *In*: V.S. Kale (Ed) *Flood Studies in India*, Geological Society of India, Bangalore, 229–256.

ale, Vishwas S., 1999. Long-period fluctuations in monsoonal floods in the Deccan Peninsula, India, *Jour. Geol. Soc. Ind.*, 53, 5–15.

ale, Vishwas, S., 2011. On the link between extreme floods and excess monsoon epochs in South Asia, *Climate Dynamics*, DOI 10.1007/s/0032-011-1251-6, available online 2 December 2011.

Kalra, I. D., 1973. Drainage problems in roads at high altitudes in western Himalayas. *Ind. Road Congress Jour.,* 35(2): 249–68.

Kalubarme, M. H. and Sahai, B., 1983. Remote sensing of waterlogged and salt-affected soils in t Mahi Command area. *Proc. Nat. Sym. Remote Sensing in Develop, and Management of Water Resources,* Ind. Soc. Photo. Int. and Remote Sensing, Dehradun, pp. 182–190.

Kamble, V. B., 2007. Living beyond our means — but how long? *Dreams 2047*, Vigyan Prasar, 10(3 35 and 31.

Karanth, K. R., 1997. On sustainable yield of infiltration gallery in valley fills of Peninsular India, *Jour. Applied Hydrology*, 10, 86–98.

Karamchandani, K.P., 1981. *Extension Components Social Forestry, Gujarat,* Forest Dept., Gujar State, Ahmadabad, 32 p.

Kasbekar, M. S., 1981. Noise pollution. *Scavenger,* 11(3): 15–18.

Kattlemann, R., 1987. Uncertainty in assessing Himalayan water resources, *Mountain Research & Development*, 7, 279–286.

Kaul, R. N. and Das, D. C, 1982. Soil and water management. In: *State of the Environment—Some Aspects,* Nat. Comm. Plann., Dept. of Environment, New Delhi, Chapter 2, pp. 1–16.

Kaul, V. and Pandit, A. K., 1979. Is the Dal Lake dying? *Science Reporter,* 16(11).

Kaul, V. and Raina, R., 1982. The interdisciplinary approach to save the dying Dal Lake in Kashm *In*: J. C. Bhatt and Desh Bandhu (Eds.), *First Nat. Env. Congr.,* (Abstracts) Ind. Env. Assoc, Ne Delhi, 81 p.

Kaushal, R. C, Dayal, S. K. N. and Misra, P. R., 1975. Study on the effect of closure with afforestati and gully control measures on runoff and peak discharge in Chandigarh. *Annual Report of Central Soil and Water Conservation Research and Training Institute,* Dehradun, pp. 57–59.

Kayastha, S. L., 2000. Water: Sustainable management of a vital resource, *Vsudha,* 3 (Association f Geographical Studies, Delhi, 36 p.).

Kayastha, S. L., 1981. An appraisal of water resources of India and need for national water polic *Geogr. Jour.,* 5-6: 563–72.

Kayastha, S. L. and Kumra, V. K., 1980. Noise as a factor in environmental pollution of Kanpur ci *Philippine Geogr. Jour.,* 24(3): 137–42.

Keilis-Borok, V. I., Kronvod, T. L. and Molchen, G. M., 1984. Seismical hazard for largest cities of t world. In: S. A. Fedov. (Ed.), *Earthquakes and Geological Hazard Prediction,* Intern. Geol. Cong. Colloquium, Moscow, pp. 12–34.

Keller, E. A., 1976. *Environmental Geology,* Bell and Howell, Columbus (Ohio), 488 p.

Khan, A. A., Haque, S., Huq, S. M., Kibria, K. Q. and Haque, M. A., 2003. Evidence of bacterial activı in the release of arsenic – A case study from the Bengal Delta of Bangladesh, *Jour. Geol. Soc. Ind* 61, 209–214.

Khattri, K. N., 1995. Observed and theoretical acceleration response spectra in the Tehri regio Implication for seismic hazard in the regions, *Current Science*, 69, 161–171.

Khattri, K. N., 1983. Epoch-dependent probabilistic seismic hazard map of the Himalayan regio *Geoscience,* 4(2): 121–30.

hattri, K. N., Rogers, A. M. and Perkins, D. M., 1983. Estimates of seismic hazard in north-eastern India and neighbourhood. *Bull. Ind. Soc. Earthquake Tech.*: 20(1), 1–22.

hattri, K. N. and Tyagi, A. K., 1982. Seismicity pattern in the Himalayan plate boundary and identification of areas of high seismic potential. *Techtonophysics,* 96: 281–97.

hattri, K. N. and Wyss, M., 1978. Precursory variation of seismicity rate in the Assam area. India. *Geology,* 6: 685–88.

hoshoo, T. N., 1982. Energy from plants. Presidential Address, *69th Session of Ind. Sci. Congr.,* Ind. Sci. Congr. Assoc, Calcutta, 100 p.

hoshoo, T. N. and Ahmad, K. T., 1981. Air pollution and plants. In: A. K. Sharma and Archana Sharma (Eds.), *Impact of the Development of Science and Technology on Environment,* Ind. Sci. Cong. Assoc, Calcutta, pp. 78–94.

hybri, M. L., 1973. Ravine reclamation in Kota district, Rajasthan. *Soil Conservation Digest,* 2(1): 57–60.

enholz, H., Hafner, H. and Schneider, G., 1984. Stability, instability and conditional instability: mountain ecosystem concepts based on a field survey of the Kakani area in the Middle Hills of Nepal. *Mountain Research and Development,* 4(1): 63–77.

ernholz, H., Hafner, H., Schneider, G. and Tamrakar, R., 1983. Mountain hazards mapping in Nepal's Middle Mountains with maps of land use and geomorphic damages (Kathmandu-Kakani area). *Mountain Research and Development,* 3(3): 195–220.

ersch, C. A., 1965. The Vaiont Reservoir disaster. *Min. Information Service,* 18(7): 129–34.

mothi, M. M. and Juyal, N., 1996. Environmental impact assessment of a few selected watersheds of the Chamoli district (Central Himalaya) using remotely sensed data, *Intern. Jour. Remote Sensing,* 17, 1391–1405.

•llmannsperger, F., 1977. Change in microclimate and biotopes leading to erosion. In: H.C. Rieger (Ed.), *Dialogue: Himalayan Mountain Ecosystem,* Max Mueller Bhawan, New Delhi, pp. 76–80.

•teswaram, P., 1974. Climate and meteorology of humid tropic area. In: *Natural Resources of Humid Tropical Asia,* UNESCO, Paris, pp. 27–85.

•vda, V. A., 1983. Loss of productive land due to salinization. *Ambio,* 12(2): 91–93.

ishna, Brahman, N. and Kanungo, D. N., 1976. Inference of granitic batholiths by gravity studies in South India. *Jour. Geol Soc. Ind.,* 17: 45–53.

ishnamurthy, K., 1980. Potential role of intense ionizing radiation sources in municipal sludge management and environmental protection. In: B. Patel (Ed.), *Management of Environment,* Wiley Eastern, New Delhi, pp. 149–69.

ishnaswamy, V. S. and Jain, M. S., 1975. A review of some of the major landslides in northern and northwestern Himalaya. *Proc. Sem. Landslides and Toe Erosion Problem with Special Reference to Himalayan Region,* Indian Soc. Eng. Geol., Calcutta, pp. 3–40.

ishnaswamy, V. S. and Shanker, R., 1982. Scope of development, exploration and preliminary assessment of geothermal resource potentials of India. *Rec. Geol. Surv. India,* 111(2): 17–38.

ishnaswamy, V. S., Sundaram, R. K., Raju, K. C. C, Seshagiri, D. N. and Agrawal, A. N., 1982. Geotechnical problems of Sipra Dam. *Ind. Jour. Eng. Geol.,* 11: 37–54.

Krynine, D. P. and Judd, W. R., 1957. *Principles of Engineering Geology and Geotechnics,* McGraw Hill, New York., 730 p.

Kukla, G. and Gavin, J., 1981. Summer ice and carbon dioxide. *Science,* 214: 497–503.

Kulkarni, V. N., 1982. Dam foundation and faults. *Proc. Sem. Accelerated Hydroelectric Development in India,* Central Board of Irrigation and Power, New Delhi, 39: 99–123.

Kulkarni, A. V., 2011. *In*: *Snow and Glaciers of the Himalaya*, Space Applications Centre, Ahmadabad.

Kumar, T., 2009. Energy security of India through optimum development of fossil fuels and alternati energy resources, *Jour. Indian Geol. Congress*, 1, 41–61.

Kumar, Adarsh, 1983. Pollution of groundwater by nitrates in U.P. *Proc. Sem. Assessment, Development and Management of Groundwater Resources,* Central Groundwater Board, New Delhi, 2: 427–3

Kumar, G. and Shome, S. K., 1970. Alaknanda tragedy—a geotechnical review. In: *Proc. River Valley Projects in North India,* Roorkee University and Society for Engineering Geology, Calcutta.

Kumar, R., Saxena, N. C. and Singh, B., 1975. Measurement of horizontal and vertical movements the surface caused by mining excavation. *Jour. Eng. Geol.,* 7: 429–36.

Lagwankar, V. G., Garde, A. K. and Patil, K. D., 1994. Trends in reservoir sedimentation in India a abroad. *In*: *Proc. Workshop on Reservoir Sedimentation*, Irrigation Department, Government Karnataka, Bangalore, pp 127–134.

Lal, Murari and 7 others, 2001. Future climate change: Implications for Indian summer monsoon and variability, *Current Science*, 81, 1196–1207.

Lamba, B. S., 1977. Effects of disturbed mountain ecosystem on the wild life in the western Himalay In: H. C. Rieger (Ed.), *Dialogue: Himalayan Mountain Ecosystem,* Max Mueller Bhawan, Ne Delhi, pp. 91–101.

Lawrence, R. D. and Yeats, R., 1979. Geological reconnaissance of the Chaman Fault in Pakistan. I A. Farah and K. A. Dejong (Eds.), *Geodynamics of Pakistan,* Geol. Surv. Pakistan, Quetta, pp. 351–58.

Legget, R. F., 1962. *Geology and Engineering,* McGraw-Hill, New York, 884 p.

Leopold, Aldo, 1941. Lakes in relation to terrestrial life pattern. In: *Proc. Hydrology,* Wisconsin Un Press, Madison, pp. 17–22.

Limaye, S. D., 1975. Utility of percolation tanks for water conservation in arid and semiarid regions. I *Water for Human Needs,* Ind. Comm. for I.W.R.A., New Delhi, 4: 205–208.

Liverson, David, 1980. *Geology and Urban Environment,* Oxford University Press, New York, 190 p

Lucas, B. A. and Friedman, S. (Eds.), 1983. *Technology Choice and Change in Developing Countries,* Tycooly, Dublin.

Lunkad, S. K. and Sharma, Anita, 2010. Combating negative impact of green revolution on groundwat soil and land in Haryana, *In*: B.S. Paliwal (Ed) *Global Groundwater Resources and Management* Scientific Publishers, Jodhpur, 511–528.

Madan, A. S., 1977. Various methods of avalanche control and their suitability for Indian condition *Proc. Intern. Workshop Snow & Avalanches,* Snow & Avalanche Studies Establishment, Mana pp. 99–116.

Madhavan, N. and Subramanian, V., 2000. Sulphide mining as a source of arsenic in the environmen *Current Science*, 78, 702–708.

adhavan, K., 1982. Tunnelling in rocks with high in-situ stresses. *Proc. Fourth Intern. Congr. Engin. Geol,* Oxford & IBH, New Delhi, 5(2): 79–84.

adson, P. H. and Natarajan, A., 2011. Challenges and prospects of wind energy to attain 20% grid penetration by 2020 in India, *Current Science*, 101, 35–42.

aingnet, M., 1983. Tentative megamorphological study of the Sahara. In: Garden, R. and Scoging, H. (Eds.), *Megageomorphology,* Clarendon, Oxford, pp. 113–33.

aithani, P. B., Massive uranium deposits found in Andhra Pradesh, *The Hindu*, March 20, 2011.

aithani, P. B., Taneja, P.C. and Rajendra, A., 1995. Sandstone-type uranium deposit at Phlangdiloin, West Khasi Hills, Meghalaya, *Jour. Atomic Minerals*, 3, 45–60.

ajumdar, S. K., Basu, B., Shivashankar, J., Arunachalam, A. and Rajagaraju, M. K., 1995 Palakollu–Pasarlapudi petroleum system, Krishna–Godavari Basin, India, *In*: *Proceedings Petrotech-5, Technology Trends in Petroleum Industry*, B.R. Publishing Corporation, New Delhi, pp. 495–508.

ajumdar, Kaushik, 2004. A study of fluctuation in radon concentration behaviour as an earthquake precursor, *Current Science*, 86, 1288–1292.

andwal, N. K., 1975. Geotechnical evaluation of some of the problems of the Ramganga project, Kalagarh, Garhwal District, U.P. *Himalayan Geology,* 5: 365–385.

albara, B. D., 2009. Tunnelling experience through Reasi Thrust (MBF) Zone for Katra–Reasi railway line sector (J & K), *Jour. Engineering Geol.*, 36, 195–201.

all, R. K., Gupta, A., Singh, R., Singh, R. S. and Rathore, L. S., 2006. Water resources and climate change: An Indian perspective, *Current Science*, 90, 1610–1626.

ani, A., 1981. *Handbook of Solar Radiations: Data for India,* Allied Publishers, New Delhi, 498 p.

ani, A. and Moolay, D. A., 1983. *Wind Energy Data for India,* Allied Publishers, New Delhi.

ann, H. S. and Prakash, I., 1982. The desert habitat and desertification. In: *State of Environment—Some Aspects,* Department of Environment, New Delhi, Chapter III, pp. 1–49.

ann, H. S. and Prakash, I., 1983. *Halting the March, Ecodevelopment in the Thar,* Department of Environment, New Delhi, pp. 1–36.

pinga, J., 1984. Out in the cold voids of nature. *Ecoforum,* 9(1): 1 and 16.

arsh, W.M. and Dozier. J., 1981. *Landscape: An Introduction to Physical Geography,* Addison-Wesley, Reading, 637 p.

athur, H. N., Babu, R., Johri, P. and Singh, B., 1976. Effect of clear-felling and reforestation on runoff and peak rates in small watersheds. *Indian Forester,* 102: 219–26.

athur, O. P., 1983. Artificial recharge studies in the Ghaggar river basin, Haryana. *Proc. Sem. Assessment, Development and Management of Groundwater Resources,* Central Groundwater Board, New Delhi, 2:317–24.

ears, B., Jr., 1970. *Essentials of Geology,* Van Nostrand, New York, 420 p.

eher-Homji, V. M., 1982. Do forests control climate ? *Science Today,* Jan., pp. 37–42.

ehra, S. R. and Natarajan, T. K., 1966. *Handbook on Landslide Analysis and Correction,* Central Road Research Institute, New Delhi, 114 p.

ehrotra, R. C. and Joshi, M. N., 1982. Environmental degradation of Himalayan region due to exploitation of minerals. *Proc. National Seminar on Minerals and Ecology,* Indian School of Mines, Dhanbad, 3(13): 1–10.

Mehta, K. K., 1983. *Reclamation of Alkali Soils in India,* Oxford & IBH, New Delhi, 280 p.

Mehta, P. N. and Dua, K. J. S., 1972. The significance of the Krol and Nahan thrust zones to t planning of the Yamuna and Giri-Bata hydel schemes, *Dr. J. B. Auden Commemorative Volume,* Ind. Soc. Engineering Geology, Kolkata, pp. 305–316.

Mehta, P. N. and Pradhan, S. D., 1972. Geologic causes for mishaps and failures of engineeri structures, *Rec. Geol. Surv. India,* 104 (2): 85–96.

Mehta, P. N. and Sanganeria, J. S., 1972. Geotechnical considerations in the selection of undergrou power house site for the Yamuna Hydel Scheme, Stage II, Phase I, Dehradun district, U.P. *Proc. Symp. Geol. and Eng. Problems of River-Valley Projects,* Ind. Soc. Eng. Geol., Calcutta, pp. 33 48.

Mehta, P. N., Shome, S. K. and Mathur, S. K., 1972. Geological problems of the location of the pow house and surge-shaft sites for the Yamuna Hydel Project, Stage II, Phase II. *Proc. Symp. Geol. and Engin. Problems of River Valley Projects,* Ind. Soc. Eng. Geol., Calcutta, pp. 349–56.

Meigh, A. C. and Skipp, B. O., 1981. Site investigation—methodology and modern practices, *Proc. Sem. Accelerated Hydroelectric Development of India,* CBI & P Publication No. 150, I: 111–23.

Menard, H. W., 1963. Some rates of regional erosion. *Jour. Geol.,* 69: 154–61.

Menon, Geeta, 1983. Silence please. *Femina,* Feb., pp. 18–22.

Middlemiss, C. S., 1890. Geological sketch of Nainital with some remarks on the natural conditio governing the mountain slopes. *Rec. Geol. Surv. Ind.,* 21: 213–34.

Miller, G.T. and Armstrong, P., 1982. *Living in the Environment,* Wadsworth International, Belmo California, 498 p.

Miller, G. and Tyler, Jr. 2007. *People and Environment*, Cengage Learning, Singapore, 696 p.

Mishra, V. B., 2005. The Ganga at Varanasi and travail to stop her abuse, *Current Science*, 89, 75 763.

Mitra, Nandini, 2007. Vanishing islands, *Down To Earth*, January, pp 25–34.

Mishra, P. S., Ramchandran, B. and Chhibber, I. B., 1982. Protection of Kuteshwar limestone depos and its significance in mine planning, Bansagar Dam Project, Rewa-Jabalpur districts, M.P. *Proc. Fourth Intern. Congr. Eng. Geol,* Oxford & IBH, New Delhi, 2(1): 345–58.

Misra, R. C, 1985. Environmental problems created by exploration of mineral resources in the Himalay In: Singh, J. S. (Ed.), *Environmental Regeneration in Himalaya,* Cen. Him. Env. Assoc, Nainital, pp. 155–68.

Mithal, R. S., 1982. The physical impact of Ramganga Dam Project on its drainage basin and catchme areas. Report-3 to Council of Scientific and Industrial Research, New Delhi, 49 p.

Mitra, A. K., 1982. Chemical characteristics of surface waters at selected gauging stations in rive Godavari, Krishna and Tungbhadra. *Ind. Jour. Env. Health,* 24(2): 165–79.

Moddie, A. D., 1977. Environmental oriented development policy for hill regions. In: H.C. Rieger (Ed *Dialogue: Himalayan Mountain Ecosystem,* Max Mueller Bhawan, New Delhi, pp. 175–78.

Moddie, A. D., 1981. Himalayan environment. In: J. S. Lall and A. D. Moddie (Eds.), *The Himalaya,* Oxford Univ. Press, New Delhi, pp. 341–50.

Mohan, J. and Tiagi, S. S., 1982. Siltation of reservoirs in Ganga-Yamuna valley. *Proc. Fourth Intern. Cong. Eng. Geol.,* Oxford & IBH, New Delhi, 7: 3–10.

ohan Rao, N. and Rangachari, N., 1983. Snow-avalanches of Himalaya—an analysis. *Proc. First Nat. Symp. Seasonal Snow Cover,* Vigyan Bhawan, New Delhi.

olnar, P. and Tapponnier, P., 1975. Cenozoic tectonics of Asia: effects of a continental collision. *Science,* 189: 419–26.

ool, Pradeep, 2001. Glacial lakes and glacial lake outburst flood events in the Hindukush–Himalaya Region, *In*: *Mountain Flash Floods*, ICIMOD, Kathmandu, pp. 7–9.

onto, Geethanjali, 2012. Nuclear India — to be not to be, *Current Science*, 102, 973–990.

oss, R. P. and Morgan, W. B., 1981. *Fuelwood and Rural Energy, Production and Supply in the Humid Tropics,* Tycooly, Dublin, 224 p.

otwani, A. V., 1981. Summary of reports of committees formed by the Govt. of India, relating to hydroelectric projects. *Proc. Sem. Accelerated Hydroelectric Development in India,* Central Water and Power Commission, New Delhi, 1: 31–38.

oullin, M. M., 1982. Hydrology and international drinking water supply and sanitation decade. In: *Water for Human Consumption,* Tycooly, Dublin, pp. 317–37.

ukherjee, S., 1983. Gandak-Ghaghara interfluvial belt, a potential repository for groundwater. *Proc. Sem. Assessment, Development and Management of Groundwater Resources,* Central Groundwater Board, New Delhi, 2: 59–73.

unn, R. E., 1975. What is the problem? In: E. R. Munn (Ed.), *Environmental Impact Assessment: Principles and Procedures,* John Wiley, New York, pp. 1–19.

urthy, Y. K. and Srinivasa, B. S., 1972. Influence of geology on dam design. In: Krishnaswamy, V. S. and Mehta, P. N. (Eds.), *Symp. Geol. and Eng. Problems of River Valley Projects,* Ind. Soc. Eng. Geol., Calcutta, pp. 1–8.

ibo, Z., 1986. Cascade development and multipurpose utilization of small rivers, *Urja*, 19(3), 179–184.

aidu, Y. Somu, and Kavitha, C., 2012. Alarming acidic nature of reainwater on the industrial zone of Visakhapatnam and its implication, *Ind. Jour. Geophys. Union*, 16, 65–68.

ain, Huang, 1982. Engineering geological analysis method for the forecast of reservoir-induced seismicity. *Proc. 4th Intern. Eng. Geol. Congr.,* Oxford & IBH, New Delhi, 7: 141–46.

ainwal, H. C., Negi, B. D. S., Chaudhary, M., Sajwan, K. S. and Gaurav, A., 2008. Temporal changes in the rate of recession: Evidence from Satopanth and Bhagirath Kharak Glaciers, Uttarakhand, Using Total Station Survey, *Current Science*, 94, 653–660.

air, V. N., Singhal, B. B. S. and Parkash, B., 1983. Hydrochemistry of groundwater in Chorwad-Madhavpur area, Junagarh district, Gujarat. Abstract, *Sem. Assessment, Development and Management of Groundwater Resources,* Central Groundwater Board, New Delhi, 74 p.

arain, H. and Gupta, H., 1968. Koyna earthquake. *Nature,* 217(5134): 1138–39.

arasimha, R. and Shrinivasa, V., 1984. Specification of design wind loads in India. *Sadhana,* Ind. Acad. Sci., Bangalore, 7(3): 259–74.

arasimhan, T. N., 2011. Water and public engagement in Science, *Current Science*, 100, 1781–1782.

arayanan, Kottilil, 2004. India's choice of an energy policy and its impact on future economic growth, *Jour. Ind. Geophys. Union*, 8(2), 79–96.

arayanan, K.G., 2011. Potential of renewable energy in meeting future needs of electricity, *Curr. Sci.*, 102, 1088–1089.

Naruse, T., 1977. A morphological study of Punjab plains and the northern part of Thar desert in Ind In: H. Isida (Ed.), *Green Revolution in Punjab: Techniques, Background & Effects,* Centre Res. World Regional Geography, University of Hiroshima, 47–66.

Nath, S. K., 1981. Flood control in Brahmaputra Valley in Assam, India. *Intern. Conf. Flood Disaster,* Ind. Nat. Sci. Acad., New Delhi.

Natarajan, T. K. and Gupta, S. C., 1980. Techniques of erosion control for surficial landslides. *Proc. Intern. Symp. Landslides,* New Delhi, pp. 413–17.

Natarajan, T. K., Murthy, A. V. S. R., Chandra, D. and Naik, T. K., 1983. Geological aspects of landslid in Nilgiri hills, *Jour. Eng. Geol,* 12(2-3): 131–46.

Natarajan, T. K. and Murthy, A. V. S. R., 1985. Geotextiles in erosion control works for landslide corre tion. *Proc. Sem. Construction of Roads in Hill Areas,* Ind. Road Congress, New Delhi, 5: 45–53.

Nautiyal, J. C. and Babor, P. S., 1985. Forestry in the Himalayas. *Interdisciplinary Sci. Reviews,* 10(27–41.

Nautiyal, M. D., Kumar, A., Singh, B. P. and Saini, D. S., 2002. Impact of water quality on human heal with special reference to selenium toxicity (Preprint).

Nawani, P. C., Sanwal, R. and Khanduri, H. C., 1991. On disproval of river-bed dislocation at Teh Dam site in the light of recent exploration, *Jour. Engin. Geol.*, 20, 98–107.

Nayak, P. N., 2004. Earthquake predictions using geoelectrical precursors, *Proc. Nat. Acad. Sci .India* 74A, 191–217.

Nayak, S. and Mandal, M., 2012. Impact of landuse and land cover changes on temperature trends ov Western India, *Curr. Sci.*, 102, 1166–1173.

Negi, G. C. S. and Joshi, Varun, 1996. Geohydrology of spring in a mountain watershed, *Current Science*, 71, 772–775.

Negi, G. C. S., Joshi, V. and Kumar, K., 1998. Spring sanctuary development to meet household wat demand in the mountains: A call for action, *In*: *Research for Mountain Developments: Some Initiatives and Acknowledgements*, Gyanodaya Prakashan, Nainital, 25–46.

Negi, S. S., 1982. *Environmental Problems in the Himalaya,* Bishan Singh Mahendrapal Sing Dehradun, 188 p.

Nikolaev, N., 1974. Tectonic conditions favourable for causing earthquakes occurring in connecti with reservoir filling. *Eng. Geol.,* 8: 171–89.

Niyas, N. T., Srivastava, A. K. and Hatwar, H. R., 2009. Variability and trend in the cyclonic storms ov north Indian Ocean, *Met. Monograph,* No. 3, Cyclone warning – 3/2009.

Niyogi, D., 1970. Geological background of beach erosion at Digha, West Bengal. *Bull. Geol. Mining, Metallur. Soc. India,* 43: 1–36.

Odum, E. P., 1959. *Fundamentals of Ecology* (2nd ed), W.B. Saunders, Philadelphia, 546 p.

Ohri, M. L. and Ahuja, V. K., 1981. *Textbook of Irrigation Engineering* (2nd edn), India Pub. Hous Delhi, 296 p.

Olson, G. K., 1981. *Soil and the Environment,* Chapman & Hall, New York, 178 p.

Omvir Singh, 2012. Rainfall, runoff and sediment transfer from a forested Lesser Himalayan watersh in Himachal Pradesh, *Ind. Jour. Geophys. Union*, 15, 207–210.

i, B. R., 1984. Utilization of sludge gas for power generation through aerogas turbines. *Sadhana,* 7(3): 223–36. Ind. Acad. Sci., Bangalore.

kiser, C. C, Eaton, J. P., Healy, J. H. and Raleigh, C. B., 1969. Earthquake predictions and control. *Science,* 166(3912): 1467–74.

l, S. C. and Das, R. C, 1952. Construction of the Shillong-Silchar road, Assam. *Ind. Road Congress Jour.,* 14(2): 57–110.

liwal, H. V., 1985. Environmental aspects of lead-zinc mining in India. *Industry & Environment,* 8(i): 13–17.

liwal, B. S. and Paliwal, S. C., 2010. Depleting groundwater resources in the Great Thar Desert of India, *In*: B.S. Paliwal (Ed) *Global Groundwater Resources and Management*, Scientific Publishers, Jodhpur, 1–38.

liwal, B. S., Vyas, A., Kaushik, S., Trivedi, S. K., Rathor, P. S. and Sisodia, M. S., 2010. Interaction of the surface water with the groundwater in some cities of northwestern Rajasthan, *Global Groundwater Resources and Management*, Scientific Publishers, Jodhpur, pp. 497–510.

lmer, L., 1976. River management criteria for Oregon and Washington. In: D. R. Coates (Ed.), *Geomorphology and Engineering,* Dowden, Hutchinson and Ross, Stroudsburg (Pennsylvania), pp. 329–46.

ncholi, D. M., 1997. Engineering geology and seismotectonic investigation, Sardar Sarovar Project, Gujarat. *Miscellaneous Publication No. 63*, Geological Survey of India, Kolkata, 95–101.

ndey, A. N., Pathak, P. C. and Singh, J. S., 1983. Water sediment and nutrient movement in forested and non-forested catchments in Kumaun Himalaya. *Forest Ecology and Management,* 7: 19–29.

ndit, M. K., Sodhi, N. S., Koh, L. P., Bhaskar, A. and Brook, B. W., 2007. Unreported yet massive deforestation driving loss of endemic biodiversity in the Indian Himalaya, *Biodiversity and Conversation*, 16, 153–163.

ndit, R. N. and Chadha, D. K., 1983. Hydrology and mathematical model study of the saline areas of lower Ghaggar basin, Haryana, India. *Proc. Sem. Assessment, Development and Management of Groundwater Resources,* Central Groundwater Board, New Delhi, 2: 27–35.

nt, G., 1972. The significance of the epigenetic gorges to the planning of river valley projects in the Uttarakhand division, U.P., *Symp. Geol. and Eng. Problems of River Valley Projects,* Ind. Soc. Eng. Geol., Calcutta, pp. 439–460.

nt, G., Narula, G. L. and Shome, S. K., 1980. *Geotechnical Studies Relating to the Planning and Design of the Tehri Dam Project* (Unpublished Report to Geological Survey of India).

nt, M. C., Sharma, A. P. and Sharma, P. C, 1980, Evidence for the increased eutrophication of lake Nainital as a result of human interference. *Environmental Pollution,* 1: 149–61.

rizek, R. R., 1973. Impact of highways on the hydrogeologic environment. In: D. R. Coates (Ed.), *Environmental Geomorphology,* Dowden, Hutchinson and Ross, Stroudsburg (Pennsylvania), pp. 151–99.

tel, V. J., 1983. *A New Strategy for High Density Agroforestry,* Jivnajbhai Patel Agrocentre, Kardej (Gujarat), 104 p.

thak, B. D., 1982. Groundwater resources of U.P., India. In: B. B. S. Singhal (Ed.), *Engineering Geosciences,* Sarita Prakashan, Meerut, pp. 111–31.

Pathak, P. C, Pandey, A. N. and Singh, J. S., 1983. Overland flow, sediment output and nutrient loss fro certain forested sites in the central Himalaya, India. *Jour. Hydrology,* 71: 239–51.

Pathak, S., 1974. Role of forests in soil conservation with special reference to Ramganga watershe *Soil Conservation Digest,* 2(1): 44–47.

Patil, A. R., 1982. Significance and concept of critical triangles in the treatment of cavernous foundatio at the Kamthikheri Dam Project in Maharashtra. *Jour. Eng. Geol.,* 1(112): 25–35.

Patil, D. N., Bhonsale, V. N. Suka, S. K. and Powar, K.B., 1984. The Khardi earthquake of 198 *Current Science,* 53(15): 805.

Patnaik, N., 1978. Soil conservation and watershed management for flood retardation and prevention review. *Indian Jour. Soil Conservation,* 6(1): 43–55.

Patnaik, N., 1981. Role of soil conservation, afforestation and salinity control in mitigating floods. *Proc. Intern. Conf. Flood Disaster,* Ind. Nat. Sci. Acad., New Delhi (preprint).

Paul, Ajay, 2010. Evaluation and implications of seismic events in Garhwal-Kumaun region of Himalay *Jour. Geol. Soc. India*, 76, 414–418.

Paul, D. K., 2004. Public policies for hazard management: Appropriate structural designs and buildi codes, *In*: K.S.Valdiya (Ed) *Coping With Natural Hazards: Indian Context*, Orient Longma Hyderabad, 71–90.

Paul, S. K., Bartarya, S. K., Rautela P. and Mahajan, A. K, 2000. Catastrophic mass movement of 19 monsoons at Malpa in Kali Valley, Kumaun Himalaya, *Geomorphology*, 35, 169–180.

Pearce, P., 1983. Landfilling: a longterm option for hazardous waste disposal. *Industry & Environment,* UNEP, Special issue No. 41983, pp. 57–63.

Perla, R. I. and Martinelli M., 1978. *Avalanche Handbook,* U.S. Dept. of Agriculture, Forest Servic Washington, D.C., 254 p.

Petak, W. J. and Atkisson, A. D., 1982. *Natural Hazard Risk Assessment and Public Policy,* Springe Verlag, New York, 489 p.

Petrov, M. P., 1976. *Deserts of the World,* John Wiley, New York, 447 p.

Petrova-Jusjunas, L. P., 1982. Some new positions of the nature and mechanism of the landsli processes and their application of water-storage shores stability. *Proc. Fourth Intern. Congr. Eng. Geol.,* Oxford and IBH, New Delhi, 7(4-5): 83–96.

Phadtare, P. N. and Bradley, E., 1983. Optimum artificial recharge potential in the alluvial aquifers of t Mehsana areas, Gujarat, *In*: *Proc. Sem. Assessment Development and Management of Groundwater Resources*. Central Groundwater Board, New Delhi, 307–316.

Philips, G., Gupta, R. P. and Manickavasagam, R. M., 1984. Studies on some geoenvironmental aspec of Singrauli belt. *Sem. Env. Aspects Geology,* (Abstract) Centre Earth Sci. Studies, Trivandru pp. 54–55.

Pichamuthu, D.V., 2011. Note on the evolution of India's mineral policy and its impact on the miner industry, *Jour. Geol. Soc.* India, 78, 285–289.

Polland, J. F., 1969. Land subsidence in western United States. In: *Geologic Hazards and Public Problems,* U.S. Govt. Printing Office, Washington, pp. 77–96.

Prajapati, M. C., Misra, D. D. and Agrawal, M. C., 1975. Dry farming practices under normal a aberrant monsoon conditions in Agra region of Yamuna ravines. *Soil Conservation Digest,* 3(37–42.

akash, V., Pain, D. J., Cunningham, A. and 6 others (2003). Catastrophic collapse of Indian white-backed *Gyps bengalensis* and long-billed *Gyps indicus* vulture population, *Biological Conservation*, 109, 381–390.

akash, V. A. and Chadha, D. K., 1983. Genetic causes of salinization of groundwater in Haryana, India. *Bull. Ind. Geol. Assoc,* 16(1): 100.

ess, F. and Brace, W. F., 1966. Earthquake predictions, *Science,* 152(3729): 1575–84.

udhvi Raju, K. N. and Krishna Bhagavan, S. V. B., 1984. Human-induced geomorphic hazards in Visakhapatnam coastal strip, A.P. *Trans Inst. Ind. Geographers,* 6(1): 79–84.

rnachandra Rao, V., Mohan Rao, K. and Raju, D.S.N., 2000. Quaternary phosphorites from the continental margin off Chennai, SE India. *Jour. Sedimentary Research*, 70, 1197–1209.

yravaud, J-P, Davidar, P. and Laurance, W. F., 2010. Cryptic destruction of India's natural forests, *Conservation Letters*, 1–5.

asim, S. Z., 1980. Endless energy from the sea, *Science Today,* September, pp. 17–23.

asim, S. Z., 1981. Bonanza from the sea, *Science Today,* June, pp. 15–24.

asim, S. Z. and Sengupta, R., 1980. Present status of marine pollution in India. In: B. Patel (Ed.), *Management of Environment,* Wiley Eastern, New Delhi, pp. 310–29.

ghunath, B., Das, D. C. and Thomas, P. K., 1970. Some results of investigations on hydrology of the subwatershed in Nilgiris. LASH UNESCO Publication, 96(3): 118.

ghunath, H. M., 1982. *Ground Water,* Wiley Eastern, New Delhi, 450 p.

i, K. L., 2000. Coal mining environment and related geohazards: The Indian scenario, *In*: A.M. Pathan and S.S. Thigale (Eds) *Contrilutions To Environmental Geosciences*, Aravali Books, New Delhi, 221–228.

i, S. P., Valdiya, K. S. and Rawat, J. S., 1998. Management of water resources: Spring sancturies, *In*: K.S.Valdiya (Ed), *The Khulgad Project: An Experiment In Sustainable Development*, Gyanodaya Prakashan, Nainital, 41–60.

ina, V. K., 2011. *Himalayan Glaciers: A State-of-the-Art Review of Glacial Studies, Glacial Retreat and Climate Changes*, G.B. Pant Institute of Himalayan Environment, Kosi-Katarmal, 56 p.

ina, V. K., Kaul, M. K., Sangewar, C. V. and Mukherjee, B. P., 1982. Investigation of avalanche for engineering structures, upper Sind Valley, Jammu and Kashmir. *Proc. Fourth Intern. Congr. Eng. Geol.,* Oxford & IBH, New Delhi, 1: 351–59.

jawat, A. S., Sastry, C. V. S. and Narain, A, 1999. Application of pyramidal processing on high resolution IRS/IC data for tracing migration of the Saraswati River in parts of Thar Desert, *In*: B.P. Radhakrishna and S.S. Merh (Eds), *Vedic Saraswati: Evolutionary Histroy of a Lost River in Northwest India*, Geol. Society of India, Bangalore, 259–272.

jendra, C. P., Rajendra, K., Srinivasalu, S., Andrade, V., Aravazhi, P. and Sanwal, J., 2011. Geoarchaeological evidence of a Chola-period tsunami from an ancient port at Kaveripattinam on the SE cost of India, *Geoarchaeology: An International Journal*, 26, 867–887.

ju, D. V. and Raju, K. C. C., 1982. Coastal erosion in Kerala, India—a critical appraisal. *Proc. Fourth Intern. Congr. Eng. Geol.,* Oxford & IBH, New Delhi, 7: 213–30.

ma, 1971. An alternative to Ganga-Cauvery link. *Science Today,* Sept. 1971.

Ramanna, Raja, 1984. A 15-year programme for nuclear power in India. *Meghnad Saha Medal Lecture,* Indian National Sci. Acad., New Delhi, p. 34.

Ramchandra Rao, M. B. (Ed.), 1953. A compilation of papers on the Assam Earthquakes of Augu 1950. Central Board of Geophysics, Pub. No. 4, Govt. of India, 112 p.

Ramchandran, B. and Gangopadhyaya, S., 1972. Engineering geological features of soft-soil areas. *Rec. Geol. Surv. India,* 104(2): 65–84.

Ramchandran, B. and Ghosh, D. K., 1980. Geotechnical appraisal and forecasts relating to river-vall project development—success and shortfalls. *Rec. Geol. Surv. India,* 112(6): 1–6.

Ramkrishnan, P. S. and Tokay, O. P., 1978. Preliminary observations on the impact of 'jhum'(shiftin cultivation on the forested ecosystem. *Proc. National Seminar on Resources Development and Environment in the Himalayan Region,* Department of Science and Technology, New Del pp. 343–54.

Ramasamy, S. M., 2002. Time transgenive tectonics and its control over groundwater regions of Tam Nadu, India, *In*: Thangarajan, M. et al., (Eds) *Sustainable Development of Groundwater Resources in Semi-arid Region*, Oxford & IBH, New Delhi, pp 411–415.

Ramasamy, S. M. and Muthukumar, M., 2008. Geospacial modeling of geosystems and landslid mapping and mitigation, the Nilgiri Mountains, South India, *Indian Landslides*, 1, 45–54.

Ramalingeswara Rao, B., 2000. Historical seismicity and deformation rates in the Indian Peninsul Shield, *Jour. Seism*., 4, 247–258.

Ramamurthy, M. V., 2011. Earthquake prediction research and pre-disaster management strategy, *Inter. Jour. Earth Sciences & Engineering*, 4, 941–946.

Ramanathan, V., Gutzen, P. J., Kiehl, J. T., and Rosenfeld, P., 2001. *Science*, 294, 2119–2124.

Raman, M. V., Ramprasad, T., Kamesh Raju, K. A. and Desa, M., 2007. Occurrence of gas hydrat along the continental margins of India, particularly the Krishna–Godavari offshore basin, *Jour. Environ. Studies*, 64, 675–693.

Ramana, M. V., Thomas, D. G. and Varghese, S., 2001. Estimating nuclear waste production in Ind *Current Science*, 81, 1458–1462.

Rane, V. S. and Kulkarni, A. W., 1984. Realignment of lower Borghat road on Bombay-Pune Nation Highway No. 4., *Ind. Road Congress Jour.,* 45(3): 111–59.

Rao, D. G., Katyal J. C., Sinha, S. K., and Srinivas, V., 1995. Impacts of climate change on *sorghum* productivity in India: Simulation study. *In*: *Climate Change and Agriculture: Analysis of Potential International Aspects*, Amer. Soc. Agronomy, Madison, 325–337.

Rao, K. L., 1975. *India's Water Wealth,* Orient Longman, New Delhi, 255 p.

Rao, K. L. and Palta, B. R., 1973. Great man-made lake Bhakra, India. In: W. C. Akermann et al. (Eds *Man-Made Lakes: Their Problems and Environmental Effects,* Amer. Geophys. Union, Washingto pp. 170–85.

Rawat, Y. S., Pandey, U., Singh, J. S. and Singh, S. P., 1983. The U.P. Himalaya: Ecological per pectives. In: O. P. Singh (Ed.), *The Himalaya: Nature, Man and Culture,* Rajesh Pub., New Del pp. 47–66.

Rawat, J. S. and Rawat, M. S., 1994. Accelarated erosion and denudation in the Nanakosi watershe Central Himalaya, *Mountain Research and Development*, 14, 25–38.

wari, Nandini, 1983. Sanitary landfill in Delhi: A landscape study. *Indian Environment* (Dec. issue), Indian Environmental Society, New Delhi.

chardson, S. D. and Reynolds, J. M., 2000. An overview of glacial hazards in the Himalaya, *Quaternary International*, 65/66, 31–47.

eger, H. C., 1977. An approach to a dynamic ecosystem for a watershed. In: H.C. Rieger (Ed.), *Dialogue: Himalayan Mountain Ecosystem,* Max Mueller Bhawan, New Delhi, pp. 155–70.

Ritcher, G., 1978. *Plant Metabolism,* Croom Helm, London, p. 475.

Robinson, 1977. *Relationship Between Soil Erosion and Sediment Delivery.* I.A.H.S., File No. 122, 159 p.

the, J. P., 1973. Geophysics report. In: Ackerman, W. C. et al. (Eds.), *Man-Made Lakes: Their Problems and Environmental Effects,* Amer. Geophy. Union, Washington, pp. 441–54.

wan, L. C. and Lathram, E. H., 1980. Mineral exploration. In: B. S. Siegal and A. R. Gillespie (Eds.), *Remote Sensing in Geology,* John Wiley, New York, pp. 553–605.

y, S. C. and Ananthkrishnam, S., 1980. Mussoorie-Dehradun mining belt, U.P. In: *A Decade of Environmental Geoscientific Studies,* Geol. Surv. India, Calcutta, pp. 75–77.

uddle, K. and Manshard, W., 1981. *Renewable Natural Resources and the Environment,* Tycooly, Dublin, 396 p.

upakheti, M. and Nakajima, T., 2011. Satellite view of particulate pollution in the Hindukush-Himalayas, *In*: *Earth Observation and Climate change*, ICIMOD, Kathmandu, 13–16.

h, S.L., 2010. Shale Gas: An important future energy resource, *Dew Journal*, December, 84–86.

ha, A. K., Saha,, A. and Das, N., 1982. A statistical analysis of air pollution data from five sampling stations in southern Bengal. *Proc. Ind. Nat. Sci. Acad.,* 48A: 144–60.

Saha, A. K. et al., 1984. *Report of the Cyclone Review Committee,* Department of Science and Technology (Unpublished), New Delhi, Vol. 2, 480 p.

hai, A. K., 1999. A brief review of climate change: Causes, observations and modelling studies, *Proc. National Acad. Sci. India*, LXIX (I), 1–17.

hu, K. C. and Mukherjee, S., 1983. Effectivity of heavy metal pollution of water on the underlying sediments along Ulhas River course. *Proc. Silver Jubilee Seminar on Environmental Management,* I.I.T., Bombay.

in, Kalachand and Gupta, H. K., 2010. Gas hydrates — future potential source of energy in India, *In*: *Glimpses of Geoscience Research in India*, Indian National Science Academy, New Delhi, 2010, pp. 244–250.

ini, D. S., Singh, S. B., Sikka, V. M. and Bhatnagar, N. C., 1983. Waterlogging in Faridkot and Ferozpur districts of Punjab as surplus-water balance problem. *Proc. Sem. Assessment, Development and Management of Groundwater Resources,* Central Groundwater Board, New Delhi, 2: 215–25.

late, E. and Vose, P. B., 1983. Depletion of tropical rain forests. *Ambio,* 12(2): 67–71.

ndhu, T. S., 1973. Pollution of groundwater by manganese in U.P. *Proc. Sem. Assessment, Development and Management of Groundwater Resources,* Central Groundwater Board, New Delhi, 2: 311–14.

stry, C. A., 1984. Groundwater contamination by tanneries in Vaniyambadi, Tamil Nadu, India, *In*: C.R.Krishnamurti, (Ed.), *Facets of Environmental Problems,* Ind. Nat. Sci. Acad., New Delhi, pp. 15–20.

Sastry, G., Juyal, G. P. and Samra, J. S., 1995. Conservation measures for sustainable developme of degraded lands in Himalaya, with special reference to Doon Valley, *Himalayan Geology*, 6(2 47–51.

Sati, S. P., Sundriyal, Y.P., Rana, N. and Dangwal, Surekha, A., 2011. Recent landslides in Uttarakhan nature's fury or human folly, *Current Science*, 100, 1617–1620.

Satyapal, A., 1981. Mineral industry in India. *Proc. National Mineral Convention,* Federation of Indi Mineral Industries, New Delhi, pp. 8–18.

Schneider, W. J., 1970. *Hydrologic Implications of Solid Waste Disposal,* U.S. Geol. Surv. Circular 6 F., 10 p.

Schultz, J. R. and Cleaves, A. B., 1955. *Geology in Engineering,* John Wiley, New York, 592 p.

Seeber, L. and Armbruster, J., 1979. Seismicity of the Hazara arc in northern Pakistan: Decollement v basement faulting. In: A. Farah and K. A. De Jong (Eds.), *Geodynamics of Pakistan,* Geol. Surv. Pakistan, pp. 131–42.

Sengupta, R., 1981. Marine and coastal environment management. In: Desh Bandhu (Eds.), *Environmental Management,* Indian Environmental Society, New Delhi, pp. 67–80.

Sengupta, B., Bahuguna, R., Kumar, S., Singh, Rajendra and Kau, R., 1991. Sandstone type uraniu deposit at Domiasiat, West Khasi Hills, Meghalaya, Northeastern India, *Current Science,* 61, 46–47.

Sen Sharma, S. B. and Chowdhury, A. K., 1972. Geological studies during the planning and constructi of Jaldhaka hydel project. *Proc. Symp. Geol. and Engin. Problems of River Valley Projects,* Ind. Soc. Engineering Geologists, Calcutta, 357–376.

Seth, S. L., Das, D. C. and Gupta, G. P., 1982. *Floods in Arid and Semiarid Areas—Rajasthan.* National Flood Commission Report, Vol. I, Chapter 10.

Shafi, Mohammad S., 1982. Human settlement and urbanization. In: *State of Environment—Some Aspects,* National Committee on Environmental Planning, Department of Environment, New Del Chapter I, pp. 1–41.

Shah, B. A., 2010. Arsenic-contaminated groundwater in Holocene sediments from parts of Midd Ganga Plain, U.P., *Current Science*, 98, 1359–1365.

Shah, S. L., 1981. Ecological degradation and future of agriculture in Himalaya. *Presidential Address, 41st All India Agricultural Economics Conference,* Indian Society of Agricultural Economist, Ne Delhi, 22 p.

Shakya, S. K. and Singh, J. P., 2010. New drainage technologies for salt-affected waterlogged areas southwest Panjab, India, *Current Science*, 99, 204–212.

Shanker, Ravi, 1988. Heat flow map of India and discussion on its geological and economic significanc *Indian Minerals*, 42, 89–110.

Sharma, A. K., 1981. Presidential Address. In: A.K. Sharma and Archana Sharma (Ed.), *Impact of the Development of Science and Technology on Environment,* Ind. Sci. Congr. Assoc, Calcutta, pp. 5–3

Sharma, Arun Kumar, 1981. *Structural Study of Nainital with Special Reference to the Hillside Instability,* (Unpublished Ph.D. Thesis), Kumaun University, Nainital, 126 p.

Sharma, A. P., Jaiswal, S., Negi, V. and Pant, M. C., 1982. Photoplankton community analysis in t lakes of Kumaun Himalaya. *Acta Hydrobiol.,* 93(2): 173–93.

arma, G. G., 1958. Landslide investigation and treatment in Assam. *Ind. Road Congess Jour.,* 23(3): 319–84.

arma, H. S., 1980. *Ravine Erosion in India,* Concept Pub., New Delhi, 96 p.

arma, K.D., 2009. Groundwater management for food security, *Current Science*, 96, 1444–1447.

arma, K. D. and Yangani, S. N., 1982. Flash flood of July 1979, in the Luni basin—A rare event in the Indian desert. *Jour. Sci. Hydrologique,* 27: 385–97.

arma, R., Nayak, S. R. and Sahai, B., 1983. Study of sedimentation in Ukai Reservoir using landsat data. *Proc. Sem. Env. Maps and Atlases,* Nat. Atlas Org., Calcutta, pp. 2–9 (mimeo).

arma, S. S., 2004. Mitigating avalanche disasters in Indian Himalaya, *In*: K.S. Valdiya (Ed) *Coping with Natural Hazards: Indian Context*, Orient Longman, Hyderabad, 124–133.

arma, S. C, 1975. Landslips in the Deccan Traps with special reference to Western Ghat. *Proc. Sem. Landslides and Toe Erosion Problem with Special Reference to Himalayan Region,* Indian Soc. Eng. Geol., Calcutta, pp. 255–62.

arma, V. K. and Sanwal, Rajendra, 1998. Disastrous rock avalanche at Malpa on 18th August 1998, *Jour. Engineering Geol.*, 27, 71–74.

elialin, N. V., 1984. Earthquake sources, hazard and disaster. In: Fedov, S. A. (Ed.), *Earthquakes and Geological Hazard Prediction,* Colloquium Intern. Geol. Congr., Moscow, pp. 3–13.

ekar, A.V., 1999. Municipal solid waste management — The Indian perspective. *Jour. Ind. Assoc. Environ. Management*, 26, 100–199.

enoi, R. S. and Srinivasan, P. B. 1982. Significance of seismotectonics in reservoir projects on Narmada river. *Proc. Fourth Intern. Congr. Eng. Geol,* Oxford & IBH, New Delhi, 7(4): 243–56.

eshagiri, D. N., Upendra, R. Lakshmikanthan, C. B. and Subramanian, K., 1982. Landslide zonation in Nilgiri plateau, Tamil Nadu, India. *Proc. Fourth Intern. Congr. Eng. Geol,* Oxford & IBH, New Delhi 1(1): 379–90.

etye, S. R., 2011. Review of the book "*Research Design*" in *Current Science*, 101, 682–683.

ivakumar, A. R., 2007. World's largest rainwater harvesting project in Karnataka, *Current Science*, 92, 161–163.

ome, S. K. (Ed.), 1983. *Geotechnical Features of Major Dams in India,* Ind. Soc. Eng. Geol., Calcutta, 114 p.

bal, V. K., Raju, S. V., Mondal, M. E. A. and Nauroz Ahmad, M., 2006. CO_2 storage in coal mines, *Current Science*, 91, 405–406.

kka, A. K., 2002. Augmentation and management of groundwater resources: Participatory watershed management as a policy and strategy. *In*: Thangarajan et al. (Eds) *Sustainable Development and Management of Groundwater Resources in Semi-arid Region*, Oxford & IBH, New Delhi, pp. 563–969.

mpson, D. W., 1976. Seismicity changes associated with reservoir loading. *Eng. Geol,* 10: 123–50.

ngh, Ajit, 1985. Reinforced earth retaining walls. *Proc. Sem. Construction of Road in Hill Areas,* Ind. Road Congress, New Delhi, III-59–73.

ngh, D. D. and Gupta, H. K., 1980. Source dynamics of two great earthquakes of the Indian subcontinent: The Bihar-Nepal earthquakes of January 15, 1934 and the Quetta earthquakes of May 30, 1935. *Bull. Seism. Soc. Amer.,* 70(3): 757–73.

Singh, Gurucharan, 1981. Experience of flood disasters of countries prone to periodic floods. *Proc. Intern. Conf Flood Disaster,* Ind. Nat. Sci. Acad., New Delhi (mimeo).

Singh, Gurbachan, 1994. Agroforestry in salt-affected soils, *In*: *Proc. Eighth ISCO Conference*, Ne Delhi, 1466–1483.

Singh, G., Joshi, R. D., Chopra, S. K. and Singh, A. B., 1974. Late Quaternary history of vegetatio *Philo. Trans. Royal Society, London,* Ser. B, 267 (889): 467–501.

Singh, G., Ram Babu, Bhushan, L. S. and Abrol, I. P., 1992. Soil erosion rates in India, *Jour. Soil & Water Conservation*, 47, 97–99.

Singh, J. S., Pandey, Uma and Tiwari, A. K., 1984. Man and forest: A central Himalayan case stud *Ambior* 13(2): 80–87.

Singh, J. S., Pathak, P. C. and Pandey, A. N., 1983. A hypothesis to account for the major pathway to s loss from Himalaya. *Environmental Conservation,* 10(4): 343–45.

Singh, J. S. and Saxena, A. K., 1980. Analysis of forest grazing and vegetation in parts of Kuma Himalaya. *Ind. Jour. Range Management,* 1(1): 13–32.

Singh, J. S., Singh, S. P., Saxena, A. K. and Rawat, J. S., 1984. India's Silent Valley and its threaten rainforests ecosystems. *Environment Conservation,* 11: 223–33.

Singh, J. S. and Kushwaha, S. P. S., 2008. Forest biodiversity and its conservation in India, *International Forest Review*, 10, 292–304.

Singh, K., 1977. Some alternative strategies for optional land and other resources in the hills of U.P. *Ind. Jour. Agri. Econ.,* 32(3): 113–20.

Singh, K. D., 2008. Balancing fuelwood production and consumption in India, *International Forest Review*, 10, 190–200.

Singh, K. P. and Bhayana, N., 1982. Environmental effects of industrial effluents on the groundwat regime in Ludhiana area, Punjab, India. *First Nat. Env. Congr.,* Ind. Env. Assoc, New Del (Abstracts), p. 5.

Singh, K.P., 2000. Development of waterlogging and salinity in southwestern parts of Punjab sta India, *In*: A.M. Pathan and S.S. Thigale (Eds) *Contributions to Environmental Geoscience*, Aravali Books, New Delhi 97–106.

Singh, Neelam, 2002. Fig leaf for the business world, *Down to Earth*, March 15, 2002, 35–40.

Singh, O. P., Khan, T. M. A. and Rahman, M. S., 2001. Has the frequency of intense tropical cyclon increased in the north Indian Ocean?, *Current Science*, 80, 575–580.

Singh, P. and Mahajan, J. L., 1982. Some geotechnical considerations on tunnelling in Himalayas. *Proc. Fourth Intern. Congr. Eng. Geol,* Oxford & IBH, New Delhi, 5(2): 65–74.

Singh, P. P., 1984. Assessment of soil deterioration due to irrigation by Saryu Canal Project, *Annual Report MAB Workshop,* Department of Environment of Environment, New Delhi (mimeo).

Singh, Samar, 1981. Soil conservation and afforestation for moderating/mitigating floods. *Intern. Conf. on Flood Disasters,* Ind. Nat. Sci. Acad., New Delhi (mimeo).

Singh, Samar, 1982. *Soil Conservation Problem, Approach and Progress in India,* Soil and Wat Conservation Division, Ministry of Agriculture, New Delhi, 40 p.

Singh, Satyajit, 1993. Political economy of large dams: Some tentative assertions, *In*: Balakrishnan, (Ed) *Environmental Problems and Prospects in India*, Oxford & IBH, New Delhi, 195–232.

ngh, Tarlochan, 1995. Artificial recharge of groundwater in Panjab, India, *In*: A.I. Johnson and D.G. Payne (Eds) *Artificial Recharge of Groundwater*, Amer. Soc. Civil Engineers, New York, p. 831–838.

ngh, V. and Ghosh, D. R., 1975. A plan for development of irrigation in drought-prone areas in India. *In*: *Water for Human Needs,* Ind. Comm. for Water Resources, New Delhi, 4: 185–95.

nghal, B. B. S., 1963. Occurrence and geochemistry of groundwater in the coastal region of Midnapur, West Bengal, India. *Economic Geology,* 58: 419–33.

nghal, B. B. S., 1973. Some observations of occurrences, utilization and management of groundwater in the Deccan Trap areas of central Maharashtra. *Proc. Intern. Symp. Development of Groundwater Resources,* Madras, 3: 75–81.

nghal, H. S. S., 1982. Sedimentation in reservoirs with special reference to Matatila Reservoir. *Proc. Fourth Intern. Congr. Eng. Geol.,* Oxford & IBH, New Delhi, 7: 11–17.

nghota, J. S. and Aggarwal, K. G., 1979. Silting of Govindsagar. *Symp. Silting of Reservoirs,* Central Board of Irrigation and Power, New Delhi, 126: 23–34.

nha, A. K., Srivastava, V. C. and Nagar, M., 1982. Land degradation around Singrauli—An observation. *Proc. Fourth Intern. Congr. Eng. Geol.,* Oxford & IBH, New Delhi, 1: 57–68.

nha, B. P. C, 1983. A critical review of the groundwater resource evaluation methodologies and estimates in India. *Proc. Sem. Assessment, Development and Management of Groundwater Resources,* Central Groundwater Board, New Delhi, 2: 13–25.

nha, Rajiv and Friend, P. F., 1994. River systems and their sediment flux, Indo-Gangetic Plains, northern Bihar, India, *Sedimentology*, 41, 825–845.

nha-Roy, S. P., 1983. Management of groundwater resources in the coastal tract of West Bengal. *Proc. Sem. Assessment, Development and Management of Groundwater Resources,* Central Groundwater Board, New Delhi, 2: 275–86.

nvhal, H., Agrawal, P. N., King, G. C. P. and Gaur, V. K., 1973. Interpretation of measured movement at a Himalayan (Nahan) Thrust. *Geophy. Jour. Royal Astr. Soc,* 34: 203–210.

ndhi, T. N.; 1983. Pollution of natural waters by manganese in U.P. *Proc. Sem. Assessment, Development and Management of Groundwater Resources,* Central Groundwater Board, New Delhi, 2: 335–45.

ni, V., Sharma, D. C., Srivastava, K. S. and Sisodia, M. S., 1999. Hydrological, geophysical and isotope study to trace the course of the buried Saraswati River in Jaisalmer district, Rajasthan, *In*: B.S. Paliwal (Ed) *Geological Evolution of Northwestern India*, Scientific Publishers, Jodhpur, 305–311.

Snil, V. 1984. *The Bad Earth: Environmental Degradation in China,* Zed Press, London, 245 p.

nivasan, J., 2008. Climate change, greenhouse gas and aerosols, *Resonance*, 13, 1146–1155.

nivasan, P. B., Bansode, R. G. and Chandra, P., 1982. Evaluation of geological and seismotectonic parameters for a design, seismic coefficient for Navagam Dam, Gujarat. *Proc. Fourth Intern. Congr. Eng. Geol.,* Oxford & IBH, New Delhi, 8(6): 97–105.

Srinivasan, P. B., Chalapati Rao, R. V. and Bansode, R. G., 1982. Engineering geological problems in hard-rock areas. *Rec. Geo I. Surv. India,* 109(2): 49–64.

nivasan, P. B., Raju, K. C. C. and Raju, D. V., 1980. Geomorphology and its influence on the erosion of Kerala coast. *Bull. O.N.G.C.,* 17(1): 277–84.

Srivastava, A. K., 1972. Geological studies relating to the alignment of the Pandoh-Baggi tunnel, Bea Sutlej link project, H.P. *Sym. Geol. Eng. Problems of River Valley Projects,* Ind. Soc. Eng. Geol., Calcutta, pp. 317–26.

Srivastava, H. N. and Dube, R. K., 1982. Seismicity studies of some important dams in India. *Proc. Fourth Intern. Congr. Eng. Geol.,* Oxford & IBH, New Delhi, 8(4): 219–27.

Srivastava, P.C., 2009. *Hind Mahasagar kee Samudri Khanij Sampada*, Shilalekh, Delhi, 296.

Srivastava, K. N., Jaitle, G. N., and Waza, J. L., 1972. Geological factors controlling the location a design of Pandoh dam, Beas-Sutlej link project, Himachal Pradesh. *Symp. Geol. and Eng. Problems of River Valley Projects,* Ind. Soc. Eng. Geol., Calcutta, pp. 209–220.

Srivastava, L. S. and Basu, S., 1982. Seismic zoning of India. *Proc. Fourth Intern. Congr. Eng. Geol.,* Oxford & IBH, New Delhi, 8: 119–27.

Srivastava, V. K. and Roy, A. K., 1982. Seismotectonics and seismic risk study in and around Delhi. *Proc. Fourth Intern. Congr. Eng. Geol.,* Oxford & IBH, New Delhi, 8: 77–86.

Stille, H., Brantmark, J., Heiner, A. and Martna, J., 1998. Evaluation of squeezing experiences in t Uri Project, India, *In*: V.D. Choubey (Ed) *Hydropower Development in the Himalayas*, Oxford IBH, New Delhi, 281–304.

Stocklin, J., 1980. Geology of Nepal and its regional frame. *Jour. Geol. Soc, London,* 137: 1–34.

Strahler, A. N., 1981. *Physical Geology,* Harper & Row, New York, 612 p.

Strahler, A. N. and Strahler, A. H., 1973. *Environmental Geosciences: Interaction Between Natural Systems and Man,* Hamilton, Santa Barbara, 511 p.

Subramanian, S. K., 1978. Biogas system in Asia: A survey. In: Andrew, B., Leo, Pyle and Subramania S. K. (Eds.), *Biogas Technology in the Third World,* Third World Development Research Cent Ottawa, pp. 97–121.

Subramanian, V., 1978. Input by Indian rivers into the world oceans. *Proc. Ind. Acad. Sci.,* 87A: 7 88.

Subramanian, V., 1979. Chemical and suspended sediment characteristics of rivers of India. *Jour. Hydrology,* 44: 37–55.

Subramanian, V., 1993. Sediment load of Indian rivers, *Current Science*, 64, 928–930.

Subramanian, V., 2000. *Water: Quantity-Quality Perspective in South Asia*, Kingston Intern. Publishers, Surrey, 256 p.

Subramanian, V. and Ramanathan, A. L., 1996. Nature of sediment load in the Ganges–Brahmaput river system in India, In: J.D. Milliman and B.U. Haq (Eds) *Sea-level Rise and Coastal Subsidence* Kluwer, Acad. Pub., Amsterdam, 151–168.

Subramanian, V. and Dalavi, R. A., 1978. Some aspects of stream erosion in the Himalaya. *Him. Geol.,* 8: 822–34.

Sudradjat, A. and Tilling, R., 1984. Volcanic hazards in Indonesia: the 1982-1983 eruption Galunggung. *Episodes,* 7(2): 13–19.

Sukant, Roy, 1975. Jaldhaka hydroelectric project stage I, West Bengal. In: *Engineering Geology Case Histories* (Sp. Pub. 29), Geol. Surv. Ind., Calcutta, pp. 199–207.

Sukhatme, S.P., 2011. Meeting India's future needs of electricity through renewable energy resource *Current Science*, 101, 624–630.

ndaraiya, E. and Wakpanjar, V. N., 1996. Reservoir created in karstic environment, *In*: C.V.J. Varma and A.R.G. Rao (Eds), *Proc. Symp. Earthen Dams*, Central Board of Irrigation & Power, New Delhi 349–362.

ndaresan, B. B., 1980. *National Air Quality Monitoring Network, 1978-1979,* National Environmental Engineering Research Institute, Nagpur, pp. 2–5.

ndaresan, B. B., 1982. Waste management. In: *State of Environment—Some Aspects,* National Committee on Environmental Planning, Department of Environment, New Delhi, Chapter IV, pp. 1–41.

resh, D. S., Babu, M., Nandakumar, V., John, B., Jayaprasad, B. K. and Pramod, S. V., 2000. Siltation analysis in Neyyar reservoir and forest degradation in its catchments: A study from Kerala State, India, *Environ. Geology*, 39, 391–397.

waminathan, M. S., 1982. Foreword. In: *Report of the Task Force for the Study of Ecodevelopment in the Himalayan Region,* Planning Commission, New Delhi.

wamy, M. L., 1981. Inadequacies in investigation of hydroelectric project in Andhra Pradesh. *Proc. Sem. Accelerated Hydroelectric Development in India,* Central Board of Irrigation and Power, New Delhi, 1(150): 37–44.

wanson, F. J. and Dyrness, L. T., 1975. Impact of clearcutting and road construction and soil erosion by landslides in the Western Cascade Range, Oregon. *Geology,* 3: 393–96.

wanston, D. N. and Swanson, F. J., 1978. Timber harvesting, mass erosion and sheepland forest geomorphology in the Pacific northwest. In: Coates, D. R. (Ed.), *Geomorphology and Engineering,* Halsted, Stroudsburg (Pennsylvania), pp. 199–221.

ngri, A. K., 2000. *Integration of Remote Sensing Data with Conventional Methodologies in Snow-melt Runoff Modelling in Bhagirathi River Basin, U.P. Himalaya* (unpublished report), Remote Sensing Application Centre, Lucknow, 601 p.

nk, R. W. (Ed.), 1973. *Focus on Environmental Geology,* Oxford Univ. Press, New York, 534 p.

jwani, K. G., 1975. Classification and reclamation of gullied lands. *Indian Jour. Soil & Water Conservation in India,* 25: 29–40.

jwani, K. G., Dayal, R., Verma, V., Godwani, H. M. and Gupta, S., 1975. Classification, control and reclamation of ravine-lands. In: Tejwani, K. G., Gupta, R. K. and Mathur, H. N. (Eds)., *Soil and Water Conservation Research 1956-1971,* Indian Council of Agricultural Research, New Delhi, pp. 254–310.

ampi, P. K., 1997. *Evaluation Study in Terms of Landslide Mitigation in Parts of Western Ghats, Kerala,* Centre for Earth Science Studies, Trivandrum, 33 p.

anikaimani, G., 1980. Mangrove formations in Indian Peninsula. In: M. Nagaraj and C. P. Mullik (Eds), *Current Trends in Botanical Research,* Kalyan Pub., New Delhi, pp. 233–335.

igale, S. S., 1983. The impact of physical determinants on groundwater occurrence in the aggraded landforms associated with Western Ghats of Maharashtra, India. *Intern. Conf. Groundwater and Man,* Sydney (Australia), pp. 319–27.

igale, S. S. and Zambre, M. K., 1982. An integrated study of hydrogeology of the Shivaganga basin, Maharashtra. *Proc. Intern. Symp. Hydrological Aspects of Mountain Watersheds,* School of Hydrology, Roorkee Univ., 1(10): 8–15.

ling, R. I., 1983. *Monitoring Active Volcanoes,* General Interest Publication, U. S. Geol. Survey, Washington, 13 p.

Thigale, S. S. and Khandge, A., 2000. Landslides in Western Ghats of Maharashtra, *In*: A.M. Path and S.S. Thigale (Eds), *Contributions to Environmental Geoscience*, Aravali Books Intern., Ne Delhi, 161–174.

Thomas, M. F. and Kale, Vishwas S., 2011. Tropical Environments, *In*: K.J. Gregory and A. Goud (Eds), *Handbook of Geomorphology*, Sage Publications, London, 447–468.

Thussu, J. L., 1999. Geoenvironmental appraisal of Haryana, *Jour. Geol. Soc. India*, 54, 621–640.

Tiwari, S. C. and Srivastava, N. K., 1983. Artificial recharge by injection method in overexploit Mehsana alluvial aquifers—A case study. *Proc. Sent. Assessment, Development and Management of Groundwater Resources,* Central Groundwater Board, New Delhi, 6: 297–305.

Tiwari, S. K., 1982. *Recent Trends in Wind Energy,* INSDOC, New Delhi.

Tolba, M. K., 1983. *Development Without Destruction,* Tycooly, Dublin, 197 p.

Tyagi, G. D. and Sharma, K. S., 1982. Adverse geology necessitating heavy support rectification wh tunnelling at Loktak Project. *Proc. Fourth Intern. Congr. Eng. Geol.,* Oxford & IBH, New Del 8(6): 97–105.

Tyler, Miller G. and Armstrong, Patrick, 1982. *Living in the Environment,* Wadsworth, Belmo California, 498 p.

Unnikrishnan, A. S., Rupa Kumar, K., Fernandes, S. E., Michael, G. S. and Patwardhan, S. K., 2006. S level changes along the Indian coast: observation and projections, *Current Science*, 90, 362–368

Upadhyaya, V. S., 1980. Prevention of slope failures in deep-cut excavations by means of anchorpa in the Tallhatta open channel, Kalinadi Hydroelectric Project, Karnataka, India. *Geotech,* 8(2 353–57.

Uprety, B. K., 1988. Impact of riverbed erosion, flood hazard and river-shifting in Nepal, *In*: *Proc. Intern. Symp. Impact of Riverbed Erosion, Flood Hazard and Problems of Population Displacement* Dhaka, April 11–13, 27 p.

Vaidyanath, L. N. and Ghosh, D. K., 1982. Evaluation of engineering properties of rocks in centr India. *Rec. Geol. Surv. Ind.,* 112(6): 1–6.

Vaidyanathan, R., 1983. Flood-plain mapping studies in India based on remote-sensing techniques—review. *Proc. Nat. Sym. Remote Sensing in Development and Management of Water Resources,* Ind. Institute Rem. Sensing, Dehradun, pp. 74–78.

Valdiya, K. S., 1977. Structural set-up of the Kumaun Lesser Himalaya. In: *Himalaya, Science de la Terre,* C.N.R.S., Paris, 268 pp. 449–62.

Valdiya, K. S., 1981. The tectonics of the central sector of the Himalaya. In: F. M. Delany and K. Gupta (Ed.), *Zagros-Hindukush-Himalaya-Geodynamic Evolution,* Amer. Geophy. Union, Washington, pp. 87-110.

Valdiya, K. S., 1984. *Aspects of Tectonics: Focus on South-Central Asia,* Tata McGraw-Hill, New Del 319 p.

Validya, K. S., 1985. Accelerated erosion and landslide-prone zones in the Himalayan region. In: J. Singh (Ed.), *Environmental Regeneration in Himalaya: Strategies and Concepts,* Cen. Him. Env. Assoc, Nainital, pp. 12–38.

Valdiya, K. S., 1985. Accelerated erosion and landslide-prone zones in the Himalayan region. *In*: J Singh (Ed) *Environmental Regeneration in Himalaya: Strategy and Concept*, Gyanodaya Prakashan, Nainital, 12–38.

ldiya, K. S., 1992. Must we have high dams in the geodynamically active Himalayan domain? *Current Science*, 62, 289–296.

ldiya, K. S., 2001. Tectonic resurgence of the Mysore Plateau and surrounding regions in cratonic southern India, *Current Science*, 81, 1068–1089.

ldiya, K. S., 2002. *Saraswati: The River That Disappeared*, Universities Press, Hyderabad, 116 p.

ldiya, K. S., 2004. *Geology, Environment and Society*, Universities Press, Hyderabad, 226 p.

ldiya, 2008. ***Facing the Hazards: Earthquakes and Landslides***, Gyanodaya Prakashan, Nainital, 87 p.

ldiya, K. S., 2010. *The Making of India: Geodynamic Evolution*, Macmillan, Noida, 816 p.

ldiya, K. S., 2012. *Geography, Peoples and Geodynamics of India in Purans and Epics*, Aryan Books International, New Delhi, 240 p.

ldiya, K. S., Joshi, D. D., Sanwal, R. and Tandon, S. K., 1984. Geomorphologic development across the Main Boundary Thrust: An example from the Nainital hills in Kumaun Himalaya. *Jour. Geol. Soc. Ind.,* 25(12): 761–74.

radarajulu, R. and Dhanalakshmi, S., 1975. Sea levels and waves along the Madras coast. *Ind. Jour. Marine Sci.,* 4: 115–23.

Varma, O. P., 1999. *Emerging Trends in Mineral Exploration and Valuation*, Indian Geological Congress, Roorkee, 21 p.

sudevaiah, R. D., Singh, J. S. P. and Guha, D. P., 1965. Runoff, soil loss determination studies at experimental station: II-Effects of annual cultivated grain crops and perennial grasses on 5 per cent slopes. *Ind. Jour. Soil and Water Conservation,* 13(39): 1–36.

rma, V. K., Dogra, S. D., Joshi, D. D. and Kumar, A., 1982. Groundwater in Pathankot piedmont. *Proc. First Intern. Symp. Soil Geol. and Landforms: Impact on Landuse Planning in the Developing Countries,* Bangkok, Thailand, 8: 1–9.

hra, B. B., 1975. Land management problems in India—A microview. *Soil Conservation Digest,* 3(1): 54–57.

hra, B. B., 1980. *A Policy for Land and Water,* Sardar Patel Memorial Lectures, Department of Environment, New Delhi, 28 p.

hra, C. P., 1978. Glacier resources of the Himalaya and their importance to environmental studies, *Proc. Nat. Sem. Resources Development and Environment in the Himalayan Region,* Department of Science and Technology, New Delhi, pp. 441–59.

hra, C. P. and Sehgal, M. N., 1977. The significance of geological factors in the Himalayan ecosystem, In: H. C. Rieger (Ed.), *Dialogue: Himalayan Mountain Ecosystem,* Max Mueller Bhawan, New Delhi, pp. 10–77.

hra, C. P., 2000. Glaciology, 25 years ago, In: O.P. Varma and T.M. Mahadevan (Eds), *Research Highlights in Earthscience*, Indian Geological Congress, Roorkee, 199–206.

as, J. N. and Parthasarathy, R., 2001. *Saket Environmental Handbook*, Saket, Ahmedabad, 455 p.

adia, D. N., 1975. *Geology of India* (4th edn.), Tata-McGraw-Hill, New Delhi, 508 p.

angpang, T. and Tangjang, S., 2012. Slash-and-burn agriculture in eastern Himalayan zone of Arunachal Pradesh, NE India, *Curr. Sci.*, 1247–1248.

are, Leslie, 1981. Energy from the sun, *Span,* Oct. 22(10): 16–18.

Warsi, W. E. K. and Molnar, P., 1977. Gravity anomalies and plate tectonics in the Himalaya. I *Himalaya, Science de la Terre,* C.N.R.S., Paris, 268: 463–78.

Watson, R. T., Zinyowera, M. C. and Moss, R. H., 1998. *The Regional Impact of Climate Change* Cambridge University Press, Cambridge, 514 p.

Weber, F., and Hoskins, M. W., 1983. *Soil Conservation Technical Sheets,* Forest, Wildlife and Ran Experiment Station, *University of Idaho* Moscow, 94 p.

Wiggins, J. H., Slosson, J. E. and Krihn, J. P., 1978. *Natural Hazard—Earthquakes, Landslide Expansive Soil-Loss Models,* J. H. Wiggins Technical Report, Redondo Beach, California, 162 p

Worthington, E. B., 1977. *Arid Land Irrigation in Developing Countries,* Pergamon, Oxford.

Ye Hong, Wen-Yu Zhang, Zhi-Shui Yu, and Qin Xia, 1981. Seismicity and recent crustal movements the Himalayan region. In: *Geological and Ecological Studies of Qinghai-Xizang Plateau,* Science Press, Beijing and Garden and Beach Science Pub., New York, 1: 65–80.

Yokoyama, I., 1984. Volcanic processes revealed by geophysical observations of the 1977-1982 activı of Uso volcano, Japan. In: S. A. Fedov (Ed.), *Proc. Colloq. Earthquakes and Geological Haza Prediction,* Intern. Geol. Congr. Moscow, pp. 136–44.

Zachar, D., 1982. *Soil Erosion,* Elsevier, Amsterdam, 547 p.

Zaruba, Q. and Mencl, V., 1969. *Landslides and their Control,* Elsevier, Amsterdam, 205 p.

Zaruba, Q. and Mencl, V., 1976. *Engineering Geology,* Elsevier, Amsterdam, 504 p.

Ziswilas, V., 1967. *Extinct and Vanishing Species,* Springer-Verlag, New York.

Zubairi, A. H. and Agarwal, P. D., 1985. Excavation and rock cutting in context of hill road constructio *Proc. Sem. Construction of Roads in Hill Areas,* Ind. Road Congress, New Delhi, 111–25–34.

Zutshi, D. P., 1984. Impact of human settlement on the ecology of rural lakes of Kashmir. *Annual Report, MAB Programme,* Department of Environment, New Delhi (mimeo).

Subject Index

Authors' Index

Author's Profile

Padma Shree K.S. Valdiya is the recipient of numerous prestigious awards including S.S. Bhatnagar ze (1976) of the Council of Scientific & Industrial Research; National Mineral Award of Excellence 997) of the Ministry of Mines; Pitambar Pant Environment Fellowship (1982–1984) of the Ministry Environment and Forests; Hindi Sevi Samman (Atmaram Puraskar, 2007) of the Ministry of Human source Development; National Lecturer (1977–1978) of the University Grants Commission; GM odi Award for Science (2012) from GM Modi Foundation; D.N. Wadia Medal (1995) of Indian ational Science Academy; and L. Rama Rao Gold Medal (1980) of the Geological Society of India. Professor Valdiya is a Fellow of all the three National Science Academies of India, the Third World Academy of Science, the Geological Society of America and The Geological Society of Nepal.

Born at Kalaw in Myanmar, he had his early education in Pithoragarh, a small town in Kumaun Himalaya, and higher education in Lucknow University. Obtaining Ph.D in 1963 from Lucknow niversity, he studied at Johns Hopkins University, USA as a Fulbright Fellow during 1965–1966.

Currently he is Honorary Professor of Geodynamics at Jawaharlal Nehru Centre for Advanced ientific Research, Bengaluru. He has been Additional Director in Wadia Institute of Himalayan ology; Professor of Geology, Dean: Science and Vice-Chancellor in Kumaun University; Golden Jubilee Research Professor of Indian National Science Academy at Jawaharlal Nehru Centre for dvanced Scientific Research; Distinguished Guest Professor at Indian Institute of Technology, umbai; and Visiting Honorary Professor at the Indian Institute of Technology, Roorkee.

He is the author of several publications including *Geology of Kumaun Lesser Himalaya* (1980), *Aspects of Tectonics: Focus on Southcentral Asia* (1984), *Environmental Geology: Indian Context* (1987), *Dynamic Himalaya* (1998), *Saraswati, the River That Disappeared* (2002), *Geology, Environment and Society* (2005), *The Making of India: Geodynamic Evolution* (2010), *Ek Thi Nadi Saraswathi* (2010) and *Geography, Peoples, Geodynamics of India in Purans and Epics* (2012). He is a prolific writer in Hindi, popularizing earth science and general science.

He specializes in tectonics, with special reference to active faults and environmental geology.

Made in the USA
Monee, IL
08 November 2026

68055350R00354